Transport of Ions
and Water
in Animals

Transport of Ions and Water in Animals

edited by

B. L. GUPTA

R. B. MORETON

TRANSPORT OF IONS AND WATER IN ANIMALS

Errata

Page 129, line 18, 160 mM should read 140 mM
and line 20, 140 mM should read 160 mM.
Page 274, line 30, it should read see p. 267.

1977

ACADEMIC PRESS
LONDON   NEW YORK   SAN FRANCISCO
A Subsidiary of Harcourt Brace Jovanovich, Publishers

ACADEMIC PRESS INC. (LONDON) LTD
24-28 Oval Road
London NW1

US edition published by
ACADEMIC PRESS INC.
111 Fifth Avenue,
New York, New York 10003

Library of Congress Catalog Card Number: 77-71820
ISBN: 0-12-307650-1

Text set by Santype International Ltd.,
Salisbury, Wiltshire.
Printed by Page Bros. (Norwich) Ltd.,
Norfolk, England.

Contributors

J. ABBOTT *Department of Physiology, Kings College, Strand, London WC2, England*

M. J. BERRIDGE *A.R.C. Unit of Invertebrate Chemistry and Physiology, Department of Zoology, University of Cambridge, Downing Street, Cambridge (CB2 3EJ,) England*

W. H. DANTZLER *Department of Physiology, University of Arizona, College of Medicine, Tucson, Arizona 85724, U.S.A.*

E. B. EDNEY *Laboratory for Nuclear Medicine and Radiation Biology, University of California, Los Angeles, California 09924, U.S.A.*

A. FINKELSTEIN *Department of Physiology, Albert Einstein College of Medicine, Bronx, New York 10464, U.S.A.*

J. D. GEE *Tsetse Research Laboratory, University of Bristol School of Veterinary Science, Langford House, Langford, Bristol BS18 7DU, England**

G. GIEBISCH *Department of Physiology, Yale University School of Medicine, New Haven, Conneticut 06510, U.S.A.*

B. L. GUPTA *Department of Zoology, University of Cambridge, Downing Street, Cambridge CB2 3EJ, England*

T. A. HALL *Biological Microprobe Laboratory, Department of Zoology, University of Cambridge, Downing Street, Cambridge CB2 3EJ, England*

A. E. HILL *Department of Physiology, Downing Site, Cambridge, England*

E. K. HOFFMANN *Institute of Biological Chemistry A, Universitetsparken 13, 2100 Copenhagen, Denmark*

C. R. HOUSE *Department of Veterinary Physiology, Royal (Dick) School of Veterinary Studies, University of Edinburgh, Summerhall, Edinburgh, Scotland*

L. B. KIRSCHNER *Department of Zoophysiology, Washington State University, Pullman, Washington 99163, U.S.A.*

W. J. LARSEN *Department of Anatomy, University of Iowa Medical School, Iowa City, Iowa 52240, U.S.A.*

C. LITTLE *Department of Zoology, University of Bristol, Bristol BS8 1UG, England*

A. M. P. LOCKWOOD *Department of Oceanography, University of Southampton, Southampton S09 5NH, England*

*Present address: Parasitology Department, Central Research, Pfizer Ltd., Sandwich, Kent, England.

J. MACHIN *Department of Zoology, University of Toronto, Toronto 5, Canada*

S. H. P. MADDRELL *A.R.C. Unit of Invertebrate Chemistry and Physiology, Department of Zoology, University of Cambridge, Downing Street, Cambridge CB2 3EJ, England*

R. B. MORETON *A.R.C. Unit of Invertebrate Chemistry and Physiology, Department of Zoology, University of Cambridge, Downing Street, Cambridge CB2 3EJ, England*

S. NEDERGAARD *Institute of Biological Chemistry A, August Krogh Institute, Universitetsparken 13, 2100 Copenhagen, Denmark*

J. NOBLE-NESBITT *Department of Biological Sciences, University of East Anglia, Norwich NR4 7TJ, England*

J. E. PHILLIPS *Department of Zoology, University of British Columbia, Vancouver, British Columbia, Canada*

W. T. W. POTTS *Department of Biology, University of Lancaster, Bailrigg, Lancaster, England*

W. T. PRINCE *A.R.C. Unit of Invertebrate Chemistry and Physiology, Department of Zoology, University of Cambridge, Downing Street, Cambridge CB2 3EJ, England*

R. D. PRUSCH *Division of Biological and Medical Sciences, Brown University, Providence, Rhode Island 12912, U.S.A.*

J. A. RIEGEL *Westfield College, University of London, Hampstead, London, NWB 75T, England*

G. SACHS *Department of Membrane Biology, University of Alabama Medical Centre, Birmingham, Alabama 35233, U.S.A.*

E. SKADHAUGE *Institute of Medical Physiology A, University of Copenhagen, Juliane Maries Vej 28, 2100 Copenhagen, Denmark*

J. E. TREHERNE *A.R.C. Unit of Invertebrate Chemistry and Physiology, Department of Zoology, University of Cambridge, Downing Street, Cambridge CB2 3EJ, England*

H. H. USSING *Institute of Biological Chemistry A, August Krogh Institute, Universitetsparken 13, 2100 Copenhagen, Denmark*

B. J. WALL *Marine Biological Laboratory, Woods Hole, Massachusetts 02543, U.S.A.*

V. B. WIGGLESWORTH *Department of Zoology, University of Cambridge, Downing Street, Cambridge CB2 3EJ, England*

I. ZERBST-BOROFFKA *Insitut fur Tierphysiologie und Augewandte Zoologie, Berlin-West, West Germany*

K. ZERAHN *Institute of Biological Chemistry A, August Krogh Institute, Universitetsparken 13, 2100 Copenhagen, Denmark*

T. ZEUTHEN *Department of Physiology, Downing Site, Cambridge, England*

Preface

In his contribution to the *Festschrift* for Sir James Gray on his retirement from the Chair in Zoology in Cambridge University, J. Arthur Ramsay recalled Gray to have said in the 1930s that "there was not — or if there was there ought not to be — any such subject as Comparative Physiology". Gray's fear was that Comparative Physiology, drawing inspiration from Comparative Anatomy, might become a futile attempt to establish phylogenetic relationships of physiological functions. However, the subject of Comparative Physiology has developed into the study of fundamental problems in General Physiology by seeking out organisms, tissues or cells where these may be best investigated. Arthur Ramsay's own research, endearingly reviewed by Sir Vincent Wigglesworth in the Introduction to this volume, played a major part in steering the comparative physiology of solute and water transport in this direction. Unlike many General Physiologists and Biophysicists, Ramsay never studied a phenomenon outside the context of the organism's ecological requirements — a need which is echoed in Ussing's Prelude. This is the approach which Ramsay adopted in his own work and implanted in the minds of several generations of aspiring young physiologists through his teaching and textbooks: this is the approach which has guided us in putting together this volume of essays to commemorate the retirement of J. Arthur Ramsay from the Chair of Comparative Physiology in the University of Cambridge.

To cover the subject fully, at all levels of organization from simple lipid bilayers to whole metazoan organisms, we have deviated from the usual practice of *Festschrifts* by also inviting workers who were not Ramsay's students and associates but whose work was nevertheless influenced by Ramsay's scientific thoughts and numerous technical innovations. A section on current approaches (Chapters 1–5) documents the present state of the art in microanalytical techniques for transport physiology, many of which were pioneered by Ramsay. A new dimension to the microanalytical approach has been added by the emerging technique of X-ray microanalysis applied to frozen-hydrated samples. Chapter 4 includes the first set of extensive quantitative data obtained by this method from many of Ramsay's "pet" tissues. This method offers most promise in helping to choose between various theoretical models for Haidenhain's *Triebkraft* to explain the phenomenon of active transport of water

(Chapters 5, 7, 21—25) — a phenomenon which provided the major theme in Ramsay's own research endeavours.

To us the preparation of this volume has been a pleasure because of the feeling that it will form a useful source of information as well as a tribute to Ramsay's own struggle with the unknown, by focusing attention on the fundamental problems in transport physiology which remain unresolved. Our task was made even more pleasant by the enthusiastic cooperation of the contributors who readily concurred with our aims. It was this enthusiasm which has resulted in a volume that is double the size of what we had planned for. We appreciate the understanding of the Academic Press in accepting this expansion in recognition of the contributor's efforts. We also thank all those publishers who generously permitted the various authors to reproduce material.

We have a special gratitude to Dr J. E. Treherne and Professor Leonard B. Kirschner for their advice and whole-hearted support in this venture from the outset. Mr R. Hughes and Miss Jean Sanderson performed the unenviable task of compiling the indexes and Mrs M. V. Clements provided the secretarial help. We thank Dr D. A. Parry for encouragement and for the use of Departmental office facilities.

August, 1977   *The Editors*

Contents

I. Current Approaches

II. Mechanisms and Control of Transport

IV. Ion Transport at Tissue Level

J. Arthur Ramsay, MBE, M.A., Ph.D. (Cantab.), F.R.S. *Retired on 30 September 1976 from the Chair of Comparative Physiology, Cambridge University.*

J. Arthur Ramsay

The work of a scientist can epitomize a period as effectively as the artistries of Beerbohm or Astair can recall theirs. For me, Arthur Ramsay epitomizes a fascinating phase in the development of the biological sciences: the emergence of Comparative Physiology as a fully fledged scientific discipline. His work, more than that of any other, provided uniquely valuable microtechniques for the quantitative analysis of the miniature tissues and organs of the invertebrates which are frequently the working material of comparative physiologists. His experimental rigour and clarity of thought, especially in his 1950s papers, provided an inspiration for a younger generation of biologists. His textbooks guided and influenced biological undergraduates for a decade or more. His skilful editing of the *Journal of Experimental Biology* influenced the standard of publication of a wealth of papers on Comparative Physiology.

Despite more than half a lifetime in Cambridge, Arthur Ramsay remains, unmistakably and unshakably, a Scot. His 40 years in the ancient English University were enlivened by his regular exercise of his national musical instrument and sustained by his native beverage. Generations of Cambridge undergraduates were diverted by his expert fingering of a pibroch on a biro pen during supervisions, or the suspicion of a few surreptitious steps of a Highland reel during perambulations in laboratory classes. Users of the Zoology workshop were astonished by the wealth and range of Celtic imprecation. Characteristically, his retirement from the Chair of Comparative Physiology and Fellowship of Queens' College, Cambridge, was followed by an abrupt retreat north of the Highland line to the braes above Loch Ness.

Arthur Ramsay is, in some ways, reminiscent of the Duke of Wellington. He shares his decisive powers of organization, eccentricities, directness and essential simplicity in mode of life. As would the Iron Duke, he will detest these biographical reminiscences. Indeed I view my involvement with this publishing venture with some trepidation, for I was specifically enjoined by him not to engage in any attempts to commemorate his retirement from the Cambridge scene.

It is, however, an adequate testimony of our regard for Arthur Ramsay that the contributors to this dedicatory volume should risk his wrath to express our admiration and gratitude for his outstanding contributions to biological sciences.

J. E. Treherne

August, 1977

Introduction – J. Arthur Ramsay and Transport Physiology

V. B. Wigglesworth

Department of Zoology, University of Cambridge, Cambridge, England

Arthur Ramsay is one of the most ascetic of scientists. He never publishes preliminary notes. He never (save under severe external pressure) discourses about his work before scientific societies at home, and never abroad. He publishes his discoveries once — and that is that. And yet in spite, or perhaps because, of this self-denying ordinance his achievements have always been most highly regarded. The fact is that he does not make clever discoveries or inventions to display his ingenuity, nor does he hasten into print to forestall others in the field. I suppose he is inspired by a certain scientific curiosity; but he recognizes only one enemy: the problem he is out to solve. He sees research as a single-minded combat with this powerful adversary. In these introductory remarks I attempt to show that this wily opponent has not escaped unscathed.

Ramsay was early concerned with measuring the evaporation of water from animals. He went carefully into the physical theory of evaporation in still air and in moving air, and particularly the effects of temperature on rates of diffusion (which had been virtually neglected); and he devised apparatus for giving controlled conditions of temperature, humidity and wind velocity. All this led to pinpointing the limitations of the law of saturation deficiency in descriptions of water loss by insects in still air (Ramsay, 1935a).

Applying these ideas and his wind tunnel methods to the living cockroach Ramsay demonstrated the great influence of air movement on rate of water loss, particularly from the spiracles; and, when a permeable surface is being studied, the reduction in water loss caused by the lowering of the surface temperature by evaporation.

1

He confirmed the abrupt increase in water loss from the cockroach when the temperature is raised above 30°C, but showed that this was not due to increased tracheal ventilation, as had been supposed, because it occurred also in the dead insect with the spiracles blocked. Moreover tiny droplets of water sprayed on the surface of the cuticle remained for hours without evaporation at 25°C, but dried up immediately at 33°C. This was found to result from a "film of fatty substance which spreads over the surface and suffers a change of state either of expansion or of melting at 30°C". He pointed out that if other insects have such a film, then the temperature of its melting point may be very important in the ecological relations of the insect to the temperature of its environment (Ramsay, 1935b).

At a meeting of the Society for Experimental Biology in 1935 Ramsay suggested that this lipid material might be similar to the "cuticulin" I had described in *Rhodnius*. I pointed out in discussion that it should be compared rather with the extractable waxes obtained by Bergmann in the silkworm. I believe I suggested that it would be interesting to compare the properties of such extracts in different insects; and I confidently expected that Ramsay would make such comparative studies. But the years went by; he published with Sydney Manton on water loss in *Peripatus* (Manton and Ramsay, 1937), with Butler and Sang (Ramsay *et al.*, 1938) on the very steep gradient of humidity at the surface of leaves, and briefly on acetylcholine in nerve-muscle preparations of the snail (Ramsay, 1940).

Then the war came and Ramsay turned his attention to higher things. Along with other studies on radiowave propagation it was he who discovered that the mysterious radar signals, which had been ascribed to passing angels, were really reflections from night-flying birds. So it fell to me in the war years to extend the transpiration studies to other insects and to show that Ramsay's predictions were indeed correct and that the waterproofing waxes of insects vary in hardness with their environments and that (as suggested to us by Danielli) the abrupt increase in transpiration rate takes place at the "transition temperature of free rotation", some five or ten degrees centigrade below the melting point.

Returning to experimental biology after the war, Ramsay first studied excretion and osmotic relations in the earthworm (Ramsay, 1946, 1949a,c). It was at this stage that he devised his now classic method of freezing point determination for small quantities (Ramsay, 1949b). This was the method which everyone who had worked in this field had dreamed about: an ultramicromethod for osmotic pressure. This was soon followed by his improved flame photometers, capable of the simultaneous determination of sodium and potassium in samples of 10 nl (Ramsay, 1950b; Ramsay *et al.*, 1951, 1953), coupled with what was of equal or even greater importance, the devising of methods of collection of minute samples of fluid from the nephridia of the earthworm and the lumen of the rectum and Malpighian tubules of mosquito larvae and other insects.

This combination of refined methods of analysis with remarkable ingenuity and dexterity in collecting samples, has enabled Arthur Ramsay to put our knowledge of osmoregulation and urine formation in invertebrates upon a new foundation. By direct measurements he was able to show that the rectum in mosquito larvae is the main site of osmotic regulation, the epithelium diluting the contents or concentrating them according to the medium in which the larva finds itself — and with appropriate adaptations in species from fresh water (*Aedes aegypti*) and brackish water (*A. detritus*) (Ramsay, 1950a).

But although the Malpighian tubules do not appear to play any part in the osmotic regulation of the urine they do assist in sodium conservation in a medium poor in salts by secreting a fluid with a lower sodium content than the haemolymph (Ramsay, 1951). Furthermore, there is a striking difference between the movement of sodium and potassium: whereas the concentration of sodium in the Malpighian tubule fluid is consistently lower than in the haemolymph, there is a circulation of potassium from haemolymph to tubule, and so to rectum or midgut, followed by reabsorption back to the haemolymph. Despite changes in the medium the sodium and potassium content of the haemolymph is held remarkably constant (Ramsay, 1953a).

Likewise in the terrestrial, blood-feeding insect *Rhodnius*, the upper portion of the Malpighian tubule can secrete a urine which contains more potassium and less sodium than the haemolymph. As the fluid passes down the tubule the differences between urine and haemolymph become reduced; and during storage in the rectum, water and sodium but not potassium are reabsorbed (Ramsay, 1952).

In a wide range of insects the concentration of potassium in the contents of the Malpighian tubules was always greater than in the haemolymph and the sodium concentration usually less. In *Dixippus*, *Pieris*, *Tenebrio*, *Dytiscus* and *Aedes* measurements of electrical potential across the wall of the Malpighian tubule showed that the interior of the tubule is always positive with respect to the haemolymph; and that in *Locusta* and *Rhodnius*, although the content of the tubule proved to be negative with respect to the haemolymph, when the concentration ratios were expressed in terms of potential difference it became clear that even in these insects there must be an active transport of potassium. On the other hand there was no such evidence in the case of sodium and the conclusion was that sodium is probably moved by passive diffusion (Ramsay, 1953b).

It was at this stage that Ramsay devised the method of studying the activities of single Malpighian tubules isolated in drops of haemolymph under liquid paraffin; measurements of osmotic pressure and ionic composition being made on haemolymph and urine (Ramsay, 1954). For these experiments the stick insect *Dixippus* was the chief subject. Once more it was found that the tubules are far more efficient in regulating potassium than sodium (Ramsay, 1955a). But sodium also can be transported against an electrochemical gradient. It is in the rectum that the greater part of the reabsorption of potassium and of water takes place (Ramsay, 1955b). As regards other inorganic ions, the concentrations of calcium, magnesium and chloride are always less in the tubule fluid than in the medium, that of phosphate is always greater. But in all cases it seemed that "potassium is the prime mover in generating the flow of urine and that in consequence of this secretion conditions are created which enable water and other constituents of the urine to follow" (Ramsay, 1956).

Turning to the behaviour of organic substances, Ramsay found, contrary to expectation, that all compounds of low molecular weight in the haemolymph, appear to enter the Malpighian tubule by passive diffusion (as in the glomerulus of the vertebrate nephron) and that the metabolically useful materials such as amino acids and sugars, along with inorganic ions and water as already shown, are reabsorbed in the rectum (Ramsay, 1958).

In more recent years Ramsay has taken up the challenge of the cryptonephric system of insect larvae, such as the mealworm *Tenebrio*, which feed on relatively dry material and produce highly desiccated excreta. It is a fact of observation that this powerful drying capacity in the excretory system is related to an intimate association between the rectum and the Malpighian tubules, such that successive compartments invest the rectal chamber, in which the fluid urine from the Malpighian tubules and the contents from the midgut are converted into dry solid faecal pellets. As Ramsay showed, these pellets are in equilibrium with a relative humidity which averages 90%, but may be as low as 75%!

Ramsay's plan of campaign has been to measure the osmotic pressures and ionic contents of the successive compartments (rectal lumen, perinephric fluid, Malpighian tubular fluid, and haemolymph) with a view to formulating a theory to account for the high gradients attained. Thus the freezing point depression of the perinephric fluid under dry conditions may reach 8°C! The freezing point depression of the fluid in the Malpighian tubules within the perinephric space is very close to that of the perinephric fluid and can be almost completely accounted for by potassium chloride which may reach a concentration of over 2M. The perinephric membrane is fairly impermeable to water. The active process involved appears to be secretion of potassium chloride from haemolymph to the perirectal tubules; and this draws water passively from the perinephric fluid and thereby decreases the osmotic work done by the rectal epithelium in removing water from the faeces (Ramsay, 1964).

In collaboration with Grimstone and Mullinger, Ramsay (1968) attempted to pin down more clearly the source of the potassium chloride supply which lies at the basis of this scheme. In a study of fine structure much attention was given to the "leptophragmata" which are points of contact, like tiny port holes, between the Malpighian tubules and the perinephric membrane. Each window is formed by the very thin cell body of a leptophragma cell (a modified cell of the Malpighian tubule) containing mitochondria and clothed internally with microvilli. The outcome of this combined structural and experimental study is a model of the rectal complex according to which the high osmolarity of the fluids concerned is brought about by the inward secretion of potassium chloride from the haemolymph, unaccompanied by water, at the leptophragmata.

This model provides a formal solution to a very tough thermodynamical problem. Indeed it has been Ramsay's insistence on actual measurements which has underlined the toughness of the problem he is trying to solve. My only hesitation about the outcome is that it lays greater weight upon the slender leptophragma cells than they look structurally qualified to bear. This doubt is strengthened by the fact that in the larvae of Lepidoptera there is a cryptonephridial system more or less like that of the mealworm but without leptophragmata — and in the larvae of clothes moths the excretory system has desiccatory powers even greater than those of the mealworm.

During recent years Arthur Ramsay has been locked in mortal combat with the cryptonephridial system of the Lepidoptera. I am not in a position to reveal the outcome — but I have reason to hope that when the ink is dry, even this latest opponent will be found to have been battered to his knees, *si deis placit*.

(In the event, Ramsay's results were published (Ramsay, 1976) just in time for me to recall this Introduction before the present volume went to press.

It turned out that in the caterpillars of *Pieris* and *Manduca* (the lepidopterous larvae used in his experiments) the cryptonephridial system does not show the powers of desiccating the faecal pellets that is seen in *Tenebrio*; but the system is actively engaged in the transfer of ions, a consequence perhaps of the high throughput of food and the massive transfer of ions that occurs in the midgut of Lepidoptera. The outstanding puzzle was that more sodium could be eliminated in the faeces than enters the rectal complex from the intestine. Evidence strongly points to the Malpighian tubules as the route of entry. No leptophragmata are present; but a new factor came to light: there is a tidal flow of Malpighian tubule fluid in and out of the cryptonephridial system, brought about by muscular compression of a dilated region in the outer cryptonephridial tubules. In this way the segments of the tubules which lie free in the body cavity can readily take up ions from the haemolymph and as the tubule contents are shuttled to and fro these ions can be excreted, if necessary, or they can build up the high osmolarity in the perinephric fluid and thus contribute to the removal of water from the faeces.

Ramsay points out that there is evidence for a similar tidal movement in the mealworm. He suggests that this may be the basal mechanism in the cryptonephridial system of *Tenebrio*, the leptophragmal mechanism being superimposed upon it.)

The history of science, which is largely a matter of elucidating priorities in conceptions and experiments, is a terribly difficult subject. But there is no doubt at all that during the past 40 years there have been great advances in our knowledge of transpiration, osmoregulation and excretion among invertebrates; and that an unduly large proportion of the key ideas and the key techniques in this field have their source in the publications which I have here briefly reviewed.

REFERENCES

Grimstone, A. V., Mullinger, A. M. and Ramsay, J. A. (1968). Further studies on the rectal complex of the mealworm *Tenebrio molitor*, L. (Coleoptera, Tenebrionidae). *Phil. Trans. R. Soc.* B. 253, 343—382.

Manton, S. M. and Ramsay, J. A. (1937). Studies on the Onychophora: III The control of water loss in *Peripatopsis*. *J. Exp. Biol.* 14, 470—472.

Ramsay, J. A. (1935a). Methods of measuring the evaporation of water from animals. *J. Exp. Biol.* 12, 355–372.

Ramsay, J. A. (1935b). The evaporation of water from the cockroach. *J. Exp. Biol.* 12, 373–383.

Ramsay, J. A. (1940). A nerve muscle preparation from the snail. *J. Exp. Biol.* 17, 96–115.

Ramsay, J. A. (1946). Role of the earthworm nephridium in water balance. *Nature*, 158, 665.

Ramsay, J. A. (1949a). The osmotic relations of the earthworm. *J. Exp. Biol.* 26, 46–56.

Ramsay, J. A. (1949b). A new method of freezing point determination for small quantities. *J. Exp. Biol.* 26, 57–64.

Ramsay, J. A. (1949c). The site of formation of hypotonic urine in the nephridium of *Lumbricus*. *J. Exp. Biol.* 26, 65–75.

Ramsay, J. A. (1950a). Osmotic regulation in mosquito larvae. *J. Exp. Biol.* 27, 145–157.

Ramsay, J. A. (1950b). The determination of sodium in small volumes of fluid by flame photometry. *J. Exp. Biol.* 27, 407–419.

Ramsay, J. A. (1951). Osmotic regulation in mosquito larvae: the role of the Malpighian tubules. *J. Exp. Biol.* 28, 62–73.

Ramsay, J. A. (1952). The excretion of sodium and potassium by the Malpighian tubules of *Rhodnius*. *J. Exp. Biol.* 29, 110–126.

Ramsay, J. A. (1953a). Exchanges of sodium and potassium in mosquito larvae. *J. Exp. Biol.* 30, 79–89.

Ramsay, J. A. (1953b). Active transport of potassium by the Malpighian tubules of insects. *J. Exp. Biol.* 30, 358–369.

Ramsay, J. A. (1954). Active transport of water by the Malpighian tubules of the stick insect, *Dixippus morosus* (Orthoptera, Phasmidae). *J. Exp. Biol.* 31, 104–113.

Ramsay, J. A. (1955a). The excretory system of the stick insect, *Dixippus morosus* (Orthoptera, Phasmidae). *J. Exp. Biol.* 32, 183–199.

Ramsay, J. A. (1955b). The excretion of sodium, potassium and water by the Malpighian tubules of the stick insect *Dixippus morosus* (Orthoptera, Phasmidae). *J. Exp. Biol.* 32, 200–216.

Ramsay, J. A. (1956). Excretion by the Malpighian tubules of the stick insect *Dixippus morosus* (Orthoptera, Phasmidae): calcium, magnesium, chloride, phosphate and hydrogen ions. *J. Exp. Biol.* 33, 697–708.

Ramsay, J. A. (1958). Excretion by the Malpighian tubules of the stick insect *Dixippus morosus* (Orthoptera, Phasmidae): amino acids, sugars and urea. *J. Exp. Biol.* 35, 871–891.

Ramsay, J. A. (1964). The rectal complex of the mealworm *Tenebrio molitor*. L. (Coleoptera, Tenebrionidae). *Phil. Trans. R. Soc.* B. 248, 279–314.

Ramsay, J. A. (1976). The rectal complex in the larvae of Lepidoptera. *Phil. Trans. R. Soc. Lond.* B. 274, 203–226.

Ramsay, J. A. and Brown, R. H. J. (1955). Simplified apparatus and procedure for freezing-point determinations upon small volumes of fluid. *J. Sci. Instrum.* 32, 372–375.

Ramsay, J. A., Butler, C. G. and Sang, J. H. (1938). The humidity gradient at the surface of a transpiring leaf. *J. Exp. Biol.* **15**, 255—265.

Ramsay, J. A., Falloon, S. W. H. W. and Machin, K. E. (1951). An integrating flame photometer for small quantities. *J. Sci. Instrum.* **28**, 75—80.

Ramsay, J. A., Brown, R. H. J. and Falloon, S. W. H. W. (1953). Simultaneous determination of sodium and potassium in small volumes of fluid by flame photometry. *J. Exp. Biol.* **30**, 1—17.

Ramsay, J. A., Brown, R. H. J. and Croghan, P. C. (1955). Electrometric titration of chloride in small volumes. *J. Exp. Biol.* **32**, 822—829.

Prelude

H. H. Ussing

Institute of Biological Chemistry, University of Copenhagen, Denmark

The Editors kindly asked me to write a "Prelude" for this book and I consented with the thought that a book in honour of Professor Ramsay would be a most appropriate thing. Later it dawned on me that I did not know what a prelude meant in this context, so I had to resort to the study of various dictionaries. These studies led me to the definition which is going to guide me through the following pages: "A prelude is a movement introducing the themes or chief subjects of a concert."

The opening bars sound peaceful and promising: Recent years have seen an enormous expansion of the interest in transport of ions and water through biological membranes. To some extent this is due to the development of electron microscopic methods which have given reality to the once elusive concept of the cell membrane based, mainly, on biophysical evidence. But progress in membranology has drawn its main force from other sources. Major ones undoubtedly have been the greatly improved analytical methods. In this context it is fair to say that few scientists have done more than Professor Ramsay to scale down analytical methods to the level where they can serve the needs of, say, insect physiologists. The development of tracer methods has been of immeasurable value to the "transport workers" as has the perfection of microelectrodes.

All these technical advances in turn have led to spectacular theoretical advances.

The period of rapid development in the transport field started toward the end of the last world war. In the first decade the most spectacular advances were of biophysical and physico-chemical nature. The concepts of active and passive transport were refined and made accessible for suitable tests. Phenomena like exchange diffusion, co-transport and counter-transport were described and the potential role of solvent drag and other types of interaction was formalized, both in the framework of kinetics and in that of irreversible thermodynamics. The Hodgkin—Huxley theory for the development of action potentials was one of the peak achievements of this era.

Most of the developments mentioned so far have now ended up in textbooks, and are not recorded in this book, but they shaped many of the themes which are now being elaborated and amplified. In the last two decades new instruments have joined the orchestra, and new themes mix with the old ones. Biochemistry is playing an increasingly important role. The chemistry of membranes, and transport ATPases are being eagerly studied, and in these studies high resolution electron microscopy endeavours to bridge the gap between chemistry and morphology.

The picture which is shaping up from the combined biochemical, biophysical and morphological efforts, seems to be the following. The bilayer of membrane lipid is surprisingly tight to nearly all biologically important substances (perhaps with the exception of water). Penetration, both active and passive takes place by way of highly specific devices (channels, carriers, etc.) which usually admit only one or a few molecular species. On the membrane level interaction between different permeating species is probably less than imagined a few years ago. Interaction will, however, take place in regions like the lateral spaces and possibly in intracellular spaces. Mechanisms of the type discussed by Diamond (see Chapters 5 and 7) under the name of the "standing gradient hypothesis" undoubtedly are important, but many problems relating to the role of lateral spaces in net movements of water are still wide open.

The mushrooming growth of interest in transport processes has had its price: Transport studies have developed more or less independently in many areas of physiology, biophysics, biochemistry, medicine, pharmacology and zoology, just to mention a few, each discipline developing its own nomenclature and its own paradigms. In the studies of transport in animals there has been a tendency to stick to one group, be it mammals, amphibia, insects or shrimps, and there can be little doubt that language barriers have developed between groups, working with different types of "pet" animals.

This book should serve as a short-cut to the ideas developed about transport processes in experimental objects different from ones own. It may be timely, now to acknowledge the existence of at least two contrasting themes in the transport studies presented here. One group of investigators, consisting mainly of general physiologists and biophysicists, more or less tacitly consider animals primarily as manufacturers of suitable tissues or organs for the study of certain general transport mechanisms. What such investigators (I am one of them) look for and what, consequently, they usually find, are phenomena and mechanisms which are common to many or all living things. That approach has been very successful, but by its very nature it tends to disregard or play down the profound differences which may exist between different types of animals.

The other group of investigators are interested primarily in the biological role of different transport processes and are therefore unwilling to perform experiments under conditions which are outside the natural range of environmental variation. The dialogue between these groups of investigators is often difficult; but, even unwillingly, they can do each other great services by designing experiments and making observations which the workers from the other camp would not have considered worth while but which nevertheless open new lines of thought.

I.   Current Approaches

1. Microsample Analysis

C. Little

Department of Zoology, University of Bristol, Bristol, England

INTRODUCTION

Early methods applied to the study of the chemical composition of animals required several ml of sample so that analyses were limited to large animals. Improvements came first in the accuracy and sensitivity of methods used for mammalian body fluids because of their relevance to human medical problems. Many of these methods were not applicable to invertebrates, and especially not to marine forms. Robertson and Webb (1939) were the first to produce an accurate series of methods for marine animals, using "micro-estimations". These required 1 ml of sample. Since then, many methods have been developed, and the scale of operations has been reduced progressively from ml to μl to nl. Two particular fields of investigation have required the development of these micromethods. Vertebrate renal physiologists, striving to understand the functioning of kidney tubules, have been forced to adapt their methods to working with nl volumes; while invertebrate physiologists studying comparable renal systems or investigating ionic regulation in small animals have had to adapt first to μl and then to nl volumes. The analytical procedures themselves are not limited by the available sample volume *per se*, but by total quantity of the substance to be measured. Nevertheless, it is least confusing in general to discuss the methods in terms of the volume required because it is this which determines the handling procedure. I am not aware of any methods which are able to operate when volumes are reduced to pl (i.e. 10^{-12} l): this is probably related to the difficulties of manipulating such small volumes without contamination.

It is not possible in this article to cover the whole range of micro-analytical techniques, and the selection must be a matter of personal choice. I shall not deal with radioactive techniques or neutron activation analysis; nor shall I consider the vast field of the analysis of organic compounds.

II. OSMOTIC PRESSURE

The earliest micro-methods to be developed were concerned with the measurement of the colligative properties of solutions, usually by the observation of depression of freezing point (Drucker and Schreiner, 1913), or by measurement of the vapour pressure of the fluid in question (Baldes, 1934). When Krogh's book was published, (1939) volumes as small as 0·1 to 1·0 μl could be used. Since this time, many variations of these methods have been used, but the greatest advance has been the development of freezing-point methods which can be used routinely with samples of 0·1 to 1 nl. Such a method was originally described by Ramsay (1949), and a similar technique was later published by Hargitay et al. (1951). The method of Ramsay was developed by Ramsay and Brown (1955) and has been so successful that several instruments based on it are now on the market. All of these depend on the determination of melting point rather than freezing point because this overcomes the problem of supercooling. While Ramsay's method is ideal because it is an equilibrium method, the commercial instruments mainly operate by allowing the temperature to rise at a fixed rate and noting the temperature at which the last part of the sample melts. This mode of operation is quicker and easier, but could produce results of doubtful validity if solutions with different protein contents and surface tensions are used. A return to equilibrium methods would surely be a change for the better.

Two further developments may now be required in this field. One is a reduction in scale to allow estimation of freezing point in yet smaller volumes. This should not present a great problem, given better optical arrangements and methods for preventing salts from leaching out of the capillary glass. Such a method might be relevant, for instance, in the investigation of contractile vacuoles, or of the "formed bodies" described by Riegel (1966). The other possible development is an increase in accuracy of measurement, such as is required in the comparison of blood and pericardial fluid in molluscs. Here it is desirable to measure differences of the order of 10 mOsm/l in a total of about 1000 mOsm/l. One approach would be to use a

Ramsay-type apparatus, which involves a large volume of cooling liquid, and to introduce a more sensitive temperature-measuring device than the usual Beckmann thermometer. Since there is reason to believe that the large thermal capacity of the cooling bath used by Ramsay and Brown is an advantageous feature, such an approach might be more valuable than one using a miniaturized bath.

Much of the work on microsample analysis has been concerned with the component cations and anions which contribute to the total osmotic pressure. It can be said that as yet there are very few methods which, at equivalent sample volumes, allow the estimation of individual ions with an ease or accuracy such as obtained with Ramsay's apparatus for freezing point. It is now possible, however, to estimate many cations and anions in volumes of the order of nl, and there appear to be further possibilities emerging from developments in inorganic chemistry.

III. CATIONS

Most physiologists wishing to estimate cations are initially satisfied if they can measure the concentrations of the four most abundant ions — sodium, potassium, calcium and magnesium. Early methods involved specific titrations for individual ions, and in 1950, when Kirk's book "Quantitative Ultramicroanalysis" was published, it was possible to determine each of the above cations by such methods in volumes of a few μl. All these methods, however, were complex and time-consuming, in particular because they each involved different techniques. In 1955 Shaw published an account of procedures which allowed calcium, magnesium, sodium and potassium all to be estimated by chloride titrations, thus greatly reducing the time and variety of techniques involved. It was the introduction of flame photometry, however, which caused a revolution in chemical analysis, because it offered the possibility of measuring the concentration of a number of ions by basically similar methods. For the monovalent ions, flame photometry soon became routine despite interference problems, and a large number of commercial flame photometers became available. Some attempts to counteract interferences were made by using internal standards such as lithium, but in general such procedures do little more than provide a spurious feeling of accuracy. Better results are to be obtained by attempting to include interfering ions in the calibrating standards. Sensitivities capable of dealing with volumes of the order of nl have not been forthcoming in normal flame photometers, which utilize much diluted samples and measure emission from a steady flow of sample through the flame. Great increases in sensitivity were, however, produced when single samples were injected into a flame and a measurement was made of the total emission from that sample. Such "integrating" flame photometers were described by Ramsay (1950) (for details see Ramsay et al., 1951), and by Müller (1958), and allowed determination of sodium and potassium in volumes of approximately 1 nl. The drawbacks to these methods were inherent instability and large overall standard errors. However, later commercial development of the electronics allowed such errors to be reduced. Interference errors still remain, and are difficult to estimate or to overcome completely.

An alternative emission method using a helium glow photometer was described by Vurek and Bowman (1965), and has been used extensively in the analysis of fluid from kidney tubules. In this method the sample is placed on an iridium wire in a stream of helium. Electrical heating of the wire generates very energetic metastable helium atoms which transfer their energy to the atoms of the sample, and thus excite them. The light emitted is then recorded by photocells and the total emission is integrated. Sensitivities for sodium and potassium are in the picomole range, i.e. of the same order as integrating flame photometers.

Atomic absorption spectrophotometry is the third widely used method for measuring cations. In the case of sodium and potassium, the sensitivity of flame absorption methods is not generally increased in comparison with flame emission, but spectral interference is to a great degree eliminated. Following the initial design of Russell et al. (1957), many varieties of flame atomic absorption spectrophotometer have become available. These allow routine analysis of samples with volumes of the order of 1 μl, after dilution to 1–2 ml, for a very wide variety of elements. A number of non-flame absorption methods have recently been described, and these allow analysis of volumes comparable to those employed by integrating flame photometers and helium glow photometers. The general principle is to use an electrically-heated carbon furnace instead of a flame (e.g. L'vov, 1961; Massmann, 1968; West and Williams, 1969; Montaser et al., 1974). The sample is placed in the furnace which is programmed to heat in three steps: first to dry the sample, then to ash it, and finally to atomize it. The transient absorption peak may be measured by a rapid-response recorder, or the area under the curve may be measured by an integrating voltmeter. Once again, sensitivities are of the order of picomoles.

All three of the above methods for estimating sodium and potassium can now be used for measuring divalent ions. Formerly calcium and magnesium were usually determined by EDTA or EGTA titrations on samples of $1-10$ μl. In these titrations (e.g. van Asperen and van Esch, 1956), total calcium and magnesium is determined using an indicator such as eriochrome black T, and calcium is determined using murexide. Magnesium is therefore estimated by difference. The advent of flame atomic absorption allowed measurement of calcium and magnesium individually in volumes of approximately 1 μl, with vastly increased speed and convenience. Provided that lanthanum is used as a "releasing agent", interference from anions such as phosphate and bicarbonate can be eliminated. Another great advantage of flame atomic absorption is that a very wide range of elements can be investigated using basically similar preparation techniques. Non-flame atomic absorption methods as described above can also be used to measure divalent ions, and sensitivities are then of the same order as those achieved for monovalent ions. Similarly, adaptations of ultramicro flame photometers (Haljamäe and Wood, 1971) and of helium-glow photometers (Vurek, 1967; Marcus and Jamison, 1972) for use with higher temperatures allow the determination of calcium and magnesium, and some other metals, in picomole quantities. As in many micro-methods, the accuracy of these techniques could be improved, and a number of significant advances have recently been made. In atomic absorption, for instance, background correctors are now available, and these can be used to eliminate much interference. More accurate results are to be obtained with carbon furnace techniques when the temperature of the furnace is accurately controlled, and this is now possible (Lundgren et al., 1974). Multi-element detectors such as the vidicon tube are now being developed (e.g. Busch et al., 1974) so that many elements can be determined simultaneously. Many other improvements and developments are likely in these fields, and the determination of picomole quantities should soon be relatively easy for a wide variety of cations. In particular, the development of non-flame methods is likely to increase because of the lower background levels associated with electrical heating as compared with flames (see, e.g. Layman and Hiefte, 1975).

An alternative approach to the direct photometric methods considered above is to form a fluorescent complex with the cation in question, and to measure this fluorescence. Such an approach has been successful with calcium in 80 nl samples using calcein (Duarte and Watson, 1967), and with magnesium in 5 nl samples using

N,N'-bis-salicylidene-2-3-diaminobenzofuran (Brunette *et al.*, 1974; Brunette and Crochet, 1975).

Thus far, only analysis of samples which have been withdrawn from organisms has been considered. In recent years microelectrodes have been developed which are capable of measuring ion activity within cells. These electrodes have been of two types: one uses ion-selective glass and has been used mainly for sodium, potassium and pH; the other uses liquid ion-exchange membranes, and potentially at least has a wider application including divalent cations. The measurement of sodium and potassium inside cells using glass electrodes is now well established. Since Hinke's original descriptions in 1959, many developments have been made (see, e.g. Thomas, 1970; reviewed by Lev and Armstrong, 1975), and although there are some problems of selectivity, there seems to be no reason why they should not be used on microsamples withdrawn from the organism. This has rarely been attempted, but measurements by Khuri (1969) show that results are comparable to those obtained by photometric methods. The present state of liquid ion-exchange electrodes for measuring calcium is not so satisfactory. Commercially available electrodes require at least 10 μl, and selectivity appears to be more of a problem than with glass electrodes. No really satisfactory microelectrodes for either calcium or magnesium have been described, although preliminary work has shown that it is technically possible to construct liquid ion-exchange electrodes on a micro scale (Orme, 1969; Walker, 1971). Interference from potassium and from magnesium, however, make these electrodes of little practical use. With regard to microelectrodes in general, it must be emphasized that they determine the activity of an ion, and normally this is what is required by physiologists. If total concentration is required, recourse must be had to sampling and analysis by some process such as atomic absorption.

Yet another approach for measuring intracellular activity has been used by Ashley and Ridgway (1970), who determined free intracellular calcium levels by injecting the calcium-activated luminescent protein aequorin into muscle cells and measuring the light emitted during contraction. A similar protein has now been extracted from the hydroid *Obelia geniculata*, and should be more widely available (Campbell, 1974). However, this type of method is likely to be restricted to the measurement of calcium.

Perhaps the ultimate in cellular analysis may be provided by electron microprobe X-ray analysis, which is capable of estimating ion concentrations in various parts of sectioned tissues and cells.

This is dealt with elsewhere in this volume (see, Chapter 4) but one important application to microsample analysis must be mentioned here. Morel *et al.* (1969), working on fluid from renal tubules, used nl samples which were deposited on a cold beryllium block and then freeze-dried. Analysis of these dried samples with an electron probe proved very satisfactory, and since electron probes are likely to become relatively common instruments, such a procedure may have much to recommend it.

IV. ANIONS AND pH

The analysis of anions in small sample volumes has shown little tendency for the development of any overall method to emerge equivalent to the photometric methods used to determine cations. The three major anions normally considered to make up a substantial fraction of the osmotic pressure are chloride, bicarbonate and sulphate. Of these, the estimation of chloride is by far the most satisfactory at present. Wigglesworth (1938) first used the Volhard titration to determine chloride in volumes of 0.3 μl, and Cunningham *et al.* (1941) described a potentiometric method requiring several μl of sample. Shaw described an electrometric titration in 1955, and in the same year Ramsay *et al.* described two methods using electrometric titrations. One of these uses approximately 1 μg of chloride (normally a volume of about 1 μl), and is a standard silver nitrate titration using a potentiometric end-point. The second method uses volumes of the order of nl. In this method silver ions are added not from silver nitrate but by passing a current through a fine silver electrode: the electricity passed is used to charge a bank of condensers, and the total is recorded when the titration has been completed. Since 1955 a major development in chloride analysis has been the design of chloride-sensitive microelectrodes (Kerkut and Meech, 1966; Neild and Thomas, 1973). These have been used effectively inside nerve cells, but have not been applied to micropuncture samples. An entirely different approach has been employed by Haljamäe and Wood (1971). Using nl volumes, they precipitated the chloride with silver, and then estimated the silver remaining in solution with an integrating micro-flame photometer. This method has the advantage that it requires procedures very similar to those used for cation analysis, so that five ions can be measured by virtually the same technique.

The measurement of bicarbonate is complicated by the equilibria between $HCO_3{}^-$, H_2CO_3, $CO_3{}^{--}$ and free CO_2, which are affected by pH. Various methods have been used, usually involving the measurement of two variables. Calculation of the remaining variables is then possible using the Henderson—Hasselbalch equation, always provided that the relevant dissociation constants have been established for the fluid in question. The variables normally measured are total CO_2, pCO_2 and pH. With sample volumes of the order of 50 μl, most routine measurements are now made by measuring pH with a glass electrode and pCO_2 with a carbon dioxide electrode such as that developed by Severinghaus and Bradley (1958). Alternatively, pH may be measured when the sample has been equilibrated with two or more known pCO_2 levels (the Astrup method). The pH of the fluid *in situ* is also measured, and the original pCO_2 can then be calculated. This equilibration method has recently been adapted for use with nl samples by Karlmark and Sohtell (1973). In this method a droplet is placed in liquid paraffin which is then equilibrated with a range of known pCO_2 values. The pH of the sample is then measured with an antimony microelectrode, and the original pCO_2 *in vivo* can be obtained. The concentration of $HCO_3{}^-$ can be calculated from the Henderson—Hasselbalch equation.

The third parameter which may be measured, namely total CO_2, has until recently been measured most conveniently by Conway's (1962) microdiffusion technique. This latter is perhaps the most widely applicable of all methods which are used to estimate anions. As originally described it needs 0·1 to 1 ml, and can be used for chloride, total CO_2, NO_3—N, NH_3, Br, and I. The principle is to place a sample in a dish which contains two compartments, to cover the dish, and then to volatilize the substance under investigation. The gas evolved is trapped in a solution in the second compartment and estimated by titration. For total CO_2, Little and Ruston (1970) devised a method of microdiffusion necessitating approximately 1 μl of sample and using a freezing-point determination instead of a titration. Little (1974) reduced the scale of operations to approximately 1 nl by drawing appropriate volumes of liquids into a capillary tube and allowing the microdiffusion to take place through liquid paraffin instead of through air. In this way the diffusion of water vapour is reduced to a minimum while the diffusion times are also reduced because of the very small distances involved. This technique is simple to use, but as in other methods which measure total carbon dioxide it is necessary to draw samples under mercury and to keep them so until analysis because of the rapid rate of equilibrium with the carbon dioxide dissolved in the liquid paraffin.

It is evident that the discovery of a transparent fluid which is impermeable to CO_2 would here be of the greatest value. A recent alternative method of measuring total CO_2 has been devised by Vurek *et al.* (1975). This method makes use of the liberation of heat when CO_2 is allowed to react with anhydrous lithium hydroxide. Samples of 10—15 nl are acidified and the CO_2 evolved is transferred by a carrier gas to a small crystal of lithium hydroxide placed between thermistors. The change in temperature of the crystal is monitored and the total heat output is then integrated. This is directly proportional to the CO_2 evolved from the sample.

The measurement of pH still presents many technical problems, although the development of glass electrodes has certainly improved the situation. One of the major problems is that if CO_2 is allowed to enter or leave the fluid, the pH will change. Measurement of pH *in situ* is therefore to be preferred, but this is not always either easy or possible. When samples of the order of 50 μl are withdrawn from an organism, pH can be measured with capillary glass electrodes since the surface : volume ratio of such samples is relatively small, and little CO_2 can diffuse in or out. Ramsay (1956) attempted to develop a capillary glass electrode using 0·5 μl samples, but with volumes of this size kept under liquid paraffin it is likely that CO_2 will diffuse in (or out) before measurements can be made. Other types of electrode, using the quinhydrone system, or a variety of metal electrodes such as tungsten and antimony, could be used, but their specificity is not as high as that of glass electrodes, and the problem of CO_2 diffusion remains. Once again, a transparent liquid impermeable to CO_2 is required.

Recently glass pH electrodes have been developed for intracellular measurements. Initial studies by Caldwell (1954, 1958) produced electrodes small enough to be used inside crab muscle fibres, although when these electrodes were used in seawater or in crab saline they showed a slow drift towards the pH as finally read. This drift is also found with the currently available glass electrodes utilizing volumes as large as 10 μl, and may be due to the weak buffering capacity of sea water. Thomas (1974) has described a glass electrode which is small enough to be used inside snail neurones, and needs to penetrate only 100 μm.

The final major anion to be investigated is sulphate, and at the present time the methods of analysis of this ion are undoubtedly far more unsatisfactory than those of the others mentioned. Titrimetric methods and gravimetric methods usually require volumes of the order of 1 ml. Conductimetric methods (e.g. Roach, 1963) require volumes of the order of 100 μl, and have a low accuracy when employed in such liquids as sea water because the changes in conductivity are small as a percentage of the total conductivity (Little, 1967). Spencer (1960) described two spectrophotometric methods for use with 10 μl samples. One of these uses the standard method of precipitation as benzidine sulphate, the other the release of chloranilic acid from barium chloranilate. These methods are of use with only a very limited number of liquids, since even low concentrations of calcium give rise to impossibly low recoveries. The recent development of sulphate electrodes, as with those for other ions such as nitrate, utilize ion exchangers held in hydrophobic membranes. As yet these electrodes require relatively large volumes — of the order of 50 μl — and suffer from interference from other anions. One method of analysis which seems likely to produce sensitive and accurate measurements is the reduction of sulphate to sulphide and subsequent analysis of the sulphide. It is unfortunate that rather violent reducing agents are required, making the operation somewhat time-consuming and hazardous (see, e.g. the method of Davis and Lindstrom, 1972), since sulphide can be determined in very small quantities (e.g. Strickland and Parsons, 1968). If a convenient reduction method were to be found, it might then be possible to volatilise the sulphide as H_2S and to use the principle of microdiffusion.

V. NITROGENOUS COMPOUNDS

Many other components of animal body fluids are often investigated, but attention has perhaps been focussed in particular on the determination of nitrogenous compounds. A wide variety of methods may be used, involving specific ion electrodes, chromatography, enzymatic methods, radio-active labelling and direct colorimetry, but the method of widest application to small volumes is the reduction of the nitrogen to ammonia, and estimation of this by micro-diffusion. A procedure for the estimation of total nitrogen and protein nitrogen, using a Kjeldahl digestion procedure, has been outlined by Shaw and Beadle (1949). This method uses fluid volumes of 1 μl or less, and employs a standard titration procedure following the microdiffusion. Shaw and Staddon (1958) introduced a conductimetric method of measuring the quantity of ammonia which was volatilized from the sample. This method can estimate down to 1×10^{-3} μg of nitrogen, and the authors suggested that with modifications the limit could probably be lowered still further. The method of Karlmark (1973), designed for use with nl samples, can measure less than 1×10^{-4} μg of nitrogen. This method employs formaldehyde to release hydrogen ions from ammonium ions, and then estimates the hydrogen ions by titration with hydroxyl ions liberated from an antimony trioxide electrode. Because formaldehyde combines with amino acids as well as with ammonium ions, the method estimates a total of ammonium and amino nitrogen.

VI. FUTURE DEVELOPMENTS

It is impossible in a short space to discuss the microanalysis of the many other substances of interest to the physiologist and biochemist, but it is relevant to speculate upon possible new developments in the whole field of microsample analysis. Evidently the microsampling physiologists and the analytical electron microscopists are coming to a common meeting ground with the development of electron microprobe X-ray analysis (see Chapter 4). Besides this, physiologists are increasingly having to absorb evermore sophisticated chemical techniques, with the expansion of such methods as atomic absorption spectrophotometry. Despite these developments in the direction of expensive and highly sophisticated equipment, there is still room for the introduction of relatively simple techniques, especially for the estimation of anions, utilizing volumes of the order of nl. Two particular aspects may be discussed. Some techniques are best suited for use with small volumes but high concentrations — for example the electrometric titration of chloride. Others are better adapted to large volumes of low concentration — for example the recently introduced technique of anodic stripping voltametry which can be used to measure some metals at concentrations of the order of 10^{-12} M/l. In order for these latter type techniques to be used by the microanalyst, volumes of the order of nl must be accurately diluted. The type of manipulation described by Ramsay (1950), utilizing fine silica capillaries in combination with a mercury reservoir can be used but is extremely fragile. It dispenses fixed volumes which are not known accurately but which are repeatable, as is the case with much micro-work, in which methods are usually calibrated using the same dispensers both for standards and for unknown solutions. Fixed-volume self-adjusting pipettes as small as 1—5 nl, constructed from silica capillaries, have been described by Prager *et al.* (1965), and in a more robust form by Riegel (1970). This type of pipette is accurate to approximately 1%. The system of Little (1974) is used for drawing up liquids and operates well at the level of several nl. It utilizes a micrometer and can therefore deal with variable volumes. A system is now required which will dispense volumes of the order of nl or less with a very high accuracy.

The development of methods which deal with the alternative approach — that is, the analysis of relatively high concentrations in small volumes is in an advanced stage with reference to cations. It is in the field of anion analysis that developments must occur. Here a number of alternatives suggest themselves. Colorimetric methods are notoriously unreliable, and are often subject to interference. There is also the problem of constructing special spectrophotometers to accommodate small samples, and although this is not impossible, it is likely to prove expensive. This type of method may also be made possible by use of fluorescent complexes, as already achieved with the cations calcium and magnesium. Perhaps a more likely approach is a more general application of the microdiffusion technique which has already been shown to be readily adaptable to small volumes. Such an application might allow a wider variety of anions to be determined by similar methods. A development on these lines, paralleling the spectrometric analyses of cations, would be a most useful one.

ACKNOWLEDGEMENT

I would like to thank Dr A. E. Dorey for his comments on this paper.

REFERENCES

Ashley, C. C. and Ridgway, E. B. (1970). *J. Physiol.* **209**, 105–130.
Asperen, K. van and Esch, I. van (1956). *Archs néerl. Zool.* **11**, 342–360.
Baldes, E. J. (1934). *J. sci. Instrum.* **11**, 223–225.
Brunette, M. G. and Crochet, M. E. (1975). *Analyt. Biochem.* **65**, 79–88.
Brunette, M. G., Vigneault, N. and Carriere, S. (1974). *Am. J. Physiol.* **227**, 891–896.
Busch, K. W., Howell, N. G. and Morrison, G. H. (1974). *Analyt. Chem.* **46**, 575–587.
Caldwell, P. C. (1954). *J. Physiol.* **126**, 169–180.
Caldwell, P. C. (1958). *J. Physiol.* **142**, 22–62.
Campbell, A. K. (1974). *Biochem. J.* **143**, 411–418.

Conway, E. J. (1962). "Microdiffusion Analysis and Volumetric Error". Crosby Lockwood and Son Ltd., London.

Cunningham, B., Kirk, P. L. and Brooks, S. C. (1941). *J. biol. Chem.* **139**, 11—19.

Davis, J. B. and Lindstrom, F. (1972). *Analyt. Chem.* **44**, 524—532.

Drucker, C. and Schreiner, E. (1913). *Biol. Zbl.* **33**, 99—103.

Duarte, C. G. and Watson, J. F. (1967). *Am. J. Physiol.* **212**, 1355—1360.

Gupta, B. L., Moreton, R. B. and Hall, T. A. (1977). Electron microprobe X-ray analysis. This volume.

Haljamäe, H. and Wood, D. C. (1971). *Analyt. Chem.* **42**, 155—170.

Hargitay, B., Kuhn, W. and Wirz, H. (1951). *Experientia* **7**, 276—278.

Hinke, J. A. M. (1959). *Nature, Lond.* **184**, 1257—1258.

Karlmark, B. (1973). *Analyt. Biochem.* **52**, 69—82.

Karlmark, B. and Sohtell, M. (1973). *Analyt. Biochem.* **53**, 1—11.

Kerkut, G. A. and Meech, R. W. (1966). *Comp. Biochem. Physiol.* **19**, 819—832.

Khuri, R. N. (1969). *In* "Glass Microelectrodes" (Eds M. Lavallée, O. F. Schanne and N. C. Hébert), J. Wiley and Sons. Inc., London.

Kirk, P. L. (1950). "Quantitative Ultramicroanalysis". Wiley, New York and London.

Krogh, A. (1939). "Osmotic Regulation in Aquatic Animals". Cambridge University Press, London.

Layman, L. R. and Hiefte, G. M. (1975). *Analyt. Chem.* **47**, 194—202.

Lev, A. A. and Armstrong, W. McD. (1975). *In* "Current Topics in Membranes and Transport" (Eds F. Bronner and A. Kleinzeller), Vol. 6. Academic Press, N.Y. and London.

Little, C. (1967). *J. exp. Biol.* **46**, 459—474.

Little, C. (1974). *J. exp. Biol.* **61**, 667—675.

Little, C. and Ruston, G. (1970). *J. exp. Biol.* **52**, 395—400.

Lundgren, G., Lundmark, L. and Johansson, G. (1974). *Analyt. Chem.* **46**, 1028—1031.

L'vov, B. V. (1961). *Spectrochim. Acta* **17**, 761—770.

Marcus, D. and Jamison, R. L. (1972). *Analyt. Chem.* **44**, 1523—1525.

Massmann, H. (1968). *Spectrochim. Acta* **23B**, 215—226.

Montaser, A., Goode, S. R. and Crouch, S. R. (1974). *Analyt. Chem.* **46**, 599—601.

Morel, F., Roinel, N. and Le Grimellec, C. (1969). *Nephron* **6**, 350—364.

Müller, P. (1958). *Exp. Cell Res. Suppl.* **5**, 118—152.

Neild, T. O. and Thomas, R. C. (1973). *J. Physiol.* **231**, 7P—8P.

Orme, F. W. (1969). *In* "Glass Microelectrodes". (Eds M. Lavallée, O. F. Schanne and N. C. Hébert), J. Wiley and Sons Inc., London.

Prager, D. J., Bowman, R. L. and Vurek, G. G. (1965). *Science* **147**, 606—608.

Ramsay, J. A. (1949). *J. exp. Biol.* **26**, 57—64.

Ramsay, J. A. (1950). *J. exp. Biol.* **27**, 407—419.

Ramsay, J. A. (1956). *J. exp. Biol.* **33**, 697—708.

Ramsay, J. A. and Brown, R. H. J. (1955). *J. sci. Instrum.* **32**, 372—375.

Ramsay, J. A., Falloon, S. W. H. K. and Machin, K. E. (1951). *J. sci. Instrum.* **28**, 75—80.

Ramsay, J. A., Brown, R. H. J. and Croghan, P. C. (1955). *J. exp. Biol.* **32**, 822—829

Riegel, J. A. (1966). *J. exp. Biol.* **44**, 379—385.

Riegel, J. A. (1970). *Comp. Biochem. Physiol.* **35**, 843—856.

Roach, D. K. (1963). *J. exp. Biol.* **40**, 613—623.

Robertson, J. D. and Webb, D. A. (1939). *J. exp. Biol.* **16**, 155—177.

Russell, B. J., Shelton, J. P. and Walsh, A. (1957). *Spectrochim. Acta* **8**, 317—328.

Severinghaus, J. W. and Bradley, A. F. (1958). *J. appl. Physiol.* **13**, 515—520.

Shaw, J. (1955). *J. exp. Biol.* **32**, 321—329.

Shaw, J. and Beadle, L. C. (1949). *J. exp. Biol.* **26**, 15—23.

Shaw, J. and Staddon, B. W. (1958). *J. exp. Biol.* **35**, 85—95.

Spencer, B. (1960). *Biochem. J.* **75**, 435—440.

Strickland, J. D. H. and Parsons, T. R. (1968). *Bull. Fisheries Res. Bd., Canada* **167**, 41—44.

Thomas, R. C. (1970). *J. Physiol.* **210**, 82P—83P.

Thomas, R. C. (1974). *J. Physiol.* **238**, 159—180.

Vurek, G. G. (1967). *Analyt. Chem.* **39**, 1599—1601.

Vurek, G. G. and Bowman, R. L. (1965). *Science* **149**, 448—450.

Vurek, G. G., Warnock, D. G. and Corsey, R. (1975). *Analyt. Chem.* **47**, 765—767.

Walker, J. L. (1971). *Analyt. Chem.* **43**, 89A—93A.

West, T. S. and Williams, X. K. (1969). *Anal. chim. Acta* **45**, 27—41.

Wigglesworth, V. B. (1938). *Biochem. J.* **31**, 1719—1722.

2. Micropuncture Techniques [*]

G. Giebisch

Department of Physiology, Yale University School of Medicine, New Haven, U.S.A.

I. INTRODUCTION

Our present state of knowledge as to how the kidney fulfils its main task of regulating the concentration and content of both ionic and non-ionic solutes in body fluids is based on two types of renal function studies. First, overall glomerular and tubular function can be evaluated by renal clearance methods (Levinsky and Levy, 1973). Secondly, experiments using micropuncture techniques can be carried out at the single nephron level. The clearance approach consists in comparing the amounts of water and solutes filtered at the glomerular level with those excreted during the same time in the final urine. Measurements of renal clearances provided the first quantitative information on the three elements of renal function, i.e. rates of glomerular filtrate formation, and rates of tubular reabsorption and secretion. Using such renal clearance methods, the amounts of solutes reabsorbed or secreted can be estimated as the difference between filtered and excreted moieties. This method depends critically on a reference substance that can be used for the accurate measurement of glomerular filtration rate and for transepithelial water movement. It is now well established that inulin fulfils these requirements since it is freely filtrable and neither reabsorbed nor secreted during its passage from the glomerular filtrate along the nephron (Marsh and Frasier, 1965).

Concerning the usefulness of renal clearance methods with respect to their incisiveness in defining specific tubular transport sites and

[*] Work in the author's laboratory has been supported by grants from the NIH and NSF.

transport mechanisms, several limitations are obvious. Since renal clearances measure overall nephron function, that is the sum of many individual epithelial transport functions operating simultaneously at specific tubular sites along the tubule, they can only rarely be used to provide detailed information on tubular transport sites and cellular mechanisms of solute and water transport.

Indeed, it has been customary to interpret clearance data with respect to tubular transport sites and mechanisms on the basis of information obtained by studies on individual functional units of the kidney, i.e. by studying single nephron function. In the following, the most frequently employed methods of studying glomerular and tubular function will be discussed and, briefly, their impact on present concepts of renal function outlined. From these considerations it will become apparent that beginning with the initiation of micropuncture methods, and the design of appropriate ultramicromethods which were pioneered by Professor Ramsay, such studies have been uniquely powerful in identifying and exploring individual transport sites along the tubule (Richards and Walker, 1937; Walker *et al.*, 1941; Windhager, 1968; Giebisch, 1972; Gottschalk and Lassiter, 1973). Also, by measuring directly the driving forces acting on individual solutes and water as they cross well-defined segments of the renal tubule, additional insights into the cellular mechanisms of renal tubular transport processes have been obtained (Giebisch and Windhager, 1973).

II. STUDIES ON SINGLE RENAL TUBULES

Figures 1 and 2 summarize some of the micropuncture methods which have been most successfully employed. In free-flow micropuncture experiments, tubular fluid can be collected from individual glomeruli and superficial tubules along the nephron. Both proximal and distal tubules can be approached from the surface but it is also possible to puncture medullary tubules, i.e. parts of Henle's loop and adjoining blood vessels in some rodents in which papillary structures are well developed (Gottschalk and Mylle, 1959; Gottschalk and Lassiter, 1973). Such samples of tubular fluid (TF) can be analysed for a large number of solutes and this information used to establish the concentration profile along the proximal and distal tubule as well as along Henle's loop and the collecting duct. By withdrawing fluid from known sites along the nephron and measuring its inulin concentration it is also possible to measure single nephron filtration

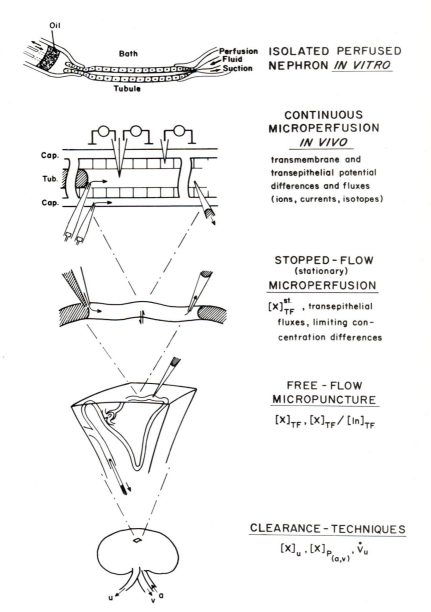

ISOLATED PERFUSED NEPHRON *IN VITRO*

CONTINUOUS MICROPERFUSION *IN VIVO*

transmembrane and transepithelial potential differences and fluxes (ions, currents, isotopes)

STOPPED - FLOW (stationary) **MICROPERFUSION**

$[x]_{TF}^{st.}$, transepithelial fluxes, limiting concentration differences

FREE - FLOW MICROPUNCTURE

$[x]_{TF}, [x]_{TF} / [In]_{TF}$

CLEARANCE - TECHNIQUES

$[x]_{u}, [x]_{P_{(a,v)}}, \dot{V}_{u}$

Fig. 1. Summary of methods for studying nephron function. (Hierholzer)

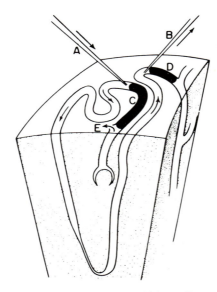

Fig. 2. Microperfusion experiment in the rat kidney. E: puncture site for escape of tubular fluid, A: perfusion pipette placed distal to the occluding oil block C, B: collecting pipette, placed proximal to the occluding oil block D. (From Marsh and Frasier, 1965.)

rate (Wright and Giebisch, 1972) and to quantify the rate of solute reabsorption or secretion, either in absolute or fractional terms, along precisely defined segments of the nephron. Such free-flow micropuncture studies have been widely used to map the site of many individual transport processes along the nephron (Gottschalk and Lassiter, 1973; Giebisch and Windhager, 1973).

Glomeruli can be easily punctured in amphibian kidneys but, with notable exceptions in some rat strains, are rarely present in sufficient numbers on the surface of adult mammalian kidneys. In studies in which single glomeruli have been successfully punctured in mammalian kidneys, the nature of the glomerular fluid as an ultrafiltrate of plasma has been firmly established, confirming earlier results obtained in studies on amphibian glomeruli (Harris et al., 1974). It has also been feasible to measure the hydrostatic pressure both in single glomerular capillaries and in the glomerular capsule (Brenner et al., 1971, 1974, 1976). It is thereby possible to define both the hydrostatic pressure difference that effects ultrafiltration as well as deduct the hydraulic permeability coefficient of the glomerular capillary filter directly from measurements of single nephron glomerular filtration rate and the hydrostatic pressure difference across the glomerulo-capillary membrane.

A number of microperfusion approaches have also gained wider use. All have in common a greater functional isolation of well-defined tubular segments. Stopped-flow, or stationary, microperfusion studies (see Fig. 1) are carried out by initially filling a nephron segment with stained oil, then splitting the oil with a test-droplet of perfusion fluid, and, in a final step, collect the perfusate and compare its composition with that of the initial perfusion fluid (Gertz, 1973; Kashgarian et al., 1963; Malnic et al., 1966b). Using this technique, net fluid reabsorption can be measured by observing the change of volume of the luminal fluid sample which gradually disappears as tubular fluid and solute reabsorption progresses. In addition, the use of special perfusion solutions has also allowed the evaluation of the steady-state distribution of ions across different segments of the nephron under conditions approximating complete cessation of net fluid movement. This is achieved by placing an initially electrolyte-free poorly penetrating non-electrolyte solution (raffinose, polyvinyl pyrrolidone) into the tubule. Subsequently, ions will diffuse into the initially electrolyte-free solution until a steady-state transepithelial concentration difference develops ("static-head" condition). The comparison of such transtubular ionic concentration differences with the transepithelial electrical potential difference provides the information necessary to decide whether a given ion species is transferred by active or passive transport mechanisms (Giebisch and Windhager, 1973; Kashgarian et al., 1963).

Microperfusion of single nephrons may also be carried out continuously as also shown in Figs 1 and 2 (Windhager et al., 1969; Frömter et al., 1973; Bank and Aynedjian, 1972). An important extension of such luminal perfusion methods is the additional simultaneous perfusion of peritubular capillaries surrounding perfused proximal, distal or collecting duct segments. This permits one to evaluate the effects of ionic substitutions and to test directly whether local electrolyte changes, drugs and hormones affect transport processes residing within well-defined tubular segments. Clearly, the measurement of electrical potential differences, measured either across the renal tubular epithelium, or across individual tubular cell membranes, is also of considerable importance in defining the transport mode of ions.

Finally, even more complete functional isolation of tubular segments has become possible by the dissection and the perfusion, *in vitro*, of mammalian (Burg and Orloff, 1973) and reptilian (Dantzler and Bentley, 1975) tubules. Above all, the development of these rather demanding techniques has made it possible to study those tubular segments which are not accessible for micropuncture or microperfusion from the kidney surface. This approach has been particularly useful in studying the transport properties of the mammalian collecting tubule and various parts of Henle's loop including its thick and thin limb. In addition to thus providing an opportunity of directly studying non-surface tubules, this technique of isolated tubular perfusion also allows one to measure cellular solute concentrations directly by tissue analysis.

Several types of nephrons have played a key role in studies using micropuncture techniques. One is the amphibian nephron, an example of which is shown in Fig. 3. Its main characteristics are its

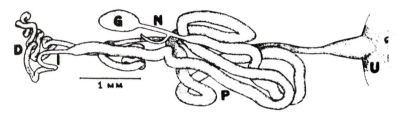

Fig. 3. Nephron of Necturus. G: glomerulus, N: neck, P: proximal tubule, I: intermediate segment, D: distal tubule, U: ureter. (From Kempton, 1937.)

relatively large size, the presence of a ciliated neck joining the glomerulus to the proximal tubule, the intermediate segment that connects the proximal tubule to the distal tubule, and collecting ducts that run at right angles to the tubules before joining the ureter. Some amphibian kidneys, such as those of Amphiuma, are character- ized by an abundance of distal convolutions on the surface (Sullivan, 1968).

In sharp contrast to amphibian and reptilian nephrons, some proximal tubules in birds and all nephrons in mammals are followed by a loop-like structure, the loop of Henle. Whereas amphibian and reptilian nephrons are mostly arranged in parallel to the kidney surface, the mammalian nephron and its blood supply is arranged in functional units which are oriented at right angles to the surface, such that the loop of Henle dips into the medullary part of the kidney (see Fig. 4). It is now well established that the hair-pin

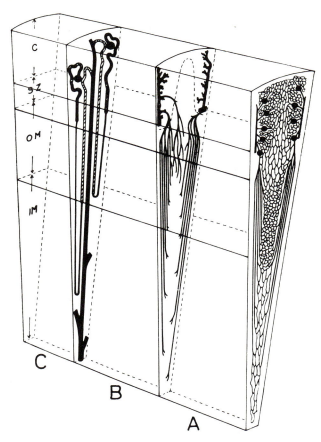

Fig. 4. Schematic representation of three sections of rat kidney in which three structural components are illustrated separately. A: the arterial vessels and the capillaries, B: the venous vessels, C: a collecting duct with a long and short loop of Henle. C = Cortex; SZ = Subcortical zone; OM = Outer zone; IM = Inner zone. (From Kriz, 1970.)

arrangement of the loop of Henle is functionally related to the ability of the kidney to produce an osmotically concentrated urine. It is interesting that some birds, particularly those endowed with the capacity of producing an osmotically concentrated urine, have kidneys containing a mixture of nephrons, consisting of some nephrons having a simple proximal and distal tubule without loops of Henle and being arranged at right angles to collecting ducts. In addition to these, there is an additional nephron population, having a

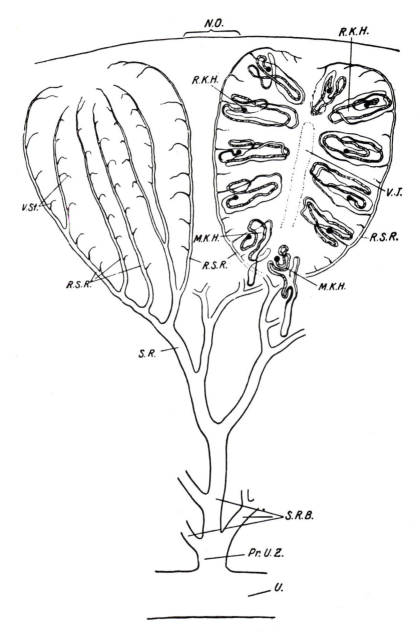

Fig. 5. Section of avian kidney showing arrangement of different types of nephrons. (Modified from Braun and Dantzler, 1975.)

loop of Henle, which is anatomically arranged in parallel to both *vasa recta* and collecting ducts (Fig. 4). Dantzler and his associates have studied the functional properties of these kidneys (see Fig. 5) and discovered interesting differences in the response of these two distinct nephron populations to water deprivation (Braun and Dantzler, 1975). Thus, in severe dehydration, only those nephrons lacking a loop structure, decreased their filtration rate. In contrast, those having a loop continued to function and were able to produce an osmotically concentrated urine.

It has become increasingly apparent, however, that even in mammalian kidneys there may exist significant differences in the behavior of various nephron populations. Figure 6 shows that the length and location within the kidney of various nephrons may vary widely. Not all superficial nephrons have a loop of Henle reaching

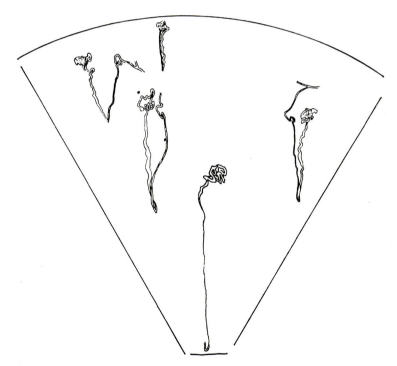

Fig. 6. Isolated nephrons dissected from the kidney of Aplodonta rufa. The nephrons are drawn in the relative position which they occupy in the kidney. Note nephrons with short loops and the absence of a thin segment in the cortical nephron. (From Nungesser and Pfeiffer, 1965.)

into the inner medulla. In accordance with such structural differences, some functional heterogeneity has begun to emerge. Deep
nephrons filter at a higher rate than superficial ones (Jamison, 1970;
Horster and Thurau, 1968), and the possibility exists that they
might, under some experimental conditions at least, also deliver
larger amounts of fluid and sodium chloride into the collecting ducts
than their superficial counterparts (Stein *et al.*, 1976).

III. RECENT APPLICATIONS OF MICROPUNCTURE AND MICROPERFUSION METHODS

The following is a brief outline of recent work at the single nephron
level which has given new insight into the function of the nephron.

A. Studies on the Determinants of Glomerular Filtration Rate

It is well known that a sizeable amount of fluid, some one-fifth to
one-third of the volume of blood plasma entering the renal
glomerulus, is filtered across its capillaries. Using micropuncture
techniques the composition of fluid collected from the glomerular
capsule has been found to be that expected from an ultrafiltrate, i.e.
one closely resembling plasma water with respect to low mol. wt.
solutes (Harris *et al.*, 1974). Filtration becomes restricted for solutes
having a greater mol. wt. than 5000, the size of inulin, and this
restriction becomes almost complete once the molecular size of
serum albumin is reached (Renkin and Gilmore, 1973).

Rate of glomerular filtrate formation is determined by the driving
forces that govern fluid exchange across capillary membranes, i.e. the
balance of forces between transglomerular hydraulic and colloid-
osmotic pressures.

Only recently has it become possible to measure and vary the
hydrodynamic driving forces acting across the mammalian glomerular
filter. Two advances have been instrumental in opening this area of
haemodynamic research. First, a mutant strain of Wistar Rats has
been discovered with fairly numerous glomeruli on the surface of the
renal cortex. This was an important advance as glomeruli are not
generally present on the surface of mammalian kidneys. A second
advance was the successful adaptation of Wiederhielm's servonulling
pressure measuring technique to the direct measurement of glomerular capillary and glomerular capsular pressure (Brenner *et al.*, 1971;
Brenner *et al.*, 1974; Deen *et al.*, 1972; Brenner *et al.*, 1972; Brenner
et al., 1976).

From these measurements, it has now become possible to evaluate the net driving force for ultrafiltration along the glomerular capillary circuit. The results of Brenner's group have led to a significant downward revision of the glomerular capillary pressure (approximately 45 mm Hg) as compared to previous, indirect estimates (Renkin and Gilmore, 1973). At the same time it has been shown that filtration pressure equilibrium can be reached during the transition from afferent to efferent end of the glomerular capillary. This observation suggests that glomerular filtration rate is importantly controlled by plasma flow rate. Using a mathematical model in which the glomerular capillary network is simplified to a single tube, and the conservation of volume and flow and Starling's hypothesis are employed, it has been possible to derive the rate of change of protein concentrations with distance along the glomerular capillary. In addition, it has now also become possible to derive values of K_f, the hydraulic filtration coefficient. The effects of specific hemodynamic perturbations on the hydrodynamic determinants of glomerular filtration have thus become the subject of direct experimental evaluation.

Another new and interesting area of research utilizing micropuncture techniques is that of possible feedback control of glomerular filtrate formation. Thus, the close anatomical proximity of the vascular pole of the glomerulus with the *macula densa* cells in the distal tubule of the same nephron as well as some functional relationship between distal tubular flow rate and glomerular filtration pressure have been taken as evidence of a feedback control system operating within single mammalian nephrons (Thurau and Mason, 1974; Schnermann *et al.*, 1970; Wright and Schnermann, 1974; Wright, 1974a).

Figures 7 and 8 depict the experimental approach used to provide supportive data for glomerular feedback control. Figure 7 shows the arrangement that permits measurement of glomerular stop-flow pressure during the perfusion, via a second micropipette, of the downstream tubular segment, including Henle's loop and the *macula densa* area. Using these microperfusion techniques, it was possible to measure proximal stop-flow pressure* while varying the rate of fluid

---

* Proximal stop-flow pressure is measured by filling a loop immediately distal to a glomerulus with viscous, stained mineral oil, and placing a pressure-measuring micropipet just proximal to the block. Pressure will rise above the oil block due to continued filtration until it stops filtration. This "stop-flow" pressure approximates glomerular capillary pressure minus colloid osmotic pressure. Since the latter is constant, tubular stop-flow pressure is an index of single nephron glomerular filtration rate.

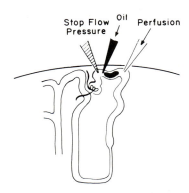

Fig. 7. Single nephron showing measurement of glomerular stop-flow pressure to study feedback control mechanism during loop perfusion. (Modified from Wright, 1974.)

delivery from the late proximal puncture site. Figure 9 depicts a more elaborate experimental microperfusion and collection arrangement that allows, during perfusion of Henle's loop and the *macula densa*, not only the collection (and analysis) of the fluid emerging from Henle's loop, but also the direct measurement of single nephron glomerular filtration rate. An additional pipette, in close proximity to the glomerulus, monitors tubular pressure and signals the occurrence of luminal pressure changes that could occur due to displacement of the separating oil column.

An extensive series of microperfusion experiments using the approaches summarized in Figs 7 and 8 have shown that an increase in flow rate through Henle's loop is followed by significant reductions in single nephron filtration rate and stop-flow pressure (Wright and Schnermann, 1974). This is demonstrated in Fig. 9 (see upper panel) in which the change of the perfusion rate from 10 to 40 ml/min is consistently accompanied by a significant fall in stop-flow pressure. This effect is not changed by the addition of either amiloride, acetazolamide or mercaptomerin to the perfusion solution. In sharp contrast, it is apparent from inspection of the lower panel of Fig. 9 that the same augmentation of tubular flow rate through Henle's loop and past the *macula densa* has become totally ineffective when sodium chloride reabsorption has been blocked by the addition, to the perfusion fluid, of (1) poorly reabsorbable salts or non-electrolytes (Na_2SO_4, mannitol) or (2) by diuretic agents which inhibit sodium chloride reabsorption along the

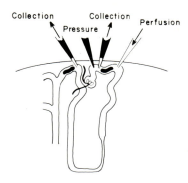

Fig. 8. Single nephron showing arrangement suitable to perfuse Henle's loop, collect early distal tubular fluid and measure single nephron filtration rate at constant tubular pressure. (Modified from Wright, 1974.)

thick ascending limb of Henle's loop, i.e. that nephron segment that constitutes the *macula densa* (Burg and Green, 1973a,b,c). These findings strongly support the hypothesis that the observed feedback regulation of glomerular filtration depends on the rate of tubular sodium chloride reabsorption by *macula densa* cells of the distal tubule. This is the most likely explanation of these perfusion experiments since, with enhanced flow rate through Henle's loop, net sodium transport increases proportionately. Since the diuretics that act at this nephron site also interfere with the feedback response, it is virtually certain that transport of sodium chloride (or some other constituent of distal tubular fluid) is a key factor in the feedback mediation (Wright, 1974a; Wright and Schnermann, 1974).

B. Studies on Various Aspects of Tubular Transport Mechanisms

A very large number of micropuncture and microperfusion studies see Fig. 10 have dealt with the mapping of transport functions along mammalian, reptilian and amphibian nephrons. Clearly, it will not be possible to discuss the results of the large number of studies in any detail within the context of this paper. Rather, an attempt will be made to highlight recent developments in which the skilful application of micropuncture and microperfusion methods has played an incisive role.

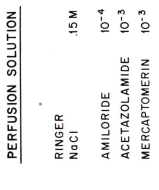

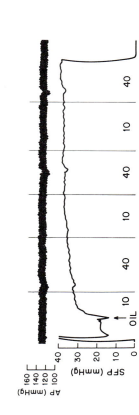

Fig. 9. Effects of loop perfusion rate, osmotic diuretics (Na_2SO_4, mannitol) and transport inhibitors on proximal stop-flow pressure (P_{SF}). Tracings show arterial pressure (AP) and proximal intratubular pressure during perfusion of the loop of Henle at 10,40 and 0 nl/min. (From Wright and Schnermann, 1974.)

Fig. 10. Principle of continuous microperfusion. After terminating collection with pipette No. 1, the tubule is punctured more proximally with micropipette No. 2 and a second sample is obtained. Note oil block proximal to the pump and small hole proximal to oil block which allows escape of glomerular fluid. (From Ullrich *et al.*, 1969.)

1. Proximal Tubule

This is the nephron site where the bulk of filtered water, salt and organic solutes are reabsorbed (Giebisch and Windhager, 1973). The key role of active sodium transport has been demonstrated in a number of coupled transport processes. Active sodium reabsorption is the primary osmotic driving force for fluid reabsorption (replacement by choline, lithium or tetraethyl ammonium as well as metabolic inhibitors abolish fluid movement) and, similar to other epithelia, it has become apparent that the interspaces between tubule cells provide both an important transport route as well as the site of osmotic coupling between active sodium and water transport (Giebisch and Windhager, 1973; Frömter, 1974; Sackin and Boulpaep, 1975; Burg and Green, 1976).

Several aspects of proximal tubular sodium transport have been clarified: they concern (1) ionic requirements of sodium transport such as its sensitivity to reduction of peritubular bicarbonate and potassium, (Ullrich *et al.*, 1971; Green and Giebisch, 1975a,b; Burg and Green, 1976); (2) its control by alterations in peritubular "physical factors", such as changes in peritubular hydrostatic and colloid-oncotic pressure (Windhager *et al.*, 1969; Windhager, 1973; Brenner and Troy, 1971); (3) the interaction of sodium during its entry across the luminal cell membrane with both glucose and amino acids, mediated by an electrogenic co-transport mechanism, where the "downhill" movement of sodium provides the energy for glucose accumulation in proximal tubule cells (Ullrich *et al.*, 1974a,b.); (4) the limiting nature of the luminal cell membrane for sodium transport as evidenced by an increase in both net rate of sodium transport and of cellular sodium content when substances known to increase membrane permeability, such as amphotericin, are given

(Stroup *et al.*, 1974; Spring and Giebisch, 1977), (5) the loose nature of coupling between tubular sodium and hydrogen ion transport at the luminal cell membrane (Green and Giebisch, 1975a,b) and a similarly loose coupling between peritubular sodium extrusion from cell to peritubular fluid and of potassium uptake in the opposite direction (Giebisch *et al.*, 1973).

Further studies have dealt with the functional heterogeneity of the proximal tubule. Recent studies dealing with some of the properties of the mammalian proximal tubular epithelium have clearly demonstrated that there are significant functional differences between "early" and "late" proximal tubular segments. Many transepithelial concentration differences, particularly those of chloride and bicarbonate (Walker *et al.*, 1941; Gottschalk, 1963; Le Grimellec *et al.*, 1975), but also those of phosphate, glucose and amino acids, are already established very early along the proximal convoluted tubule and stress the greater functional capacity of the earliest segment of the proximal tubule.

Functional differences have also been observed between the convoluted and straight sections of the proximal tubule. The proximal tubular convolution has a higher sodium than chloride permeability (Boulpaep and Seely, 1971; Frömter *et al.*, 1971; Frömter, 1974); the reverse is true with respect to the straight descending section of the proximal tubule (Schafer *et al.*, 1974). These two tubular sections also differ with respect to their bicarbonate-sensitivity to sodium transport. Deletion of bicarbonate from luminal and peritubular perfusion medium blocks a sizeable fraction of sodium and fluid reabsorption across the convoluted segment but not across the straight, descending part of the proximal tubule (Schafer *et al.*, 1974; Ullrich *et al.*, 1971; Green and Giebisch, 1975a,b). It is also the straight, descending portion of the proximal tubule that is endowed with the higher capacity to secrete organic acids like para-amino hippurate into the lumen (Burg and Orloff, 1973). In the studies dealing with the functional assessment of the straight portion of the proximal tubule, the isolated tubule preparation played again a key role since this nephron segment cannot be punctured from the surface.

2. *Loop of Henle*

The function of the loop of Henle has also been successfully explored by micropuncture and microperfusion techniques. Thus, many aspects of the countercurrent operation of osmotic urine concentration are based on the comparison of the composition of fluid collected from the descending and ascending limbs of Henle's loop, the collecting duct as well as of the two limbs of the vasa recta. This topic has been extensively reviewed (de Rouffignac, 1972; Wirz and Dirix, 1973; Jamison, 1974; Berliner, 1976) but it should be realized that some aspects of the concentrating process are still unsettled. This concerns particularly the role of active sodium chloride transport in the thin ascending limb of Henle's loop and the importance of solute addition versus water abstraction in the thin, descending limb of Henle's loop (Kokko and Rector, 1972; Stephenson, 1972). It is very likely that a careful analysis of solute and water movement in different segments of the loop structure will ultimately resolve this problem.

Another recent advance in our knowledge of the function of Henle's loop is based on studies on the isolated thick ascending portion of Henle's loop. Active chloride transport plays a major role in this nephron segment. In contrast to many other tubule segments, the lumen of the thick, ascending limb of Henle's loop is electrically positive with respect to the peritubular fluid and it can be shown to pump chloride ions actively from lumen to blood (Burg and Green, 1973a; Burg and Stoner, 1974). It is also of great interest that the presently most powerful diuretics such as furosemide, ethacrynic acid and mercurial components exert their diuretic action, from the lumen, by inhibiting active chloride reabsorption (Burg and Green, 1973a,b; Burg and Stoner, 1974) at this tubule site.

3. Distal Tubule and Collecting Duct

The function of the distal tubule has also been successfully explored by free-flow micropuncture. This nephron segment reabsorbs normally some 10% of the filtered sodium load (Giebisch and Windhager, 1964, 1973; Khuri *et al.*, 1975a) but responds to the increased delivery of sodium chloride with a powerful enhancement of salt reabsorption. It thus curtails sodium chloride loss whenever more proximally located nephron sites fail to transport sodium at their normal rate. The distal tubule is also the main site of tubular potassium secretion (Giebisch, 1971; Brenner and Berliner, 1973; Wright, 1974b; Giebisch, 1975; Grantham, 1976), and metabolic stimuli such as changes in potassium, sodium and acid-base balance as well as changes in the level of circulating mineralo—corticoids are translated at the late distal tubular level into the appropriate secretory or reabsorptive transport rate (Malnic *et al.*, 1964, 1966a,

1971; Giebisch, 1971; Wright *et al.*, 1971, 1974b; Brenner and Berliner, 1973; Khuri *et al.*, 1975b; Giebisch, 1975; Grantham, 1976). Thus, from micropuncture studies of the distal tubule this nephron site has clearly emerged as the main control site of potassium transport in the kidney.

On the basis of studies on the isolated, perfused cortical collecting tubule we have now a fairly extensive body of information concerning the transport properties of this terminal nephron segment. The cortical collecting tubule secretes potassium, continues the reabsorption of sodium and is exquisitely sensitive to the action of anti-diuretic hormones (Grantham *et al.*, 1970; Burg *et al.*, 1973). It is also an important site of control of sodium transport (Stein and Reineck, 1975) and is an additional site of action of mineral-ocorticoids (Hierholzer and Lange, 1974). It is of interest that there are significant differences in the urea permeability of the cortical collecting tubule and the papillary collecting ducts. Only the latter permit urea to leak from lumen into the interstitium. This process is under the control of antidiuretic hormones and is of significance for the appropriate supply of urea to the medullary countercurrent operation of osmotic urine concentration (Grantham, 1974).

C. *Acidification of the Urine*

Our knowledge of the role of various tubular segments in the renal regulation of acid-base balance has also been greatly advanced by application of micropuncture and microperfusion methods, in conjunction with the development of micro pH electrodes (Malnic and Giebisch, 1972; Malnic, 1974; Karlmark, 1974; Rector, 1973) and methods to measure titratable acid and ammonia in nanoliter samples. Figure 11 shows the main features and applications of the antimony electrode system. As shown at the top, the intratubular placement of the pH-sensitive and of a glass reference electrode allows for the instantaneous *in situ* measurement of the pH of tubular fluid. Measurement of pH values of a collected sample which had been equilibrated under oil at known CO_2-tensions allows calculations of tubular bicarbonate concentrations. Finally, it is also possible to record the pH of buffer samples such as bicarbonate and phosphate which had been isolated by oil from the remainder of the luminal fluid (Malnic *et al.*, 1971, 1972, 1974). This latter approach has been particularly useful to obtain quantitative information of the rate of acidification, and of steady-state pH differences across chosen

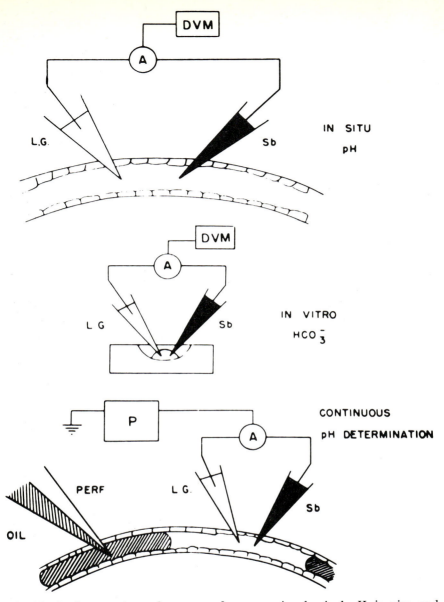

Fig. 11. Antimony electrode system for measuring luminal pH *in vivo* and bicarbonate concentrations *in vitro*. Top: *in situ* measurement. The voltage difference between the Sb microelectrode (Sb) and a glass micro-electrode serving as reference electrode is measured by a high-impedance voltmeter (DVM), Middle: measurement of bicarbonate concentration by pH determination of tubular sample under oil at known pCO2, Bottom: continuous pH measurement during stopped-flow micro-perfusion. (From Malnic, 1974.)

nephron segments in different experimental conditions. Advances have been made by measuring several kinetic aspects of tubular acidification in stationary microperfusion experiments. We now have a fairly comprehensive picture of the general properties of the hydrogen ion transport system, in terms of the topography of bicarbonate reabsorption, of ammonia secretion, of the capacity of segments of the proximal and distal tubule to secrete hydrogen ions and of the active transport and permeability components of tubular acidification (Malnic and Giebisch, 1972; Malnic, 1974; Rector, 1973).

C. Electrophysiological Methods

Electrophysiological methods have been used extensively during the last decade to study the behaviour of single nephrons and impressive progress has been made in this area (Boulpaep, 1971, 1976; Frömter *et al.*, 1971, Frömter, 1974). A large body of data now permits a precise definition of the electrochemical potential gradients across individual membranes of tubule cells. Figure 12 summarizes some useful approaches. It is possible to measure both peritubular transmembrane potential differences across single tubule cell mem-

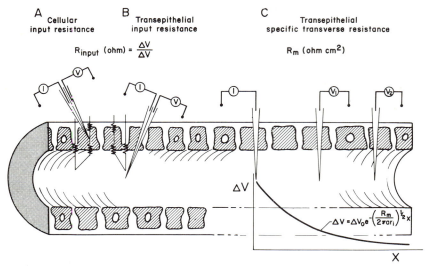

Fig. 12. Summary of electrophysiological methods used on single Nephrons. (From Grandchamp and Boulpaep, 1974.)

branes as well as transepithelial potential differences across the nephron. This is achieved by the application of standard Ling-Gerard type electrodes, although axial electrodes have also been employed (Spring, 1972). Clearly, the precise definition of electrical potential gradients across individual membranes of the tubule cell is crucial to the assessment of active and passive transport processes.

Electrophysiological methods have also been uniquely powerful in defining low-resistance coupling between the tubular lumen and the peritubular space. For such studies electrical resistance measurements have been made either by measuring effective resistance across cell membranes and the epithelium as a whole, or by measuring the cable properties of the tubular epithelium (Boulpaep, 1971, 1972, 1976a; Giebisch and Windhager, 1973). Specific transmembrane and transepithelial potential and resistance measurements during various ionic substitutions have also provided quantitative information dealing with the permeability of the tubule to ions (Boulpaep, 1971, 1976b; Frömter, 1974). Such an arrangement is depicted in Fig. 13 in which luminal and peritubular perfusion methods are combined with measurements of transepithelial potential differences. From the magnitude and direction of the diffusion potentials generated by selective ion substitutions such potential measurements have yielded information on tubular ion permeabilities. Such information is now available on almost all nephron segments (Boulpaep, 1976a,b).

Recent observations on electrophysiological effects of luminally adminstered glucose and amino acids have shown that the stimu-

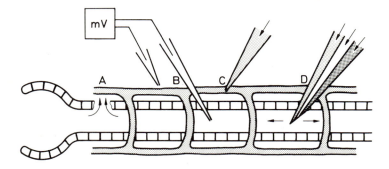

Fig. 13. Perfusion technique and electrical measuring technique as applied to single nephrons. A: drainage of glomerular filtrate, B: microelectrodes, C: peritubular perfusion pipette, D: triple-barreled luminal perfusion pipette. (From Frömter and Gessner, 1974.)

lation of sodium transport that results from the interaction of glucose and sodium at the luminal cell membrane is electrogenic, i.e. it contributes directly to the polarization of the proximal tubular epithelium (Frömter, 1974). Addition of either amino acids or glucose leads to a prompt and reversible potential shift making the lumen more negative. These sugar and amino acid-induced electrical potential changes are best explained by co-transport of sodium with glucose or amino acids across the luminal membrane of proximal tubule cells. It is assumed that translocation of sodium depolarizes the luminal cell membrane. Although current spread along low-resistance intercellular pathways also depolarizes the peritubular cell membrane, there is a significant shift of the transepithelial potential difference that renders the lumen electrically more negative. As a matter of fact, it now appears that in the proximal tubule only that fraction of sodium transport associated with sugar or amino-acid transport is "electrogenic", and renders the lumen electrically negative.

Evidence obtained from both transepithelial and transmembrane potential measurements has shown that several other tubular ionic transport processes are also directly electrogenic. Stated differently, stimulation of specific ion pumps can be shown to directly polarize tubular cell membranes. Thus, stimulation of sodium transport in distal tubules activates peritubular sodium-potassium exchange such that the flux ratio of sodium/potassium exceeds unity and that the cell becomes more negative (Wiederholt and Giebisch, 1974). It has also been shown by perfusion studies of isolated collecting tubules *in vitro* that both potassium (Boudry *et al.*, 1976) and hydrogen ion secretion (Burg and Stoner, 1974; Stoner *et al.*, 1974) are directly electrogenic. Activation of these transport processes leads to a shift of the luminal potential toward more positive values. These effects become apparent when reabsorptive sodium transport had been abolished. It is the latter transport process which normally plays a key role in the generation of the normally predominating luminal negativity. Hence, the overall transepithelial potential difference across the collecting tubule is generated by the complex interplay of directionally opposing electrogenic transport processes.

A method for electrophoretic injection of cations into single proximal tubule cells has also been developed and has provided some insight into the electrical properties of sodium transport (Tadokoro and Boulpaep, 1972). Figure 14 shows schematically the use of double-barreled microelectrodes for electrophoretic injection of various ion species and for recording the resulting electrical potential changes. Injection of sodium ions by accurately timed current

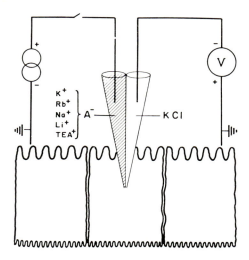

Fig. 14. Iontophoresis in Kidney Cells. Method for electrophoretic injection of cations into single proximal cells of Necturus kidney. The left barrel of the microelectrode contains a concentrated solution of a salt of the cation to be injected and is connected to a constant current source. The right barrel is filled with 3M KCl and serves to record the peritubular membrane potential difference. (From Tadokoro and Boulpaep, 1972.)

application has been shown to depolarize the membrane potential by increasing the cellular sodium and by decreasing cellular potassium. The amplitude of this depolarization is proportional to the duration of current injection. With time the potential repolarizes to the initial potential level. From the sensitivity of the repolarization process to metabolic inhibitors and cooling it is virtually certain that this repolarization is due to the activity of a sodium-potassium exchange pump that restores the intracellular potassium concentration. Under special conditions, i.e. when the energy barrier against which sodium transport takes place is reduced, it can again be shown that the peritubular cell membrane of single tubule cells hyperpolarizes, and is thus capable of electrogenic sodium extrusion.

REFERENCES

Bank, N. and Aynedjian, H. S. (1972). *Yale J. Biol. and Med.* **45**, 312–317.
Berliner, R. W. (1976). *Kidney Int.* **9**, 214–222.
Boudry, J. F., Stoner, L. C. and Burg, M. B. (1976). *Am. J. Physiol.* **203**, 239–244.

Boulpaep, E. L. (1971). *In* "Electrophysiology of Epithelial Cells" (Ed. G. Giebisch), pp. 119—146. F. K. Schattauer Verlag, Stuttgart.

Boulpaep, E. L. (1972). *Am. J. Physiol.* 222, 517—531.

Boulpaep, E. L. (1976a). *Ann. Rev. Physiol.* 38, 20—36.

Boulpaep, E. L. (1976b). *Kidney Int.* 9, 88—102.

Boulpaep, E. L. and Seely, J. F. (1971). *Am. J. Physiol.* 221, 1084—1096.

Braun, E. J. and Dantzler, W. H. (1975). *Am. J. Physiol.* 229, 222—228.

Brenner, B. M. and Berliner, R. W. (1973). *In* "Handbook of Physiology" (Eds J. Orloff and R. W. Berliner), Section 8, pp. 497—520. American Physiological Society, Washington, D.C.

Brenner, B. M. and Troy, J. L. (1971). *J. Clin. Invest.* 50, 336—349.

Brenner, B. M., Troy, J. L. and Daugharty, T. M. (1971). *J. Clin. Invest.* 50, 1776—1780.

Brenner, B. M., Troy, J. L., Daugharty, T. M., Deen, W. M. and Robertson, C. R. (1972). *Am. J. Physiol.* 223, 1184—1190.

Brenner, B. M., Deen, W. M. and Robertson, C. R. (1974). *In* "MTP International Review of Science" (Ed. K. Thurau), Physiology Series 1, Vol. 6, pp. 335—356. Butterworths, University Park Press.

Brenner, B. M., Deen, W. M. and Robertson, C. R. (1976). *Ann. Rev. Physiol.* 38, 9—19.

Burg, M. B. and Green, N. (1973a). *Kidney Int.* 4, 245—251.

Burg, M. B. and Green, N. (1973b). *Am. J. Physiol.* 224, 659—668.

Burg, M. B. and Green, N. (1973c). *Kidney Int.* 4, 301—308.

Burg, M. B. and Green, N. (1976). *Kidney Int.* 9, 189—197.

Burg, M. B. and Orloff, J. (1973). *In* "Handbook of Physiology" (Eds J. Orloff and R. W. Berliner), Section 8, pp. 145—159. American Physiological Society, Washington, D. C.

Burg, M. B. and Stoner, L. (1974). *Fed. Proc.* 33, 31—36.

Burg, M. B., Stoner, L., Cardinal, J. and Green, N. (1973). *Am. J. Physiol.* 225, 119—124.

Dantzler, W. H. and Bentley, S. K. (1975). *Am. J. Physiol.* 229, 191—199.

Deen, W. M., Robertson, C. R. and Brenner, B. M. (1972). *Am. J. Physiol.* 223, 1178—1183.

de Rouffignac, Chr. (1972). *Kidney Int.* 2, 297—303.

Frömter, E. (1974). *In* "MTP International Rev. of Science" (Ed. K. Thurau), Physiology Series I, Vol. 6, pp. 1—38. Butterworths, University Park Press.

Frömter, E. and Gessner, K. (1974). *Arch. ges. Physiol.* 351, 85—98.

Frömter, E., Müller, C. W. and Wick, T. (1971). *In* "Electrophysiology of Epithelial Cells" (Ed. G. Giebisch), pp. 119—146. F. K. Schattauer Verlag, Stuttgart.

Frömter, E., Rumrich, G. and Ullrich, K. J. (1973). *Pflügers Arch. ges. Physiol.* 343, 189—220.

Gertz, K. H. (1973). *Arch. ges. Physiol.* 276, 336—356.

Giebisch, G. (1971). *In* "The Kidney, Morphology, Biochemistry, Physiology" (Eds Ch. Rouiller and A. Muller), pp. 329—382. Academic Press, New York and London.

Giebisch, G. (1972). *Yale J. Biol. and Med.* **45**, 187−456.

Giebisch, G. (1975). *Yale J. Biol. and Med.* **48**, 315−336.

Giebisch, G. and Windhager, E. E. (1964). *Am. J. Med.* **36**, 643−669.

Giebisch, G. and Windhager, E. E. (1973). *In* "Handbook of Physiology" (Eds J. Orloff and R. W. Berliner), Section 8, pp. 315−376. American Physiological Society, Washington, D.C.

Giebisch, G., Sullivan, L. P. and Whittembury, G. (1973). *J. Physiol.* **230**, 51−74.

Gottschalk, C. W. (1963). *Harvey Lect. Ser.* **58**, 99−123.

Gottschalk, C. W. and Lassiter, W. E. (1973). *In* "Handbook of Physiology" (Eds J. Orloff and R. W. Berliner), Section 8, pp. 129−143. American Physiological Society, Washington, D.C.

Gottschalk, C. W. and Mylle, M. (1959). *Am. J. Physiol.* **196**, 927−936.

Grandchamp, A. and Boulpaep, E. L. (1974). *J. Clin. Invest.* **54**, 69−82.

Grantham, J. J. (1974). *In* "MTP International Review of Science" (Ed. K. Thurau), Physiology Series I, Vol. 6, pp. 247−272. Butterworth, University Park Press.

Grantham, J. J. (1976). *In* "The Kidney" (Eds B. M. Brenner and F. C. Rector), Vol. 1, pp. 299−317. W. B. Saunders. Philadelphia, London and Toronto.

Grantham, J. J., Burg, M. B. and Orloff, J. (1970). *J. Clin. Invest.* **49**, 1815−1825.

Green, R. and Giebisch, G. (1975a). *Am. J. Physiol.* **229**, 1216−1226.

Green, R. and Giebisch, G. (1975b). *Am. J. Physiol.* **229**, 1206−1215.

Harris, C. A., Baer, P. G., Chirito, E. and Dirks, J. H. (1974). *Am. J. Physiol.* **227**, 972−976.

Hierholzer, K. and Lange, S. (1974). *In* "MTP International Review of Science" (Ed. K. Thurau), Series I, Vol. 6, pp. 273−334. Butterworth and Co-London, Unversity Park Press, Baltimore.

Horster, M. and Thurau, K. (1968). *Arch. ges. Physiol.* **301**, 162−181.

Jamison, R. L. (1970). *Am. J. Physiol.* **218**, 46−55.

Jamison, R. L. (1974). *In* "MTP International Review of Science" (Ed. K. Thurau), Physiology Series I, Vol. 6, pp. 199−245. Butterworth, University Park Press.

Karlmark, B. (1974). *Acta Physiol. Scand.* **91**, 243−256.

Kashgarian, M. H., Stoeckle, H., Gottschalk, C. W. and Ullrich, K. (1963). *Arch. ges. Physiol.* **277**, 89−106.

Kempton, R. T. (1937). *J. Morphol.* **61**, 51−62.

Khuri, R. N., Wiederholt, M., Strieder, N. and Giebisch, G. (1975a). *Am. J. Physiol.* **228**, 1249−1261.

Khuri, R. N., Wiederholt, M., Strieder, N. and Giebisch, G. (1975b). *Am. J. Physiol.* **228**, 1262−1268.

Kokko, J. P. and Rector, F. C., Jr. (1972). *Kidney Int.* **2**, 214−228.

Kriz, W. (1970). *In* "Urea and the Kidney" (Ed. B. Schmidt-Nielsen), pp. 342−357. Excerpta Medica Foundation, Amsterdam.

Le Grimellec, C., Poujeol, P. and de Rouffignac, Chr. (1975). *Arch. ges. Physiol.* **354**, 133−150.

Levinsky, N. G. and Levy, M. (1973). *In* "Handbook of Physiology" (Eds J. Orloff and R. W. Berliner), Section 8, pp. 103—117. American Physiological Society, Washington, D.C.

Malnic, G. (1974). Tubular handling of H. *In* "MTP International Review of Science" (Ed. K. Thurau), Series I, Vol. 6, pp. 79—106. Butterworth and Co., London.

Malnic, G. and Giebisch, G. (1972). *Kidney Int.* 1, 280—296.

Malnic, G. and de Mello Aires, M. (1971). *Am. J. Physiol.* 220, 1759—1767.

Malnic, G., Klose, R. M. and Giebisch, G. (1964). *Am. J. Physiol.* 206, 647—686.

Malnic, G., Klose, R. M. and Giebisch, G. (1966a). *Am. J. Physiol.* 211, 529—547.

Malnic, G., Klose, R. M. and Giebisch, G. (1966b). *Am. J. Physiol.* 211, 548—559.

Malnic, G., de Mello Aires, M. and Giebisch, G. (1971). *Am. J. Physiol.* 211, 1192—1208.

Malnic, G., de Mello Aires, M., de Mello, G. B., and Giebisch, G. (1972). *Arch. ges. Physiol.* 331, 275—278.

Malnic, G., de Mello Aires, M. and Cassola, A. C. (1974). *In* "Ion selective microelectrodes" (Eds H. J. Berman and N. C. Hebert), pp. 89—108, Plenum Press.

Marsh, D. and Frasier, C. (1965). *Am. J. Physiol.* 209, 283—287.

Nungesser, W. C. and Pfeiffer, E. W. (1965). *Comp. Biochem. Physiol.* 14, 289—297.

Rector, F. C., Jr. (1973). *In* "Handbook of Physiology" (Eds J. Orloff and R. W. Berliner), Section 8, pp. 431—454. American Physiological Society, Washington, D.C.

Renkin, E. M. and Gilmore, J. P. (1973). *In* "Handbook of Physiology" (Eds J. Orloff and R. W. Berliner), Section 8, pp. 185—248. American Physiological Society, Washington, D.C.

Richards, A. N. and Walker, A. M. (1937). *Am. J. Physiol.* 118, 111—120.

Sackin, H. and Boulpaep, E. L. (1975). *J. Gen. Physiol.* 66, 671—733.

Schafer, J. A., Troutman, S. L. and Andreoli, T. E. (1974). *J. Gen. Physiol.* 64, 582—607.

Schnermann, J., Wright, F. S., Davis, J. M., Stackelberg, W. V. and Grill, G. (1970). *Arch. ges. Physiol.* 318, 147—158.

Spring, K. R. (1972). *Yale J. Biol. and Med.* 45, 426—431.

Spring, K. R. and Giebisch, G. (1977). *J. Gen. Physiol.* in press.

Stein, J. H. and Reineck, H. J. (1975). *Physiol. Rev.* 55, 127—141.

Stein, J. H., Osgood, R. W. and Kunau, R. T. (1976). *J. Clin. Invest.* 58, 767—773.

Stoner, L. C., Burg, M. B. and Orloff, J. (1974). *Am. J. Physiol.* 227, 453—459.

Stroup, R. F., Weinman, E., Hayslett, J. R. and Kashgarian, M. (1974). *Am. J. Physiol.* 212, 1341—1349.

Sullivan, W. J. (1968). *Am. J. Physiol.* 214, 1096—1103.

Tadokoro, M. and Boulpaep, E. L. (1972). *Yale J. Biol. and Med.* 45, 432—435.

Thurau, K. and Mason, J. (1974). *In* "MTP International review of Science" (Ed. K. Thurau), Series I, Vol. 6, pp. 357—389. Butterworth, London.

Ullrich, K. J., Frömter, E. and Baumann, K. (1969). *In* "Laboratory Techniques in Membrane Biophysics" (Eds H. Passow and R. Stämpfli), pp. 106—129. Springer.

Ullrich, K. J., Radtke, H. W. and Rumrich, G. (1971). *Pflüg. Arch. ges. Physiol.* **330**, 149—161.

Ullrich, K. J., Rumrich, G. and Kloess, S. (1974a). *Arch. ges. Physiol.* **351**, 35—48.

Ullrich, K. J., Rumrich, G. and Kloess, S. (1974b). *Arch. ges. Physiol.* **351**, 49—60.

Walker, A. M., Bott, P. A., Oliver, J. and MacDowell, M. C. (1941). *Am. J. Physiol.* **134**, 580—595.

Wiederholt, M. and Giebisch, G. (1974). *Fed. Proc.* **33**, 387.

Windhager, E. E. (1968). Micropuncture techniques and nephron function. Butterworth, London.

Windhager, E. E. (1973). *In* "Transport Mechanisms in Epithelia" (Eds H. H. Ussing and N. A. Thorn), pp. 596—606. Academic Press, London.

Windhager, E. E., Lewy, J. E. and Spitzer, A. (1969). *Nephron* **6**, 247—259.

Wirz, H. and Dirix, R. (1973). *In* "Handbook of Physiology" (Eds J. Orloff and R. W. Berliner), Section 8, pp. 415—430, American Physiological Society, Washington, D.C.

Wright, F. S. (1974a). *New England J. Med.* **291**, 135—141.

Wright, F. S. (1974b). *In* "MTP International Review of Science" Series I, Vol. 6, Kidney and Urinary Tract Physiology (Ed. K. Thurau), pp. 79—106. Butterworth, London.

Wright, F. S. and Giebisch, G. (1972). *Kidney Int.* **1**, 201—209.

Wright, F. S. and Schnermann, J. (1974). *J. Clin. Invest.* **53**, 1695—1708.

Wright, F. S., Strieder, N., Fowler, H. B. and Giebisch, G. (1971). *Am. J. Physiol.* **221**, 437—448.

3. *In Vitro* Microperfusion

W. H. Dantzler

Department of Physiology, University of Arizona, Arizona, U.S.A.

I. INTRODUCTION

The study of transport in small tubular epithelial structures, such as vertebrate renal tubules, has posed many problems. Micropuncture techniques (see Chapter 2) have permitted sampling of the luminal fluid at exposed sites in these structures *in vivo* and *in vitro*. Professor Ramsay has pioneered the use of these techniques with invertebrates (see for example, Ramsay, 1949, 1951, 1952, 1953). Variations in micropuncture techniques for vertebrate renal tubules have also permitted simultaneous perfusion of the lumen and peritubular blood vessels at some sites, but they have not allowed complete control of the environment on both sides of the epithelium. In recent years, however, a technique has been developed for the isolation and perfusion of vertebrate renal tubules (Burg *et al.*, 1966). This technique has made possible the study of transport in various tubular segments under circumstances in which the environment on both sides of the epithelium is controlled.

The present chapter concerns this *in vitro* perfusion technique. In it, I consider the basic method and the types of information that it can provide with selected examples. I also describe a number of modifications in the basic technique which have been developed to study specific problems. Although the general technical considerations are discussed, no attempt is made to describe in detail how to perform each aspect of the technique. Moreover, no attempt is made to include all studies that have been performed with this technique. Instead, certain types of studies are selected to illustrate the range of applications of the technique and some of the specific advantages and problems involved in these applications.

II. BASIC METHOD

In principle, the basic technique is very simple. Briefly, a segment of renal tubule is teased from fresh tissue and transferred to a special bathing chamber on a microscope stage. The two ends of the tubule are held between glass micropipettes and the tubule is perfused through an inner micropipette (Fig. 1). In practice, the technique is very difficult to use successfully. It generally requires about a year of continual practice to obtain consistently reproducible data.

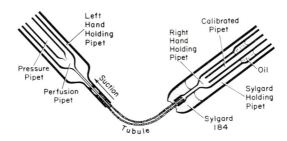

Fig. 1. Arrangement for perfusing isolated renal tubules. See text for details.

A. Dissection of Tubules

One of the most critical steps in this technique is that of teasing tubular segments from fresh tissue. During the dissection process, the tissue is maintained in an artificial Ringer medium or in blood serum from the same species. The dissection is performed under a stereomicroscope with fine foreceps and finely ground metal needles. It must be done without the aid of enzymatic agents, such as collagenase, since these substances weaken the tubules and cause them to rupture during perfusion (Burg *et al.*, 1966). Therefore, only species whose renal connective tissue permits dissection of the tubules without enzymes can be studied successfully. Table I lists the species and tubular segments that have been studied to date. Although human nephrons can be teased from fresh tissue and perfused (Abramow and Dratwa, 1974), they are not readily available for experimental work. The only other mammals from whose kidneys nephrons can be teased readily are rabbits. Most studies with isolated, perfused renal tubules have involved nephron segments from rabbits or snakes of the genus *Thamnophis*. The nephrons from these animals, like those from rabbits, can be

TABLE I

Examples of renal tubule segments perfused

Species	Tubule segment	Length perfused mm	Outside diameter μm	References[a]
MAMMALS				
Man	Medullary collecting duct	0.4	43	Abramow and Dratwa (1974)
Rabbit	Proximal tubule			
	convoluted portion	1–2	50.2	Burg and Orloff (1968)
	straight portion	2–4	44.8	Burg and Orloff (1968)
				Tune et al. (1969)
	Loop of Henle			
	thin descending limb			Kokko (1970)
				Imai and Kokko (1974a)
	thin ascending limb			Rocha and Kokko (1973a)
	thick ascending limb	0.6–2.0		Burg and Green (1973)
	Distal convoluted tubule	0.3–0.8		Gross et al. (1975)
	Collecting duct			
	cortical	0.5–4.5	35–40	Schafer and Andreoli (1972b)
				Grantham et al. (1970)
	medullary		35–40	Schafer and Andreoli (1972a)
REPTILES				
Snake	Proximal–proximal tubule	1.1–2.1	55.6	Dantzler (1974a)
(Thamnophis spp.)	Distal–proximal tubule	0.9–2.2	55.6	Dantzler (1974a)
	Thin intermediate segment	0.5–2.0	20	Dantzler and Bentley (unpublished observations)
	Distal convoluted tubule	1.2–1.5	40	Dantzler (1975)
AMPHIBIANS				
Bullfrogs	Proximal tubule	0.2–1.7	75	Irish and Dantzler (1976)
(Rana catesbiana)	Distal diluting segment	—	—	Stoner (1975)
Salamanders	Distal diluting segment	—	—	Stoner (1975)
(Ambystoma tigrinum)				
FISH				
Flounder	Proximal tubule	1.0–3.3		Burg and Weller (1969)
(Pseudopleuronectes americanus)				

[a] When a number of studies have been performed on a tubule segment, those references showing the dimensions of the perfused segments have been selected.

dissected free of the surrounding renal tissue with relative ease. Although nephrons can be teased from the kidneys of some amphibian species, they are much more difficult to dissect. Because of these difficulties, studies with nephrons from these species may have to be limited to a few comparisons of common transport processes and to more detailed studies of transport processes unique to these species.

The dissection must be performed with great care to avoid damaging the walls of the tubules or crushing shut the ends of the segments when cutting them. When the latter occurs, the distal end of the tubule may not open during the attempt to establish perfusion. It should also be pointed out that for any species being studied, the quality of the renal tissue and the ease with which nephrons can be dissected varies markedly from animal to animal. This is true whether the animals are bred under controlled conditions (e.g. New Zealand white rabbits) or obtained from a wild population (e.g. snakes of the genus *Thamnophis*). However, in those species from whose kidneys nephrons can be teased, most of the anatomically different nephron segments can be studied. The fact that transport processes can be studied in nephron segments inaccessible to other techniques is a major reason for the importance of the technique despite its difficulty.

B. Basic Perfusion System

The basic perfusion system is shown diagramatically in Fig. 1. A number of variations on this system have been used in different laboratories and some important ones will be considered below in regard to the need to obtain certain specific types of information. However, the principal features of the perfusion system are the same in all circumstances and were first outlined by Burg and his colleagues (1966). The pipettes are pulled on a vertical pipette puller, and the tips are shaped with a microforge. Tip diameters are chosen that are appropriate for the tubule segments being perfused (Table I). The perfusion pipette moves independently of the holding pipette so that its tip can be positioned in the lumen of the tubule after the tubule is drawn into the holding pipette. For the perfusion pipette to move smoothly into the lumen of the tubule without rupturing the cells, its tip must be centered in that of the holding pipette. This is often aided by set screws and by constrictions in the thin portion of the holding pipette which also act to squeeze the tubule tightly between the wall of the two pipettes (Fig. 1). Perfusion pressure can be applied with gravity, a closed air pressure system, or a perfusion pump.

For most studies, it has become important to change the perfusion fluid rapidly during the course of the perfusion. Although it is possible to do this in several ways, we and others (Burg and Green, 1973) have found it most convenient to maintain a permanent concentric pipette within the perfusion pipette (designated as the pressure pipette in Fig. 1). The perfusion fluid can be changed through this pipette in a few seconds without disrupting the perfusion.

The tubule is sealed into the collection pipette with the non-polymerized form of Sylgard 184 (Dow Corning) (Burg *et al.*, 1970). This is a biologically inert encapsulating resin which forms an excellent seal between the tissue and the glass. The tubule can be fitted into a single pipette with a Sylgard seal or a double pipette as shown in Fig. 1. In either case, the Sylgard seal reduces the need for a close fit between the tubule and the glass and eliminates leaks at this end of the system. The double pipette system shown makes it easier to maintain a layer of mineral oil on top of the fluid that is accumulating to prevent evaporation (Fig. 1). If the oil comes in contact with the Sylgard, it will dissolve it. However, it should be noted that we (Randle and Dantzler, unpublished observations) and others (Schafer *et al.*, 1974a) have found that evaporation from the collection pipette, even in the absence of an oil layer, is negligible. The perfusion fluid that accumulates in the collection pipette is removed with a calibrated constant bore capillary or a volumetric constriction pipette.

C. Bathing and Perfusion Media

Natural serum and an ultrafiltrate of serum have been used for the bathing and perfusion media, respectively, in some studies with rabbit proximal tubules (Burg and Orloff, 1968). Artificial media have been used in studies with nephrons from human kidneys (Abramow and Dratwa, 1974) and from kidneys of non-mammalian vertebrates (Burg and Weller, 1969; Dantzler, 1973), as well as in many studies with nephrons from rabbit kidneys (Grantham and Burg, 1966; Kokko, 1972a, 1973; Irish and Dantzler, 1976). Any of these media may be varied in the course of an experiment. This potential for controlling the media on both sides of the tubule during the course of an experiment is another feature of the technique which makes it particularly valuable for detailed study of the transepithelial transport processes. In general, experiments with nephrons from kidneys of mammals have been performed at $37°C$ and those with nephrons from kidneys of non-mammalian vertebrates have been performed at $25°C$.

III. TYPES OF INFORMATION OBTAINED

A. Fluid Absorption

One important function that can be studied with isolated, perfused renal tubules is net fluid absorption. Factors in the bathing medium and perfusion fluid which may influence such transport can be controlled and carefully evaluated. In order to measure net fluid absorption, most investigators have used some type of volume marker in the perfusion fluid. The ideal marker should not be lost from the tubule lumen by active absorption or passive diffusion through the tubule wall. Also, it should not be lost by adherence to the glassware used in the perfusion and collection systems. Some of the markers that have been used by various investigators are: ^{125}I-albumin (Kokko, 1970); ^{131}I-albumin (Burg and Orloff, 1968); ^{125}I-iothalamate (Rocha and Kokko, 1973a); ^{125}I-polyvinylpyrolidone (Rocha and Kokko, 1973b); sucrose-[14]C (Schafer and Andreoli, 1972a); inulin-carboxyl-[14]C (Imai and Kokko, 1974a); and inulin-methoxy-[3]H (Schafer and Andreoli, 1972b).

Although albumin has been used successfully, it tends to stick to glass and the investigator must use great care to obtain complete recovery. We and some others (Schafer and Andreoli, 1972b) have found inulin-methoxy-^{3}H to be a consistently reliable marker. However, before using it, we dialyze it in tubing with a 3,500 M.W. cut-off to elminate any small mol. wt fragments. Inulin-carboxyl-^{14}C appears to penetrate or bind to the epithelium (Dantzler, unpublished observations), and ^{125}I-iothalamate, although it often works satisfactorily, may occasionally cross the proximal tubule epithelium (Kokko, 1972a; Dantzler, unpublished observations).

Some comparative values for proximal tubule fluid absorption rates in several species are shown in Table II. As can be seen, fluid absorption is much greater in the proximal convoluted portion than in the proximal straight portion of rabbit nephrons whereas it is the

TABLE II
Representative fluid absorption rates in proximal tubules[a]

Species	Proximal tubule segment	Fluid absorption $nl \cdot mm^{-1} \cdot min^{-1}$	References
Rabbits	Convoluted portion	1.18 ± 0.07 (40)	Tune *et al.* (1969)
	Straight portion	0.47 ± 0.06 (15)	Tune *et al.* (1969)
Snakes (*Thamnophis* spp.)	Proximal portion	0.87 ± 0.04 (127)	Dantzler (unpublished observations)
	Distal portion	0.90 ± 0.03 (142)	Dantzler (unpublished observations)
Bullfrog (*Rana catesbiana*)	Proximal portion	0.34 ± 0.07 (29)	Irish and Dantzler (1976)

[a] Numbers are means ± SE. Values in parentheses indicate numbers of tubules.

same throughout the snake proximal tubule. These data further illustrate the advantage of the isolated perfusion technique in determining differences in transport in various segments of nephrons which would not be accessible to *in vivo* techniques.

Factors which might control fluid absorption rates in rabbit proximal tubules have been studied with this preparation. For example, with rabbit blood serum in the bath and an ultrafiltrate of rabbit blood serum for the perfusion fluid, fluid absorption rate appears to be independent of perfusion rate, at least at perfusion rates above $10 \, nl \, min^{-1}$ (Burg and Orloff, 1968; Kokko, 1972a). Thus, the balance between glomerular filtration rate and fluid absorption, observed *in vivo*, does not appear to be maintained *in vitro*. This indicates that the control system for the maintenance of glomerulotubular balance in mammals is not intrinsic to the tubule. These findings also illustrate another advantage of the isolated tubule technique. It permits differentiation between control mechanisms which reside in the tubules and those which require factors present only in the intact organ or animal.

Furthermore, it is possible, by altering the bathing and perfusion media to determine specific factors which may alter fluid absorption. It has been possible, for example, to demonstrate that the colloid osmotic pressure in the bathing medium can influence directly the rate of fluid absorption in rabbit proximal tubules in the absence of a capillary bed (Imai and Kokko, 1972; Grantham *et al.*, 1972; Imai and Kokko, 1974b). By examining the transepithelial movements of urea, which presumably penetrates cells, and sucrose, which presumably moves between cells, Imai and Kokko (1974b) suggested that this effect might involve extracellular pathways between cells. On the other hand, factors, such as the chloride concentration difference between lumen and bath, which may influence the transepithelial potential difference (see below), may not influence the rate of fluid absorption (Cardinal *et al.*, 1975).

Some modifications in the basic perfusion technique have also helped to define characteristics of fluid movement in rabbit proximal tubules. Burg and Orloff (1968) used an inner pipette on the collection side to compress the end of the perfused tubule, thereby increasing outflow resistance and causing the tubule to dilate. Using this technique, they found that fluid absorption was not directly related to tubule diameter or volume. Grantham and his colleagues (Grantham *et al.*, 1972; Grantham *et al.*, 1973) completely occluded the distal end of the tubule by sucking a knuckle of the tubule into the holding pipette. Since the distal end was occluded, they assumed that at a constant hydrostatic pressure net fluid absorption was equal to the rate of movement of fluid out of the perfusion pipette. This was measured by the movement of tritiated water into the bathing medium (Grantham *et al.*, 1972), or by the movement of an oil column along a portion of the perfusion pipette (Grantham *et al.*, 1973). This last approach first permitted them to recognize that the presence of para-aminohippurate (PAH) in the bathing medium could lead to net fluid secretion into the lumen when they observed that the oil drop moved up the perfusion pipette (Grantham *et al.*, 1974).

Another important example of the use of the basic technique for the measurement of fluid absorption involves the study of the effects of antidiuretic hormone (ADH) on mammalian collecting ducts. Several of the volume markers discussed above have been used to assess changes in net water absorption. When the perfusion fluid was hypo-osmotic to the bathing medium, both vasopressin and cyclic 3', 5'-AMP added to the bathing medium markedly enhanced net water absorption in rabbit cortical collecting ducts (Grantham and Burg, 1966). In the absence of an osmotic gradient, these agents also increased the diffusional permeability to water measured with tritiated water (Grantham and Burg, 1966). The agents were ineffective when applied to the luminal surface of the cortical collecting ducts (Grantham and Burg, 1966). Abramow and Dratwa (1974) have also been able to use the same technique to demonstrate directly that vasopressin increases the osmotic water permeability of a segment of human medullary collecting duct. These results are in general agreement with observations made in studies with anuran epithelia (Leaf, 1967; Orloff and Handler, 1967). Moreover, they directly confirmed suppositions concerning the mammalian collecting ducts which had been derived indirectly from studies of renal function in intact animals (Wirz, 1956).

These initial studies with isolated collecting ducts did not permit any additional conclusions concerning the mechanisms by which vasopressin enhanced osmotic flow. However, recent studies with the same techniques, but more rapid and complete collections, have permitted more detailed analyses of both osmotic and diffusional permeabilities and a more direct evaluation of the changes in permeability of the luminal membrane to water induced by ADH (Schafer and Andreoli, 1972a; Schafer *et al.*, 1974a,b). These studies have suggested that ADH increases the water solubility in luminal plasma membranes (Schafer *et al.*, 1974b). It has also been possible to combine studies of isolated, perfused rabbit collecting ducts with electron microscopic analyses of the changes in the tubules during ADH treatment (Grantham *et al.*, 1969). These studies indicated that the osmotic flow of water in response to ADH might enter the cells across the luminal membrane but exit into the lateral intercellular spaces as well as across the basal membrane.

B. Electrical Measurements

1. Transepithelial Potential Difference

The basic perfusion system can be modified rather simply for the measurement of transepithelial potential differences. In general, this is done by connecting the bath and perfusion pipette via equivalent bridges of Ringer solution in 4% agar to a saturated potassium chloride solution containing calomel half cells. The circuit is then completed with an electrometer and a voltage reference source. Since the measurements of transepithelial potential differences are made through the perfusion pipette, the difficulties of localizing an electrode in the lumen during *in vivo* micropuncture experiments (Hegel and Frömter, 1966; Frömter *et al.*, 1967) are avoided. However, in these experiments, it is important that good electrical insulation exist at the perfusion end of the tubule. This can be accomplished with Sylgard 184 as already described for the distal end of the tubule (Fig. 1). The most convenient method of maintaining a Sylgard seal at the perfusion end is with a separate concentric outer pipette (Burg and Green, 1973).

The potential difference across rabbit proximal convoluted tubules with rabbit blood serum in the bathing medium and an ultrafiltrate of rabbit blood serum as perfusate averaged −4 to −5 mv, lumen negative (Burg and Orloff, 1970; Kokko, 1973). A similar value has been found for frog and salamander proximal tubules (Stoner, 1975). This potential disappeared with the addition of ouabain to the bathing medium and returned to control levels when ouabain was removed from the bath, suggesting that it might be secondary to active sodium transport.

It was also observed by Kokko and Rector (1971) that the transepithelial potential difference in rabbit proximal convoluted tubules decreased with decreasing perfusion rates at rates below 10 nl min^{-1}. They postulated that, at slow perfusion rates, some constituents of the perfusion fluid which influence the potential difference might be completely absorbed. To examine this possibility, Kokko (1973) perfused rabbit proximal convoluted tubules with artificial solutions of various compositions. The potential difference decreased significantly when either glucose or amino acids were removed from the perfusion and decreased virtually to zero when both glucose and amino acids were removed. When, in the absence of glucose and amino acids, the bicarbonate concentration in the perfusate was reduced from 26 mEq l^{-1} to 5.4 mEq l^{-1} and replaced with chloride, the potential changed at about 3 mV, lumen positive. This positive potential difference appeared to be a chloride diffusion potential secondary to the increased chloride concentration in the lumen since it did not occur when bicarbonate was replaced with methyl sulphate. Absorption of glucose, amino acids and bicarbonate with a concomitant rise in the intraluminal chloride concentration does occur early in the proximal tubule *in vivo*. These *in vitro* data suggested that the potential might be negative in the early proximal convoluted tubule *in vivo* and might become positive further along the nephron. Although the earliest portions of the proximal convoluted tubule are accessible to direct micropuncture *in vivo* only under special circumstances, these studies have now been performed and the predicted pattern does appear to prevail in rats (Barratt *et al.*, 1974; Frömter and Gessner, 1974). Thus, the *in vitro* perfusion technique permitted an earlier and more direct evaluation of factors involved in the generation of the transepithelial potential than *in vivo* measurements.

It has been suggested that the absorption of glucose, amino acids, and bicarbonate might facilitate the entry of sodium into the cell and its access to an electrogenic pump on the peritubular side (Kokko, 1973). This possibility has yet to be examined directly. However, as noted above, other studies of fluid absorption in isolated, perfused rabbit proximal convoluted tubules have indicated that the rate of fluid absorption, which apparently depends on sodium absorption, may be independent of the magnitude and direction of the transepithelial potential (Cardinal *et al.*, 1975). The process of fluid and ion absorption in the proximal renal tubule is clearly complicated and not yet well understood, but perfusion of isolated tubules permits investigators to manipulate directly some of the factors involved.

2. *Transepithelial Resistance*

It is also possible to measure transepithelial resistance in isolated perfused tubules (Helman *et al.*, 1971; Lutz *et al.*, 1973), and this permits further analysis of the transport properties. Current can be injected through the perfusion pipette, the voltage measured at both the perfusion and collection ends, and the transepithelial resistance determined by cable analysis. In these experiments, it is essential that both ends of the tubules be well insulated with Sylgard.

In measurements of resistance in rabbit cortical collecting ducts, a single perfusion pipette with a single bridge circuit can be used for the injection of current and the measurement of voltage (Helman *et al.*, 1971). This is possible because the electrical resistance is very high (138,000 ohm cm) and, thus, is the same order of magnitude as that of the perfusion pipette. This high electrical resistance is consistent with the fact that the cortical collecting ducts can maintain large transepithelial ion concentration gradients.

In measurements of electrical resistance in rabbit proximal tubules, however, separate circuits for injecting current and recording voltage are necessary because the resistance (864 and 1188 ohm cm for the convoluted and straight portions, respectively) is much lower than that of the distal tubule (Lutz *et al.*, 1973). Nevertheless, it is possible to use a single perfusion pipette for the injection of current and the recording of voltage (Lutz *et al.*, 1973). The low electrical resistance appears to reflect the high permeability of these segments to electrolytes. Similar low resistance values have also been found for the thick ascending limb of Henle's loop (Burg and Green, 1973). Cable analysis seems to be applicable to these tubule segments and the resistance measurements, although difficult, avoid some of the problems of localization of the electrode tips and non-uniform current spread involved in attempts to measure resistance by *in vivo* micropuncture methods.

C. Transport of Inorganic Ions

Methods for studying transport of inorganic ions are similar to those employed with other isolated epithelial structures, but are considerably more direct than those that can be employed with renal tubules *in vivo*. For example, isotopic sodium has been used in the bathing medium and perfusion fluid to measure unidirectional and net sodium fluxes in rabbit proximal convoluted tubules (Kokko *et al.*, 1971). The net absorptive flux of sodium was 20% of the unidirectional flux and compared closely with the calculated short-circuit current (Lutz *et al.*, 1973). These data indicated that sodium was actively absorbed. Moreover, when a poorly absorbed solute, such as raffinose, was present in the perfusion fluid, the sodium concentration, which normally did not change during absorption, decreased to a level $33-35$ mEq l^{-1} less than that in the bath (Kokko *et al.*, 1971).

Another, and quite different example of the measurement of active transport of inorganic ions in isolated renal tubules involves chloride transport in the thick ascending limb of Henle's loop. This segment of the mammalian nephron is inaccessible to direct *in vivo* micropuncture, but indirect evidence clearly indicated that sodium and chloride were absorbed (Walker *et al.*, 1941; Wirz, 1956; Gottschalk *et al.*, 1963; Clapp and Robinson, 1966). This segment of the rabbit nephron has been isolated and perfused *in vitro* (Burg and Green, 1973; Rocha and Kokko, 1973a). Sodium and chloride fluxes were measured isotopically, fluid absorption was assessed with a volume marker, and the potential difference and transepithelial resistance were measured as described above. Net sodium chloride absorption occurred with the resultant lowering of the concentration in the lumen below that in the bath. In contrast to the proximal tubule, however, the potential difference across the epithelium was about 3 to 9 mv, lumen positive. This indicated that chloride absorption was active because chloride was transported against both an electrical and a chemical gradient. Moreover, the transepithelial potential difference appeared to be caused by active chloride transport since it approached zero when chloride in the perfusion fluid and bath was replaced by sulphate. The possibility of some active sodium absorption was not completely ruled out, but was not clearly supported by the equilibrium distribution or the flux ratios. The water permeability in this segment was also found to be very low, as predicted from *in vivo* studies (Wirz, 1956).

The isolated, perfused tubule technique made possible for the first time direct exploration of some of the transport processes in the thick ascending limb of Henle's loop where significant dilution of the tubular fluid takes place (Burg and Green, 1973). The studies directly confirmed the net absorption of sodium chloride and the low permeability of this segment to water. They also demonstrated that, in contrast to other segments of the nephron, sodium chloride absorption in this segment consists primarily of the active absorption of chloride with sodium following passively along the resulting electrical potential gradient.

This technique has recently permitted the identification of a functionally similar region in the distal renal tubules of both frogs and salamanders, animals without loops of Henle (Stoner, 1975). In this segment, the potential difference is about 12 mV, lumen positive, and rapid sodium chloride absorption occurs as a result of active chloride transport. Since water permeability is low, the rapid absorption of sodium chloride results in dilution of the tubule fluid.

D. Transport of Organic Molecules

Transport of organic substances that are secreted by renal tubules, such as para-aminohippurate (PAH), can be studied with isolated perfused tubules by placing the substance, labelled radioactively, in the bathing medium and observing its appearance in the lumen (Tune *et al.*, 1969; Dantzler, 1974a; Irish and Dantzler, 1976). For those substances that are actively transported into the lumen (Tune *et al.*, 1969; Dantzler, 1973, 1974a), a concentration develops in the lumen which is greater than that in the bath. Therefore, if the concentration in the bathing medium is kept low enough, it is possible to demonstrate transport against a concentration gradient from bath to lumen at most perfusion rates. It is generally unnecessary to measure unidirectional fluxes, to have the substance in the perfusion fluid initially, or to measure fluid absorption simultaneously.

By perfusing different segments of isolated nephrons, the site of such secretory transport can be identified. For example significant PAH transport was found to occur only in the straight portion of rabbit proximal tubules (Tune *et al.*, 1969) and the distal portion of snake proximal tubules (Dantzler, 1974a). However, transport occurs throughout the frog proximal tubules (Irish and Dantzler, 1976). Also, uric acid secretion occurs throughout snake proximal tubules and does not appear to involve the same mechanism or site of transport as PAH (Dantzler, 1973, 1974a, 1975).

The study of the active absorption of organic molecules, such as glucose, is somewhat more complicated than the study of active secretion. The volume of the bathing medium is generally large (50 μl to 2 ml, depending on the chamber design) compared with the volume in the tubule lumen (about 1.5 nl). Therefore a substance placed in the lumen alone will not develop a concentration in the bathing medium greater than that in the lumen. Moreover, the measurement of unidirectional fluxes is always complicated by the rapid transport of the substance that has moved from the bath into the lumen out of the lumen again. The most effective approach involves the measurement of net absorption from the lumen when the concentration of the substance is identical in the perfusion fluid and bath. In this case, a volume marker must be included in the perfusion fluid to permit corrections for water absorption. Moreover, if an isotopically labelled substance is used to measure transport, the labelled as well as the unlabelled concentrations must be indentical in the perfusion fluid and bathing medium to avoid problems of independent isotopic flux (Barfuss and Dantzler, 1975). Measurements of glucose absorption in isolated rabbit proximal renal tubules have indicated that transport is greater in the convoluted than in the straight portion (Tune and Burg, 1971). In snake proximal renal tubules, however, glucose transport is greater in the distal than in the proximal portion (Barfuss and Dantzler, 1975).

In studies of the transepithelial transport of isotopically labelled organic molecules, the tubules can be recovered during perfusion for measurements of the concentration of the transported substance in the cell water. The procedure involves removing the tubules rapidly from the perfusion pipettes with a minimum of contamination by bathing medium and perfusion fluid (Tune *et al.*, 1969; Dantzler, 1973, 1974a). In these studies, mineral oil can be added to the perfusion pipette behind the perfusion fluid. Samples are collected in the usual fashion until the volume of fluid in the perfusion pipette is exhausted and oil fills the tubule lumen. The tubule then is pulled rapidly from the pipettes with a glass needle and transferred through an oil layer to a drop of trichloroacetic acid for precipitation of protein and extraction of the labelled substrate. The removal of the tubule from the pipettes and the transfer to the trichloroacetic acid can be accomplished in less than five seconds. Following this extraction, the tubule is placed in chloroform to remove the oil. Finally, it is dried and weighed on a quartz fiber ultramicrobalance (Bonting and Mayron, 1961). Appropriate corrections are made in the dry tissue weight for weight lost during the extraction procedures. The radioactivity in the trichloroacetic acid extract is then determined. For transported substances which may be metabolized by the cells, this extract is usually analysed with some form of chromatography to determine how much of the radioactivity is still in the form of the transported substance.

In some experiments it is not feasible to end the perfusion by filling the tubule with oil. In these cases, the amount of the transported substance remaining in the lumen can be determined from the luminal volume (calculated from the luminal dimensions measured during perfusion) and the average concentration of the substance in the luminal fluid (Tune and Burg, 1971). If a marker for volume absorption has been used, the amount of this marker extracted from the tubule can be used instead of the luminal dimensions to determine the volume of fluid in the lumen (Barfuss and Dantzler, unpublished observations). In these last two situations, the amount of the transported substance in the lumen must be subtracted from the amount extracted from the tubule.

In order to determine the concentration of the transported substance in the cell water, the average cell water content of the tubules must also be determined. This can be done by equilibrating a series of tubules with an appropriate medium containing tritiated water (Burg *et al.*, 1966; Dantzler, 1973). The average value for these tubules can then be used to calculate the cell water in each subsequent tubule. The concentration of the transported substance in the cell water is then determined for each tubule from the amount of extracted substance and the cell water content.

With these data, it is possible to compare the concentration of the transported substance in the tubule cell water with that in the bath and lumen during the transport process. For example, this comparison has been made in studies on the transport of PAH from bath to lumen by the straight portion of rabbit proximal tubules (Tune *et al.*, 1969), the distal portion of snake proximal tubules (Dantzler, 1974a), and all portions of frog proximal tubules (Irish and Dantzler, 1976). In each case, the concentration of PAH was greater in the lumen than in the bath and greater in the cells than in either the lumen or the bath. These data suggested that transport from bath to lumen was active and that it occurred by active transport into the cells on the peritubular side and passive movement from the cells into the lumen. A similar pattern was observed for uric acid in all portions of snake proximal tubules (Dantzler, 1973).

In any of these studies in which the substance transported can be ionized, the electrical potential differences across the epithelium and the cell membranes must be considered in evaluating the concentration in the cells and lumen. PAH and urate, for example, exist almost entirely in the anionic form at physiological pH. However, the interior of the cells is probably significantly negative compared to the bath and the lumen is slightly negative compared to the bath (Burg and Orloff, 1970; Dantzler, unpublished observations). Therefore, luminal and cellular concentrations greater than those in the bath further support the concept of active transport.

When intracellular concentrations of a transported substance are high, a determination must also be made about whether the substance is free in the cell water or bound to intracellular structures. Some information on this point can be obtained by determining the cell water concentrations, as described above, during studies of passive unidirectional flux across the epithelium. For example, during the passive flux of PAH and urate from lumen to bath the cellular concentrations in rabbit and snake tubules never approached those in the lumen (Tune *et al.*, 1969; Dantzler, 1973, 1974a). Since the amounts crossing the epithelium in these studies were many times those found in the cells, these data suggested that there was no significant tissue binding and that the concentrations measured were representative of free concentrations in the cell water.

E. Passive Properties of Epithelium

1. Transepithelial Permeabilities

The apparent passive transepithelial permeabilities to inorganic ions such as sodium and chloride, and organic molecules such as urea, can be determined in isolated perfused tubules from the transepithelial fluxes. If a substance that moves passively from bathing medium to tubule lumen is added to the bathing medium only (usually in an isotopically labelled form), the passive permeability from bath to lumen can be calculated from the transepithelial flux and the concentration difference. In general, the area of the epithelium used in these calculations is that of the luminal membrane determined from the diameter of the lumen and the tubule length measured during perfusion. If the perfusion rate is kept sufficiently rapid that the concentration in the collected fluid remains well below that in the bath, half the final concentration in the collected fluid is generally a reasonable approximation of the mean concentration in the tubule lumen during perfusion. If the concentration in the collected fluid approaches that of the bath, the probable exponential increase in the concentration within the tubule lumen during perfusion should be considered (Imai and Kokko, 1974a).

If the substance moving passively from bath to lumen is actively transported out of the lumen again, this will be a source of error in these measurements. However, this problem can be reduced by maintaining a rapid perfusion rate and, if possible, blocking the active absorption with an appropriate inhibitor (Barfuss and Dantz-ler, unpublished observations). It is also possible to compare the permeabilities in both directions (bath-to-lumen and lumen-to-bath) in a single tubule and correct for differences caused by active transport (Imai and Kokko, 1974a). These calculations, however, do not take into account the effect of any potential difference across the epithelium on the permeability of ionized substances. For these reasons, the permeabilities are probably best considered as apparent permeabilities only.

Measurements of apparent passive transepithelial permeabilities from lumen to bath can be more complicated if there is significant fluid absorption from the tubule lumen. If a substance moves passively from lumen to bath, it can be added to the perfusion fluid alone and the transepithelial permeability calculated from its rate of disappearance from the lumen. In the presence of net fluid absorption, it appears reasonable to assume that the change in flow velocity is linear with respect to tubule length. If this assumption is made, the transepithelial permeability can be calculated from the following equation (Grantham and Burg, 1966):

$$P = \frac{V_i - V_o}{A}\left[\frac{\ln C_i/C_o}{\ln V_i/V_o} + 1\right]$$

In this equation, P is the permeability coefficient; V_i and V_o are the initial perfusion rate and collection rate, respectively; A is the area of the luminal membrane; and C_i and C_o are the concentrations of the substance whose permeability is being measured in the initial perfusate and collected fluid, respectively. This equation can be modified to include the reflection coefficient for the substance whose permeability is being studied (Schafer and Andreoli, 1972a). However, this modification has only a small effect on the value obtained with the equation above and, in the absence of accurate measurements of the reflection coefficient (see below), the above expression gives a reasonable value for the permeability.

If there is no fluid absorption, the above equation for trans-epithelial permeability reduces to the following (Grantham and Burg, 1966):

$$P = \frac{V_i}{A} \ln \frac{C_i}{C_o}$$

Since this simplifies the determination of passive permeabilites from lumen to bath, it is desirable to prevent fluid absorption during these measurements. This is relatively easy to do in many nephron segments, but it is difficult to achieve an equilibrium solution which eliminates all net transepithelial fluid movement in any individual proximal tubule (Kokko *et al.*, 1971).

Some examples of permeabilities of various segments of rabbit nephrons to water, sodium, chloride and urea, determined with these techniques, are shown in Table III. The very high permeabilities of the thin ascending limb of the loop of Henle to sodium and chloride and of the descending limb to water have given impetus to the re-evaluation of the counter-current multiplier system in the mammalian kidney (Kokko and Rector, 1972; Stephenson, 1972).

2. Hydraulic Conductivities

The hydraulic conductivity (Lp) of the tubular epithelium can be determined by measuring net fluid absorption, as described above, in response to an imposed osmotic gradient (Kokko *et al.*, 1971). In addition, the perfusion system has recently been modified to determine the hydraulic conductivity directly from measurements of the hydrostatic pressure across the tubular epithelium (Horster and Larsson, 1975). In this modification, a sampling pipette is fitted tightly against a ring of hardened Sylgard in the collecting pipette so that the pressure delivered through the tubule lumen is transmitted to it. This pressure is then measured with a manometer and used for the determination of hydraulic conductivity.

TABLE III

Permeability coefficients (P) for water (dw), sodium (Na), chloride (Cl) and urea in rabbit nephron segments[a]

Nephron segment	P_{dw}	P_{Na}	P_{Cl}	P_{urea}	References[b]
	cm sec^{-1} × 10^{-5}				
Proximal convoluted tubule		9.33 ± 1.44 (6)	3.8	5.3 ± 0.6 (6)	Kokko et al. (1971) Imai and Kokko (1974b) Kokko (1972b)
Descending limb of Henle	446.0 ± 113.5 (6)	1.61 ± 0.27 (6)		1.51 ± 0.47 (6)	Imai and Kokko (1974a) Kokko (1970) Kokko (1972b)
Thin ascending limb of Henle	45.3 ± 5.4 (6)				
Thick ascending limb of Henle		25.5 ± 1.8 (20)	117.0 ± 9.1 (13)	6.74 ± 1.04 (10)	Imai and Kokko (1974a)
Cortical collecting duct		6.27 ± 0.38 (9)	1.06 ± 0.12 (4)		Rocha and Kokko (1973a)
Without ADH	40.1 ± 4.2 (3)	0.06		0.16 ± 0.010 (3)	Burg et al. (1970) Frindt and Burg (1972)
With ADH	77.2 ± 6.2 (3)			0.15 ± 0.018 (3)	Burg et al. (1970) Burg et al. (1970)

[a] Values are means or means ± SE. Numbers in parentheses indicate number of experiments.

[b] When the permeabilities for any tubular segment come from more than one source, the references are listed in the same order as the permeabilities to which they refer when one reads from left to right.

3. Reflection Coefficients

The reflection coefficient (σ) for various solutes in relation to the epithelium of different nephron segments can also be determined with this perfusion technique. This measurement indicates the degree to which the osmotic effect exerted by a solute across a membrane deviates from the theoretical osmotic effect predicted by the van't Hoff relationship. It is defined as the ratio of the experimentally observed osmotic effect to the theoretical osmotic effect (Staverman, 1951). If a solute penetrates a membrane without any restrictions, it will exert no osmotic force and will have a reflection coefficient equal to zero. If the solute does not penetrate the membrane at all, it will exert its full theoretical osmotic effect and have a reflection coefficient of 1.0.

In experiments to determine the reflection coefficient, the osmotic effect across a membrane of the solute to be tested is compared with the osmotic effect of a reference solute which is known not to penetrate the membrane. Two different approaches to these measurements have been used with isolated, perfused renal tubules, (Kokko *et al.*, 1971). In the first approach the perfusion fluid contains the reference solute (generally raffinose) at some previously measured osmolality. The concentration of test solute added to the bathing medium which results in no net fluid movement across the epithelium is then determined. The reflection coefficient is the ratio of the osmolality of the reference solution to the osmolality of the test solution which results in no net flow. The second method involves a comparison of the increment in osmotic flow induced by equal concentrations of test and reference solutes. A known concentration of a reference solute is added to the bathing medium and the induced increase in fluid absorption, measured as described earlier, is determined. The bathing medium is then replaced by one containing a concentration of test solute known to have the same osmolality in free solution as the concentration of reference solute. The increment in fluid absorption from the control level produced by this solution is then measured. The ratio of the increment in fluid absorption produced by the test solute to that produced by the reference solute equals the reflection coefficient.

The accuracy of these methods for determining σ is limited by the movement of either the test solute or water across the epithelium during the measurements. Yet, such movements must occur, so that the values can only be considered approximate. The sources of errors in these measurements have been well reviewed in a series of recent publications (Kokko *et al.*, 1971; Kokko, 1970, 1972a). Some representative values for reflection coefficients in various segments of rabbit nephrons are shown in Table IV.

TABLE IV

Reflection coefficients (σ) for sodium chloride and urea in rabbit nephron segments[a]

Segment	σ Sodium chloride	σ Urea
Proximal convoluted	0.71 ± 0.06 (6)	0.91 ± 0.05 (6)
Tubule	0.68 ± 0.06 (6)	
Descending limb		
of Henle	0.96 ± 0.01 (6)	0.95 ± 0.04 (6)

[a] Values are means $\pm$ SE. Numbers in parentheses indicate numbers of experiments. Values for sodium chloride are from Kokko (1970) and Kokko *et al.* (1971). Values for urea are from Kokko (1972b).

F. Passive Properties of Individual Membranes

1. Permeabilities with Intact Cells

When the active transepithelial transport of organic molecules is being studied in isolated, perfused tubules, it is possible to determine the passive permeabilities of the luminal and peritubular membranes individually as well as the permeability of the entire epithelium (Tune *et al.*, 1969; Dantzler, 1973, 1974a,b; Irish and Dantzler, 1976). The luminal membrane permeability for substances that are transported from bath to lumen with an active step at the peritubular membrane can be determined if the concentration of the substances in the cell water is measured during the transport studies. The permeability be calculated from the net transepithelial transport rate and the concentration difference across the membrane (Tune *et al.*, 1969; Dantzler, 1973, 1974a).

For these same substances, such as PAH or uric acid, that are transported from bath to lumen, the permeability of the peritubular membrane can be determined from the passive efflux of the substance from tubules with oil-filled lumens (Dantzler, 1974b). This technique was developed to limit efflux to movement from the cells across the peritubular membrane only. The tubules are teased from fresh tissue, attached to the perfusion system, and the lumens are then filled with mineral oil. The tubules are removed from the

D

perfusion system and incubated in a bath with an appropriate concentration of the substrate whose transport is being studied. They are allowed to accumulate the substance, which is transported into the cells across the peritubular membrane, until a steady-state concentration is reached. Since substances such as PAH and uric acid have no significant lipid solubility, they do not enter the oil in the tubule lumen (Dantzler, 1974b). Therefore, the efflux of these substances can be measured across the peritubular membrane alone.

Following the incubation, the tubules are washed in bathing medium without substrate to remove extracellular fluid and transferred rapidly through a series of baths without substrate to measure the efflux. At the end of the efflux periods, the tubules are analysed for any remaining substrate in the cell water. The concentration of substrate in the tubule cell water during each efflux period is determined from the efflux data. The efflux of the substrate from cells to bath is proportional to the concentration difference between the cells and the bath, and the constant of proportionality is the permeability coefficient.

Table V shows some representative permeabilities of luminal and peritubular membranes to PAH in snake and frog proximal renal tubules. A luminal membrane permeability much larger than the peritubular membrane permeability favors the passive movement of substrate, which has been transported into the cells on the peritubular side, into the lumen rather than back into the bath (Dantzler, 1974a,b; Irish and Dantzler, 1976).

For organic substances which are transported from lumen to bath, the passive movement from bath to lumen can be used to determine the permeabilities of the luminal and peritubular membranes as well

TABLE V

Apparent permeabilities of luminal and peritubular membranes to PAH in bullfrogs and garter snakes[a]

	P_L[b] cm sec^{-1} × 10^{-5}	P_P[c] cm sec^{-1} × 10^{-5}
Bullfrog (*Rana catesbiana*)	3.80 ± 0.53 (21)	0.66 ± 0.077 (8)
Garter snake (*Thamnophis* spp.)	3.50 ± 0.91 (15)	0.50 ± 0.065 (5)

[a] Values are means ± SE. Figures in parentheses indicate number of tubules.
[b] P_L indicates apparent permeability of luminal membrane.
[c] P_P indicates apparent permeability of peritubular membrane. Values for bullfrogs are from Irish and Dantzler (1976). Those for garter snakes are from Dantzler (1974b).

as the transepithelial permeability described above. In these experiments, the cell water concentration of the substance is determined and the permeabilities are calculated from the net fluxes and the concentration difference across each membrane. As described earlier, precautions must be taken to minimize transport of the substrate from lumen back into the bath. In all these studies, if the substrate is ionized, electrical potential differences may influence the permeabilities and it is best to consider them apparent, rather than absolute, permeabilities.

2. Hydraulic Conductivities with Isolated Basement Membranes

Some of the physical properties of the basement membrane alone can also be measured with a modification of the basic perfusion technique (Welling and Grantham, 1972). In these studies, the tubule segments are first perfused in the usual fashion. Then the bathing medium is replaced with one containing sodium deoxycholate, Triton X-100, or with distilled water. These treatments cause the cells to loosen from the basement membrane so that they can be washed out of the distal end of the tubule. Welling and Grantham (1972) found that sodium deoxycholate produced the cleanest preparation of basement membrane.

By using the perfusion technique, described earlier, in which the distal end of the basement membrane preparation is completely occluded, it is possible to measure hydraulic conductivity from the rate of transmural flow in response to a difference in colloid osmotic pressure. For example, Welling and Grantham (1972), found that the hydraulic conductivity of basement membranes from rabbit proximal straight, proximal convoluted, and cortical collecting tubules was 300–800 times greater than that of the intact epithelium. This hydraulic conductivity was similar to that of peritubular and glomerular capillaries *in vivo*.

These investigators also observed that the hydraulic conductivity determined from a difference in hydrostatic pressure was greater than that determined from a difference in colloid osmotic pressure. This appeared to result from the fact that the basement membrane was permeable to the albumin used for colloid osmotic pressure. In addition, they found that the basement membrane was a relatively tough, elastic structure with a Young's modulus similar to that of pure tendon collagen. These direct measurements of the passive properties of basement membranes alone permit more quantitative estimates of the roles of hydrostatic and colloid osmotic pressures in controlling fluid absorption than can be obtained with less direct measurements.

IV. CONCLUSION

This discussion has concerned the basic method for perfusing and studying isolated renal tubules *in vitro*. It has included examples of some types of information that can be obtained with the technique and modifications of the basic technique to permit measurements that are not possible with other methods. Since the technique was developed for use with isolated renal tubules and has not been used with other structures, the discussion has been limited to studies with renal tubules. It should be obvious, however, that it could be used with other small tubular biological structures. The only requirement is that these structures can be removed from surrounding tissue without significant damage to their walls. For example, among the vertebrates, some arterioles or capillaries may be accessible for such studies. Among small invertebrates, larval intestinal segments and some portions of nephridia could possibly be studied. I have not attempted to include all conceivable approaches, even for renal tubules, in the present discussion. The imaginative investigator should be able to devise other experiments and variations in the technique which will enable him to study more directly the fundamental problems of biological transport.

ACKNOWLEDGMENT

The personal research and the preparation of this chapter was supported in part by National Science Foundation Research Grant BMS75-09918.

REFERENCES

Abramow, M. and Dratwa, M. (1974). *Nature* **250**, 492—493.
Barfuss, D. W. and Dantzler, W. H. (1975). *Physiologist* **18**, 130.
Barratt, L. J., Rector, F. C. Jr., Kokko, J. P. and Seldin, D. W. (1974). *J. Clin. Invest.* **53**, 454—464.
Bonting, S. L. and Mayron, B. R. (1961). *Microchem. J.* **5**, 31—42.

Burg, M. B. and Green, N. (1973). *Am. J. Physiol.* **224**, 659—668.

Burg, M. B. and Orloff, J. (1968). *J. Clin. Invest.* **47**, 2016—2024.

Burg, M. B. and Orloff, J. (1970). *Am. J. Physiol.* **219**, 1714—1716.

Burg, M. B. and Weller, P. F. (1969). *Am. J. Physiol.* **217**, 1053—1056.

Burg, M. B., Grantham, J., Abramow, M. and Orloff, J. (1966). *Am. J. Physiol.* **210**, 1293—1298.

Burg, M. B., Helman, S., Grantham, J. and Orloff, J. (1970). *In* "Urea and the Kidney" (Eds B. Schmidt-Nielsen and D. W. S. Kerr), pp. 193—199. Excerpta Medica Foundation, Amsterdam.

Cardinal, J., Lutz, M. D., Burg, M. B. and Orloff, J. (1975). *Kidney Int.* **7**, 94—102.

Clapp, J. R. and Robinson, R. R. (1966). *J. Clin. Invest.* **45**, 1847—1853.

Dantzler, W. H. (1973). *Am. J. Physiol.* **224**, 445—453.

Dantzler, W. H. (1974a). *Am. J. Physiol.* **226**, 634—641.

Dantzler, W. H. (1974b). *Am. J. Physiol.* **227**, 1361—1370.

Dantzler, W. H. (1975). *In* "International Symposium on Amino Acid Transport and Uric Acid" (Eds P. Deetjen, S. Silbernagl, R. Greger, and F. Lang), Georg Thieme, Stuttgart (in press).

Frindt, G. and Burg, M. B. (1972). *Kidney Int.* **1**, 224—231.

Frömter, E. and Gessner, K. (1974). *Pflügers Arch.* **351**, 69—83.

Frömter, E., Wick, T. and Hegel, U. (1967). *Pflügers Arch.* **294**, 265—273.

Gottschalk, C. W., Lassiter, W. E., Mylle, M., Ullrich, K. J. Schmidt-Nielsen, B., O'Dell, R. and Pehling, G. (1963). *Am. J. Physiol.* **204**, 532—535.

Grantham, J. J. and Burg, M. B. (1966). *Am. J. Physiol.* **211**, 255—259.

Grantham, J. J., Ganote, C. E., Burg, M. B. and Orloff, J. (1969). *J. Cell Biol.* **41**, 562—576.

Grantham, J. J., Burg, M. B. and Orloff, J. (1970). *J. Clin. Invest.* **49**, 1815—1826.

Grantham, J. J., Qualizza, P. B. and Welling, L. W. (1972). *Kidney Int.* **2**, 66—75.

Grantham, J. J., Irwin, R. L., Qualizza, P. B., Tucker, D. R. and Whittier, F. C. (1973). *J. Clin. Invest.* **52**, 2441—2450.

Grantham, J. J., Qualizza, P. B. and Irwin, R. L. (1974). *Am. J. Physiol.* **226**, 191—197.

Gross, J. B., Imai, M. and Kokko, J. P. (1975). *J. Clin. Invest.* **55**, 1284—1294.

Hegel, U. and Frömter, E. (1966). *Pflügers Arch.* **291**, 121—128.

Hegel, U., Frömter, E. and Wick, T. (1967). *Pflügers Arch.* **294**, 274—290.

Helman, S. I., Grantham, J. J. and Burg, M. B. (1971). *Am. J. Physiol.* **220**, 1825—1832.

Horster, M. and Larsson, L. (1975). VIth Internat. Congress Nephrology. Abstracts of Free Communications. **91**.

Imai, M. and Kokko, J. P. (1972). *J. Clin. Invest.* **51**, 314—325.

Imai, M. and Kokko, J. P. (1974a). *J. Clin. Invest.* **53**, 393—402.

Imai, M. and Kokko, J. P. (1974b). *Kidney Inst.* **6**, 138—145.

Irish, J. M., III and Dantzler, W. H. (1976). *Am. J. Physiol.* **230**, 1509—1516.

Kokko, J. P. (1970). *J. Clin. Invest.* **49**, 1838—1846.

Kokko, J. P. (1972a). *Yale J. Biol. Med.* **45**, 332—338.

Kokko, J. P. (1972b). *J. Clin. Invest.* **51**, 1999—2008.

Kokko, J. P. (1973). *J. Clin. Invest.* **52**, 1362—1367.

Kokko, J. P. and Rector, F. C. (1971). *J. Clin. Invest.* **50**, 2745—2750.

Kokko, J. P. and Rector, F. C., Jr. (1972). *Kidney Int.* **2**, 214—223.

Kokko, J. P., Burg, M. B. and Orloff, J. (1971). *J. Clin. Invest.* **50**, 69—76.

Leaf, A. (1967). *Am. J. Med.* **42**, 745—756.

Lutz, M. D., Cardinal, J. and Burg, M. B. (1973). *Am. J. Physiol.* **225**, 729—734.

Orloff, J. and Handler, J. (1967). *Am. J. Med.* **42**, 757—768.

Ramsay, J. A. (1949). *J. Expl Biol.* **26**, 65—75.

Ramsay, J. A. (1951). *J. Expl Biol.* **28**, 62—73.

Ramsay, J. A. (1952). *J. Expl Biol.* **29**, 110—126.

Ramsay, J. A. (1953). *J. Expl Biol.* **30**, 358—369.

Rocha, A. S. and Kokko, J. P. (1973a). *J. Clin. Invest.* **52**, 612—623.

Rocha, A. S. and Kokko, J. P. (1973b). *Kidney Int.* **4**, 326—330.

Schafer, J. A. and Andreoli, T. E. (1972a). *J. Clin. Invest.* **51**, 1264—1278.

Schafer, J. A. and Andreoli, T. E. (1972b). *J. Clin. Invest.* **51**, 1279—1286.

Schafer, J. A., Patlak, C. S. and Andreoli, T. E. (1974a). *J. Gen. Physiol.* **64**, 201—227.

Schafer, J. A., Troutman, S. L. and Andreoli, T. E. (1974b). *J. Gen. Physiol.* **64**, 228—240.

Staverman, A. J. (1951). *Rec. Trav. Chim. Pays-Bas.* **70**, 344—352.

Stephenson, J. L. (1972). *Kidney Int.* **2**, 85—94.

Stoner, L. (1975). *Fed. Proc.* **34**, 392.

Tune, B. M. and Burg, M. B. (1971). *Am. J. Physiol.* **221**, 580—585.

Tune, B. M., Burg, M. B. and Patlak, C. S. (1969). *Am. J. Physiol.* **217**, 1057—1063.

Walker, A. M., Bott, P. A., Oliver, J. and MacDowell, M. C. (1941). *Am. J. Physiol.* **134**, 580—595.

Welling, L. W. and Grantham, J. J. (1972). *J. Clin. Invest.* **51**, 1063—1075.

Wirz, H. (1956). *Helv. Physiol. Pharmacol. Acta* **14**, 353—362.

4. Electron Probe X-ray Microanalysis

B. L. Gupta, T. A. Hall and R. B. Moreton

Department of Zoology, University of Cambridge, Cambridge, England

I. INTRODUCTION

Our understanding of the mechanisms of ion and water transport by cells and tissues depends on accurate measurements in microvolumes corresponding to physiological compartments. The compartments of interest range from intracellular organelles to a whole organism. Techniques of microsample analysis, micropuncture and microperfusion (Chapters 1, 2 and 3) have now been applied to measure the ionic and organic composition of hitherto inaccessible spaces. Ramsay initiated many of these methods and applied them to the physiology of excretion and osmoregulation in many small invertebrate species (see Introduction).

Parallel advances in the electron microscopy of the fine structure of cells and tissues have revealed a complexity with which the existing concepts of transport mechanisms must be reconciled (Berridge and Oschman, 1972; Oschman *et al.*, 1974). Most of the transporting epithelia are now known to include an intricate system of intercellular spaces and extracellular channels not accessible to direct extraction by any microsampling technique. Such sub-microscopic compartments within tissues form an essential structural basis of all current hypotheses to explain the transport of isotonic, hypotonic or hypertonic fluids (Chapters 5 and 7).

The cells are subdivided into a complex system of membrane-bound organelles, each with its own ionic composition different from the surrounding cytosol. The ionic composition of organelles is important not only in the regulation of their own specific functions, but also in that mitochondria and endoplasmic reticulum are said to maintain the intracellular homeostasis of ions

83

like Ca^{2+}. This cation, in turn, is implicated in the general regulation of ion and water fluxes across the cell and hence the tissues (Rasmussen, 1970; Berridge, 1975 and Chapter 9). The sub-microscopic complexity of cell structure also necessitates a reconsideration of the belief that free ions and water are more or less uniformly distributed in the cytosol (Bernal, 1965; Gupta, 1976). Studies with ion-selective intracellular electrodes (Chapter 20) are beginning to reveal gradients of free ionic Na, K and Cl across the cells. However, the interpretation of results from microelectrodes and from the measurement of radioisotope fluxes remains limited by the lack of reliable information on the total ionic concentrations in localized microvolumes of intracellular milieu.

The method of electron probe X-ray microanalysis has the potential of reliable quantification of elemental concentrations in microvolumes with a simultaneous observation of fine structure in cell and tissue slices. The feasibility of this approach has been long established (Andersen, 1967; Hall, 1971) and the technique has been applied extensively to non-biological materials for a long time (Reed, 1975). Its use in biology has been delayed by the limited instrumental sensitivity, which makes the measurement of the normally low elemental concentrations at best arduous, by problems of specimen stability during analysis, and by the lack of suitable preparatory procedures. Most, but not all, of the technical difficulties now seem to be resolved. The application of the technique to biology has been extensively reviewed in recent years (Hall et al., 1974; Echlin and Galle, 1975).

The capability of this method to determine the concentrations of many elements in one microvolume enormously extends the range of microsample analysis (Chapter 1). But clearly, for the in situ resolution of problems in transport physiology such information can be relevant only if the fine structure and the distribution of diffusible materials in the specimen remain characteristic of a defined physiological state, without being altered during preparation or analysis. Not all approaches to electron probe microanalysis satisfy this requirement. The method has its maximum potential when applied to biological materials prepared and analysed in a deep-frozen and hydrated state (Moreton et al., 1974; Gupta, 1976; Echlin, 1975a). This is the approach which is described in the following pages, together with some results obtained so far.

For readers who are not familiar with the method, brief sections on physical principles, instrumentation and quantification are given first.

II. PHYSICAL PRINCIPLES

In the Rutherford—Bohr model of atomic structure, an atom has a positively charged nucleus and one or more electrons moving in "orbits" around the nucleus, with the number of electrons matching the nuclear charge. The orbits are grouped into "shells", with the "K" shell orbits closest to the nucleus and lowest in energy, the "L" shell orbits further away on average and somewhat higher in energy, etc. The orbits within a shell may differ slightly in energy, and are specified by subscripts such as L_I, L_{II}, L_{III}, etc. If an electron in an incident beam (for example, in an electron microscope) has sufficient energy, it may remove an orbital electron from its shell and the atom is then said to be in an excited state or ionized. An electron from an outer orbit may then fall down into the vacated orbit, and an X-ray photon may be emitted with a quantum energy given precisely by the difference in energy between the two electron orbits. For X-rays emitted in this way, there is a conventional labelling scheme which corresponds to the initial and the final orbits of the transition; for example, the radiation designated $K_{\alpha 1}$ arises from an electron transition from orbit L_{III} to a K-shell orbit (Fig. 1).

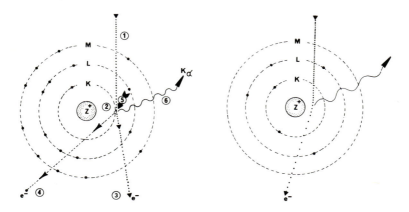

Fig. 1. Physical basis of X-ray generation in the electron microprobe. Left. Generation of characteristic X-rays. (1) An electron in the beam impinges on an atom. (2) An electron in an inner atomic orbit is "knocked out". (3) and (4) The incident and ionized electrons leave the atom. (5) An electron from an outer orbit "drops" into the vacated orbit, accompanied by (6) the emission of a characteristic X-ray quantum. Right. Generation of X-ray continuum. Deceleration of an incident electron is accompanied by the emission of a quantum in the X-ray continuum.

The nuclear charge defines the identity of an atom as belonging to one or another chemical element. Since the energies associated with the orbits are also determined primarily by the nuclear charge, it follows that each element has its own set of orbital energy levels, and hence its own X-ray emission spectrum with its own set of X-ray quantum energies. The X-rays which are emitted as a consequence of ionization are therefore aptly described as "characteristic" radiation, each emission being characteristic of the emitting element. The specific nature of this radiation provides the basis of X-ray spectroscopy and X-ray microanalysis.

As with any other part of the electromagnetic spectrum, each X-ray photon has a specific quantum energy and also displays certain wave-like properties with a specific wave length. The quantum energy and the wave length are related by the equation:

$$\lambda E = 1239.6$$

where the quantum energy E is expressed in units of electron volts (eV) and the wave length λ is expressed in nanometers (nm).

An electron in an incident beam may also suffer deceleration by interacting with the electrical field produced by the nucleus of an atom in a specimen, and the energy lost by the incident electron in this interaction may appear as an X-ray photon. The quantum energy of such a photon can be anything from zero up to the energy of the incident electron. Hence such interactions produce an additional X-ray spectrum called the "continuum", "white" radiation, or "Bremsstrahlung" (which translates as "braking radiation"). The continuum is well known as an unwanted background which can greatly complicate the observation of weak characteristic radiations, but it can also serve as a measure of specimen mass (cf. Section IV below).

Both characteristic and continuum X-radiations are automatically generated during all electron microscopy. Thus if the electron optical instrument is fitted with the means to reduce the area irradiated by the incident beam, and suitable X-ray spectrometers are fitted, the instrument becomes an electron probe X-ray microanalyser. The basic procedure in almost all types of electron probe X-ray microanalysis is to form some kind of electron microscope image of the specimen, and then to limit the irradiating electron beam to a microarea selected within the image and to analyse the emitted X-rays.

III. DIFFERENT INSTRUMENTAL APPROACHES

For a person entering the field of biological X-ray microanalysis, the commercially available systems offer a bewildering set of choices. The specimen images may be formed by "conventional" electron microscopy (i.e. with a non-scanning electron beam and with image-forming electron lenses) or by scanning electron microscopy (i.e. by irradiating the specimen point-by-point sequentially with a finely focussed beam and forming the image on a synchronously scanned cathode ray tube). The signal for a scanning image may be based on secondary, backscattered, absorbed or transmitted electrons. There is also a choice between two quite different kinds of X-ray spectrometers: energy-dispersive systems utilizing solid-state ("SiLi") detectors, and wave-length selective systems based on diffracting crystals. Finally one must decide whether the specimens should be in bulk form, or in the form of thin or ultra-thin sections.

The characteristics of the available instrument types have been described generally by several authors, for example Reed (1975), and the subject has recently been reviewed briefly from the standpoint of the biologist (Hall, 1975a). Here a general recapitulation would be too lengthy, but we shall discuss the choice of conditions specifically for the analysis of fully or partially hydrated specimens in physiology.

1. Spatial Resolution of the X-ray Analysis

This parameter depends on the scattering and consequent spread of the incident electrons within the specimen. We believe that the size of the spaces of interest in transport physiology requires an analytical spatial resolution at least as good as 0.5 μm, preferably better. Bulk specimens are therefore precluded on the grounds of excessive electron spread; one wants sections no thicker than approximately 1 μm and high beam voltages (preferably 50 kV or higher).

2. Image Contrast

The intrinsic image contrast in unstained, highly hydrated sections of soft tissue is very low. Image formation by a scanning beam is preferable in order to take advantage of the contrast control and enhancement available in the electronic processing of the scanning signal.

For the observation of internal structure, the transmission scanning mode (Echlin, 1975b) is probably best.

3. Spatial Resolution in the Image

In the transmission scanning image, this parameter also depends on electron scatter and spread, and the section should therefore be as thin as is consistent with other requirements.

4. X-ray Spectrometry

Energy-dispersive spectrometers have the advantages of ease of operation, simultaneous detection of many elements and high X-ray collecting power. But their peak-to-background performance, inferior to that of diffracting spectrometers, is inadequate for the assay of important elements in hydrated specimens — for example, sodium at a concentration of 20 mM. It is highly desirable to use both energy-dispersive and diffracting spectrometers. The latter have a poor X-ray collecting power and therefore, in order to gain sufficient X-ray intensity, the sections should be as thick as is consistent with the required spatial resolution.

5. Low-temperature Facilities

A section will remain hydrated in vacuum under the electron beam only at low temperature. For analysing hydrated sections of cryo-preserved soft tissues, the instrument must have a cold stage maintained at below $-150°C$ and a colder anti-contamination plate. There must also be a system for keeping the specimen at low temperature without condensation from the atmosphere while it is inserted into the instrument.

For the microanalysis of frozen-hydrated sections in our Laboratory, we use a JEOL JXA-50A scanning microanalyser operated at 50 kV. The instrument includes a full range of scanning microscopy facilities, two diffracting spectrometers, a Kevex energy-dispersive spectrometer, a low-temperature transmission stage cooled by liquid nitrogen to $-170°C$, an anti-contamination cap cooled by liquid nitrogen to $-190°C$, and an airlock which is evacuated during loading of the specimen. (The energy-dispersive detector is not part of the standard instrument, and the cold stage and cold cap have been modified from the manufacturer's design.) The tissue sections are most often cut at a thickness of 1 μm.

IV. QUANTIFICATION OF THE X-RAY DATA

Most of the many methods of obtaining quantitative results in the X-ray microanalysis of biological specimens have been reviewed recently (Hall, 1975b). Here we shall summarize the approach used in our Laboratory for 1-μm thick, hydrated sections of soft tissue mounted on ultra-thin supporting films.

A. What the Microprobe Measures

Microprobe measurements of the "concentration" of an element a actually determine the local mass fraction f_a, which is defined as the mass of element a in the analysed microvolume divided by the total mass of specimen in the same volume. The physiologist may prefer to express such a mass fraction in terms of mM kg^{-1}, i.e. millimoles of element a per kg of specimen, denoted here as C_a. The conversion between f and mM kg^{-1} is direct and unambiguous. For example with sodium, mol. wt 23, a value of C_{Na} = 100 mM kg^{-1} amounts to 2.3 g of Na per kg of specimen, or f_{Na} = 2.3/1000 = 0.23%.

In order to interpret the microprobe measurements, the physiologist must realise that the results refer to the specimen as it exists at the time of analysis (e.g. if the specimen is dehydrated, the "total mass" will be the local dry mass) and they refer to the total amount of an element locally present, bound plus unbound. It is essential to establish clearly the relationship between a measurement of mM kg^{-1} and a measurement (by another method) of "mM l^{-1}" (millimoles per litre).

In a simplified model, suppose that a microvolume is fully hydrated and contains a mass of water M_H, a "dry" mass M_D, a mass M_{aH} of element a dissolved in the water, and a mass M_{aD} of element a bound to the "dry" material. A microprobe measurement in this volume will give (if for the moment we neglect the contribution of dissolved elements to the total mass):

$$f_a = \frac{M_{aH} + M_{aD}}{M_H + M_D}. \tag{1}$$

If the specimen were then dehydrated and the only effect was to remove the water (no movement of element a or of dry mass into or out of the microvolume), the measurement would give:

$$f_{a1} = \frac{M_{aH} + M_{aD}}{M_D}. \tag{2}$$

On the other hand, if there is another method which measures all of element a in the aqueous phase and which samples only that phase, the result will give:

$$f_{a2} = \frac{M_{aH}}{M_H}. \tag{3}$$

(Extraction by micropipette, followed by atomic absorption analysis, might ideally perform in this way.) Finally, a measurement by microelectrode in the aqueous phase *in situ* will give a result corresponding to:

$$f_{a3} = \frac{\alpha M_{aH}}{M_H}. \tag{4}$$

where the activity factor α (less than unity) takes account of the fact that the dissolved element is not completely ionized. In the case of measurements on the aqueous phase alone (equations 3 and 4), there is a direct conversion between amount per kilogramme and amount per litre, since one is speaking of the amount of element per given amount of water. But clearly, in order to compare microprobe measurements (equations 1 and 2) with measurements on the aqueous phase alone (equations 3 and 4), one must be able to make some judgement on the relative amounts of the element in bound and unbound form and one must have some idea of the local dry-mass fraction. In many extracellular spaces the dry-mass fraction may reasonably be taken to be zero and equations (1) and (3) become the same. This is not in general valid for intracellular compartments.

B.   Microprobe Measurement of Mass Fraction or Millimoles Per Kilogramme

In a thin specimen, the intensity of emission of a characteristic radiation of an element a is a measure of the amount of element a per unit area. To go from amount of element per unit area to mass fraction one clearly needs a measure of total specimen mass per unit area, and the intensity of continuum radiation can serve as such a measure. The ratio of characteristic to continuum intensities is proportional to the mass fraction. This approach leads to the equation:

$$f_a(sp) = \frac{(n_a/W)_{sp}}{(n_a/W)_{st}} f_a(st) \frac{g(sp)}{g_{st}}$$

$$= \frac{R_a(sp)}{R_a(st)} f_a(st) \frac{g(sp)}{g(st)}. \tag{5}$$

Here $f_a(sp)$ and $f_a(st)$ are respectively the mass fractions of element a in an analysed microregion of a thin specimen and in a thin standard; sp and st identify specimen and standard; n_a is the number of characteristic quanta detected and W is the simultaneously observed count of continuum photons; and the factors g take account of the fact that standard and specimen may not generate the same continuum intensity per unit mass per unit area. The signal W is best observed in a quantum energy band above the main characteristic lines. The choice of energy band is somewhat arbitrary and must come from a compromise. At low energies there is a relatively large extraneous continuum background from bulk material near the specimen, while at high energies the continuum signal from the specimen is weak. Most often we use a 50-kV beam and count continuum in the quantum energy band 10–22 keV.

In the second line of equation (5) we have introduced the notation $R_a = n_a/W$ for the sake of convenience later on.

The quantities n_a and W must be the background-corrected values. The characteristic count n_a may be obtained with either an energy-dispersive or a diffracting X-ray spectrometer. W must be corrected for continuum generated outside the specimen. The main extraneous contributions come from supporting films, grids or specimen holders and the magnitudes of the corrections must be determined by suitable background runs.

To obtain the quantity n_a one may also have to correct for the absorption of characteristic X-rays within the specimen. For thin specimens of soft tissue this correction is usually negligible except for sodium. In our instrument, in order to correct for absorption in a fully hydrated 1-μm section the recorded sodium count should be multiplied by approximately 1.3. A simple correction scheme has been detailed elsewhere (Hall, 1971). To perform the correction, one needs a value for the local specimen mass per unit area; this can be obtained from a mass standard (cf. Section 4 on standards below).

In equation (5), according to a simple theory of the generation of X-ray continuum, the quantity g for specimen or standard is given by:

$$g = \sum_r (f_r Z_r{}^2 / A_r) \qquad (6)$$

where the sum is taken over all constituent elements, Z is atomic number and A is atomic weight. (Equation 6 is the most convenient form for the calculation of g when the composition of the material is directly known only in terms of the mass fractions f. When the chemical formula is known, it is more convenient to use the entirely equivalent form:

$$g = \frac{\sum_r (N_r Z_r{}^2)}{\sum_r (N_r A_r)} \qquad (7)$$

where N_r is the number of atoms of element r in the chemical formula.) In the application of equation (5) to microanalysis in general, the factor $g(sp)$ poses a difficulty since it involves the mass fractions which one seeks to measure. Therefore in general, in order to evaluate $g(sp)$, for the assayed elements the observed X-ray counts must be introduced in some way, either iteratively or by an algebraic equivalent, and some estimation must be made of the composition of the microvolume apart from the assayed elements (see Hall, 1971). However, in soft tissues which consist mainly of the elements H, C, N and O, $g(sp)$ can be estimated reasonably well. Typical values of g are 3.3 for protein, 3.0 for lipid, 3.4 for carbohydrate, 3.4 for nucleic acid (aside from the phosphorus content which should be brought into the formulation separately when the phosphorus radiation is recorded), and 3.67 for water. The limited spread of these values makes it practical in the case of the analysis of soft tissues to insert a simple average into equation (5). Furthermore, with appropriate standards, the factor $g(sp)/g(st)$ in equation (5) can usually be kept close to unity.

Equation (5) can be put just as well in terms of C_a, mM kg^{-1}:

$$C_a(sp) = \frac{R_a(sp)}{R_a(st)} \, C_a(st) \, \frac{g(sp)}{g(st)}. \qquad (8)$$

We have already noted that the measurement of C_a must refer to the specimen as it exists at the time of analysis. How can one obtain a value for mM kg^{-1} wet weight from a measurement on a section which is partially or fully dehydrated? Unfortunately this question requires careful, extensive consideration.

In one method, the tissue is surrounded by a medium containing salts and organic matter, and it is assumed that the medium has the same dry weight fraction and undergoes dehydration to the same degree as the tissue itself. This medium, located at the periphery of the tissue section, can then serve as a standard. If we define f as the ratio of local mass during measurement to the fully hydrated local mass and we use H to denote the state of full hydration and D for the conditions during analysis, then we may write (neglecting the corrections associated with the factors g):

$$\frac{R_a(sp, D)}{R_a(st, D)} = \frac{R_a(sp, D) \cdot f}{R_a(st, D) \cdot f} = \frac{R_a(sp, H)}{R_a(st, H)} = \frac{C_a(sp, H)}{C_a(st, H)}, \qquad (9)$$

from which it follows that:

$$C_a(sp, H) = \frac{R_a(sp, D)}{R_a(st, D)} C_a(st, H). \qquad (10)$$

Equation (10) gives the desired wet weight fraction in terms of the X-ray counts on the dehydrated fields and the known "wet" composition of the medium. However this equation cannot give precise results because a) one cannot really be sure that the degree of dehydration is the same in the standard and locally in the specimen (unless they are *totally* dehydrated) and more importantly, b) it is impossible to have equal dry weight fractions in general in the standard and locally in the specimen because the local dry weight fraction in the specimen varies from point to point. In sum, the factor f in equation (9) really cannot be accurately taken as the same in the standard and locally in the specimen. Equation (10) is therefore not used in practice for largely dehydrated sections except as a check on the reasonableness of a result.

Another method can be based on the same kind of peripheral standard. If it is assumed that the standard and the analysed region of the specimen have the same mass per unit area prior to dehydration, then it is necessary to monitor only the characteristic radiation as a measure of elemental amount per unit area, and we may write:

$$C_a(sp, H) = \frac{n_a(sp)}{n_a(st)} C_a(st, H). \tag{11}$$

This is an approach used by Dörge et al. (1975). It is assumed that the amount of element a per unit area is not affected by dehydration, or at least is affected equally in standard and specimen, an assumption which may be inadequate especially because of the possibility of non-uniform shrinkage. But the main difficulty is in the assumption of uniform mass per unit area prior to dehydration. In our specimens at least, the accuracy of this assumption is limited because of non-uniformity in the thickness of the section when it is cut, and because of further distortions as the section is deposited onto its support.

None of these difficulties is pertinent to the analysis of fully hydrated sections, but the image contrast in such sections is too poor for microanalysis. The most promising approach seems to be to analyse after a minimal dehydration which is just enough to give an adequate image (in practice a mass loss of perhaps 10% or less). In material which is so slightly dehydrated, one may use equation (10) with only a slight error associated with the underlying assumptions.

The remaining question in this scheme is how to establish that there has been only slight dehydration at the time of analysis. The best and most direct index of degree of hydration is the ratio of a characteristic signal to continuum in the peripheral medium. With a stable analytical system and a medium of fixed composition this ratio should be consistent for fully hydrated material and only slightly elevated for slightly dehydrated material; a substantial increase above the value for fully hydrated medium indicates substantial dehydration.

The degree of hydration can also be judged from changes in mass per unit area as manifested in the continuum signal. The slight initial dehydration which is imposed for the sake of image contrast is usually performed in our practice before the specimen enters the microanalyser, so it is not possible to compare continuum signals before and after this step. However, one may fully dehydrate after analysis, within the analyser, and record the effect on the continuum intensity from a selected region. In going from slight to full dehydration, one might expect a change in continuum signal corresponding to the dry weight fraction. (For example, in material with a dry weight fraction of 20%, the full dehydration might be expected to reduce the continuum intensity by a factor of approximately five.) Unfortunately, two other factors may also affect the change in mass per unit area during dehydration: shrinkage and latent radiation damage.

A simple shrinkage affects characteristic and continuum counts in the same way. Hence one may compensate for shrinkage by using the ratio of characteristic counts before and after dehydration as a correction factor.

Latent radiation damage caused by the electron beam is more difficult to handle. When dehydration is carried out after analysis by raising the specimen temperature, it is believed that there is a loss in organic mass, which is due to radiation damage but does not actually occur until the temperature rises (Stenn and Bahr, 1970; Hall and Gupta, 1974). In areas which are to be used to estimate degree of hydration or dry mass fractions, it may be possible to keep the radiation dose well below a damaging level, although this is very tedious in practice; it may also be possible in the future to establish dehydration temperatures at which the loss of organic mass remains latent. However when we interpret dehydration data at present, we have to recognise that some of the loss of mass accompanying dehydration may be a loss of organic mass due to radiation damage, typically approximately 30% of the organic mass in many organic polymers. Consequently, at this time we can use loss of mass on dehydration as a good qualitative check to confirm an initially high level of hydration, but we are not ready to use the mass loss data routinely in a quantitative way.

While the ratio of characteristic to continuum radiation from the medium is currently the clearer way of establishing a state of nearly full hydration, the observation of continuum change during dehydration offers a further important possibility. If the problem of latent radiation damage can be mastered, the measurement of local mass loss on dehydration should give a direct way of measuring dry weight fractions in individually analysed microregions.

In the work reported below we have analysed both slightly dehydrated and fully dehydrated sections, using the approaches embodied in equations (10) and (11).

C. Measurement of Elemental Ratios

In thin specimens the relative amounts of two elements a and b may be measured by means of the equation:

$$\left(\frac{C_a}{C_b}\right)_{sp} = \frac{(n_a/n_b)_{sp}}{(n_a/n_b)_{st}}\left(\frac{C_a}{C_b}\right)_{st}, \qquad (12)$$

where $(C_a/C_b)_{st}$ is the ratio of the elemental amounts in a thin standard, and the quantities n are corrected for background and absorption just as mentioned above.

In the measurement of elemental ratios there is no need to measure mass per unit area or to obtain the continuum count $W*$, or to estimate the factors g, and the degree of hydration is irrelevant. Thus measurements of elemental ratios are simpler and more reliable than measurements of mass fractions.

D. Standards

In conjunction with the methods described above, the standards must be thin (approximately 1 μm or thinner), homogeneous and of known composition. It should be stressed that there is no need to know the thickness since all of these methods involve ratios which are independent of thickness.

For absolute measurements one may use sections of minerals (Hall and Peters, 1974), but the corrections are simpler and more accurately made if one uses preparations of the relevant elements in sections of organic material. Many types of organic standards have been described in recent reviews (Spurr, 1975; Hall, 1975b). One of the most stable, now coming into widespread use, is the section of epoxy resin containing the relevant elements in soluble complexes (Spurr, 1974; Jessen et al., 1974).

For the measurement of elemental ratios, again it is possible to use minerals (Cliff and Lorimer, 1975; Rowse et al., 1974) but the simplest preparations are dried droplets from an aerosol spray of a solution of mixed salts (Morgan et al., 1975).

---

* Except that specimen mass per unit area must enter into the calculation of any required absorption correction.

For the measurement of the absolute mass fractions of several elements it may be convenient to use an absolute standard for one element and to determine the others absolutely by ratio. Thus for example, potassium may be assayed by means of a standard of potassium in epoxy, and the other elements can then be determined from ratio standards containing potassium, via equation (12).

The standard which we use most often is the medium on the periphery of the tissue section. Besides the great convenience of immediate access, such standards have substantial advantages. As described above, in slightly dehydrated sections the parallel dehydration of standard and tissue gives some compensation so that one may estimate mM kg^{-1} wet weight by means of equation (10). Also, the g factors for tissue and standard can be favourably matched. If the tissue and the medium have the same dry weight fraction, then they will have approximately the same g values when fully hydrated and the g values will change in parallel through slight dehydration in spite of the substantial difference between g for water (3.67) and g for protein or carbohydrate (3.3 or 3.4), so that the g factors need not be estimated. Finally, the medium as standard is essential to the method of analysis based on equation (11).

For the calculation of absorption corrections and for quick estimates of section thickness it is important to have an additional standard to measure mass per unit area. Such a standard should be a thin, uniform film of known composition and thickness, again preferably organic to minimize effects needing correction. At present we are using 2-μm films of polycarbonate, $C_{16}H_{14}O_3$, density 1.2 g cm^{-3} ("Makrofol", supplied by Siemens). (These films have a thin coating of aluminium, which can be measured and taken into account.) The mass per unit area in a specimen, M_{sp}, can be measured by reference to such a standard through the equation:

$$M_{sp} = \frac{W_{sp}}{W_{st}} M_{st} \frac{g_{st}}{g_{sp}}. \tag{13}$$

Here the quantities W_{sp} and W_{st} must be obtained from separate runs on specimen and standard; since they are proportional to probe current and detection time, these parameters must be held constant or allowance should be made for any variation. To minimize beam damage, the irradiated "microregion" in the mass standard should be large (say, 100 x 100 μm) and the integrated dose to the standard should be kept well below 10^{-10} Coulomb per μm^2.

V. ANALYSIS OF FROZEN-HYDRATED SECTIONS

A general scheme for the fully quantitative analysis of ions in tissue sections kept hydrated below $-80°C$ has been published earlier (Moreton *et al.*, 1974; Gupta, 1976). The current version of the method is summarized in Figs 2 and 3. A brief evaluation of each step is given below.

A. State of Cells and Tissues

The first step involves the transfer of tissue samples to minichucks, silver or copper pins or silver tubes for quench-freezing and transfer to a microtome. Even if the cells and tissues have been previously assessed for physiological function, the possible hazards during this manipulation cannot be overstated. A point emphasized by Ramsay (1953) and sometimes overlooked is the rapid change in small samples by evaporation of water in the dry atmosphere of a laboratory. Where bathing medium is to be used as a standard for quantification of data, the change in concentration in the thin film of defined solution surrounding the tissue can significantly alter the final analysis. Osmotic effects generated by this rapid evaporation may also alter ionic compositions in delicate but rapidly transporting tissues like insect salivary glands and Malpighian tubules within the microdrops on pinheads. In contrast to physiological experiments *in vitro*, where tissues can rapidly recover, such altered conditions will be preserved by freezing and will be reflected in the results of analysis. We routinely manipulate our living samples in an atmosphere saturated with water vapour. For larger vertebrate tissues, more elaborate methods to minimize the lag between the removal of tissue in a certain functional state and the rapid freezing have been devised for muscle (Sjöström *et al.*, 1974; Bacaner *et al.*, 1973) and frog skin (Dörge *et al.*, 1974).

B. Processing and Preservation of Samples

To determine the distribution of ions faithfully reflecting the physiological state, all diffusion processes in cells and tissues must be *instantly* stopped. Chemical methods of tissue fixation, with or without the addition of ion precipitating agents, must cause drastic redistribution and loss of water-soluble substances. They are not

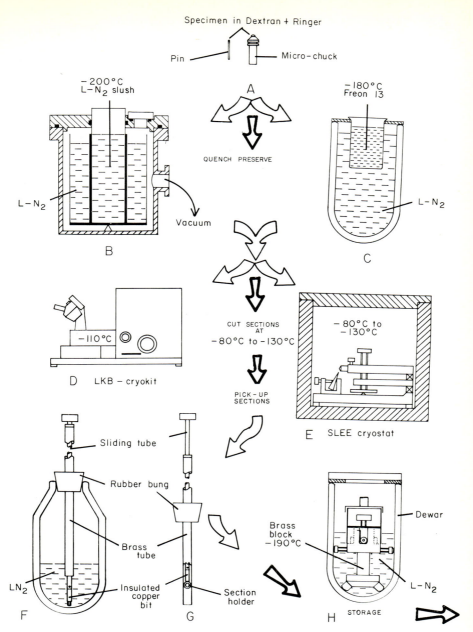

Fig. 2. Preparation and storage of frozen-hydrated sections for X-ray microanalysis. The tissue is mounted on pin or micro-chuck (A), quenched in a cryogen (B or C) and sectioned in a cryokit (D) or cryostat (E). The sections, mounted on special collars, are transferred from cryokit or cryostat into storage (H) by means of a transfer rod (G) which has been pre-cooled (F).

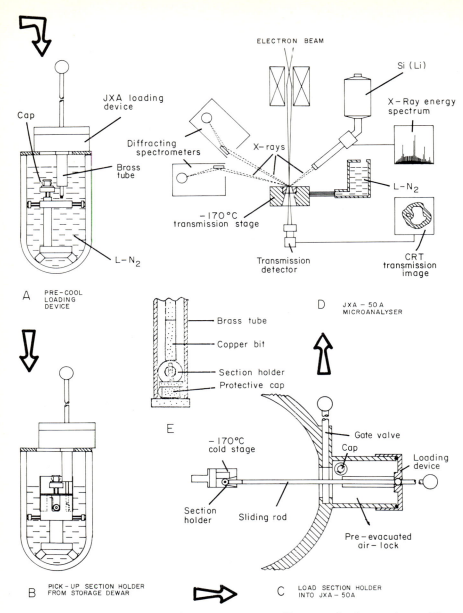

Fig. 3. Transfer of sections from storage to cold-stage of microanalyser. The JXA loading device is cooled (A) and then used to transfer sections from storage (B) to the airlock, and thence to the cold-stage of the analyser after airlock evacuation (C). The device is designed to keep the section holder cold and protect it from the atmosphere during transport (E). The microanalyser facilities shown in (D) are all operated simultaneously.

suitable to preserve tissues for quantitative microanalysis (Sjöström, 1975; Sjöström *et al.*, 1974). Ions can diffuse at an appreciable rate even in ice, and only below the eutectic temperature is the redistribution arrested. For most ionic species this temperature is thought to be $-40°C$ although ions like Ca may retain mobility down to $-55°C$ (Meryman, 1966). Furthermore, in most animal epithelia, water constitutes up to 80% of the volume in cells and may form 99% of the fluids in lumens, intercellular spaces and serosal cavities. Crystallographic studies on the structure of ice indicate that bulk water cannot be frozen to form an amorphous or vitreous phase even if cooled below $-180°C$. The significance of crystalline structure in frozen water at various temperatures depends on the mode of study. In general a microcrystalline structure, resolved only by X-ray and electron diffraction, seems to exist below $-130°C$. As the temperature rises, the crystal size grows. Above $-80°C$ coarse hexagonal crystals begin to form and may continue to grow by fusion. Such transitions in crystalline phases are slow, occurring over hours. If microcrystals are surrounded by organic matrices, as in cells, the thermotropic change to larger crystals may be vastly slowed or even prevented. Once the water is frozen to form larger crystals, a change to microcrystalline structure by lowering the temperature apparently cannot be achieved. The formation of large ice crystals obviously destroys the fine structure of cells and tissues. Moreover, all molecules other than water are excluded from such pure crystals and become concentrated at crystal boundaries, thereby causing serious osmotic effects. Ideally, the formation of resolvable ice crystals should be prevented if the samples can be instantly and uniformly cooled to well below $-130°C$. In practice, cooling rates of the order of $10^4°C\ s^{-1}$ may be needed for a good preservation of ultrastructure and hence ionic distribution (Riehle, 1968).

Microcrystals certainly form during the freezing step, and it is clear that the mean diameter of the crystals must be kept smaller than the required microanalytical spatial resolution.

1. Rapid Freezing

The simplest method of rapidly freezing cells and tissue samples is by plunging them into clean liquid nitrogen. However the boiling caused by the warm sample forms an insulating envelope of gaseous nitrogen (Leydenfrost) around the block, reducing the cooling rate and increasing the ice-crystal damage. This difficulty should be overcome by the use of nitrogen near its melting point in the form of a solid-liquid slush (Umrath, 1974; Roomans and Seveus, 1976), although Hodson and Williams (1976) obtained less satisfactory freezing of vertebrate tissues with melting nitrogen than with fluorocarbons. In our samples of insect epithelia, droplets of whole chick blood and Dextran-containing salt solutions the results so far have also been less satisfactory with melting nitrogen. The most reproducible results appear to be achieved with cryogens with relatively high boiling points, cooled by liquid nitrogen to form a mixture of solid and liquid phases (Echlin and Moreton, 1976).

The cryogens in common use are fluorocarbon refrigerants (dichlorodifluoromethane, monochlorodifluoromethane, monochlorotrifluoromethane), pentane, isopentane, etc. (Pearse, 1968; Rebhun, 1972; Clark et al., 1976). The results described in this chapter were obtained by freezing the tissues in monochlorotrifluoromethane (Freon 13, melting point $-181°C$, boiling point $-81°C$), in the arrangement illustrated in Fig. 2C. The dispersion of heat from the specimen can be further improved by the use of a generous volume of the cryogen and by mounting the specimens in preparation for quenching on thin metallic pins, tubes or microchucks. With microchucks one also avoids a later intermediate and possibly hazardous step of mounting a frozen specimen onto the microtome chuck.

Possibly, a common cause of coarse ice-crystal formation in biological samples may be the slow freezing which begins during transition through the cold atmosphere above the cryogen-containing vessels. This effect is probably reduced or avoided by mechanically propelling the specimen into the cryogen by devices like those described among others by Rebhun (1972) and Echlin and Moreton (1976). However, the damaging effects of high mechanical forces on impact with the cryogen and during deceleration are difficult to evaluate. Small droplets of whole chick blood, or delicate tubular epithelia carefully oriented to obtain transverse sections, are often distorted into shapes less suitable for sectioning. Cracks may also develop in the block of frozen tissue on mechanical impact.

Since the frozen sections are not cut below $-140°C$, any residual liquid or solid nitrogen adhering to the tissue block boils off, but cryogens with high boiling points do not. Liquid films of such cryogens can cause serious difficulties in low-temperature sectioning. Good sections become contaminated by smeared cryogens which may refreeze when the sections are transferred to a specimen stage at much lower temperature for microanalysis. We have frequently found elevated levels of chlorine in frozen-hydrated sections contaminated with residual monochlorotrifluoromethane. This problem is alleviated when sections are analysed in a dry state, although adsorption of halide molecules onto the tissue during sublimation could add extraneous chlorine to the specimen.

In principle the cleanest method of quench-freezing samples is by pressing them against a highly polished block of pure copper cooled to liquid-nitrogen temperature (Eränkö, 1964; Christensen, 1971) or a silver block at liquid-helium temperature (Van Harrevald et al., 1965). Suitably shaped recesses in the surface of the metal can also help to shape the block into a pyramid — a more favourable geometry for sectioning than a hemispherical droplet (Seveus and Kindel, 1974). Although the cooling rates are faster due to the much higher thermal conductivity of the metals (405 Joule m^{-1} s^{-1} $°K^{-1}$ for copper) as compared to liquid cryogens (0.1 Joule m^{-1} s^{-1} $°K^{-1}$, Echlin and Moreton, 1976), superior freezing is confined to a depth of $10-20$ μm at the surface of the sample. We have had a limited success with this method because our tissues are often enclosed by a film of bathing medium which usurps the benefits of rapid cooling.

After quench-freezing, if the specimen holders are not already in liquid nitrogen, they should be immediately transferred to clean liquid nitrogen in a small vessel. The whole vessel is then transferred to the chamber of the cryomicrotome. If transferred through the laboratory atmosphere, the sample would rapidly gain heat and moisture. All implements which come into contact with the specimen-bearing metal, such as forceps, must be precooled to liquid nitrogen temperatures.

More extensive discussion on the quench-freezing of biological samples may be found in various recent symposia and reviews (Meryman, 1966; Roth and Stumpf, 1969; Mazur, 1970; Moore, 1972; Plattner et al., 1972; Rebhun, 1972; Hall et al., 1974; Echlin and Galle, 1975; Echlin and Moreton, 1976).

2. The Role of Cryoprotectants

A large number of low mol. wt substances, many of which penetrate the cells, have been added to biological samples to prevent ice-crystal formation during freezing (Meryman, 1966; Echlin and Moreton, 1976). Glycerol and dimethylsulphoxide (DMSO) are used as cryoprotective agents for the preparation of freeze-fracture replicas for electron microscopy. Both substances are known to increase greatly the permeability of cell membranes and hence cannot be suitable for a physiologically meaningful localization of ions. Larger molecules like Dextran and polyvinyl pyrolidone (PVP), which do not penetrate the cells but do seem to prevent the formation of ice crystals, are more acceptable as cryoprotectants in microanalytical studies. Serum albumin, gelatine (Meryman, 1966) and sucrose (Roomans and Seveus, 1976) have also been used. Before the employment of any such agents in nonphysiological concentrations, their effect on the proper function of the tissue must be assessed *in vitro*. For example we found that 10—20% PVP (W/V), when included in the bathing solution for *in vitro* studies of *Rhodnius* Malpighian tubules, virtually blocked the rapid fluid secretion which normally follows stimulation with 5-hydroxytryptamine (5-HT). On the other hand, the inclusion of up to 20% Dextran (Pharmacia-Sigma, approximate MW 237,000) had no significant effect on normal function. Apart from a cryoprotective role, Dextran-containing bathing salines of defined composition are very suitable standards for the quantification of the analytical data (cf. Section IV above; also Gupta *et al.*, 1976b); and the added macromolecules also seem to improve the sample mechanically, greatly reducing the danger of shattering during sectioning. How the macromolecular protectants prevent freezing damage remains speculative. Conceivably they improve cooling rates by increasing the thermal conductivity of the fluids surrounding the cells, or perhaps they retard the freezing of this fluid and thus delay the effect of the latent heat of formation of ice as the medium freezes.

C. Cryomicrotomy

No matter how well the tissue blocks are quench-frozen, changes can still occur if the temperature of the samples is later allowed to rise much above $-130°C$. Until recently, cryosections for the study of water-soluble components of tissues were cut at temperatures between $-60°C$ and $-85°C$ (Roth and Stumpf, 1969; Appleton, 1968; Echlin *et al.*, 1973; Moreton *et al.*, 1974). In retrospect, the lower limit seems to have been largely due to technical difficulties in maintaining cryostat microtomes at lower temperatures and perhaps due to a general belief that fresh-frozen tissues are too brittle to section below $-80°C$ (Appleton, 1974a). Nevertheless Hodson and Marshall (1970) claimed to have obtained ultra-thin sections at

−198°C. More recent experience has shown that it is feasible to obtain frozen sections with the tissue block down to −140°C and the knife a few degrees warmer (Seveus and Kindel, 1974; Hodson and Williams, 1976; Appleton, 1974a). At such low temperatures, the sections have usually been cut at a thickness of 50−100 nm and analysed after they have been allowed to dry.

1. Cryomicrotomes

There are essentially two instrumental approaches to sectioning at low temperatures. First, there are the conventional cryostat cabinets containing a microtome (Pearse, 1968) fitted with improved refrigeration to maintain temperatures down to −70°C (Stumpf and Roth, 1965; Dörge *et al.*, 1974) or −85°C (Moreton *et al.*, 1974). The use of liquid-nitrogen cooling rather than two-stage cascade refrigeration units now enables the temperatures of the cabinets to be maintained down to −140°C (SLEE, London). The cabinets have also been adapted to take ultramicrotomes for ultrathin sections (Appleton, 1974a). Most of the operations with these cryostats are controlled from outside and the spacious chamber allows for additional fittings such as anti-roll plates, devices to pick up sections (Appleton, 1974a) and storage of sections mouted on EM grids or metal collars. For the cutting of sections with a thickness of 1−2 μm, like ours, freshly sharpened, carbon-steel knives with a low bevel angle (12−15°) (Roth and Stumpf, 1969) are convenient and perhaps superior to the commonly used but less suitable glass and diamond knives (Hodson and Marshall, 1976). If the knives can be laterally displaced by means of an external control, a very large part of the cutting edge can be used without the danger of frost. Alternatively, adaptors for the use of either glass or diamond knives are available. Stereomicroscopes with a long working distance are fitted outside the observation windows to scrutinize the section-cutting operation. The great advantages of the cryostat approach are that the tissue block, knife and sections are maintained at the relatively uniform temperature of a large chamber and that there is space for warm, moist air to rise, so that the tissue should be well buffered from moist currents by a substantial layer of cold, dry gas. However certain manual operations such as specimen loading, recovery of sections and some adjustments have to be performed through the opening at the top, thereby increasing the risk of frost deposition onto the cut face of the specimen and onto the sections.

This risk can be minimized by the use of dry, long-sleeved rubber gloves and by maintaining a "curtain" of flowing, cold, dry nitrogen at the window (Appleton, 1974a).

The second approach is to adapt one of the various models of ultramicrotomes by building a small cryochamber around the cutting area (Christensen, 1971; Hodson and Marshall, 1970). Such "cryokit" attachments are now sold for several commercial brands of ultramicrotomes (Sorvall-Porter-Blum, Cambridge-Huxley, Reichert, LKB). In most of the models, thermostatically controlled liquid-nitrogen feeds provide independent regulation of the temperatures of the tissue block and the knife down to −140°C. But to our knowledge, no such control is available for the temperature of the cryochamber in most models. In Christensen's device for the Porter-Blum ultramicrotome (Sorvall), the cryochamber is cooled by feeding liquid nitrogen from the bottom. The cutting area is therefore equilibrated with the chamber temperature of around −75°C. Other authors, using either an LKB cryokit (Seveus and Kindel, 1974; Roomans and Seveus, 1976) or a Cambridge-Huxley cryoattachment (Hodson and Marshall, 1976), emphasize the low temperatures of tissue block and knife but do not mention the temperature in the chamber.

In addition to a SLEE cryostat, we also use an LKB cryokit with Ultratome III to cut relatively thick (1–2 μm) and ultra-thin (c. 100 nm) sections for microanalysis. With the cryokit as delivered, thermocouple measurements revealed that when the block holder and the knife were maintained in the range of −140°C to −120°C, the temperature in different areas of the chamber ranged from −70°C to −40°C. Under such conditions turbulent currents are generated, making sectioning exceedingly difficult. Cut sections exposed to the higher temperatures in the chamber can grow coarse ice crystals and also rapidly dry by sublimation. By reinforcing the thermal insulation of the plastic walls and introducing a continuous flow of cold (c. −180°C) dry nitrogen, together with other modifications, we can now maintain the temperature inside the cryochamber at −110°C. This has enabled us to obtain 1–2 μm thick sections of insect epithelia and other samples in a fully hydrated state. More details of our modifications to the LKB cryokit are described elsewhere (Gupta and Cooper, unpublished), and a similar modification to the LKB cryokit has been described recently by Kirk and Dobbs (1976). We have no experience with other makes of cryokit, but in our model the small distance between the access ports on the top and the cutting area in the chamber vastly increases the problem of frosting, which becomes progressively worse with time as one works at a very low temperature.

2. Trimming

A truncated pyramid is the most suitable shape for cutting thin sections. Trimming of the tissue blocks has been recommended either by razor blade under liquid nitrogen (Saubermann and Echlin, 1975), by a suitable concavity in steel knives (Dörge *et al.*, 1974) or by a knife-rotating facility on the ultramicrotome (Appleton, 1974a). As a rule, we avoid this step in order to decrease the risk of damage to the tissue.

3. The Fidelity of Frozen Sections

Artifacts generated in the cutting of thin sections are well known (Glauert and Phillips, 1965) and are even more troublesome in the case of unfixed, frozen sections because there are no means to relieve them harmlessly. Compression and "ripples" (Hodson and Marshall, 1976) or fractures resulting in cracks (Echlin, 1975a) degrade the image and complicate the interpretation of analytical data. Very slow cutting speeds and the use of either an anti-roll plate (Pearse, 1968) or a vacuum suction device (Appleton, 1974a) may reduce this problem. More fundamentally, the mechanics of the sectioning process at low temperature are not fully established. Kirk and Dobbs (1976) have examined the replicas of cut surfaces from tissue.blocks sectioned either at $-70°C$ or below $-110°C$. From the evidence they conclude that while true sections with a smooth surface are cut at $-70°C$, microfracturing occurs at $-110°C$ resulting in the splitting of the two halves of cell membranes. If true cutting involves thawing at the interface of the block and the knife, the resultant redistribution of electrolytes by diffusion in the melted zone must pose a primary objection to the entire approach. The complexity of the parameters involved in the process of cutting a small block at the low temperature of a large cryochamber precludes any realistic assessment of the actual volume which may thaw and refreeze within the section. Often quoted estimates of a 30-nm thick "thaw" zone on each surface of the section (Thornberg and Menger, 1957; Hodson and Marshall, 1970) have not been experimentally established. Appleton (1974a) has argued that the adhesion of sections to form a ribbon suggests that some melting must occur at least at the initial line of impact in each cycle of the microtome. Do sections of epoxy resin blocks also stick to each other by melting? The nature of the phenomenon is clearly not known. Appleton's own evidence from a test model (NaCl in carboxymethyl cellulose) and other published information on the microanalysis of frozen sections from a range of unfixed tissues do not suggest that such thawing and refreezing during cutting cause any serious redistribution of electrolytes within the range of the available analytical resolution. However the hazard remains.

Another difficulty in sectioning is shattering in large aqueous areas, such as extracellular spaces or vacuoles. Material may actually drop out, leaving a fragmented section. The danger is probably greatest when coarse ice crystals have formed.

In practice, the most serious damage both to the structure and to the distribution of water soluble substances is likely to occur not during sectioning but during the formation of ice crystals. As discussed above, crystallization is not confined to the initial step of quench-freezing, but may occur in any of the subsequent steps. The low contrast in the electron microscopical images of sections examined in a fully hydrated state generally masks the appearance of ice crystals, but they become readily visible in partially or fully dried sections. However, as long as the ice crystals are not larger than the minimum microvolumes which can be analysed, they need not in themselves affect the utility of sections.

In spite of the claims of "routine success", to cut sections at low temperatures remains a hazardous step in the microanalysis of unfixed, frozen tissues. A good quality frozen section free from artifacts is delicately translucent and "glassy" in appearance. With due care, such sections can be obtained with sufficient frequency. However, in microanalysis the question posed by Appleton (1974b), "Can we trust the frozen section?", must always be asked.

D. Transfer of Sections to the Holder

"Conventional" transmission electron microscopes only accept 2—3 mm diameter grids or screens with a choice of mesh shapes or sizes. Several workers using scanning electron microscopes for microanalysis have also used conventional EM grids to pick up sections (Hutchinson *et al.*, 1974). Such grids severely restrict the area of the section which can be observed in transmission modes. Scattered electrons hitting the grid-bars in close proximity greatly increase the extraneous background in X-ray spectroscopy. For the analysis of hydrated sections, the very low mass of the conventional grid and the problem of obtaining adequate thermal contact between the grid and the specimen holder of the microscope greatly increase the risk of artifacts due to temperature rise and dehydration. The relative roominess of the stages in scanning microanalysers allows one to mount sections on bulkier holders with larger holes. In our approach, we have from the beginning (B. L. Gupta, T. A. Hall and T. Weis-Fogh, unpublished, 1970) used aluminium or Duraluminium collars as shown in Fig. 4. The section holder can be directly inserted into the cold stage.

To pick up sections, the holders are first thoroughly cleaned and the top surface polished, and they are then covered with a taut film of nylon, 100–200 nm thick, which has been pre-cast on specially designed metal frames (Fig. 5). When dry, the film is nicked with a sharp instrument around the edges to expose metal and allow better thermal contact with a 10–15 nm thick aluminium layer deposited in a vacuum evaporator on the top. Such coated holders are stored in a desiccator with clean silica gel as the desiccant (not $CaCl_2$ if one intends to analyse for Ca or Cl). Just before use, the coated holders are "flashed" in a vacuum evaporator to render the aluminium films hydrophilic (Dubochet *et al.*, 1971).

If the sections are cut in a cryostat with steel knives, the holder can be mounted on a spring-loaded metal jig (Fig. 6) to guide it while it is pressed against the section resting on the knife. This method allows the sections to be picked up with minimum disturbance and contamination. Contact and adhesion between the sections and the film can be improved by carefully pressing between two copper blocks kept in the cryostat. (Dörge *et al.*, 1974, collect their sections on collodion film mounted at the end of a carbon tube.)

The very small working area around the glass knife in a cryokit does not allow the use of our jig for directly pressing large annular holders onto the sections on the knife, although Hutchinson *et al.* (1974) used this approach for EM grids. Appleton has used a vacuum device to stretch a ribbon of ultrathin sections onto a pre-positioned EM grid. Other workers have recommended the use of thin wire probes (Christensen, 1971) and mouse whiskers (Saubermann and Echlin, 1975); they manipulate the sections onto horizontally oriented holders. To transfer sections in the LKB cryokit, we use

E

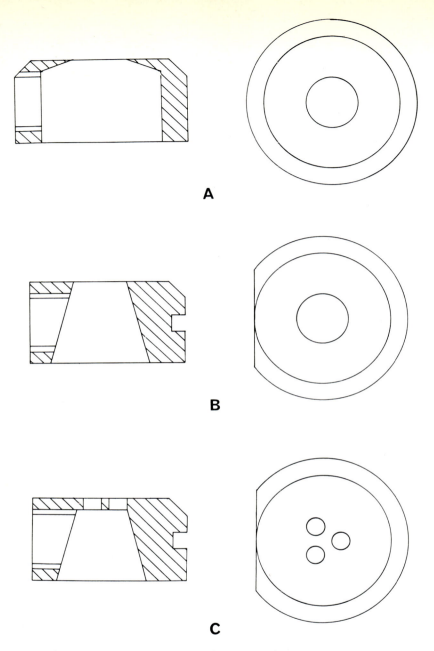

Fig. 4. Varieties of the section holder (or "collar"). A. 3-mm hole on top, giving minimum contribution of bulk metal to the X-ray continuum. B. Like A, but with groove around the side to fit a jig in the cryostat (cf. Fig. 6). C. Three 1-mm holes on top, for more efficient cooling of the sections. The threaded hole in the side is engaged by transfer rods.

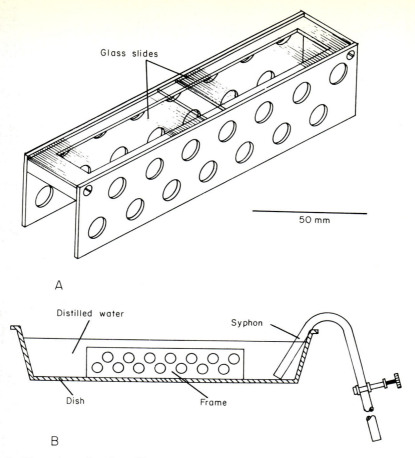

Fig. 5. Mounting of nylon film on the holders. Nylon film is cast on a water surface (B). After the water is drained and the film has dried on the metal frame, the frame (A) is disassembled and the film is transferred to the section holders by passing the holders through the holes.

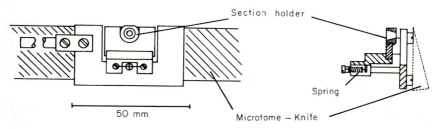

Fig. 6. A jig mounted on the steel knife of the SLEE cryostat microtome to pick up sections by pressing the coated holders directly onto the sections on the knife.

either an eyelash mounted at the end of an applicator stick, or an adaptation of Appleton's vacuum device (Gupta and Cooper, unpublished).

High electrostatic charge often develops in the very dry atmosphere of a cryochamber. A few bursts directed at the cutting area from an anti-static gun, sold for use with gramophone records, can be helpful (Fraser, 1976). The earthing of knife holders and metallic coatings on eyelash probes is also beneficial. Frozen sections appear to bear a negative charge. The application of a slight positive charge onto the vacuum device vastly improves the formation of ribbons and their retention (Gupta and Cooper, unpublished). Some initial experiments suggest that the adhesion of sections to the metallic coating on the holders could also be improved if the latter are made to bear a slight positive charge. Organic glues such as polylysine (Mazia *et al.*, 1974) and N-butyl benzene (Saubermann and Echlin, 1975) used to improve the adhesion of specimens to substrates may be permissible for purely morphological studies. In our experience such adhesives vastly degrade microanalysis by extraneous and indeterminate contribution to the X-ray continuum. Further, if they are applied only at the edge of a section, they do not provide the required intimate contact between the entirety of the section and the conducting supporting film.

E. Top Coating (and the Question of Electrical and Thermal Stability)

A uniform contact of the section with the substrate film is of vital importance a) for the electrical stability of sections under the relatively high beam currents used for microanalysis (especially with diffracting X-ray spectrometers) and b) to prevent dehydration of the irradiated microareas, which must occur if there is not an efficient dissipation of the heat generated locally by the interaction of the beam with the specimen. We have often found that small flat regions of a section, in good uniform contact with the aluminium-coated nylon substrate, are very stable even under an intense static probe (10 nA delivered into a diameter of 100 nm, which amounts to a current "density" of 10^3 nA per μm^2). This is not surprising; a calculation by one of us has shown that for our operating conditions, one should expect a local temperature rise of the order of only $10°C$ (Saubermann and Echlin, 1975). On the other hand, sections sometimes dehydrate rapidly and even disintegrate under a 10-nA beam even when it is distributed over a much larger area, such as $10 \mu m^2$. This instability is due primarily to the fact that 1-μm thick, deep-frozen sections are not uniformly flat but are compressed or "corrugated" in spite of the use of an anti-roll plate or the other devices mentioned above. Even when flattened by metal stampers (Christensen, 1971; Appleton, 1974a; Saubermann and Echlin, 1975; Gupta and Cooper, unpublished), the sections do not attain a uniform contact with the substrate.

(The presence of near-liquid substances, like fluorocarbons or N-butyl benzene, greatly increases the instability of sections under the beam, especially if they come between the section and the supporting film.)

One well known technique to alleviate instability under the beam is the prior application of a metallic top coating, thus providing another conducting path besides the film under the specimen. In our laboratory a coating of aluminium or carbon, $10-15$ nm thick, may be applied while the section holder is kept at $-170°C$ on the special cold table of the vacuum evaporator (Echlin and Moreton, 1974; Saubermann and Echlin, 1975). However, this step itself requires the loading of the section holder into the high vacuum of the coating unit, and the parts of the section which are not in good thermal contact with the substrate often dry before the section is protected by the top coat. We believe that this technical problem will be overcome, and it should be possible to improve stability by top-coating without dehydration. At present we do not use top coating in the study of frozen-hydrated sections; we simply cut the sections as small as possible and try to mount them with maximum flatness.

F. Storage and Transfer of Sections to the Microanalyser

In most published work on frozen sections of unfixed tissues, the sections are allowed to dry by sublimation at low temperature ($c.$ $-70°C$) in the cryochamber for several hours before electron microscopic examination (Christensen, 1971; Appleton, 1974a; Sjöström, 1975; Hodson and Marshall, 1976). Such dried sections are very vulnerable to destruction by the absorption of moisture from the atmosphere. Elaborate procedures have therefore been devised to protect the sections during transfer to the electron microscope for either morphology or microanalysis.

If sections are to be analysed fully hydrated, they must not be left in the cryochamber for a long time because they would slowly dehydrate. In order to prevent recrystallization of ice, sublimation or condensation from the atmosphere, at all times from the moment of removal from the cryochamber through the analysis of the section in the hydrated state, the holder should be kept below $-130°C$ and kept shielded from the laboratory air. Our transfer system is based on the shroud device (Steere, 1957) where the frozen specimen is withdrawn inside a metal tube pre-cooled with liquid nitrogen. A pre-cooled cap seals the end of the tube during the few moments when it is exposed to the warm air of the laboratory, on the way from one cold, dry environment to another. Similar systems are used by Hutchinson et al. (1974), and Gullasch and Kaufmann (1974).

Since the microanalyser stage accepts only one holder at a time, other section-bearing holders are "parked" in an intermediate transfer and storage Dewar (Fig. 2H). After receiving sections in the cryochamber, each holder is transferred to this "parking Dewar" by means of the device described by Saubermann and Echlin (1975). Thermocouple measurements have shown that as long as liquid nitrogen is maintained in the bottom of the Dewar, the temperature of the brass block and the specimen remains below $-190°C$. We have repeatedly stored specimens in this Dewar for several days without evident loss or gain of moisture.

Unlike the commercial Dewars for storing biological samples (supplied, for example, by Union Carbide), the specimens in this arrangement are kept cold without immersion in liquid nitrogen. We have previously noted that direct immersion of the section holders may break the films or otherwise cause mechanical disturbance (Echlin et al., 1973).

For study in the JXA-50A microanalyser, the sections on their holders are transferred to the cold stage through an air-lock chamber. The arrangement for this transfer is depicted in Fig. 3A—D. The entire procedure takes about two and a half minutes. Tests on a holder fitted with a thermocouple have shown that the temperature of the holder remains below $-150°C$ throughout.

The efficacy of the method in preserving the hydration of a sample has been confirmed additionally by the examination of sections of pure ice as well as saline solutions.

G. Study of the Section in the Microanalyser

During microanalysis the section holders are in intimate contact with the specimen stage, which remains below $-170°C$ (Taylor and

Burgess, in press). The temperature of the sections depends on many factors but given a good contact with the substrate film, they may be only 5–10°C warmer.

1. Morphological Image

An incident-light optical microscope is incorporated in the microanalyser to bring the specimen precisely to the focus of the diffracting spectrometers. This microscope is invaluable for initial search and assessment of the quality of the sections and to see the degree of contamination by frost. Some deposition of frost is unavoidable during the mounting, storage and transfer of sections if the conditions have been established to avoid dehydration. In practice, the frost forms mostly on the metal rim of the holder, and to a lesser extent on the film over the hole. Frost particles c. 1 μm in diameter are more numerous near the rim than towards the centre of the hole. This suggests that the substrate film near the centre of the 3-mm hole is slightly warmer and the frost settles preferentially at the colder periphery. (Such evidence of a temperature gradient is not apparent on section holders with 2-mm holes or with three 1-mm holes, as depicted in Fig. 4C). Frequently there are also irregularly distributed fluffy aggregates of frost presumably picked up during manipulation in the cryochamber. Such areas are of course avoided during routine microanalysis. The retention of microdroplets of frost on the substrate film during several hours of examination and microanalysis is a convenient indication that the specimen has not grossly dried by warming.

Among the various modes of visualization (Echlin, 1975b) of the section under the electron beam, the scanning transmission mode is generally most suitable. The secondary electron image of a fully hydrated section merely reveals the outlines and gross irregularities. Any details of the tissue in this image are clear indications of partial or complete dehydration. The scanning transmission image of a fully hydrated tissue section also reveals very little detail of the structure. The range of contrast in transmission can be somewhat improved by the use of a dark-field geometry, i.e. by directing the axis of the unscattered beam away from the electron detector, although this tactic is not as effective for 1-μm sections as it is for ultra-thin specimens because of the effects of multiple scattering. As always with dark-field illumination, holes and completely opaque fields both appear as black in the image and there is a maximum of signal at some intermediate level of electron opacity, but in practice the general effect is of positive contrast with dense or thick areas of tissue appearing relatively dark.

The visualization of structural details in scanning transmission images of hydrated sections is vastly improved if they are allowed to lose about 10% of their water content by leaving the section-bearing holders in an atmosphere of cold dry nitrogen in the cryochamber of the microtome (e.g. Fig. 7C). The duration of such treatment and the resulting degree of dehydration depend on several conditions and have to be determined empirically. In our cryokit chamber with temperatures below $-100°C$, a storage of 15–30 minutes seems best.

Further improvement in the structural details of the image occurs with progressive drying, either by longer storage in the cryochamber or by raising the temperature of the specimen inside the column of the microanalyser. Inside the column, the specimen can be warmed by heating the specimen stage (Moreton et al., 1974; Saubermann and Echlin, 1975). At a partial pressure of water vapour of around 10^{-5} Torr, rapid dehydration of the section occurs at stage temperatures in the range -100 to $-90°C$. Continuous monitoring of the image on the display tube during this process invariably shows dramatic changes in the intrinsic contrast of the tissue structure, and much increased detail in both the secondary and transmission modes, heralding complete dehydration.

The rate of sublimation increases steeply as the stage is warmed. Around the time when the process of dehydration reaches its end point, the rate of release of water vapour from specimen, holder and cold-stage is enough to produce a sudden deterioration in the column vacuum (Moreton et al., 1974). Such a deterioration is generally manifest if there is no cold anti-contamination plate in the specimen chamber. (The modified cold plate in our JXA microanalyser seems to trap the water vapour from stage and specimen so well that there is no change in the reading of the vacuum gauge when the stage is heated, as long as the plate is kept cold.)

Dehydration by warming of the stage is not practical if the section is to be analysed after drying, since analysis should be carried out below $-150°C$ to minimize radiation damage and the processes of warming and re-cooling the stage would be very time-consuming. Therefore, in order to compare X-ray analytical data from hydrated and dehydrated states of a section, we routinely dehydrate by means of the specimen-loading device of the JXA analyser. The gate valve to the evacuated air-lock chamber is opened and the specimen holder is withdrawn from the stage at the end of the warm loading rod and kept there in the column for 1–3 minutes at c. 10^{-5} Torr; it is then returned to the stage for further study. A point of drying can be visually established through the observation window by the sudden loss of the slightly frosty appearance of the specimen holder.

Saubermann and Echlin (1975) have observed a mass-18 peak from a quadrupole mass spectrometer when they dehydrate by warming the stage to c. −90°C, and they suggest that this peak provides one way to demonstrate that the specimen was hydrated. We do not observe such a peak, using the same mass spectrometer, when we dehydrate much more rapidly with the JXA specimen transfer-rod, even with the chamber vapour trap uncooled. A calculation of the total water content of a tissue section, and the effect which might be expected from the release of all of this water into the column, suggests that the peaks seen during warming of the stage are due to release of vapour not from the section but from other surfaces.

It has been claimed on the basis of high-resolution transmission microscopy that structure is not preserved when deep-frozen sections are allowed to dry rapidly under vacuum between 10^{-5} and 10^{-2} Torr (Sjöström, 1975; Appleton, 1974a). Appleton believes that there is gross distortion due to surface tension during rapid sublimation. But no mention is made of the temperatures at which the dehydration was accomplished. It may be that in these studies, drying occurred at relatively high temperatures, above −70°C. At the resolution of our scanning transmission image (100−200 nm) we have not generally found any gross disruption of structures in sections of several insect epithelia, maturing erythrocytes of chick embryos, HeLa cells or other tissues. Nor as a rule do we find any significant changes in the distribution of electrolytes or other heavy elements. However some shrinkage of structures does occur on drying, particularly in areas which are not flat on the substrate film. Tissues suffering bad ice-crystal damage during initial quenching also show greater distortion and shrinkage on drying, as do specimens contaminated with residues of cryogens. The change in dimensions has to be kept in mind when comparing X-ray data from specimens before and after dehydration (see Section IV B).

The lack of structural detail in the images of fully hydrated sections in our experience seems to be in contradiction with the detailed images obtained in some other published works. Bacaner et al. (1974) and Hutchinson et al. (1974) reported on the spatial distribution of ions in 100−200 nm thick sections of a vertebrate muscle analysed at −100°C on the cold stage of a Cambridge Stereoscan microscope. The specimens were mounted on standard

EM grids. Saubermann and Echlin (1975) have published detailed transmission images of mouse liver and *Thermobia* rectum, from 1–2 µm thick frozen sections analysed with the instrumental arrangement described by Moreton *et al.* (1974). The sections were top-coated with evaporated aluminium. By the criteria of image detail and of the highly favourable peak to background ratios in the published X-ray spectra, the specimens in these studies would appear to be substantially dehydrated.

The reason for the poor detail in the image of a fully hydrated section is that the interaction of the electron beam with water (atomically H and O) is not very different from the interaction with the H, C, N and O atoms which constitute the bulk of organic matter. Only phosphorus in the membranes and in nucleic acids (e.g. in condensed chromatin and nucleolus) may impart a substantially higher electron-scattering coefficient to the organic material. Hence the electron beam undergoes virtually the same interaction throughout a fully hydrated section, and the contrast between different areas after dehydration corresponds to the differing amounts of water which they have lost. However, with scanning images, the chain of signal amplification enables one to enhance those differences in contrast which do exist, so that even a slight dehydration can produce ample contrast on the display tube. For instance, if an intense beam produces a local dehydration in a small area in poor thermal contact with the substrate film and this field is then viewed within its surround at lower magnification, the contrast can be made to correspond to the full brightness range of the display tube even if the X-ray continuum count shows that there is a difference of only 5–10% in mass per unit area across the boundary.

2. X-ray Spectroscopy

Once the structures of interest have been identified in the transmission image at low magnification (500–2,000x), relatively stable fields are selectively viewed at higher magnifications (3,000–20,000x). Depending on the size of the structure to be analysed, the probe is localized to the area of interest either in the static mode (probe area *c.* 10^4 nm^2) or scanning a raster of suitable dimensions. Localization has to be based on the after-image on the display tube. However it is essential to be aware that in most scanning instruments, including ours, there is a shift in image registration as one goes from large-area scan to small-area scan and to static probe. We cope with this problem by identifying the profile of the area of interest on the image wave-form-monitor. In this way one can also deal with drifts of specimen or beam during a measurement by adjusting the probe to keep the same profile on the monitor.

For most analyses we use an electron beam with an accelerating voltage of 50 kV and a probe current of 10 nA (measured with a Faraday cup built into the specimen stage). A complete X-ray spectrum is recorded with a Kevex Si(Li) retractable spectrometer fitted with a 30-mm^2 detector (resolution 160 eV) and an 8-μm Be window. Permanent magnets are mounted in front of the detector window to remove backscattered electrons. In typical runs, the Si(Li) spectrometer is preset for a live time of 40 sec and the dead-time fraction is in the neighbourhood of 50%, giving a real running time of approximately 80 sec, and the diffracting spectrometers are preset to run concurrently for a real time of 80 sec.

The energy-dispersive X-ray spectra from fully hydrated sections show low peak to background ratios (Fig. 7). This is in accord with the observations of Gullasch and Kaufmann (1974) on whole red blood cells, Dörge et al. (1975) on blocks of kidney and Fuchs and Lindemann (1975) on bulk samples, all examined in the frozen-hydrated state. Fuchs and Lindemann also noted that the peak to background ratios increased as their specimens dehydrated. We have repeatedly seen the same effect as sections dehydrate. Neither Bacaner et al. (1974) nor Saubermann and Echlin (1975) have discussed this sensitive criterion of hydration in relation to their spectra, although it need not affect the elemental distributions which they have reported.

Because of the low peak to background ratios, the energy-dispersive spectra of hydrated specimens are not generally useable for the study of Na and Mg, or for calcium in the presence of 100 mM K due to interference with the potassium K_β X-ray line. For these elements we use fully focussing diffracting spectrometers, with an RAP diffracting crystal for Na and Mg and a PET diffracting crystal for Ca and also K. The potassium X-ray signals are generally monitored simultaneously by both energy-dispersive and diffracting spectroscopy to give a check on both systems.

Background counts in the diffracting spectrometers are occasionally determined by offsetting the linear-tracking spectrometers by about 4 mm from peak position, but the backgrounds are invariably low (generally less than 10 counts per run) and need not be measured accurately or often.

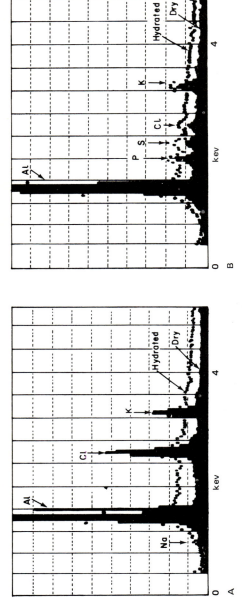

Fig. 7. Effects of partial and full dehydration on X-ray spectra and images. A and B. X-ray spectra from a section of maturing red blood cells of 6 day old chick embryos. A: Peripheral medium, saline solution plus 10% w/v Dextran. B: Cytoplasm of a cell. Dotted spectra: Frozen-hydrated. Bar spectra: Same fields after dehydration in the column. Dehydration greatly increases peak to background ratios by reducing the continuum background. (Gupta, Hall, Johnson, Schor and Northfield, unpublished.) C. and D. Scanning transmission images from 1-μm thick sections. C: Malpighian tubules of *Rhodnius*, frozen-hydrated. The change of continuum on subsequent dehydration indicated that the section had retained 80—90% of its original water when the image was obtained. D: Rectal papillae of *Calliphora* (cf. Fig. 12), as seen after full dehydration in the column. When fully hydrated, sections show very little detail in the image. *bb*: brush border; *cut*: cuticle; *cyt*: cytoplasm; *lu*: lumen; *ms*: lateral membrane stacks; *nu*: nucleus; *sal*: saline solution. Arrows delineate intercellular channels.

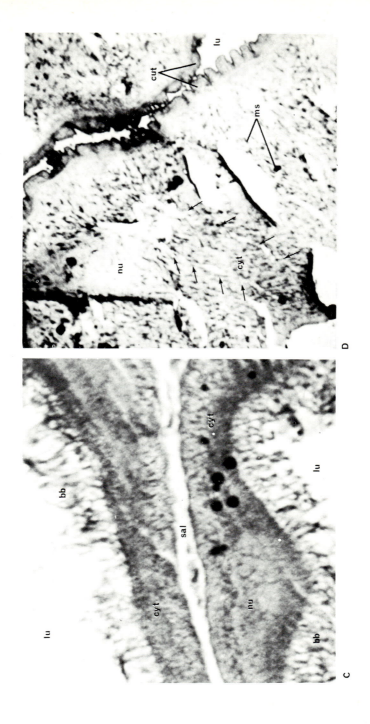

3. *Control of Conditions during Analysis*

a   *Degree of hydration*

We have already discussed above, in Section IV B, the basis for the quantitative analysis of hydrated sections, and we derived equations for the determination of wet weight mass fractions from X-ray observations on slightly dehydrated or fully dehydrated sections. We also noted that degree of hydration may be judged from the X-ray spectra (especially from the peripheral standard), and from the extent of mass loss accompanying a final, deliberate dehydration. Here we should describe more fully how the deliberate dehydration is used to judge the average level of hydration of a section. The procedure is as follows:

A fairly large field of Dextran-saline standard in the section, free of visible frost, is selected and analysed. A similarly large field from the tissue is also analysed. Analysis is then carried out in the various microareas of interest. When all the needed data have been recorded, the specimen is dried by warming in the column and analysis is repeated on fields close to the initial pair. The loss of mass, as reflected in the continuum count, provides an estimate of the initial degree of hydration of the section.

The applicability of this approach to the measurement of local dry weight fraction, and the complications due to shrinkage and latent radiation damage, have been discussed above (Section IV B).

b   *Contamination*

Measurements of local mass may be falsified by two kinds of contamination within the microanalyser: condensation of water vapour, and the deposition of pump-oil fractions "cracked" by the electron beam. The latter is especially difficult to manage when ultra-fine probes are used for the analysis of ultrathin specimens, because the rate of deposition is then especially high and the true tissue mass is relatively low; ultra-high vacuum systems may be needed for such work. Our conditions of analysis are less demanding. We have a liquid-nitrogen baffle above the oil diffusion pump, a molecular sieve trap in the forepump line of the rotary pump, and a cold-trap much colder than the stage and fitted closely above it. Chamber pressure is in the range $10^{-6}-10^{-5}$ Torr. Mass per unit area in our specimens is not significantly affected by contamination as long as the chamber cold-trap is in use.

Certain heavy elements, notably sulphur and chlorine, may be present in the residual gases in the column. (They probably come from pump oils, O-rings and various plastics.) We have occasionally seen contamination of a section or supporting film with these elements, but never under the protection of the chamber anti-contamination plate. The danger of chlorine contamination must be kept in mind when anomalously high levels are found in the microprobe analysis of tissues.

c   Loss of elements

It is well known that certain elements may be volatilized or otherwise removed by the electron probe. Such effects have been noted for Na, S, Cl and K. The effects depend greatly on the particular conditions — nature of the specimen and the binding of the element, and the intensity of the beam — so that stability in any given case is difficult to predict. However the effect is generally slow enough to be manifest as a decline from an initially higher counting rate. In our specimens we have noted such losses only with chlorine, and that only occasionally with currents of 10 nA or more focussed into areas of the order of $0.1 \, \mu m^2$ or less.

VI. BIOLOGICAL RESULTS

The ultimate test of any physical technique applied to cells and tissues must lie in the biological relevance of the information produced. In spite of the reservations expressed above on the various preparatory and analytical steps, we have been able to measure the absolute elemental concentrations of Na, K, Cl, P, S, Ca and Mg in several fluid transporting epithelia of insects (Gupta, 1976; Gupta *et al.*, 1976a) and in maturing erythrocytes of 3—20 day old chick embryos. These results are briefly described in the following pages.

A. Malpighian Tubules and Salivary Glands of Adult *Calliphora*

These simple tubular epithelia, when bathed *in vitro* in a drop of normal Na-rich saline, will secrete a fluid in the lumen which is nearly iso-osmotic with the bathing fluid. However, the Na and K concentrations in the secreted fluid are essentially reversed (Berridge, 1968; Berridge and Oschman, 1969; Oschman and Berridge, 1970, 1971; Maddrell, 1971 and Chapter 21). In both tissues it has been suggested that the movement of water is coupled to the concentration gradients of ions (mainly K and Cl) actively transported into the cytoplasmic processes of basal infolds and the extracellular channels in the apical microvillate brush-border. It has also been suggested that the isotonicity of the secreted fluid is achieved by a "standing gradient osmotic flow" hypothesized by Diamond and Bossert (1967, 1968). No information has been available on the actual concentration of ions either in the cells or in the extracellular channels.

1. Malpighian Tubules

a   Ion distribution in and around the cells

Initial results (Gupta, 1976) from 1—2 µm thick sections analysed in the hydrated state demonstrated that, as expected from physiological data, the lumen contents contained 138 mM of K and 26 mM of Na per kg wet weight. The average concentration of these major cations inside the cells was also about the same but there was somewhat more K (157 mM) and less Na (13 mM) in the brush border area. In the basal part of the cells (which included the basal lamina) both the K and Na concentrations were high (125 and 87 mM respectively). These results established that the technique did allow retention and measurement of the authentic distribution of ions in major tissue compartments, and provided *in situ* information on the intracellular concentrations (see Fig. 10 in Gupta, 1976). Subsequently, a more detailed analysis for Na, K, Cl, P, Ca and Mg was performed on this tissue, and the results are summarized in Fig. 8 (Gupta and Hall, unpublished).

The higher phosphorus and lower chlorine content of the lumen as compared with the bathing medium (Fig. 8, column 5) confirm a well established feature of the fluids secreted by the Malpighian tubules of a variety of insects (Maddrell, 1971 and in this volume).

A new feature is the analysis of the ion distribution in the basal infolds where electron microscopy (Berridge and Oschman, 1969) shows extracellular channels 20—30 nm wide, alternating with cytoplasmic processes. In scanning transmission images of partially

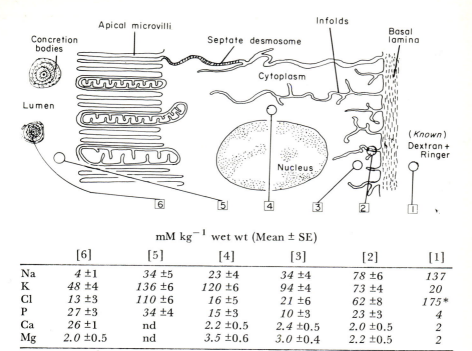

mM kg^{-1} wet wt (Mean ± SE)

	[6]	[5]	[4]	[3]	[2]	[1]
Na	4 ±1	34 ±5	23 ±4	34 ±4	78 ±6	137
K	48 ±4	136 ±6	120 ±6	94 ±4	73 ±4	20
Cl	13 ±3	110 ±6	16 ±5	21 ±6	62 ±8	175*
P	27 ±3	34 ±4	15 ±3	10 ±3	23 ±3	4
Ca	26 ±1	nd	2.2 ±0.5	2.4 ±0.5	2.0 ±0.5	2
Mg	2.0 ±0.5	nd	3.5 ±0.6	3.0 ±0.4	2.2 ±0.5	2

* Extra Cl is from Tris-HCl buffer.   nd = not detected.

Fig. 8. Results of microprobe analyses of frozen-hydrated sections of Malpighian tubules of *Calliphora*. Details of the sketch are based on "conventional" electron microscopy of fixed and embedded tissues (cf. Berridge and Oschman, 1969). The numbers in boxes refer to the columns in the table, reading from right to left.

hydrated sections, the membrane infolds appear as dark reticular striations separated by "clear" cytoplasm. If the width of the extracellular channels, as given by Berridge and Oschman, represents the true *in vivo* dimension, it is clear that even the finest electron probe used so far in our study (about 100 nm in diameter) is too large to be confined purely to such a channel. Thus, static probes centred on the infolds ([2] in Fig. 8) indicate concentrations of Na, K and Cl midway between those in the bathing medium [1] and in the cytoplasm [3], and a phosphorus content of 23 mM probably contributed by phospholipids in the membranes. However, the analysis does resolve the intervening cytoplasmic processes (about 300 nm in width), as indicated by the lower concentrations of Na, Cl and P but higher K.

Analysis of the basal cytoplasm suggests that a significant

concentration gradient of K (from about 80 mM to 120 mM per kg wet weight) may exist in this part of the cell. The direction of such a gradient is, however, the reverse of that postulated by Berridge and Oschman (1969) when applying the "standing gradient" hypothesis to these tissues. The somewhat complex ultra-structural geometry of *Calliphora* Malpighian tubules has precluded more detailed analysis at the resolution presently available.

b   Mineralized concretions

Spherical or sub-spherical bodies containing organic and inorganic salts have been observed as concentrically lamellate structures in the electron microscopic study of Malpighian tubules and other tissues of insects (Wigglesworth, 1965; Smith, 1968). Electron microprobe analysis of ultrathin sections from fixed tissues (Sohal *et al.*, 1976) has suggested that they can bind cations.

Dense, spherical objects $1-3\,\mu m$ in diameter were commonly found in our sections of tubules from 8–12 day old female flies. These bodies (area [6] in Fig. 8), both in the cytoplasm and in the lumen, contain much less Na and K but more Ca than their surroundings. Moreover, their total cation complement (108 mequivalent kg^{-1} wet weight) exceeds that of anions (67 mequivalent, taking P as divalent phosphate). The anion deficit could be made up either by carbonate (Eastham, 1925) or by various organic anions (Wigglesworth, 1965). The exact role of such "concretion bodies" in the excretory function of Malpighian tubules remains to be elucidated (see Maddrell, Chapter 21).

2. Salivary Glands

The secretory portion of *Calliphora* salivary glands also produces a K-rich fluid isotonic with the bathing medium, at a rate which can be increased 60-fold *in vitro* by stimulation with 10^{-8} M 5-hydroxy-tryptamine (5-HT) in the bathing medium. In this case, Cl is the only major anion in the secretion, and its concentration is balanced by the sum of Na and K (see Berridge, Chapter 9). The apical microvillate membrane of these cells is deeply invaginated to form canaliculi (Oschman and Berridge, 1970), which are believed to be the primary sites of K and Cl secretion. Water movement occurs passively, to maintain isotonicity of the secretion. Microprobe measurements are of particular interest here, since electrical potentials recorded across the apical surface of the gland (Berridge *et al.*, 1975) may well not reflect accurately the events occurring at such deeply invaginated canaliculi (Keynes, 1969), especially when each opening into the main lumen is guarded by a fibrous intima decorated with patches resembling epicuticle (Oschman and Berridge, 1970). The possibility that chloride movements may take place partly by shunting through the septate junctions between the cells has not yet been excluded from physiological events in this tissue (Berridge *et al.*, 1976).

Figure 9 is a diagrammatic summary of microanalytical results from the secretory portion of the unstimulated gland (Gupta, Berridge and Hall, unpublished). In the main lumen the measured concentrations of Na, K and Cl are very close to the values from flame-photometric analysis of bulk secretory fluid (cf. Oschman and Berridge, 1970; Prince and Berridge, 1973; Berridge *et al.*, 1976).

In the canaliculus the ion concentration of the fluid near the blind base is different than that of the fluid near the opening into the main lumen. In the canalicular base the concentration of K is about 40 mM higher than either in the main lumen or in the main cytoplasm. The Cl concentration is about 70 mM lower than in the lumen although it is about 60 mM higher than that of the cytoplasm. Since the measured level of P was less than 5 mM, the Cl deficit may be compensated by fixed negative charges on the surface coat or glycocalyx of the microvilli. Alternatively, the anion deficit could be made up by other anions such as HCO_3, which are not detected in the microprobe. The concentrations of K, Cl and Na in the fluid near the canalicular opening are however very similar to those in the luminal fluid. Presumably, the other anions, if present, have already been replaced by Cl, possibly entering through a paracellular route of intercellular junction.

Inside the cells, the average concentrations of Na, K and Cl appear to be respectively about 20, 120 and 40 mM kg^{-1} wet weight. The figure for potassium may be compared with recent estimates using ion selective electrodes which indicate activities corresponding to concentrations in free solution of 145 mM K in the resting gland and 150 mM K (range 120—180 mM) under 5-HT stimulation (Berridge and Schlue, personal communication). As discussed in Section IV A above, for such a comparison one needs to know the amount of free water and the fraction of ions firmly bound. A reasonable estimate of the local free water can be made by comparing mass counts before and after drying the section (see p. 95). In the main cytoplasmic area of the salivary gland section, the residual dry mass was

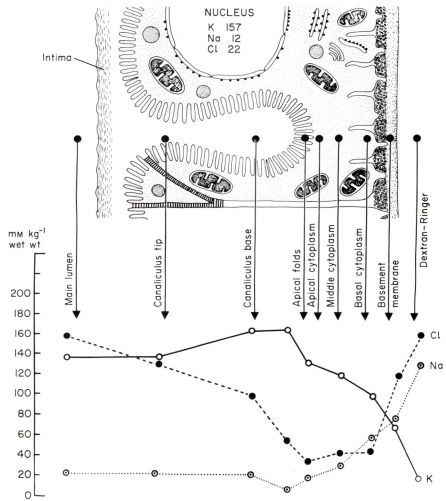

Fig. 9. Results of microprobe analyses of frozen-hydrated sections of unstimulated salivary glands of *Calliphora* adults. Details of the sketch are based on the account of Oschman and Berridge (1970). Arrowed circles show the approximate positions but not the size of the electron probe.

approximately 30%. Estimates of the fraction of bound potassium in cells vary widely (cf. Hodgkin and Keynes, 1953; Hays *et al.*, 1968). Taking a figure from cecropia mid-gut of 15—20% (Zerahn, Chapter 15), of the 120 mM K determined by microprobe analysis, some 100 mM remains to be distributed in cell-water estimated as 70% of the total mass, giving a free concentration of about 140 mM — quite close to the microelectrode figure for unstimulated glands. More accurate estimates of the freely exchangeable K could be obtained by replacement of intracellular K with a different ion, e.g. Rb or Na, which have been shown by Berridge *et al.* (1976) to maintain fluid secretion in this tissue.

It is also noteworthy that the highest concentration of K (167 mM) and the lowest concentration of Na (10 mM) occur over the apical membrane infolds. These values are remarkably similar to K 157 mM and Na 13 mM first found in the apical brush border of the Malpighian tubules in *Calliphora* (Gupta, 1976 and section VI, 1a above). In both tissues the apical membrane is believed to be the site of a potassium secreting mechanism.

As indicated above for the Malpighian tubules of *Calliphora*, ion concentrations in the basal cytoplasm of the salivary gland cells are not uniform. In fact, we have not observed a step-change across the basal cell membrane, but rather the concentrations change progressively over about 5 μm of cytoplasm towards the base of the canaliculi. Thus the Na concentration gradually falls from 140 mM in the bathing medium to about 20 mM while the K concentration rises from 20 mM to about 120 mM in the apical cytoplasm and 140 mM in the canaliculus. The concentrations of Cl in the cell do not show such steep gradients (Fig. 9). As in Malpighian tubules, the levels of both Na and K in the basal lamina are high, but in this tissue there is also a high level of chloride. If ions are bound to any considerable extent in the cytoplasm, then concentration gradients may simply reflect different binding capacities of different regions, and variations in the amount of free water. Also, so far we have not completely ruled out the possibility that in the basal region of the cell the characteristic X-ray signals may include a variable contribution from the extracellular fluid in the narrow infolds, even when the electron probe is visually localized over the cytoplasm. The alternatives can only be resolved by definitive experimental analysis of tissues in different physiological states. However, intracellular gradients in the activity of Na, K and Cl have been demonstrated with ion-selective microelectrodes between the basal and the apical membranes in the 100-μm high columnar cells of the newt gall bladder (Zeuthen, 1975 and Chapter 20).

B. Malpighian Tubules of *Rhodnius* (Fig. 7C)

The general properties of fluid secretion by the upper secretory part of *Rhodnius* Malpighian tubules are similar to those of the salivary glands in *Calliphora* but the rate of secretion after stimulation by a hormone in *vivo* and by the hormone or by 5-HT *in vitro* can be increased up to ×1000. Moreover, *Rhodnius* tubules are unusual because the Na/K ratios in the stimulated secretion normally approach unity. The simple fine-structure of this tissue makes it very suitable for microanalysis. Figs. 10 and 11 (Gupta *et al.*, 1976b) summarize the results from microprobe analysis of this epithelium both in the resting condition and after stimulation with 5-HT.

The general distribution of ions measured in these tubules is comparable with the *Calliphora* tissues discussed above. However, no physiological information had previously been available on the fluid slowly secreted by unstimulated tubules. Our results clearly bring out the major differences in the concentration of ions both in the cells

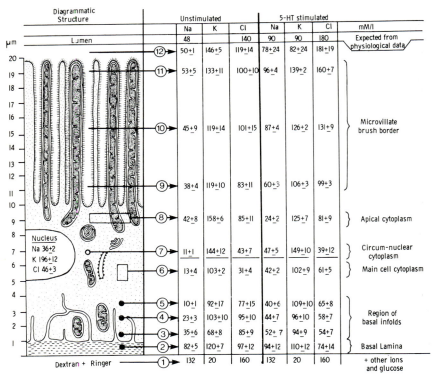

Fig. 10. A comparison of electrolyte distributions in unstimulated and stimulated Malpighian tubules (upper portion) of *Rhodnius prolixus*. (From Gupta *et al.*, 1976b.)

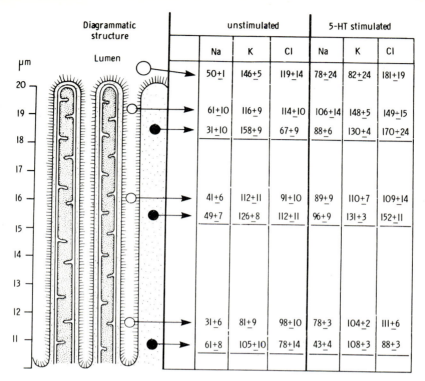

Fig. 11. As in Fig. 10, but restricted to the microvillar region. The values are averages of static-probe measurements from different microvilli and channels. (From Gupta *et al.*, 1976b.)

and in the secreted fluid under two different functional states. The secreted fluid in the lumen of unstimulated tubules contains more K but less Na and Cl than in the lumen of 5-HT stimulated tubules. It also contains about 20–30 mM of phosphorus. In terms of ionic composition the fluid secreted by unstimulated tubules of *Rhodnius* resembles the average composition of the fluid secreted by the Malpighian tubules of a number of insects including *Calliphora*. K, Na and Cl contents of 5-HT stimulated secretion in our microprobe results agree with the known physiological data (see Maddrell, Chapter 21). The phosphorus contents of this fluid were less than 2 mM.

The much lower Na content of the secretion of unstimulated tubules is paralleled by a very low Na concentration in the general cytoplasm. On 5-HT stimulation the average intracellular concentration of Na rises from about 20 mM to 45 mM but there is no similar change in the K content. This suggests that at least one of the actions of 5-HT in this tissue is to promote entry of Na ions into the cells and thus increase the intracellular pool of Na^+. A similar change in intracellular Na^+ as part of the hormonal stimulation of fluid transport has been postulated for many vertebrate epithelia but the evidence has remained equivocal because of the difficulties in accurately measuring the intracellular concentrations of ions by the techniques used so far (Handler and Orloff, 1973; Finn, 1976; Gupta *et al.*, 1976b).

To gain more information on the brush-border as the site of solute-solvent coupling, we have attempted to measure the concentration of ions along the lengths of microvilli and the intervening extracellular channels. The diameter of both these structures is about 200—400 nm — much less than the thickness of a hydrated section (about 1 μm). This causes difficulties in X-ray resolution due to: (a) the spread of the electron beam discussed above (Section III.3) and (b) the possible overlap of structures within the thickness of the section. Significantly different measured concentrations of Na, K and Cl between the microvilli and the adjoining channels at different sites along their length (Fig. 11) suggest, however, that one is able to obtain a better X-ray resolution than expected. In the extracellular channels as well as the microvilli, in both functional states of the tubules, the concentration of Na and K at the basal end is much lower than at the apical end. The channel fluid at the apical end is also different from the fluid in the lumen. The pattern of Cl distribution is intricate; a general deficiency of Cl over Na + K in the extracellular channels may be compensated by the fixed anionic charges of the glycocalyx (cf. canaliculi of *Calliphora* salivary glands). As in the basal area, the direction of the concentration gradients of ions is opposite to that needed to support the "standing gradient" hypothesis. Perhaps it is premature to put exact functional interpretations on the ionic concentrations measured in the brush border since the theories of solute-solvent coupling are still a subject of debate (see Chapters 5, 7).

C. Rectal Papillae of *Calliphora*

An important function of the rectal complex in terrestrial insects is to remove water and essential solutes from faecal matter (Wigglesworth, 1932; Ramsay, 1971; Wall, Chapter 23). The ability to remove water varies in different species but in some insects such as *Thermobia* water can be removed from faecal contents in equilibrium with an atmosphere of 45% relative humidity (see Noble-Nesbitt, Chapter 22). In *Calliphora* the removal of water can occur against a gradient of 300 mosmol between the lumen and the haemolymph (Phillips, 1969). From a fine-structural analysis it has been proposed that the mitochondria-associated stacks of highly infolded lateral cell membranes are sites of solute secretion into intercellular channels. The high concentration of solutes in these channels then provides the osmotic force for the removal of water from the lumen via the cells (Gupta and Berridge, 1966; Berridge and Gupta, 1967). The major ion transported was thought to be K. The existence of a Mg-ATPase was histochemically demonstrated on the stacked membranes while a K-Mg-ATPase was found on the apical membrane-leaflets (Berridge and Gupta, 1968).

The complexity of fine-structure in the rectal papillae (Fig. 7D) has so far not permitted a detailed microprobe analysis in fully hydrated sections. Analyses of partially or fully dried sections of rectum under experimental conditions similar to those used earlier by Berridge and Gupta (1967) and by Phillips (1969) have revealed a much higher concentration of Na in the stacked lateral infolds and in narrow intercellular channels than elsewhere in the cytoplasm (Fig. 12, Gupta, Wall and Hall, in preparation). It appears therefore that Na may be the major cation used to generate fluid transport even when K is the major cation present in the lumen. A lower overall concentration of ions in the long narrow intercellular channels may be taken as evidence of solute reabsorbtion (Berridge and Gupta, 1967; Wall, 1971 and Chapter 23). If the fluid in these channels is iso-osmotic with the cytosol, the ion-deficiency may be compensated by organic molecules (Wall, Chapter 23). Alternatively, the fluid may well be hypo-osmotic. Unfortunately, we have not so far been able to obtain reliable measurements on the ionic composition of fluid in the infundibular spaces which open into the haemolymph. No other information is available on the composition of the absorbate produced by the rectal papillae, Phillips' observations being on the entire rectal complex. The details of microanalytical results from this tissue are to be published elsewhere (Gupta, Wall and Hall, in preparation).

D. Basal Laminae

All tissues have a sheet of glycoprotein, frequently containing collagen, on the serosal surface (facing the body fluids) which is

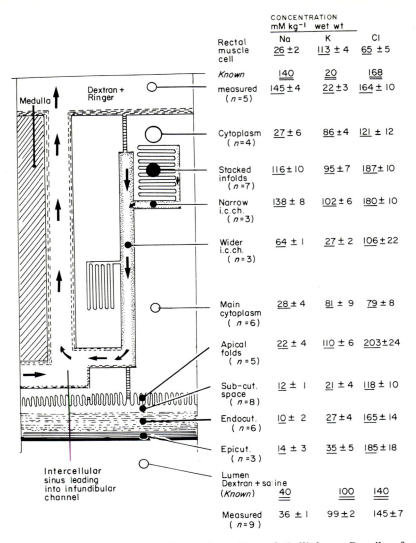

CONCENTRATION
mM kg⁻¹ wet wt

	Na	K	Cl
Rectal muscle cell	26 ±2	113 ± 4	65 ±5
Known	140	20	168
measured (n=5)	145±4	22±3	164± 10
Cytoplasm (n=4)	27 ± 6	86 ±4	121 ± 12
Stacked infolds (n=7)	116± 10	95 ±7	187± 10
Narrow i.c.ch. (n=3)	138 ± 8	102 ± 6	180± 10
Wider i.c.ch. (n=3)	64 ± 1	27 ± 2	106 ± 22
Main cytoplasm (n=6)	28 ± 4	81 ± 9	79 ± 8
Apical folds (n=5)	22 ± 4	110 ± 6	203±24
Sub-cut. space (n=8)	12 ± 1	21 ± 4	118 ± 10
Endocut. (n=6)	10 ± 2	27 ±4	165 ± 14
Epicut. (n=3)	14 ± 3	35 ± 5	185 ± 18
Lumen Dextran + saline (*Known*)	40	100	140
Measured (n=9)	36 ± 1	99±2	145±7

Fig. 12. Microprobe analyses of rectal papillae of *Calliphora*. Details of the highly schematic diagram are based on Gupta and Berridge (1966) and Berridge and Gupta (1967, 1968). The arrows indicate the pathway which the absorbed water is believed to follow. The flies were kept without food and water for two days and the rectal lumen was injected with saline-Dextran a few minutes before quenching; the "measured" values in the saline-Dextran on the serosal side are based on the saline injected into the lumen as standard, and *vice versa*. Very high concentrations of Na and K in the stacked infolds and in the narrow intercellular channels were also found in recta from normally fed flies as well as in recta where the lumen had been injected with 150 mM of KCl. i.c.ch.: intercellular channel.

histologically referred to as basement membrane, basal lamina etc. The exact role of the basal lamina in ion and water transport has remained enigmatic. Very little information is available on the permeability properties of this structure although some elegant attempts have been made to study them in isolated form *in vitro* (see Dantzler, Chapter 3). From such studies and from the use of macromolecular and colloidal tracers (Kessel, 1970; Oschman and Berridge, 1971) it is commonly believed that the basement membranes do not in general present any barrier to the free diffusion of small molecules, ions and water but may selectively exclude larger molecules.

In most of the sections from insect epithelia (e.g. Fig. 10) we have found that the basement lamina reveals concentrations of Na, K and Cl which are quite different from those either in the bathing solution on one side or in the cytoplasm on the other. Potassium levels are usually higher than in either of the adjoining fluids while Na and Cl are much lower than in the bathing fluid. These ionic concentrations in the basal laminae are not average values of ionic concentrations in the two adjacent compartments and therefore cannot be explained purely by the limited spatial resolution of the X-ray analysis. Furthermore, the ionic concentrations show tissue-specific differences.

It thus appears that basal laminae in the tissues we have examined may sequester K with the exclusion of Na and Cl. The high levels of phosphorus and sulphur which we have also found in these structures may indicate the presence of phosphate- and sulphate-containing proteo-glycans, which could be responsible for such discrimination. Since this is the first definitive indication of such ion-discriminating properties in basement membranes, their importance in transport physiology, especially in determining the ionic composition of the fluid reaching the basal cell membrane will have to be re-examined. A restrictive hydraulic and ionic conductivity of basal laminae covering the openings of intercellular channels in fluid-absorbing tissues will have important implications in the choice of different theories for solute-solvent coupling (see Chapters 5, 7).

E. Nuclei of Maturing Red Blood Cells in Chick Embryos

Lowestein and Kanno (1963) first demonstrated that the nuclear envelope in some cells could have a high resistance, and maintain an electrical potential difference. The intranuclear ionic environment therefore may be different from that of the surrounding cytoplasm. There are also suggestions that intranuclear binding as well as free ionic concentrations of Na, K, Ca, Mg, and Zn have important roles in the regulation of nuclear function during the cell cycle (Jung and Rothstein, 1966; Brennan and Lichtman, 1973; Berger and Skinner, 1974). Very high levels of intranuclear Na (up to 500 mM kg^{-1} dry weight) are said to be involved in the dissociation of DNA from histones to allow replication and transcription. A change in Na levels is also believed to mediate hormonal induction of gene activity (Kroeger, 1967). However, the evidence for such high concentrations of Na and other ions comes from the chemical analysis of nuclei isolated under anhydrous conditions in inorganic solvents or in material lyophilized at sub-zero temperatures, usually $-30°C$ (Naora et al., 1962; Siebert and Langendorf, 1970; Kirsch et al., 1971).

We have examined nuclei suspended in glycerol, after being similarly isolated (Kirsch et al., 1971) from the red cells of 3—16 day old chicks at 24 hour intervals. Microprobe analysis of the whole nuclei revealed very high concentrations of Na and K (up to 500 mM kg^{-1} dry weight) as well as Ca and Mg (up to 100 mM kg^{-1} dry weight) in nuclei from 3—8 day old chicks, with a steep decline in these concentrations in the older cells. After 12 days no ions could be detected in the condensed nuclei.

However, studies of nuclei not subjected to anhydrous isolation procedures have totally failed to confirm these results: repeated analyses of nuclei in quench-frozen and sectioned whole red blood cells from 5—12 day old embryos have been made, with the sections both in the hydrated state, and after drying in the microscope column. The nuclei always showed a higher concentration of P and K and a lower concentration of Fe and Cl than in the cytoplasm, but the Na levels were approximately the same in both. No evidence of unusually high Na concentrations was found. When the ion pumps on the cell membranes were blocked by treating for 2—12 hours with either 10^{-4}M ouabain or 10^{-4}M ouabain + 10^{-4}M ethacrynic acid before freezing, the Na/K ratio in the cytoplasm rose, to exceed unity. In the nuclei of such cells the Na concentration increased in parallel with that in the cytoplasm, in some cases even exceeding it. It would therefore seem that intra-nuclear Na is not being lost in our analyses of the sectioned material, and that the elevated levels recorded in isolated nuclei are an artifact of the fractionation procedures. A detailed description of these results is to be published elsewhere (Gupta, Hall, Johnson, Schor and Northfield, unpublished).

F. Results From Other Laboratories

Although there is a considerable volume of literature on the localization of ions in sectioned biological tissues by microprobe analysis (Hall *et al.*, 1974; Echlin and Galle, 1975) most of the work has been carried out on chemically treated and embedded materials. Very few studies have been performed on quench-frozen specimens sectioned at low temperatures. Notable amongst such studies is the work of Bacaner *et al.* (1974) on rabbit psoas muscle, Kriz *et al.* (1972) on kidney slices, Dörge *et al.* (1974) on frog skin (all using a scanning microscope), and by Appleton (1974b), Sjöstrom (1974) and Somlyo, Shuman and Somlyo (1976) using a TEM-type analytical instrument. With the exception of Dörge *et al.* and Somlyo *et al.*, all other work is mostly qualitative or semiquantitative in terms of ionic concentrations and will not be discussed here.

Dörge *et al.* (1974) have used 2 μm thick sections of frog skin cut and dried by sublimation at $-70°$C. Quantitative localization of Na and K under normal conditions and with skins treated with ouabain and amiloride demonstrated that 20–30 mEquivalent kg^{-1} wet weight of Na in the cells of the *stratum granulosum* is a part of the Na-transport pool.

Somlyo *et al.* (1976) have analysed approximately 100 nm thick sections of frog's toe muscle, cut at $-130°$C, dried at 10^{-5} Torr and analysed at $-110°$C, using a Philips EM300 microscope with a computer-programmed Kevex X-ray spectrometer. The measured intracellular concentrations of Na, K, Mg, P and Cl expressed as mM per kg dry weight were consistent with the known values from the chemical analysis of the whole muscle. They also found that in relaxed muscle the terminal cisternae had a very high concentration of Ca. In fibres treated with hypertonic (2.2X) Ringer solutions, cisternae had high concentrations of Cl (up to 1500 mM kg^{-1} dry weight) but very little Ca. Both these studies also support the observation that the ions are quantitatively retained in major cell compartments in sections cut and dried at low temperatures. (Also see Addendum.)

VII. CONCLUSIONS

Although the technique of electron microprobe X-ray analysis is not new, in the last few years great advances have been made in its application to biological problems. In this chapter we have described a method for quantitative microanalysis of about 1 μm thick, frozen-hydrated sections cut from quench-frozen tissues in known physiological states. By the choice of tissues whose structure is reasonably simple, and for which adequate confirmatory data are available in the form of quantitative analysis carried out by other means, it has been possible to demonstrate that by these newly developed procedures the electrolyte distributions are well-preserved and the spatial resolution in analysis is adequate to measure the ionic concentrations in microareas of the order of 100 nm and to reveal the existence of ionic gradients over a distance of a few μm. With this capability a wide variety of new information on the transport of ions and water in cells and tissue is now accessible, as illustrated by the following observations selected from the preceeding section.

1. In the unstimulated Malpighian tubules of *Rhodnius* the ionic composition in the lumen is similar to that in the Malpighian tubules of *Calliphora* (high K, low Na) and the Na level in the cytoplasm is similarly low. Stimulation with 5-HT raises the Na concentration of the fluid in the lumen to match the K concentration, as was already known from physiological studies. Microprobe analysis has now revealed that on 5-HT stimulation the average level of Na in the cytoplasm rises from about 20 mM to about 40 mM, indicating that the dramatic increase in fluid secretion depends on an elevation of cytoplasmic Na.

2. The concentrations of Na, K and Cl within the cytoplasm of secretory cells are not uniform. In *Calliphora* salivary glands, there is a gradual increase in K and a gradual decrease in Na over about 5 μm in the cytoplasm from the basal cell membrane to the apical folds at the base of the secretory canaliculi. K/Na ratios determined separately show a similar increase, with the highest ratio of 15 occurring over the area of the apical folds.

3. In the rectal papillae of *Calliphora*, which can absorb fluid from the lumen against a considerable osmotic gradient, the concentration of Na in the stacked lateral infolds is 3—4 times higher than elsewhere in the cytoplasm although the lumen contained 40 mM Na, 100 mM K and 140 mM Cl. It therefore seems that Na is the major cation used to generate the fluid absorption even when K is the major cation present in the lumen. The total cation concentration in the large intercellular channels is much lower than either in the stacks or in the cytoplasm, thus suggesting the possibility of recycling of ions.

4. In several insect epithelia the microprobe analysis of the basement laminae reveals K concentrations higher than either in

the bathing medium or in the adjacent cytoplasm while Na and Cl are much lower than in the bathing fluid. This lamina may not therefore be freely permeable to all ions and it could affect the composition of the extracellular fluid at the basal cell membrane.

5. Microprobe analyses of nuclei, isolated from maturing red blood cells of the chick embryo under anhydrous conditions at low temperatures, confirmed the presence of Na at high concentrations, but in sections of quench-frozen cells no such concentrations of Na were found. This strongly suggests that very high levels of Na in isolated nuclei are a consequence of fractionation procedures.

The tissues studied so far have been relatively simple. In this laboratory, in addition to further studies on ion and water transport related to the problems posed in Chapters 7, 20, 22 and 25, we hope to examine the accessibility to ions of neuronal elements in the nervous system (see Chapter 19). It is known that under favourable conditions better spatial resolution can be obtained by the use of thinner sections. Both the image quality and resolution of analysis can also be improved by the controlled dehydration of frozen sections, once it has been established that this does not significantly alter the ionic distributions.

ACKNOWLEDGEMENTS

The work reported here was supported by a grant from the Science Research Council of the United Kingdom originally awarded to (the late) Professor T. Weis-Fogh and Drs P. Echlin, B. L. Gupta, T. A. Hall, and R. B. Moreton. We are grateful to Messrs N. G. B. Cooper, P. G. Taylor, A. J. Burgess, D. J. Tyler, G. G. Runnalls, R. Northfield and M. J. Day for technical assistance. Many of the drawings are by Mr A. J. Burgess. We also thank our collaborators for allowing us to summarize unpublished results and Dr D. A. Parry, the Head of the Zoology Department, Cambridge University, for his keen interest and an unfailing support for the project.

REFERENCES

Andersen, C. A. (1967). *Methods biochem. Anal.* 15, 147–270.
Appleton, T. C. (1968). *Acta Histochem. Suppl.* 8, 115–133.
Appleton, T. C. (1974a). *J. Microscopy* 100, 49–74.

Appleton, T. C. (1974b). *In* "Electron Microscopy and Cytochemistry" (Eds E. Wisse, W. Th. Daems, I. Molenaar and P. van Duijn), pp. 229—241. North Holland/American Elsevier, Amsterdam and New York.

Bacaner, M., Broadhurst, J., Hutchinson, T. and Lilley, J. (1973). *Proc. natn. Acad. Sci. U.S.A.* **70**, 3423—3427.

Berger, N. A. and Skinner, A. M. (1974). *J. Cell Biol.* **61**, 45—55.

Bernal, J. D. (1965). *Symp. Soc. exp. Biol.* **19**, 17—32.

Berridge, M. J. (1968). *J. exp. Biol.* **48**, 159—174.

Berridge, M. J. (1969). *J. exp. Biol.* **50**, 15—28.

Berridge, M. J. (1975). *Adv. Cyclic Nucleotide Res.* **6**, 1—98.

Berridge, M. J. and Gupta, B. L. (1967). *J. Cell Sci.* **2**, 89—112.

Berridge, M. J. and Gupta, B. L. (1968). *J. Cell Sci.* **3**, 17—32.

Berridge, M. J. and Oschman, J. L. (1969). *Tissue and Cell* **1**, 247—272.

Berridge, M. J. and Oschman, J. L. (1972). "Transporting Epithelia." Academic Press, New York.

Berridge, M. J., Lindley, B. D. and Prince, W. T. (1975). *J. exp. Biol.* **62**, 629—636.

Berridge, M. J., Lindley, B. D. and Prince, W. T. (1976). *J. exp. Biol.* **64**, 311—322.

Brennan, J. K. and Lichtman, M. A. (1973). *J. Cell Physiol.* **82**, 101—112.

Christensen, A. K. (1971). *J. Cell Biol.* **51**, 772—804.

Clark, J., Echlin, P., Moreton, R., Saubermann, A. and Taylor, P. (1976). *In* "Scanning Electron Microscopy (Part I). Proc. 9th Ann. Scanning Electron Microscope Symposium" (Ed. O. Johari), pp. 83—90. IIT Research Institute, Chicago.

Cliff, G. and Lorimer, G. W. (1975). *J. Microscopy* **103**, 203—207.

Diamond, J. M. and Bossert, W. H. (1967). *J. gen. Physiol.* **50**, 2061—2083.

Diamond, J. M. and Bossert, W. H. (1968). *J. Cell Biol.* **37**, 694—702.

Dörge, A., Gehring, K., Nagel, W. and Thurau, K. (1974). *In* "Microprobe Analysis as Applied to Cells and Tissues" (Eds T. Hall, P. Echlin, and R. Kaufmann), pp. 337—349. Academic Press, London and New York.

Dörge, A., Rick, R., Gehring, K., Mason, J. and Thurau, K. (1975). *J. Microscopie Biol. Cell.* **22**, 205—214.

Dubochet, J., Ducommun, M., Zollinger, M. and Kellenberger, E. (1971). *J. Ultrastruc. Res.* **35**, 147—167.

Eastham, L. (1925). *Q. Jl. Microsc. Sci.* **69**, 385—398.

Echlin, P. (1975a). *J. Microscopie Biol. Cell.* **22**, 215—226.

Echlin, P. (1975b). *J. Microscopie Biol. Cell.* **22**, 129—136.

Echlin, P. and Galle, P. (1975). "Biological Microanalysis". Soc. Francaise Microscopie Electronique, Paris.

Echlin, P. and Moreton, R. (1974). *In* "Microprobe Analysis as Applied to Cells and Tissues" (Eds T. Hall, P. Echlin and R. Kaufmann), pp. 159—174. Academic Press, London and New York.

Echlin, P. and Moreton, R. (1976). *In* "Proceedings of 9th Annual Scanning Electron Microscopy Symposium" (Ed. O. Johari), pp. 753—762. IIT Research Institute, Chicago.

Echlin, P., Gupta, B. L. and Moreton, R. B. (1973). *In* "Scanning microscopy: Systems and applications" (Ed. Nixon, W. C.), pp. 242—245. Institute of Physics, London.

Eränkö, O. (1954). *Acta Anat.* **22**, 331—336.

Finn, A. L. (1976). *Physiol. Rev.* **56**, 453—464.

Fraser, T. W. (1976). *J. Microscopy* **106**, 97—99.

Fuchs, W. and Lindemann, B. (1975). *J. Microscopie Biol. Cell.* **22**, 227—234.

Glauert, A. M. and Phillips, R. (1965). *In* "Techniques for Electron Microscopy" (Ed. Kay, D. H.), pp. 213—253. Blackwell, Oxford.

Gullasch, J. and Kaufmann, R. (1974). *In* "Microprobe Analysis as Applied to Cells and Tissues" (Eds. T. Hall, P. Echlin and R. Kaufmann), pp. 175—190. Academic Press, London and New York.

Gupta, B. L. (1976). *In* "Perspectives in Experimental Biology" (Ed. P. Spencer-Davies), pp. 25—42. Pergamon Press, Oxford and New York.

Gupta, B. L. and Berridge, M. J. (1966). *J. Morph.* **120**, 23—82.

Gupta, B. L., Hall, T. A., Maddrell, S. H. P. and Moreton, R. B. (1976a). *J. Cell Biol.* **70**, 176a.

Gupta, B. L., Hall, T. A., Maddrell, S. H. P. and Moreton, R. B. (1976b). *Nature, Lond.* **264**, 284—287.

Hall, T. A. (1971). *In* "Physical Techniques in Biological Research" (Ed. G. Oster), 2nd ed. Vol. IA, pp. 157—275. Academic Press, New York and London.

Hall, T. A. (1975a). *J. Microscopie Biol. Cell.* **22**, 163—168.

Hall, T. A. (1975b). *J. Microscopie Biol. Cell.* **22**, 271—282.

Hall, T. A. and Gupta, B. L. (1974). *J. Microscopy* **100**, 177—188.

Hall, T. A. and Peters, P. D. (1974). *In* "Microprobe Analysis as Applied to Cells and Tissues" (Eds T. Hall, P. Echlin, and R. Kaufmann), pp. 229—238. Academic Press, London and New York.

Hall, T. A., Echlin, P. and Kaufmann, R. (1974). "Microprobe Analysis as Applied to Cells and Tissues". Academic Press, London and New York.

Handler, J. S. and Orloff, J. (1973). *In* "Handbook of Physiology" (Eds J. Orloff and R. W. Berliner), Section 8, pp. 791—814. American Physiological Society, Washington.

Hays, E. A., Lang, M. A. and Gainer, H. (1968). *Comp. Biochem. Physiol.* **26**, 761—792.

Hodgkin, A. L. and Keynes, R. D. (1953). *J. Physiol. Lond.* **119**, 513—523.

Hodson, S. and Marshall, J. (1970). *J. Microscopy* **91**, 105—117.

Hodson, S. and Williams, L. (1976). *J. Cell Sci.* **20**, 687—698.

Hutchinson, T. E., Bacaner, M., Broadhurst, J. and Lilley, J. (1974). *In* "Microprobe Analysis as Applied to Cells and Tissues" (Eds T. Hall, P. Echlin and R. Kaufmann), pp. 191—200. Academic Press, London and New York.

Jessen, H., Peters, P. D. and Hall, T. A. (1974). *J. Cell Sci.* **15**, 359—377.

Jung, C. and Rothstein, A. (1966). *J. gen. Physiol.* **50**, 917—932.

Kessel, R. G. (1970). *J. Cell Biol.* **47**, 299—303.

Keynes, R. D. (1969). *Q. Rev. Biophys.* **2**, 177—281.

Kirk, R. G. and Dobbs, G. H. (1976). *Sci. Tools.* **23**, 28—31.

Kirsch, W. M., Schulz, D., Nakane, P., Lasher, R. and Tandami, Y. *J. Neurosurg.* **34**, 301—309.

Kriz, W., Schnermann, J., Höhling, H. J., von Rosenstiel, A. P. and Hall, T. A. (1972). *In* "Recent advances in Renal Physiology" (Eds H. Wirz and F. Spinelli), pp. 162—171.

Kroeger, H. (1967). *Mem. Soc. Endocr.* **15**, 55—66.

Lowenstein, W. R. and Kanno, Y. (1963). *J. gen. Physiol.* **46**, 1123—1140.

Maddrell, S. H. P. (1971). *Adv. Insect Physiol.* **6**, 199—331.

Mazia, D., Sale, W. S. and Schatten, G. (1974). *J. Cell Biol.* **63**, 212a.

Mazur, P. (1970). *Science, Washington* **168**, 939—949.

Meryman, H. T. (1966). "Cryobiology". Academic Press, London and New York.

Moor, H. (1972). *J. Microscopie* **13**, 159—160.

Moreton, R. B., Echlin, P., Gupta, B. L., Hall, T. A. and Weis-Fogh, T. (1974). *Nature, Lond.* **247**, 113—115.

Morgan, A. J., Davies, T. W. and Erasmus, D. A. (1975). *J. Microscopy* **104**, 271—280.

Naora, H., Naora, H., Izawa, M., Allfrey, V. G. and Mirsky, A. E. (1962). *Proc. natn. Acad. Sci. U.S.A.* **48**, 853—859.

Oschman, J. L. and Berridge, M. J. (1970). *Tissue and Cell* **2**, 281—310.

Oschman, J. L. and Berridge, M. J. (1971). *Fedn. Proc. Fedn. Am. Socs. exp. Biol.* **30**, 49—56.

Oschman, J. L., Wall, B. J. and Gupta, B. L. (1974). *Symp. Soc. exp. Biol.* **28**, 297—341.

Pearse, A. G. E. (1968). "Histochemistry. Theoretical and Applied". Churchill, London.

Phillips, J. E. (1969). *Can. J. Zool.* **47**, 851—863.

Plattner, H., Fisher, W. M., Schmitt, W. W. and Bachmann, L. (1972). *J. Cell Biol.* **53**, 116—126.

Prince, W. T. and Berridge, M. J. (1973). *J. exp. Biol.* **58**, 367—384.

Ramsay, J. A. (1953). *J. exp. Biol.* **30**, 358—369.

Ramsay, J. A. (1971). *Phil. Trans. R. Soc. Lond.* B. **262**, 251—260.

Rasmussen, H. (1970). *Science, Washington* **170**, 404—412.

Rebhun, L. I. (1972). *In* "Principles and Techniques of Electron Microscopy. Biological application" (Ed. M. A. Hayat), Vol. 2, pp. 3—52. Reinhold, New York.

Reed, S. J. B. (1975). "Electron Microprobe Analysis". Cambridge University Press, Cambridge.

Riehle, W. (1968). "Ueber die Vitrifizierung verdünnter wässriger Lösungen". Ph.D. Thesis, Zurich.

Roomans, G. M. and Sevéus, L. A. (1976). *J. Cell Sci.* **21**, 119—127.

Roth, L. J. and Stumpf, W. E. (1969). "Autoradiography of Diffusible Substances". Academic Press, New York and London.

Rowse, J. B., Jepson, W. B., Bailey, A. T., Climpton, N. A. and Soper, P. M. (1974). *J. Physics* E **7**, 512—514.

Saubermann, A. J. and Echlin, P. (1975). *J. Microscopy* **105**, 155—191.

Sevéus, L. and Kindel, L. (1974). *Proc. 8th Int. Congr. Electron Microsc., Canberra* 2, 52–53.

Siebert, G. and Langendorf, H. (1970). *Naturwissenschaften* 57, 119–124.

Sjöström, M. (1975). *J. Microscopie Biol. Cell.* 22, 415–424.

Sjöström, M., Johansson, R. and Thornell, L. E. (1974). *In* "Electron Microscopy and Cytochemistry" (Eds E. Wisse, W. Th. Daems, I. Molenaar and P. van Duijn), pp. 387–391. North-Holland/American Elsevier, Amsterdam and New York.

Smith, D. S. (1968). "Insect Cells. Their Structure and Function". Oliver and Boyd, Edinburgh.

Sohal, R. S., Peters, P. D. and Hall, T. A. (1976). *Tissue and Cell* 8, 447–458.

Somlyo, A. V., Shuman, H. and Somlyo, A. P. (1976). *J. Cell Biol.* 70, 336a.

Spurr, A. R. (1974). *In* "Microprobe Analysis as Applied to Cells and Tissues" (Eds T. Hall, P. Echlin and R. Kaufmann) pp. 213–227. Academic Press, London and New York.

Spurr, A. R. (1975). *J. Microscopie Biol. Cell.* 22, 287–302.

Steere, R. K. (1957). *J. biophys. biochem. Cytol.* 3, 45–60.

Stenn, K. and Bahr, G. F. (1970). *J. Ultrastruc. Res.* 31, 526–550.

Stumpf, W. E. and Roth, L. J. (1965). *Nature, Lond.* 205, 712–713.

Taylor, P. and Burgess, A. J. (1977). *J. Microscopy* in press.

Thornberg, W. and Menger, P. E. (1957). *J. Histochem. Cytochem.* 5, 47–52.

Umrath, W. (1974). *J. Microscopy* 101, 103–106.

Van Harrevald, A., Crowell, J. and Malhotra, S. K. (1965). *J. Cell Biol.* 25, 117–137.

Wall, B. J. (1971). *Fedn. Proc. Fedn. Am. Socs. exp. Biol.* 30, 49–56.

Wigglesworth, V. B. (1965). "The Principles of Insect Physiology". 6th ed. Methuen, London.

Zeuthen, T. (1975). *J. Physiol. Lond.* 256, 32P.

ADDENDUM

After this article had been sent to press the following papers were published reporting new information relevant to the subject matter discussed in this chapter.

Appleton, T. C. and Newell, P. F. (1977). *Nature, Lond.* 266, 854–855.

Cameron, I. L., Sparks, R. L., Horn, K. L. and Smith, N. R. (1977). *J. Cell Biol.* 73, 193–199.

Marshall, A. T. (1977). *Microscopica Acta* 79, 254–266.

Sauberman, A. J., Riley, W. and Echlin, P. (1977). *In* "Scanning Electron Microscopy." (Ed. O. Johari), Vol. I, pp. 347–356, IITRI, Chicago.

Shuman, H., Somlyo, A. V. and Somlyo, A. P. (1977). *In* "Scanning Electron Microscopy." (Ed. O. Johari), Vol. I, pp. 663–672, IITRI, Chicago.

Attention is also drawn to a number of papers presented at the 1st International Meeting on "Low Temperature Microscopy" held at Cambridge in April 1977 (for abstracts see *Proc. R. microsc. Soc.* 12 (2), 65–79).

5. Analysis of Computer Models

E. Skadhauge

Institute of Medical Physiology A, University of Copenhagen, Copenhagen, Denmark

I. INTRODUCTION

The young physiologist who becomes interested in microanalysis will sooner or later discover that we owe many analytical methods to J. A. Ramsay. In numerous laboratories around the world one finds micro-osmometers, micro-chloride titrators and micro-flame photometers which are copies of Ramsay's instruments (Ramsay and Brown, 1955; Ramsay *et al.*, 1955; Ramsay *et al.*, 1953). The modern development in renal micropuncture is unthinkable without Ramsay's work.

If the physiologist becomes more curious and asks what Ramsay used his instruments for: his publications, (particularly Ramsay, 1964; Grimstone *et al.*, 1968), reveal highly interesting studies of renal and intestinal osmoregulation concerning mechanisms with intriguing interactions of salt and water transport. I have always admired Ramsay's studies, his experimental design, his honest writing and the clarity of the language. Finally the reader will be inspired by the good quantitative treatment Ramsay gives for his data. When I got the right to suggest guest speakers at the University of Copenhagen, the first person I invited was J. A. Ramsay. I had thus several good reasons to be pleased when I was asked to write a short contribution to his Festschrift.

The title "Computer Models" should not be too narrowly interpreted: the right word is "quantitative". Computer simulation can be regarded only as a convenient tool for the treatment of models of a certain complexity. Often good approximations or dimensional analysis can provide analytical solutions. This survey will include three types of models of interaction of salt and water flow across and along epithelia.

The first type is the "black box" model. The epithelium is treated either as a true "black box", to which thermodynamic coefficients are assigned, or as two rate-limiting membranes with an unstirred layer of free diffusion — generally the cell — between.

The second type of model treats diffusion and flow in intercellular spaces and through "tight" junctions, or considers intercellular transport together with that across the cell.

In the third type of model, flow along a tube with transport across its wall is treated. Generally these models are macroscopic in the sense that diffusion need not be considered in the flow equations. The wall of the tube — the lining epithelium — is treated as a "black box" to which measured or estimated transport parameters are assigned.

Common to all these models is their aim of quantitatively explaining the coupling of solute and solvent flows. In most cases the particular goal is to understand the apparently active transport of water i.e. water following solutes in the absence of or against a trans-epithelial osmotic pressure gradient. This survey will be limited to epithelial transport; membrane phenomena have been excluded and so have models explaining integration at a higher level.

A. What can Models do — and what not?

In a general sense quantitative model treatments may "be of aid in interpreting experiments, evaluating hypotheses, predicting behaviour, suggesting critical experiments and understanding abnormal processes" (Palatt et al., 1970).

In the physiology of epithelial transport, theoretical model calculations are of particular value in assessing quantitative limits of performance, so that it can be judged whether the proposed mechanism is compatible with the majority of biological findings. If some of the system parameters are regulated, for example by hormones, the parameter sensitivity and mechanism of regulation can be elucidated.

It should be remembered that a model is never proved right even if it survives quantitative scrutiny: it can never rule out alternative explanations. A quantitatively correct model is, however, of great value even if it does not include the most detailed information, for it provides a frame-work for further elucidation. This has most clearly been demonstrated by Hodgkin and Huxley (1952) model of the nerve axon membrane. Its "highly satisfactory" (Hodgkin, 1965) agreement with experimental findings — it predicted correctly the shape of the action potential and the conduction velocity — has allowed further physiological work to centre around the properties of the sodium and potassium channels.

II. CELLULAR MODELS OF COUPLING OF SALT AND WATER FLOW

These models may be called thermodynamic or "black box" models. Although they may depend upon two barriers in series, and may even have a shunt pathway, they are characterized by having no special anatomical structure assigned to the model parts. They can therefore work over a large range of physical properties, including size, of the components.

Such models were introduced to explain the proportionality, often isotonicity, observed between solute (largely NaCl) and water absorption across various epithelia. Curran (1960) suggested that the NaCl transported into the cell made the cell hyperosmotic to the mucosal and serosal fluid, so that water flowed passively from the mucosal lumen to the cell interior. The mucosal membrane of the cell was assumed to have a high reflection coefficient to solutes of the cell, the serosal membrane a low reflection coefficient, which would reduce the effective osmotic pressure and allow solute-linked water flow across the serosal membrane under the slight hydrostatic pressure created by the flow of water into the cell. Curran and McIntosh (1962) showed this mechanism to work in a non-biological model system. Kaye et al. (1966) have visualized the model for the gall bladder (Fig. 1).

Patlak et al. (1963) subjected Curran's model to analysis by irreversible thermodynamics for a single, neutral solute. They showed that the model could yield isotonic flow and also, for certain membrane characteristics, give solvent flow without a net solute flow. The model was not applied to any particular epithelium. Whitlock and Wheeler (1964) found all the observed transport characteristics of the gall bladder in agreement with the model.

Clarkson (1967) also applied irreversible thermodynamics to experiments on rat ileum in vitro. When currents were applied across the tissue a weak electro-osmotic water flow was observed, and as in other preparations there was proportionality between active sodium flow and water flow in the absence of an osmotic gradient. His model

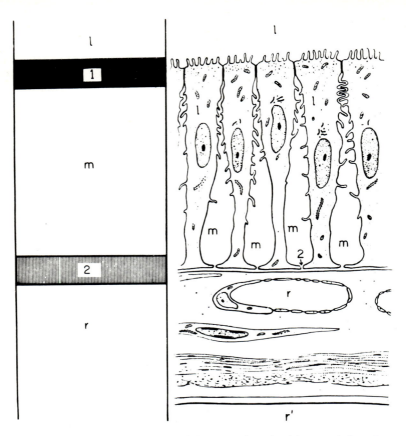

Fig. 1. The three-compartment model. Reproduced with permission from the authors and the publisher (Kaye *et al.*, 1966). The middle compartment (*m*) may be localized to either the cells or the lateral intercellular spaces or both. The serosal compartment begins *in vivo* in the capillaries (*r*), *in vitro* (*r'*) at the peritonaeum.

is interesting because, before the leakiness of tight junctions was known, it predicted the presence of both a channel for active transport of ions, and a passive shunt with extracellular properties.

Another thermodynamic approach has been to calculate the flow of water following the passive ion flux, by so-called co-diffusion. This is related to solvent drag by the Onsager reciprocity as electro-osmosis is related to streaming potentials (see later). Codiffusion may result in solute-linked water flow and even uphill water transport (Diamond, 1962). This mechanism is, however, unattractive in biological tissues since organs in the vertebrate kingdom which regularly receive hyperosmotic solutions with sodium chloride concentrations higher than on the other side of the epithelium, such as the chicken coprodeum (Bindslev and Skadhauge, 1971a) and the intestine of euryhaline fish (Skadhauge, 1969, 1974a) are almost totally sodium impermeable. Diamond (1962) extended the description to the water following active sodium transport and assigned the same reflection coefficient to this water as to that following passive sodium diffusion. There is, however, no experimental or theoretical justification for this procedure.

Two interesting calculations concerning the osmolality of the fluid transported across an isolated epithelium where the absorbate was allowed to drip off the serosal side have been made. Diamond (1964) found isotonicity of the transported fluid of the dripping gall bladder over a wide range of mucosal osmolalities. The problem was whether this isotonicity was due to formation of an isosmotic fluid in the epithelial tissue itself, or to osmotic equilibration occurring while the scanty hyperosmotic secretion moved along the serosal surface before dripping off. In this case equilibrium might be reached by general osmosis. Diamond calculated according to certain assumptions that the passive water permeability was insufficient to give isotonicity. Marro and Germagnoli (1966), however, in another calculation showed that equilibration could occur when no restricting assumptions concerning the passage across the epithelium were made. Great care should therefore be taken in the interpretation of coupling of solute and solvent flows in unilateral preparations.

Recently Hill (1975b and Chapter 7), unsatisfied with the local osmosis model (see later), revived the principle of electro-osmosis. If an osmotic or hydrostatic force, P, creates a water flow across a membrane with fixed charges, and the membrane is bathed in an electrolyte solution, a potential difference or "streaming potential", PD, will be set up. If a current, I, is passed across the membrane, a water flow, Jv, follows. As is well known from the Onsager reciprocal relations these flows and forces are related by $PD/P = Jv/I$.

The problem is whether electro-osmosis can create the necessary driving force. This can be attacked experimentally by measurement of streaming potentials, and theoretically by calculation of the maximal thermodynamic driving force. The gall bladder has a high rate of solute-linked water flow and a low transmural electric potential difference. This organ might therefore be regarded as suitable for the investigation of electro-kinetic phenomena. Apparent streaming potentials were demonstrated (Diamond and Harrison,

1966) although the documentation for the water flow was thin (Diamond, 1966). The origin of the apparent streaming potential was later interpreted as polarization effects in the unstirred layers adjacent to the epithelium (Wedner and Diamond, 1969). With hindsight it is therefore not surprising that electro-osmosis in this organ was found insufficient in magnitude and in the wrong direction to explain the solute-linked water flow (Diamond, 1964). In the chicken coprodeum, however, calculations based on water flow and electric potential difference (Skadhauge, 1973) have demonstrated that electro-osmosis can produce water movements of the same magnitude as the observed solute-linked water flow. If streaming potentials could be measured, electro-osmotic water flow could safely be calculated from the reciprocity of the Onsager relation, which has been shown to hold in experiments with clay plugs (Saxén, 1892) and calculations for narrow tubes (Sørensen and Koefoed, 1974). However, as pointed out by Hill (1975b), the *PD*'s measured across the epithelium may not be those determining the electro-osmosis. The transmural *PD* may be reduced by shunting, or a streaming potential occurring across the "tight" junction might be dissipated along the lateral cellular space. This leads to the conclusion that electro-osmosis is likely to be partly or wholly a local cell membrane phenomenon. Hill showed that the cell membrane potential can provide a sufficient driving force, calculating that the likely intracellular *PD* of −50 mV is equivalent to an osmotic driving force of 100 mOs.

Hill went on to develop a model with electrogenic sodium pumps at each membrane and electro-osmotic coupling of a variable fraction of the hydraulic permeability at both membranes. The lateral intercellular space was a variable shunt without electro-osmosis. He used a coupling coefficient for electro-osmosis developed from straightforward irreversible thermodynamics, and ranging from zero to unity. A weak point is the assumption of an average solute concentration for the membrane equal to that of the luminal bathing fluid. The result of the model is that the transported fluid can be near isosmotic even for coupling coefficients of 0.3 to 0.4, provided the shunt conductance is low. The assumption of electrogenic sodium transport in both membranes needs close scrutiny. An interesting prediction was that the serosal side would have to be bathed with the secretion in order to give isotonicity under all conditions. If impermeant solute was added to the luminal side the secretion became hypertonic; if added to the serosal side it became hypotonic. This would seem to be a natural consequence of the general transmural osmotic driving force. Other logical predictions were that the osmolality rose if the shunt became chloride permeable, so that chloride could not drive current through the electro-osmotic pathway. Conversely, changing the sodium permeability of the shunt did not change osmolality, but only flow rate. (Also see Chapter 7.)

III. STRUCTURAL MODELS OF COUPLING OF SALT AND WATER FLOW

Structural or "microscopic" models describe flow in intercellular spaces, and through the tight junctions. They depend heavily on interaction of diffusion and flow, and are very sensitive to size, particularly length of the channels, and diffusion coefficients.

The first models of this type were introduced by Diamond and Bossert (1967, 1968), as summarized by Diamond (1971). They involved diffusion, osmosis and flow in intercellular spaces. These were assumed closed at the "tight" junctions since the existence of leaky "tight" junctions was not then appreciated (Fig. 2a). The aim was to explore whether realistic channel properties could result in isosmotic, "solute-linked" water flow, as observed experimentally. The independent variables were the length, L, of the channel, its radius, r, osmotic permeability of the wall, P_{os}, the diffusion coefficient of the solute, D, the cellular osmolality, C_o, and the solute transport rate into the channel, N.

The important dependent parameters calculated as functions of the distance, x, from the closed end of the channel, were the osmolality of the lateral intercellular space, $C(x)$, the linear velocity, $V(x)$, and, most important, the osmolality of the emerging fluid, $C(x = L)$. This is the integrated solute flow into the channel divided by the water flow at L. Since no analytical solution existed to the non-linear differential equation describing the system, the authors used digital computation. Hempling (1968) has presented an analogue computer calculation, and Segel (1970) found approximate solutions using an interesting dimensional analysis, introducing dimensionless variables.

The main conclusion from this early stage of modelling was that active transport of solute into the intercellular space could produce a near isosmotic outflow or a fixed hypertonicity nearly independent of solute flow when the ultrastructural and diffusional parameters were within biological limits. Segel's analysis demonstrated that

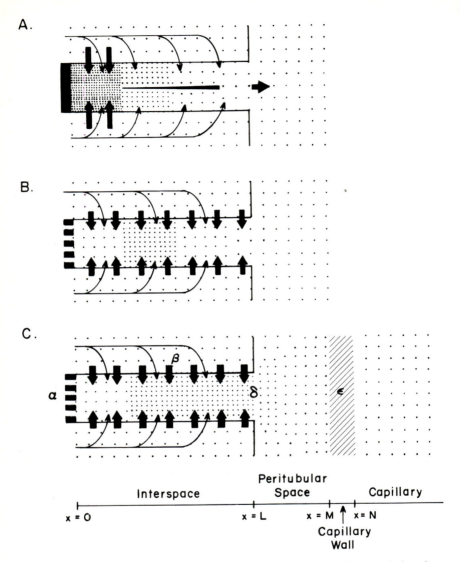

Fig. 2. The interspace hypertonicity model. Reproduced with permission from Sackin and Boulpaep (1975). Solid arrows indicate solute pumps, thin arrows water flow. A shows near equilibrium due to localization of solute pumps at the closed tight junction end. B shows a permeable tight junction and solute pumps distributed along the length of the channel. This leads to strong hyperosmotic flow. If, however, as in C, the osmolality of the fluid emerging from the interspace may be larger than outside, and equilibration occurs in the pericellular space, isotonic flow may be obtained.

constant emerging osmolality would result if the parameters were related by:

$$\frac{C_o \cdot P \cdot L^2}{D \cdot r} = \text{constant}.$$

r is now effective radius $r = 2(A/S)$, where A is area, and S circumference of the channel. Note the similarity between this expression and that derived by Stender *et al.* (1974) (see later). A high sensitivity to channel length is apparent from the relation. Diamond and Bossert used a channel length of 100 μm in most of their calculations. This is rather unfortunate since most epithelia which transport isotonically have shorter cells.

The major weakness in Diamond and Bossert's analysis was, however, a rather unrealistic restriction of solute pumps to the first 10% of the channel adjacent to the closed tight junction end. This was necessary in order to create a sufficiently "flat" concentration profile towards L to prevent diffusion of solute at the open end from "running ahead" of the bulk flow and thus making the emerging fluid hypertonic. Other authors (Hill, 1975a, Sackin and Boulpaep, 1975) have dealt with this problem (see later).

The early models, furthermore, considered the cellular osmolality, C_o, to be equal to that of the luminal (and serosal) fluid. This simplification is obviously mathemathically and physically impossible. Later models have taken the osmotically driven water flow through the cell into account. This problem was first studied by Lindemann, but only published in abstract form (Lindemann and Pring, 1969). The cell will be hyperosmotic to the surroundings. This implies that some water will circulate from the serosal side through the cell into the intercellular space, making the transmural solute-linked water flow of higher osmolality than that emerging from the space. Only if the luminal membrane has a low reflection coefficient for the transported solute will isosmotic transport be created by osmotic equilibrium in the space. There is nothing to indicate, however, that this assumption is tenable.

The aforementioned difficulties have led to further considerations of the intercellular space models. The discovery that many "tight" junctions are leaky, has led to models which take account of the flow across them. Since Ussing and Windhager (1964) and Ussing (1965) drew attention to the importance of the paracellular shunt, and the leakiness of the "tight" junctions was demonstrated (Ussing, 1971) the centre of interest has thus moved from the lateral intercellular space itself to the whole paracellular route including "tight" junctions, foot processes and subepithelial unstirred layers.

A. Recent Interspace Hypertonicity Models

Working from the assumption of uniformly distributed solute pumps and an interspace length of 5—30 μ, Hill (1975a) concluded that to achieve near-isotonic secretion, the osmotic permeability of the channel walls would have to be $10^{-2}-10^{-1}$ cm/sec osm., some 2—3 orders of magnitude larger than Diamond and Bossert's highest estimate. This would seem to confine the interspace hypertonicity model to special long channels with localized solute pumps, such as might occur in layers of several cells. It is, however, possible that the transmural osmotic permeabilities as determined experimentally are too low, due to unstirred layer effects.

Another approach was taken by Sackin and Boulpaep (1975). They abandoned the requirement of $C(L) = C_o$, under which the hypertonicity of emerging fluid only depends upon the value of $dC/dx(x = L)$, and assumed $C(x) = C_o$ only in the capillaries under the epithelium. Figure 2c shows the model according to Sackin and Boulpaep's assumptions. The main conclusion is that the local osmosis model can still yield isotonic flow with a low interspace hypertonicity and no major deviation from isotonicity towards the tight junction, provided osmotic equilibration is achieved in the space beneath the epithelial cell layer. It may be necessary to assume some restriction to solute movement in this unstirred layer in order to prevent too rapid dissipation of the gradient. A real transcellular osmolality difference may even contribute to a general osmotic water flow. Although Sackin and Boulpaep's work has confirmed the viability of the local osmosis model they were unable to solve completely the problem of the required osmotic permeability coefficient being larger than that observed experimentally, in order to yield near-isotonic flow. Water flow across the tight junction must also be considered.

B. Kidney Tubule

In three models special aspects of the control and regulation of salt and water flow across the tubule have been considered. These models will be referred to only briefly. In one the effects of applied electric current (Spring, 1973), in another the effect of hydrostatic pressure (Huss and Marsh, 1975), and in a third the interaction of diffusion of two anions across the tight junction (Schafer et al., 1975) have been calculated. The physiological background of these model treatments is the pronounced leakiness of the tight junction in this epithelium and the isosmotic character of the absorbate under all transport conditions.

Spring's model assumed that the majority of current would pass across the tight junctions resulting in electro-osmosis. He could explain the volume flow (and electric currents) observed experimentally.

Huss and Marsh expanded the Diamond-Bossert model considerably to include "tight" junctions and the basement membrane, and particularly hydrostatic pressure. The most important finding was that the osmotic and the hydrostatic pressure were of the same order of magnitude. This may, at least in part, explain the dependence of resorption in the tubule upon hydrostatic pressure. In general it aids in understanding the large effects of hydrostatic pressure differences on solute-linked water flow in leaky epithelia. The pressure sensitivity was determined by the "tight" junction parameters.

Schafer et al. used their model to simulate solute and solvent flow across the "tight" junction; the solutes entered the intercellular space passively. Different reflection coefficients were assumed for chloride and bicarbonate. The background for the model was the conclusion of several workers that NaCl may be transported across the epithelium by active $HCO_3 =$ transport into the intercellular space from the cell, dragging Na^+ passively across the "tight" junction. Chloride is shunted across the "tight" junction which is more permeable to this ion than to HCO_3^-. The water flow results from codiffusion with NaCl. The model indicated that hypertonicity of the intercellular space was unlikely as a driving force for water, but a quantitatively sufficient flow could result from the oppositely directed gradients for HCO_3^- and chloride, provided they had, as postulated, different reflection coefficients.

C. Frog Skin

An important problem in epithelial transport is whether the flow of one species is linked to that of another by direct solute-solute interaction or through entrainment in a water flow (solvent drag). To find a quantitatively tenable explanation of some unusual results on frog skin, Stender *et al.* (1973) developed a model Fig. 3 of solvent drag by solute-linked water flow. The unusual or unexpected finding was that if the outside of the frog skin bathed in Ringer's solution was made hypertonic with urea and if a tracer solute (sucrose) was present in equal concentration on both sides an increased permeability to the tracer solute in both directions was observed. Furthermore there was a net transport in the inward direction, opposite to that of the net osmotic water flow (Ussing and Johansen, 1969).

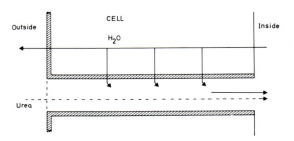

Fig. 3. The anomalous solvent drag model. Reproduced with premission from Stender *et al.* (1974). Urea diffuses through the "tight" junction to the lateral intercellular space dragging water osmotically from the cell. This water escapes predominantly through the foot processes augmenting diffusion of molecules in the direction of flow, impeding diffusion in the opposite direction.

In the so-called anomalous solvent drag model Ussing and Johansen (1969) suggested that this deviation of the sucrose flux ratio from unity might be due to entrainment of sucrose molecules in a local water flow between the cells in the inward direction, created by urea diffusion into the intercellular spaces. The anomalous solvent drag model was formulated mathematically and simulated on an analogue computer. The hypertonicity in the intercellular space, created by urea diffusion through the tight junction, led to water flow from the interior through the cell and back to the inside. In this flow other solute molecules could be driven towards the serosal side. The movement of tracer molecules entering from the outside would be speeded up, but that from the inside would be impeded. The concentration profiles of a typical case are shown in Fig. 4.

The problem was solved by dimensional analysis, since every

Concentration of NaCl and Driving Species

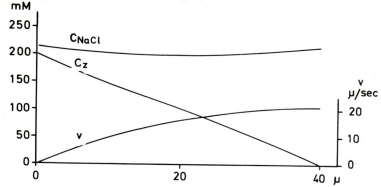

Relative Concentration of Driven Species

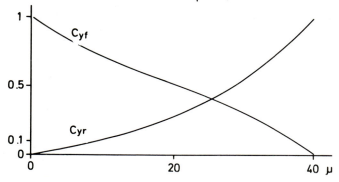

Length of Intercellular Space

Fig. 4. Anomalous solvent drag model. Concentration profiles through the lateral interspace. Reproduced with permission from Stender *et al.* (1974). The driving species (C_z) is urea. Driven species, y, is sucrose; f, indicates flow in outside–inside direction, r, flow in the opposite direction. V, is velocity of water flow. The Ringer's solution is assumed to behave as NaCl.

dimensionless expression, particularly the flux ratio, could be fully described by only two dimensionless parameters:

$$\alpha = \frac{D_z}{D} \quad \text{and} \quad \beta = \frac{P_{os} \cdot S \cdot C_z \cdot L^2}{A \cdot D_z}$$

C_z is the concentration of driving species on the outside, D_z its diffusion coefficient, D is the diffusion coefficient of the driven species, P_{os} is osmotic permeability, S and A and L circumference, area, and length of the channel. The Ringer fluid is assumed to be NaCl with its diffusion coefficient in free solution, and to bathe the epithelium on both sides. The concentration of the driven species does not appear osmotically since it is only added in tracer dose. By analogue computation $\ln R$ (R = flux ratio) was, within the relevant biological range, found to be lineary related to β, so that:

$$\ln R = 0.290 \cdot \beta \cdot \frac{D_z}{D}.$$

Calculation of β and R from this equation allows the estimation of anomalous solvent drag for any epithelium. The main result of applying the model to frog skin data was that the calculated flux ratios of driven species were in agreement with most experimental findings. Due to the sensitivity to length (L^2 is a factor in β) deviation from unity was calculated to be too small to measure in short, epithelia such as toad bladder (10 μm), as was confirmed by experiment (Urakabe *et al.*, 1970). Other calculations showed that the alternative explanation, direct solute-solute coupling, was far from being quantitatively sufficient.

IV. MODELS OF SALT AND WATER FLOW ALONG AND ACROSS TUBES

These models may be called "macroscopic", since they have been applied to structures of lengths from mm to cm with linear velocities which allow diffusion to be neglected. Problems of kidney tubules, and especially of intestinal transport have been dealt with.

A. Retrograde Flow of Urine in Birds

Skadhauge and Kristensen (1972) studied an intriguing problem in the regulation of salt and water excretion in birds. The problem was how in the dehydrated state they manage to pass the hyperosmotic ureteral urine retrogradely into the coprodeum and large intestine without counteracting the water conserving work of the kidney. The transmural flow of NaCl and water occurring during the retrograde flow of ureteral urine was estimated by analogue computation. The values for urine flow, urine osmolality, osmotic permeability coefficients of the epithelium, the net NaCl absorption rate as governed by "Km" (concentration at half maximal absorption rate) and "V_{max}" (maximal absorption rate), and the solute-linked water flow were used; experimental parameters being determined by *in vivo* perfusion studies in the domestic fowl. The retrograde passage of the hyperosmotic ureteral urine was found to result in a net water gain, but at the expense of a hyperosmotic NaCl absorption. With the parameters of the dehydrated bird a 14% fractional water absorption occurred, but 78% of the NaCl was absorbed at a concentration of 835 mequivalent l^{-1}. The model was further used to evaluate the quantitative influence of the parameters upon the fractional water absorption, and the effect of the change in transport parameters from the hydrated to the dehydrated state. Three conclusions were reached. Firstly, since water absorption was more sensitive to urine osmolality than to urine flow the post-renal water absorption would seem to increase if the kidney concentrated the urine less well. If the fall in glomerular filtration rate (which also leads to a decreased urine flow), as observed in the dehydrated state, was taken into account the water conservation would be maximal when the urine osmolality was maximal in the dehydrated state. The model has thus aided in understanding teleological adaptation. Secondly, that water absorption occurred only when NaCl was absorbed. This — together with the previously mentioned strong dependence on urine osmolality — led to the conclusion that a bird with a good renal concentrating ability such as the budgerigar (Krag and Skadhauge, 1972) could not survive on a diet of natural seeds (which are low in NaCl). When such a discrepancy arises the question is whether the model is insufficient in its description of the biological phenomenon, or whether the numerical values assigned to the parameters are wrong. A study on the Australian parrot, or galah, a seed eating bird with a good renal concentrating ability from a salt-depleted desert, showed the latter to be the case (Skadhauge, 1974b). Thirdly, the adaptation of absorption parameters to dehydration was predicted to result in a 20% augmentation of fractional water absorption, thus quantifying the role of retrograde transport in osmoregulation.

B. Drinking Rate and Intestinal Absorption in Fish

There is in the vertebrate kingdom one other case of hyperosmotic fluids running into the intestinal tract: marine teleosts (bony fish) drink the surrounding sea water. They absorb salt and water through the wall of the gut, and excrete NaCl through the gills. This saves water to make up for osmotic losses to the surroundings. (see Chapters 17 and 18).

When euryhaline teleosts have been adapted to waters of a new salinity, the drinking rate, the gill NaCl turnover, and the intestinal transport parameters change. In order to obtain insight into the interaction of the parameters of the system Kristensen and Skadhauge (1974) simulated the flow along and across the gut on an analogue computer. Data obtained from the yellow european eel, the rainbow trout and the Cyprinodont, *Aphanius dispar*, were used. The model of the intestine is shown in Fig. 5.

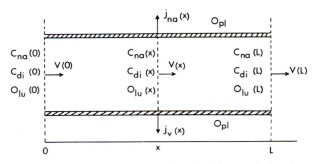

Fig. 5. Model of the fish intestine. Reproduced with permission from Kristensen and Skadhauge (1974). The intestine is simplified to a straight, rigid tube with sea water taken in by drinking entering at one end ($x = 0$) and leaving through the other end ($x = L$). The fluid contains NaCl of concentration C_{Na}, and divalent ions, $C_{di} V(O_{lu}) O_{lu}$ indicates the osmolality. The transmural transport parameters for sodium, j_{Na}, and water, j_v, are considered of equal value for the whole area.

When the experimental values for drinking rate, maximal NaCl absorption rate, concentration at half maximal absorption rate, osmotic permeability coefficients, solute-linked water flow, and concentrations in the gut were used in the calculations, good consistency was achieved and predictions could be made. There was a close linkage between drinking rate and maximal NaCl absorption rate. A large water absorption was only possible close to an optimal drinking rate for each value of maximal NaCl absorption rate. Water absorption had little sensitivity to the osmotic permeability coefficient of the intestinal wall.

When exposed to waters of higher salinity the intestinal NaCl absorption would have to increase roughly with the square of the drinking rate, which would be very costly for energetic reasons, if the osmotic permeability of the gills was unchanged. In waters hyperosmotic to sea water drinking rate was found to be unchanged, showing that intestinal NaCl absorption went up only in proportion to salinity. This demonstrated indirectly that the effective osmotic permeability of the gills must decrease. Simulation has thus pinpointed the crucial elements of adaptation to high salinity and given clues to new investigations.

A somewhat similar problem of salt and water flow was studied by analogue computation by Bentzel *et al.* (1974). These authors sought an analysis of the time course of volume expansion of a droplet of isomotic sodium-free solution injected into an oil-filled proximal tubule of the *Necturus* kidney. Due to diffusion of Na^+ (and anions) and water into the droplet, it expands and its Na concentration increases. The distance between the oil columns can be followed under the microscope. As a result of NaCl diffusion into the drop the usual tubular NaCl (and water) absorption begins, and after some minutes a state of (practically) zero volume flow and constant NaCl concentration is reached. The authors assumed the volume change to be caused by the difference between influx and efflux of sodium and water across a common, single barrier. The volume and therefore the transmural water flow was exclusively a dependent parameter. Neither hydrostatic, osmotic nor local solute-linked water flow entered the equations. With this simple pump-leak system (for sodium) it was possible to simulate the time course of volume expansion, predict the observed steady-state sodium concentration and the passive sodium permeability and active absorption rate. This analysis is of particular interest because it allows calculation of bidirectional fluxes (assuming pump-leak) without isotope measurements. Considering the problems of exchange diffusion and calculations of passive fluxes from electric and isotope measurements this independent analysis deserves interest.

V. GENERAL DISCUSSION

In the years after the cellular and intercellular osmotic models of solute-solvent linkage were introduced, attempts have been made to demonstrate experimentally the local areas of hypertonicity required by the models. In most cases these efforts have not been successful, because methods of measuring average osmolality or of localizing ions are not yet quantitatively sufficient at the sub-cellular level (see Chapter 4). In only one case has intercellular hypertonicity been demonstrated directly (Wall *et al.*, 1970) and in another inferred indirectly (Machen and Diamond, 1969). The latest development in modelling has led us to expect only a slight hypertonicity, either inter or intracellularly which makes direct demonstration even more elusive. Furthermore, even demonstration of intercellular hypertonicity will not distinguish between cellular or intercellular mechanisms, since the osmolality of the cell will in any case follow that of its surroundings. Other approaches will be needed.

One requirement of an open-end local osmosis model is that the volume flow stops immediately after cessation of the net solute flow, because the solute gradient will dissipate within one second. The author is not aware of any experiment deliberately designed to test this necessary condition. In one report, however, a separation in time over several minutes between cessation of net NaCl flow and net water flow was observed (Schilb and Brodsky, 1970). These authors observed in the turtle urinary bladder *in vitro* that water flow continued several minutes after net NaCl absorption stopped following poisoning with ouabain. During this period the cellular sodium concentration increased. Their experiments can be explained by a Curran model, but not by local osmosis in lateral spaces. The conclusion of recent model-building seems, however, to make the distinction irrelevant; isotonic water flow seems necessarily dependent on interaction of cell and intercellular space.

It is still not possible to distinguish between local hypertonicity mechanisms with water moving osmotically across the membranes, and electro-osmotic systems. The possibility of electro-osmosis should be studied more intensively, particularly since two findings will readily be explained by electro-osmosis:

First, as pointed out by many authors, epithelia with more leaky "tight" junctions have isotonic absorbates, whereas epithelia with very high osmolalities of the absorbate (2—3 M NaCl) such as the frog skin have truly "tight" junctions. Second, in the secretion from the proximal intestine induced by cholera toxin, the change from sodium absorption to chloride secretion is associated with H_2O secretion without electron microscopic change.

Finally, too little attention, both in model building and in experiments, has been paid to the situation where lumen and plasma are anisotonic. In some cases of vertebrate osmoregulation the maximal ability to transport water against an osmotic gradient in solute-linked coupling is of great physiological importance, e.g. bird cloaca (Skadhauge, 1973) and fish intestine (Skadhauge, 1969, 1974a).

A combination of two oppositely directed bulk flows of water which operates when the lumen is anisotonic to the plasma was suggested on kinetic grounds by Skadhauge (1969, 1974a). This is particularly important when solute-linked water flow occurs against an osmotic gradient. Recently Berry and Boulpaep (1975) have suggested that the solute-linked water flow proceeds through the intercellular space, the reverse flow through the cell. The general osmotic water flow must therefore always be considered when lumen and plasma are anisotonic. In the only epithelium where a systematic test of water flow over a wide range of mucosal osmolalities has been made in the presence or absence of sodium ions, the coprodeum of the fowl (Bindslev and Skadhauge, 1971a,b), the solute-linked water flow was only measurable in a ±100 mOsmol range around plasma osmolality. At higher and lower osmolalities solute-linked water flow disappeared completely in the general osmotic water flow in the two directions. It should also be remembered that in many epithelia as little as 20—60 mOsmol higher osmolality on the luminal side will prevent absorption of water, thus inhibiting or overcoming solute-linked water flow. The general osmotic water flow will accordingly in many cases be of overwhelming importance for osmotic loss or gain.

The "macroscopic" models of flow along and across intestine may be applied to absorption in mammalian colon and in ducts of salivary glands and pancreas where regulation and parameter sensitivity needs clarification.

VI. SUMMARY

The coupling of salt and water flow across epithelia in the absence of or even against an osmotic gradient can be quantitatively accounted for by local hypertonicity or electro-osmosis models. Such mechanisms may cause isosmotic water flow if slight hypertonicity exists in the intercellular spaces and water and salt moves across the tight junctions or if a part of the equilibrium occurs in the sub-epithelial unstirred layers. When the osmolalities on the mucosal and serosal side are not identical two separate bulk-flows of water are likely to occur, one through the cell, and one through the intercellular space and "tight" junction. The transport across this extracellular shunt explains why epithelia which are electrically leaky and solute-permeable produce absorbates closer to isotonicity than tight epithelia. Several of the models have strong predictive value concerning interaction of bulk flow, diffusion and electrical events.

Macroscopic models of flow along and across the intestinal tract have elucidated the role of the intestine in osmoregulation, and pinpointed fruitful areas for further studies of adaptation mechanisms. These models can readily be expanded to study the quantitative interaction of ion and water flow in mammalian colon, and along the ducts of secretory glands.

REFERENCES

Bentzel, C. J., Spring, K. R., Hare, O. K. and Paganelli, C. V. (1974). *Amer. J. Physiol.* **226**, 127—135.

Berry, C. A. and Boulpaep, E. L. (1975). *Am. J. Physiol.* **228**, 581—595.

Bindslev, N. and Skadhauge, E. (1971a). *J. Physiol.* **216**, 735—751.

Bindslev, N. and Skadhauge, E. (1971b). *J. Physiol.* **216**, 753—768.

Clarkson, T. W. (1967). *J. gen. Physiol.* **50**, 695—727.

Curran, P. F. (1960). *J. gen. Physiol.* **43**, 1137—1148.

Curran, P. F. and McIntosh, J. R. (1962). *Nature, Lond.* **193**, 347—348.

Diamond, J. M. (1962). *J. Physiol.* **161**, 503—527.

Diamond, J. M. (1964). *J. gen. Physiol.* **48**, 15—42.

Diamond, J. M. (1966). *J. Physiol.* **183**, 58—82.

Diamond, J. M. (1971). *Fed. Proc.* **30**, 6—13.

Diamond, J. M. and Bossert, W. H. (1967). *J. gen. Physiol.* **50**, 2061—2083.

Diamond, J. M. and Bossert, W. H. (1968). *J. Cell Biol.* **37**, 694—702.

Diamond, J. M. and Harrison, S. C. (1966). *J. Physiol.* **183**, 37—57.

Grimstone, A. V., Mullinger, A. M. and Ramsay, J. A. (1968). *Phil. Trans. Roy. Soc. Lond.* B **253**, 343—382.

Hempling, H. G. (1968). *J. gen. Physiol.* **51**, 273—276.

Hill, A. E. (1975a). *Proc. Roy. Soc. Lond.* B **190**, 99—114.

Hill, A. E. (1975b). *Proc. Roy. Soc. Lond.* B **190**, 115—134.

Hodgkin, A. L. (1965). "The conduction of the nervous impulse". Liverpool University Press, England.

Hodgkin, A. L. and Huxley, A. F. (1952). *J. Physiol.* **117**, 500—544.

Huss, R. E. and Marsh, D. J. (1975). *J. Membr. Biol.* **23**, 305—347.

Kaye, G. I., Wheeler, H. O., Whitlock, R. T. and Lane, N. (1966). *J. Cell. Biol.* **30**, 237—268.

Krag, B. and Skadhauge, E. (1972). *Comp. Biochem. Physiol.* **41A**, 667—683.

Kristensen, K. and Skadhauge, E. (1974). *J. exp. Biol.* **60**, 557—566.

Lindemann, B. and Pring, M. (1969). *Arch. ges. Physiol.* **307**, 55.

Machen, T. E. and Diamond, J. M. (1969). *J. Membr. Biol.* **1**, 194—213.

Marro, F. and Germagnoli, E. (1966). *J. gen. Physiol.* **49**, 1351—1353.

Palatt, P. J., Saidel, G. M. and Macklin, M. (1970). *J. theor. Biol.* **29**, 251—274.

Patlak, C. S., Goldstein, D. A. and Hoffman, J. F. (1963). *J. theor. Biol.* **5**, 426—442.

Ramsay, J. A. (1964). *Phil. Trans. Roy. Soc. Lond.* B **248**, 279—314.

Ramsay, J. A. and Brown, R. H. J. (1955). *J. Sci. Instrum.* **32**, 372—375.

Ramsay, J. A., Brown, R. H. J. and Falloon, S. W. H. W. (1953). *J. exp. Biol.* **30**, 1—17.

Ramsay, J. A., Brown, R. H. J. and Croghan, P. C. (1955). *J. exp. Biol.* **32**, 822—829.

Sackin, H. and Boulpaep, E. L. (1975). *J. gen. Physiol.* **66**, 671—733.

Saxén, U. (1892). *Ann. Phys. Chem.* **47**, 46—68.

Schafer, J. A., Patlak, C. S. and Andreoli, T. E. (1975). *J. gen. Physiol.* **66**, 445—471.

Schilb, T. P. and Brodsky, W. A. (1970). *Am. J. Physiol.* **219**, 590—596.

Segel, L. A. (1970). *J. theor. Biol.* **29**, 233—250.

Skadhauge, E. (1969). *J. Physiol.* **204**, 135—158.

Skadhauge, E. (1973). *Dan. Med. Bull.* **20**, suppl. 1, 1—82.

Skadhauge, E. (1974a). *J. exp. Biol.* **60**, 535—546.

Skadhauge, E. (1974b). *J. Physiol.* **240**, 763—773.

Skadhauge, E. and Kristensen, K. (1972). *J. theor. Biol.* **35**, 473—487.

Spring, K. R. (1973). *J. Membr. Biol.* **13**, 323—352.

Stender, S., Kristensen, K. and Skadhauge, E. (1973). *J. Membr. Biol.* **11**, 377—398.

Sørensen, T. S. and Koefoed, J. (1974). *J. Chem. Soc., Faraday Trans* **II**. **70**, 665—675.

Urakabe, S., Handler, J. S. and Orloff, J. (1970). *Amer. J. Physiol.* **218**, 1179—1187.

Ussing, H. H. (1965). *Acta physiol. Scand.* **63**, 141—155.

Ussing, H. H. (1971). *Phil. Trans. Roy. Soc. Lond.* B **262**, 85—90.

Ussing, H. H. and Johansen, B. (1969). *Nephron* **6**, 317—328.

Ussing, H. H. and Windhager, E. E. (1964). *Acta physiol. Scand.* **61**, 484—504.

Wall, B. J., Oschman, J. L. and Schmidt-Nielsen, B. (1970). *Science* **167**, 1497—1498.

Wedner, A. J. and Diamond, J. M. (1969). *J. Membr. Biol.* **1**, 92—108.

Whitlock, R. T. and Wheeler, H. O. (1964). *J. clin. Invest.* **43**, 2249—2265.

II.   Mechanisms and Control of Transport

6. Water, Nonelectrolyte and Ion Permeability of Lipid Bilayer Membranes*

A. Finkelstein

Departments of Physiology, Neuroscience and Biophysics, Albert Einstein College of Medicine, New York, U.S.A.

Permeability of lipid bilayers has been studied in two different systems. One involves a single planar (or spherical) bilayer separating two macroscopic aqueous compartments that can be stirred and sampled and which may contain electrodes for passing current and recording potential differences across the membrane (Finkelstein, 1974b). The other consists of a dispersion in water of single walled or multiwalled vesicles from which the leakage of solutes can be measured by various techniques (Kinsky, 1974). Since virtually all of the quantitative data comes from studies on the planar bilayers, I shall consider only that system in this article.

A discussion of the permeability of lipid bilayers divides itself naturally into two parts: the unmodified and the modified membrane. The former may consist of a single, chemically defined lipid such as dioleophosphatidylcholine or contain the entire pantheon of phospholipids, neutral lipids, cerebrosides, etc. that can be extracted from brain. The latter is an unmodified membrane to which has been added some exogenous agent that alters its permeability properties.

I. UNMODIFIED MEMBRANES

To a first approximation the permeability of lipid bilayers is what one would expect for a thin (~50 Å) layer of liquid hydrocarbon (Finkelstein and Cass, 1968). That is, the barrier to permeation is the

* Supported by NSF grant number BMS 74-01139.

hydrocarbon interior of the bilayer formed from the hydrophobic tails of the lipids; the polar head groups of the lipids serve merely to anchor this hydrocarbon region between two aqueous phases and do not themselves offer significant resistance to water and solute permeation. Molecules (including water) cross the membrane by a solubility-diffusion mechanism; that is, they partition into the hydrocarbon region of the membrane at one interface, diffuse through the hydrocarbon interior, and then partition out of the membrane at the other interface. "Lipophilic" molecules, i.e. those that are reasonably soluble in hydrocarbon (e.g. butanol, unionized forms of weak acids), are very permeant; "hydrophilic" molecules (e.g. urea, sugars, amino acids) are poorly permeant, and the most hydrophilic of all, small ions, are essentially impermeant. Within this simple model, however, there exist interesting features of great relevance to the mechanism of transport (particularly water transport) across plasma membranes.

A. Water Permeability

Through changes in the degree of unsaturation and length of the fatty acid tails of phospholipids, the amount of cholesterol and the temperature, the water permeability of lipid bilayers has been varied several hundred fold, and undoubtedly could be varied even more (Finkelstein, 1976a). The range of values for the water permeability coefficient $[P_d(H_2O)]$, which extends from 1×10^{-2} cm/sec for egg lecithin membranes at $36°$ (Huang and Thompson, 1966) to 2×10^{-5} cm/sec for sphingomyelin:cholesterol membranes at $14.5°$ (Finkelstein, 1976a), encompasses almost the entire spectrum of values reported for plasma membranes and tissues. Thus, there is no problem in accounting for the magnitude of the water permeability of cells and tissues simply from the known properties of lipid bilayers and the fact that all plasma membranes have a lipid bilayer as their basic structural element.* There are other considerations, however, that have led physiologists to believe that for many cells and tissues, the major route of water transport is through aqueous pores. These are: (1) ratios of the osmotic permeability coefficient,

---

* If there is any problem at all, it is not in accounting for why $P_d(H_2O)$ is so high for such membranes as that of the red blood cell, but rather why it is so low for the membranes of certain eggs such as *Fundulus* (Dunham *et al.*, 1970), in which $P_d(H_2O)$ is about 20 fold smaller than the lowest value so far obtainable with lipid bilayers.

P_f, to $P_d(H_2O)$ significantly greater than one; (2) solvent drag of solutes accompanying osmotic flow; and (3) graded permeability to small molecules (molecular sieving).

With the possible exception of the red cell, unstirred layers make it impossible to establish the true existence of the first two criteria in almost all cases (Dainty, 1963; Hays, 1972). The third criterion, molecular sieving, has convincingly established the existence of pores in some cases, but interestingly, this phenomenon is most pronounced in multicellular systems such as capillary endothelia (Pappenheimer *et al.*, 1951), and certain "loose" epithelial tissues (Wright and Pietras, 1974). It thus appears that large pores may be restricted to intercellular regions, and that if pores exist in plasma membranes, they have relatively small radii. It would therefore be useful to have additional criteria for determining whether water moves across a plasma membrane primarily by a solubility-diffusion mechanism or by passage through narrow, aqueous pores penetrating the bilayer. One such criterion has recently emerged from the study of water and nonelectrolyte permeability of lipid bilayers.

It appears that although the absolute values of nonelectrolyte and water permeabilities can be varied over a several hundred fold range, through changes in lipid composition and temperature, the relative values of these permeabilities remain constant to within about a factor of two (Finkelstein, 1976a). Therefore, if water traverses the lipid bilayer of a cell or tissue primarily by a solubility-diffusion mechanism, P_d (water)/P_d (solute) should approximate the value found in artificial lipid bilayers, where "solute" is a molecule, such as n-butyramide or 1,6 hexanediol, that crosses the cell membrane by a solubility-diffusion mechanism without the aid of a special transporting system. Conversely, if P_d (water)/P_d (solute) greatly exceeds the value found in artificial lipid bilayers, strong evidence exists that water crosses the membrane by an alternative pathway; i.e. through pores. This criterion has been applied to the action of antidiuretic hormone (ADH) on toad urinary bladder and the conclusion reached that water crosses the mucosal membrane of that tissue through narrow aqueous pores (Finkelstein, 1976b). This same test can be applied to a solute suspected of being transported through pores or by "carriers". If the ratio of its permeability coefficient to that of a solute such as 1,6 hexanediol, that almost certainly crosses the membrane by a solubility-diffusion mechanism, greatly exceeds the value found in artificial lipid bilayers, it is most likely that the molecule in question is transported by a specialized pathway through the cell membrane.

B. Ion Permeability

I have already indicated that lipid bilayers are virtually impermeable to small ions such as Na^+, K^+, and Cl^-. This is primarily a consequence of the enormous electrostatic energy (Born energy) required to move a charge of approximately 2Å radius from water, with its high dielectric constant (≈ 80), into the low dielectric constant (≈ 2), hydrocarbon interior of the bilayer. On the other hand, bilayers are substantially permeable to large lipophilic ions such as tetraphenylboron ($TPhB^-$) and tetraphenylarsonium ($TPhAs^+$), both because the electrostatic energy required to move these large ions into hydrocarbon is much less than for small ions, and because of the intrinsic lipid solubility of the phenyl groups.* The virtual impermeability to small ions is manifested electrically by the very low conductance ($\approx 10^{-8}$ mho/cm^2) of lipid bilayers bathed by NaCl or KCl.

C. Summary

One of the major motivations for studying lipid bilayers is that they are the basic structural element of cell membranes. Their permeability properties, reviewed above, reveal that they share with plasma membranes a high permeability for lipophilic molecules and have a comparable water permeability. On the other hand, they show none of the ion permeability and selectivity manifested by cell membranes (their conductance being about 10^5 fold less than a "typical" cell membrane), nor do they discriminate, as can cell membranes, among nonelectrolytes either on the basis of size or steriochemistry.* Furthermore, we noted that although values for P_d (water) span those of cell membranes, the route of water transport in at least some cells and tissues is not by a solubility-diffusion mechanism.

Physiologists have long believed that these special permeability characteristics of cell membranes result from the presence of carriers or pores in these membranes. It is interesting that some of these same

---

* $TPhB^-$ is much more permeant than $TPhAs^+$, because the interior of the bilayer is several hundred millivolts positive with respect to the aqueous solutions. This interesting feature of lipid bilayers that makes them intrinsically more permeable to anions than to cations will not be considered in this article (see, for example, Andersen *et al.*, 1976).
* For those nonelectrolytes (such as 1,4 butanediol and n-butryamide) that are not transported by special mechanisms, P_d's comparable in value to those found in cells and tissues are obtainable in artificial lipid bilayers (Finkelstein, 1976b).

characteristics can be reproduced in artificial lipid bilayers by the introduction of "modifiers" that function in either one of these capacities. Although most of the modifiers so far discovered are of small (~1000) mol. wt, whereas the natural modifiers of cell membranes are presumably larger mol. wt proteins, there is reason to believe that many of the physical principles pertaining to the model system are applicable to biological systems. With this in mind, we now turn to a consideration of the properties induced in lipid bilayers by some modifiers and their mechanism of action.

II. MODIFIED MEMBRANES

A. Carriers

Certain cyclic antibiotics such as the depsipeptides (e.g. valinomycin and the enniatins) and the macrocyclic tetalides (e.g. nonactin and monactin) function as alkali cation carriers (Fig. 1). When added in micromolar or smaller quantities to the aqueous phases on one or both sides of the membrane, they increase membrane conductance by many orders of magnitude, and at these conductances, the membrane is ideally cation selective. In fact, the membrane markedly discriminates among the alkali cations. A valinomycin-treated membrane is at least 300 times more permeable to K^+ than to Na^+ (Mueller and Rudin, 1967). Conductance increases linearly with both antibiotic concentration and cation concentration (McLaughlin $et\ al.$,

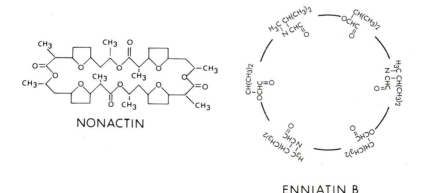

NONACTIN

ENNIATIN B

Fig. 1. Structural formulae for the cation carriers nonactin and enniatin B.

G

1970).* These results are consistent with the view that the current-carrying species is of the form:

$$(antibiotic\text{-}cation)^+.$$

Clearly, this is a carrier system. The antibiotic complexes with the cation at one interface, diffuses across the membrane, and discharges the cation at the other interface.

These antibiotics are cyclic molecules that have many carbonyl or ether oxygens. The oxygens form the lining of a cavity in which the naked alkali cation sits (Fig. 2), having exchanged the H_2O oxygens

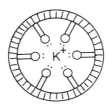

Fig. 2. Schematic diagram of a cation carrier, such as valinomycin. The carbonyl oxygens provide a polar environment for the cation, and in fact substitute for the first hydration shell normally surrounding the cation in water. The exterior aspect of the molecule is non polar, thus making it compatible with the hydrocarbon interior of the membrane.

in its first hydration shell in free solution for these carbonyl and ether oxygens. Although these molecules are cyclical, the ion does not sit in the hole of a donut. Instead, the molecule enfolds the ion to create a spherical cavity (Dobler *et al.*, 1969; Kilbourn *et al.*, 1967; Pinkerton *et al.*, 1969). This brings the polar oxygens inward to surround the cation, and they in turn are "buried" within the complex. Thus, the external aspect of the complex is completely non polar, and this, together with its large size (which greatly reduces the Born energy), makes it very lipid soluble. The order and degree of selectivity among alkali cations reflects the magnitude of the difference in free energy between the ion in its hydrated form in water and the ion "hydrated" by the oxygens of the antibiotic.

The ability of these antibiotics to pull alkali cations into an organic phase is not confined to lipid bilayers. The same phenomenon is observed in bulk partitioning experiments between water and

* The linear dependence of conductance on cation concentration is obtained on membranes made with lipids having no net charge. We shall not consider the complications arising from surface potentials associated with charged lipid membranes (see, for example, McLaughlin *et al.*, 1970).

organic solvents (Eisenman *et al.*, 1969; Pressman *et al.*, 1967). Thus, the complexing and partitioning properties of these antibiotics is an inherent property of the molecules themselves, and they do not require a bilayer structure for their action to be manifest. The action of the channel formers, however, that we shall now take up, requires a bilayer.

B. Channel Formers

The conductances induced by the carriers are relatively ohmic, although their nonlinear aspects have attracted attention because of information they yield both about energy barriers in the membrane and about the kinetics of transport processes (Hall *et al.*, 1973; Läuger, 1972). Among the channels, however, there are two broad classes. Members of the first class have, like the carriers, relatively linear I-V characteristics. Members of the second class, on the other hand, display dramatic voltage-dependent conductances analogous to those observed in excitable cells such as nerve and muscle (see Mueller and Rudin, 1969). We shall restrict our attention to two representatives of the former class, as a proper discussion of voltage dependence is beyond the scope of this article.

1. Nystatin and Amphotericin B

These molecules are so similar both in their structure and in their action on bilayers that we shall not distinguish between them in our discussion. They differ drastically in chemical structure from the molecules just considered (Fig. 3). They have a lactone ring, many

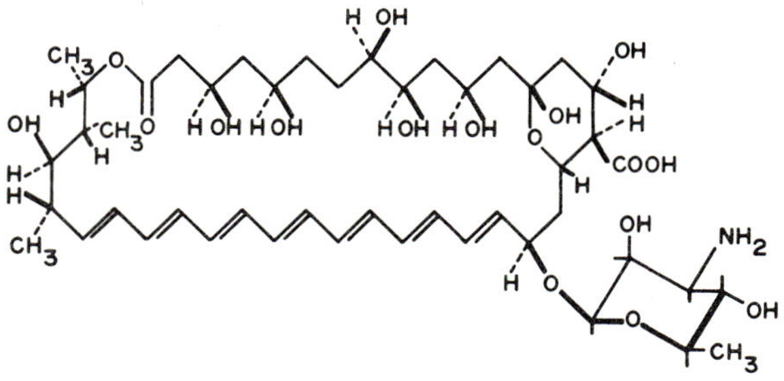

Fig. 3. Structural formula of amphotericin B. Nystatin differs from amphotericin B in that it contains a tetraene and a diene in place of the heptaene chromophore; in other respects the molecules are very similar.

hydroxyl groups, one amino sugar and carboxyl group (which make them zwitterions between pH 5 and 8), and a polyenic chromophore. The most salient physical chemical difference between these molecules and the carriers is that the latter are very lipophilic and therefore quite soluble in hydrocarbon, whereas nystatin and amphotericin B are not very soluble in either water or hydrocarbon. They are amphipathic molecules, with the hydrophilic hydroxyl groups, amino sugar, and carboxyl group segregated from the hydrophobic polyene chain (Fig. 4).

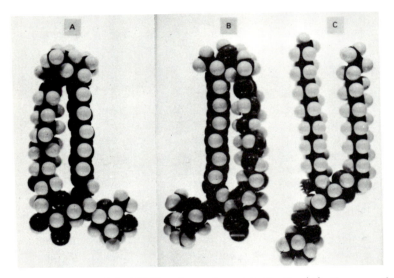

Fig. 4. Molecular model (CPK) of amphotericin B. In (A) the completely hydrophobic face of amphotericin B is seen; in (B) the molecule has been rotated 180° about its long axis to reveal the opposite face with its many hydroxyl groups. Note that the molecule consists of two chains: a polyene chain (seen on the right in (A) and on the left in (B)) and an amphipathic chain. The hydrophobic and hydrophilic faces of the amphipathic chain are seen in (A) and (B) respectively. At the bottom of the figures are the polar amino sugar and carboxyl groups; at the top is a single hydroxyl group (seen most clearly in (B)). In (C), a CPK model of lecithin is shown for comparison with amphotericin B. (From Finkelstein and Holz, 1973).

a. Effect on ion permeability

When added in micromolar or less amounts to both sides of the membrane, nystatin greatly increases the conductance of sterol-containing (usually cholesterol) bilayers (Cass *et al.*, 1970). The conductance increase is primarily due to increased univalent anion permeability.* In contrast to the linear dependence of conductance on antibiotic concentration observed with carriers, nystatin-induced conductance is proportional to a large power of the antibiotic concentration (6th to 12th power, depending on the lipid (Cass *et al.*, 1970)), implying that many nystatin molecules are involved in forming a conductance site.

b. Effect on water and nonelectrolyte permeability

Concomitant with the conductance increase produced by nystatin is an increase in membrane permeability to water and small nonelectrolytes (Holz and Finkelstein, 1970). There are two salient features of this permeability increase (Holz and Finkelstein, 1970). (a) $P_f/P_d(H_2O)$ (where P_f is the permeability coefficient for water as measured in an osmotic experiment) has an approximate value of 3, and (b) the permeability to nonelectrolytes decreases with increasing Stokes-Einstein radius of the molecule; molecules larger than glucose (radius $\approx 4Å$) are virtually impermeant. Both of these features are attributes of water and nonelectrolyte transport through aqueous pores. In fact these permeability data combined with the amphipathic nature of nystatin and the dependence of permeability on a large power of its concentration make it virtually certain that nystatin forms aqueous pores in bilayer membranes.

c. Model of the pore

There is a remarkable topological similarity between amphotericin B and a phospholipid (Fig. 4). The major differences are that the two chains in the antibiotic are rigidly fixed with respect to each other, and one of the chains has a polar face. By analogy to the phospholipids, amphotericin B can be oriented with its polar amino sugar-carboxyl end anchoring it to the aqueous phase and the two chains extending through half the bilayer. Although this orientation would be energetically unfavourable for a single molecule, because of all the OH's that end up contacting the hydrocarbon tails of the phospholipids, several molecules could pack together to bring the OH's into proximity with each other and away from the lipid hydrocarbon.

---

* When added at much larger concentrations to one side of the membrane, nystatin induces a cation permeability. For possible explanations for the discrepancy between the one-sided and two-sided effects, see Marty and Finkelstein, (1975).

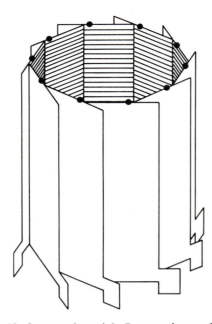

Fig. 5. Diagram of half of an amphotericin B-created pore. Each amphotericin B molecule is schematized as a plane with a protuberance and a solid dot (●). The shaded portion of each plane represents the hydroxyl face of the amphipathic chain, the protuberance represents the amino sugar, and ● represents the single hydroxyl group at the non polar end of the molecule. The aqueous phase is at the bottom of the figure and the middle of the membrane is at the top. We see that the interior of the pore is polar, whereas the exterior is completely non polar; there is also a wedge in the exterior of the pore, between each pair of amphotericin B molecules, that can accommodate a sterol molecule. Note the ring of hydroxyl groups in the middle of the membrane that can hydrogen bond with an identical structure from the other side to form a complete pore. (From Finkelstein and Holz, 1973.)

One such packing is the cylindrical arrangement shown in Figure 5 as the model for the amphotericin B pore (Finkelstein and Holz, 1973).

There are several interesting features to this structure: (1) The exterior aspect is nonpolar, thus making it compatible with the bilayer interior, whereas the interior is copiously lined with hydroxyl groups, providing a favourable polar environment for water, polar nonelectrolytes and ions.* (2) The complete pore consists of two

---

* The anion selectivity presumably results either from the partial substitution of these hydroxyls for the anion hydration shells, or from orientation of the —OH dipoles to make the channel interior positive.

"half pores", hydrogen bonded together in the center of the bilayer by a ring of OH groups formed from the single hydroxyl groups at the nonpolar end of each molecule. Each half pore spans half of the bilayer, and this provides a rationale for the greater effectiveness of the antibiotic when present on both sides of the membrane (Cass *et al.*, 1970). (3) The internal radius of the pore is of the appropriate dimension (4–7 Å, depending on the number of molecules packed together to form the structure) to account for the molecular sieving data. (4) Cholesterol packs snugly in the wedge between each pair of amphotericin B molecules, accounting for the sterol requirement for activity.

Despite the complex, multimolecular structure of the channels, they are not permanently fixed in the membrane. Removal of the antibiotic from aqueous solutions causes the conductance to decline with time to the low value characteristic of unmodified membrane. Clearly, there is a dynamic equilibrium between antibiotic molecules in the membrane and those in solution, and channels must continuously be forming and breaking up.

2. Gramicidin A

The evidence that this pentadecapeptide is a channel former comes not from measurements of water and nonelectrolyte permeability, but from observation of fluctuations in the conductance of membranes treated with minute amounts of the material (Hladky and Haydon, 1972). There appears to be a unit fluctuation of about 5×10^{-12} mho in 0.1 M NaCl; larger fluctuations are multiples of this size (Fig. 6). The clear implication is that the unit conductance step is the conductance of a single site, and from the magnitude of this conductance, it is almost certain that this site is a channel and not a carrier (Hladky and Haydon, 1972).* The channel is perfectly selective for univalent cations (Myers and Haydon, 1972), but unlike the carriers, it discriminates poorly among the alkali cations, the order and degree of selectivity being approximately the same as the sequence obtained from the mobilities of the ions in free solution (Hladky and Haydon, 1972).

There have been two models proposed for the channel structure, both of which are dimers with the gramicidin A molecules in helical

---

* The argument is simply that the rate of charge movement is incompatible with a bulky complex diffusing back and forth across the membrane, picking up and discharging an ion at each interface, whereas it is perfectly consistent with a small ion moving through a channel.

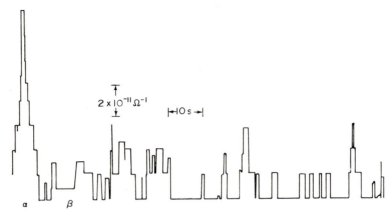

Fig. 6. Conductance fluctuations for a membrane treated with a small amount of gramicidin A. (One is actually seeing the current fluctuations in the presence of a constant voltage across the membrane.) The membrane separates 0.5 M NaCl solutions at 23°C. The events marked α and β on the left hand side of the record occurred very infrequently. (From Hladky and Haydon, 1972.)

configuration (Urry *et al.*, 1971; Veatch *et al.*, 1974). The details of these will not concern us. The points to be stressed about them are: (1) the channel passes through the molecules themselves; (2) it is lined by carbonyl oxygens to form the polar pathway for cations; these oxygens completely or partially substitute for the inner hydration shell of the cations. (This is the channel analogue for the cation carriers discussed earlier.) (3) The outer aspect of the helix is hydrophobic, allowing it to sit in the bilayer, and (4) the interior radius of the channel is about 2Å. This last feature of the model is supported by the observation that gramicidin A channels are permeable to water but not to urea (Finkelstein, 1974a). They are, thus, "tighter" than the amphotericin B channels. Indeed, their permeability for water but impermeability to urea mimics the characteristics of the ADH-stimulated cortical collecting tubule (Grantham and Burg, 1966), which I have suggested contains similar sized pores (Finkelstein, 1976b).

3. Comparison of Gramicidin A Channels with those of Nystatin and Amphotericin B

In spite of the total difference in the chemical nature of gramicidin A and the polyenes, there are interesting physico-chemical similarities between the structures of their respective channels. In both cases the channel interior is lined by polar groups (either carbonyl or hydroxyl), whereas the exterior is non polar. By this arrangement a parallel polar pathway permitting the movement of ions and other polar molecules is inserted through the hydrophobic interior of the bilayer. It is very likely that the naturally occurring channels in cell membranes have a similar segregation of polar interior and hydrocarbon exterior. Notice, however, that this segregation is achieved in two different ways. With the polyenes, the channel is formed by a "barrel stave" arrangement of many molecules, whereas with gramicidin A, the channel is created in the interior of an individual molecule. (Even though two molecules are apparently necessary to make a gramicidin A channel, it is quite clear that such a structure could be formed by a single, larger molecule.) *A priori*, either type of channel could be made by proteins in cell membranes. The former would be built up from subunits, whereas the latter would result from the proper folding of a single peptide chain.

C. Summary

We have seen that the permeability of lipid bilayers can be modified by introducing into them molecules that function as carriers or channel formers. Although we have discussed several of these molecules individually, there is nothing that precludes having more than one of these transport systems present in the same membrane, and indeed this is easily achieved experimentally (Finkelstein and Cass, 1968). Such a membrane is a mosaic structure that molecules may cross either through unmodified lipid regions or through one or more of the parallel, specialized transport systems inserted in the bilayer. This situation is phenomenologically the same as that which physiologists have long assumed to exist for plasma membranes. Of course, the transporters in plasma membranes have not yet been chemically isolated and identified. One of the major endeavours of workers dealing with lipid bilayers is to extract these transporters from cell membranes, insert them into planar lipid bilayers, and thereby reconstitute biological transport activity.

REFERENCES

Andersen, O. S., Finkelstein, A., Katz, I. and Cass, A. (1976). *J. Gen. Physiol.* **67**, 749–771.

Cass, A., Finkelstein, A. and Krespi, V. (1970). *J. Gen. Physiol.* **56**, 100–124.

Dainty, J. (1963). *Advan. Botan. Res.* **1**, 279.

Dobler, M., Dunitz, J. D. and Krajewski, J. (1969). *J. Mol. Biol.* **42**, 603—606.

Dunham, P. B., Cass, A., Trinkaus, J. P. and Bennett, M. V. L. (1970). *Biol. Bull.* **139**, 420—421.

Eisenman, G., Ciani, S. and Szabo, G. (1969). *J. Memb. Biol.* **1**, 294—345.

Finkelstein, A. (1974a). *In* "Drugs and Transport Processes" (Ed. B. A. Callingham), pp. 241—250. MacMillan, London.

Finkelstein, A. (1974b). *In* "Methods in Enzymology" (Eds S. Fleischer and L. Packer), Vol. 32, pp. 489—501. Academic Press, New York.

Finkelstein, A. (1976a). *J. Gen. Physiol.* **68**, 127—135.

Finkelstein, A. (1976b). *J. Gen. Physiol.* **68**, 137—143.

Finkelstein, A. and Cass, A. (1968). *J. Gen. Physiol.* **52**, 145s—172s.

Finkelstein, A. and Holz, R. (1973). *In* "Membranes. Lipid Bilayers and Antibiotics" (Ed. G. Eisenman), Vol. 2, pp. 377—408. Marcel Dekker Inc., New York.

Grantham, J. J. and Burg, M. B. (1966). *Amer. J. Physiol.* **211**, 255—259.

Hall, J. E., Mead, C. A. and Szabo, G. (1973). *J. Memb. Biol.* **11**, 75—97.

Hays, R. M. (1972). *In* "Current Topics in Membranes and Transport". (Eds F. Bonner and A. Kleinzeller), Vol. 3, pp. 339—366. Academic Press, New York.

Hladky, S. B. and Haydon, D. A. (1972). *Biochim. Biophys. Acta* **274**, 294—312.

Holz, R. and Finkelstein, A. (1970). *J. Gen. Physiol.* **56**, 125—145.

Huang, C. and T. E. Thompson, (1966). *J. Mol. Biol.* **15**, 539—554.

Kilbourn, B. T., Dunitz, J. D., Pioda, L. A. and Simon, W. (1967). *J. Mol. Biol.* **30**, 559—563.

Kinsky, S. C. (1974). *In* "Methods in Enzymology. Biomembranes." (Eds S. Fleischer and L. Packer), Vol. 32, pp. 501—513. Academic Press, New York.

Läuger, P. (1972). *Science* **178**, 24—30.

Marty, A. and Finkelstein, A. (1975). *J. Gen. Physiol.* **65**, 515—526.

McLaughlin, S. G. A., Szabo, G., Eisenman, G. and Ciani, S. M. (1970). *Proc. Natl. Acad. Sci. U.S.* **67**, 1268—1275.

Mueller, P. and Rudin, D. O. (1967). *Biochem. Biophys. Res. Comm.* **26**, 398—404.

Mueller, P. and Rudin, D. O. (1969). *Current Topics in Bioenergetics* **3**, 157—249.

Myers, V. B. and Hayden, D. A. (1972). *Biochim. Biophys. Acta.* **274**, 313—322.

Pappenheimer, J. R., Renkin, E. M. and Borrero, L. M. (1951). *Am. J. Physiol.* **167**, 13—46.

Pinkerton, M., Steinrauf, L. K. and Dawkins, P. (1969). *Biochem. Biophys. Res. Comm.* **35**, 512—518.

Pressman, B. C., Harris, E. J. Jagger, W. S. and Johnson, J. H. (1967). *Proc. Natl. Acad. Sci. U.S.* **58**, 1949—1957.

Urry, D. W., Goodall, M. C., Glickson, J. D. and Mayers, D. F. (1971). *Proc. Natl. Acad. Sci. U.S.* **68**, 1907—1911.

Veatch, W. R., Fossel, E. T. and Blout, E. R. (1974). *Biochemistry* **13**, 5249—5256.

Wright, E. M. and Pietras, R. J. (1974). *J. Memb. Biol.* **17**, 293—312.

7. General Mechanisms of Salt-Water Coupling in Epithelia

A. E. Hill

The Physiological Laboratory, University of Cambridge, Cambridge, England

I. INTRODUCTION: SOLUTE-LINKED WATER FLOWS

Of the many situations in which water moves across cell membranes in response to various thermodynamic gradients, I shall discuss here one of the most important but least understood: that of water flow coupled to salt transport across epithelia. The general features of this flow in an "ideal" case are (1) that the water flow is obligatorily linked to salt transport in the same direction; (2) the water flow is essentially "non-osmotic" i.e. it can occur when there is no osmotic or hydrostatic pressure across the tissue; (3) the osmolarity of the transported fluid is proportional to that of the bathing medium adjacent to the proximal membrane.* This proportionality holds over quite large concentration ranges, and in several systems the transported fluid is virtually isotonic with the bathing medium, for example in pancreatic epithelium (Case *et al.*, 1968), Malpighian tubule (Maddrell, 1971), and gall-bladder (Diamond, 1964b). This is not to say that non-osmotic flow cannot occur in addition to the solute-linked fraction, but in most situations within the animal body there are no appreciable osmotic gradients and it is the solute-linked flow which is predominant *in vivo*; in this respect isotonic flows are probably of cardinal importance in that they do not lead directly to

---

* There is no classification of epithelial membranes to please everyone. I have adopted a functional one here, defining the flows as occurring across the epithelium from the proximal to the distal membrane. In absorptive epithelia the distal membrane is serosal and supported by a "basement membrane", while the proximal membrane is mucosal. The reverse is the case in secretory epithelia such as salivary glands and Malpighian tubules.

any internal osmotic gradients or accumulations of salt which in turn would lead to swelling, or which would have to be dissipated by other means. When epithelia produce concentrated secretions or absorbates they are usually mediating exchanges between the animal body and its environment.

Originally it was supposed by several workers (e.g. Visscher *et al.*, 1944) that the water was secreted by the epithelial cells, and the salts accompanied it; with time however, the opposite view has come to prevail, that active pumping of salt is the prime mover, and that the water flow is passive down its own potential gradient. As I shall argue further on, Visscher's view may not be wholly wrong in the final analysis; first however, it is important to pursue the development of subsequent ideas.

In reviewing the problem, Diamond (1965) went through the various mechanisms which might be operative, rejecting all of them except local osmosis; this concept was later elaborated into the standing-gradient osmotic theory (Diamond and Bossert, 1967). I shall do the same here, but in greater detail and to no firm conclusions.

II. CO-DIFFUSION

As discussed by Kedem (1965), and Diamond (1965), co-diffusion is a rather vague concept, but it seems to be the flow of salt down its concentration gradient, accompanied by an osmotic flow of water in the same direction.

If we assume as a starting-point that there are no gradients of osmotic or hydrostatic pressure driving the water flows across the epithelium, then there can still be differences in chemical composition between proximal and distal solutions; if the epithelium has different reflection co-efficients for the various species in solution, then osmotic flow will result. In Fig. 1A a simple situation is shown where the basal epithelial face is bathed by non-electrolytes and a diffusive salt flow of magnitude $P_s \Delta\pi/RT$ is accompanied by an osmotic flow $L_p \Delta\pi$ in the same direction. In this naive form it could not possibly be taken seriously as a mechanism of epithelial water transfer, but it should be remembered that most fluid transporting epithelia can produce isotonic fluids with different composition to that of the proximal bath, and recently Schafer *et al.* (1975) have put forward a scheme for the fluid transport process in proximal tubule in a far more sophisticated form (Fig. 1B) in which small ionic concentration differences are operative in driving the water flows.

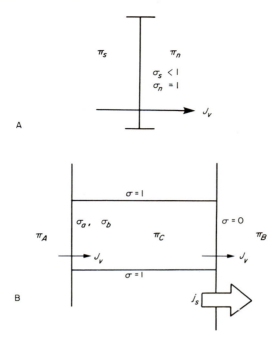

Fig. 1A. A simple co-diffusive system. In the absence of any osmotic gradient across the membrane $(\pi_s = \pi_n)$ the salt s diffuses towards the non-electrolyte bath n, accompanied by a volume flow driven by the force $(\pi_n - \sigma_s \cdot \pi_s)$. B. Osmotic flow across a system with ionic concentration differences, but no overall osmotic gradient. Active ion transport j_s creates a difference in concentration of two ions a and b between proximal and distal baths, which are isotonic with a channel space C, i.e. $\pi_A = \pi_B = \pi_C$. The proximal membrane has different reflective properties for the two ions $(\sigma_a \neq \sigma_b)$ whereas the lateral membranes are impermeable $(\sigma = 1)$ and the distal membrane is non-selective $(\sigma = 0)$. The osmotic volume flow J_v is generated at the proximal membrane as in A above. As applied to absorptive epithelia, the channel is visualized as an intercellular space opening by a junctional complex onto the proximal (apical) bath.

It is very difficult to discuss such a scheme at present because it is so isolated; in any event, it must be capable of being visualized in osmotic terms, and these run as follows: active transport sets up ionic concentration differences between two isotonic regions which cause a volume flow due to the thermodynamic driving force $(\sigma_a \Delta \pi_a - \sigma_b \Delta \pi_b)$ where a and b are the two ions. This is formally equivalent to a concentration gradient established by an active transport system, driving water osmotically across a membrane whose reflection co-efficient is less than 1.0. The electrical potentials introduced into this analysis do not affect the osmotic properties in any way, and this scheme can be regarded as a particular case of a double-membrane system, discussed below.

The above scheme introduces a more important point for absorptive epithelia; what controls the tonicity of the distal bath? This is a cardinal point which will crop up again in this discussion for the following reason: if either the *in vivo* conditions, or the experimental set-up, requires that this basal solution be created (or influenced) by the epithelium itself, then it will not do to assume isotonicity of the baths as a prerequisite for any model — this must itself be generated by the model.

III. DOUBLE-MEMBRANE COUPLING

The double-membrane theory of osmotic coupling was introduced by Curran (1960) to explain the intestinal absorption of water linked to active salt uptake. Essentially it comprises two membranes in series with a compartment in between. Salt is transported into the central compartment (which might be a cell, for example), raising its osmotic pressure and consequently its hydrostatic pressure (Fig. 2).

The volume transfer in such a system is given by:

$$J_v = \frac{\dfrac{L_{p_1} \cdot L_{p_2}}{L_{p_1} + L_{p_2}} \cdot (\sigma_1 - \sigma_2) \cdot js}{\omega_1 + \omega_2 + \dfrac{L_{p_1} \cdot L_{p_2}}{L_{p_1} + L_{p_2}}(\sigma_1 - \sigma_2)^2 \cdot C} \tag{1}$$

and it can be seen that volume flow in the direction of active transport is set up when $\sigma_1 > \sigma_2$. The behaviour of a double-membrane system such as this is in fact very complicated and gives rise to many properties which are of great interest; Curran and McIntosh (1962) showed experimentally that such a system would transfer volume but never dealt with the problem of predicting the osmolarity of the fluid transferred. Before going on to discuss the phenomena of double-membrane coupling, it is perhaps the place to mention the models of House (1964), and Schilb and Brodsky (1970). These are similar to the scheme of Schafer *et al.* (1975) in that active transport is used to create an ionic concentration difference, this time between the central compartment and the baths; this is envisaged as being a sodium pump, creating a potassium-rich central compartment bounded by two sodium-rich baths. Differences

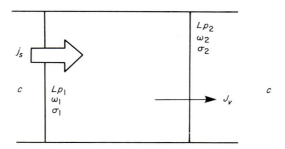

Fig. 2. Volume flow in a simple two-membrane system. Active transport across the first membrane at a rate j_s raises the concentration and pressure in the central compartment, creating a volume flow across the system given by equation (1). L_p is the hydraulic conductivity, ω the solute permeability, and σ the reflection coefficient.

in selectivity to the two ions, where membrane 1 has a higher P_{Na} than P_K, and membrane 2 is the reverse, possessing in addition an outwardly directed sodium pump, will give rise to volume flow in the direction of sodium pumping. There are similarities here to the epithelial conditions, applied by House (1964) to frog skin and Schilb and Brodsky (1970) to turtle urinary bladder, in which the central compartment is the cell interior and membranes 1 and 2 are the proximal (mucosal) and distal (serosal) membranes, respectively; there has been no real attempt to calculate the tonicity of the absorbates however, and as far as water equilibrium is concerned these schemes are similar in their essential nature to other double-membrane models. They differ in organisation, but the coupling between the active salt transport and the water flow is that of osmosis at a planar junction.

A comprehensive double-membrane model has been developed by Patlak *et al.* (1963), who use the equations for solute and volume flow developed by Kedem and Katchalsky (1958); one should perhaps note that the solute flow equation:

$$Js = \omega RT \Delta c + (1 - \sigma)\bar{c}J_v \tag{2}$$

only really holds for a *homogenous* membrane, i.e. one in which there is no possibility of circulatory flow occurring, and this is a proviso. However, after integrating the flow equations across a membrane element, the equation:

$$C_0 = C_i e^{(1-\sigma)J_v/P} + \frac{Js}{(1-\sigma)J_v} \cdot (1 - e^{(1-\sigma)J_v/P}) \tag{3}$$

is obtained which relates the concentration on one side, c_0, to that on the other c_i from which the two flows of solute J_s, and volume J_v, emanate. This equation is then applied to two membranes in series, eliminating the internal pressure terms. The resultant equations are:

$$\frac{J_s}{J_v} = \frac{c[e^{-\alpha_2} - e^{\alpha_1}] + \frac{j_s}{J_v}\left[\frac{1 - e^{\alpha_1}}{1 - \sigma_1}\right]}{\left[\frac{e^{-\alpha_2} - 1}{1 - \sigma_2}\right] + \left[\frac{1 - e^{\alpha_1}}{1 - \sigma_1}\right]} \tag{4}$$

and:

$$J_v = \frac{L_{p_1} \cdot L_{p_2}}{L_{p_1} + L_{p_2}} \cdot RT(\sigma_1 - \sigma_2)(e^{-\alpha_2} - 1)\left[c - \frac{(J_s/J_v)}{(1 - \sigma_2)}\right] \tag{5}$$

where $\alpha_i = (1 - \sigma_i)J_v/P$ and c is the concentration which is indentical for both proximal and distal baths. These equations define the absorbate osmolarity J_s/J_v and the volume transfer J_v as a pair of non-linear simultaneous equations which have to be solved numerically for particular cases. They encompass the models of Curran (1960), House (1964), Whitlock and Wheeler (1964), and Schilb and Brodsky (1970), all of which can be arrived at by either (i) manipulating equations (4) and (5), or (ii) deriving similar ones in which there are two solutes, or the active transport is moved to the second membrane.

Now a property of equations (4) and (5) is that the absorbed osmolarity, J_s/J_v, can be hypertonic, isotonic, or even hypotonic to the proximal medium, c; this medium *in vitro*, it should be remembered, bathes both sides of the system. It would thus seem that virtually any system could be described by double-membrane coupling if the central compartment is capable of being identified, and Schilb and Brodsky (1970) regard this as the cell interior, whilst Kaye *et al.* (1966) consider it to be the intercellular space. How is the distal bath maintained at c? Some mechanism must either add (or subtract) salt to solute-clamp the distal bath at the concentration c, and this mechanism is not part of the model. In virtually every case where a fluid transporting epithelium can be studied *in vitro*, it will function under conditions where the distal membranes are bathed by the emergent fluid itself, or at least, the solution adjacent to these membranes must be profoundly affected by that of the emergent fluid (usually due to diffusion through a basement membrane). Equations (4) and (5) can be therefore amended by an additional constraint which defines the basal concentration as:

$$c_{basal} = Js/J_v. \tag{6}$$

When equations (4) and (5) are derived anew using equation (6), hypertonic solutions only are obtained, becoming isotonic in the limit. According to double-membrane theory, hypotonic solutions cannot therefore be produced by unilateral preparations, i.e. preparations bathed by their own secretions.

By numerical solution we can explore the behaviour of the model, and using approximate values for the solute pumping rate and the osmotic permeability of the membranes, the emergent osmolarity of the transported fluid can be calculated. In Fig. 3 are shown some results of calculations involving solute pumping rates well within the spectrum of these found in epithelial systems, in the context of varying hydraulic conductivities and proximal bath concentrations.

The first thing to be noted is that osmotic equilibrium of the transported fluid with the proximal bath to within the limits observed in nature i.e. a few % in many systems, is not achieved without enormous hydraulic conductivities; it can be seen from equation (5) that the overall hydraulic conductivity given as:

$$L = \frac{L_{p_1} \cdot L_{p_2}}{L_{p_1} + L_{p_2}} \tag{7}$$

enters as a simple parameter, and this is the conductivity observable in practice. It is possible to mitigate somewhat the requirement for high hydraulic conductivity by raising the solute permeability of the membranes, as can be seen in the graph; but this has the effect of making the pumping system very inefficient. At the higher permeability values shown in Fig. 3 for example, the fraction of salt crossing the epithelium is less than 1% of that pumped at the first membrane due to the fact that most of it recirculates by leaking back again into the proximal bath. From what we know of pumping efficiencies in various epithelia (Keynes, 1969), they are nothing like as low as that, quite apart from the fact that the overall solute permeability is closer to the lower values given in Fig. 3.

The double-membrane system we have so far considered is one without a shunt pathway, and is therefore maximally efficient; but we should bear in mind that the overall osmotic and solute permeabilities of many epithelia may partially represent those of shunt

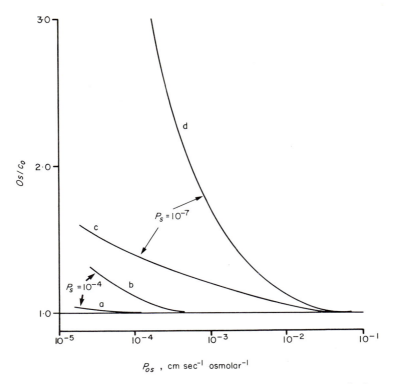

Fig. 3. A graph of the effect on O_s, the emergent osmolarity, of changes in osmotic permeability of the membranes in a unilateral double-membrane system, where the transported fluid forms the distal solution. The pumping rate is 6.0×10^{-10} osm sec^{-1} cm^{-2}, which is not high for transporting epithelia. At higher proximal bath osmolarities a and c, ($c_o = 0.2$) the emergent osmolarities are reasonably isotonic, but at lower c_o-values b and d, ($c_o = 0.02$) they become very hypertonic and only fall to within 3% hypertonicity above osmotic permeabilities of 3.0×10^{-2}. The effect of raising the solute permeability from 10^{-7} to 10^{-4} cm sec^{-1} is to facilitate equilibration, but at the higher value the emergent solute flux is only 1% of the pumping rate.

pathways, so that the true transcellular conductivities will be lower still. It has recently been suggested that due to an unstirred layer, represented by the serosal layer in many systems, the measured conductances could be lower than the true values by a substantial fraction (Wright *et al.*, 1972), but Fig. 3 suggests that a correction of up to three orders of magnitude, from 10^{-5} to 10^{-2} cm sec^{-1} osmolar^{-1}, would be required.

   Another feature of double-membrane coupling pointed out by

Diamond (1965), is that of the "uniqueness" of isotonic flow in this model. In fact it is misleading to use the term "osmotic equilibration" when referring to fluid production in the double-membrane model, because strictly there is none: water and salt flows are coupled by hydrostatic, osmotic and diffusive forces, and the ratio of the two flows changes with the concentration of the proximal bath. At low concentrations the transported fluid is hypertonic; at high concentrations it becomes hypotonic. In the unilateral case, where the model is modified by equation (6), the hypotonicity is not observed. In Fig. 4 I have plotted one against the other, where it can be seen that the deviation from a line of constant tonicity (in this case the isotonic line) is quite marked, and in the physiological range of solute permeabilities and hydraulic conductivities it is really very severe.

A final complication arising from double-membrane theory is that of the internal pressures generated, which in general are quite high, and which would rarely be found because the cells would swell (or

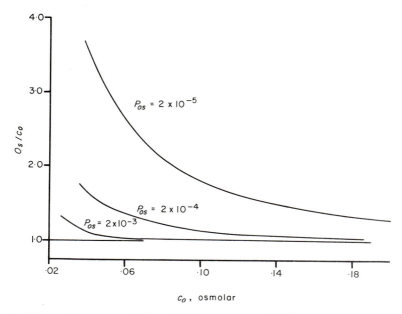

Fig. 4. The relative osmolarity of the emergent fluid across a symmetrical double-membrane system as a function of the proximal bath osmolarity c_o. The solute permeability is 10^{-7} cm sec^{-1}. The three curves represent different osmotic permeabilities, P_{os}. The secretion constitutes the distal bath according to equation (6).

shrink, if the solute pump were sited at the basal membrane). In the models of House (1964) and Shilb and Brodsky (1970) pressure gradients are not considered, but this does not cause them to vanish; there will always be pressure differences between the central compartment and the baths, as can be rigorously proved (Patlak *et al.*, 1963) unless there exists a cellular mechanism for overcoming them — an unspecified mechanism which is outside the model itself. In the example described by House (1964) the pressure gradient is abolished when a particular relationship holds between the reflection co-efficients and the hydraulic conductivities of the two membranes, and the concentrations in baths and the central compartment. This is a very special and highly restrictive condition which I doubt could ever be attained in practice without some form of sophisticated feedback control between the parameters.

It thus appears that double-membrane coupling of water flow to active salt transport is a very poor candidate indeed, unless some way can be devised to get around the requirement for very high hydraulic conductivities which are not an observed feature of most epithelial systems; there also exist good theoretical reasons for doubting their existence, as we shall see later.

The local osmotic theory of Diamond (1964b) differs from a double membrane system in that it represents the osmotic equilibrium as occurring across a single membrane. In this case no pressure gradient is set up (by definition) as the local osmotic space opens onto the distal bath. There are also no real unstirred layer problems, a point with which it is perhaps convenient to deal now as we shall require to consider it later in another form. If the passive solute permeability of the membrane shown in Fig. 5 is low, such that the rate of salt flow across it is given almost entirely by the active pumping term, js, then the combined effects of diffusion and convection are constant, and

$$js = vc - D \frac{dc}{dx} \qquad (8)$$

everywhere basal to the membrane, v being the convective velocity due to volume transfer. This equation integrates to give:

$$c = \frac{js}{v} + k e^{vx} \qquad (9)$$

where k is a constant. There is only one possible solution for (9) in the steady state and that is:

$$c = \frac{js}{v} \qquad (10)$$

the constant being zero. Thus there is no gradient of concentration and the distal bath in the unilateral preparation is created by the emergent fluid.

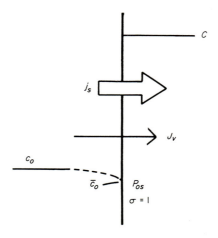

Fig. 5. Equilibration across a single membrane whose osmotic permeability is P_{os} and whose reflection coefficient is 1.0. Solute is pumped at a rate j_s across the membrane causing a steady-state concentration difference, $c - c_o$. $\bar{c}_o$ is the concentration adjacent to the membrane, which in general may be either higher or lower than c_o. Side 1 is therefore unstirred, but on side 2 stirring is irrelevant in the unilateral preparation i.e. when the transported fluid creates the distal bath.

The convective velocity due to osmosis (assuming zero solute volume transfer) is:

$$v = P_{os}(c - \bar{c}_o) \qquad (11)$$

where P_{os} is the osmotic permeability, and $\bar{c}_o$ the proximal bath osmolarity adjacent to the membrane. In general $\bar{c}_o$ will differ from c_o, the bulk osmolarity of the bath, due to an unstirred layer on the proximal side, described by a similar equation to (9). If the ratio of the overall secretion rates of salt and water is equal to their concentration ratio in the proximal bath, as they are in isotonic transfer, then:

$$\bar{c}_o \simeq \frac{js}{v} \simeq c_o \tag{12}$$

and equations (10) and (11) then give a quadratic in c:

$$2c = c_o + \sqrt{c_o^2 + \frac{4js}{P_{os}}} \tag{13}$$

This equation is also derivable from equations (4) and (5) of Patlak et al. (1963) when modified by equation (6). Calculations of the osmotic permeability required to produce secretions hypertonic to the proximal bath by 3% (perfectly isotonic are not obtainable by an osmotic mechanism) again show that enormous values are required, when suitable values are used for the solute pumping rates. In rabbit gall-bladder for example, the production of absorbates hypertonic by a few % at observed rates in 1/10 Ringer's solution requires an osmotic permeability of more than 10^{-2} cm sec^{-1} osmolar^{-1}, which is three orders of magnitude greater than the transepithelial permeability. Similarly high values can be calculated for other epithelia. This requirement was noted by Diamond (1965) who argued that special channels (inaccessible to passive transepithelial flow) must mediate the local osmosis; the very high values required for P_{os} however, make it virtually impossible for any membrane to mediate it. The production of isotonic secretion by simple equilibration over the whole epithelium has been seriously argued by Marro and Germagnoli (1966), but these authors seem to have confused an equilibrium with steady-state transfer maintained by pumping.

This concludes the discussion on double-membrane coupling, including single-membrane equilibration. All the systems considered are ones in which water flow and the active salt pumping occur normal to planar membrane surfaces, and consequently the flow vectors are normal to the membranes at all points of the system. If this is not true, then the compartments must be internally stirred for the expressions derived above to hold. I think it is apparent that none of these systems offers a satisfactory description of quasi-isotonic flow. Simple local-osmosis perhaps comes nearest to acceptability in that no internal pressure differences are required, and within the isotonic region there is a fixed relationship between the tonicity of the transported fluid and that of the proximal bath. The osmotic permeabilities required to bring the system well within the isotonic region are too high, however.

IV. STANDING-GRADIENT FLOW

When the membranes conducting the active salt transfer and osmotic equilibrium are not planar but convoluted, there will be gradients of all the variables set up i.e. of pressure, concentration, osmotic pressure, fluid velocity, and where ion pumping is involved, of electric potential. Essentially, active transport of solute into a space bordered by a membrane creates a rise in osmotic pressure, and this pulls in water as in a simple local-osmotic system. A hydrostatic gradient is therefore created which drives fluid along the system, depending on the disposition of the hydraulic conductivities in different areas. Diamond and Bossert (1967) considered this system (a system with "standing-gradients" in the steady-state) to apply to the intercellular spaces of epithelia, to membrane infolds, or possibly to the spaces between microvilli on the apical cell-surface (Fig. 6).

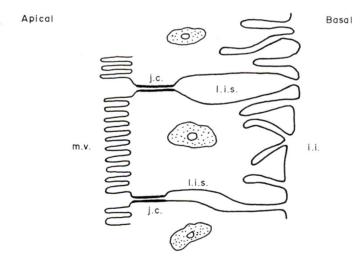

Fig. 6. Typical spaces which might serve as osmotic coupling channels for salt and water flow; all the components shown here are rarely present in this well-developed form together in one epithelium. Usually either the basal or the apical surface membranes show extensive folding, and occasionally the junctions are wide open. The direction of fluid transport is apical → basal in absorption and basal → apical in secretion. m.v. = microvilli; l.i.s. = lateral intercellular space; j.c. = junctional complex; i.i. = intracellular basal infoldings.

For absorbing epithelia, they examined a model representing a channel of constant cross-section closed at one end by an impermeable junction (Fig. 7). Salt pumping into this channel at a

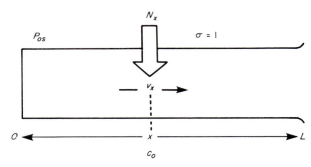

Fig. 7. A standing-gradient osmotic channel into which solute is pumped at a rate N. The raised concentration draws in water from the bath at c_o, creating a volume flow down the channel. The walls are assumed to be ideally semi-permeable, the channel length is L, and the separation $2r$.

rate N, over lateral membranes whose reflection co-efficient is unity, and whose osmotic permeability is P_{os}, creates a concomitant osmotic water flow into the channel. The flow pattern is not laminar in fact, due to volume entry along the walls, but there is an axial velocity v which increases with distance from the closed end given by:

$$N_{(x)} + \frac{Dr^2}{4P_{os}} \frac{d^3 v}{dx^3} - \frac{r^2 v}{4P_{os}} \frac{d^2 v}{dx^2} - \frac{c_o r}{2} \frac{dv}{dx} - \frac{r^2}{4P_{os}} \left(\frac{dv}{dx}\right)^2 = 0 \quad (14)$$

This differential equation is difficult to solve in the range of physiological parameters, but it can be done by various numerical techniques (Diamond and Bossert, 1967; Weinbaum and Goldgraben, 1972; Hill, 1975a; Huss and Marsh, 1975; Sackin and Boulpaep, 1975), and if the lateral membranes are assumed to be impermeable the emergent osmolarity of the system can be calculated according to the expression:

$$o_s = \frac{c}{a} \cdot v_L^{-1} \int_0^L N_{(x)} \cdot dx \quad (15)$$

where c/a is the ratio of circumference to area of any cross-section of the channel whose length is L.

If this model applies to the spaces bounded by membranes that are a feature of all the volume transporting epithelia (in fact of almost all epithelia), then these systems could produce quasi-isotonic secretions by adjusting r, the effective radius of the channels.

Segel (1970) has derived an analytical approximation to equations (14) and (15), which assumes that approximate osmotic equilibration has been achieved within the channel and thus defines the dependence of O_s, the emergent osmolarity, on P_{os}, L, r and D, in the isotonic region:

$$\frac{O_s}{c_0} = \left[1 - \frac{\lambda \sin \kappa \lambda^{-1}}{\kappa \cosh \kappa} \right]^{-1} \tag{16}$$

where:

$$\kappa = \sqrt{\frac{c_0 P_{os} L^2}{rD}} \tag{17}$$

Considering the parameter κ we can see that a great enough reduction in r will ensure that $O_s/c_0 \to 1.0$. The active pumping rate N does not enter equations (16) and (17) because of a further assumption in the derivation: that solute pumping is confined to the "blind" end of the channel. The solute then has to transverse a length of channel into which only osmotic water flow occurs, and as long as λ^{-1}, the fractional length given up to pumping, is less than about one-half, the equilibration is not affected by N (within the spectrum of physiological pumping rates) provided κ is large enough. This lends great support to standing-gradient theory, because there are many systems where the pumping rate can be altered by inhibitors, hormones, or temperature, and only the rate of fluid transport is affected but not the tonicity.

When we come to examine the geometry of various intercellular spaces, it is apparent that most of them are best approximated by a cleft between two parallel membrane surfaces, the lateral membranes. Thus $c/a = 1/r$, where $2r$ is the separation. As we have previously noted, the emergent osmolarity for several systems can be isotonic to about 2% and in this case equations (16) and (17) show that κ^2 must then exceed 20. Taking an osmotic permeability for the lateral membranes of 10^{-5} cm sec^{-1} osmolar^{-1}, a diffusion coefficient (NaCl) of 1.5×10^{-5} cm^2 sec^{-1}, and a proximal bath osmolarity of 0.3 osmolar, it appears that the geometrical factor L^2/r must be about 100 cm. Many of these epithelia can function isotonically in medium of much lower osmolarity that 0.3, and in many cases L^2/r must approach 1000 cm. When we come to examine approximate values for L^2/r taken from published ultrastructural studies however, we see that a mean value of 0.06 cm characterizes the epithelia chosen (Table I).

TABLE I

Values of the geometrical function L^2/r for various systems, (Hill, 1975a).

System	L^2/r (cm)	Emergent fluid
Reptilian proximal tubule	0.028	isotonic
Rat proximal tubule	0.115	isotonic
[a]Malphigian tubule (*Rhodnius*)	0.28	isotonic
[a]Avian salt gland	0.035	hyperosmotic
Amphibian urinary bladder	0.01	near-isotonic
[a]Salivary gland		
rat	0.016	isotonic
insect	0.022	isotonic
Mammalian gall bladder	0.059	isotonic
[a]Exocrine pancreas	0.008	isotonic
Small intestine	0.04—0.09	isotonic

[a] Secretory epithelia where the channels may be formed by basal infolds and apical microvilli.

Although in general there is a dearth of good structural information, it appears that L^2/r is three to four orders of magnitude too small to encompass standing-gradient flow coupling within the lateral or basal spaces. In some epithelia there is considerable choice as to what spaces one might choose to fit to standing-gradient theory for there are often microvilli on the apical surface, lateral intercellular spaces and basal membrane invaginations; most basal and apical convolutions are impossibly short however, and I have usually taken the lateral space as the only obvious candidate.

There is no evidence that solute pumping is really confined to the blind ends of epithelial spaces, in which case we have to use the full differential equation (14) to investigate channels with evenly distributed pumps. Obviously the solute injected into a channel near the open end will contribute to the emergent tonicity, and this has been calculated for a few systems where all the parameters required for the calculation can be assembled (Hill, 1975a). From the dependence of the emergent osmolarity on the osmotic permeability of the channel membranes of the four epithelia shown in Table II, we

TABLE II

The relative emergent osmolarity from four epithelial systems at a proximal bath osmolarity of 0.3, calculated from equations (14) and (15), (Hill, 1975a).

Osmotic Permeability cm sec^{-1} osmolar^{-1}	Emergent osmolarity relative to proximal bath (0.3 OsM)			
	Proximal tubule (rat)	Small intestine (rat)	Mammalian gall-bladder	Malpighian[a] tubule
10^{-5}	61.6	136.5	130.0	35.6
10^{-4}	7.25	14.74	14.09	4.65
10^{-3}	1.79	2.55	2.49	1.52
10^{-2}	1.18	1.28	1.27	1.13
10^{-1}	1.09	1.08	1.08	1.07

[a] Direction of fluid transport is from the basal to the apical surface.

can see that to reach 2% hypertonicity the osmotic permeability must approach 10° i.e. 1.0; it therefore seems that the standing-gradient osmotic theory cannot be applied in any simple way to these spaces without postulating enormous osmotic permeabilities.

In several systems there is the additional factor that the solute pumps on the basal membrane, which are not part of a lateral system or which are part of very short basal convolutions, will also contribute hypertonic solution to the overall epithelial secretion.

Why could not these high osmotic permeabilities be a feature of fluid transporting epithelia? I think there are three reasons that make this more or less impossible.

(i) An osmotic permeability of this magnitude would reflect itself as a high overall permeability of the epithelium. The transepithelial permeability P_{epi} is due to the proximal and distal membrane, P_a and P_b in series or:

$$P_{epi} = \frac{P_a \cdot P_b}{P_a + P_b} \tag{18}$$

and this overall value is approximately the same as that of a single membrane, 10^{-5} cm sec^{-1} osmol^{-1}. The osmotic permeabilities of almost all single biological membranes and phospholipid bilayers studied to date (see Chapter 6) fall within a range of 10^{-8} to 10^{-5} cm sec^{-1} osmolar^{-1} (P_{os} for the red cell membrane and for occasional lipid films is 10^{-4}) and this means that if the osmotic permeability of the distal (latero-basal) membrane were very high, the proximal membrane might mask this by bringing P_{epi} down to approximately 10^{-5}.

The proximal membrane however, must also be very permeable indeed to osmotic water flow, due to the fact that water has to enter the cell down a water potential gradient that cannot exceed a few % of that of the proximal bath (Hill, 1975a). This is only true if water crosses both membrane systems by osmotic flow, in which case it follows from a simple calculation that the proximal membrane must have an osmotic permeability of more than 10^{-2} cm sec^{-1} osmolar^{-1}. The transepithelial permeability given by equation (18) would therefore have to possess a P_{epi} — value of at least 10^{-2} cm sec^{-1} osmolar^{-1} if standing-gradients were responsible for the fluid flow. This enormous permeability is not observed, and indeed this pinpoints one of the greatest difficulties in this osmotic theory of fluid production, which is that water flow over the proximal membrane has to be accounted for too, and this is just as great a problem as fluid production at the distal membrane.

(ii) In addition to the above argument, we should note the effect of closing the spaces by reverse osmosis, which induces changes in P_{epi} which are attributable to the occlusion of latero-basal membrane (Wright *et al.*, 1972; Smulders *et al.*, 1972); I have calculated from this occlusion in mammalian gall-bladder that the lateral membrane must have a P_{os} of about 10^{-5} cm sec^{-1} osmolar^{-1} by ignoring the possibility of flow over the terminal junction, which only tends to increase the estimate (Hill, 1975a). Two things which complicate this calculation are: (a) that the osmotic conductance of the channel is almost certainly "distributed" rather than "lumped" (a hyperbolic function), and (b) the transepithelial conductance may be underestimated due to unstirred layer effects. Neither of these effects could mask an osmotic conductance of 10^{-1} cm sec^{-1} osmolar^{-1} however, and similar calculations could be made on several other epithelia.

(iii) Conductivities as high as those required might be achieved in theory with membrane pores, but these would have to be so large and so numerous that their calculated reflection co-efficient would be zero (Hill, 1975a); on hydrodynamic grounds therefore, the osmotic permeability σL_p would seem to be limited for small solutes.

In summary then, it seems quite impossible to reconcile standing-gradient flow coupling with our present knowledge of cell ultra-structure. This may come as a surprise to many who accept the theory as essentially plausible and natural, based upon the simple idea of local osmotic action. It represents an advance upon local osmotic theory however, in that the geometry was not considered until Diamond and Bossert drew attention to it, and there is no obvious reason why it could not work except for the fact that most epithelia do not possess suitably narrow channels; that is, presuming the channel dimensions seen in electron-micrographs are not seriously expanded by the fixation, or that there are not myriads of minute channels too small to be yet seen. Perhaps *in vivo* the dimensions of the spaces are too imprecise and too variable for the important phenomenon of isotonic flow (or near-isotonic flow) to be entrusted to the vagaries of baso-lateral geometry. There is one experimental fact which could rule out simple standing-gradient osmotic theory overnight, and that is a demonstration that the cell interior is hypertonic to the proximal bath during isotonic transport. In this case osmotic equilibrium could never bring the emergent osmolarity down to that of the proximal bath. Having said that however, it might be that some scheme of ion-recirculation across the channel membranes is in operation, such that salt pumped into the space is reabsorbed further down the channel, thus causing the osmolarity to fall towards the channel mouth. I don't wish to be pessimistic, but there are great difficulties here; to name but a few: reverse pumps have to be established further down the same cell membrane, but as intercellular spaces vary widely in size these would have to be distributed in a complex way to counter-balance the effect of variable geometry; if the membrane osmotic permeabilities are about 10^{-5} cm sec^{-1} osmolar^{-1}, a fairly normal value, then at least 90% of the salt initially pumped into the channel would have to be removed, and this would require very delicate balancing; if the cell interior is not stirred, then the osmotic gradient would presumably reverse, and the lower end of the channel would function like a "backward" standing-gradient system (Diamond and Bossert, 1968), and so on. A reabsorptive theory is a possibility but it seems to me, an improbability.

Before leaving standing-gradients we should note that Maddrell (1971) has suggested that in extensively folded membrane systems, as seen in Malpighian tubules, both "forward" and "backward" channels can be regarded as lying adjacent to one another in which case osmotic equilibration might be assisted. This interesting idea has not been formally investigated as yet.

V. JUNCTIONAL OSMOTIC EQUILIBRIUM

The junctional complex ("tight junction" or *zonula occludens* of Farquhar and Palade, 1963) may have an osmotic permeability which is not so easily discussed as that of a "unit" membrane, due to its fine structure being unknown; in addition, the fact that it always lies in series with a lateral intercellular space makes it difficult to determine its intrinsic conductance. Up until now we have treated water flow as being a membrane phenomenon, but if it appears that the standing-gradient theory (as outlined above) cannot operate in practice, then perhaps water can enter the lateral space via the tight junction, as shown in diagrammatic form in Fig. 8; here the osmotic permeability of the channel walls is set at zero so that all the equilibrium is via the junction.

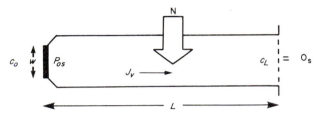

Fig. 8. Osmotic equilibration over a junctional complex. The junctional osmotic permeability is wP_{os} and active solute pumping extends from the complex for a length L. At the end of the pumping region there is no further change in concentration and so c_L is equal to the emergent osmolarity O_s. The lateral walls are impermeable to salt and water.

The analytical approximation describing the emergent osmolarity O_s in this system is:

$$2O_s = c_o + \sqrt{c_o^2 + \frac{8NL}{wP_{os}}} \tag{19}$$

Which is analogous to the equation for local osmosis (equation 13), but in this case O_s is the osmolarity at the end of the solute input region, after which the secretion does not change (Hill, 1975c). wP_{os} seems difficult to assess, but it can be dealt with in the following way: the osmotic permeability of the whole epithelium is divided by the linear extent of junctional complex (per cm^2 epithelium) to give a value for the permeability of the junction per cm length; this is taken as equal to wP_{os}, w being essentially unknown. w and P_{os} enter the calculation as a product and therefore it is unnecessary to separate them. The same is true of N and L; if we decide that all the pumping takes place in 10% of the channel nearest to the junctions, then N will be ten times higher to accommodate the known pumping rate. Assignment of N is accomplished by dividing the transepithelial pumping by the area of the cell membranes distal to the junctions; it is difficult to do accurately, and is based on the assumption that pumping is evenly distributed over the basal and lateral membranes (which are continuous of course). This is probably true, but if all the pumping is confined to the intercellular channels then N must be higher, which makes osmotic equilibrium down to c_o more difficult. Basically the cell circumference has to be determined, and if this is not estimated correctly then the same error occurs in NL and wP_{os} which cancel out as can be seen in equation 19. Table III contains calculated values of the emergent osmolarity at different proximal osmolarities from both lateral intercellular spaces (equation 19) and from the whole epithelium due to addition of the salt efflux from the basal membrane (Hill, 1975c). It is apparent that the secretion is never really isotonic, but gets near it at the higher apical osmolarities when the basal salt flux is neglected.

The junctional permeability wP_{os} is obtained by assuming that all the transepithelial permeability is due to the junctions alone. This is certainly untrue, due to the parallel contribution of the membranes, and it should be remembered that an epithelium possessing membranes of osmotic permeability $\sim 10^{-5}$ cm sec^{-1} osmolar^{-1}, and with the degree of membrane convolution usually seen, could account for most of the overall epithelial permeability even if there were no junctional complexes at all. We do not as yet know the hydraulic conductivity of a junctional complex, but if it is a simple opening or slit then its reflection coefficient is likely to be very small indeed and hence so is P_{os}. I don't think that junctional equilibration is a very promising idea, because the junction would have to manifest such a high osmotic conductance to encompass isotonic fluid flow at the range of proximal tonicities used in most experiments that it would be apparent in the overall osmotic permeability of the tissue, and there is no real evidence that any substantial fraction of this permeability is due to the junctions at all. The junctional permeability wP_{os} can be estimated on hydrodynamic grounds by use of the expression:

$$wP_{os} = \frac{w^3}{12\eta d} \tag{20}$$

which defines the hydraulic conductivity per unit length of a slit, where w is the separation, and d is the depth. For calculated values

TABLE III

Selected data and emergent relative osmolarities for four epithelial junctions mediating osmotic equilibration (Hill, 1975c). The epithelial secretion represents the sum of the contributions from both the channels and the basal membranes, assuming that the salt pumping is uniformly distributed (see text).

Epithelium	$wP \times 10^8$ cm^2 sec^{-1} osmolar^{-1}	$NL \times 10^{10}$ mosm. cm^{-1} sec^{-1}	Tonicity of Luminal bath osmol/l	Relative tonicity of lateral space fluid	Relative tonicity of epithelial absorbate
Gall-bladder (rabbit)	1.1	7.5	0.5	1.39	a
			0.05	7.90	a
Proximal tubule (Necturus)	9.1	0.75	0.5	1.01	1.43
			0.05	1.30	1.86
Proximal tubule (rat)	130.0	6.4	0.5	1.01	5.04
			0.05	1.46	7.27
Small intestine (rat)	4.25	3.4	0.5	1.06	a
			0.05	3.08	a

a Similar to that of the lateral spaces.

of NL in equation (19) it is apparent that wP_{os} cannot be lower than 10^{-7} (cm^2 sec^{-1} osmolar^{-1}) for reasonable equilibration (Hill, 1975c); if however, we assume a separation of 2×10^{-7} cm (20 Å) and a junctional depth of 0.5×10^{-4} cm, wP_{os} comes out to be 3×10^{-10} which is very small indeed, and which depends upon the reflection co-efficient being unity. Not only is this unlikely to be true for sodium or chloride ions traversing a 20 Å junction, but the junction may not be an open slit; it may contain material. It seems that the junctional permeability, on simple theoretical grounds, should be insufficient to mediate osmotic equilibration to any extent. In addition, some epithelia such as amphibian proximal tubules, have large distal membrane convolutions which are not coupled to the proximal bath via junctions. If these are not pumping salt at all, then much higher values of N must be used to calculate the equilibration in their lateral spaces. If they are contributing to the transepithelial salt flow then they are too short to produce anything but strongly hypertonic solutions by the analysis of the previous section.

I think it is a quite natural question to ask why both junctional and lateral equilibration could not be acting together, perhaps in conjunction with a putative "unstirred layer" in the serosal region or in distal convolutions, which would pull water osmotically across the epithelium through the cells. I have represented this in Fig. 9 as an osmotic circuit whose output is the emergent osmolarity, O_s.

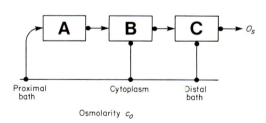

Fig. 9. An optimal circuit of osmotic equilibrators operating in series to produce the emergent osmolarity O_S. A represents the junctional equilibration, B the intercellular channel, and C the final (planar) equilibration over the whole epithelium. In the optimal case all the solute pumping is confined to the channel ends near the junctions (A); if pumping occurs in B across the channel membranes, or in C across the distal membranes, then O_S will be higher. B "sees" the osmolarity of the cytoplasm, which may be different from that of the proximal bath. The functioning of the three equilibrators A to C is described by equations (19), (14–17), and (22) respectively.

H

The simple answer is, of course, that if the separate components will not create a quasi-isotonic secretion by themselves, neither will their combination. The standing-gradient osmotic system represented by the intercellular space (B) has the property that solution passing into it at one end (or any point, in fact) undergoes partial equilibration before reaching the channel end. Let us assume that the whole channel can produce 50% equilibration, such that if solute was injected into the end near the junction, the space would produce a secretion of osmolarity $2\,c_o$. If there is now a junctional equilibration (A) replacing a blind end, producing also a fluid 50% equilibrated with an osmolarity $2\,c_o$ this will undergo 50% further equilibration on passing down the channel, to give a fluid 75% equilibrated of osmolarity $1.33\,c_o$. To produce an emergent osmolarity of $1.02\,c_o$, which would be experimentally acceptable, the channel would have to possess 96% equilibration properties, i.e. (B) would have to be able to function near the isotonic limit itself, without help from the junction equilibration, which we have seen is virtually impossible.

If we now add an unstirred layer represented by (C) there will be further equilibration, but as it turns out this is very slight. In the unilateral case where the distal membrane of the epithelium is bathed by its own secretion, we have seen that there is no concentrated layer adjacent to the membrane (equations 8 to 13) due to the fact that $dc/dx = 0$. Even when the distal bath is not similar in osmolarity to the secretion, the serosal layer is unstirred and so dc/dx is zero there at all but minute volume flows. Dainty and House (1966), have applied an equation to this situation in which two concentration zones within unstirred layers adjacent to the membranes mediate the transepithelial water flow, but this equation is derived by treating unstirred layer widths as though they are independent of the convective and solute velocities, which is untrue. The situation at (C) in Fig. 9 is one in which active salt input into the unstirred layer is accompanied by water which has been picked up in (A) and (B) i.e. by fluid pumping into the layer. If the solute transfer output of (B) is N osmoles sec^{-1} per cm^2 epithelial surface, as a solution of osmolarity mc_o where c_o is that of the proximal bath, then the volume transfer from (B) is N/mc_o. The emergent osmolarity O_s from (C) is then given by:

$$O_s = \frac{N}{P_{os}(o_s - c_o) + N/mc_o} \tag{21}$$

which re-arranges to give:

$$2o_s = \left(c_o - \frac{N}{P_{os}mc_o}\right) + \sqrt{\left(c_o - \frac{N}{P_{os}mc_o}\right)^2 + \frac{4N}{P_{os}}} \qquad (22)$$

where P_{os} is now the transepithelial osmotic permeability. At values of P_{os} around 10^{-5} (cm sec^{-1} osmolar^{-1}) with a mean solute input rate N of 10^{-9} (osmoles cm^{-2} sec^{-1}) the equilibration properties of (C) as defined by equation (22) are very poor indeed; there is scope for variation in both N and P of course, but the osmotic permeability is the dominant parameter and has to reach very high values before equilibration is substantially improved. These permeabilities are those of whole epithelia and are not observed, nor could they be masked by underestimated values due to unstirred layer phenomena. In our example we might consider the output, $1.33\, c_o$ of the stage (B). With a proximal bath osmolarity of 0.1 (osmolar) the emergent osmolarity from (C) and thus into the distal bath, is $1.28\, c_o$.

The above is merely an example chosen to illustrate the nature of osmotic flow in the coupled elements (A) to (C). In reality the situation is far worse for such a serial model of fluid production due to the much poorer equilibration properties of the individual elements. I cannot see how any judicious arrangement of these elements can get around the basic fact that neither the junctions nor the cell membranes seem to have anything like the required osmotic permeability, which in some cases would have to exceed observed values for cell membranes and epithelia by several orders of magnitude. The problem has been treated in a very detailed manner by Sackin and Boulpaep (1975) who have solved equations representing standing-gradient osmosis in a channel with an osmotically permeable end. They show, as in Table 3, that *Necturus* proximal tubule could function near-isotonically if virtually all the osmotic conductance of the epithelium were concentrated at the junctions, but they do not consider how well the system should function at lower bath osmolarities, and they confine the salt pumping to the lateral channels instead of distributing it over the whole of the baso-lateral membranes, which is an assumption similar to that made by Diamond and Bossert (1967) in confining it to the channel ends, and which serves to lower the osmolarity of the emergent fluid. An interesting aspect of their treatment is that the hypertonicity of the emergent fluid can be lowered by placing a salt-reflective barrier after the epithelium; this could be a capillary wall, although its reflection coefficient is very low, and it should be remembered that in excised epithelia the capillaries are non-functional. A reflective barrier serves to convert the system (functionally) into a double-membrane system with internal pressure gradients. In short, the osmotic permeability of the junctions is the controlling parameter, and this is completely unknown, nor can it be meaningfully calculated from existing data; as we have seen, although the hydraulic conductivity may sometimes be quite high, the osmotic permeability may be very small indeed.

It is very difficult to show that a concept such as local osmosis is inapplicable in these circumstances, and indeed it may play some role albeit a minor one, but I think the onus is now upon us to examine alternative models of fluid transfer in epithelia.

VI. ELECTRO-OSMOSIS

I have suggested this as a means of fluid transfer (Hill, 1975b) although the idea has been considered before and rejected (Diamond, 1965; House, 1974). If water is unable to move fast enough by osmotic gradients then some mechanism in which water is coupled to the solute flux directly seems worth considering. In electro-osmosis the transmembrane potential can move water, because there is specific frictional interaction between water and one of the ions in a special pore, which excludes the counter-ion. If such a specific model is not considered, but merely a generalized salt-water friction, then the water has to be driven by salt gradients, which is undesirable from many standpoints (Hill, 1975b).

There are basically two objections to electro-osmosis (i) that transepithelial potentials are of the wrong sign and are too small, and (ii) that the efficiency of electro-osmosis, in terms of molecules of water transferred to ions transported, is in general far too small to account for the osmolarity of the emergent secretion. The first point arises from streaming-potential experiments where an osmotic or hydrostatic gradient is applied across the epithelium. In most epithelia the sign of the streaming potential indicates that the normal positive transepithelial secretory potential, such as is usually observed in the absence of gradients, could only drive the water slowly and in the wrong direction (Diamond, 1965). Coupled water flow might take a different route however, and there exists a substantial negative potential (30–80 mv) from the cell interior to the distal bath, which would be quite enough and in the right direction. In addition, it is now clear that there are probably diffusion potentials of opposite sign superimposed on the streaming potentials, due to concentration polarization (Wedner and Diamond, 1969), and that the streaming-potential is probably due to paracellular flow via the junctional complex, whereas fluid transfer during secretion could well be transcellular. For these reasons it must be clear that the organization is too complicated to allow any simple generalizations from streaming potential experiments or indeed to infer that there should be a strict stoichiometry between transepithelial potential and water flow rates.

The second point is not so easily discussed or dismissed, and must involve future studies. It has been shown (Hill, 1975b) that from a system with only thermodynamic limitations, it is possible to construct a double-membrane electro-osmotic model in which isotonic (or near-isotonic) fluid transfer is generated without high osmotic permeabilities of the membranes, and without involving large intracellular pressures or concentrations. Moreover, the model is self-regulating, in that the osmolarity of the emergent fluid follows that of the proximal bath when this is changed without any readjustment of the basic parameters. Encouraging as this might seem, there are physical limitations which must come into play. Membrane pores of the largest diameter likely to allow efficient electro-osmosis could only contain a maximum of about 150—200 water molecules, and if one ion traversed such a pore it could only carry out this number and no more. This is about half the ratio of water to ions involved in the transfer of an isotonic physiological saline; most epithelia however, can function quasi-isotonically in about one-tenth this osmolarity, and therefore require couplings of up to 4000 water molecules per ion.

If these couplings are to be achieved by an electro-osmotic mechanism in epithelial membranes there must be substantial ionic recirculation involved, presumably involving the junctional complexes. In addition, the recirculated ions must be pumped at some stage as no passively distributed species can recirculate without violating thermodynamic principles. In relation to this requirement there are two observations which are topics of current interest and about which our knowledge is at present rudimentary. The first concerns the sodium pump in these systems, which is generally supposed to be the primary active transport mechanism that powers the fluid transport (Van Os and Slegers, 1971). If this is an identical pump to that of nerve, frog skin, erythrocyte, and other cells, it presumably involves pumping of potassium ions into the cytoplasm together with sodium extrusion; potassium ions are not transported in a reverse direction to the overall fluid flow however, and so there should be substantial potassium recirculation at the basolateral membranes at a rate similar to that of sodium pumping. The second involves the junction and paracellular pathway. Schultz (1972) has pointed out that the absence of large secretory potentials in "leaky" epithelia may be due to ionic shunting via paracellular pathways, and this could only take place across the junctions. In several fluid-transporting epithelia the paracellular pathway is of quite low resistance (Frömter and Diamond, 1972) and the junctions are known to be cation-permeable (Moreno and Diamond, 1974), in which case there is probably a cation current, presumably of sodium ions, back to the proximal bath. The amount of recirculation needed to accommodate isotonic flow at plasma osmolarities is about 60% but if dilute secretions can be produced then this rises to a very high percentage. Such a recirculatory scheme would not help to lower the emergent osmolarity in an osmotic equilibration theory because osmosis relies upon existing concentration gradients, not upon ion flows, and the values of N (the ion pumping rates) are calculated as being those which leave the epithelium into the distal bath.

If electro-osmosis is to explain fluid transfer across epithelia, then obviously a great deal more experimental work is needed to establish whether the electro-osmotic permeability of the constituent membranes is in fact measurable, and these permeabilities must then be embedded in a model in which the ion flows are known with greater accuracy than at present. At the moment we do not have a sufficiently well-developed model of intercellular spaces, in which both concentration and potential gradients have been calculated together, in the presence of possible electrogenic pumps and a selective junction i.e. a full standing-gradient ionic model. Electro-osmosis thus stands as a possible mechanism to be more fully investigated, both experimentally and theoretically.

VII. FLUID PUMPING

A final model of fluid transfer we might consider is one in which solution is transferred in bulk from bath to bath, possibly by pinocytosis. This has been suggested by Grim (1963), by Frederiksen and Leyssac (1969) for gall-bladder, and by Leyssac (1966) for proximal tubule epithelium. The experiments in support of this hypothesis are not very convincing to my mind, and have to meet the objection that a whole range of electrolytes and organic compounds present in the proximal bath are not transported across the epithelium at all (Diamond, 1964a, 1964b); obviously simple pinocytosis is not really an acceptable theory, but it could be argued that sub-microscopic regions of the proximal membrane might invaginate as vesicles with little enclosed volume but with absorbed salt (NaCl) which is then released into the vesicle lumen; the vesicle would then osmotically equilibrate with the cytoplasm and cross the cell to fuse with the distal membrane and discharge NaCl into the extracellular infoldings by reverse-pinocytosis. This idea is certainly feasible when one considers what cells can do if they put their minds to it; it obviously relies on the cytoplasm being iso-osmotic to the proximal bath, a fact which is by no means established as yet.

However, there are certain simple calculations which can be made on the basis of the fluid transfer rates which make pinocytosis highly unlikely. In a tissue such as mammalian gall-bladder electron microscopy does not reveal the cells to be filled with numerous vesicles, in fact it is hard to see any, but let us suppose that there are vesicles with a diameter of 0.1 μm which are large enough to be clearly seen. To accommodate a fluid transfer rate of about 50 microlitres cm^{-2} hr^{-1} there would have to be more than 10^{14} vesicles crossing 1 cm^2 of epithelium per hour. These vesicles would possess a membrane area of 3×10^4 cm^2, which must fuse with the baso-lateral membrane. If we allow a factor of 30 for convolution of this membrane, then to preserve constant cell dimensions basal to the junctions in the steady-state, it must turn over about 1000 times per hour. This membrane material must be removed from the apical membrane at the same rate. I think this amount of membrane could never be metabolized so fast, but if the lipid could rotate around the cell in an amoeboid fashion, somehow traversing the junctions, this problem might conceivably be overcome. With traversal times of 1 sec across the cell however, there should be 10^4 vesicles per cell in transit, as there are 10^6 cells per cm^2 epithelium, and I feel that these could not escape notice. With a cytoplasmic transit time of 1 minute, which is still short compared with cytoplasmic streaming velocities, the vesicles should occupy the whole cell volume. Reducing the vesicle size does nothing but increase the membrane turnover rate, as their surface to volume ratio increases. A similar argument has been advanced by Parsons (1963).

VIII. SUMMARY

The greater part of this review has been devoted to a consideration of osmosis as the driving force for water flow during epithelial fluid secretion, with rather pessimistic conclusions. I cannot see how the standing-gradient osmotic theory can be reconciled with the available data on osmotic permeability and cell geometry, and the study of fluid production in *Necturus* proximal tubule by Sackin and Boulpaep (1975) and that of Lim and Fischbarg (1976) on corneal endothelium, both come to a similar conclusion. Measurements of fluid secretion in *Necturus* gall bladder at low osmolarity, of which a preliminary report has appeared (Hill, 1976), are also inexplicable in terms of standing gradients.

Recently Diamond (1977) has argued that epithelial osmotic permeabilities have been grossly underestimated owing to unstirred layer effects; correct values, up to a thousand times greater, are now being claimed, to put the theory on a secure footing again. However, the intercellular spaces are separated from the cell contents by single membranes, not by leaky epithelia, and so the osmotic permeability of the latter is really somewhat irrelevant. In addition, there is no new experimental data which would support such high P_{os} values, and unstirred layer corrections support a revision of epithelial permeabilities of about two times, not a thousand.

It is perhaps inevitable that interest will shift to the role of junctions in mediating osmotic flow during secretion, although as I have tried to show, there are serious problems in this interpretation. The greatest is the magnitude of the reflection coefficient of the junctions. If this is low then osmosis cannot occur across the junctions, and already we know that many non-electrolytes traverse epithelia by the paracellular route, including even inulins. It must surely be apparent that the junctional reflection coefficient for ions is of central importance in any theory of paracellular water flow. If this eventually turns out to be low, "fluid pumping" will have to be taken very seriously.

ACKNOWLEDGEMENTS

I should like to acknowledge the great help of Bruria S. Hill and Guillermo Whittembury in discussing and reading this contribution.

REFERENCES

Case, R. M., Harper, A. A. and Scratcherd, T. (1968). *J. Physiol. Lond.* **196**, 133–149.
Curran, P. F. (1960). *J. gen. Physiol.* **43**, 1137–1148.

Curran, P. F. and McIntosh, J. R. (1962). *Nature. Lond.* **193**, 347—348.
Dainty, J. and House, C. R. (1966). *J. Physiol. Lond.* **182**, 66—78.
Diamond, J. M. (1964a). *J. gen. Physiol.* **48**, 1—14.
Diamond, J. M. (1964b). *J. gen. Physiol.* **48**, 15—42.
Diamond, J. M. (1965). *Symp. Soc. exp. Biol.* **19**, 329—347.
Diamond, J. M. (1977). *The Physiologist* **20**, 10—18.
Diamond, J. M. and Bossert, W. H. (1967). *J. gen. Physiol.* **50**, 2061—2083.
Diamond, J. M. and Bossert, W. H. (1968). *J. Cell Biol.* **37**, 694—702.
Farquhar, M. G. and Palade, G. E. (1963). *J. Cell Biol.* **17**, 375—412.
Frederiksen, O. and Leyssac, P. P. (1969). *J. Physiol. Lond.* **201**, 201—224.
Frömter, E. and Diamond, J. M. (1972). *Nature New Biology.* **235**, 9—13.
Grim, E. (1963). *Am. J. Physiol.* **205**, 247—254.
Hill, A. E. (1975a). *Proc. R. Soc. Lond.* B **190**, 99—114.
Hill, A. E. (1975b). *Proc. R. Soc. Lond.* B **190**, 115—134.
Hill, A. E. (1975c). *Proc. R. Soc. Lond.* B, In press.
Hill, A. E. (1976). *J. Physiol. Lond.* **263**, 201.
Hogben, C. A. M. (1960). *Physiologist* **3**, 56—62.
House, C. R. (1964). *Biophys. J.* **4**, 401—416.
House, C. R. (1974). "Water Transport in Cells and Tissues". pp. 407—408. Arnold, London.
Huss, R. E. and Marsh, D. J. (1975). *J. Membr. Biol.* **23**, 305—347.
Kaye, G. I., Wheeler, H. O., Whitlock, R. T. and Lane, N. (1966). *J. Cell. Biol.* **30**, 237—268.
Kedem, O. (1965). *Symp. Soc. exp. Biol.* **19**, 61—73.
Kedem, O. and Katchalsky, A. (1958). Thermodynamic analysis of the permeability of biological membranes to non-electrolytes. *Biochim. biophys. Acta* **27**, 229—246.
Keynes, R. D. (1969). *Quart. Rev. Biophys.* **2**, 177—281.
Leyssac, P. P. (1966). *Acta Physiol. Scand.* **70**, suppl. 291.
Lim, J. J. and Fischbarg, J. (1976). *Biochim. biophys. Acta* **443**, 339—347.
Maddrell, S. H. P. (1969). *J. exp. Biol.* **51**, 71—97.
Maddrell, S. H. P. (1971). *Phil. Trans. R. Soc. London.* B **262**, 197—207.
Marro, F. and Germagnoli, E. (1966). *J. gen. Physiol.* **49**, 1351—1353.
Moreno, J. H. and Diamond, J. M. (1974). *In* "Membranes — A series of Advances "(Ed. G. Eisenman), Marcel Dekker, New York.
Parsons, D. S. (1963). *Nature, Lond.* **199**, 1192—1193
Patlak, D. S., Goldstein, D. A. and Hoffman, J. F. (1963). *J. theoret. Biol.* **5**, 426—442.
Sackin, H. and Boulpaep, E. L. (1975). *J. gen. Physiol.* **66**, 671—733.
Schafer, J. A., Patlak, C. S. and Andreoli, T. E. (1975). *J. gen. Physiol.* **66**, 445—471.
Schilb, T. P. and Brodsky, W. A. (1970). *Am. J. Physiol.* **219**, 590—596.
Schultz, S. G. (1972). *J. gen. Physiol.* **59**, 794—798.
Segel, L. A. (1970). *J. theoret. Biol.* **29**, 233—250.
Smulders, A. P., Tormey, J. M. and Wright, E. M. (1972). *J. Membr. Biol.* **7**, 164—197.

Van Os, C. H. and Slegers, J. F. G. (1971). *Biochim. biophys. Acta.* **241**, 89–96.

Visscher, M. B., Fetcher, E. S., Carr, C. W., Gregor, H. P., Bushey, M. S. and Barker, D. E. (1944). *Am. J. Physiol.* **142**, 550–575.

Wedner, H. J. and Diamond, J. M. (1969). *J. Membr. Biol.* **1**, 92–108.

Weinbaum, S. and Goldgraben, J. R. (1972). *J. Fluid Mech.* **53**, 481–512.

Whitlock, R. T. and Wheeler, H. O. (1964). *J. Clin. Invest.* **43**, 2249–2265.

Wright, E. M., Smulders, A. P. and Tormey, J. M. (1972). *J. Membr. Biol.* **7**, 198–219.

8. Cation and Anion Transport Mechanisms

G. Sachs

Laboratory of Membrane Biology, University of Alabama, Birmingham, Alabama, U.S.A.

Ion transport occurs across membranes of bacteria, animal and plant cells as well as across membranes of intracellular organelles such as mitochondria and chloroplasts. The transport of ions can be passive, with the ions moving down electrochemical gradients or it may be active due to the presence of ion pumps.

I. CATION PUMPS

These pumps are now regarded as reversible machines capable of interchanging chemical energy and ion gradients. At this stage it seems that we can classify these pumps in at least two ways; according to the chemical energy utilized, or according to the ions transported. Thus the ion pump of mitochondria can utilize energy generated by an oxidation-reduction reaction oriented appropriately (Fig. 1). In the scheme illustrated, the oxidation of AH_2 results in a

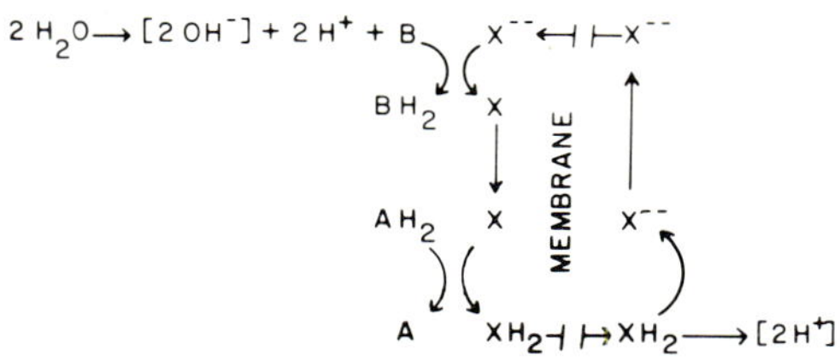

Fig. 1. A schematic diagram illustrating an electrogenic H^+ transport across a membrane by a redox reaction.

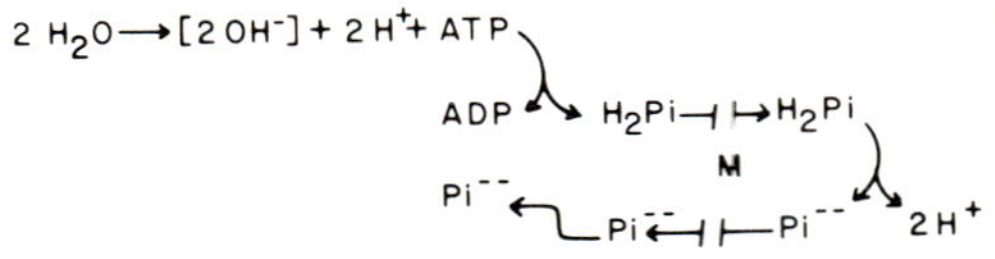

$$2\,H_2O \longrightarrow [2\,OH^-] + 2\,H^+ + ATP$$

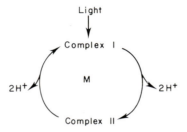

Fig. 2. A schematic diagram illustrating the electrogenic development of a H^+ gradient by mitochondrial ATPase.

transfer of $2H^+$ across the membrane and the development of a H^+ ion gradient. An H^+ gradient can also develop in mitochondria from the hydrolysis of ATP (Fig. 2) or in the purple membrane of halophile bacteria from light activation of lumirhodopsin (Fig. 3). All

Light

Complex I

M

$2H^+$ $2H^+$

Complex II

Fig. 3. A simplified illustration of the electrogenic H^+ movement by the action of light on lumirhodopsin.

these pumps, as illustrated, are electrogenic and the electrochemical gradient for H^+ can be written as:

$$\Delta \bar{\mu}_H + = \Delta \psi \frac{2 \cdot 3RT}{F} \qquad \Delta pH = \text{proton motive force}$$

In the case of mitochondria (or in chloroplasts or in bacteria) these pumps are reversible so that an electrochemical H^+ gradient ($\bar{\mu}_{H^+} = 220$ mv or more) can result in the synthesis of ATP. This gradient develops from the potential and H^+ gradient properties of the oriented redox reactions across the mitochondrial membrane. This gradient is therefore the coupling mechanism between two forms of chemical reactions, oxidation-reduction (redox) reactions and ATP (phosphorylation) reactions. It took many years of work to realize that this H^+ gradient was the key to the mechanism of oxidative phosphorylation (Mitchell, 1966).

The details of the H^+ translocation across the mitochondrial membrane are as yet not clear since this requires the participation of an ATPase (F_1) consisting of several subunits and of other hydrophobic proteins (e.g. F_0) which are responsible for the H^+ conductance or transport across the membrane.

The H^+ ion gradient can also be directly coupled to solute transport across bacterial or organelle membranes (Fig. 4) and this H^+ cotransport seems to be a major mechanism for solute transport (Kaback, 1974). From these considerations, H^+ pumps occupy a central role in membrane biology.

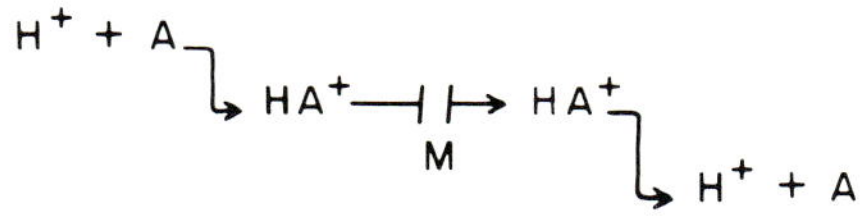

$$H^+ + A \searrow HA^+ \longrightarrow\!|\!\longmapsto HA^+ \searrow H^+ + A$$
$$M$$

Fig. 4. A simple illustration of an electrogenic H^+ solute cotransport across a membrane.

The most specialized H^+ pump, that of stomach, appears to operate by a different mechanism, a mechanism similar to other ion pumps such as the $Na^+ + K^+$ (Skou, 1965) or Ca^{++} (Martonosi, 1972) ATPases present in plasma membranes of eukaryotic cells. These ATPases form a covalently linked phosphate ester during transport in contrast to the H^+ ATPase of mitochondria which simply binds the phosphate. The H^+ pump of stomach depends on the presence of a $H^+ + K^+$ ATPase (Ganser and Forte, 1973) which, like the $Na^+ + K^+$ ATPase, can be thought of as having the mechanism shown in Fig. 5.

In this scheme a subunit of the protein X is phosphorylated by ATP but only following the binding of the obligatory cation A^+. The conformation (X_1)Pi is then converted to (X_2)Pi where the affinity for A^+ declines and A^+ is released. The affinity for the form (X_2)Pi for B increases and the movement of B^+ through the transport complex facilitates the conversion of (X_2)Pi to (X_1)Pi. In the presence of B^+, (X_1)Pi breaks down to (X_1) and Pi. Thus ATP hydrolysis is coupled to the exchange of cations A^+ and B^+ by a conformational change of the phosphorylated subunit. The energization is for transport of A^+ and this reaction can be modelled as an electrogenic EMF and series conductance for A^+ and a parallel conductance for B^+. An anion C^- could be substituted for B^+ and

ATP
ADP
(X)₁ Pi A⁺
(X)₂ Pi A⁺
(X)₁ A⁺
A⁺
M
A⁺
(X)₁
Pi
B⁺
(X)₁ Pi
(X)₂ Pi
B⁺
B⁺
B⁺

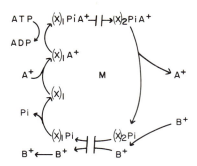

Fig. 5. An illustration of an ATP driven cation exchange pump operating via a phosphorylated intermediate of the ATPase.

this would result in cotransport of A^+ and C^-. The potential depends on the relative conductances of the two limbs.

It appears however, that the $Na^+ + K^+$ ATPase and the $H^+ + K^+$ ATPase exist as dimers of the transporting subunit. The transport reaction can be diagrammed as in Fig. 6. This diagram shows that the dimer exists only in the form $(X_1 Y_2)$ or $(X_2 Y_1)$ but not in $(X_1 Y_1)$ or $(X_2 Y_2)$ forms. Hence the translocation of A^+ is coupled to the (X_1)Pi to (X_2)Pi conversion as for the monomer. But the conversion of (X_1)Pi to (X_2)Pi is coupled to the conversion of (Y_2)Pi to (Y_1)Pi and hence strictly to the translocation of B^+.

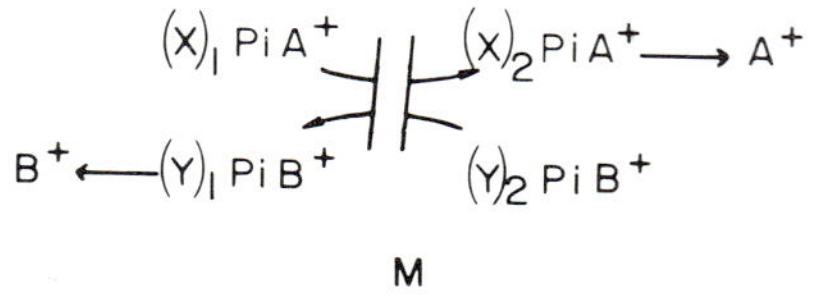

(X)₁ Pi A⁺
(X)₂ Pi A⁺
A⁺
B⁺
(Y)₁ Pi B⁺
(Y)₂ Pi B⁺
M

Fig. 6. A dimer model for a cation translocating ATPase where the dimer form is obligatorily asymmetric.

The difference then between the monomer and dimer situation is that in the latter the electrogenicity of the pump will disappear. Ion transport across plasma membranes of mammalian cells has therefore variable electrogenicity, perhaps as a function of the state of aggregation of pump subunits.

This type of pump is structurally simpler than the H^+ ATPase of mitochondria. The phosphorylating (and presumably transporting) subunit for all three types of this ATPase has a mol. wt of 100,000 or so. Other subunits may also be associated but the function of these subunits is not known.

In the case of the $H^+ + K^+$ ATPase the only known function of this ATPase is to generate an H^+ gradient. The $Na^+ + K^+$ ATPase plays a broader role, to maintain a high cellular K^+, and hence cell potential, and to maintain an inward electrochemical gradient of Na^+.

As for the H^+ cotransport process in yeast or bacteria, many mammalian cells (e.g. kidney, intestine, ascites cells) utilize Na^+ cotransport for solute uptake. This is electrogenic and as for H^+, a sodium motive force across a membrane can be described by:

$$\bar{\mu}_{Na} = \Delta \psi \frac{2 \cdot 3RT}{F} \Delta pNa^+$$

This sodium motive force then determines the organic solute (e.g. sugar, amino acids) concentration gradient achieved across plasma membranes of animal cells. The $Na^+ + K^+$ ATPase determines this electrochemical gradient a) by creating an Na^+ gradient, b) a potential due to a K^+ diffusion potential, and c) due to the electrogenicity of the pump.

In evolution therefore, the primary transport was that of H^+ by an electrogenic mechanism and the electrochemical H^+ gradient was utilized for chemical synthesis and accumulation of metabolites. This type of pump has been related to the organelles of higher cell types. These cells have developed a different type of plasma membrane pump, as for Na^+ or H^+, which are now coupled to specific cell functions such as secretion, or solute accumulation, or generation of a potential as the need may be. During this evolutionary process the hydrolytic and transport functions have been combined into one protein, but in lower organisms and in cell organelles they are maintained separately.

This brief summary of cation transport processes shows that our level of understanding of cation movement across cell membranes is relatively well advanced. In contrast, anion movement across membranes is less well understood.

II. ANION PUMPS

Passive movement of anion such as Cl^- is extremely rapid across the human red blood cell membrane and requires the participation of a specific protein. Anion transport blocking agents have been synthesized (Cabantchik and Rothstein, 1974) and have proved very useful in delineating the protein component responsible.

An understanding of an active anion transport has been much more difficult in the sense that no membrane-bound enzymer system has been shown to transport anions and, except for limonium (Hill and Hill, 1973) no ATPase has been shown to have a requirement for an anion such as Cl^-. Mitochondrial ATPase (Mitchell and Moyle, 1971) and perhaps plasma membrane ATPases in kidney (Kinne et al., 1975) or in salivary gland have been shown to be stimulated by HCO_3^- and by other oxyanions, but the relationship of this stimulation to the transport of Cl^- or other anions, including HCO_3^- is obscure.

By standard criteria the gastric mucosa contains a Cl^- pump. Thus there is a net active transport of Cl^- against an electrochemical gradient (Hogben, 1951). Removal of Cl^- abolishes the potential across the epithelium (Durbin and Heinz, 1957). Short-circuiting the mucosa does not abolish net Cl^- movement, and Ussing equation for flux ratio is not obeyed under these circumstances. Metabolic inhibition of the mucosa also abolishes net flux of Cl^- from serosa to mucosa.

The transport of Cl^- is dependent on the presence of Na^+ in the bathing solutions (Sachs et al., 1966) and is inhibited in high K^+ solutions (Hogben, 1968). Inhibition of Cl^- transport by amytal (which blocks NADH dehydrogenase activity and ATP synthesis in gastric mitochondria) is reversed by bypassing the amytal block with menadione and ascorbate (Hersey, 1974); this also restores 80% of the ATP synthesis.

In a tissue such as rabbit colon (Frizzell et al., 1976), Cl^- movement appears to be absorptive and non-electrogenic (i.e. in exchange for HCO_3^- secretion or with H^+ absorption). However, the addition of cAMP elicits an active electrogenic Cl^- secretion, at the same time reducing the exchange mechanism. Considering other mammalian tissues such as cornea, ileum or seminal vesicles, it has been suggested that a primary effect of cAMP on epithelia is on Cl^- transport mechanisms (Frizzell et al., 1976; also see Berridge, in this volume).

Antral tissue (Flemström and Sachs, 1975) seems quite similar to colon, in that although there is Cl^- secretion, this secretion is non-electrogenic and the short circuit current of antrum can be accounted for by Na^+ absorption.

Thus, the Cl^- transport by gastrointestinal epithelia can be electrogenic in the direction of serosa to mucosa (gastric fundus, cAMP treated colon), non-electrogenic from mucosa to serosa (colon), or non-electrogenic from serosa to mucosa (antrum). The non-electrogenic Cl^- transport in colon or antrum is independent of Na^+. In contrast, the electrogenic transport in gastric fundus does appear to be Na^+ dependent (Sachs *et al.*, 1966).

Data from red blood cells and from gastric fundus (Sanders *et al.*, 1972) strongly suggest the presence of an obligatory $HCO_3^- $-$Cl^-$ exchange mechanism. Gastric fundus appears to have an insignificant HCO_3^- conductance whereas antrum has a significant HCO_3^- conductance, as well as apparently an active HCO_3^- secretion; antrum in this respect resembles pancreas and salivary glands. Thus, an additional anion transport mechanism is $HCO_3^- $-$Cl^-$ exchange which could account for the net absorption of Cl^- found in the unstimulated colon (Frizzell *et al.*, 1976).

In an attempt to reconcile many of these data (Field, 1976) it has been recently postulated that active Cl^- transport could be understood without postulating an active anion pump.

The main features of this scheme is shown in Fig. 7. In Fig. 7A, the mucosal surface of the colon contains a $HCO_3^- $-$Cl^-$ exchange carrier. The HCO_3^- is derived from metabolic CO_2. The Cl^- exists across the serosal membrane, accompanied by H^+, by a non-electrogenic cotransport path. All the current is carried by Na^+, pumped by the $Na^+ + K^+$ ATPase.

With the addition of cAMP, a neutral path for NaCl entry is present on the basal surface (Fig. 7B). The Na^+ is also transported by the electrogenic Na^+ pump, but does not contribute to the current. Exiting with Na^+ is HCO_3^- and the accumulated Cl^- diffuses down a conductance path across the luminal membrane. Hence, HCO_3^- available for $HCO_3^- $-$Cl^-$ exchange across this surface is depleted and an inward Cl^- flux is inhibited along with the appearance of electrogenic outward flow. Short-circuiting the tissue will show that Cl^- flux under either 7A or 7B will violate the Ussing equation for the flux ratio.

In Figure 7C, a possible situation is depicted in gastric mucosa. The NaCl path with electrogenic outward Na^+ accounts for the Cl^- current observed in the resting mucosa. Stimulation of acid secretion results in an increase of HCO_3^- in the cell and a $HCO_3^- $-$Cl^-$ exchange across the basal surface of the tissue. The Cl^- exits across the apical surface, accompanied by K^+. The $H^+ + K^+$ ATPase results in $H^+ : K^+$ exchange. The net effect is secretion of HCl and an electrogenic transport of Cl^-.

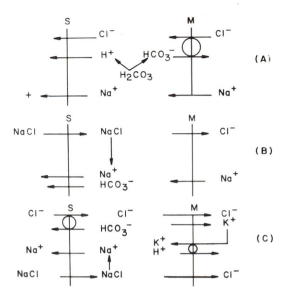

Fig. 7. A. An illustration of a mechanism resulting in a non-electrogenic Cl⁻ transport, driven by a diamox sensitive HCO_3^--Cl^- exchange mechanism. B. An illustration of a mechanism resulting in the development of an electrogenic Cl⁻ transport which depends on Cl⁻ accumulation due to NaCl entry followed by an electrogenic extrusion of Na^+. C. A scheme for Cl⁻ transport in gastric mucosa where electrogenic Cl⁻ is present under all conditions, as in B above, and the stimulation of secretion results in HCO_3^--Cl^- exchange at the opposite membrane surface from A above and in KCl cotransport across the luminal surface, followed by an active ATPase dependent K^+ : H^+ exchange across this surface.

Many variations on this type of scheme are possible. The key features are either a non-electrogenic Cl⁻ transport that is a function of a HCO_3^--Cl^- exchange carrier, or an electrogenic transport that is a function of a neutral NaCl path and a Cl⁻ conductance on the opposite membrane. This eliminates the need for postulating an active Cl⁻ pump. However, it is possible to increase the complexity of the scheme by postulating an energized HCO_3^--Cl^- exchange. Thus, while there is clear evidence for active cation pumps, the anion transport in various tissues may indeed be indirect.

REFERENCES

Cabantchik, Z. I. and Rothstein, A. (1974). *J. Membr. Biol.* **15**, 207—225.

Durbin, R. P. and Heinz, E. (1957). *J. gen. Physiol.* **41**, 1035—1047.

Field, M. (1976). *In* "Pathophysiology of Cell Membranes". (Eds T. E. Andreoli, D. Faustil, and J. F. Hoffmann), Plenum Press, New York. In press.

Flemström, G. and Sachs, G. (1975). *Am. J. Physiol.* **228**, 1188—1198.

Frizzell, R. A., Koch, H. S.. and Schulz, S. G. (1976). *J. Membr. Biol.* **27**, 297—316.

Ganser, A. L. and Forte, J. G. (1973). *Biochim. biophys. Acta* **307**, 169—180.

Hersey, S. J. (1974). *Biochim. biophys. Acta.* **244**, 157—203.

Hill, B. S. and Hill, A. E. (1973). *In* "Ion Transport in Plants". (Ed. W. P. Anderson), pp. 379—384. Academic Press, London and New York.

Hogben, C. A. M. (1951). *Proc. Nat. Acad. Sci. U.S.A.* **37**, 393—395.

Hogben, C. A. M. (1968). *J. gen. Physiol.* **51**, 240—249.

Kaback, H. R. (1965). *Science, Washington* **186**, 882—892.

Kinne, R., Murer, H., Kinne-Saffran, E., Thees, M. and Sachs, G. (1975). *J. Membr. Biol.* **21**, 375—395.

Martonosi, A. (1972). *In* "Metabolic Transport". (Ed. L. E. Hokin), pp. 317—350. Academic Press, New York and London.

Mitchell, P. (1966). *Biol. Rev.* **44**, 445—502.

Mitchell, P. and Moyle, J. J. (1971). *Bioenerg.* **2**, 1—11.

Sachs, G., Shoemaker, R. and Hirschowitz, B. I. (1966). *Proc. Soc. exp. Biol. Med.* **123**, 47—52.

Sanders, S. S., O'Callaghan, J., Butler, C. F. and Rehm, W. S. (1972). *Am. J. Physiol.* **222**, 1348—1354.

Skou, J. C. (1965). *Physiol. Rev.* **45**, 596—617.

9. Cyclic AMP, Calcium and Fluid Secretion

M. J. Berridge

Department of Zoology, University of Cambridge, Cambridge, England

I. INTRODUCTION

Ramsay made two important technical contributions to the study of insect epithelia. He introduced and perfected numerous micro-techniques for measuring the ionic concentration and osmotic pressure of minute volumes of liquid (Ramsay *et al.*, 1953; Ramsay and Brown, 1955; Ramsay *et al.*, 1955). He also began working with organs *in vitro* (Ramsay, 1954) and his technique for studying isolated Malpighian tubules has been utilized and adapted by many subsequent workers (Berridge, 1966a,b, 1970; Irvine, 1969; Maddrell, 1969; Maddrell *et al.*, 1971; Coast, 1969; Pilcher, 1970a,b; Farquarson, 1974; Kaufman and Phillips, 1973; Gee, 1975; Maddrell and Phillips, 1975; Wall *et al.*, 1975). This technique has been particularly useful for unravelling the control of fluid secretion in an insect salivary gland.

There were two main reasons for selecting *Calliphora* salivary glands as a model system for studying both the mechanism and the control of fluid secretion. The epithelium is composed of a single cell type (Oschman and Berridge, 1970) which facilitates the interpretation of both pharmacological and physiological experiments; and since the epithelial cells are organized into a long tube, they can be studied *in vitro* using Ramsay's Malpighian tubule technique (Ramsay, 1954). For this, a portion of the secretory region of the gland is removed from the fly and placed in a drop of saline maintained under liquid paraffin in a watch glass (Fig. 1). The cut end of the gland is ligated with a silk thread which is used to pull this end out into the liquid paraffin. A pair of tungsten cutting

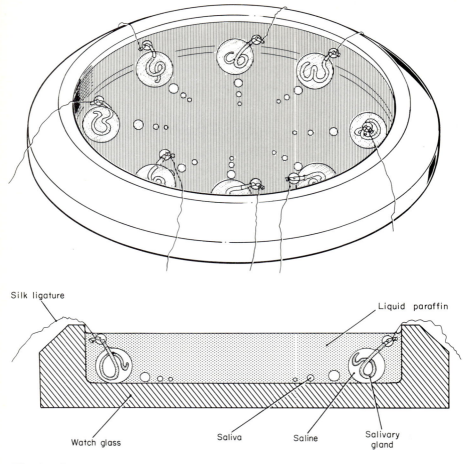

Fig. 1. The Ramsay technique for studying the secretory activity of isolated organs. In this case the tubular salivary glands of *Calliphora* are shown secreting into liquid paraffin. The fluid collecting around the ligatured end can be removed at periodic intervals and by measuring the volume of these drops of saliva it is possible to estimate the rate of secretion.

forceps are used to nick the gland immediately behind the ligature to provide an opening for saliva to escape into the liquid paraffin. This saliva can be collected at periodic intervals and by measuring the volume it is possible to monitor the rate of secretion. Drops can also be removed from under the paraffin and analysed by different techniques to determine their composition.

II. THE CONTROL AND MECHANISM OF SECRETION

The secretion of watery saliva by *Calliphora* salivary glands is regulated by 5-hydroxytryptamine (5-HT) (Berridge, 1970, 1972). The question we have to consider, therefore, is how a simple molecule like 5-HT can interact with the secretory cells to induce a dramatic acceleration of ion and water transport (Fig. 2). It is a problem of information transduction: the chemical configuration of 5-HT contains information which must be recognized by the cell and somehow translated into internal messages which will accelerate the relevant transport systems. In the case of *Calliphora* salivary glands, both 3,5′ cyclic adenosine monophosphate (cyclic AMP) and calcium seem to function as internal signals. Before considering how these two second messengers mediate the action of 5-HT it is necessary to describe some basic features of the secretory mechanism itself.

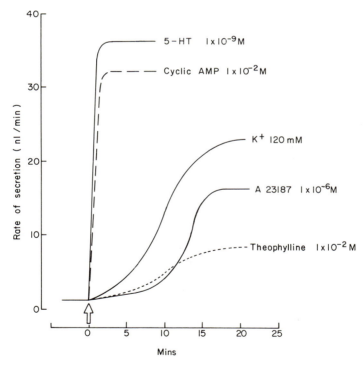

Fig. 2. The ability of various agents to stimulate fluid secretion by isolated salivary glands of *Calliphora*. The information for these curves was taken from Berridge (1970); Prince *et al.* (1973) and Berridge *et al.* (1975b).

Fluid secretion is driven by an electrogenic potassium pump which creates the gradient for a passive flow of chloride and water (Oschman and Berridge, 1970; Prince and Berridge, 1973; Berridge *et al.*, 1976a). Despite an enormous excess of sodium in the bathing medium, the glands much prefer potassium which is selectively secreted into the lumen (Berridge *et al.*, 1976a). In the complete absence of external potassium, the glands can switch over to using sodium but there is a reduction in the rate of secretion. There appears to be a relatively non-specific electrogenic pump capable of transporting cations into the lumen with anions following passively. Chloride is the major anion in saliva (Oschman and Berridge, 1970; Prince and Berridge, 1973).

In order to find out more about these ionic mechanisms it was necessary to monitor various electrical parameters. Transepithelial potential and resistance can be measured using a modified sucrose-gap technique (Berridge and Prince, 1972; Berridge *et al.*, 1975a). Intracellular potentials can be recorded by inserting a microelectrode into the cells lying in the perfusion bath of the perspex chamber (Prince and Berridge, 1972). At rest the lumen of the gland is approximately 15 mV positive with respect to the bathing medium but this potential is reduced almost to zero during treatment with 5-HT (Fig. 3). This rapid depolarization is caused by a large decrease in resistance (as indicated by the decrease in the height of the potential deflexions which result from passing constant current pulses across the gland), resulting from a dramatic increase in chloride permeability (Berridge *et al.*, 1975a). The change in resistance is abolished when chloride is replaced with the impermeant anion isethionate, but is more or less independent of the cation composition of the bathing medium. Current clamp experiments have also revealed that chloride movement is passive (Berridge *et al.*, 1975a). An interesting feature of the resistance measurements is that there is a simultaneous decrease in resistance across both the basal and apical surfaces of the cell (Berridge *et al.*, 1975a). During the action of 5-HT, therefore, there is an increase in the movement of both potassium (by means of an active pump) and chloride (by an increase in passive permeability).

There are several ways of dissociating the effects of 5-HT on potassium and chloride movements, which suggests that these two transport mechanisms are regulated independently of each other. During the action of 5-HT the transepithelial potential depolarizes (Fig. 3) despite the large increase in potassium transport which is necessary to drive the higher rate of fluid secretion (Fig. 2). Any increase in positive potential arising from this acceleration of

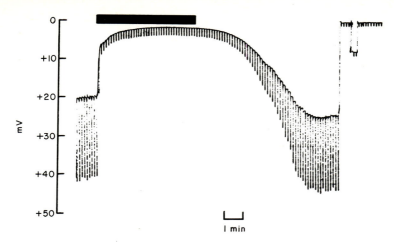

Fig. 3. The effect of 1×10^{-8} M 5-HT (solid bar) on transepithelial potential. The height of the downward potential deflexions, which were the result of passing constant current pulses across the gland, provide an indirect measure of resistance. Note that when the potential suddenly depolarized after addition of 5-HT, there was a simultaneous decrease in resistance. At the end of the record the microelectrode was withdrawn from the lumen.

potassium transport is short-circuited by chloride ions which can move freely because of the large increase in chloride permeability. However, it is possible to unveil the enhanced activity of the potassium pump by reducing the flow of chloride. When chloride in the bathing medium is replaced with isethionate (an impermeant anion) the transepithelial potential hyperpolarizes instead of depolarizing (Berridge and Prince, 1972). Although this hyper-polarization has a time course ($t_{1/2}$ = 30 sec) which is remarkably similar to that for the onset of fluid secretion ($t_{1/2}$ = 35 sec), it is very much slower than the increase in chloride movement which has a half-time of five secs (Berridge and Prince, 1972). This temporal separation of the effect of 5-HT on chloride and potassium transport may account for the biphasic potential response observed during brief 5-HT treatments (Berridge and Prince, 1972; Berridge et al., 1975a). The initial depolarization is caused by the rapid increase in chloride movement but this is not sustained and is followed by a marked hyperpolarization before the potential returns to the resting level (the first response in Fig. 4). The ability of 5-HT to increase potassium and chloride transport, apparently by independent mechanisms, must be uppermost in our minds as we go on to consider the role of intracellular second messengers.

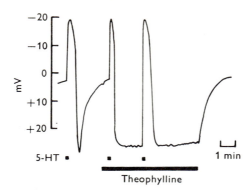

Fig. 4. The effect of short treatments with 1×10^{-8} M 5-HT on the trans-epithelial potential in the presence or absence of theophylline (1×10^{-2} M). (Taken from Berridge and Prince, 1972).

III. THE ROLE OF SECOND MESSENGERS

Both cyclic AMP and calcium seem to function as second messengers during stimulus-secretion coupling in these salivary glands. The chemical information contained in the 5-HT molecule is recognized by specific receptors which are capable of translating the information into an increase in the level of these two internal signals which then activate secretion. The exact method of information transduction is still unclear and will be considered in detail later. All the agents which can activate secretion (see Fig. 2) depend on their ability to elevate the intracellular level of either cyclic AMP (e.g. the addition of exogenous cyclic AMP or the phosphodiesterase inhibitor theophylline) or calcium (elevated potassium or the divalent cation ionophore A 23187).

The ability of exogenous cyclic AMP, or a number of closely related cyclic nucleotides, to stimulate fluid secretion indicates that this nucleotide may have an important second messenger role (Berridge, 1970, 1973). During the action of 5-HT there is an increase in the intracellular level of cyclic AMP (Prince et al., 1972). The increase in fluid secretion observed with theophylline (Fig. 2) can be explained by the ability of this methyl xanthine to inhibit the enzyme phosphodiesterase (PDE) which normally degrades cyclic AMP. An increase in the intracellular level of cyclic AMP during the action of theophylline would also explain why this drug can potentiate the action of both 5-HT and exogenous cyclic AMP (Berridge, 1970).

If cyclic AMP functions as a second messenger, how does it activate secretion? When cyclic AMP is applied to salivary glands the transepithelial potential hyperpolarizes which is completely opposite to the depolarization seen with 5-HT (Berridge and Prince, 1971, 1972). The increase in positivity induced by exogenous cyclic AMP is very similar to the hyperpolarization seen in chloride-free media after stimulation with 5-HT. It was argued earlier that this hyperpolarization was caused by an increase in potassium transport in the absence of a parallel flow of chloride. Cyclic AMP may thus be responsible for stimulating the potassium pump, with little or no effect on chloride permeability. The increase in lumen positivity apparently creates a sufficient electrical gradient to drag across enough chloride to maintain a flow of saliva. Any reduction in the availability of chloride in the bathing medium (by replacing a certain percentage of chloride with the impermeant anion isethionate) causes a marked reduction in the rate of secretion when glands are stimulated with cyclic AMP but there is no effect when 5-HT is the stimulant (Prince and Berridge, 1973). Cyclic AMP is thus able to duplicate the increase in cation transport but is clearly incapable of producing the large increase in chloride permeability normally induced by 5-HT.

The effect of 5-HT on chloride permeability is apparently mediated by calcium. Some evidence for the role of calcium has come from studies with the ionophore A 23187 (Prince et al., 1973; Berridge, 1975; Berridge and Prince, 1976) which is capable of stimulating secretion as long as there is calcium in the bathing medium (Fig. 2). Similarly, the ability of excess potassium to activate secretion depends on the presence of external calcium (Berridge et al., 1975b). One of the problems in trying to understand the role of calcium as a second messenger is to determine the source of activator calcium. 5-HT can stimulate the uptake of ^{45}Ca suggesting that an influx of external calcium may be responsible for generating the calcium signal (Prince et al., 1972). However, the glands may also utilize calcium stored in intracellular reservoirs because fluid secretion can continue for a considerable period either in the complete absence of external calcium (Prince and Berridge, 1972; Berridge et al., 1974) or when calcium entry is blocked by lanthanum (Berridge et al., 1975c). The extent of these reservoirs seems to vary with age (Berridge et al., 1975c). Newly emerged flies, whose salivary glands have just completed differentiation (Berridge et *al.*, 1976b), cannot secrete in the absence of external calcium; independence of external calcium develops during the first few days of adult life. The source of this internal supply of calcium has not been determined, but X-ray microprobe analysis of glands treated with lead to stabilize internal deposits suggest that the mitochondria are potential suppliers of internal calcium (Berridge *et al.*, 1975c).

When glands are stimulated by 5-HT in a calcium-free medium there is an initial normal depolarization and decrease in resistance, but as the calcium reservoir depletes and calcium becomes limiting the potential hyperpolarizes and the resistance returns to its resting value (Berridge *et al.*, 1975a). Since the removal of calcium has no effect on the increase in the intracellular level of cyclic AMP (Prince *et al.*, 1972), we can once again account for this hyperpolarization on the basis of cyclic AMP stimulating the potassium pump in the absence of the calcium-dependent increase in chloride permeability. The picture which is beginning to emerge is that during the action of 5-HT there is an increase in the intracellular level of both cyclic AMP and calcium; the former activates the electrogenic potassium pump whereas the latter increases chloride permeability (Fig. 5). The use of calcium as a second messenger would certainly account for the increase in chloride permeability which occurs simultaneously across both the basal and apical membranes (Berridge *et al.*, 1975a).

The presence of these two second messengers acting independently of each other can account for the biphasic change in potential mentioned earlier. It is postulated that the calcium signal is switched on very rapidly to produce the early depolarization. However, during a short stimulation this calcium signal will be rapidly removed and replaced with a pulse of cyclic AMP which develops more slowly to cause the large hyperpolarization. The recovery from this hyperpolarization would then represent the destruction of the cyclic AMP signal by phosphodiesterase. If this enzyme is inhibited by the addition of theophylline, the potential is clamped in the hyperpolarized condition and recovers only when theophylline is finally removed (Fig. 4).

Although the current model in which the action of 5-HT is mediated by two separate second messengers fits most of the observations, there are certain contradictions which require further comment. When fluid secretion is stimulated by either high potassium or by the ionophore A 23187, there is little or no increase in the intracellular level of cyclic AMP (Prince *et al.*, 1973; Berridge *et al.*, 1975b). These experiments seem to indicate that secretion can be activated independently of cyclic AMP. However, such results must

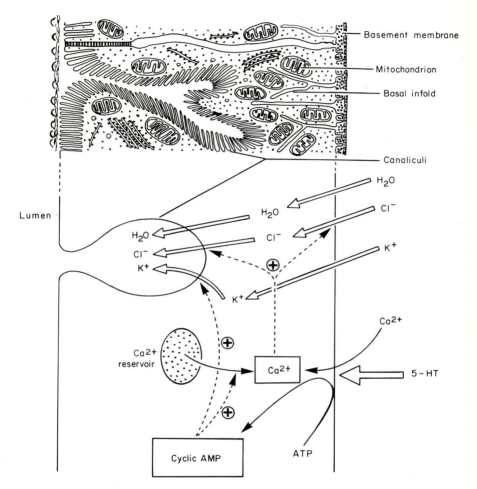

Fig. 5. Control of fluid secretion by the secretory region of adult *Calliphora* salivary glands. The main structural feature of the cells is the presence of long canaliculi formed by large infoldings of the apical plasma membrane. The basal plasma membrane has short infoldings some of which come very close to the canaliculi. The ability of 5-HT to stimulate fluid secretion seems to be mediated by both cyclic AMP and calcium. There is still some doubt concerning the action of these second messengers within the cell. Electrophysiological experiments suggest that calcium increases the chloride permeability of both the apical and basal membranes whereas cyclic AMP seems to activate an electrogenic potassium pump located on the apical membrane. Cyclic AMP may also act by releasing calcium from an internal reservoir.

be accepted with caution because treatment with ionophore or with high potassium solutions may induce additional effects leading to the onset of secretion which are unrelated to the normal course of events during the action of 5-HT. The onset of secretion induced by these two forms of stimulation is much slower than that observed with 5-HT (Fig. 2) which further suggests that they may not mimic the normal course of events. However, we cannot simply dismiss these results on the basis of unknown pharmacological effects and we must seriously consider the possibility that cyclic AMP is not directly concerned with the onset of secretion. There seems little doubt that cyclic AMP plays some role during the action of 5-HT but we must question whether or not cyclic AMP acts directly on the potassium pump or whether it acts indirectly by means of yet another intermediary. One possibility, which requires serious consideration, is that cyclic AMP acts indirectly by speeding up the release of internal calcium. Calcium will then be responsible for stimulating both potassium and chloride movement. In order to account for the apparent separation of second messenger effects mentioned earlier, it is necessary to postulate a hierarchy of calcium-requiring effects. Potassium transport is least sensitive and can make do with the supply of internal calcium released through the action of cyclic AMP. The increase in chloride permeability requires a higher concentration of calcium which is reached by the additional input of calcium when 5-HT increases the influx of external calcium. Although some of these suggestions are clearly very speculative, they focus attention on an important characteristic of second messengers — that is, their ability to interact with each other during cell activation.

IV. THE INTERACTION OF CYCLIC AMP AND CALCIUM

There is growing evidence that second messengers are not separate entities within a cell but that they interact with each other by means of various feedback loops (Berridge, 1975). Calcium is capable of influencing the intracellular level of cyclic AMP through effects on both adenyl cyclase and phosphodiesterase. In some cases, as in the mammalian salivary gland (Butcher, 1975) or turkey erythrocyte (Steer and Levitzki, 1975), calcium inhibits adenyl cyclase. There are other examples, however, where calcium interacts with a calcium-dependent receptor protein to form a complex which can stimulate adenyl cyclase (Brostom et al., 1975). A similar calcium-protein complex is also responsible for activating phosphodiesterase, in particular, the form which degrades cyclic GMP (Kakiuchi *et al.*, 1975). Perhaps of greater significance is the ability of cyclic AMP to modulate the intracellular level of calcium. In the mammalian heart cyclic AMP can have profound effects on the normal ebb-and-flow of calcium which drives each contraction-relaxation cycle. It can increase the influx of calcium during the slow phase of the action potential (Reuter, 1974) responsible for contraction. The rate of relaxation is also modulated by cyclic AMP which can stimulate the Ca-ATPase on the sarcoplasmic reticulum (Kirchberger *et al.*, 1972; Tada *et al.*, 1975). Cyclic AMP activates a protein kinase to phosphorylate a small protein (phospholamban) which then stimulates the Ca-ATPase. In the heart, therefore, cyclic AMP not only promotes calcium entry across the sarcolemma during contraction, but it can also modulate relaxation by speeding up the ability of the sarcoplasmic reticulum to sequester calcium.

A similar effect of cyclic AMP on both internal and external cell membranes has been noted in several other systems. For example, in lymphocytes cyclic AMP can reduce the uptake of calcium induced by the plant-lectin concanavalin A (Freedman *et al.*, 1975). Cyclic AMP seems to exert a similar inhibitory effect on the antigen-induced uptake of calcium by peritoneal mast cells (Foreman *et al.*, 1975). An effect of cyclic AMP on calcium transport may account for the increase in calcium efflux which has been observed during cell activation in many systems such as the liver (Friedmann, 1972), toad bladder (Thorn and Schwartz, 1965), insulin-secreting β-cells (Brisson and Malaisse, 1973) and *Calliphora* salivary gland (Prince *et al.*, 1972). There are at least two explanations to account for this increased efflux of calcium. Cyclic AMP may stimulate the extrusion of calcium from the cell, or it may stimulate the release of calcium from intracellular reservoirs. There is some evidence to favour the latter. Borle (1974) reported that isolated mitochondria from liver, kidney or heart will release calcium when treated with a low concentration of cyclic AMP. However, this observation cannot be repeated (Borle, 1976; Scarpa *et al.*, 1976), and the source of internal calcium in many cells remains to be established.

There is indirect evidence suggesting that cyclic AMP may be responsible for releasing calcium stored within the salivary glands of *Calliphora*. Like 5-HT, exogenous cyclic AMP can induce an increase in calcium efflux (Prince *et al.*, 1972). Both 5-HT and cyclic AMP are capable of depleting the internal reservoir of calcium when glands are stimulated in calcium-free media (Prince *et al.*, 1972). Other experiments have shown that there is very little depletion of the reservoirs when glands are not stimulated (Berridge *et al.*, 1974). The calcium which is released from these internal reservoirs seems to be functional. When a comparison is made between glands from newly emerged flies, which lack an internal reservoir of calcium, with older flies it is clear that the latter are less susceptible to the inhibitory effect of lanthanum which blocks the influx of external calcium. Similar experiments have shown that when glands are depleted of their internal reservoir they become much more susceptible to the inhibitory action of the local anaesthetic procaine (Berridge and Prince, 1976). The glands with a functional reservoir can release enough internal calcium to compensate for the absence of an influx of external calcium so as to maintain secretion (Berridge *et al.*, 1975c). There is still much to learn about how cyclic AMP and calcium interact with each other but we already know enough to realize that these feedback interactions are of central importance during cell activation.

V. CONCLUSIONS

The salivary glands of *Calliphora* have provided a unique system for studying not only the mechanism but also the control of secretion. Indeed the ability to manipulate the rate of secretion with various drugs has greatly facilitated the study of how ions are transported. The current model envisages an electrogenic potassium pump on the apical membrane as the "prime mover" (a term originally introduced by Ramsay) during fluid secretion. Chloride follows passively and together with potassium it creates the osmotic gradient to draw water out of the cell. The removal of potassium, chloride and water from the cytoplasm is then replenished by a continuous entry of these components across the basal plasma membrane (Fig. 5). The ability of 5-HT to accelerate this secretory mechanism is apparently mediated by both cyclic AMP and calcium. Calcium seems to be responsible for regulating the passive permeability of both the basal and apical membranes to chloride which thus greatly facilitates the transepithelial flow of chloride. The exact function of cyclic AMP is still not clear. It may act either directly on the electrogenic potassium pump or indirectly to stimulate the release of internal calcium (Fig. 5). In order to find out more about how these second messengers act we need to concentrate our efforts on two main problems. Firstly, we need to know more about the mechanisms of ion transport in particular the properties of the potassium pump. By defining the nature of the pump it will be easier to assess which second messenger is responsible for regulating its activity. Secondly, it will be important to find out more about the feedback interactions which operate between cyclic AMP and calcium. It is clear from studies on other cells that these second messenger feedback loops are of fundamental importance in the regulation of cellular activity. (Also see Chapters 4 and 25 and Gupta *et al.*, 1977.)

REFERENCES

Berridge, M. J. (1966a). *J. exp. Biol.* **44**, 553—566.

Berridge, M. J. (1966b). *J. Insect Physiol.* **12**, 1523—1538.

Berridge, M. J. (1970). *J. exp. Biol.* **53**, 171—186.

Berridge, M. J. (1972). *J. exp. Biol.* **56**, 311—321.

Berridge, M. J. (1973). *J. exp. Biol.* **59**, 595—606.

Berridge, M. J. (1975). *Adv. Cyclic Nucleotide Res.* **6**, 1—98.

Berridge, M. J. and Prince, W. T. (1971). *Phil. Trans. R. Soc.* B **262**, 111—120.

Berridge, M. J. and Prince, W. T. (1972). *J. exp. Biol.* **56**, 139—153.

Berridge, M. J. and Prince, W. T. (1976). *In* "Regulation of function and growth of eurkaryotic cells by intracellular cyclic nucleotides". (Eds J. E. Dumont, B. L. Brown and N. J. Marshall), pp. 591—607. Plenum Press, New York.

Berridge, M. J., Lindley, B. D. and Prince, W. T. (1974). *In* "Alfred Benzon Symposium VII. Secretory Mechanisms of Exocrine Glands." (Eds N. A. Thorn and O. H. Petersen), pp. 2101—2109, Munksgaard, Copenhagen.

Berridge, M. J., Lindley, B. D. and Prince, W. T. (1975a). *J. Physiol.* **244**, 549—567.

Berridge, M. J., Lindley, B. D. and Prince, W. T. (1975b). *J. exp. Biol.* **62**, 629—636.

Berridge, M. J., Oschman, J. L. and Wall, B. J. (1975c). *In* "Calcium Transport in Contraction and Secretion." (Eds E. Carafoli, F. Clementi, W. Drabikowski and A. Margreth), pp. 131—138. North-Holland Publishing Company, Amsterdam.

Berridge, M. J., Lindley, B. D. and Prince, W. T. (1976a). *J. exp. Biol.* **64**, 311—322.

Berridge, M. J., Gupta, B. L., Oschman, J. L. and Wall, B. J. (1976b). *J. Morphol.* **147**, 459—482.

Borle, A. B. (1974). *J. Membrane Biol.* **16**, 221—236.

Borle, A. B. (1976). *J. Membrane Biol.* **29**, 209—210.

Brisson, G. R. and Malaisse, W. J. (1973). *Metabolism* **22**, 455—465.

Brostrom, C. O., Huang, Y-C., Breckenridge, B. McL. and Wolff, D. J. (1975). *Proc. Natl. Acad. Sci. U.S.A.* **72**, 64—68.

Butcher, F. R. (1975). *Metabolism* **24**, 409—418.

Coast, G. M. (1969). *J. Physiol. Lond.* **202**, 102P.

Farquharson, P. A. (1974). *J. exp. Biol.* **60**, 13—28.

I

Foreman, J. C., Hallett, M. B. and Mongar, J. L. (1975). *Br. J. Pharmacol.* **55**, 283–284P.

Freedman, M. H., Raff, M. C. and Gomperts, B. (1975). *Nature, Lond.* **255**, 378–382.

Friedmann, N. (1972). *Biochim. biophys. Acta* **274**, 214–225.

Gee, J. D. (1975). *J. exp. Biol.* **63**, 391–401.

Gupta, B. L., Hall, T. A. and Moreton, R. B. (1977). *J. exp. Biol.* in press.

Irvine, H. B. (1969). *Am. J. Physiol.* **217**, 1520–1527.

Kakiuchi, S., Yamazaki, R., Teshima, Y., Uenishi, K. and Miyamoto, E. (1975). *Advances in Cyclic Nucleotide Res.* **5**, 163–178.

Kaufman, W. R. and Phillips, J. E. (1973). *J. exp. Biol.* **58**, 537–547.

Kirchberger, M. A., Tada, M., Repke, D. I. and Katz, A. M. (1972). *J. mol. cellular Cardiol.* **4**, 673–680.

Maddrell, S. H. P. (1969). *J. exp. Biol.* **51**, 71–97.

Maddrell, S. H. P. and Phillips, J. E. (1975). *J. exp. Biol.* **62**, 367–378.

Maddrell, S. H. P., Pilcher, D. E. M. and Gardiner, B. O. C. (1971). *J. exp. Biol.* **54**, 779–804.

Oschman, J. L. and Berridge, M. J. (1970). *Tissue and Cell* **2**, 281–310.

Pilcher, D. E. M. (1970a). *J. exp. Biol.* **52**, 653–665.

Pilcher, D. E. M. (1970b). *J. exp. Biol.* **53**, 465–484.

Prince, W. T. and Berridge, M. J. (1972). *J. exp. Biol.* **56**, 323–333.

Prince, W. T. and Berridge, M. J. (1973). *J. exp. Biol.* **58**, 367–384.

Prince, W. T., Berridge, M. J. and Rasmussen, H. (1972). *Proc. Natl. Acad. Sci. U.S.A.* **69**, 553–557.

Prince, W. T., Rasmussen, H. and Berridge, M. J. (1973). *Biochim. biophys. Acta* **329**, 98–107.

Ramsay, J. A. (1954). *J. exp. Biol.* **31**, 104–113.

Ramsay, J. A. and Brown, R. H. J. (1955). *J. scient. Instrum.* **32**, 372–375.

Ramsay, J. A., Brown, R. H. J. and Falloon, S. W. H. W. (1953). *J. exp. Biol.* **30**, 1–17.

Ramsay, J. A., Brown, R. H. J. and Croghan, P. C. (1955). *J. exp. Biol.* **32**, 822–829.

Reuter, H. (1974). *Circulation Res.* **34**, 599–605.

Scarpa, A., Malmstrom, K., Chiesi, M. and Carafoli, E. (1976). *J. Membrane Biol.* **29**, 205–208.

Steer, M. L. and Levitzki, A. (1975). *J. biol. Chem.* **250**, 2080–2084.

Tada, M., Kirchberger, M. A. and Katz, A. M. (1975). *J. biol. Chem.* **250**, 2640–2647.

Thorn, N. A. and Schwartz, I. L. (1965). *Gen. comp. Endocrin.* **5**, 710.

Wall, B. J., Oschman, J. L. and Schmidt, B. A. (1975). *J. Morph.* **146**, 265–306.

10. Amino Acid Transport

S. Nedergaard

Institute of Biological Chemistry A, August Krogh Institute, Copenhagen, Denmark

I. INTRODUCTION

Amino acid uptake has mainly been studied in vertebrates, in intestine, reviewed by Schultz and Curran (1970), Schultz *et al.* (1973), and in single cells, Christensen (1973). The general hypothesis for the amino acid uptake mechanism is a co-transport or a counter transport, where the amino acid together with sodium forms a ternary complex with a carrier — Na-gradient hypothesis.

This paper will only deal with amino acid transport in invertebrates and not all the literature will be covered. Amino acid uptake in single celled animals, protozoa, is reviewed by Dunham and Kropp (1973) and will not be discussed here. Investigations on amino acid uptake are described for animals representing the following phyla: Vermes, Arthropoda, Mollusca and Echinoderma.

II. VERMES

Amino acid uptake has been studied in Polychaeta-marine worms belonging to the class of Annelida. Some of the Polychaeta are adaptable to brackish water. At high salinities they are osmoconformers, but they are able to regulate their ionic composition at low salinities (Potts and Parry, 1964).

Some Polychaete worms are able to take up free amino acids through the integument from the seawater in which they live. The concentration of free amino acids in oceanwater is rougly 5% of the total dissolved organic material which corresponds to about 5×10^{-7} moles per litre (with seasonal variations). *Nereis limnicola* and *N. succinea* are able to take up glycine from seawater (Stephens, 1972), but when the salinity is reduced to 150 mEq Cl⁻/l the amino acid uptake stops. The cessation is not due to reduced concentration of some ions in the bathing medium but due to the drop in osmolarity as the worms are capable of taking up glycine from a solution where the osmolarity is made up to 200 mEq/l by galactose. When sodium is replaced by choline or lithium, uptake rates decrease. In the brittle star, *Ophiactes*, the amino acid uptake is influenced by the salinity of the adaptation medium. At 50% normal salinity the amino acid uptake drops to about half. The free amino acids in the tissue drop to about half whereas the incorporation of amino acids into proteins etc. increases about ten fold.

The uptake of one amino acid by Annelid worms can be inhibited by other amino acids in the bathing medium, but only within the same group of amino acids; for example, neutral amino acids inhibit the uptake of other neutral amino acids (also see Gomme *et al.*, 1976).

It may be supposed that the uptake of amino acids from the very dilute concentrations found in seawater through the body surface of the marine Annelid worms may serve as a means to help the animals regulate their osmolarity.

III. ARTHROPODA

The arthropods form the largest animal phylum with about 75% of all animal species. They live in all oecological niches, from the top of the mountains to the deep sea.

The arthropods have an exoskeleton which is periodically replaced during development from larva to adult. Some arthropods such as some crustaceans have a cardiovascular system. In others such as most insects this system may either be slightly developed or absent.

Aspects of amino acid transport have been investigated in species from two classes of arthropods: Crustacea and Insecta.

A. Crustacea

Most crustaceans live in water. The species in which amino acid uptake has been investigated are all marine. Their internal milieu is closely related to seawater in ionic composition. The amount of free amino acids in the blood of crustacea is as low as that of vertebrates. The pattern of the free amino acid pool in the blood seems to be widely different from species to species.

1. Midgut

Uptake of glycine by the midgut of the shrimp, *Penaeus marginatus*, has been investigated by Ahearn (1974). The midgut tissue is lined with columnar epithelium with microvilli on the apical side. The epithelium is attached to a thin basal lamina. Beneath the basal lamina circular and longitudinal muscle layers, connective tissue fibers and mesothelium are present.

The experiments were performed with midguts which were dissected out of the animals, opened longitudinally, and incubated in an oxygenated saline solution containing labelled glycine at final concentrations from 0.05 mM to 10 mM. First the amount of glycine taken up by the total midgut tissue was measured and then the amount taken up from the serosal side was estimated by ligating both ends of an intact midgut and incubating as before. The amount of glycine taken up by the mucosa was then found by subtraction. By using the subtraction method it is assumed that the amino acid is entering and going through the tissue from the two sides by the same route, which might not necessarily be the case.

Ahearn found that the total uptake (mucosa plus serosa) reached steady-state after incubation for one hour with 1 mM glycine in the artificial seawater as the bathing solution. The glycine concentration of the midgut tissue was found to be about eight times that of the bathing solution. The uptake via the serosal side was much slower and after one hour it was only slightly higher than the glycine concentration in the medium.

When sodium was replaced by choline in the bathing media the uptake of glycine was markedly reduced but mostly at low glycine concentrations in the medium. It appeared that only glycine uptake from the mucosal side was affected by lack of sodium in the bathing solution. The glycine uptake was blocked by metabolic inhibitors, the most potent being dinitrophenol. Ouabain also inhibited the glycine uptake. Glycine uptake into the midgut epithelium was also reduced by the presence of other amino acids in the bathing medium.

The author concludes that the midgut of the marine shrimp has an active glycine transport mechanism in mucosa, which has an absolute sodium dependency and is of similar nature as the mechanism in mammalian intestine.

2. Nerves

The uptake of amino acid into peripheral nerves has been studied in the spider crab, *Maia squinado*, by Baker and Potashner (1971, 1973) and in the shore crab, *Carcinus maenas*, by Evans (1973).

The concentrations of free amino acids in the blood from *C. maenas* are shown in Table I (Evans, 1972). It is seen that the concentration is very low; for glutamate it is only 0.01—0.06 mM. The value for free glutamate present in the spider crab has not been measured accurately, but is less than 1 mM, whereas the concentration of glutamate in the walking leg nerves is 40—50 mM. For the walking leg nerves of *C. maenas* this value is about 56 mM. So there is an extreme concentration difference across the neuronal membranes for glutamate: in *C. maenas* the ratio is about 10^3.

The uptake of glutamate has been investigated in crabs because this amino acid is thought to act as an excitatory synaptic transmitter at the invertebrate neuro-muscular junction.

Glutamate is accumulated inside the crab nerve, not only against a steep concentration gradient, but also against a potential gradient of −60 mV as glutamate is negatively charged at physiological pH values.

Baker and Potashner (1971) found that the uptake of glutamate into the spider crab nerve was divided in a sodium sensitive and a sodium insensitive uptake. It was found that the sodium sensitive uptake requires two sodium ions for each molecule of glutamate. Potassium in the bathing solution acts as a competitive inhibitor of the stimulating effect of sodium. Glutamate uptake is inhibited by the presence of other amino acids; strongest effect is produced by L-cysteic acid and L-aspartic acid, and the least by L-lysine. In a later paper, Baker and Potashner (1973) describe the effect of metabolic poisons on the sodium dependent glutamate uptake by the spider crab nerve and find that it is inhibited by iodoacetamide, and that the uptake generally is reduced in parallel with the reduction in ATP. In the presence of external sodium, the glutamate influx increases with the increased concentration of intercellular ATP, but in the absence of external sodium, changes in ATP have no effect on the glutamate influx. Evans (1973) has investigated the uptake of glutamate by the peripheral nerve of the shore crab, *C. maenas*. Like Baker and Potashner (1973), he finds that the glutamate uptake can be divided in a sodium sensitive and a sodium insensitive uptake (Fig. 1). The sodium sensitive uptake shows the typical saturation kinetics of a carrier-mediated process. The sodium activated uptake is

TABLE I

The composition of the free amino acid pools from various tissues of Carcinus[a]

Sample / Amino acid	Whole blood	Plasma	Cells	Peripheral nerve	Muscle
Taurine	2.090 ± 0.310	0.325 ± 0.050	209.200 ± 33.700	117.30 ± 6.87	50.53
Aspartate	0.039 ± 0.010	0.017 ± 0.005	1.780 ± 0.250	307.90 ± 13.84	5.21
Glutamine	1.180	0.633	55.700	36.90	
Serine	0.140	0.083	6.330	4.66	
Glutamate	0.223 ± 0.009	0.055 ± 0.016	18.930 ± 2.930	56.20 ± 3.00	48.13
Proline	0.720 ± 0.138	0.446 ± 0.117	25.990 ± 4.310	37.90 ± 2.06	131.60
Glycine	0.613 ± 0.100	0.350 ± 0.014	26.510 ± 8.100	11.37 ± 2.46	134.60
Alanine	0.819 ± 0.140	0.426 ± 0.047	53.800 ± 10.680	33.65 ± 2.46	27.40
Valine	0.056 ± 0.007	0.042 ± 0.011	2.170 ± 0.050	1.06 ± 0.02	4.41[b]
Methionine	0.029 ± 0.001	0.018	0.739 ± 0.020	1.01 ± 0.06	
Isoleucine	0.020 ± 0.003	0.015 ± 0.003	0.723 ± 0.112	0.42 ± 0.05	2.27
Leucine	0.037 ± 0.006	0.023 ± 0.004	1.098 ± 0.553	0.72 ± 0.05	3.48
Tyrosine	0.026 ± 0.003	0.019 ± 0.006	1.306 ± 0.196	0.81 ± 0.08	0.40
Phenylalanine	0.016 ± 0.001	0.013 ± 0.004	0.613 ± 0.055	0.26 ± 0.02	0.67
Lysine	0.033 ± 0.001	0.022 ± 0.003	1.699 ± 0.434	0.92 ± 0.06	2.54
Histidine	0.038 ± 0.006	0.022 ± 0.002	1.783 ± 0.289	0.75 ± 0.05	1.33
Arginine	0.318 ± 0.039	0.164 ± 0.028	18.400 ± 3.035	13.99 ± 0.33	48.40
Total	6.392	2.672	426.800	626.80	460.90

[a] The results are expressed as μmoles/ml blood ± S.E. for the whole-blood and plasma fractions and as μmoles/ml cell water ± S.E. for the cell fraction, peripheral nerve and muscle. The values for muscle are all calculated from Duchâteau et al., (1959), except for the taurine value which comes from Chaplin et al., (1970), using a correction factor obtained from Florkin and Schoffeniels (1969).
[b] Present but not estimated. From Evans (1972).

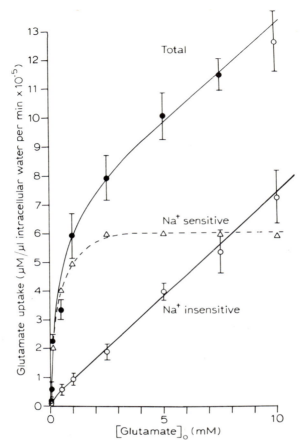

Fig. 1. The rate of uptake of glutamate is plotted against the concentration of glutamate in the bathing medium. The total uptake represents the uptake from normal saline (494 mM Na+) and the Na+-insensitive uptake represents the influx from a Na+-free (choline) medium. The linear portions of the plot were obtained by a linear regression analysis. The Na+-sensitive plot was obtained by the subtraction of the Na+-insensitive component from the total uptake at each glutamate concentration. The bars represent 2 S.E. and n = 10. (From Evans, 1973)

the major part of the glutamate uptake but, contrary to the spider crab nerve, Evans finds that only one sodium ion is needed for one molecule of glutamate. He finds that the sodium sensitive glutamate uptake is inhibited by other dicarboxylic amino acids, mainly cysteic acid and aspartic acid, while the sodium insensitive transport is not affected much by other amino acids. Metabolic inhibitors had very little effect on the two types of glutamate uptakes and that too only after preincubation.

Evans concludes by suggesting that the peripheral nerve of *Carcinus* possesses an active uptake mechanism for glutamate, which apparently is dependent on energy provided by a combination of the sodium ion gradient across the cell membranes and metabolic processes. The energy providing source for the sodium insensitive glutamate uptake is not clear.

B. Insecta

The insects live in all kinds of habitat all over the world. They are able to adapt to extreme living conditions. The environmental conditions are therefore very different from one species to the next. The feeding habits are variable, some eat an extremely specialized diet while others feed on many different types of food. Many have one diet when they are larvae and another when they are adults.

The composition of insect blood varies between species both with respect to total amino acid concentration and the presence of individual amino acids, Table II (Jeuniaux, 1971). The amino acid concentration in the blood varies even within the same animal during development, and is further influenced by changes in diet (for references see Wigglesworth, 1965).

The free amino acid concentration is much higher in insect blood than in vertebrate blood, and in most species the greater part of the total osmotic pressure comes from free amino acids. Ussing (1946) measured 300–400 mg % non-protein amino nitrogen in blood from *Oryctes* and *Melolontha*, and also found that the concentration of free amino acids in the gut lumen was smaller than in the blood of *Melolontha* and therefore suggested that amino acids are transported actively across the gut wall.

1. Absorption

a   Midgut

J. A. Ramsay found in some preliminary experiments (1958) that in the gut contents of the stick insect, *Dixipus morosus*, free amino acids were detectable in the foregut, present in quantity in the

TABLE II

Distribution and concentration of the principal free amino acids in the hemolymph of some insects

(mg/100 ml Hydrolysed plasma[a])

Amino acids	Exopterygotes				Hymenoptera	Coleoptera[b]		Diptera	Lepidoptera			
	Aeschna sp larvae	Carausius morosus adults	Periplaneta americana adults[i]	Locusta migratoria nymphs	Apis mellifera larvae	Hydrophilus piceus adults	Popillia[h] japonica larvae	Gasterophilus larvae	Euproctis crysorrhoea larvae[c]	Saturniidea[e] pupae	Sphyngidae[f] pupae	Papilio machaon pupae[d]
Alanine	46	10–60	7	34	58	60	146–187	—	33	7–300	16–250	103–213
Arginine	19–27	17–19	19	24	50–74	7–11	48–81	8	44–58	107–243	59–576	126–127
Aspartic acid (total)	4–13	6–14	2	13	32–33	17–18	42–47	14	9–22	4–36	5–55	14–19
Glutamic acid (total)	32–63	50–77	24	166	308–347	131–195	309–526	314	302–343	83–468	62–240	202–226
Glycine	22–54	23–31	53	97	72–84	17–26	288–325	5	48–94	20–82	4–57	48
Histidine	7–21	55–58	23	30	17–30	8–12	169–225	1	107–161	23–196	3–127	71–89
Isoleucine	16–18	7–13	6	21	20–24	8–25	36–54	8	15–32	14–83	20–65	40–56
Leucine	22–29	10–14	7	21	25–30	7	20–25	7	13–23	15–108	40–73	56–80
Lysine	6–14	20–28	11	47	74–104	20–24	29–94	8	50–105	113–471	64–433	325–401
Methionine	4–13	9–13	4	6	19–23	3	3–12	7	1–13	11–148	25–81	122–163
Phenylalanine	5–11	6–9	6	11	8–12	6–7	13–17	7	8–15	7–72	8–49	24–43
Proline	12–41	10–16	42	62	368–418	122–283	264–507	16	129–157	62–478	28–230	146–256
Threonine	12–23	29–40	8	20	27–49	12–17	11–29	23	30–54	1–136	20–82	47–57
Tyrosine	3–13	5–8	25	28	3	2–9	11–37	22	0–5	2–76	8–146	4–5
Valine	23–29	22–25	12	48	58–59	11–20	94–150	15	29–49	34–127	22–105	101–120
Total	399.0	293–424	248	636.0	1239.0	445–721	1723–2162	465[g]	870–1164	1124–1989	515–1819	1575–1769
Serine	24		14	49		22–35						

[a] The values have been rounded to unity. (From Duchâteau and Florkin, 1958; Stevens, 1961; Shotwell et al., 1963).

[b] Other species studied: Leptinotarsa decemlineata.

[c] Other species studied: larvae of Cossus cossus (sum of the 15 amino acids: 938 mg/100 ml); Amathes xanthographa (1027 mg/100 ml), Triphaena pronuba (1352 mg/100 ml), Imbrasia macrothyris (497 mg/100 ml), Pseudobunaea seydeli (709 mg/100 ml), Smerinthus ocellatus (700 mg/100 ml).

[d] Other pupae studied: Lasiocampa quercuss (sum of 15 amino acids: 2317 to 2430 mg/ml), Euproctis chrysorrhoea (1066 mg/100 ml) and Smerinthus ocellatus (1645 mg/100 ml).

[e] 15 species belonging to the genus Citheronia, Eacles, Saturnia, Antheraea, Actias, Hyalophora, Philosamia.

[f] Species studied: Deilephila elpenor, Sphinx ligustri, Celerio euphorbiae, Laothoe populi, L. austanti, and L. populi × austanti.

[g] Without alanine.

[h] Shotwell et al., 1963.

[i] Stevens, 1961.

midgut, and absent from the hindgut. When a gut from a well fed animal was dissected out and placed in a Ringer's solution after it had been tied off at both ends, amino acids were detected in the bathing solution. When the experiment was done with separate portions of the gut it was found that nearly all the amino acids came from the midgut.

Treherne (1959) found that the concentration of glycine, serine, and glutamate was higher in the haemolymph of this insect than was normally found in insect blood. Treherne investigated the uptake of glycine and serine from the lumen of the alimentary canal into the haemolymph in the locust, *Schistocerca gregaria*. The uptake of these amino acids was measured by injection into the gut lumen of ^{14}C labelled amino acid in a solution as close as possible to the composition of the haemolymph, so the amino acid concentration on the two sides of the intestinal tissue was the same. The osmotic pressure was also kept the same on the two sides. It was found that the labelled amino acids disappeared from the lumen of the alimentary canal and fastest from the caeca. At the same time, however, water was disappearing from the caeca measured by the addition of a dye, Amaranth, or by ^{131}I-labelled albumin to the solution injected into the lumen. It was found that water disappeared faster from the lumen of the caeca than the amino acids, therefore the amino acid concentration in the lumen increased and became higher than in the haemolymph. The amino acid could then be absorbed passively along a concentration gradient (Fig. 2). The author points out that this way of absorption of amino acids by an established concentration gradient may not be the only way by which amino acid are absorbed in the intestinal canal of this insect.

According to Shyamala and Bhat (1966), Bhatnagar (1962) found that absorption of alanine, glycine, and glutamic acid was rapid in the gut of *Oxycarenus hyalimipennis*, but that absorption of arginine, leucine, methionine and tryptophan was much slower.

Bragdon and Mittler (1963) found that the aphid, *Myzus persicae*, did not excrete ingested amino acids in the same proportion as they were fed to the animals. Of the six amino acids investigated the uptake by the animal was increasing in the following order: asparagine, lysine, leucine, threonine, valine and methionine.

Shyamala and Bhat (1966) studied uptake from a mixture of amino acids by the isolated midgut of the fifth instar larva of *Bombyx mori* and found that tyrosine, alanine and the basic amino acids were absorbed at a greater rate than glycine, aspartate, leucine, valine, methionine; the slowest absorbed being glutamate. 10^{-4} M

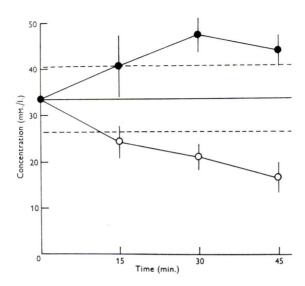

Fig. 2. The changes in concentration of total glycine (closed circles) and ^{14}C-labelled glycine (open circles) in the caeca following the injection of experimental fluid into the gut lumen. The vertical lines through the points represent twice the standard error of the mean. The three horizontal lines represent the mean and twice the standard error of the initial glycine concentration in the haemolymph. (From Treherne, 1959)

dinitrophenol and 5×10^{-6} M cyanide were found to have no effect on the amino acid absorption.

The amount of L-glutamate in the haemolymph of the locust is low compared to the amount of other amino acids. L-glutamate, however, is very abundant in plant proteins and therefore present in the lumen of the digesting animal in high concentrations. It could be expected that L-glutamate was metabolized by the cells of the alimentary tract. This was investigated by Murdock and Koidl (1972a). They used a perfusion technique where they flushed the haemolymph out of the living animal and established a constant rate of perfusion of the haemolytic space with saline. ^{14}C-labelled L-glutamate was injected into the gut as described by Treherne (1959). The amount of ^{14}C in the perfusate was measured with time and about 30% appeared during the first hour of perfusion. 80% of the radioactivity in the perfusate was found to be CO_2/bicarbonate, so an extensive metabolism of the substance occurs in the gut wall. A similar experiment was also performed with alanine and glycine and

it was found that alanine crosses the gut about four times faster than
L-glutamate but much less of the alanine was converted to CO_2/
bicarbonate. Glycine rapidly reached maximum absorption faster
than alanine but was not — or hardly — metabolized by the gut wall.
The animals used were *S. gregaria* and *Locusta migratoria*. Murdock and Koidl (1972b) found that the major part of the ^{14}C-L-glu-
tamate infused into the locust gut, not converted to $CO/_2$/bicar-
bonate, was found in the blood as glutamine.

Orlowski and Meister (1970) have proposed that γ-glutamyltranspeptidase is involved in amino acid transport in vertebrates. Bodnaryk *et al.* (1974) have found by a histochemical method that
γ-glutamyltranspeptidase is present in the larva of the housefly,
Musca domestica, and the blowfly, *Sarcophaga bullata*. The enzyme
is found in the proximal half of the Malpighian tubules and in the
midgut epithelium, and located in the brushborder. The enzyme is
absent from the hindgut. Bodnaryk and Skillings (1971) found that
γ-L-glutamyl-L-phenylalanine is present in growing housefly larvae.
Bodnaryk *et al.* (1974) therefore suggested that the membrane
bound enzyme γ-glutamyltranspeptidase is involved in translocation
of amino acids across the insect midgut epithelium.

Nedergaard (1972) found that α-aminoisobutyric acid (AIB), a
non-metabolized amino acid, can be transported actively from the
lumen side to the blood side by the isolated midgut of fifth instar
larvae of *Hyalophora cecropia*. The AIB flux ratio was between 10
and 60 with equal AIB concentrations on the two sides of the
midgut, but became close to one when the metabolism was inhibited
with nitrogen (Fig. 3). So it was concluded that AIB is transported
across the gut wall by an active process. The isolated Cecropia
midgut maintains a potential difference of about 100 mV, lumen side
positive, when bathed with identical potassium containing solutions
on both sides. This is generated by the active transport of potassium
from blood to lumen, which is independent of sodium (Harvey and
Nedergaard, 1964). Amino acid transport in the vertebrate intestine
is connected to the sodium transport. Here the active amino acid
transport could be related to the active potassium transport; the
effect on the amino acid transport was therefore measured after
short-circuiting the potential difference across the midgut. In short-
circuited midgut the active potassium transport is increased by about
100% (Harvey and Zerahn, 1971). Instead of an increase in amino
acid transport with the increased potassium transport, a considerable
decrease occurred in the active amino acid flux. The decrease in
amino acid transport might either be caused by the increased

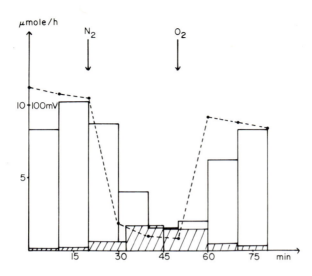

Fig. 3. Effect of anoxia on α-aminoisobutyric acid fluxes across the Cecropia midgut. The white columns are α-aminoisobutyric acid fluxes from lumen to blood, the hatched columns are fluxes from blood to lumen. The dotted line is the potential difference across the midgut. The two fluxes are not measured on the same midgut. The potential difference shown belongs to the lumen to blood experiment, but the variations in the potential difference of the other experiment were very similar. The arrows indicate the time for the change in aerating gas. (From Nedergaard, 1973)

potassium transport or by the lack of potential difference. Experiments were therefore performed where all potassium in the bathing solutions was substituted by sodium (sodium is not transported by the midgut when Ca^{++} is present in the bathing solution, and the potential difference is therefore close to zero). An artificial potential gradient was established across the tissue and the amino acid flux from lumen to blood side was measured and found to increase with increasing potential difference, lumen side positive (Fig. 4). This shows that the active potassium transport is not involved directly in the AIB transport. In experiments with potassium in the bathing solutions where the natural potential difference was more or less abolished by short-circuiting, it was found that the AIB uptake increased with increasing potential difference (Fig. 5). The effect of the potential difference might mean that it is the passive flux of the potassium ion from lumen to blood that is driving the amino acid uptake, but in experiments with very little potassium or other ions

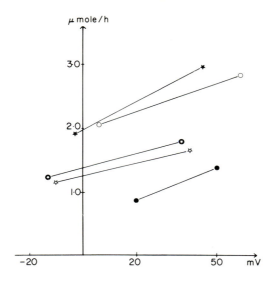

Fig. 4. Effect of an applied midgut potential on α-aminoisobutyric acid flux from lumen to blood in the absence of potassium. Abscissa: potential difference in mV. Ordinate: μmole α-aminoisobutyric acid flux from lumen to blood per hour. Each symbol is one midgut, i.e. one experiment. (From Nedergaard, 1973)

on the luminal side there seemed to be no decrease in the AIB transport (Nedergaard, in preparation).

The effect of the potential difference on the active amino acid transport could be via the influence on the charge of the transported amino acid. The pH of the bathing solutions in experiments so far described here was about eight. At this pH the experimental amino acid, AIB, is negatively charged and it will therefore be moving against an electrochemical gradient when it is transported from lumen to blood. To test if an overall negative charge on the amino acid was necessary for the transport, the fluxes of lysine in both directions was measured at pH 7.5, where lysine bears a positive charge. The ratio between the two fluxes was found to be about 10, so lysine is also transported from lumen to blood. When pH on the lumen side was raised to about 10, lysine was still transported into the blood although lysine has no net charge at this pH. The overall charge on the transported amino acid seems therefore not to be involved in the transport mechanism (Nedergaard, 1973).

The amount of AIB, which accumulated in the tissue was measured both during active transport from lumen to blood and

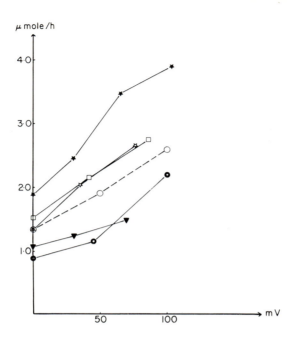

Fig. 5. Effect of the midgut potential on α-aminoisobutyric acid flux from lumen to blood.

Abscissa: potential difference in mV. Ordinate: μmole α-aminoisobutyric acid flux from lumen to blood per hour. Each symbol is one midgut, i.e. one experiment. The large open circles are the mean values of all the experiments of the figure. The highest point is the amino acid flux at the natural spontaneous potential difference and the point on the ordinate is the flux from lumen to blood when the midgut potential is short-circuited. (From Nedergaard, 1973)

during passive flux in the opposite direction. It was found that during active transport, AIB accumulated in the tissue at concentrations from about 10 mM to about 40 mM, directly proportional to the size of the active flux. Inhibition of the active amino acid transport for instance by reducing the potential difference or inhibiting the metabolism reduced the amount of AIB in the tissue. It was also found that the inhibition did not effect the amino acid flux right away, the flux continued with unchanged speed for the first 15 minutes after which it started to drop off (Fig. 3). This indicates that the active part of the AIB transport is located at the entry into the tissue. If we view the tissue as a black box, the active transport site is located on the luminal side. This also indicates that the active potassium pump is placed on the luminal side of the midgut tissue (Nedergaard, in preparation).

The uptake of AIB into the tissue from the blood side increased with increasing amino acid concentration of up to 10 mM AIB on the blood side. The concentration in the tissue was then about three times that of the bathing solution. Saturation occurred at 10 mM AIB in the bathing solution. This accumulation in the tissue was independent of the passive AIB flux from blood to lumen. Contrary to the fluxes across the midgut tissue, which reached steady state within 20—30 minutes, the uptake into the tissue from the blood side did not reach steady-state before almost an hour. The uptake into the tissue is an exchange process because the amino acid in the tissue cannot be washed out again except with an amino acid present in the washing solution. The uptake is not very specific because acidic, neutral and basic amino acids can all exchange. This amino acid exchange from the blood into the tissue cannot be inhibited with metabolic inhibitors such as nitrogen and iodoacetate, or a mixture of the two. This is in contrast to the amino acid uptake from the lumen into the tissue which is almost abolished by nitrogen (Nedergaard, in preparation). The amino acid exchange uptake from the blood side is found to be independent of either the active transport of potassium or the presence of these ions in the bathing medium.

We have now discussed three types of transports into and across the midgut tissue: 1. the active AIB flux from lumen to blood, 2. the passive AIB flux from blood to lumen, and 3. the exchange of AIB from the blood into the tissue. How do these three fluxes work together in the same tissue? Firstly, if the two transports from each side of the midgut tissue went into the same pool, it would be expected that in steady-state the pool size would be about the same whether it was loaded from one or the other side of the tissue. But the AIB concentration in the midgut tissue is proportional to the active AIB transport from lumen to blood. Therefore the AIB concentration in the tissue in steady-state varies over a wide range, from about 10 mM to about 40 mM, while the variation in the tissue concentration in steady-state when AIB is entering from the blood side via the exchange mechanism is only from about 25 mM to 30 mM, and is independent of the flux. (Variations in tissue concentration are due to individual differences in animals.) Secondly, the exits from these two pools in the tissue are different. The AIB taken up by the exchange mechanism does not leave the tissue again except when AIB, or another amino acid, is present in the bathing solution, while the AIB coming from the lumen via the active transport mechanism does appear on the blood side of the midgut whether or not there is an amino acid present in the bathing solution. So these two fluxes do not pass the midgut tissue through the same route.

The midgut of the Cecropia larva is a one cell-layer thick epithelium consisting of two types of cells, columnar cells and goblet cells, with only a thin, discontinuous muscle layer on the blood side. Each of the two mechanisms could therefore use one cell type; in this case the active transport of AIB would most likely be through the columnar cells because the goblet cells typically secrete in the direction opposite to the amino acid transport. This would fit with the idea of uptake of AIB from the blood side, because the amino acid uptake could be used for enzyme production for the digestion. Unfortunately the relative amount of goblet cells in the total tissue volume is not known. Wood (1972) estimates it to be between one eighth and one tenth of the total volume; this would give an AIB concentration of 200–300 mM in the cells, a surprisingly high value. The larva of *Simbex femorata*, a sawfly, has no goblet cells in the midgut, but lives under similar conditions as Cecropia larva (Turbeck, private communication). The amino acid transport in this gut might be worth investigating.

It seems therefore realistic to assume that the amino acid exchange transport from the blood side goes into all the cells and is presumably used for protein synthesis. This means that the amino acid from the active transport will travel through the cells in a compartment that does not mix with the rest of the cell, and also does not mix with the amino acid coming from the opposite side into the same cell. A special transport passage through cells has been proposed by Voute *et al.*, (1975) for the active sodium transport in the frog skin through the endoplasmic reticulum.

Lastly, at least some of the small passive flux from blood to lumen goes backwards through the active transport route as the flux is altered by a change in potential difference. The flux could also go between the cells; or both routes may exist. But it does not go into the cells via the amino acid exchange mechanism because the passive flux reaches steady-state in 20–30 minutes, whereas the exchange is not in steady-state before about one hour. It also looks as though the amino acid entering the cells through the exchange mechanism does not appear on the lumen side of the midgut, or does so only in very small amounts, because the passive flux is constant from about 30 minutes to one hour.

2. Excretion

a   Malpighian tubules

Ramsay (1958) measured the uptake of amino acids from the haemolymph into the lumen of the Malpighian tubules of *Dixippus morosus*. Glycine is present in the haemolymph in a concentration of 10 mM 1^{-1} but is not present in normal urine secreted by the Malpighian tubules. Only when the glycine concentration in the blood is raised to about 70 mM 1^{-1} or more does it start to appear in the urine.

Isolated Malpighian tubules were placed in a droplet of Ringer solution under paraffin oil with the cut end of the tubules pulled out in the paraffin oil outside the drop (see Berridge, in this volume). The urine secretion by the tubules then formed a new droplet in the paraffin oil. This way the two fluids, the one bathing the outside and the one bathing the inside of the secreting epithelium, were very elegantly separated (Ramsay, 1954).

The uptake into the urine was measured for six amino acids, with increasing concentrations on the blood side. The results are shown in Table III. It is seen that all of these metabolically useful amino acids appear in the urine. Furthermore, the ratio between the amount in the urine and the amount in the bathing medium (U/P) is virtually independent of the concentration of the substance in the medium. In the case of glycine, the range of concentration in the bathing medium is varied by a factor of 37, whereas the corresponding range of U/P only varies by a factor of 3.2. For all the other amino acids the contrast is even more pronounced. It is also seen that there seems to be no limit to the capacity of the tubule to remove the amino acids from the medium.

In order to see if there was any competition or interaction between amino acids, the uptake into the urine from a mixture of three amino acids were compared to the uptake of separate amino acids. The results are given in Table IV. In the case of the mixtures, the U/P ratio is generally higher than for the amino acids presented separately. This could be taken as an evidence for interaction but it may equally well be related to the observation shown in Table IV; that is the U/P ratio tends to increase with increasing amino acid concentration in the medium. There is therefore no evidence for competitive inhibition affecting the excretion of one amino acid in the presence of another.

Since the concentration in the urine for all amino acids studied

TABLE III

Amino acid	P	U/P	R
DL-Alanine-I-^{14}C	4.4	0.23 ± 0.029 (9)	1.69 ± 0.24 (9)
	56	0.25 ± 0.048 (10)	1.45 ± 0.41 (10)
	109	0.32 ± 0.041 (6)	1.11 ± 0.43 (6)
	161	0.37 ± 0.061 (9)	1.19 ± 0.30 (9)
L-Arginine-^{14}C(G)	0.34	0.39 ± 0.07 (4)	1.34 ± 0.14 (4)
	53	0.35 ± 0.09 (4)	1.29 ± 0.21 (4)
	105	0.40 ± 0.19 (4)	1.09 ± 0.29 (4)
	158	0.62 ± 0.29 (4)	0.40 ± 0.01 (4)
Glycine-^{14}C(G)	4.5	0.14 ± 0.03 (4)	Not recorded
	43	0.17 ± 0.05 (7)	Not recorded
	82	0.22 ± 0.03 (6)	Not recorded
	121	0.27 ± 0.05 (5)	Not recorded
	166	0.45 ± 0.07 (4)	Not recorded
L-Lysine-^{14}C(G)	0.15	0.13 (1)	Not recorded
L-Proline-^{14}C(G)	0.16	0.22 ± 0.03 (4)	1.22 ± 0.54 (4)
	26	0.55 ± 0.08 (4)	2.25 ± 0.07 (4)
	52	0.59 ± 0.04 (4)	1.43 ± 0.54 (4)
	105	0.58 ± 0.13 (7)	1.94 ± 0.57 (7)
	157	0.63 ± 0.09 (4)	1.21 ± 0.75 (4)
DL-Valine-1-^{14}C	6.3	0.24 ± 0.04 (8)	2.72 ± 0.42 (8)
	59	0.24 ± 0.04 (8)	2.36 ± 0.64 (7)
	111	0.23 ± 0.04 (8)	2.44 ± 0.61 (8)
	163	0.25 ± 0.05 (8)	2.40 ± 0.53 (7)

(Mean ± standard deviation (no. of observations). Amino acids as specified in catalogue of Radiochemical Centre, Amersham. P, concentration in medium, mM/1; U, concentration in urine, mM/1; R, rate of urine flow, mμ 1/min.) (From Ramsay, 1958).

here is less than in the medium, it is not nescessary to postulate an active transport for the neutral molecules. In contrast to the neutral amino acids, arginine carries a positive charge at physiological pH but as its U/P ratio is not very different from the other amino acids it seems that the same transport mechanism is operating in all cases. The simplest and the most likely way by which the amino acids enter into the urine is by a passive, non-specific process.

b   Hindgut

Ramsay (1958) found that when the separate regions of the gut from *Dixipus morosus* were filled with Ringer containing 70 mM 1^{-1}

TABLE IV

	U/P(mixture)	U/P(separate)
Group 1. Arginine	0.30	0.25
Proline	0.76	0.56
Valine	0.29	0.20
2. Alanine	0.22	0.24
Glycine	0.32	0.19
Valine	0.24	0.21

(From Ramsay, 1958).

glycine, the greatest output was from the hindgut, and was associated with an obvious decrease in the volume of the hindgut contents. Ramsay therefore suggested that amino acids which have entered into the urine are reabsorbed when the urine passes through the hindgut.

Wall and Oschman (1970) measured the free amino acid concentration in the different compartments of the rectum of the hydrated cockroach, *Periplaneta americana*. They found that the concentration in the sinus of the rectal pad was higher than in the haemolymph, which again was higher than the concentration in the lumen of the rectum. The free amino acids therefore seem to be moving against a concentration gradient.

Balshin and Phillips (1971) studied the uptake of glycine by the isolated rectum from the desert locust, *Schistocerca gregaria*. The isolated rectum was mounted as a sac with the lumen side facing the outside (everted sac). The two sides of the isolated rectum were bathed with identical Ringer solutions, except that sucrose was added to the lumen side in a concentration of 400 mosmole to prevent movement of fluid from lumen to blood. The tissue was equilibrated for one hour with ^{14}C-glycine in the medium. Then the bathing solutions were renewed and an equal concentration of ^{14}C labelled glycine, was again added to both sides. Samples were taken from both sides at regular intervals for up to six hours. The potential difference and the net water flow was measured during the experiment. The flux of glycine from lumen to blood side was measured with either 1, 10, or 50 mM glycine in the bathing solutions. The results are shown in Table V, which also shows the effect of metabolic poisons and the effect of the presence of sodium in the bathing solutions. It is seen that there is a net uptake of glycine across the rectum into the blood side. This uptake is abolished by metabolic inhibitors and by the absence of sodium. It is therefore assumed that transport of glycine from lumen to blood across the

TABLE V

Composition of medium		Net rates (mean s.e.)	
Glycine (mmol/1)	External sucrose (mosm/1)	Glycine (nm h^{-1} cm^{-2})	Water (μl h^{-1} cm^{-2})
1	400	9 + 2	−1.0 ± 0.5
10	400	106 ± 11	0.4 ± 0.9
50	400	202 ± 31	−1.0 ± 1.1
10	0	171 ± 14	7.2 ± 0.6
10	50	−2 ± 4	0.2 ± 1.3
with 10^{-2}M KCN + IAA			
10	400	1 ± 2	0.3 ± 0.1
with Na replaced by K			

(From Balshin and Phillips, 1971).

tissue is active and dependent on the presence of sodium in the bathing solutions. The uptake mechanism thereby resembles that in mammalian intestine.

3. Nerves

a Central nervous system

Evans (1975) has measured the uptake of one amino acid, glutamate, into the central nervous system (CNS) of *Periplaneta americana*. L-glutamate was chosen for the study because of its important metabolic roles. It is the precursor of γ-aminobutyric acid, which is an important neuro-transmitter in insects. The concentration gradient for glutamate between the CNS and the blood in this insect is about thirty to one. Evans measured the uptake of glutamate into the abdominal nerve cord at varying glutamate concentrations in the bathing medium and found that, as in the peripheral nerves of crab, the uptake could be divided into two components: a sodium dependent and a sodium independent uptake (Evans, 1973). The sodium insenstive influx is linearly related to the glutamate concentration in the bathing medium between 0.005 and 1 mM. The sodium sensitive glutamate uptake was obtained by subtracting the sodium insensitive uptake from the total uptake of glutamate from normal saline containing 150 mM sodium. The sodium sensitive glutamate uptake showed saturation kinetics with a K_m value of 0.33 mM. Metabolic poisons slightly inhibited the sodium sensitive glutamate uptake. The effect of other amino acids on the glutamate uptake was not very pronounced; the best inhibitor being 2-amino-adipic acid which inhibited the sodium sensitive glutamate uptake about 40%. The sodium insensitive glutamate uptake was hardly inhibited at all by any of the tested amino acids.

b   Peripheral nerve

Glutamate uptake by isolated cockroach nerve-muscle preparation has been investigated by Faeder and Salpeter, (1970). The uptake is enhanced by nerve stimulation. Leucine uptake is significantly lower than glutamate and is not affected by nerve stimulation. However, both glutamate and leucine incorporation into protein is increased by nerve stimulation. Radioautographic experiments showed that glutamate uptake was highest at the neuromuscular junctions, but it was also high in the nerve sheath.

IV. MOLLUSCA

The molluscs have a well developed cardiovascular system. The alimentary canal has a distention in the midgut into which a big liver, the hepatopancreas, secretes.

Amino acid transport has been investigated in animals from two classes of this phylum: Lamellibranchia and Cephalopoda. They all live in water, most of them in the sea and are osmoconformers over a wide range of salinity.

A. Lamellibranchia

Amino acid uptake by the isolated gills of a mussel, *Mytilus californianus* Conrad has been studied by Wright *et al.*, (1975). They measured the uptake of the model amino acid, cyclo-leucine, and found a linear uptake with time, which reached maximum at a steady-state concentration in the tissue of about 2000 times that of the medium (0.05 mM). This uptake was inhibited only by other neutral amino acids and not affected by basic or acidic amino acids in the medium.

The effect of metabolic poisons has not been measured. The authors conclude that this amino acid uptake through the gills of the animal is used to provide food for the animal because the dimension of the uptake is of the same order of magnitude as the energy supply for the whole animal, and because it is shown by Péquignat (1973) that the gills contain chymotrypsin and other digestive enzymes. May be this amino acid uptake could also be participating in the volume regulation as in other marine softbodied animals (see also Hoffman, in this book).

B. Cephalopoda

The uptake of glutamate into the giant axon of the squid, *Loligo forbesi*, has been studied by Baker and Potashner (1973). They find that the accumulation of glutamate is similar to that in periferal leg nerves of the spider crab, *Maja squinado*, as the sodium sensitive influx of glutamate needs two sodium ions for each glutamate molecule. The giant axon from the squid differs from the spider crab nerve in that potassium does not inhibit the stimulation of the glutamate uptake by sodium.

The glutamate uptake is blocked by metabolic inhibitors. This is not a secondary inhibition due to changes either in the membrane potential or in sodium, potassium, or calcium concentrations.

V. ECHINODERMATA

The echinoderma have a rather simple digestive canal and blood vascular system with no heart. The internal cavities of the animals are filled with a fluid with an ionic composition close to that of seawater and with numerous free cells.

A. Echinoidea

Bramford and James (1972) have studied amino acid uptake in the isolated gut from a seaurchin, *Echinus esculentus*. The gut tissue was cut into small pieces and incubated for ten minutes in artificial seawater containing a radioactive amino acid. The amount of radioactivity in the tissue was then measured, and for L-alanine was found to be about three times the concentration in the bathing medium. With increasing alanine concentration in the medium the amount in the tissue increased and reached a maximum at about 10 mM alanine in the medium. The authors suggest that an active, carrier-mediated process is involved in the alanine absorption, but they were not able to inhibit the absorption by agents such as dinitrophenol and iodoacetate. However, starfish tissues are known to be insensitive to traditional metabolic inhibitors (Ferguson, 1966) and Bromford and James therefore assume that this might also be the case for sea-urchins. Leucine is also accumulated in the gut tissue and does compete with alanine. The uptake system prefers the L-isomers of the amino acids, but can accumulate D-alanine. Aspartic acid appears not to be accumulated.

The authors suggest that the amino acid uptake by the isolated gut of the seaurchin has the same characteristics as that from the mammalian intestine.

B. Holothurioidea

From the class of Holothurioidea the sea cucumber, *Chiridota rigida*, has been used for investigation of amino acid uptake through the outer surface of the animal.

The holothurian body wall is a complex structure consisting of a superficial cuticle, a single layer of surface epithelial cells, a thick collagenous dermal region (often deliminated into specific zones) a muscular layer and an inner peritoneal lining of the coelomic cavity.

Ahearn and Townsley have found that glycine and L-amino-isobutyric acid, AIB, are accumulated in the integument of the sea cucumber, *Chir. rigida*, when the whole animal is bathed in a medium containing these amino acids. The glycine concentration in the integuments showed saturation in about 60 minutes at about two to three times the concentration in the bathing medium. The glycine uptake into the integument was dependent on the presence of sodium in the bathing solution as there was no uptake at all when all sodium in the artificial seawater was replaced by choline.

Glycine and α-aminoisobutyric acid were taken up from the bathing medium into the tissue, and not through the epithelium into the body cavity. Only very small amounts of radioactive amino acids could be found in the blood, and here they reached steady-state already after five to ten minutes, while the radioactive AIB concentration in the tissues did not reach steady-state before about two hours. Glycine was used by the tissue; some was metabolized to CO_2 and some built into non-diffusable compounds.

Only a minor part of the radioactive amino acids were taken up by the animal's digestive system so the authors conclude that the uptake through the body wall is the main source of energy provision for these animals. The AIB uptake was inhibited by the presence of other amino acids in the bathing medium. Alanine, glycine and proline inhibited the AIB uptake about 50%.

VI. CONCLUSIONS

Amino acid transport has been reviewed here in species representing four invertebrate phyla. Several transport mechanisms are found: passive, passive with an established concentration gradient, amino acid exchange, active transport as co-transport with either one or two sodium ions, active transport dependent upon the potential gradient across the tissue, and transport by means of the enzyme α-glutamyltranspeptidase. The most common mechanism appears to be a co-transport with one sodium ion; it is present in all four phyla and seems to resemble the one known from vertebrates. It should be noted that only a few species have been investigated and that too not very thoroughly. The insects are so far the best investigated of the invertebrate classes and show the greatest variety of amino acid uptake mechanisms.

Active transport mechanisms play an important role throughout the animal kingdom. Apart from the active transports of amino acids mentioned here, there are active transport mechanisms for sodium, potassium, calcium, magnesium, etc (see Sachs, in this volume). There may, however, be only a single active transport principle taking care of it all. Diversity could come from, say, changes in the configuration around the active site or by some kind of filter placed in the canal through the cell membrane, either near the active site or further away, etc.

The configuration around the active site may account for the stereospecificity in the case of amino acid transport; usually only the L-forms are transported.

The larvae of some insects have an active potassium transport mechanism in the midgut epithelium (Harvey and Nedergaard, 1964, Zerahn, in this volume). When calcium is left out of the bathing medium the midgut is able not only to transport potassium but also sodium, lithium and caesium in the same direction as potassium and with the same characteristics. Calcium is in this case responsible for the specificity of the active transport mechanism (Harvey and Zerahn, 1971).

Caesium, however, can also be transported when calcium is present but only when the experimental conditions are very much in favour of caesium, i.e. very little or no potassium present in the bathing solutions.

Calcium also plays another role in the same tissue; it appears to regulate the net uptake of α-aminoisobutyric acid. Without calcium in the bathing solutions the passive amino acid flux from blood to lumen is much higher than when calcium is present. This means that the net uptake is very small as the passive flux then is increased to nearly the same magnitude as the active amino acid uptake. When calcium is present in the bathing solutions the passive flux is small, the result being a high netflux into the blood. Calcium has no effect on the active transport of amino acid (Nedergaard, preliminary experiments).

These examples are all from insects. Many insects have to adapt to drastic changes during their life when both their internal and external milieu change. It would seem unlikely that the animals should change their active transport mechanisms completely to adapt to their new life, it would be much simpler if they just rearranged the "filters" or whatever it is that determines the specificity. So the genetic determination of the transport mechanisms in insects might be different and less rigid from that in vertebrates, as the transport mechanisms are meant to be changeable. Therefore when we alter the conditions for isolated tissue from an insect in an experiment, we might be able to see some of these changes and may get a little closer to understanding the transport mechanism.

ACKNOWLEDGEMENTS

I am indebted to H. H. Ussing, S. O. Andersen, K. Zerahn and I. Gøthgen for reading and discussing the manuscript.

REFERENCES

Ahearn, G. A. (1974). *J. exp. Biol.* **61**, 677—696.

Baker, P. F. and Potashner, S. J. (1971). *Biochim. biophys. Acta* **249**, 616—622.

Baker, P. F. and Potashner, S. J. (1973). *Biochim. biophys. Acta* **318**, 123—139.

Balshin, M. and Phillips, J. E. (1971). *Nature* **233**, 53—55.

Bhatnagar, P. (1962). *Indian J. Ent.* **24**, 66—67.

Bodnaryk, R. P. and Skillings, J. R. (1971). *Insect Biochem.* **1**, 467—479.

Bodnaryk, R. P., Bronskill, J. F. and Fetterly, J. R. (1974). *J. Insect Physiol.* **20**, 167—181.

Bragdon, J. C. and Mittler, T. E. (1963). *Nature* **198**, 209—210.

Bramford, D. R. and James, D. (1972). *Biochem. Physiol.* **42A**, 579—590.

Chaplin, A. E., Huggins, A. K. and Munday, K. A. (1970). *Int. J. Biochem.* **1**, 385—400.

Christensen, H. N. (1972). In "Role of Membranes in Secretory Processes, International Conference on Biological Membranes" (Eds L. Bolis, R. D. Keynes, and W. Wilbrandt), pp. 433—447. North-Holland, Amsterdam.

Duchâteau, G. and Florkin, M. (1958). *Arch. Int. Physiol. et Biochim.* **66**, 573—591.

Duchâteau, G., Florkin, M. and Jeuniaux, C. (1959). *Arch. Int. Physiol. et Biochim.* **67**, 173—181.

Dunham, P. B. and Kropp, D. L. (1973). "Regulation of Solutes and Water in Tetrahymena". (Ed. A. M. Elliot), Dowden Hutchinson and Ross, Inc.

Evans, P. D. (1972). *J. exp. Biol.* **56**, 501—507.

Evans, P. D. (1973). *Biochim, Biophys. Acta* **311**, 302—313.

Evans, P. D. (1975). *J. exp. Biol.* **62**, 55—67.

Faeder, I. R. and Salpeter, M. M. (1970). *J. Cell Biol.* **46**, 300—7.

Ferguson, J. C. (1966). *Comp. Biochem. Physiol.* **19**, 259—266.

Florkin, M. and Jeuniaux, C. (1974). "Hemolymph: Composition. The Physiology of Insecta." (Ed. N. Rockstein), Vol. V. Academic Press.

Florkin, M. and Schoffeniels, E. (1969). *In* "Molecular Approaches to Ecology." pp. 100. Academic Press, London and New York.

Gomme, J., Schrøder, C. and Heister, A. M. (1976). FEBS symposium on the Biochemistry of Membrane transport, Switzerland.

Harvey, W. R. and Nedergaard, S. (1964). *Proc. Nat. Acad. Sci. U.S.A.* **51**, 757—765.

Harvey, W. R. and Nedergaard, S. (1964). *Proc. Nat. Acad. Sci. U.S.A.* **51**, 757—765.

Harvey, W. R. and Zerahn, K. (1971). *J. exp. Biol.* **54**, 269—274.

Harvey, W. R., Haskell, J. A. and Zerahn, K. (1967). *J. exp. Biol.* **46**, 235—248.

Jeuniaux, C. (1971). *In* "Chemical Zoology, Arthropoda." Vol. VI, pp. 64—118. Academic Press, London and New York.

Murdock, L. L. and Koidl, B. (1972a). *J. exp. Biol.* **56**, 781—794.

Murdock, L. L. and Koidl, B. (1972b). *J. exp. Biol.* **56**, 795—808.

Nedergaard, S. (1972). *J. exp. Biol.* **56**, 167—172.

Nedergaard, S. (1973). "Transport Mechanisms in Epithelia". (Eds H. H. Ussing, and N. A. Thorn), pp. 372—391. Munksgaard, Academic Press.

Orlowski, M. and Meister, A. (1970). *Proc. Nat. Acad. Sci. U.S.A.* **67**, 1248—1255.

Péquignat, E. (1973). *Marine Biol.* **19**, 227—44.

Potts, W. T. W. and Parry G. (1964). "Osmotic and Ionic Regulation in animals." Pergamon Press.

Ramsay, J. A. (1954). *J. exp. Biol.* **31**, 104—13.

Ramsay, J. A. (1958). *J. exp. Biol.* **35**, 871—891.

Schultz, S. G. and Curran, P. F. (1970). *Physiol. Revs.* **50**, 637—711.

Schultz, S. G., Frizzell, R. A. and Nellans, H. N. (1974). *Ann. Rev. Physiol.* **36**, 51—91.

Shotwell, O. L., Bennett, G. A., Hall, H. H., Van Etten, C. H. and Jackson, R. W. (1963). *J. Insect Physiol.* **9**, 35—42.

Shyamala, M. B. and Bhat, J. V. (1966). *J. Insect. Physiol.* **12**, 129—135.

Stephens, G. C. (1972). *In* "Symposium on Nitrogen Metabolism and the Environment." (Eds J. W. Cambell and C. Goldstein), pp. 155—184. New York, Academic Press.

Stevens, T. M. (1961). *Compt. Biochem. Physiol.* **3**, 304—9.

Treherne, J. E. (1959). *J. exp. Biol.* **36**, 533—545.

Ussing, H. H. (1946). *Acta Physiol. Scand.* **11**, 61—84.

Voute, C. L., Møllgård, K. and Ussing, H. H. (1975). *J. Membrane Biol.* **21**, 273—289

Wall, B. J. and Oschman, J. L. (1970). *Am. J. Physiol.* **218**, 1208—1215.

Wigglesworth, V. B. (1965). "The Principles of Insect Physiology." Methuen and Co. London.

Wright, S. H., Johnson, T. L. and Crow, J. H. (1975). *J. exp. Biol.* **62**, 313—326.

Wood, J. L. (1972). Ph.D. Thesis. Cambridge University, Cambridge, England.

11. The Hormonal Control of Excretion

J. D. Gee

Tsetse Research Laboratory, University of Bristol School of Veterinary Science, Bristol, England

I. INTRODUCTION

The hormonal control of excretion includes a multitude of subjects far too numerous to be covered in a short article such as this. So, to be in keeping with the title of this volume I have restricted my account to some aspects of the control of the excretion of water and ions. I have further restricted my theme to the hormonal control of excretion in insects, for it is in that subject that my own interest lies and it is one in which recent investigation owes much to the pioneering work of Professor Ramsay.

In following the course of the hormones concerned with the control of excretion in insects from their source through to their effect at their target tissues and their subsequent destruction, I have attempted to compare the hormonal control of excretion in insects with the control of excretion by vasopressin in mammals. This type of comparison is not entirely new, for in 1944 E. and B. Scharrer pointed out the similarities between the intercerebralis-cardiacum-allatum system of the insects (which we shall see secretes hormones controlling excretion in several insect species) and the hypothalamo-neurohypophysial system of the vertebrates (which in mammals secretes vasopressin). A generalized comparison of the hormonal control of excretion by these two systems is shown in Fig. 1.

It must be stressed at the outset that the antidiuretic effect of vasopressin is only one aspect of the hormonal control of excretion in mammals. However, it is hoped that the comparison of the control of excretion by hormones in insects with the control of mammalian excretion by vasopressin may be of interest to both insect and mammalian physiologists.

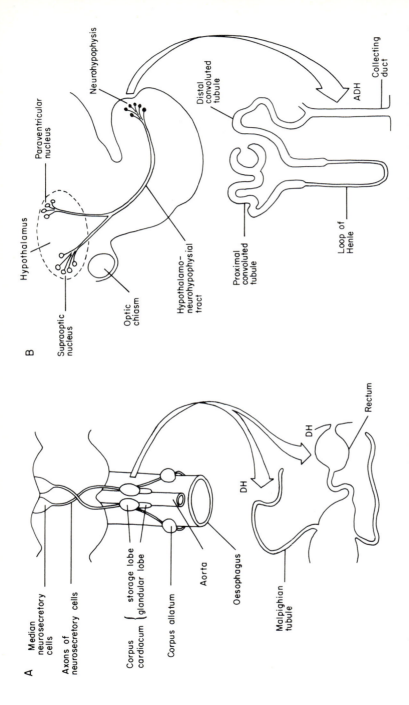

Fig. 1. A diagramatic comparison of the hormonal control of excretion in insects and mammals, showing control of Malpighian tubule and rectal function by diuretic hormone (DH) in *Schistocerca* (A) and control of kidney tubule permeability by vasopressin (ADH) in the dog (B). (Further explanation see text).

II. SOURCE

In the insects investigated to date, the hormones which control excretion by the Malpighian tubules and rectum are produced by neurosecretory cells. These cells have the properties of nerves but are also able to synthesize, transport and release hormones (Maddrell, 1974). The neurohormones are synthesized in the cell bodies, where they are packaged into neurosecretory granules (NSG) which are transported down the axons to the sites of storage and release — the neurohaemal areas. At present there is no unifying concept of the source of the hormones controlling excretion, for, as we shall see, neurosecretory cells containing factors which alter the rate of excretion have been identified almost throughout the insect nervous system — from the brain to the terminal abdominal ganglion.

The diuretic hormones (DHs) of *Anisotarsus* (Núñez, 1956), *Dysdercus* (Berridge, 1966), *Schistocerca* (Mordue, 1969) and *Carausius* (Pilcher, 1970) are synthesized by neurosecretory cells in the brain and released into the haemolymph from the corpus cardiacum — the neurohaemal organ of these cells (Fig. 1). Material with diuretic activity can also be detected in the brains of *Rhodnius* (Maddrell, 1963) and *Glossina* (Gee, 1975), though in these insects there is no evidence that this material is ever normally released into circulation. From cytological observations Nayar suggested that the brains of *Iphita* (1960) and *Periplaneta* (1962) produced an antidiuretic hormone (ADH). This was confirmed in *Periplaneta* by Wall and Ralph (1964), who proposed that this hormone was released from the corpus allatum. In some instances the evidence of the nature of the hormones is contradictory. In the brain of *Schistocerca*, Mordue (1969) identified a DH effective on the Malpighian tubules and rectum, whereas in the brain of *Locusta* Cazal and Girardie (1968) detected both a DH effective on the Malpighian tubules and an ADH which stimulated rectal reabsorption. There is also confusion about the nature of the hormones released from the corpus cardiacum of *Carausius*. Unger (1965) proposed that Neurohormone D from this organ was a DH, whereas Vietinghoff (1967) suggested that it was an ADH and attributed a diuretic effect to Neurohormone C. As we shall see (p. 274), care must be taken in identifying not only the source of the hormone but also its function.

Besides acting as the neurohaemal organ for neurosecretory cells in the brain, the corpus cardiacum synthesizes its own intrinsic hormones. In *Schistocerca*, for example, Mordue (1970) demonstrated that the glandular lobe produces an ADH which controls the rectal glands, and in extracts of whole corpora cardiaca of *Locusta*, Cazal and Girardie (1968) found an ADH effective on the Malpighian tubules, which they had not detected in the brain. Mordue (1970), however, found no evidence of this hormone, and his experiments demonstrate the importance of examining the storage and glandular lobes of the corpus cardiacum separately, for they may, as in the locust, contain different hormones having opposite effects on the same organ. To complete the list of insects in which antidiuretic activity has been attributed to the corpus cardiacum we must add *Apis* (Altmann, 1956) and *Gryllus, Periplaneta* and *Clitumnus* (in which de Bessé and Cazal, (1968) have also demonstrated the presence of ADH in the *organs périsympathiques*).

Excretion in *Apis* is further controlled by a DH which originates from the corpus allatum (Altmann, 1956). In *Periplaneta* also the corpus allatum contains a DH (Mills, 1967), besides the potent ADH synthesized by the brain (Wall and Ralph, 1964). Extracts of the corpora allata of *Schistocerca*, on the other hand, contain neither DH nor ADH (Mordue and Goldsworthy, 1969).

The intercerebralis-cardiacum-allatum system is not the only site of hormone synthesis and release in insects. In *Rhodnius* neurosecretory cells in the fused mesothoracic ganglionic mass (MTGM) synthesize a DH (Maddrell, 1963) which is subsequently released from neurohaemal areas on the abdominal nerves (Maddrell, 1966). DH is also present in the thoracic ganglion of *Glossina* (Gee, 1975) and is released from neurosecretory axon endings in the abdomen (Maddrell and Gee, 1974). In *Corethra* (Gersch, 1967) the thoracic ganglion contains an ADH, but a DH is present in the first three abdominal ganglia. There is evidence that ADHs may be present in the ventral ganglia of *Schistocerca* (Delphin, 1963, 1965) and *Periplaneta* (Wall, 1965, 1967), and in the latter the terminal abdominal ganglion also contains a DH (Mills, 1967).

It is clear from the above review that we have abundant if sometimes contradictory evidence that in insects excretion is controlled by neurosecretory hormones. As a result of the work of Maddrell on *Rhodnius* and the work of Mordue on *Schistocerca*, in two insects we have excellent evidence of the source of the hormones concerned. However, in other insects, where hormonal activity has been detected almost throughout the nervous system, some of the observed effects of tissue extracts are probably pharmacological in nature, and further investigations are necessary to establish the exact sources of the hormones controlling excretion.

Our knowledge of the synthesis, release and effect of vasopressin is much more precise. This is not surprising since investigations of the hormonal control of excretion in insects date back only to 1956, whereas the antidiuretic effect of posterior pituitary extracts was demonstrated as long ago as 1913. Reviews of the history of research into the hypothalamo-neurohypophysial system may be found elsewhere (see, for example, Heller, 1974). However, it would be a serious omission if mention were not made of the work of E. and B. Scharrer who, in investigating this system developed the concept of neurosecretion (see Scharrer and Scharrer, 1954). They proposed that neurosecretory material originates in the supraoptic and paraventricular nuclei of the hypothalamus and passes by way of the hypothalamo-neurohypophysial tract to the neurohypophysis, in which it is stored and from which it is released (Fig. 1).

The ability of the hypothalamus to synthesize vasopressin was demonstrated by Sachs, who, together with his co-workers (for references see Sachs, 1969), provided much of our present knowledge of the synthesis of vasopressin. He proposed that vasopressin is formed at ribosomes in the cell bodies of the hypothalamus as a part of a larger, biologically-inactive precursor molecule. The active hormone is cleaved from the precursor, probably at the Golgi apparatus. Here vasopressin and precursor molecules are packaged into NSG within which the process of liberation of active hormone continues as the NSG is transported from the cell body to the neurohypophysis.

III. AXONAL TRANSPORT

In comparison with the number of attempts to find the source of the hormones which control excretion in insects, few investigations have followed their progress once they leave the site of synthesis. Goldsworthy *et al.* (1972) showed that sectioning the nerves connecting the brain to the corpus cardiacum in the locust prevented the release of DH. However, the cut end of the nerve nearest the brain regenerated its neurohaemal area and this *de novo* corpus cardiacum could then secrete normal levels of DH. This technique demonstrates not only that DH is transported along the nervi corpora cardiaci but also that the brain and not the corpus cardiacum is the site of synthesis of DH.

The rate of axonal transport of insect DH or ADH has not been measured specifically. However, by monitoring the time taken for

K

^{35}S-cysteine to pass from the median neurosecretory cells to the corpus cardiacum, Highnam and Mordue (1970) determined that axonal transport of neurosecretory material over this distance (approximately 1.5 mm) took about one hour — indicating a rate of transport of about 1.5 mm h^{-1}.

Hormones produced at sites other than the brain may also need to be transported to their neurohaemal areas. The DH of *Rhodnius*, for example, is transported from cell bodies in the MTGM to the neurosecretory axon endings on the abdominal nerves (Maddrell, 1966). The DH of *Glossina* is transported an even greater distance from the thoracic ganglion to the sites of release, which are found in the abdomen close to the Malpighian tubules (Maddrell and Gee, 1974). Axonal transport in these insects enables the neurohaemal areas to be close to the target organs.

By following the incorporation of ^{3}H-tyrosine, Pickering *et al.*, (1971) demonstrated that in the rat radioactive vasopressin is transported within NSG from the hypothalamus to the neuro-hypophysis at a rate of approximately one to three mm h^{-1}. Vasopressin and insect neurosecretory materials therefore appear to be transported at similar rates. These are too rapid to be accounted for by axonal flow and must require the active transport of the NSG (see review by Heslop, 1975).

As we saw when considering the site of vasopressin synthesis, maturation of the precursor molecule into vasopressin and its carrier protein (neurophysin) may continue within the NSG. The time course of the synthesis and axonal transport of the two molecules is the same and they are probably formed from a common precursor molecule (Pickering *et al.*, 1975). Similar events may occur during the transport of insect DHs and ADHs. In *Schistocerca* there is evidence for the presence of carrier proteins (Mordue and Goldsworthy, 1969) and in *Periplaneta* (Wall, 1967) and *Schistocerca* (Mordue, 1969) hormones extracted from their neurohaemal areas have a greater potency than when extracted from their sites of synthesis. In *Rhodnius* (Maddrell, 1963) and *Dysdercus* (Berridge, 1966), on the other hand, large amounts of DH may be extracted from the cell bodies of the neurosecretory cells, indicating that in these insects maturation of the hormone either takes place within the cell body or does not occur at all. More recently, however, Aston and White (1974) have extracted from the MTGM of *Rhodnius* high mol. wt material possessing diuretic activity. This material may represent an immature form of the DH which, when released, has a much lower mol. wt.

Clearly there is some evidence to suggest that similar events may occur during axonal transport of insect and mammalian hormones. However, much more information is required about the events occurring within the NSG during the axonal transport of the hormones which control excretion in insects before more precise comparisons may be drawn.

IV. STORAGE AND RELEASE

There is evidence that the DHs of *Schistocerca* (Mordue and Goldsworthy, 1969), *Carausius* (Pilcher, 1970) and *Rhodnius* (Maddrell and Gee, 1974) are stored at their respective neurohaemal areas before release. This storage of neurosecretory material at the release site may be of considerable value to the insect, for it enables the rapid release of hormone in response to a stimulus without requiring greatly increased hormone synthesis and axonal transport.

In mammals the neurohypophysis may contain several hundred times more vasopressin than is found in the hypothalamus, and this may be sufficient to maintain water balance for several weeks, even under conditions requiring continuous antidiuresis. Within the neurohypophysis the most recently arrived hormone appears to be most readily released (Wong and Pickering, 1976). Hormone not immediately released is stored — that which has been stored the longest being the least available for release. This mechanism may be of interest when considering the release of insect hormones. In haematophagous insects, for example, diuresis requires the rapid release of large amounts of hormone, probably necessitating the recruitment of hormone which was synthesized and stored during the period since the previous meal.

Release of neurohormones occurs in response to the arrival of nerve impulses at the neurosecretory axon endings. Douglas (1963) demonstrated that elevating the potassium concentration of the medium bathing an isolated neurohypophysis would mimic the depolarizing effect of these impulses and would induce the release of vasopressin if Ca was present in the incubation medium. Further experiments determined the concentrations of K and Ca necessary to stimulate hormone release (Douglas and Poisner, 1964). The release of the DHs of *Rhodnius* and *Glossina* can also be induced by K-rich solutions, and the concentrations of K and Ca necessary for their release are very similar to those established for the release of vasopressin (Maddrell and Gee, 1974). At the commencement of depolarization by K-rich solutions very high rates of release of both vasopressin and these insect DHs occur, but with prolonged depolarization the rates of release rapidly decline (Daniel and Lederis, 1967; Maddrell and Gee, 1974). The ultrastructural events accompanying hormone release may also be common to both mammals and insects, in that, in both, release of neurohormones takes place by exocytosis (Maddrell, 1974). There is considerable evidence that this process accounts for the release of vasopressin from the neurohypophysis (Douglas, 1974). In the insects, however, only in *Rhodnius* has a connection been observed between the release of DH and the presence of the omega-shaped profiles characteristic of exocytosis (Maddrell, 1966).

In insects and mammals similar mechanisms appear to operate at the cellular level during the release of the neurohormones concerned with the control of excretion. Impulses arriving down the neurosecretory axons cause depolarization of the axonal endings, this promotes the influx of Ca into the cells and the hormones are released by exocytosis.

V. STIMULI PROMOTING HORMONE RELEASE

Present evidence suggests that insect diuretic and antidiuretic hormones are released in response to stimuli associated both with feeding and with the osmotic pressure of the haemolymph. In *Rhodnius* (Maddrell, 1963) and *Glossina* (Gee, 1975) DH is released in response to the ingestion of blood meal. In *Schistocerca* feeding results in an increased rate of excretion by the Malpighian tubules, a response which can be mimicked by the injection of extracts of corpora cardiaca (Mordue, 1969). There is also excellent evidence that in *Dysdercus* (Berridge, 1966) and *Carausius* (Pilcher, 1970) the titre of DH in the haemolymph rises in response to feeding. In *Periplaneta* there is evidence of the control of hormone release both by feeding stimuli and by stimuli from osmoreceptors. Mills (1967) demonstrated the necessity of feeding stimuli, for DH was released only when insects were allowed to drink their fill. Release of the DH may also be controlled by inhibitory impulses from osmoreceptors in the haemolymph (Penzlin, 1971; Penzlin and Stölzner, 1971) and Wall and Ralph (1964) suggested that the ADH of *Periplaneta* was released in response to a hyperosmotic haemolymph — presupposing the presence of haemolymph osmoreceptors.

Our knowledge of the nature of the receptors which monitor these stimuli is very limited. Only in *Rhodnius*, where stretch receptors probably in the tergosternal muscles of the abdomen are stimulated as the midgut swells with blood during feeding (Maddrell, 1964a), has a specific receptor been identified. In mammals our knowledge of both the stimuli and their receptors is much more extensive (see recent reviews by Moses and Miller, 1974; Share, 1974) and it is clear that the release of vasopressin results from changes in the volume and osmolality of the blood.

Verney demonstrated in 1947 that increases in the osmolality of the extracellular fluid are detected by osmoreceptors in the hypothalamus which stimulate the release of vasopressin from the neurohypophysis. Vasopressin reduces urine flow and increases urine osmolality, thereby conserving water. This control mechanism is both sensitive and efficient, for increasing the osmolality of the extracellular fluid by only 2% causes a 90% reduction in urine flow (Verney, 1947).

Control of the release of vasopressin by changes in blood volume is also clearly of advantage to the mammal, for decreases in blood volume due to both dehydration and haemorrhage can be rectified by reducing urine output. Small reductions in blood volume are detected by stretch receptors in the left atrium and greater losses of blood, resulting in a fall in arterial blood pressure are detected by the arterial baroreceptors in the aortic arch and carotid sinus. Increased activity of these receptors results in a reduction in vasopressin release and an increase in urine flow.

It is of interest to the insect physiologist that in mammals the feedback mechanisms which control vasopressin release use receptors which monitor the osmolality and volume of the blood. These parameters may be of greater importance in the control of excretion in insects than we realise at present, for increases in the osmotic pressure of the haemolymph and decreases in haemolymph volume must occur during periods of starvation and dehydration when the release of ADH is required. As Mordue *et al.*, (1970) point out, at present we have no concrete information about the stimuli which promote the release of ADH in insects.

VI. RESPONSE OF THE TARGET TISSUES

The investigation of the hormonal control of the insect excretory system owes much to the work of J. A. Ramsay, who, in his studies of the isolated Malpighian tubules of *Carausius*, demonstrated that the functioning of the insect excretory system could be monitored *in vitro* (Ramsay, 1954). Such *in vitro* preparations form excellent bioassays for the hormones which control excretion, for they enable the direct effects of the hormones on the excretory system to be distinguished from indirect effects due to the action of hormones on other aspects of the metabolism of the insect.

The isolated Malpighian tubules of *Rhodnius* (Maddrell, 1966) and *Glossina* (Gee, 1975) respond to the presence of their respective DHs by a thousand-fold increase in their rates of fluid secretion — a response which clearly reflects the events occurring within these insects during rapid diuresis. In *Dysdercus* (Berridge, 1966) and *Carausius* (Pilcher, 1970) there is also clear evidence that the DHs stimulate fluid secretion by the Malpighian tubules (also see Chapter 21).

The response of the insect excretory system to DHs and ADHs has also been monitored by following the excretion of dyes such as amaranth (Mordue, 1969), indigo carmine (Wall and Ralph, 1964) and neutral red (Unger, 1965). The relationship between the rate of transport of fluid and of dye by insect Malpighian tubules was investigated by Maddrell *et al.*, (1974), who concluded that only in insects with tubules which transport dyes very rapidly and fluid relatively slowly and which have highly permeable walls, could the rate of dye transport form a suitable indicator of the rate of fluid transport. Where this is not the case, for example in *Rhodnius* and *Carausius*, the rate of dye secretion is independent of the rate of fluid secretion. Unfortunately, in *Carausius* both fluid transport and dye secretion have been used as a measure of the rate of excretion by the Malpighian tubules. Unger (1965) observed that Neurohormone D caused an increase in the rate of dye secretion and therefore proposed that Neurohormone D was a DH (see p.261). Vietinghoff (1967), on the other hand, monitored the rate of fluid secretion and found that Neurohormone D had an antidiuretic effect. Though at present there is no direct evidence that dye secretion is hormonally controlled, it is possible that in this insect a single hormone stimulates dye secretion while at the same time inhibiting the movement of fluid.

Care must be taken in the evaluation of experiments in which the rate of dye secretion is used as a guide to the response of the excretory system to insect DHs or ADHs. Only when the functioning of the Malpighian tubules and rectum of the insect under investigation is understood is it possible to relate changes in the rate of excretion of dye to changes in the rate of excretion of fluid (see Maddrell *et al.*, 1974). This subject is clearly one which requires further investigation for at present a hormone which stimulates either fluid or dye secretion may be termed diuretic — the danger of such a loose definition is illustrated above.

At present we have little information about the mode of action of insect ADH. Wall (1967) postulated that in *Periplaneta* a single hormone could exert an antidiuretic effect on both Malpighian tubules and rectum by reducing the passive permeability of the cells to water. It would therefore restrict water movement generated by ion transport across the Malpighian tubules and would reduce the amount of water leaking back across the rectal glands after it had been reabsorbed from the faeces.

The mammalian ADH, vasopressin, acts on the kidney to increase the permeability of the collecting ducts to water (see Wirz, 1957; Berliner and Bennet, 1967; Sawyer, 1974). This enables the fluid in this section of the tubule to equilibrate with the renal interstitium, across which an osmotic gradient is established by the action of the loops of Henle (Wirz, 1957). The highest osmotic pressure is found in the papillary interstitium, ensuring that the fluid leaving the kidney under conditions of antidiuresis is highly concentrated. The collecting ducts can be isolated and the addition of vasopressin to the medium bathing the basal (serosal) side of such isolated preparations increases their permeability to water; perfusion of the lumenal side, however, is without effect (Grantham and Burg, 1966; Handler and Orloff, 1973).

The mechanism by which the stimulus of the hormone at the receptor site causes changes in the permeability of the membranes at the opposite side of both Malpighian and kidney tubules is, unfortunately, outside the scope of this article. However, it is interesting to note that *in vitro* cyclic AMP can mimic the effects of the hormones controlling fluid movement across both insect Malpighian tubules (Maddrell *et al.*, 1971; Gee, 1976) and mammalian collecting ducts (Grantham and Burg, 1966). The response to the hormones controlling excretion may therefore be mediated by similar intracellular messengers in both insects and mammals. (also see Chapter 9.)

VII. FATES OF THE HORMONES AND THEORIES OF HORMONAL CONTROL OF EXCRETION IN INSECTS

It is important not only that the hormones controlling excretion should be released into circulation at the correct time in response to the correct stimulus but also that their effect on the target tissue should continue only for as long as it is required. This is amply demonstrated in the tsetse fly, *Glossina*, where the high rate of excretion necessary to rid the fly of the unwanted water in the blood meal would be sufficient to desiccate the fly within but a few minutes if it were to continue after sufficient water was excreted (Gee, 1975).

Destruction of DH by the Malpighian tubules has been demonstrated in *Rhodnius* (Maddrell, 1964b) *Dysdercus* (Berridge, 1966), *Carausius* (Pilcher, 1970) and *Glossina* (Gee, 1975) and has been postulated in *Schistocerca* (Mordue, 1972). In *Glossina* there also appears to be an enzyme in the haemolymph which destroys the DH and which is possibly released together with the DH (Maddrell and Gee, 1974; Gee, 1975). The presence of a similar enzyme has been postulated in *Rhodnius* (Aston and White, 1974) and *Periplaneta* (Mills, 1967). There is no evidence for the excretion of DH in the urine, though this has only been investigated in *Rhodnius* (Maddrell, 1964b) and *Carausius* (Pilcher, 1970).

The presence of an efficient mechanism for the destruction of DH has resulted in several authors proposing a similar theory of the control of excretion by the Malpighian tubules (see Maddrell, 1964b; Berridge, 1966; Pilcher, 1970; Mordue, 1972; Gee, 1975). In response to the correct stimulus a burst of DH release occurs, followed by a rate of release equal to the rate of destruction of the hormone. When the stimulus is removed, DH release ceases immediately, its titre in the haemolymph is rapidly reduced by the degradative mechanism and the rate of excretion declines equally rapidly.

In *Schistocerca* this control theory has been extended to cover the rectum as well (Mordue *et al.*, 1970). During water loading DH from the storage lobes of the corpora cardiaca both increases secretion by the Malpighian tubules and reduces rectal water reabsorption. When water conservation is required, rectal reabsorption is increased by an ADH secreted by the glandular lobes of the corpora cardiaca. Though we have little information about them at present, such ADHs may prove important in other insects. Neurosecretory axon endings have been found close to the basal surface of the insect rectum (see Gupta and Berridge, 1966; Oschman and Wall, 1969), in haematophagous insects we know little about the control of excretion during the period of days or weeks between meals, when the function of the excretory system must be the conservation of water.

The fate of the neurohypophysial hormones — vasopressin and

oxytocin — has recently been reviewed by Lauson (1974). In contrast to the situation in insects, and possibly as a consequence of the greater permeability of the glomerulus than the Malpighian tubule, vasopressin is excreted in the urine of mammals (Heller and Urban, 1935). However, as excretion cannot account for the fate of all the vasopressin released, to maintain a steady-state concentration in the blood during antidiuresis the remainder must be destroyed elsewhere in the animal. Heller and Urban (1935) demonstrated that the blood destroyed antidiuretic activity only very slowly, but *in vitro* tissue suspensions, especially of the liver and kidneys, would cause its rapid destruction. Inactivation by the liver has since been demonstrated in both perfused whole organs and in cell-free homogenates, and the site of hormone inactivation and the biochemical pathway of its degradation are known (see Lauson, 1974). In the kidney it is apparent that the tubules and not the glomeruli are responsible for the destruction of the hormone. In both the mammals and the insects, therefore, the receptors may possibly participate in the destruction of the hormone.

VIII. NATURE OF THE HORMONES

Much of our knowledge of the control of mammalian excretion by vasopressin stems from the characterization of its structure and its subsequent synthesis by du Vigneud and his colleagues (for references see Acher, 1974). The structure of vasopressin, which is the same in all the mammalian species so far investigated, with the exception of the pig, is shown below:

$$
\begin{array}{ccccccccc}
1 & 2 & 3 & 4 & 5 & 6 & 7 & 8 & 9 \\
\text{Cys} & \text{Tyr} & \text{Phe} & \text{Gln} & \text{Asn} & \text{Cys} & \text{Pro} & \text{Arg} & \text{Gly} - \text{NH}_2
\end{array}
$$

Cys—Tyr—Phe—Gln—Asn—Cys—Pro—Arg—Gly—NH$_2$

A. Arginine Vasopressin

Arginine vasopressin consists of nine amino acids with a disulphide bridge between the cysteine residues at positions one and six. It has a mol. wt of 1,084. (In vasopressin from the pig, lysine replaces arginine at position 8 and in vasotocin — the ADH of birds, reptiles and fish — isoleucine replaces phenylalanine at position 3).

Knowledge of the structure of vasopressin has enabled the synthesis of radioactively labelled hormone and the use of radioimmunoassay techniques. Such advances have assisted in the determination of the cellular origin of vasopressin and the events occurring during the axonal transport and release of the hormone. The ability to synthesize analogues of vasopressin has enabled us to study the way in which changes in the chemical structure of the hormone affect its biological activity. This has provided much information about the structure of the receptor site on the target cell and the hormone-receptor interaction (Pickering, 1970).

We have much less information about the properties and structures of the hormones controlling excretion in insects. The diuretic and antidiuretic hormones of *Schistocerca* are heat stable and soluble in methanol (Mordue and Goldsworthy, 1969) as is the DH of *Glossina* (Gee, 1975). The DH of *Dysdercus* is similarly heat stable (Berridge, 1966); that of *Rhodnius*, however, is stable only at low temperatures and is only slightly soluble in methanol (Aston and White, 1974).

It is proposed that the DH of *Rhodnius* (Aston and White, 1974) and the diuretic and antidiuretic hormones of *Schistocerca* (Mordue and Goldsworthy, 1969) and *Periplaneta* (Mills, 1967; Goldbard *et al.*, 1969) are peptides or polypeptides. *Glossina* DH also contains a peptide but this appears to be part of a more complex structure containing glucose and sialic acid residues (Gee, 1975). Aston and White (1974) have estimated the mol. wt of *Rhodnius* DH to be less than 2,000, though it may occur in high mol. wt complexes during storage. The occurrence of such complexes may account for the apparently high mol. wt (30,000) of the DH of *Periplaneta* determined by Goldbard *et al.* (1969). These authors estimated the mol. wt of *Periplaneta* ADH to be in the region of 8,000.

Recently, indirect evidence has suggested that the hormones controlling excretion in insects may be more similar in size and structure to vasopressin. Stone *et al.* (1976) have succeeded in purifying the adipokinetic hormone of *Schistocerca*, which appears to be a peptide with 9 amino acid residues and a mol. wt of about 1,100. As Mordue and Goldsworthy (1969) have demonstrated that the DH and ADH of *Schistocerca* have similar chromatographic properties to the adipokinetic hormone, these hormones may also be similar in composition and mol. wt.

The purification and characterization of the hormones which control excretion in insects is of paramount importance in any future investigation of this subject. Pure preparations would enable the diuretic and antidiuretic hormones to be distinguished from pharmacological agents which also affect the excretory organs, and the ability to synthesize the hormones would enable us to apply to the investigation of the hormonal control of excretion in insects many of the techniques developed in the study of the hypothalamo-neurohypophysial system in mammals.

IX. CONCLUSIONS

What is to be gained from this sort of comparative approach to the hormonal control of excretion? Clearly there is a considerable disparity between our depth of understanding of these two systems, yet what we do know of the hormonal control of excretion in insects demonstrates that, especially at the cellular level, similar events may be occurring in both insects and mammals. For the insect physiologist many aspects of this subject are ripe for investigation, and it is hoped that the comparative approach adopted here has indicated points at which future investigations may most profitably commence.

ACKNOWLEDGEMENTS

I would like to express my gratitude to Dr S. H. P. Maddrell and Dr B. T. Pickering for their helpful advice and constructive criticism of the manuscript and would like to thank the Overseas Development Administration of the Foreign and Commonwealth Office for financial support.

REFERENCES

Acher, R. (1974). *In* "Handbook of Physiology. Endocrinology." (Eds E. Knobil and W. H. Sawyer), Vol. 4, Part 1, pp. 119–130. Am. Physiol. Soc. Washington, D.C.

Altmann, G. (1956). *Insectes soc.* 3, 33–40.

Aston, R. J. and White, A. F. (1974). *J. Insect Physiol.* 20, 1673–1682.

Berliner, R. W. and Bennet, C. M. (1967). *Am. J. Med.* 42, 777–789.

Berridge, M. J. (1966). *J. exp. Biol.* 44, 553–566.

de Bessé, N. and Cazal, M. (1968). *C. r. hebd. Séanc. Acad. Sci., Paris* 266, 615–618.

Cazal, M. and Girardie, A. (1968). *J. Insect Physiol.* 14, 655–668.

Daniel, A. R. and Lederis, K. (1967). *J. Physiol. Lond.* 190, 171–187.

Delphin, F. (1963). *Nature Lond.* **200**, 913–914.

Delphin, F. (1965). *Trans. R. ent. Soc. Lond.* **117**, 167–214.

Douglas, W. W. (1963). *Nature Lond.* **197**, 81–82.

Douglas, W. W. (1974). *In* "Handbook of Physiology. Endocrinology." (Eds E. Knobil and W. H. Sawyer), Vol. 4, Part 1, pp. 191–224. Am. Physiol. Soc. Washington, D.C.

Douglas, W. W. and Poisner, A. M. (1964). *J. Physiol. Lond.* **172**, 1–18.

Gee, J. D. (1975). *J. exp. Biol.* **63**, 391–402.

Gee, J. D. (1976). *J. exp. Biol.* **64**, 357–368.

Gersch, M. (1967). *Gen. comp. Endocr.* **9**, 453.

Goldbard, G. A., Sauer, J. R. and Mills, R. R. (1970). *Comp. gen. Pharmac.* **1**, 82–86.

Goldsworthy, G. J., Johnson, R. A. and Mordue, W. (1972). *J. comp. Physiol.* **79**, 85–96.

Grantham, J. J. and Burg, M. B. (1966). *Am. J. Physiol.* **211**, 255–259.

Gupta, B. L. and Berridge, M. J. (1966). *J. Morph.* **120**, 23–81.

Handler, J. S. and Orloff, J. (1973). *In* "Handbook of Renal Physiology." pp. 791–814. Am. Physiol. Soc. Washington, D.C.

Heller, H. (1974). *In* "Handbook of Physiology. Endocrinology." (Eds E. Knobil and W. H. Sawyer), Vol. 4, Part 1, pp. 103–118. Am. Physiol. Soc. Washington, D.C.

Heller, H. and Urban, F. F. (1935). *J. Physiol. Lond.* **85**, 502–518.

Heslop, J. P. (1975). *Adv. comp. Physiol. Biochem.* **6**, 75–163.

Highnam, K. C. and Mordue, A. J. (1970). *Gen. comp. Endocr.* **15**, 31–38.

Lauson, H. D. (1974). *In* "Handbook of Physiology. Endocrinology." (Eds E. Knobil and W. H. Sawyer), Vol. 4, Part 1, pp. 287–394. Am. Physiol. Soc., Washington, D.C.

Maddrell, S. H. P. (1963). *J. exp. Biol.* **40**, 247–256.

Maddrell, S. H. P. (1964a). *J. exp. Biol.* **41**, 459–472.

Maddrell, S. H. P. (1964b). *J. exp. Biol.* **41**, 163–176.

Maddrell, S. H. P. (1966). *J. exp. Biol.* **45**, 499–508.

Maddrell, S. H. P. (1974). *In* "Insect Neurobiology." (Ed. J. E. Treherne), pp. 307–357. North Holland, Amsterdam.

Maddrell, S. H. P. and Gee, J. D. (1974). *J. exp. Biol.* **61**, 155–172.

Maddrell, S. H. P., Pilcher, D. E. M. and Gardiner, B. O. C. (1971). *J. exp. Biol.* **54**, 779–804.

Maddrell, S. H. P., Gardiner, B. O. C., Pilcher, D. E. M. and Reynolds, S. E. (1974). *J. exp. Biol.* **61**, 357–377.

Mills, R. R. (1967). *J. exp. Biol.* **46**, 35–41.

Mordue, W. (1969). *J. Insect Physiol.* **15**, 273–285.

Mordue, W. (1970). *J. Endocr.* **46**, 119–120.

Mordue, W. (1972). *Gen. comp. Endocr.* Supp. 3, 289–298.

Mordue, W. and Goldsworthy, G. J. (1969). *Gen. comp. Endocr.* **12**, 360–369.

Mordue, W., Highnam, K. C., Hill, L. and Luntz, A. J. (1970). *Mem. Soc. Endocr.* **18**, 111–136.

Moses, A. M. and Miller, M. (1974). *In* "Handbook of Physiology. Endocrinology." (Eds E. Knobil and W. H. Sawyer), Vol. 4, Part 1, pp. 225—242. Am. Physiol. Soc., Washington, D.C.

Nayar, K. K. (1960). *Z. Zellforsch. mikrosk. Anat.* **51**, 320—324.

Nayar, K. K. (1962). *Mem. Soc. Endocr.* **12**, 371—378.

Núñez, J. A. (1956). *Z. vergl. Physiol.* **38**, 341—354.

Oschman, J. L. and Wall, B. J. (1969). *J. Morph.* **127**, 475—510.

Penzlin, H. (1971). *J. Insect Physiol.* **17**, 559—573.

Penzlin, H. and Stölzner, W. (1971). *Experientia* **27**, 390—391.

Pickering, B. T. (1970). *In* "Pharmacology of the Endocrine System and Related Drugs: The Neurohypophysis." (Eds H. Heller and B. T. Pickering), Vol. 1, pp. 59—80. Pergamon, Oxford.

Pickering, B. T., Jones, C. W. and Burford, G. D. (1971). *In* "Neurohypophysial Hormones." (Eds G. E. W. Wolstenholme and J. Birch), pp. 58—69. Churchill Livingstone, Edinburgh and London.

Pickering, B. T., Jones, C. W., Burford, G. D., McPherson, M., Swann, R. W., Heap, P. F. and Morris, J. F. (1975). *Ann. N.Y. Acad. Sci.* **248**, 15—35.

Pilcher, D. E. M. (1970). *J. exp. Biol.* **52**, 653—665.

Ramsay, J. A. (1954). *J. exp. Biol.* **31**, 104—113.

Sachs, H. (1969). *Adv. Enzymol.* **32**, 327—372.

Sawyer, W. H. (1974). *In* "Handbook of Physiology. Endocrinology." (Eds E. Knobil and W. H. Sawyer), Vol. 4, Part 1, pp. 443—468. Am. Physiol. Soc. Washington, D.C.

Scharrer, E. and Scharrer, B. (1944). *Biol. Bull. mar. biol. Lab. Woods Hole* **87**, 242—251.

Scharrer, E. and Scharrer, B. (1954). *In* "Handbuch der mikroskopischen Anatomie des Menschen." (Ed. W. Bargmann), Bd. 6, TL. 5. Springer, Berlin.

Share, L. (1974). *In* "Handbook of Physiology. Endocrinology." (Eds E. Knobil and W. H. Sawyer), Vol. 4, Part 1, pp. 243—256. Am. Physiol. Soc., Washington, D.C.

Stone, J. V., Cheeseman, P. and Mordue, W. (1976). *Gen. comp. Endocr.* **29**, 290—291.

Unger, H. (1965). *Zool. Jb. (Physiol.)* **71**, 710—717.

Verney, E. B. (1947). *Proc. R. Soc. London.* B **135**, 25—105.

Vietinghoff, U. (1967). *Gen. comp. Endocr.* **9**, 503.

Wall, B. J. (1965). *Zool. Jb. (Physiol.)* **71**, 702—709.

Wall, B. J. (1967). *J. Insect Physiol.* **13**, 565—578.

Wall, B. J. and Ralph, C. L. (1964). *Gen. comp. Endocr.* **4**, 452—456.

Wirz, H. (1957). *In* "The Neurohypophysis." (Ed. H. Heller), pp. 157—169. Butterworths, London.

Wong, T. M. and Pickering, B. T. (1976). *Gen. comp. Endocr.* **29**, 242.

III. Transport at Cell Level

12. Control of Cell Volume

E. K. Hoffmann

Institute of Biological Chemistry A, August Krogh Institute, Copenhagen, Denmark

I. INTRODUCTION

The volume of animal cells is precisely regulated. This regulation is a fundamental cellular function and malfunctions often lead to cell-swelling and lysis. Four physical principles are involved in the determination of cell volume:— (1) water is in thermodynamic equilibrium in the system i.e. in *osmotic equilibrium*; (2) the products of activities of the permeable ions on both side of the membrane are equal at equilibrium i.e. in *Donnan equilibrium*; (3) *electroneutrality* is maintained on both sides of the membrane; (4) *hydrostatic pressure differences* between the inside and the outside of the cell are *negligible*, since membranes of animal cells are easily distensible and cannot maintain pressure gradients.

Water can move readily through the plasma membrane of most cells whereas the cell solid contains the "fixed anions" that are too large to pass through the membrane. Besides these, the system contains small ions like Na^+, K^+ and Cl^- to which the membrane is permeable. Such a system should undergo colloid osmotic swelling and cytolysis (see Wilbrandt, 1941; Hoffman, 1958). This is prevented in living cells by active adjustment of the intracellular content of the diffusible cations Na^+, and K^+. Amongst the proposed theories are the conventional pump-leak hypothesis as well as possible alternative energy dependent mechanisms. Moreover, most cells can continue to regulate their volume when transferred to anisotonic media. This adaptation appears mainly to involve very selective changes in the permeabilities to Na^+, K^+, Cl^- and amino acids induced by the changes in cell volume.

II. CONTROL OF CELL VOLUME AT THE STEADY-STATE

A. Gibbs-Donnan Equilibrium

The Gibbs-Donnan rule tells us that the products of diffusible ions on both sides of the membrane are equal at equilibrium (see e.g. Helfferick, 1962). As a simple example let us consider the situation

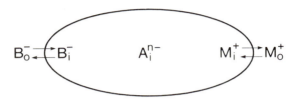

Fig. 1. A simple cell model. The cation M^+ and the anion B^- can penetrate the cell membrane. A_i^{n-} is an impermeable anion. Subscript i and o refer to inside and outside respectively.

in Fig. 1. From the Gibbs-Donnan rule we obtain (activity coefficients being neglected):

$$[M_i^+] \times [B_i^-] = [M_o^+] \times [B_o^-] \qquad (1)$$

Since there must be electroneutrality in each compartment:

$$[M_o^+] = [B_o^-] \qquad (2)$$

and

$$[M_i^+] = [B_i^-] + n[A_i^{n-}] \qquad (3)$$

Mathematically, if two positive quantities are related so as to have a constant product (as specified for $[M_i^+]$ and $[B_i^-]$ in equation (1), then their sum is smallest when they are equal. From equation (3) it thus follows that:

$$[M_i^+] + [B_i^-] > [M_o^+] + [B_o^-] \qquad (4)$$

Hence the total osmotic concentration will be greater inside than outside. Water enters the cell and tends to level out the difference of osmotic pressure, but at the same time the salt will redistribute according to the Gibbs-Donnan rule. This will go on until all the diffusible ions are inside or until the cells burst (colloid osmotic lysis).

Such a system can avoid this colloid osmotic swelling in different ways:

1. *Contractile Mechanisms*

A sufficiently high internal hydrostatic pressure would prevent the penetration of water. This seems to be the mechanism in most bacteria and plant cells (see e.g. Rothstein, 1959).

Kleinzeller (1965) suggested that the hydrostatic pressure also contributed to maintenance of the volume of some animal cells devoid of a solid cell wall. This suggestion was based on experiments which examined the effect of ouabain on the electrolyte and water transport in the slices of kidney cortex and liver (Kleinzeller and Knotkova, 1964a). Although the concentrations of ouabain should be sufficient to completely block the action of the Na/K pump (Whittam and Willis, 1963), leached cells reincubated aerobically in the presence of ouabain could still extrude NaCl and water as an isotonic solution, the water extrusion depending on metabolic energy. Referring to a myosin-like protein isolated from membranes of some cells (Ohnishi, 1963), Kleinzeller (1965) thus pointed out that a contractile mechanism might be active in water extrusion. Later actin-like or actomyosin-like contractile proteins were indeed extracted from various non-muscle cells (see e.g. Bettex-Galland *et al.*, 1972, Komnick *et al.*, 1972 and Shibata *et al.*, 1972). On the basis of experiments on some effects of calcium and of K^+ free media, Kleinzeller *et al.* (1968) again hinted at the possible elastic or contractile mechanisms being active in rabbit kidney cortex cells. Passive cation fluxes were increased in both Ca^{++} and K^+ free media, with no change in cell volume. However it is questionable whether there was sufficient time for the cells to reach a new steady-state. Finally, a marked extrusion of water and NaCl from previously swollen cells was observed in the absence of external K^+ (see Kleinzeller, 1972). Rorive *et al.* (1972) showed that the Na^+-independent and ouabain-insensitive control of the cell volume worked solely in the lower pH range (6.2—7.2). Ca^{++} and ATP were involved. Again the results were compatible with the mechanochemical hypothesis and it was suggested that cell (membrane) ATP and Ca^{++} were implicated in determining the physical properties of the membrane. Working with rat diaphragm, Rorive and Kleinzeller (1972) found similar results. For further discussion see Kleinzeller (1972).

More studies are, however, needed to elucidate whether the apparent Na-independent, ouabain insensitive mechanism is of a mechanochemical nature. At least that part of the theory which implies high hydrostatic pressures across the cell membrane seems unlikely for most animal cells. In erythrocytes the hydrostatic pressure gradient is only $1-2$ mm H_2O (see p. 299). For a less fragile cell like the sea urchin egg a hydrostatic pressure gradient of 5 mm H_2O was found (Mitchison and Swann, 1955). These values are several orders of magnitude lower than the osmotic pressure due to the fixed anions in a Donnan-system (up to 1 atm.). The surface tension, which is directly related to the capacity of the membrane to withstand hydrostatic pressure, and which may therefore play a role in stress situations, seems to depend on metabolic processes. Thus ATP depletion in the erythrocyte produced a 15-fold increase in surface tension, correlated with a marked increase in intracellular Ca^{++} (Weed et al., 1969, Weed and LaCelle, 1973). These effects are reversible on regeneration of cell ATP or removal of Ca^{++}.

Rangachari et al. (1973) postulated a Ca^{++} dependent contractile mechanism for volume control for rat myometrium. They suggested a contractile mechanism located in the membrane of micro-pinocytotic vesicles which could squeeze out the vesicular content.

2. "Active" Transport of Water. Colloid Osmotic Pressure

It would be an obvious course for controlling cell volume to pump out the water which diffuses into the cell. Robinson originally proposed such a mechanism (1960). However, when he examined the swelling of chilled slices of kidney cortex, he realized that the cells took up electrolytes and water in the same proportions as isotonic saline (1961). This showed that the swelling was not due to a difference in the osmotic pressure but rather to charge differences caused by intracellular fixed anions. Furthermore Robinson (1966) showed that such charge differences were substantial factors in controlling cell volume, in that he could inhibit the swelling process with a high mol. wt solute, polyethylene glycol. He therefore gave up the idea of water pumps. He even found that polyethylene glycol was more effective than NaCl, osmole for osmole, because NaCl entered the cell (Robinson, 1971). This effect was independent of the nature of the monovalent cation used (McLver and Raine, 1972). Thus an effect of added electrolytes on the colloid osmotic pressure should not be neglected in the interpretation of osmotic experiments.

Certain arthropods absorb water from atmospheres having a relative humidity as low as 45% (Edney, 1966; also see Chapters 22, 26). In the case of the mealworm, Ramsay (1964) showed that it was the cryptonephric complex which had such remarkable capacity for absorbing water. This seems to be an example of active water transport (see Ramsay, 1971 for discussion). There are indications, however, that extracellular non-electrolytes may be involved in such water vapour absorption (see Oschman et al., 1974, and Wall, in this volume, for discussion). Furthermore, a luminal injection of hypotonic solution in the blowfly rectal papillae seems to be followed by secretion of impermeant organic molecules into the intracellular spaces, apparently representing a method of drawing water from the cells (Berridge and Gupta, 1967). This seems to be a rather special case of regulating cell volume.

3. Double Donnan Equilibrium

The plasma membrane could also be impermeable to the small cations (or at least to those larger than K^+, e.g. the dominating cation on the outside, Na^+). The cell would then be in a "Double Donnan equilibrium". Van Slyke et al., (1925) and Boyle and Conway (1941) published equations based on such double Donnan equilibrium systems for red blood cells and frog sartorius muscle, respectively. Although the assumption of simple impermeability to sodium cannot now be accepted for either the red blood cell or the muscle cell, the equations have been very useful since both cell types are "functionally" impermeable to sodium (see below).

B. Pump-leak Concept

This concept states that the membrane contains an active transport system for Na^+ and K^+ which renders the cell membrane "functionally" impermeable to these ions in that the active pump fluxes are equal to and oppositely directed to the leak fluxes. For discussion of this "pump and leak" concept see Leaf (1959), Ussing (1960), Tosteson (1964) and Whittam (1964). From the pump and leak concept it follows that alteration in either the leakage permeabilities for the ions or the active transport rates will affect the total cation content of the cell and thus the cell volume. If for example, the leakage permeabilities are increased or the active transport rates decreased, the cell is no longer "functionally" impermeable to cations. The Donnan effect thus will be expressed in a redistribution of ions and a consequent change in volume (see above).

1. Cellular Swelling and Shrinkage

Cook, (1965) developed an explicit equation for the rate of change in cell volume due to an increased leakage as a function of cellular and extracellular concentrations of Na, K and Cl, and the rate coefficient k for the passive movement of one of the ions. His two assumptions were that the flux ratio equation is valid, and that the membrane potential can be determined from the chloride ratio. He measured the rate of volume change in human red blood cells by increasing the permeability to cations by simultaneously exposing the cells to ultra-violet light and inhibiting the pump with 10^{-6}Ml^{-1} strophanthin. The agreement with the postulated relation was satisfactory and the process was inhibited by sucrose, conforming to the criteria for the colloid osmotic mechanism.

Shrinkage of cells after a decrease on the sodium leak flux was seen in the ciliate *Tetrahymena* after the addition of a very low concentration of ethacrynic acid (0.2 mM) (Hoffmann and Kramhøft, 1974).

Cell swelling due to inhibition of the pump has been studied by many investigators (see e.g. Tosteson, 1964). This colloid osmotic swelling is caused by an uptake of a fluid which is not very different from physiological saline (see e.g. Leaf, 1959). The effect is important in pathology and medicine, since many tissues (brain, heart, kidney etc.) are irreversibly damaged by anoxia because cell swelling causes a collapse of the circulatory supply (Leaf, 1970; Flores *et al.*, 1972).

2. The Pump and Leak System in Brain Cortex

After the demonstration of colloid osmotic swelling in rat brain cortex (Leaf, 1956) further studies on brain slices have elucidated the pump and leak system in this tissue. Inhibition of the Na pump produced a considerable swelling of the cells (Pappius, 1964; Bourke and Tower, 1966b). The uptake of Na^+ and Cl^- during swelling was followed by a loss of K^+ (Frank *et al.*, 1968). The water uptake by cortex slices was inversely related to the concentration of ATP found in the tissues (Okamoto and Quastel, 1970a). Fluxes of electrolytes accounted for the net water fluxes in high K^+ media, which could be prevented by the presence of a relatively non-diffusible anion such as isothionate (Bourke and Tower, 1966a). Decrease in the leak permeability to sodium by the neurotrophic drug, tetrodotoxin, depressed the uptake of water (Okamoto and Quastel, 1970b). For a review of ion transport in brain tissue see Marchbanks (1970).

3. The Pump and Leak System in HK and LK Sheep Red Cells

An explicit statement of the mechanism by which cell volume control is accomplished has been given by Tosteson and Hoffman (1960) and Tosteson (1964) for the high and low K^+ (HK and LK) sheep red cell. This is now a textbook example and may be found both in Stein (1967) and Davson (1970). The equations in Tosteson and Hoffman (1960) apply only to the steady-state in a given external medium and give no information on the changes in cell composition expected when cells are placed in a different environment. Tosteson (1964) increased the range of application by choosing membrane parameters which are independent of the composition of the medium. A given membrane is characterized by six such parameters. These are: the rate constants for inward movements by leak pathways for K^+ and Na^+, ${}^{i}k_K{}^{l}$ and ${}^{i}k_{Na}{}^{l}$, the "Michaelis-Menten" constants $K_m{}^{K}$ and $K_m{}^{Na}$, and the maximum velocities ${}^{l}M_K{}^{pmax}$ and ${}^{o}M_{Na}{}^{pmax}$ for the active transport of K^+ into and Na^+ out of the cells. For the differential equations which characterize the time dependence of the cell composition in this model see Tosteson (1964).

In this treatment it was assumed that the rate coefficients for leakage of K^+ and Na^+ are invariant with the external composition. We know that the leak permeabilities actually depend both on the potential (see e.g. Donlon and Rothstein, 1969) and on the cell volume. The active role of the passive fluxes in cell volume control is discussed in Section IV. For a further discussion see Hoffmann (1978).

C. Ouabain-insensitive Na^+ (and Water) Extrusion

In agreement with the previously mentioned observations of Kleinzeller and Knothova (1964a), MacKnight (1968), Willis (1968) and Whittembury (1968) found that kidney cells extrude sodium and chloride, even in the presence of ouabain sufficient to inhibit exchange of sodium for potassium. Whittembury (1968), Whittembury and Fishman (1969) and Whittembury and Proverbio (1970) interpreted these findings as indicative of two modes of sodium extrusion from the tubule cells. The proposed model is as follows: Pump A is an electrogenic pump which extrudes sodium followed by chloride and water. This type of ion and fluid movement is inhibited by low temperature, is independent of the presence of potassium in the medium, is sensitive to ethacrynic acid, but is insensitive to ouabain. Pump B extrudes sodium in exchange for potassium. This type of ion movement also occurs at low temperature, is stimulated when the potassium concentration in the bathing medium is raised, and is sensitive to ouabain but quite insensitive to ethacrynic acid. For discussion of this work, see Giebish et $al.$ (1971). Proverbio et $al.$ (1970) concluded that pump B seemed to be related to the ATP'ase system while this was not the case for pump A.

Proverbio et $al.$ (1975) demonstrated a Mg^{++}-dependent Na^+ ATPase activity in microsomes from guinea-pig kidney cortex. This Na^+-stimulated ATPase was inhibited by ethacrynic acid or furosemide (0.1 mM). The authors attempted to relate this Na^+-ATPase with the K^+-independent, ouabain-insensitive mode of Na^+ extrusion from kidney cortex slices. Whittembury and Proverbio (1970) did not consider their ouabain-insensitive Na^+ pump inconsistent with Kleinzeller's mechanochemical model for volume regulation in kidney cells.

Also, in the medulla-cells from rat kidney an ethacrynic acid sensitive water-extruding mechanism, which was not ouabain-sensitive, seems to be involved in cell volume control (Law, 1976). MacKnight et $al.$ (1974) likewise reported that liver cells from rat, rabbit and guinea pig seemed to possess a metabolically dependent, ouabain-insensitive, potassium-independent mechanism linked to the regulation of the cellular volume. A similar system was also proposed for rat diaphragm by Kleinzeller and Knotkova (1964b), for rat uterine smooth muscle cells by Daniel and Robinson (1971) and for epithelial cells of toad urinary bladder by MacKnight et $al.$ (1975).

III. OSMOTIC BEHAVIOUR OF THE CELLS

How is the volume of a cell affected by a change in the osmotic pressure of its environment?

The water permeability of most animal cells is very high. An exception is found in the eggs of freshwater fish (Prosser, 1973) and fertilized eggs of marine fish (Holliday, 1969) in which the water permeability can be exceedingly low (also see Chapter 6).

Table I shows a few representative values of the hydraulic (osmotic) permeability P_{osm} for certain cells. Additional values have

TABLE I

Hydraulic (osmotic) permeability, P_{osm}, of some animal cells.

Cell type	$P_{osm} \times 10^4$ cm sec^{-1}	Reference
Cat erythrocyte	335	Rich *et al.* (1967)
Frog muscle fibres	128	Zadunaisky *et al.* (1963)
Human erythrocyte	122	Sha'afi *et al.* (1967)
Human erythrocyte	116	Barton and Brown (1964)
Crab muscle fibres	98.7	Sorensen (1971)
Ehrlich ascites tumour, A strain	89	Hempling (1967)
Chaos chaos	0.37	Prescott and Zeuthen (1953)
Salmon eggs, water hardened	0.004	Loeffler and Lovtrup (1970)

been published by Dick (1966) and Stein (1967). It can be seen that the hydraulic permeability falls within a surprisingly wide range of values for animal cells. It may be that the hydraulic permeability of most animal cells, except the smallest ones such as erythrocytes, has been underestimated to some extent because no account has been taken of the slow rate of intracellular diffusion of water (for discussion see Dick, 1970 and House, 1974).

A wide variety of other units have been used for the hydraulic permeability. Note that a P_{osm} of 1 cm sec^{-1} at 20°C corresponds to a hydraulic conductivity (filtration coefficient) L_p of 0.737×10^{-9} cm^3 dyne^{-1} sec^{-1} or 7.47×10^{-4} cm atm^{-1} sec^{-1}. L_p-values for a variety of cells may be found in House (1974). The high hydraulic permeability in the case of the human red cell means that if a 1 cm H_2O pressure gradient was maintained across a single red cell, it would swell to double its volume in 0.54 sec (Sidel and Solomon, 1957). It certainly seems that cells are adapted to permit a rapid flow of water through their membranes, which means that the cell and its environment always maintain osmotic equilibrium. Permeability to the major ions K^+, Na^+ and Cl^- is generally much lower than the water permeability. In the case of the red blood cell,

Na^+ and K^+ permeabilities are more than seven orders of magnitude lower than the water permeability. Other cells are, however, much more permeable to cations. As an example, Ehrlich ascites cells have a permeability of 7.8×10^{-8} cm sec^{-1} and 6.9×10^{-8} cm sec^{-1} for K^+ and Na^+ respectively (calculated from Hendil and Hoffmann, 1974). This is still, however, five orders of magnitude lower than the water permeability. This means that the membrane in most cells can be regarded as semipermeable.

A. Equations

If a cell behaves as a perfect osmometer with a semipermeable membrane, its volume should depend on the osmotic pressure of the surrounding medium according to the equation:

$$\pi(v - b) = RT \sum_j \Phi_j Q_j \tag{5}$$

derived from van't Hoff's expression (van't Hoff, 1887):

$$\pi = RT \sum_j \Phi_j m_j$$

where π is the external osmotic pressure, m_j is the molality, Q_j the amount of solute and Φ_j the osmotic coefficient of the jth solute in the cell, v is the cell volume and b is the non-solvent volume of the cell.

If $\Sigma\Phi_j Q_j$ does not change when cells are placed in solutions of different tonicities, the right hand side of equation (5) is constant and therefore:

$$\pi(v - b) = \pi_o(v_o - b) \tag{5a}$$

or

$$\frac{v - b}{v_o - b} = \frac{\pi_o}{\pi} \tag{5b}$$

where π_o and v_o are the original or isotonic external osmotic pressure and the original cell volume. This classic equation was reported by Lucké and McCutcheon, (1932), who assumed that the isotonic volume v_o may be subdivided into two components, osmotically inactive volume b, and an osmotically active volume $(v_o - b)$. The equation is usually applied in one of the following forms:

$$v = (v_o - b)\frac{\pi_o}{\pi} + b \tag{5c}$$

or

$$\frac{v}{v_o} = \frac{v_o - b}{v_o}\left(\frac{\pi_o}{\pi} - 1\right) + 1 \tag{5d}$$

Under certain experimental conditions we know π' instead of π where:

$$\pi' = \pi \frac{V - v}{V - v_o}$$

is the new external osmotic pressure before the cell volume is changed, V is the total extracellular plus cellular volume. In this case van't Hoff's law will get the following form:

$$\frac{v}{v_o} = \frac{v_o - b}{v_o}\left(\frac{a - (\pi'/\pi_o)a}{a(\pi'/\pi_o) + 1}\right) + 1 \tag{5e}$$

where

$$a = \frac{V - v_o}{v_o - b}$$

As a becomes very large equation (5e) becomes identical to equation (5d). It can be seen that a small a (0.3 in mammals) prevent wide fluctuations in cell volume as a response to changes in osmolarity of the extracellular fluid (see e.g. Olmstead, 1963).

Ponder (1948) described volume-osmotic pressure relationships in erythrocytes by an equation in a form similar to equation 5d, but he found that the degree of swelling and shrinking was often smaller than for a perfect osmometer. The volume v was given by:

$$\frac{v}{v_o} = R\frac{v_o^{H_2O}}{v_o}\left(\frac{\pi_o}{\pi} - 1\right) + 1 \tag{6}$$

where $v_o^{H_2O}$ is the measured volume of cell water at isotonic external osmotic pressure. Ponder's R value is thus the ratio between the osmotically active volume $(V_o - b)$ and the directly measured isotonic water content $(v_o^{H_2O})$.

$$R = \frac{v_o - b}{v_o^{H_2O}}$$ (7)

From (6) we get in a form resembling equation 5c:

$$v = R \cdot v_o^{H_2O} \cdot \frac{\pi_o}{\pi} + b$$

and in a form resembling 5b:

$$\frac{v^{H_2O}}{v_o^{H_2O}} = R \cdot \frac{\pi_o}{\pi} + (1 - R)$$

Thus a plot of the relative water content against the reciprocal relative osmolarity is linear with a slope R. If R is unity we have a perfect osmometer.

B. Erythrocyte Ghosts

One of the few examples of such a perfect osmotic behaviour is reported by Kwant and Seeman (1970) for erythrocyte ghosts freshly prepared and with membranes sealed at 37°C. Previous studies with erythrocyte ghosts (Teorell, 1952; Weed et al., 1963) indicated that the cells shrank predictably only if one assumed a high intracellular pressure or a high inactive volume. Kwant and Seeman (1970) considered the possibility that a considerable leakage of ions took place in the rather long interval between the time of addition of hypertonic solution and the time of measurement of the cell volume. They therefore reduced the time interval between addition of hypertonic solution and measurement of cell size to 30—60 sec. Moreover, they haemolysed in one step without any washing of the ghost, ensuring that the ghosts had an intracellular composition identical to that of the haemolysing medium. These erythrocyte ghosts seemed to behave quantitatively as perfect osmometers, since the experimental points fell on the osmometer curve predicted from the Boyle-van't Hoff law in a form similar to equation (5c) using measured values of osmolarities, cell volume and cell water. Colombe and Macey (1974) also reported that the osmotically inactive cell volume in their experiments with erythrocyte ghosts was negligible.

C. Intact Erythrocytes

The intact erythrocyte shows a linear relationship between volume and the reciprocal of the osmotic pressure, but it has an R value (equation 7) different from 1. In nearly all cases R is less than 1, since the observed volume changes are less than predicted by van't Hoff's law. This is shown in Fig. 2, which is reproduced from Kwant and Seeman (1970). The solid line for the perfect osmometer is drawn in accordance with equation (5c) with H_2O measured to 50 μl per cell and π_o equal to 0.333 osmol/l^{-1}. Many other reports

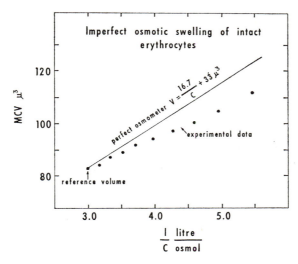

Fig. 2. Mean cell volume (MCV equal to v) of intact erythrocytes as a function of the reciprocal osmolarity $\left(\dfrac{1}{C}\ \text{equal to}\ \dfrac{1}{\pi}\right)$ in hypotonic solutions. The solid line for the perfect osmometer is:

$$v = v_o^{H_2O} \cdot \frac{\pi_o}{\pi} + b\ \mu^3 \qquad\qquad \text{(equation 5c)}$$

where $v_o^{H_2} = 5o\ \mu^3$, $\pi_o = 0.333$ Osm. and $b = 33\ \mu^3$.
The osmotic swelling is less than predicted by Boyle-van't Hoff's law. (From Kwant and Seeman, 1970.)

showed that the intact erythrocyte did not behave as a perfect osmometer (Orskov, 1946; Ponder and Barreto, 1957; Lefevre, 1964; Savitz *et al.*, 1964; Cook, 1967). They all found an R value about 0.8 in the hypertonic range. Other values for mammalian erythrocytes are given by Dick (1966) and by Ponder (1948).

In an attempt to explain this discrepancy from the Boyle-van't Hoff law, various authors have considered the following factors: (1) leakage of ions from the cells to the hypotonic medium or from the hypertonic medium into cells, (2) elasticity of the cell membrane, (3) "bound water", (4) concentration dependence of the osmotic coefficient of haemoglobin and (5) decrease in the net charge on the haemoglobin molecule with increased concentration.

1. Leakage of Ions

In the short time interval experiments like those discussed here (the time interval between addition of hypertonic medium and measurement of cell size was around 30 sec in Fig. 2) the net ion movement across the membrane is likely to be negligible. Savitz *et al.* (1964) found no significant loss of solute occurring either after shrinkage or after swelling.

An exception is the extreme case in the haemolytic process, where a prelytic loss of potassium from red cells takes place after the cell has reached its maximal volume (Seeman *et al.*, 1969). Jay and Rowlands (1975) showed, using an inverted microscope and cine camera, that for a total haemolysis time in water of 6.5 sec, the swelling time was only 1.2 sec. This was followed by a K^+ leak period of 1.5 sec during which the cell shrank a little, and then a period of 3.8 sec during which haemoglobin left the cell. For a slower haemolysis (with glycerol) the K^+ leak period was longer. Seeman *et al.* (1969) also reported that the prelytic release of K^+ was increased for slow haemolysis and that this quantitatively explained the reduced osmotic fragility in gradual haemolysis.

2. Effects of Cell Membrane Elasticity

Rand and Burton (1964), Hochmuth *et al.* (1973) and Evans (1973) all found a surface tension in the red cell membrane of $0.01-0.02$ dynes cm^{-1}. From Laplace's law we then get:

$$\Delta P = \gamma \left(\frac{1}{R_1} + \frac{1}{R_2} \right)$$

where ΔP is the pressure difference across the membrane (dynes cm^{-2} or atm), γ is the surface tension (dynes cm^{-1}) and R_1 and R_2 are the radii of curvature. For R_1 and R_2 equal to 4 μm and 1.25 μm respectively we find a pressure difference of only 1.02 mm H_2O. For further discussion of the elasticity of the red cell membrane see Evans and Hochmuth (1976a,b) and for a mathematical approach to the problem of cell membrane elasticity see Mela (1968).

Thus, any theory which implies significant hydrostatic pressure within the erythrocyte is unlikely. This is probably true for all "naked" animal cells, but it is quite different from the situation in "walled" cells, micro-organisms and plants. In plant cells the osmotic pressure in the protoplasm is usually much higher than that of environment, so that the protoplast is always pressing outward against the wall with a hydrostatic pressure — the turgor pressure. The difference in osmotic pressure may be very large, more than 100 m H_2O, compared to the few mm H_2O in the erythrocyte (for discussion see Rothstein, 1964).

3. "Bound Water"

The hypothesis of bound water implies that part of the water is unable to act as solvent for the salts in the cell, so that the effective concentration within the cell is higher than indicated by the number of moles per litre of cell water. Thus the value $v_o^{H_2O}$ in equation 6 would have to be decreased by a factor representing the fraction of bound water and R (equation 7) could be written as:

$$1 - \frac{v_o^{n\,sw}}{v^{H_2O}}$$

where $v_o^{n\,sw}$ *indicates* the osmotically inactive water. Savitz *et al.* (1964) found that 20% of the cell water was apparently unable to participate in osmotic phenomena and accounted for the phenomenon in terms of bound water associated with haemoglobin equal to 0.4 g H_2O g^{-1} haemoglobin. Drabkin (1950) found that human haemoglobin crystals contained 0.34 g bound water g^{-1} protein, and the numerical agreement was of course persuasive. Later studies indicated, however, that the amount of water bound to haemoglobin in solution would be considerably less than that of crystalline haemoglobin (see Schwan, 1965). There is other evidence against the bound water hypothesis. Miller (1964) showed that all red cell water served as solvent for glucose, and Gary-Bobo (1967) obtained similar results for various non-electrolytes. Moreover Cook (1967) found that non-solvent water was less than 4.3% of the measured water volume in an isotonic medium, with the non-solvent water being defined as the amount of water which was not available for dissolving chloride. The amount of non-solvent water appeared to increase a little in hypertonic solution but still it could not at all account for the R value of 0.77. Gary-Bobo and Solomon (1968) measured the distribution of three readily diffusible non-electrolytes. They found that even though the osmotically effective cell water was only 80% of the total cell water as reported by Savitz et al. (1964), the non-electrolytes could nevertheless dissolve in all of the cell water. They also measured the osmotic properties of a haemoglobin solution placed on one side of a cellophane membrane, after adding equal concentrations of potassium chloride to the solutions on both sides of the membrane. If the potassium chloride on the haemoglobin side of the membrane had been excluded from some of the water, the resulting osmolarity should be higher and there would have been an immediately detectable osmotic effect. There was, however, no such effect. All of the water on both sides of the membrane appeared to be available for dissolving potassium chloride. Taken together, these findings rule out the bound water hypothesis. The non-solvent water is quantitatively too small to be important in this respect.

4. The Osmotic Coefficient of Haemoglobin

Adair (1929) found that the osmotic coefficient of haemoglobin increased from 1.5 to 3.4 as the concentration increased from 2 to 5 mM. His data have been fitted to a second degree polynomial by Dick and Lowenstein (1958), and later by Gary-Bobo and Solomon (1968), giving nearly the same equation. McConaghey and Maizels (1961), investigating the osmotic coefficient of haemoglobin in the red cell, obtained a reasonable conformity with the coefficients extrapolated from Adair, though with a slightly higher value at higher concentrations. This means that Φ_j (equation 5) was not constant but increased as the cell shrank and decreased as the cell swelled. It would therefore be expected that the change in volume would be less than that predicted by equation 5a. This has been used by Dick and Lowenstein (1958) as an explanation for the variation in R. In the hypotonic range Dick and Lowenstein were able to predict the actually measured R value of 0.95 from Adair's data (see Dick,

1966 for further discussion). For the hypertonic range the predicted values were considerably larger than the experimental values (see Dick, 1966). Gary-Bobo and Solomon (1968) found the change in osmotic coefficient insufficient to account for their results and proposed a new hypothesis (see below). Dick (1969) later pointed to the possibility that the osmotic coefficient of haemoglobin increased at a greater rate at high ionic strength. The change in osmotic coefficient could then still provide an explanation for the large fraction of apparent non-osmotic water in the shrunken erythrocyte. That this cannot be the whole explanation can be seen from the experiment of Gary-Bobo and Solomon (1968), reproduced in Fig. 3. They measured the apparent osmotic water $(v_o - b)$ (equation 5c) and the cell water $(v_o^{H_2O},)$ as a function of pH. It is seen that below the iso-electric point of haemoglobin, the osmotically active volume (from equation 5c) is higher than the actual water volume. This clearly cannot be explained by any of the mechanisms proposed above, and some other mechanism must therefore be operative. Solomon (1971) has reviewed these problems.

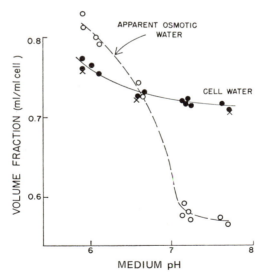

Fig. 3. Effect of medium pH on the volume fraction of measured cell water and apparent osmotic water. The filled circles are points obtained from wet and dry weight ratio, and the crosses are the distribution volume of ethanol. The open circles are the volume fractions of apparent osmotic (osmotically active) water computed from an equation similar to equation 5d. (From Gary-Bobo and Solomon, 1968).

5. Change in Net Charge on the Haemoglobin Molecule with Concentration

If $\Sigma_j \phi_j Q_j$ changes when cells are placed in solutions of different tonicities, then equation (5a) does not apply and the Boyle-van't Hoff law must be recast in another form. Since water is at thermodynamic equilibrium across the cell membrane, the value of π is the same inside and outside the cell. From equation (5) we therefore get:

$$\sum_j \phi_j^i m_j^i = \sum_j \phi_j^0 m_j^0$$

NaCl is the dominating component extracellularly, while KCl dominates in the cells. In the range of 100 mM—200 mM these electrolytes have osmotic coefficients of 0.92—0.93, independent of concentration. Thus, as a reasonable approximation we can use 0.92 for all electrolytes and treat it as a constant Φ. We then get (see Sachs et al., 1975):

$$\phi(m_+^i + m_-^i) + \phi_H m_H^i = \phi(m_+^o + m_-^o)$$

$$\phi \frac{Q_+^i + Q_-^i}{(v - b)} + \phi_H \frac{Q_H^i}{(v - b)} = \phi(m_+^o + m_-^o)$$

$$v - b = \frac{\phi(Q_+^i + Q_-^i) + \phi_H Q_H^i}{\phi(m_+^o + m_-^o)}$$

the symbols used are the same as for equation 5.

Applying the condition of electroneutrality we get:

$$m_-^i = m_+^i + z_H m_H^i$$

$$Q_-^i = Q_+^i + z_H Q_H^i$$

$$v - b = \frac{2Q_+^i + (z_H + \phi_H/\phi)Q_H^i}{m_+^o + m_-^o} \tag{8}$$

from which it is seen that the cell volume at a given osmolarity is influenced both by a change in the osmotic coefficient for haemoglobin Φ_H and in the charge on the haemoglobin molecule z_H.

The charge on haemoglobin was found (Gary-Bobo and Solomon, 1968, 1971) to be a function of concentration. This is shown at pH 6.6 in Fig. 4. The decrease in z_H occurred whether z_H was

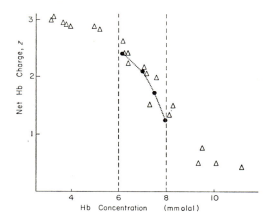

Fig. 4. Decrease of charge (z_H) with increasing concentration of haemoglobin (Hb) is shown for experiments carried out both *in vivo* (solid circles) and *in vitro* (open triangles). The vertical dashed lines indicate the attainable excursions of osmolality within intact human red blood cells (From Gary-Bobo and Solomon, 1971).

positive (pH <6.9) or negative (pH >6.9). The rate of change of z_H with concentration was greatest at the higher haemoglobin concentration, which correlated well with the observation that R deviated more from 1 in hypertonic than in hypotonic solutions.

This mechanism implies that as cells shrink they gain Cl^-. A Cl^- gain corresponding to the change in z_H was found by Gary-Bobo and Solomon (1968), but Cook (1967) found a much smaller gain in chloride under similar circumstances.

Dalmark (1975) used a different technique for changing the cell volume, based on water movement occurring when the cellular chloride was altered by titration of the cells with acid or base. The movement of chloride did not satisfactory explain the osmotic phenomena under his experimental conditions. The cause of these differences is not apparent. The reason for the deviation of R from unity is as yet unclear, but the variations in Φ_H and z_H (equation 8) are probably the main factors, with the change in Q_j resulting from the change in z_H probably being the most important.

D. Other Cells that Behave as "Non-ideal Osmometers"

Since the apparent osmotic behaviour of the red cell seems to be representative of a variety of cells, and since the osmotic behaviour of the human erythrocyte has been studied in far greater detail than that of other cells, the discussion above should serve as a useful basis for comparison with other cell types.

1. Muscle Fibres

Dydyǹska and Wilkie (1963), Reuben *et al.* (1963) and Blinks (1965) showed that striated muscle cells from the frog did not behave as ideal van't Hoff osmometers, either in the hypo- or in the hypertonic range. Figure 5 is reproduced from Blinks (1965) and shows the relative cell volume as a function of the reciprocal osmotic pressure (equation 5d). The thin line is the theoretical osmometer curve drawn from Blinks' own estimates. The *R* value was found to be

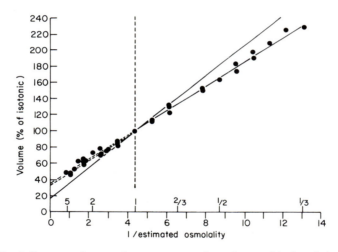

Fig. 5. The influence of osmotic pressure on the volume of isolated single fibres from *Rana temporaria* anterior tibial muscle. All volumes expressed as percentage of volume of same fibre in isotonic solution (v/v_o × 100)%. The abcissa represents the reciprocal of the external osmotic pressure ($1/\pi$); figures above the horizontal axis indicate relative tonicity (π/π_o). Separate regression lines for points in hypotonic and in hypertonic solutions; the vertical dotted line identifies isotonic solution ($\pi = \pi_o$). The thin line for the perfect osmometer is:

$$\frac{v}{v_o} \times 100 = \frac{v_o^{H_2O}}{v_o} \left(\frac{\pi_o}{\pi} - 1 \right) \times 100 \qquad \text{(equation 5d)}$$

where $\dfrac{v_o^{H_2O}}{v_o}$ × 100 is measured as 80%. (From Blinks 1965.)

0.83. This value does not seem to depend on the existence of bound water (Dydyǹska and Wilkie, 1963) but it could be due to the fixed charge on the macromolecules, as described above for the erythrocyte.

Reuben *et al.* (1964) found an R value around 0.8 in crayfish muscle fibres both in hypo- and hypertonic media.

Single isolated muscle fibres from the blue crab, *Callinectes sapidus* also showed a linear relationship between the relative cell volume and the relative osmotic pressure in hypertonic media (Lang and Gainer, 1969a). On the other hand, cells placed in hypotonic media rapidly swelled to initial peak volumes and then the volume slowly declined (see later). If the peak value of the relative volumes in hypotonic media was plotted against relative osmolarity, the points did fall on the same straight line as for the hypertonic solutions. The relative osmotically active volume was found to be 0.65, but as the actual water volume was not measured the R value cannot be assessed.

2. Mammalian Cell Lines

Lucké *et al.* (1956) observed a linear relationship between cell volume and osmolarity (equation 5c) for Gardner mouse ascites tumour cells in hypertonic solutions, and Hempling (1950) found the same for Ehrlich ascites tumour cells. From Lucké *et al.* (1956) an R value of 0.80 can be calculated. From the osmometer curve given by Hempling we get:

$$\frac{v_O - b}{v_O} = 0.69$$

(using 0.32 osmolar medium as isotonic). Assuming $v^{H_2 O} = 0.83$ (Lassen *et al.*, 1971), an R value of 0.83 for Ehrlich ascites tumour cells in hypertonic media is found. Hendil and Hoffmann (1974) confirm the linear relationship between cell volume and osmolarity in the hypertonic range but the R value seems to be higher than 0.9.

Roti Roti and Rothstein (1973) reported that mouse leukaemic cells exposed to hypertonic media for one hour shrank linearly with *decreasing* π, but they did not measure the water content, and it is thus impossible to estimate an R value. The rat erythroblastic leukaemic cell, an immature erythrocyte, also obeyed van't Hoff's law in the hypertonic range with an R value of 0.87 (Hempling and Wise, 1975).

Kleinzeller (1972) calculated an R value of 0.945 for the slices of rat kidney cortex at $0°C$ from data of Kleinzeller *et al.* (1967). Thus at least at $0°C$ the cells apparently acted as very good osmometers. No initial values were given for cell volume after transfer to different osmolarities at $25°C$, but the steady-state values at $25°$ were much lower than at $0°C$, indicating a volume regulation. The readily diffusible non-electrolyte, propylene glycol, was found to dissolve in about 90% of the cell water (Kleinzeller and Knotkova, 1967).

In renal medullary cells from rats, cell volume changes were linearly related to the reciprocal of the medium osmolarity (Law, 1975). As $v_o^{H_2O}$ was not known, it is not possible to assign an R value in this case. No secondary volume regulation like that in cortical cells seems to occur.

3. Avian Erythrocytes

The response of duck erythrocytes to hypotonic media can be divided into two phases (Kregenow, 1971a): an initial rapid phase of osmotic swelling, and a second, more prolonged phase, of shrinkage (volume-regulatory phase) (see Section IV). The increase in cell volume in the osmotic phase, regardless of the tonicity used, was less than expected from the perfect osmometer curve. Kregenow found an R value of 0.66. There was no loss of cation during this initial phase, but there was an initial net chloride loss. This suggests that a chloride shift may be responsible for part of the "non-ideal behaviour" of the duck erythrocyte, as was the case for the human erythrocyte (see above). The loss of chloride could, however, only explain approximately a third of the discrepancy.

Duck erythrocytes behave as "non-ideal osmometers" in hypertonic solutions as well (Kregenow, 1971b). Here the R value is as low as 0.55. The deviation from "ideality" is greater in hypertonic than in hypotonic media, in agreement with results from human erythrocytes. However, both R values are lower for duck erythrocytes than for human erythrocytes.

IV. CELL VOLUME REGULATION IN ANISOTONIC MEDIA

Aquatic animals which are restricted to a narrow range of salinity are called *stenohaline*, while those which tolerate a wider range of salinities are known as *euryhaline*. There are two patterns of response to osmotic stress. In some animals, body-fluid osmolarity may change with the medium. These are the *osmoconformers*. In other animals the internal osmolarity remains relatively constant even when the medium changes. These are the *osmoregulators*. Many gradations exist between the two extremes. Osmoconformers tolerate a wider variation in internal osmoconcentration than do osmoregulators (Prosser, 1973).

Many euryhaline aquatic animals are osmoconformers; this means that the osmolarity of the body fluids which are isosmotic with the environment must change a lot. Figure 6 shows the osmotic concentration of the internal medium plotted against that of the external medium for some osmoconformers as well as for some osmoregulators. For the sake of clarity only a few examples are shown. Many more are listed by Prosser (1973).

The cells of animals such as the polychaete worm *Nereis diversicolor* (Fig. 6) should thus swell and shrink in proportion to the different salinities, in accordance with the principles described in

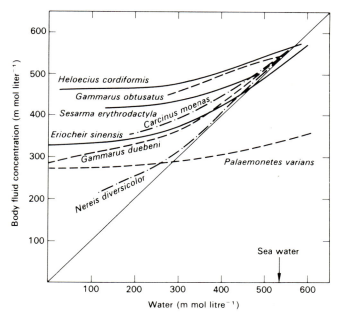

Fig. 6. The osmotic concentration of body fluids (internal medium) as a function of that of the surrounding water (external medium) for various brackish water animals. Full strength sea water is indicated by an arrow. Diagonal line indicates equal concentrations in body fluid and surrounding water. (From Schmidt-Nielsen, 1975b).

Section III. Even the relatively stable osmoregulators alter their plasma osmolarity upon exposure to a new environment, and swelling and shrinkage of cells should result. Are the tissues of some animals *de facto* capable of working at a high water content in a swollen state? Conversely, are some cells in the shrunken state able to endure a high concentration of salts? Ramsay (1954) considered these questions, but since very little was known about cellular responses at that time, he could only conclude: "The moral of all this is, that these are problems for the cell physiologist". Shortly after, he paved the way to the requisite micromeasurements by devising a method for the measurement of osmotic pressure of microsamples by freezing point determination (Ramsay and Brown, 1955), and by a method for the measurement of Cl in small volumes by potentiometric titration (Ramsay *et al.*, 1955). In the following years many experiments have actually shown that most cells seem to possess mechanisms for regulating their volume as well as ionic concentrations.

A. The Response of Cells to Hypotonic Media

Figure 7 shows the volume changes in a variety of cells following exposure to hypotonic media.

It is seen that the response of all cells shown can be divided into two phases: an initial phase of osmotic swelling and a second, more prolonged, phase of cell shrinkage, the volume-regulatory phase (Kregenow, 1971a). In the osmotic phase the cells swell more or less as perfect osmometers as has been discussed in Section III. This swelling is usually rapid, but there are variations caused by differences in permeability to water (see p. 293). During the volume-regulatory phase the volume gradually decreases towards a new steady-state at a value usually a little larger than the original. Thus, the readjustment is not perfect. The time needed for the volume regulation varies in different cell types from around five minutes up to about six hours.

In bathing media of lower osmolarity, the volume readjustment is more prolonged, although it initially proceeds more rapidly (Kregenow, 1971a; Hendil and Hoffmann, 1974).

1. The Mechanism of Regulatory Volume Decrease in normally High-K$^+$ Cells

As discussed above (p. 293), movement of water is so rapid, that the activity of water in the intra- and extracellular phase is probably the

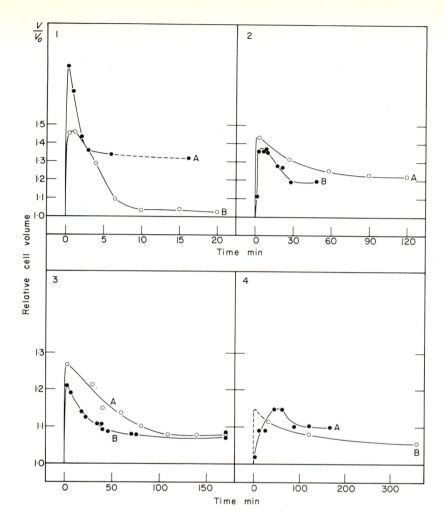

Fig. 7. Volume regulation of various cell types in hypotonic media. The relative cell volume (v/v_0) as a function of time after the cells were transferred to a hypotonic medium. The relative osmolarities of the respective hypotonic media are indicated below. The data are from the following papers: Rabbit renal tubule cells (Dellasega and Grantham, 1973); L5178Y mouse lymphocytic leukaemia cells (Roti Roti and Rothstein, 1973); *Pseudopleuronectes americanus*, red blood cells (Schmidt-Nielsen, 1975); frog (*Rana temporaria*) skin epithelial cells (McRobbie and Ussing, 1961); *Callinectes sapidus* muscle fibres (Lang and Gainer, 1969a); Ehrlich ascites mouse tumour cells (Hendil and Hoffman, 1974); frog (*Rana temporaria*) oocytes (Sigler and Janáček, 1971a); human red blood cells (Poznansky and Solomon, 1972). 1 A Rabbit renal tubule cells $\pi/\pi_0 = 0.50$, B L5178Y mouse lymphoblasts $\pi/\pi_0 = 0.45$; 2 A *Pseudopleuronectes* red blood cells $\pi/\pi_0 = 0.57$, B Frog skin epithelial cells $\pi/\pi_0 = 0.50$; 3 A *Callinectes* muscle fibres $\pi/\pi_0 = 0.65$, B *Ehrlich ascites* tumour cells $\pi/\pi_0 = 0.75$, 4 A Frog oocytes $\pi/\pi_0 = 0.57$, B Human red blood cells $\pi/\pi_0 = 0.69$.

same at any given moment and that the shift in cell size is osmotic in nature (see Kregenow, 1973).

The over-all response of the cells shown in Fig. 7 is thus considered an example of "isosmotic intracellular regulation", a term used by Florkin (1962) to describe a form of volume regulation which is achieved by adjusting the number of effective intracellular osmotic particles. During the volume-regulatory phase potassium is the major cation and chloride the major anion lost from the cell in most of the examples shown. Shrinkage seems to be a consequence of a nearly isosmotic loss of KCl and water from the cell. This has been shown in the nucleated duck erythrocyte (Kregenow, 1971a), in the frog oocyte (Sigler and Janacek, 1971a,b), in epithelial cells of frog skin (McRobbie and Ussing, 1961), in L 5178 Y mouse leukaemic cells (Roti Roti and Rothstein, 1973), in Ehrlich ascites tumour cells (Hendil and Hoffmann, 1974) and in flounder red cells (Cala, 1974). Rosenberg et al. (1972) and Shank et al. (1973) also demonstrated a loss of potassium during the volume-regulatory phase in mammalian cultured cells, presumably likewise followed by chloride. This loss of KCl was blocked by raising K_o^+ to a level which would be expected to eliminate the electrochemical gradient, which is thus almost certainly the driving force for K movement (Kregenow, 1971a). Doljanski et al. (1974) reported similar results for chick blood lymphocytes transferred to hypotonic solutions of varying cation composition. In high-Na^+ media the volume reached a maximum two minutes after transfer to a hypotonic medium and then began to decrease, but this regulatory volume decrease was obstructed by raising the external K^+ concentration.

In water-intoxicated rats the brain swelled half as much as muscle, and a drop in brain potassium content could be measured (Rymer and Fishman, 1973).

In the blue Crab (Fig. 7), which has a plasma osmotic concentration of 10^3 mOsm, similar to the surrounding seawater, volume regulation seems to involve potassium only to a very small extent, depending almost exclusively on a loss of amino acids (Lang and Gainer, 1969b). The same seems to be the case in the marine ciliate *Miamiensis avidus* (Kaneshiro et al., 1969a,b; see also p. 321). We now consider possible mechanisms for the adjustment of cellular potassium.

a   Increased K^+-permeability

That potassium loss during the regulatory shrinkage results from an increased potassium ion permeability was shown by unidirectional

flux measurements in duck red cells (Kregenow, 1971a), in human
red cells (Poznansky and Solomon, 1972a,b), in frog oocytes (Sigler
and Janacek, 1971b) and in Ehrlich ascites tumour cells (Hendil and
Hoffmann, 1974) (Table II). Cala (1974) measured the permeability

TABLE II

K^+ permeability coefficients at various osmolarities in the hypotonic range

Medium osmolarity[a] (mOsM)	$P_{K^+} \times 10^6$ (cm sec^{-1})	
	frog oocytes[b]	Ehrlich ascites[c]
100	4.96	—
175	1.52	—
225	—	0.159 ± 0.005
300	—	0.078 ± 0.004
340	0.95	—

[a] The osmolarity is adjusted by manitol and sucrose in the oocytes and the ascites cells
respectively.
[b] Sigler and Janáček (1971b).
[c] From Hendil and Hoffmann (1974).

to potassium during the regulatory-volume decrease following osmotic swelling of flounder erythrocytes, and found it to be increased
11-fold over the control value. Roti Roti and Rothstein (1973)
measured $^{42}K^+$-efflux at various times after suspending mouse
leukaemic cells in diluted media. Immediately after transfer to the
diluted medium the rate constant for efflux was five to six-fold
higher than in the control, but 10 minutes later it approached the
control value. In the cells transferred to 185 mOsm medium, the
efflux returned almost to the normal rate within five hours, in
140 mOsm it remained rather elevated even after five hours. In cells
grown for many generations in 185 mOsm, the K^+ concentration
remained reduced, the K^+ permeability seemed to be about the same
as in the control, and the pumping rate was reduced.

Ussing (1965) demonstrated that under conditions of cell-swelling
the rate of active sodium transport in the frog skin, measured as
short-circuit current, increased. According to the two-membrane
theory (Koefoed-Johnsen and Ussing, 1958) the current is carried by
potassium across the inward facing cell membrane, whereas sodium
goes via a Na^+/K^+ exchange pump. Therefore we should expect such
a stimulation of the Na^+-current if potassium permeability increases
under conditions of cell-swelling.

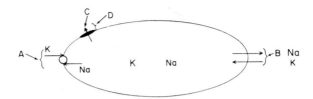

Fig. 8. Depiction of some of the membrane events that might occur when an enlarged cell (here a duck red blood cell) regulates its volume. (From Kregenow, 1974).

It thus seems obvious that the regulatory volume decrease in many cells is a consequence of a temporary increase in the permeability to potassium. Whether or not the potassium permeability remains somewhat increased when the new steady-state is reached probably depends on the experimental conditions employed and the membrane characteristics of the cell type. Since the increase in K^+ permeability is at least partly transient, the cell volume *per se* may be the sensitive indicator. That the K^+ flux is dependent on cell volume has been demonstrated in cat erythrocytes by Davson (1940), in dog erythrocytes by Parker and Hoffman (1965) and Romualdez *et al.* (1972), and in human erythrocytes by Poznansky and Solomon (1972b). Reversibility of the effect was confirmed by Poznansky and Solomon (1972a). The volume effect is independent of solute in all these species. Figure 8 shows some of the membrane events which occur in enlarged cells. As described above, findings in different cells have shown that under nonsteady-state conditions a membrane pathway, designated C in the figure, appears. It is through this pathway that the volume-controlling mechanism, D, controls K^+ loss from enlarged cells. C has been drawn as part of the volume-controlling mechanism, in this case as part of the effector portion, although at present we cannot exclude the possibility that pathway C is identical to the diffusion pathway, B, for K^+

b   Functional Separation of the Na^+-K^+ Exchange Pump from the Volume-controlling Mechanisms

The volume-regulatory phase in duck erythrocytes was not inhibited by ouabain (Kregenow, 1971a, 1974) except in enlarged cells with high Na^+ content. In these cells the shrinkage resulted from the action of the cation pump (A in Fig. 8) responding to an increase in internal Na^+ (Kregenow, 1974). In this case the process was no longer regulatory as the cells continued to shrink until they reached a minimal volume. This kind of cell shrinkage was blocked by removal of extracellular K^+. It seems that an uncoupling occurred in such high-Na^+ cells so that more Na^+ was removed than K^+ was taken up, as has been shown in human red blood cells (Post and Jolly, 1957).

Enlarged cells containing intermediate amounts of Na^+ and K^+ lost electrolytes through both the cation pump (A, Fig. 8) and the volume-controlling mechanism (D, Fig. 8) which seem to be functionally separate entities (Kregenow, 1974) using different pathways. The volume-regulatory pathway can only be used effectively by K^+.

Rosenberg et al. (1972) and Shank et al. (1973) reported that the shrinking phase in cultured mouse lymphoblasts was inhibited by ouabain. These authors concluded that volume regulation was achieved by pumping Na^+, gained during the osmotic swelling, out of the cells, followed by Cl^-. Shank and Smith (1974) suggested that increased intracellular Na^+ increased the pumping rate, since Na^+, K^+-ATPase activity was not increased and there was no enhanced pumping rate in the new steady-state after growth in hypotonic medium. Studies on the same cell type by Roti Roti and Rothstein (1973) revealed that the volume regulatory shrinkage was insensitive to ouabain and that the cation transport system (A, Fig. 8) was not directly involved. Apparently these findings contradict those of Rosenberg et al. A possible reason for the difference is that swollen cells in the study by Rosenberg et al. (1972) had a higher Na_i^+ concentration due to differences in experimental conditions. In the work by Rosenberg et al., reduction of medium osmolarity was achieved by dilution with water, whereas Roti Roti and Rothstein reduced the NaCl. Furthermore ouabain was added 10 minutes before dilution instead of at time zero. Thus the result of Rosenberg et al. can probably also be explained by the mechanism indicated in Fig. 8.

In Ehrlich ascites cells, Hendil and Hoffmann (1974) found a small (20%) increase in pumping rate (A, Fig. 8) under the new steady-state conditions (after 50 minutes), but the pumping rate in the readjustment phase is still unknown. McRobbie and Ussing (1961) also found an inhibition of the regulatory-volume decrease in frog skin pretreated with ouabain on the inside. As ouabain also inhibited the swelling in K-Ringer, they suggested that both effects might result from a decrease in the permeability to either K^+ or Cl^-. Since they observed the usual potential drop in K-Ringer with ouabain, they suggested that P_{Cl} was rate limiting and was reduced by ouabain. This possibility should be further investigated in other systems.

c   On the possible role of Ca^{++} and filaments in the membrane.

What is the volume-sensitive indicator that affects D in Fig. 8? As the cell contents are diluted, numerous intracellular substances could function in this capacity. The large selective increase in K^+ permeability might conceivably be brought about by an increase in the free intracellular Ca^{++} concentration due to dilution of ATP. This would be analogous to the well documented effect of Ca^{++} on potassium permeability in human red cells, first described by Gardos (see Gardos, 1967; Romero and Whittam, 1971; Lew, 1971a,b; Kregenow and Hoffman, 1972 and Riordan and Passow, 1973), in liver cells (van Rossum, 1970) and in snail neurons (Meech and Strumwasser, 1970; Meech, 1972; Meech and Standen, 1975). No difference in ATP concentration or ^{45}Ca influx per litre cells was, however, demonstrated in Ehrlich ascites cells suspended in either isotonic or hypotonic media (Hendil and Hoffmann, 1974). The two phenomena also differ in that the presence of absence of extracellular Ca^{++} has no effect on the potassium loss in the regulatory volume decrease (Kregenow, 1971a). The Ca-induced increase in K^+ permeability in human red cells was slightly greater in hypotonic media (Kregenow and Hoffman, 1972). It is possible that hypotonicity acts to increase the K^+ permeability by expanding the membrane matrix and increasing the accessibility of the inside of the membrane to Ca^{++}. The similarity between the volume regulatory increase in K^+ permeability and the Ca^{++} affected increase in K^+ permeability is of course persuasive, but whether or not there is any relation between them must await further analysis.

It is a tempting possibility that as the cell swells, changes in membrane elasticity develop in localized areas of the membrane and somehow act to increase K^+ permeability. This possibility is supported by the finding of actomyosin-like contracile proteins (see p. 287) in the membrane of various cells. Rosenthal *et al.* (1970) isolated such a group of fibrillar proteins and described the Ca^{++}-dependent ATPase activity associated with them. Also a Mg^{++} and Ca^{++}-stimulated ATPase is present at the outer surface of intact Ehrlich ascites cells (Ronquist and Ågren, 1975). This could be connected with local, contractile processes in the membrane for the regulation of cation permeability and thus cell volume. It is interesting that exogenous ATP together with Mg^{++} and Ca^{++} reversibly altered the volume of Ehrlich cells suspended in a Mg^{++} and Ca^{++}-free medium (Gasic and Stewart, 1968; Stewart *et al.*, 1969; Hempling *et al.*, 1969). These authors suggested that a contractile protein was present in the plasma membrane which responded to external ATP (see also p. 287). Actin-like proteins have been isolated from the membrane of Ehrlich ascites tumour cells (Kristensen *et al.*, 1973). Both in the Ehrlich cells (Hempling *et al.*, 1969) and in Hela cells (Aiton and Lamb, 1975) external ATP stimulated a large transient increase in K^+ efflux as well as an increased K^+ influx. In red cell ghosts there seems to be evidence that the cell configuration and osmotic properties may be controlled by membrane "contractile" proteins located on the internal side of the membrane and possessing Ca^{++} ATPase activity, (Palek *et al.*, 1971a,b). Posnansky and Solomon (1972a,b) suggested that the mechanism may be ascribed to a conformational change in a membrane protein which controls cation transport. They argued against localized changes in cation permeability caused by local changes in membrane elasticity because they did not think that the model sufficiently explained the observation that an increase in K^+ permeability is followed by a decrease in Na^+ permeability and *vice versa*. Moreover, they pointed out that an apparent linkage exists between Na^+ and K^+ fluxes in the same direction. In the case of dog red cells, Romualdez *et al.* (1972) also considered the primary event to be a conformational change affecting the selectivity between Na and K, possibly by an allosteric mechanism.

2. *Cells normally having a High-Na Concentration*

Studies by Parker (1972 and 1973a,b) indicated that the volume-dependent permeability changes in dog and cat red blood cells known from earlier studies (Davson, 1940; Parker and Hofman, 1965 and Sha'afi and Hajjar, 1971), may, in fact, represent a volume-regulatory mechanism (see also Parker and Hoffman, 1976). However, there are marked differences between normally high-K cells (described above) and normally high-Na cells (dogs and cats), as they regain their original volume. In dog red cells Na^+ is the important ion in volume regulation after transfer to anisotonic media (Parker *et al.*, 1975). Swollen cells regulate their volume by a calcium-dependent sodium extrusion, against an electrochemical gradient (Parker, 1973b). In such cases calcium accumulation was observed. At present most evidence suggests that a calcium-sodium exchange takes place (Parker *et al.*, 1975) so that energy for the volume-regulatory sodium extrusion is derived from passive calcium movements. It is, however, not possible at present to exclude the hypothesis that it is the calcium concentration at some critical site, rather than the movement of calcium across the membrane, which is the important determinant of volume-regulatory sodium extrusion in dog red blood cells.

3. Anion Permeability

Sulphate flux in cat red blood cells decreases with decreasing cell volume and increases with increasing cell volume (Sha'afi, 1971; Sha'afi and Pascoe, 1972). In Ehrlich ascites cells Simonsen et al. (1976) found an increase in Cl^- flux with increasing cell volume. Such changes in anion permeability are likely to play a role in cell shrinkage if net salt loss is limited by the Cl^- permeability. The Cl^- conductance in both human red cells (Hunter, 1971) and Amphiuma red cells (Lassen et al., 1975) is $10^4 - 10^6$ times smaller than that calculated from the tracer exchange (see Ussing et al., 1974). In Ehrlich ascites cells the chloride conductance was also less than expected from the exchange flux (Hoffmann et al., 1975; Simonsen et al., 1976). Because of the close relationship between conductance and permeability it thus seems very likely that the net salt loss might be limited by the Cl permeability in certain cases.

B. The Response of Cells to Hypertonic Media

The cell responds to increased tonicity by a rapid shrinkage and, in contrast to the response to decreased tonicity, many cells tested in vitro. show no apparent readjustment afterwards. This was found in Ehrlich ascites cells (Hempling, 1960; Hendil and Hoffmann, 1974), in mouse leukaemic cells (Roti et al., 1973), in crab muscles (Lang and Gainer, 1969a), in frog sartorius muscle fibres (Blinks, 1965; Bozler, 1959; 1961; Dydyńska and Wilkie, 1963; Simon et al., 1957) and in duck erythrocytes in low-K media (Kregenow, 1971b).

Osmotic experiments performed in the presence of a permeant solute are characterized by transient changes in cellular volume. Such a change is seen in red blood cells placed in a hypertonic solution containing urea (Sha'afi et al., 1970). It would be wrong, in this case, to conclude that the cell is responsible for the regulation of its volume, since the apparent regulation stems from passive entry of the added solute. On the other hand, when oocytes were transferred to different hypertonic solutions of mannitol, Sigler and Janáček

(1971b) found that a slow reswelling occurred because mannitol and water entered from the concentrated external solution. From their analysis they obtained estimates of both the hydraulic conductivity and the mannitol permeability. The water permeability was decreased and the mannitol permeability increased in the hypertonic medium. In this case a kind of regulation takes place. A discussion of experiments in which transient osmotic responses are used to evaluate the permeability characteristics of cell membranes for certain solutes may be found in a monograph by House (1974).

From the duration of such volume changes one can conclude that the water permeability and permeability and reflexion coefficient for the solute are important in dictating the nature of the osmotic response. In hypertonic solutions an increase in the permeability for the dominating external cation, Na^+, would function as a volume-regulating factor just as did the increase in K^+ permeability in hypotonic media.

1. Increased Na^+ Permeability

The Na^+ permeability of the cell membrane is increased in hypertonic media for several cell types investigated. Flounder red blood cells (Schmidt-Nielsen, 1975) returned to normal volume following shrinkage in (what was, at that stage) a hyperosmotic medium. Cala (1974) found a Na^+-permeability of 1.59×10^{-9} cm sec^{-1} in the shrunken cells compared to 0.63×10^{-9} cm sec^{-1} in the control cells. The permeability to Na^+ in the oocyte (Sigler and Janáček, 1971b) was likewise increased from 1.54×10^{-6} cm sec^{-1} in the control to 2.31×10^{-6} cm sec^{-1} in the new steady-state in hypertonic solution. Also in Ehrlich cells Pna is increased in shrunken cells (Hoffmann, 1978).

Passive movements of sodium in the high-Na^+ red blood cell were strongly influenced by cell volume both in dog red blood cells (Parker and Hoffman, 1965; Parker, 1973a; Romualdez et al., 1972) and in cat red blood cells (Sha'afi and Hajjar, 1971; Sha'afi and Pascoe, 1973). This dependence on cell volume is controlled by the metabolic state of the cell (Hoffman, 1966; Elford and Solomon, 1973; Sha'afi and Pascoe, 1973).

In the case of the dog red cell, Elford and Solomon (1974) argued that the response of the Na^+ flux to cell shrinkage was mediated by phosphoglycerate kinase, which is probably a membrane-bound enzyme (Parker and Hoffman, 1967). Shrinkage of the cells could cause a conformational change in the enzyme which, in turn, affects the conformation of the Na^+ transport site and endows it with the properties of a Na^+ channel (Elford and Solomon, 1974). Elford (1975) suggested that the changes he found in the slope of the Arrhenius plots of Na^+ and K^+ influxes were associated with the temperature dependence of these conformational changes which were catalysed by a Mg^{++}-stimulated enzyme.

The acceleration of Na^+ influx in cat red cells also depends on both the integrity of the metabolic chain and the presence of Mg^{++} (Sha'afi and Pascoe, 1973). The mechanism was likely to be the same in both cell types.

The gain in water which occurred in dog red blood cell in hyper-osmotic sodium media could be explained by the high sodium permeability of shrunken cells (Parker et al., 1975). This process was not influenced by calcium, in contrast to the regulatory volume decrease (p. 315).

Human red cells respond to volume changes in a similar manner though to a much smaller extent (Posnansky and Solomon, 1972b). Also here the cation shifts lead towards restoration of the original cell volume.

2. The Role of an Active Uptake of K^+

If the extracellular concentration of K^+ was increased, duck red blood cells transferred to hypertonic media (Fig. 9) re-established the isotonic volume (Kregenow, 1971b), suggesting that the volume-regulatory mechanisms were latent at physiological K^+ concentration in the hypertonic range. The same condition might prevail in the other cell types which failed to show any volume regulatory increase. Potassium was accumulated against an electro-chemical gradient during the volume-regulatory increase, and was therefore transported actively. This process required extracellular Na^+, and there was a general increase in both Na^+ and K^+ permeabilities. In the enlarging cells the K^+ efflux, although elevated, was initially less than the K^+ influx. Both the increase in influx and efflux were transient. Changes in cell size during the volume-regulatory phase were not altered by 10^{-4} M ouabain, and only a third of the K^+ influx seems to be ouabain sensitive. Thus a Na_o^+-sensitive but ouabain-insensitive, active accumulation of potassium must be involved.

Schmidt and McManus (1974) described a similar Na^+ and K^+ uptake which was inhibited by furosemide. These results suggest a

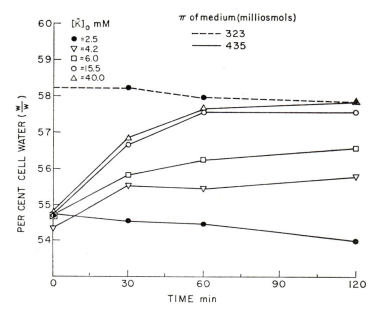

Fig. 9. The increase in the cell H_2O of duck erythrocytes incubated in hypertonic media (π = 435 mosmols) in response to raising the $[K_O]$. Duck erythrocytes were incubated in the isotonic or moderately hypertonic media, with $[K_O]$ varying between 2.5 and 40.0 mM. (From Kregenow, 1971b).

ouabain-insensitive, furosemide-sensitive, co-transport of Na^+ and K^+ which facilitates net cation movements during regulatory volume increase.

Also in human red blood cells external K^+ stimulated a furosemide sensitive net gain of Na^+ in agreement with the idea of a Na : K co-transport (Beaugé, 1975).

Another phenomenon in duck red cells seems to be associated with this regulatory volume increase. Duck red cells shrank in an artificial salt solution isosmotic to their plasma, but the cells returned to their original volume if norepinephrine or dibutyryl cyclic AMP was added to the solution (Riddick et al., 1971; Kregenow, 1973). This reswelling was associated with an uptake of K^+ which had the characteristics of the K^+ uptake brought about after shrinkage in hypertonic media. This suggested that the response brought about by norepinephrine via the adenyl cyclase system was similar to the change in cells which were exposed to a hypertonic medium.

Ørskov (1954, 1956) showed that pigeon, hen and frog red blood cells took up potassium from the plasma and swelled when the latter was made hypertonic. The similarity between this observation and the volume regulatory phase of duck erythrocytes seems obvious (Kregenow, 1971b).

Similar results were found in *Tetrahymena pyriformis* (Kramhöft, 1970). The readjustment phase was faster and more complete in a high-K^+ than in a low-K^+ medium. As indicated by Kregenow (1971b) the changes in K^+ content and permeability during the regulatory volume increase are parallel in some respects to the active accumulation of amino acids, depending on extracellular sodium. Amino acids play the same role in the isosmotic intracellular regulation in many invertebrates as K^+ does in the present examples (see next paragraph). In *Tetrahymena* both mechanisms seem to be operative (Stoner and Dunham, 1970). In this context it is of interest that the Na^+-dependent amino acid transport system demonstrated in *Tetrahymena* (Hoffmann and Kramhöft, 1969) was inhibited by ethacrynic acid, in concentrations which did not decrease the Na^+ gradient (Hoffmann and Rasmussen, 1972), and that K^+ and Na^+ were simultaneously lost from the cells, which thus shrunk (Hoffmann and Kramhöft, 1974).

There seem to be certain similarities between this apparent co-transport system and the furosemide inhibited $Na : K$ co-transport in duck red cells. It should be mentioned, however, that an increased uptake of amino acids during hyper-osmotic stress has only been suggested in very few cases (see next paragraph), and that this comparison between the role of K^+ and that of amino acids in the isosmotic intracellular regulation should therefore not be over-emphasized. On the other hand, the lack of a regulatory-volume increase in many *in vitro* systems could very well represent a difference from the *in vivo* situation partly because of the lack of the requisite extracellular concentrations of K^+ and amino acids.

C. Regulation of the Intracellular Amino Acid Pool

1. Amino Acids as Osmotic Effectors

In recent years many studies have dealt with the free amino acid content in the cells of marine euryhaline species exposed to variations in salinity. For the invertebrates the field can be followed in a series of monographs and reviews (Potts and Parry, 1964; Florkin and Schoffeniels, 1965; Florkin and Schoffeniels, 1969; Gilles, 1974b; Gilles, 1975). In addition, reviews concentrating on certain phyla or classes are available; for crustaceans, Schoffeniels and Gilles (1970); for molluscs, Schoffeniels and Gilles (1972); for annelids, Oglesby (1969), and Florkin (1969), and especially for polychaetes, Clark (1968). The role of amino acids in cell volume control has also been documented for euryhaline cyclostomes and teleosts under osmotic stress (Cholette *et al.*, 1970; Cholette and Cagnon, 1973; Lasserre and Gilles, 1971; Huggins and Colley, 1971; Lange and Fugelli, 1965; Fugelli, 1967) as well as for protozoans such as the marine ciliate *Miamiensis avidus* (Kaneshiro *et al.*, 1969b).

The participation of amino acids in isosmotic regulation of intracellular-fluid has also been investigated in freshwater and brackish water species, e.g. the freshwater ciliate *T. pyriformis* (Stoner and Dunham, 1970), the freshwater crayfish, *Astacus astacus* (Duchâteau-Bosson and Florkin, 1963), the chinese crab, *Eriocheir sinensis* (Bricteux-Grégoire *et al.*, 1962; Gilles and Schoeffeniels, 1969; Vincent-Marique and Gilles, 1970a,b) the brackish water eryhaline frog, *Rana cancrivora* (Gordon and Tucker, 1968) and the toad *Bufo viridis* (Gordon, 1965).

All these findings suggest the possibility that amino acid participation in isosmotic regulation of intracellular fluid is a general mechanism present in all organisms. It is interesting that amino acids are also active in volume regulation in a mammalian cell line, the Ehrlich ascites mouse tumour cells (Hoffmann and Hendil, 1976) where no special tolerance towards changes in environmental salinity seems to be needed.

a   The small non-essential amino acids

As stated by Gilles (1975) it seems that acclimation of euryhaline animals to a new salinity affects the concentration of all amino acids. The largest changes are, however, recorded in the small non-essential amino acids: alanine, aspartic and glutamic acids, glycine and proline (Florkin and Schoeffeniels, 1965). An exception to this is arginine, which cannot be synthesized by many of the species in which it exhibits large changes (Gilles, 1975). Alanine and glycine were dominant in most cases (Gilles, 1975). In the adductor muscle of the mollusc *Mya sp.*, glycine and alanine concentrations were proportional to the salinity of the medium (Virkar and Webb, 1970), and the concentration of alanine decreased from 159.1 mmoles kg^{-1} tissue water in $30\%_{00}$ salinity to 5.14 mmoles kg^{-1} tissue water in $2\%_{00}$ salinity. In the muscle of *Mytilus. sp.* the concentration of glycine decreased from 132.7 mmoles kg^{-1} wet weight in normal sea water to 58.9 mmoles kg^{-1} wet weight in animals acclimated to twice diluted sea water (Hoyaux *et al.*, 1976). Also in the euryhaline shrimp *Crangon crangon* glycine was the dominating amino acid in isosmotic intracellular regulation (Weber and Marrewijk, 1972).

In the osmoconforming hagfish, *Myxine sp.*, 80% of the osmotic change in the muscle was due to changes in amino acids, particularly alanine and proline but also valine, leucine and threonine (Cholette *et al* 1970). Trimethylamine oxide also played a role in volume regulation (Cholette and Gagnon, 1973). In some teleosts examined either glycine (Lassere and Gilles, 1971; Huggins and Colley, 1971) or glutamic acid (Vislie and Fugelli, 1975) were the dominating amino acids.

In the freshwater ciliate *T. pyriformis*, alanine, glycine and glutamic acid constituted about 60% of the amino acid pool of 35 mmoles l^{-1} cells (Hoffmann and Rasmussen, 1972). After exposure to 100 mM sucrose for 20 minutes, total intracellular osmolarity increased by 89 mOsm. Of this increase 38 mOsm were accounted for principally by alanine, glycine and glutamic acid, (Dunham and Kropp, 1973).

The small non-essential amino acids were also responsible for the variations seen in Ehrlich ascites cells after transfer to a hypotonic medium. The greatest change was found in alanine, for which the content in the 225 mOsm solution was only $(67 \pm 4)\%$ of the content in the 300 mOsm solution, the difference in concentration being even larger (Hoffmann and Hendil, 1976).

b   Taurine

Taurine seems to play a special role in cell volume regulation in many marine invertebrates (Allen and Garnet, 1971), particularly in the molluscs (Lange, 1963; Schoffeniels and Gilles 1972; Gilles, 1975). In the echinoderm *Asteria rubens*, the sea-star, approximately half the osmotic concentration in the tissues is due to free amino acids and taurine. These compounds change in concentration much more than any of the inorganic compounds as the animal goes from one salinity to another (Jeunieux *et al.*, 1962; Lange, 1964).

Taurine also participates in the isosmotic intracellular regulation in some euryhaline fish such as the thick-lipped mullets, *Crenimugil labrosus*, the southern flounder *Paralichthys lethostigma* (Lasserre and Gilles, 1971) and the flounder, *Platichtys flesus* (Fugelli, 1970). In the case of the latter, the taurine effect was demonstrated both in the erythrocytes (Fugelli, 1970; Zachariassen, 1972) and in the heart muscle (Vislie and Fugelli, 1975). The same was found in the muscle of the euryhaline toad *Bufo viridis* (Gordon, 1965).

Taurine is also active in the volume regulation in Ehrlich ascites mouse tumour cells (Hoffmann and Hendil, 1976). The relative change in taurine concentration after transfer to a hypotonic medium was found to be larger than the changes in all other inorganic and organic substances investigated. The taurine concentration changed from 16.3 mM in a 300 mOsm salt solution to 7.6 mM in a 225 mOsm solution. Further studies may disclose a more general role for taurine in isosmotic intracellular regulation.

2. Mechanisms of Regulation of the Intracellular Amino Acid Pool

In various euryhaline invertebrates under hypo-osmotic stress, an increase in the blood level of various amino acids occurred simultaneously with the decrease in the tissue amino acid content (see Gérard and Gilles, 1972a; Vincent-Marique and Gilles, 1970a,b and Clark, 1968). These results indicated that at least part of the amino acids were released from the tissue cells.

These conclusions, based on experiments with whole animals, have been confirmed using isolated tissues but the mechanisms responsible are still unknown. Several studies conducted on isolated tissues revealed that ninhydrin positive substances (NPS) are released into the ambient medium during the volume readjustment following hypo-osmotic stress. This was the case in *Pleuronectes flesus* erythrocytes (Fugelli, 1967), *Callinectes sapidus* muscle fibres (Lang and Gainer, 1969b), in the isolated heart muscle of *Modiolus modiolus* (Pierce and Greenberg, 1970) and in Ehrlich ascites tumour cells (Hendil and Hoffmann, 1974; Hoffmann and Hendil, 1976).

An uptake of amino acids from haemolymph into cells during hyper-osmotic stress has been suggested for *Carcinus maenas* by Siebers *et al.* (1972).

Gérard and Gilles (1972b) followed the outflux of radio-active ^{14}C-alanine from axons isolated from *Callinectes sapidus*. During hypo-osmotic stress there was an increase in the efflux of radioactive material, while during hyper-osmotic stress no significant change in the efflux occurred (Gérard and Gilles, 1972b). Gérard (1975) now has evidence that hypo-osmotic stress induces a partial inhibition of the active alanine influx and an increase in the passive efflux. It must be mentioned that the hypo-osmotic medium used was also a low Na^+ medium. The regulation was both Ca^{++} and energy-dependent and Gérard suggested that a Ca^{++}-ATPase may control the membrane permeability to amino acids.

Studies on Ehrlich ascites cells indicated that the passive leakage of amino acids, independent of the amino acid pump, was increased in a hypotonic medium (Hoffmann and Hendil, 1976). In this system isotonic and hypotonic media only differed in sucrose concentration. The ion gradient hypothesis (Morville et al., 1973) fits these data, provided we introduce a leakage pathway for the amino acids, independent of the amino acid pump, and varying in permeability with cell volume. This calculation was, however, only applied to the NPS ratio, and separate analysis should obviously be carried out using the gradients of those amino acids participating in the isosmotic regulation.

It thus seems likely that an outward movement of amino acids during hypo-osmotic stress is controlled by permeability changes. The increase in the amino acid concentration during hyper-osmotic stress is however, mainly under metabolic control (Gilles, 1972; Gilles and Gérard, 1974 and Gilles, 1974a). Discussion of this topic may be found in several reviews (Schoffeniels, 1967; Schoffeniels and Gilles, 1970; Gilles, 1974, and Gilles, 1975) and will therefore not be given here. Thus it appears that two mechanisms could exist for isosmotic regulation. The main factor in the reduction in amino acid content in cells swollen in a hypotonic medium is the increased leakage of the amino acids. The increase in amino acids found in cells exposed to hypertonic media mainly reflects a decreased catabolism of amino acids. On the other hand it is not likely that we are dealing with separate mechanisms for isosmotic regulation in hypo- or hypertonic media. It is more reasonable to consider that both mechanisms — transport control and metabolic control — are operative in both conditions but with different effectiveness.

REFERENCES

Adair, G. S. (1929). *Proc. Roy. Soc.* A **126**, 16–24.
Aiton, J. F. and Lamb, J. F. (1975). *J. Physiol.* **248**, 14–15P.
Allen, J. A. and Garrett, M. R. (1971). *Adv. Mar. Biol.* **9**, 205–253.

Barton, T. C. and Brown, D. A. J. (1964). *J. Gen. Physiol.* 47, 839—849.

Beaugé, L. (1975). *Biochim. Biophys. Acta* 401, 95—108.

Berridge, M. J. and Gupta, B. L. (1967). *J. Cell. Sci.* 2, 89—112.

Bettex-Galland, M., Probst, E. and Behnke, O. (1972). *J. mol. Biol.* 68, 533—535.

Blinks, J. R. (1965). *J. Physiol.* 177, 42—57.

Bourke, R. S. and Tower, D. B. (1966a). *J. Neurochem.* 13, 1071—1097.

Bourke, R. S. and Tower, D. B. (1966b). *J. Neurochem.* 13, 1099—1117.

Boyle, P. J. and Conway, E. J. (1941). *J. Physiol.* 100, 1—63.

Bozler, E. (1959). *Am. J. Physiol.* 197, 505—510.

Bozler, E. (1961). *Am. J. Physiol.* 200, 656—657.

Bricteux-Grégorie, S., Duchâteau-Bosson, G., Jeuniaux, C. and Florkin, M. (1962). *Archs. Int. Physiol. Biochim.* 70, 273—286.

Cala, P. M. (1974). "Volume regulation by flounder (*Pseudopleuronectes americanus*) red blood cells in anisotonic media." Doctoral Dissertation, Case Western Reserve University, Cleveland, Ohio.

Cholette, C., Gagnon, A. and Germain, P. (1970). *Comp. Biochem. Physiol.* 33, 333—346

Cholette, C. and Gagnon, A. (1973). *Comp. Biochem. Physiol.* 45A, 1009—1021.

Clark, M. E. (1968). *Biol. Bull.* 134, 252—260.

Colombe, B. W. and Macey, R. I. (1974). *Biochim. Biophys. Acta* 363, 226—239.

Cook, J. S. (1965). *J. Gen. Physiol.* 48, 719—734.

Cook, J. S. (1967). *J. Gen. Physiol.* 50, 1311—1325.

Dalmark, M. (1975). *J. Physiol.* 250, 65—84.

Daniel, E. E. and Robinson, K. (1971). *Canad. J. Physiol. Pharmacol.* 49, 178—204.

Davson, H. (1940). *J. cell. comp. Physiol.* 15, 317—330.

Davson, H. (1970). "A textbook of general Physiology." Fourth edition. Vol. I. Part 3. J. and A. Churchill, London.

Dick, D. A. T. (1966). "Cell water." Butterworths, London.

Dick, D. A. T. (1969). *J. gen. Physiol.* 53, 836—837.

Dick, D. A. T. (1970). *In* "Membranes and ion transport" (Ed. E. E. Bittar), Vol. III, pp. 211—250. Wiley-Interscience, London.

Dick, D. A. T. and Löwenstein, L. M. (1958). *Proc. Roy. Soc.* B.148, 241—256.

Doljanski, F., Ben-Sasson, S., Reich, M. and Grover, N. B. (1974). *J. Cell. Physiol.* 84, 215—224.

Donlon, J. A. and Rothstein, A. (1969). *J. Membrane Biol.* 1, 37—52.

Drabkin, D. L. (1950). *J. biol. Chem.* 185, 231—245.

Duchâteau-Bosson, G. and Florkin, M. (1963). *Comp. Biochem. Physiol.* 3, 245—247.

Dunham, P. B. and Kropp, D. L. (1973). *In* "Biology of Tetrahymena" (Ed. A. M. Elliot), pp. 165—197. Dowden, Hutchinson and Ross. Inc.

Dydyňska, M. and Wilkie, D. R. (1963). *J. Physiol.* 169, 312—329.

Edney, E. B. (1966). *Comp. Biochem. Physiol.* 19, 387—408.

Elford, B. C. (1975). *J. Physiol.* **246**, 371—395.

Elford, B. C. and Solomon, A. K. (1973). *J. Physiol.* **231**, 37—39P.

Elford, B. C. and Solomon, A. K. (1974). *Biochim. Biophys. Acta* **373**, 253—264.

Evans, E. A. (1973). *Biophys. J.* **13**, 941—954.

Evans, E. A. and Hochmuth, R. M. (1976a). *Biophys. J.* **16**, 1—11.

Evans, E. A. and Hochmuth, R. M. (1976b). *Biophys. J.* **16**, 13—26.

Flores, J., Di Bona, D. R., Beck, C. H. and Leaf, A. (1972). *J. Clin. Invest.* **51**, 118—126.

Florkin, M. (1962). *Bull. Acad. R. Belg. Cl. Sci.* **48**, 687—694.

Florkin, M. (1969). *In* "Chemical Zoology" (Eds M. Florkin and B. T. Scheer), Vol. IV, pp. 147—163. Academic Press, London.

Florkin, M. and Schoffeniels, E. (1965). *In* "studies in Comparative Biochemistry" (Ed. K. A. Munday), pp. 6—40. Pergamon Press, New York.

Florkin, M. and Schoffeniels, E. (1969). "Molecular approaches to ecology" pp. 89—111. Academic Press, New York.

Franck, G., Cornette, M. and Schoeffeniels, E. (1968). *J. Neurochem.* **15**, 843—857.

Fugelli, K. (1967). *Comp. Biochem. Physiol.* **22**, 253—260.

Fugelli, K. (1970). *Experientia* **26**, 361.

Gárdos, G. (1961). *In* "Membrane transport and metabolism" (Eds A. Kleinzeller and A. Kotyk), pp. 553—558. Academic Press, New York.

Gary-Bobo, C. M. (1967). *J. gen. Physiol.* **50**, 2547—2564.

Gary-Bobo, C. M. and Solomon, A. K. (1968) *J. gen. Physiol.* **52**, 825—853.

Gary-Bobo, C. M. and Solomon, A. K. (1971). *J. gen. Physiol.* **57**, 283—289.

Gasic, G. and Steward, C. (1968). *J. cell. Physiol.* **71**, 239—242.

Gérard, J. F. (1975). *Comp. Biochem. Physiol.* **51A**, 225—229.

Gérard, J. F. and Gilles, R. (1972a). *J. exp. mar. Biol. Ecol.* **10**, 125—136.

Gérard, J. F. and Gilles, R. (1972b). *Experientia* **28**, 863—864.

Giebisch, G., Boulpaep, E. L. and Whittembury, G. (1971). *Phil. Trans. Roy. Soc. Lond.* B. **262**, 175—196.

Gilles, R. (1972). *Life Sci.* **11**, 565—572.

Gilles, R. (1974a). *Life Sci.* **15**, 1363—1369.

Gilles, R. (1974b). *Arch. Int. Physiol. Biochim.* **82**, 423—589.

Gilles, R. (1975). Mechanisms of iono and osmoregulation. *In* "Marine Ecology" (Ed. O. Kinne), Vol. II, Part 1, pp. 259—347. Wiley Interscience, New York.

Gilles, R. and Gérard, J. F. (1974). *Life Sci.* **14**, 1221—1229.

Gilles, R. and Schoffeniels, E. (1969). *Comp. Biochem. Physiol.* **21**, 927—939.

Gordon, M. S. (1965). *Biol. Bull. mar. biol. Lab.* **128**, 218—229.

Gordon, M. S. and Tucker, V. A. (1968). *J. exp. Biol.* **49**, 185—193.

Helfferich, F. (1962). "Ion exchange." New York: McGraw-Hill.

Hempling, H. G. (1960). *J. gen. Physiol.* **44**, 365—379.

Hempling, H. G. (1967). *J. cell. Physiol.* **70**, 237—256.

Hempling, H. G. and Wise, W. C. (1975). *J. cell. Physiol.* **85**, 2, 195—208.

Hempling, H. G., Stewart, C. C. and Gasic, G. (1969). *J. cell. Physiol.* **73**, 133—140.

Hendil, K. B. and Hoffmann, E. K. (1974). *J. cell. Physiol.* **84**, 115–125.

Hochmuth, R. M., Mohandas, N. and Blackshear, P. L., Jr. (1973). *Biophys. J.* **13**, 747–762.

Hoffmann, E. K. (1978). *In* "Alfred Benzon Symposium XI." (Eds C. B. Jargensen, E. Skadhauge and J. Hess Thaysen) Munksgård, Copenhagen. in press.

Hoffmann, E. K. and Hendil, K. B. (1976). *J. comp. Physiol.* **108**, 279–286.

Hoffmann, E. K. and Kramhöft, B. (1969). *Expt. Cell. Res.* **56**, 265–268.

Hoffmann, E. K. and Kramhöft, B. (1974). *Compt. Rend. Trav. Lab. Carlsberg* **37**, 343–358.

Hoffmann, E. K. and Rasmussen, L. (1972). *Biochim. Biophys. Acta* **266**, 206–216.

Hoffmann, E. K., Simonsen, L. O. and Sjöholm, C. (1975). *Abstracts 5th, Int. Biophys. Congress Copenhagen* p. 328.

Hoffman, J. F. (1958). *J. gen. Physiol.* **42**, 9–28.

Hoffman, J. F. (1966). *Am. J. Med.* **41**, 666–680.

Holliday, F. G. T. (1969). *In* "Fish Physiology" (Eds W. S. Hoar and D. J. Randall), Vol. I, pp. 293–311. Academic Press, New York, London.

House, C. R. (1974). "Water transport in cells and tissues." Edward Arnold, London.

Hoyaux, J., Gilles, R. and Jeuniaux, Ch. (1976). *Comp. Biochem. Physiol.* **53A**, 361–365.

Hunter, M. J. (1971). *J. Physiol. Lond.* **218**, 49P–50P.

Huggins, A. K. and Colley, L. (1971). *Comp. Biochem. Physiol.* **38B**, 537–549.

Jay, A. W. L. and Rowlands, S. (1975). *J. Physiol.* **252**, 817–832.

Jeuniaux, C., Bricteux-Grégoire, S. and Florkin, S. (1962). *Cah. Biol. Mar.* **3**, 107–113.

Kaneshiro, E. S., Dunham, P. B. and Holz, G. G., Jr. (1969a). *Biol. Bull. marine biol. Lab.* **136**, 63–75.

Kaneshiro, E. S., Holz, G. G., Jr. and Dunham, P. B. (1969b). *Biol. Bull. marine biol. Lab.* **137**, 161–169.

Kleinzeller, A. (1965). *Arch. Biol. (Liège)* **76**, 217–232.

Kleinzeller, A. (1972). *In* "Metabolic Pathways" (Ed. L. E. Hokin), Vol. VI. pp. 91–131. Academic Press, New York.

Kleinzeller, A. and Knotková, A. (1964a). *J. Physiol.* **175**, 172–192.

Kleinzeller, A. and Knotková, A. (1964b). *Physiol. Bohem.* **13**, 317–326.

Kleinzeller, A., Nedvidková, J. and Knotková, A. (1967). *Biochim. Biophys. Acta* **135**, 286–299.

Kleinzeller, A. and Knotková, A. (1967). *Physiol. Bohem.* **16**, 214–226.

Kleinzeller, A., Knotková, A. and Nedvidková, J. (1968). *J. gen. Physiol.* **51**, 326–334.

Koefoed-Johnsen, V. and Ussing, H. H. (1958). *Acta Physiol. Scand.* **42**, 298–308.

Komnick, H., Stockem, W., Wohlfarth-Bottermann, K. E. (1972). *In* "Fortschritte der Zoologie" (Ed. M. Lindauer), Vol. 21, pp. 1–74. Stuttgart: Fischer.

Kramhöft, B. (1970). *Compt. Rend. Trav. Lab. Carlsberg* **37**, 343—358.

Kregenow, F. M. (1971a). *J. gen. Physiol.* **58**, 372—395.

Kregenow, F. M. (1971b). *J. gen. Physiol.* **58**, 396—412.

Kregenow, F. M. (1973). *J. gen. Physiol.* **61**, 509—527.

Kregenow, F. M. (1974). *J. gen. Physiol.* **64**, 393—412.

Kregenow, F. M. and Hoffman, J. F. (1972). *J. gen. Physiol.* **60**, 406—429.

Kristensen, B. I., Simonsen, L. O. and Pape, L. (1973). *Virchows, Arch. Abt. B. Zellpath* **3**, 103—112.

Kwant, W. O. and Seeman, Ph. (1970). *J. gen. Physiol.* **55**, 208—219.

Lang, M. A. and Gainer, H. (1969a). *J. gen. Physiol.* **53**, 323—341.

Lang, M. A. and Gainer, H. (1969b). *Comp. Biochem. Physiol.* **30**, 445—456.

Lange, R. (1963). *Comp. Biochem. Physiol.* **10**, 173—179.

Lange, R. (1964). *Comp. Biochem. physiol.* **13**, 205—216.

Lange, R. and Fugelli, K. (1965). *Comp. Biochem. Physiol.* **15**, 283—292.

Lassen, U. V., Nielsen, A.-M. T., Pape, L. and Simonsen, L. O. (1971). *J. Membrane Biol.* **6**, 269—288.

Lassen, U. V., Pape, L. and Vestergård-Bogind, B. (1975). *Abstracts 5th Int. Biophys. Congress, Copenhagen* p. 322.

Lasserre, P. and Gilles, R. (1971). *Experientia* **27**, 1434—1435.

Law, R. O. (1975). *J. Physiol. Lond.* **247**, 55—70.

Law, R. O. (1976). *J. Physiol. Lond.* **254**, 743—758.

Leaf, A. (1956). *Biochem. J.* **62**, 241—248.

Leaf, A. (1959). *Ann. N. Y. Acad. Sci.* **72**, 396—404.

Leaf, A. (1970). *Am. J. Med.* **49**, 291—295.

Lefevre, P. G. (1964). *J. gen. Physiol.* **47**, 585—603.

Lew, V. L. (1971a). *Biochim. Biophys. Acta* **249**, 236—239.

Lew, V. L. (1971b). *Biochim. Biophys. Acta* **233**, 827—830.

Loeffler, C. A. and Løvtrup, S. (1970). *J. exp. Biol.* **52**, 291—298.

Lucké, B. and McCutcheon, M. (1932). *Physiol. Rev.* **12**, 68—139.

Lucké, B., Hempling, H. G. and Makler, J. (1956). *J. cell. comp. Physiol.* **47**, 107—123.

McConaghey, P. D. and Maizels, M. (1961). *J. Physiol. Lond.* **155**, 28—45.

McIver, D. J. L. and Raine, A. E. G. (1972). *J. Physiol.* **225**, 555—564.

MacKnight, A. D. C. (1968). *Biochim. Biophys. Acta* **150**, 263—270.

MacKnight, A. D. C., Pilgrim, J. P. and Robinson, B. A. (1974). *J. Physiol. Lond.* **238**, 279—294.

MacKnight, A. D. C., Civan, M. M. and Leaf, A. (1975). *J. Membrane Biol.* **20**, 387—401.

McRobbie, E. A. C. and Ussing, H. H. (1961). *Acta Physiol. Scand.* **53**, 348—365.

Marchbanks, R. M. (1970). *In* "Membrane and ion transport" (Ed. E. E. Bittar), Vol. II, pp. 145—184. John Wiley and Sons Ltd., London.

Meech, R. W. (1972). *Comp. Biochem. Physiol.* **42A**, 493—499.

Meech, R. W. and Standen, N. B. (1975). *J. Physiol. Lond.* **249**, 211—239.

Meech, R. W. and Strumwasser, F. (1970). *Fedn. Proc. Fedn. Am. Soc. exp. Biol.* **29**, 834.

Miller, D. M. (1964). *J. Physiol. Lond.* **170**, 219—225.

Mitchison, J. M. and Swann, M. M. (1955). *J. exp. Biol.* **32**, 734—750.

Morville, M., Reid, M. and Eddy, A. A. (1973). *Biochem. J.* **134**, 11—26.

Oglesby, L. C. (1969). *In* "Chemical Zoology" (Eds M. Florkin and B. T. Scheer), Vol. IV, pp. 211—311. Academic Press, London.

Oshnishi, T. (1963). *J. Biochem. Tokyo* **53**, 238—241.

Okamoto, K. and Quastel, J. H. (1970a). *Biochem. J.* **120**, 25—36.

Okamoto, K. and Quastel, J. H. (1970b). *Biochem. J.* **120**, 37—47.

Olmstead, E. G. (1963). *Life Sci.* **10**, 745—750.

Oschman, J. L., Wall, B. J. and Gupta, B. L. (1974). *In* "Transport on the cellular level" Symp. Soc. exp. Biol. Number XXVIII. Cambridge University Press, Cambridge.

Palek, J., Curby, W. A. and Lionetti, F. J. (1971a). *Am. J. Physiol.* **220**, 19—26.

Palek, J., Curby, W. A. and Lionetti, F. J. (1971b). *Am. J. Physiol.* **220**, 1028—1032.

Pappius, H. M. (1964). *Can. J. Biochem.* **42**, 945—953.

Parker, J. C. (1972). 2nd Int. Symp. on metabolism and membrane permeability of Erythrocytes, Trombocytes and Leukocytes, Wienna 1972.

Parker, J. C. (1973a). *J. gen. Physiol.* **61**, 146—157.

Parker, J. C. (1973b). *J. gen. Physiol.* **62**, 147—156.

Parker, J. C. and Hoffman, J. F. (1965). *Fedn. Proc. Fedn. Am. Soc. exp. Biol.* 589.

Parker, J. C. and Hoffman, J. F. (1967). *J. gen. Physiol.* **50**, 893—916.

Parker, J. C. and Hoffman, J. F. (1976). *Biochim. Biophys. Acta* **433**, 404—408.

Parker, J. C., Gitelman, H. J., Glosson, P. S. and Leonard, D. L. (1975). *J. gen. Physiol.* **65**, 84—96.

Pierce, S. K. and Greenberg, M. J. (1970). *Am. Zool.* **10**, 518.

Ponder, E. (1948). "Hemolysis and related phenomena." Grune and Stratton, New York.

Ponder, E. and Barreto, D. (1957). *Blood.* **12**, 1016—1027.

Posnansky, M. and Solomon, A. K. (1972a). *J. Membrane Biol.* **10**, 259—266.

Posnansky, M. and Solomon, A. K. (1972b). *Biochim. Biophys. Acta* **274**, 111—118.

Post, R. L. and Jolly, P. C. (1957). *Biochim. Biophys. Acta* **25**, 118—128.

Potts, W. T. W. and Parry, G. (1964). "Osmotic and ionic regulation in animals." Pergamon Press, Oxford.

Prescott, D. M. and Zeuthen, E. (1953). *Acta Physiol. Scand.* **28**, 77—94.

Prosser, C. L. (1973). *In* "Comparative Animal Physiology" (Ed. C. L. Prosser), third edition, pp. 1—78. W. B. Saunders Comp., Philadelphia and London.

Proverbio, F., Robinson, J. W. L. and Whittembury, G. (1970). *Biochim. Biophys. Acta* **211**, 327—336.

Proverbio, F., Condrescu-Guidi, M., Pérez-Gonzáles, M. and Whittembury, G. (1975). Abstracts 5th Int. Biophys. Congress. Copenhagen. p. 5.

Ramsay, J. A. (1954). *In* "Active transport and secretion." (Eds R. Brown and J. F. Danielli), pp. 1—15. *Symp. Soc. exp. Biol.* Number VIII. Cambridge University Press, Cambridge.

Ramsay, J. A. (1964). *Phil. Trans. R. Soc. Lond.* **B. 248**, 279—314.

Ramsay, J. A. (1971). Insect rectum. *Phil. Trans. Soc. Lond.* **B. 262**, 251—260.

Ramsay, J. A. and Brown, R. H. J. (1955). *J. Sci. Instrum.* **32**, 372—375.

Ramsay, J. A., Brown, R. H. J. and Croghan, P. C. (1955). *J. exp. Biol.* **32**, 822—829.

Rand, R. P. and Burton, A. C. (1964). *Biophys. J.* **4**, 115—135.

Rangachari, P. K., Paton, D. M. and Daniel, E. E. (1972). *Am. J. Physiol.* **223**, 1009—1015.

Rangachari, P. K., Daniel, E. E. and Paton, D. M. (1973). *Biochim. Biophys. Acta* **323**, 297—308.

Reuben, J. P., Lopez, E., Brandt, P. W. and Grundfest, H. (1963). *Science* **142**, 246—248.

Reuben, J. P., Girardier, L. and Grundfest, H. (1964). *J. gen. Physiol.* **47**, 1141—1174.

Rich, G. T., Sha'afi, R. I., Barton, T. C. and Solomon, A. K. (1967). *J. gen. Physiol.* **50**, 2391—2405.

Riddick, D. H., Kregenow, F. M. and Orloff, J. (1971). *J. gen. Physiol.* **57**, 752—766.

Riordan, J. R. and Passow, H. (1973). *In* "Comparative Physiology" (Eds B. Bolis, K. Smith-Nielsen and S. H. P. Maddrell), North-Holland Publishing Company.

Robinson, J. R. (1960). *Physiol. Rev.* **40**, 112—149.

Robinson, J. R. (1961). *J. Physiol.* **158**, 449—460.

Robinson, J. R. (1966). *Fedn. Proc. Fedn. Am. Soc. exp. Biol.* **25**, 1108—1111.

Robinson, J. R. (1971). *J. Physiol. Lond.* **213**, 227—234.

Romero, P. J. and Whittam, R. (1971). *J. Physiol. Lond.* **214**, 481—507.

Romualdez, A., Sha'afi, R. I., Lange, Y. and Solomon, A. K. (1972). *J. gen. Physiol.* **60**, 46—57.

Ronquist, G. and Ågren, G. K. (1975). *Cancer Research* **35**, 1402—1406.

Rorive, G. and Kleinzeller, A. (1972). *Arch. Biochem. Biophys.* **152**, 876—881.

Rorive, G., Nielsen, R. and Kleinzeller, A. (1972). *Biochim. Biophys. Acta* **266**, 376—396.

Rosenberg, H. M., Shank, B. B. and Gregg, E. C. (1972). *J. cell. Physiol.* **80**, 23—32.

Rosenthal, A. S., Kregenow, F. M. and Moses, H. L. (1970). *Biophim. Biophys. Acta* **196**, 254—262.

Rothstein, A. (1959). *Bact. Rev.* **23**, 175—204.

Rothstein, A. (1964). *In* "The cellular Functions of Membrane Transport" (Ed. J. F. Hoffman), pp. 23—41. Prentice Hall, Inc., Englewood.

Roti Roti, L. W. and Rothstein, A. (1973). *Expt. Cell. Res.* **79**, 295—310.

Rymer, M. M. and Fishman, R. A. (1973). *Arc. Neurol.* **28**, 49—54.

Sachs, J. R., Knauf, P. A. and Dunham, P. B. (1975). *In* "The red blood cell" (Ed. D. Mac N. Surgenor), Vol. II, pp. 613—703. Academic Press, New York.

Savitz, D., Sidel, V. W. and Solomon, A. K. (1964). *J. gen. Physiol.* **48**, 79—94.

Schoffeniels, E. (1967). *In* "International Series of Monographs in Pure and Applied Biology". Div "Modern Trends in Physiological Sciences" (Eds P. Alexander and Z. M. Bacq), Vol. 28, pp. 55—61; pp. 157—185.

Schoffeniels, E. and Gilles, R. (1970). *In* "Chemical Zoology" (Eds M. Florkin and B. Scheer), Vol. V, pp. 255—286. Academic Press, New York.

Schoffeniels, E. and Gilles, R. (1972). *In* "Chemical Zoology" (Eds M. Florkin and B. Scheer), Vol. VII, 393—420. Academic Press, New York.

Schmidt-Nielsen, B. (1975). Comparative physiology of cellular ion and volume regulation. *J. exp. Zool.* **194**, 207—220.

Schmidt-Nielsen, K. (1975). "Animal Physiology. Adaption and Environment." Cambridge University Press, Cambridge.

Schmidt, W. F. and McManus, T. J. (1974). *Fedn. Proc. Fedn. Am. Soc. exp. Biol.* **33**, (5), 1457.

Schwan, H. P. (1965). *Ann. N.Y. Acad. Sci.* **125**, 344—354.

Seeman, P., Sauks, T., Argent, W. and Kwant, W. O. (1969). *Biochim. Biophys. Acta* **183**, 476—489.

Sha'afi, R. I. (1971). *First Europ. Biophys. Congress Proc. Vienna* 1971, pp. 377—381.

Sha'afi, R. I. and Hajjar, J. J. (1971). *J. gen. Physiol.* **57**, 684—696.

Sha'afi, R. I. and Pascoe, E. (1972). *J. gen. Physiol.* **59**, 155—166.

Sha'afi, R. I. and Pascoe, E. (1973). *J. gen. Physiol.* **61**, 709—726.

Sha'afi, R. I., Rich, G. T., Sidel, V. W., Bossert, W. and Solomon, A. K. (1967). *J. gen. Physiol.* **50**, 1377—1399.

Sha'afi, R. I., Rich, G. T., Mikulecky, D. C. and Solomon, A. K. (1970). *J. gen. Physiol.* **55**, 427—450.

Shank, B. B. and Smith, N. E. (1974). *Biochim. Biophys. Acta* **367**, 59—66.

Shank, B. B., Rosenberg, H. M. and Horowitz, C. (1973). *J. cell. Physiol.* **82**, 257—266.

Shibata, N., Tatsumi, N., Tanaka, K., Okamura, Y. and Senda, N. (1972). *Biochim. Biophys. Acta* **256**, 565—576.

Sidel, V. W. and Solomon, A. K. (1957). *J. gen. Physiol.* **41**, 243—257.

Siebers, D., Lucu, C., Sperling, K. R. and Eberlein, K. (1972). *Mar. Biol.* **17**, 291—303.

Sigler, K. and Janáček, K. (1971a). *Biochim. Biophys. Acta* **241**, 528—538.

Sigler, K. and Janáček, K. (1971b). *Biochim. Biophys. Acta* **241**, 539—546.

Simon, S. E., Shaw, F. H., Bennet, S. and Muller, M. (1957). *J. gen. Physiol.* **40**, 753—777.

Simonsen, L. O., Hoffmann, E. K. and Sjöholm, C. (1976). "Biochemistry of membrane transport." Abstracts F.E.B.S. Symp. Zürich, Switzerland, 18—23. p. 225.

Solomon, A. K. (1971). *Sci. Amer.* **224** (2), 88—96.

Sorensen, A. L. (1971). *J. gen. Physiol.* **58**, 287—303.

Stein, W. D. (1967). *In* "Theoretical and Experimental Biology" (Ed. J. F. Danielli), Vol. VI. Academic Press, New York and London.

Stewart, C. C., Gasic, G. and Hempling, H. G. (1969). *J. cell. Physiol.* **73**, 125—132.

Stoner, L. C. and Dunham, P. B. (1970). *J. exp. Biol.* **53**, 391—399.

Teorell, T. (1952). *J. gen. Physiol.* **35**, 669.

Tosteson, D. C. (1964). *In* "The cellular functions of membrane transport" (Ed. J. F. Hoffman), pp. 3—23. Prentice-Hall, Inc., Englewood Cliffs, New Jersey.

Tosteson, D. C. and Hoffman, J. F. (1960). *J. gen. Physiol.* **44**, 169—194.

Ussing, H. H. (1960). *In* "The alkali metal ions in biology" (Eds H. H. Ussing, P. Kruhöffer, J. Hess Thaysen and N. A. Thorn), p. 67. Springer Verlag, Berlin, Göttingen and Heidelberg.

Ussing, H. H. (1965). *Acta Physiol. Scand.* **63**, 141—155.

Ussing, H. H., Erlij, D. and Lassen, U. (1974). *Ann. Rev. Physiol.* **36**, 17—49.

Van't Hoff, J. H. (1887). *Z. Physik. Chem.* **1**, 481—508.

Van Rossum, G. D. V. (1970). *Nature, Lond.* **225**, 638—639.

Van Slyke, D. D., Hastings, A. B., Murray, C. D. and Sendroy, J. (1925). *J. biol. Chem.* **65**, 701—728.

Weber, R. E. and Marrewijk, W. J. A. van (1972). *Life Sci.* **11**, 589—595.

Weed, R. I. and LaCelle, P. L. (1973). *In* "Red cell membrane" (Eds G. A. Jamieson and T. J. Greenwalt), pp. 318—338. J. B. Lippincott Company, Philadelphia.

Weed, R. I., Redd, C. F. and Berg, G. (1963). *J. Clin. Invest.* **42**, 581.

Weed, R. I., La Celle, P. L., Merrill, E. W., Craib, G., Gregory, A., Karch, F. and Pickens, F. (1969). *J. Clin. Invest.* **48**, 795—809.

Whittam, R. (1964). "Transport and diffusion in red blood cells." The Williams and Wilkins Co., Baltimore.

Whittam, R. and Willis, J. S. (1963). *J. Physiol. Lond.* **168**, 158—177.

Whittembury, G. (1968). *J. gen. Physiol.* **51**, 303s—314s.

Whittembury, G. and Fishman, J. (1969). *Pflügers Archiv. Ges. Physiol.* **307**, 138—153.

Whittembury, G. and Proverbio, F. (1970). *Pflügers Arch. Ges. Physiol.* **316**, 1—25.

Wilbrandt, W. (1941). *Arch. Ges. Physiol.* **245**, 22—52.

Willis, J. S. (1968). *Biochim. Biophys. Acta* **163**, 506—515.

Vincent-Marique, C. and Gilles, R. (1970a). *Life Sci.* **9**, 509—512.

Vincent-Marique, C. and Gilles, R. (1970b). *Comp. Biochem. Physiol.* **35**, 479—485.

Virkar, R. A. and Webb, K. L. (1970). *Comp. Biochem. Physiol.* **32**, 775—783.

Vislie, T. and Fugelli, K. (1975). *Comp. Biochem. Physiol.* **52A**, 415—418.

Zachariassen, K. E. (1972). *Acta Physiol. Scand.* **84**, 31A—32A.

Zadunaisky, J. A., Parisi, M. N. and Montoreano, R. (1963). *Nature* **200**, 365—366.

Ørskov, S. L. (1946). *Acta Physiol. Scand.* **12**, 202—212.

Ørskov, S. L. (1954). *Acta Physiol. Scand.* **31**, 221—229.

13. Gap Junctions and Hormone Action

W. J. Larsen

Department of Anatomy College of Medicine, University of Iowa, Iowa, U.S.A.

I. INTRODUCTION

Hormones regulate the transport of water and ions as well as many other cellular processes. It is widely accepted that the primary sensory system of the cell consists of receptor molecules located at the cell surface. Hormone receptors are similar to carrier molecules in that they recognize specific solutes in a highly selective way.

A variety of new techniques have been developed to study the cell surface and we are beginning to resolve specializations of the membrane that may be involved in the exchange of information and of matter between adjacent cells. In this chapter I wish to discuss a particular membrane specialization, the gap junction. As a site of cell-cell contact, this junction has been implicated in the flow of molecules and information from cell to cell. I will summarize the literature on this topic, and then discuss a new hypothesis, that gap junctions may also be intimately involved in hormone action.

A. Cell-to-cell Transfer of Small Molecules

In 1952, Weidman demonstrated apparent cytoplasmic continuity between Purkinje fibres by recording potential changes induced in the limiting membrane at varying distances from a current injecting electrode. Furshpan and Potter (1957), using a similar technique, demonstrated the presence of electrical synapses in the lateral giant axon of the crayfish. Since that time, electrical synapses have been found in a variety of nerve cells and in cardiac muscle cells (Hagiwara and Morita, 1962; Bennett, 1966; Bennett *et al.*, 1967a; Bennett *et*

M

al., 1967b; Nicholls and Purves, 1970; Asada and Bennett, 1971; Weidman, 1952).

It seems likely that ionic coupling in excitable cells may allow the electrical transmission of action potentials from cell to cell. It has been more difficult to explain the low resistance pathways that have been found in many non-excitable cells (Loewenstein and Kanno, 1964; Loewenstein *et al.*, 1965; Potter *et al.*, 1966; Sheridan, 1966; Furshpan and Potter, 1968; O'Lague *et al.*, 1970; Sheridan, 1970; Michalke and Loewenstein, 1971; Johnson and Sheridan, 1971; Sheridan, 1971a, 1971b).

B. Cell-to-cell Pathways

The presence of intercellular bridges or fusions between the cell membranes of adjacent cells has been considered since the elaboration of the Cell theory by Schwann in 1839 (Wood, 1959). The search for permeable aqueous pathways connecting cell interiors was stimulated by the finding that in many normal tissues ions and small dye molecules pass from cell to cell. On the basis of largely circumstantial evidence it is suspected that the intercellular channels are located at the gap junctions.

1. Structure of the Gap Junction

Gap junctions have been described with transmission electron microscopy of stained thin sections (Fig. 1), sections of tissue infiltrated with lanthanum (Fig. 2), platinum carbon replicas of freeze fractured membrane (Fig. 3), and negatively stained isolated junctions (Benedetti and Emmelot, 1968; Goodenough and Revel, 1970; Goodenough and Stoeckenius, 1972). The three-dimensional structure of the gap junction emerging from these observations is represented in Fig. 4. Since intramembranous particles of gap junctions apparently make contact in the intermembrane space (Chalcroft and Bullivant, 1970; McNutt and Weinstein, 1973; Peracchia, 1973a, 1973b; Goodenough and Gilula, 1974), the gap junction fulfills the first prerequisite of a structure providing channels joining the interiors of adjacent cells; it appears that paired particles may span both membranes and the intercellular space.

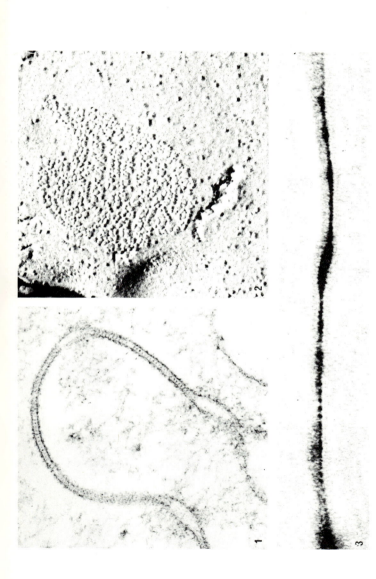

Fig. 1. Stained section of gap junction of A/B$_m$-5 cells. Note periodic densities between cells at gap junction with spacing of approximately 8 nm. These cells were also in block stained with 0.5% uranyl acetate. x119,400. Reproduced with permission of the author and Rockefeller University Press.

Fig. 2. Unstained section of gap junction infiltrated with lanthanum. Unstained intercellular bridges have a periodicity of approximately 8 nm. x236,400. Reproduced with permission of the author and Rockefeller University Press.

Fig. 3. Replica of P face of freeze fractured A/B$_m$-5 plasma membrane showing aggregation of particles characteristic of vertebrate gap junctions. The minimum center-to-center particle spacing is about 8 nm. x98,900. Reproduced with permission of the author and Rockefeller University Press.

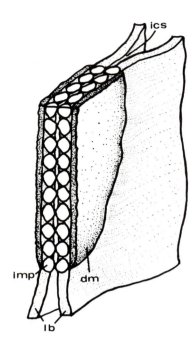

Fig. 4. Approximately one-quarter of a typical button shaped gap junction is illustrated in this diagram. Apposed membranes (lipid bilayers: lb) are separated by only 2—4 nm in these regions where intramembranous gap junctional particles (imp) apparently make contact in the intercellular space (ics). A recent study of glutaraldehyde-calcium fixed gap junctions (Larsen, 1975; see Fig. 10) suggests that some of the apparent density of the gap junctional membrane may be contributed by a cytoplasmic dense material (dm).

2. Correlation of Electrophysiological and Structural Studies

More circumstantial evidence for the existence of aqueous intercellular pathways through the gap junction particles comes from tracer and electron microscopic studies. In many cases, the existence of gap junctions has been demonstrated in tissues also shown to pass ions or small dye molecules between their constituent cells (Gilula *et al.*, 1972; Bennett and Trinkaus, 1970; Johnson and Sheridan, 1971;

Azarnia *et al.*, 1974; Larsen *et al.*, 1976). Conversely, gap junctions appear to be absent from several cell types in which the results of tracer experiments have also been negative (Gilula *et al.*, 1972; Azarnia *et al.*, 1974; Larsen *et al.*, 1976). In addition, it also seems probable that the gap junction is the only junctional specialization in animal cells which is as widely distributed throughout the phylogenetic spectrum as the coupling phenomenon demonstrated with electrophysiological and tracer experiments.

Based upon this evidence, studies of the cell-to-cell movement of ions and small dye molecules have been interpreted to indicate the presence of gap junctions in these tissues. It now appears, however, that in some tissues electrical coupling can be demonstrated while the intercellular movement of dyes such as fluorescein or procion yellow may not be detected. This is the case for at least two clones of intermediate revertants derived from somatic cell hybrids produced by fusing a tumorigenic mouse fibroblast with a human Lesch-Nyhan cell (Azarnia *et al.*, 1974; Larsen *et al.*, 1976). Most of the cells within colonies of a cloned Morris hepatoma in culture appear to be electrically coupled but fluorescein moves from cell-to-cell only occasionally and even then only to one or two adjacent cells (Azarnia and Larsen, 1976). In addition, the early blastomeres of several embryos appear to be coupled electrically, but do not show cell-to-cell passage of fluorescein (Bennett and Spira, 1971; Slack and Palmer, 1969; Bennett *et al.*, 1972; Tupper and Saunders, 1972). These observations could be explained by either quantitative or qualitative structural differences (Bennett, 1973; Bennett, 1974; Sheridan, 1976), but definitive answers await coordinated physiological and morphological studies.

Although some cells from different species and tissues may electrically couple with one another (Michalke and Loewenstein, 1971), other heterogeneous cell pairs appear to be unable to form permeable intercellular pathways. For example the Morris hepatoma cells are unable to couple with normal liver fibroblasts in culture (Azarnia and Loewenstein, 1971) although they are apparently capable of ionic coupling and limited fluorescein transfer between themselves. In addition, electron microscopy has revealed gap junctions in clones of these cells grown in artificial medium (Fig. 5).

These data suggest that structures identified as gap junctions may potentially allow the cell-to-cell movement of small molecules, but electrical coupling does not always signify the presence of gap junctions.

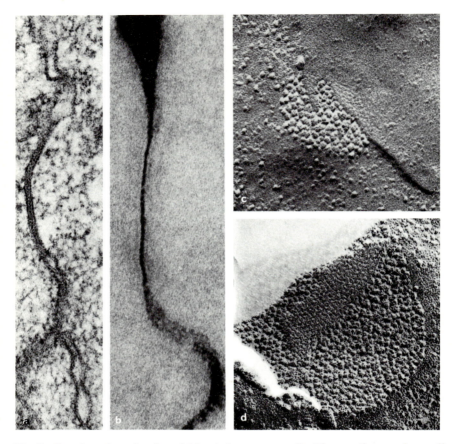

Fig. 5. Gap junctions in cloned Morris hepatoma cells. These cells grow in small colonies and form microvilli, tight junctions and occasionally canuliculi similar to those appearing in normal liver cells. a) stained thin section ×142,300. b) thin section of unstained, lanthanum infiltrated junction, ×225,100. c, d) freeze fracture replicas of gap junctions (c, ×141,700; d, ×156,400).

II. THE LIFE HISTORY OF GAP JUNCTION

The major assumption underlying most of what follows in this review is that something can be learned about the physiological significance of a cell organelle by a study of its genesis and degradation and of conditions which affect its frequency, size and form. Details of the formation and degradation of gap junctions have only recently been studied, but observations of freeze fracture replicas and stained thin sections indicate that the gap junction is much more dynamic than has been inferred from previous studies.

A. Formation

As stated above, freeze fracture studies have indicated that the gap junctions are composed of aggregates of intramembranous particles paired across intramembranous space in closely apposed membranes. There is some evidence that as membranes come into close apposition these particle aggregates enlarge by the accretion of single particles or small groups of particles. Johnson *et al.* (1974) have described this phenomenon in some detail in Novikoff hepatoma cells which were first separated and then reaggregated. They demonstrated that small particle aggregates appeared within 30—60 minutes after the cells were mixed together. The aggregation of gap junction particles was preceeded by the appearance of "formation plaques" composed of small areas of particle free or particle poor membrane in regions where adjacent cells are closely apposed. During the next two hours progressively larger aggregates of particles were observed, suggesting growth of the junction by the coalescence of single particles or small groups of particles. Similar patterns of apparent gap junctional growth have been observed in developing neural epithelium (Decker and Friend, 1974) and ependymoglial cells (Decker, 1976a) of the amphibian embryo, in rabbit granulosa cells (Albertini *et al.*, 1975), and in the cells of regenerating liver (Yee, 1972). Progressively larger particle aggregates are also observed in a line of tissue culture cells as they grow to densities leading to cell-to-cell contact and confluence (Fig. 6).

These studies suggest the possibility that gap junction formation involves the insertion of particles into the membrane, followed by their movement within the lipid bilayer to sites of membrane apposition where they may then crosslink with particles across the intercellular space. However, it is not known whether insertion of the junctional particles into the membrane occurs at some distance from or in close proximity to the growing particle-aggregate. It has been suggested that these particles may be inserted directly into the "formation plaque" (Decker, 1976a).

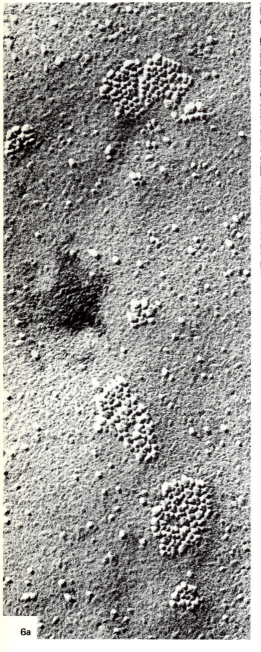

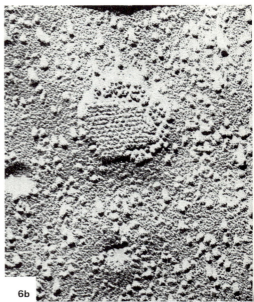

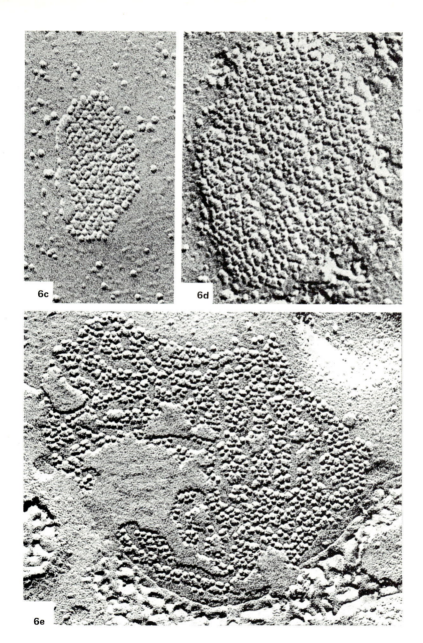

6c

6d

6e

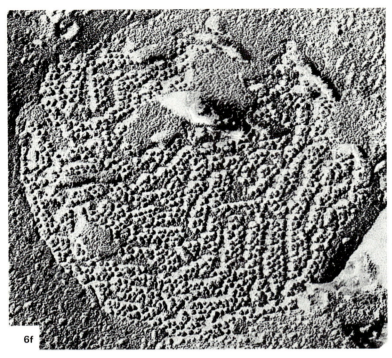

Fig. 6. Gap junctions in A/B$_m$-5 cells. These replicas were obtained during a three day period following heavy seeding of tissue culture dishes with A/B$_m$-5 cells. All junctional forms represented here are found in cultures after three days growth while the large junctions (e, f) were observed only after three days. It seems possible then, that these structures may represent various stages of gap junctional growth and maturation. Note the close packing of particles in smaller junctions (a, b, c, d), and the apparent organization of particles into distinct columns separated by particle-free aisles in the larger junctions (e, f). This latter organization has also been reported in the relatively large granulosa cell, (Albertini *et al.*, 1975) and ciliary epithelial (Kogon and Pappas, 1975) gap junction. ×168,900.

B. Destruction

Clues to the mechanism of gap junction degradation have only been obtained recently. It appears that gap junctions may be degraded after internalization into the cell (Fig. 7). This possible sequence of events has been suggested by observations of blebs formed by the invagination of one cell into another at gap junctions, and of circular gap junctional profiles apparently detached from the limiting membrane (Garant, 1972; Espey and Stutts, 1972; Merk *et*

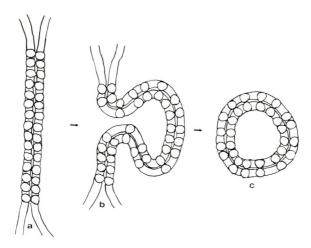

Fig. 7. Possible mechanism of gap junction interiorization. a) cross section of typical intercellular gap junction. b) one cell invaginates into the other at the gap junction forming a bleb and c) the bleb pinches off disconnecting the gap junction from the limiting plasma membranes.

al., 1973; Bjersing and Cajander, 1974; Albertini *et al.*, 1975). These circular profiles have been referred to as annular gap junctions (Merk *et al.*, 1972). In some cases annular gap junctions may enclose one or more other annular gap junctions (Fig. 8) and may also occur with great profusion in some cell types (Garant, 1972; Merk *et al.*, 1973). The separation of some of these profiles from the limiting membrane of the cell is suggested by lanthanum tracer studies (Garant, 1972) and by serial sectioning (Merk *et al.*, 1972; Espey and Stutts, 1972).

The internalization of gap junctions may lead to their degradation, since the cytoplasm enclosed within the internalized gap junction vesicles often appears degenerate in electron micrographs of stained thin sections. Mitochondria often appear washed out and possess broken membranes while enclosed ribosomes appear diffuse and swollen (Garant, 1972; Merk *et al.*, 1973). Cytoplasm on the concave surface of invaginating junctions or enclosed within gap junctional vesicles is often particularly rich in ribosomes (Garant, 1972; Merk *et al.*, 1973; Orwin *et al.*, 1973), and it has also been shown that ribosomes are specifically attracted to gap junctions in hepatocytes undergoing autolysis (David-Ferreira and David-Ferreira, 1973). In one semi-quantitative study of granulosa cells from preovulatory follicles it was suggested that the number of these gap junctions first increases and then decreases during one short period, suggesting that

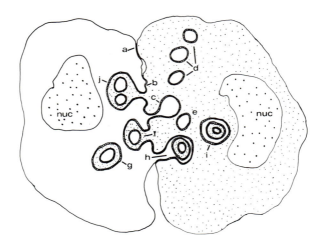

Fig. 8. Diagramatic composite of various gap junctional profiles observed in a variety of tissues. a) typical intercellular gap junctional profile is relatively small and straight or slowly curving b) longer profiles are often sinusoidal c) profile of probable gap junctional bleb d) annular gap junctional profiles are often observed in close proximity or e) in isolation f) annular profiles may occur within possible gap junctional blebs g) a single annular profile may be enclosed by another h) this latter configuration sometimes occurs within a gap junctional bleb i) three concentric annular gap junctional profiles have also been observed j) two annular profiles may occur side-by-side within a gap junctional bleb. Stippling indicates the probable origin of cytoplasm enclosed within each respective junctional profile.

these internalized gap junctional vesicles are degraded after they are interiorized (Bjersing and Cajander, 1974).

C. Turnover

These studies support the conclusion that gap junctions are rather dynamic structures capable of rapid turnover in some cell types. The rapid growth of gap junctional particle aggregates after the reaggregation of novikoff hepatoma cells (Johnson et al., 1974) and the enclosure of one or more internalized gap junctional vesicles within others is consistent with this view. A diagramatic representation of these events appears in Fig. 9.

D. Control of Turnover

The effects of hormones upon the intercellular movements of ions and small dye molecules have not been generally considered in

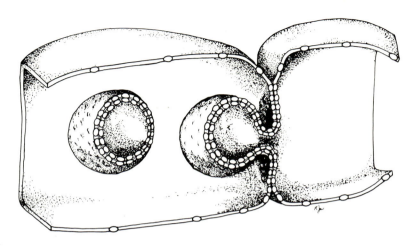

Fig. 9. Three-dimensional representation of gap junction turnover. Intramembranous particles in non-junctional membrane may aggregate within the lipid bilayer to form a gap junction. Gap junctional blebs formed by invagination of one cell into another at the gap junction may pinch off to form an interiorized gap junctional vesicle.

discussions of intercellular coupling. A number of recent studies, however, suggest that gap junction turnover is profoundly affected by a variety of hormones.

Several recent investigations have implicated several different protein hormones to stimulate interiorization of gap junction: one semi-quantitative study has demonstrated that an ovulating dose of human chorionic gonadotropin (hCG) may induce the interiorization of gap junctions in the granulosa cells of mature ovarian follicles (Bjersing and Cajander, 1974). In this study the frequency of interiorized gap junctions following an hCG injection was shown to first increase and then decrease.

Annular gap junctions appear to be particularly frequent in a variety of other tissues which are probable targets of protein hormones, including luteal cells of several species (Enders, 1973), Leydig cells of the dog (Connell and Christensen, 1975), estrogen induced adenocarcinoma of the proximal convoluted tubule of the Syrian hamster (Letourneau et al., 1975), and the pigmented retina of the human embryo (Fisher and Linberg, 1975). Extensive and highly involuted gap junctions have been described in the pigmented ciliary epithelium of the rabbit (Kogon and Pappas, 1975). Gap junctions of the adrenal cortex have been reported to be extensive (Friend and Gilula, 1972), and it has also been demonstrated that ACTH stimulates the dissociation of adrenalcortical cells in primary culture (Ramachandran and Suyama, 1975). (Dissociation of granulosa cells in mature ovarian rabbit follicles also occurs after ovulating injections of hCG (Bjersing and Cajander, 1974)). In addition, it has recently been reported that ACTH stimulates the formation of gap junctions in a line of adrenocortical Y-1 tumor cells in culture (Decker, 1976b). Ependymal cells may also possess extensive and interiorized gap junctions (Brightman, 1965) and it has recently been demonstrated that some of the ependyma may be a target of the peptide hormone angiotensin II (Phillips and Hoffman, 1976). Annular gap junctions have also been reported in several tissues not yet characterized as peptide hormone targets including the oenocytes of an insect (Locke, 1969), wool follicles (Orwin *et al.*, 1973), and mature ameloblasts of the rat enamel organ (Garant, 1972). Some of these studies are discussed in greater detail below.

Gap junctional growth and internalization may also be stimulated by estrogen, vitamin A and thyroxine. Estrogen may induce both the formation and interiorization of gap junctions in granulosa cells of immature rat follicles (Fletcher and Robertson, 1975; Merk *et al.*, 1972). Injection of estrogen restores both electrical excitability and gap junctions in rat myometrium after their disappearance following ovariectomy (Bergman, 1968). Annular gap junctions appear in adenocarcinomas of the proximal convoluted tubule of the Syrian hamster arising in response to prolonged estrogen administration and in their abdominal metastases (Letourneau *et al.*, 1975), whereas topical applications of vitamin A induce the formation of normal intercellular and annular gap junctions in a keratoacanthoma of the rabbit ear (Prutkin, 1975). The growth of gap junctions in normal embryonic chick shank epidermis in organ culture is rapid and dramatic in response to vitamin A added to the medium (Elias and Friend, 1976). Recently, Decker (1976a) has demonstrated the rapid growth of ependymoglial gap junctions in hypohysectomized *Rana pipiens* embryos injected with thyroxine.

III. FUNCTIONAL SIGNIFICANCE OF THE GAP JUNCTION

As previously discussed, in non-excitable cells there seems to be no direct evidence specifically implicating the gap junction in any particular cell function. Major hypotheses have nevertheless been based on the possibility that these structures may provide the cell-to-cell channels implied by electrophysiological and tracer studies. On the basis of equally circumstantial evidence it has been suggested that gap junctions themselves may be aggregates of hormone receptors (Albertini et al., 1975). I will first discuss several studies with possible relevance to this latter suggestion and then briefly consider other hypotheses already discussed in detail by others.

A.   Gap Junctions and Hormone Receptors in Granulosa Cells

Albertini et al. (1975) have proposed that interaction of lutinizing hormone (LH) with its receptor in the membrane of granulosa cells of mature ovarian follicles results in the aggregation of hormone-receptor complexes into gap junctions, and suggested that ovulating doses of hCG (which binds to LH receptors) induce interiorization of these receptor aggregates. Of relevance to this hypothesis are a series of studies which suggest that the presence or absence of LH receptors on granulosa cell membranes is correlated with the presence or absence respectively of gap junctions in these cells during different developmental periods.

Granulosa cells appear to acquire gap junctions and LH receptors during follicular development. Albertini and Andersen (1974b), utilizing both thin section and freeze fracture electron microscopic techniques, reported that granulosa (follicle) cells in small, immature, preantral follicles do not possess gap junctions whereas granulosa cells from early antral follicles have gap junctions which seem to enlarge as development of the follicle continues. In studies of radioactive hCG binding to porcine granulosa cells removed from follicles of different sizes, Channing and Kammerman (1973) demonstrated that granulosa cells from large follicles may bind 10–10,000 times more radioactive hCG than granulosa cells from small follicles. Amsterdam et al. (1975) have directly demonstrated large numbers of hCG binding sites on the granulosa cells of large rabbit ovarian follicles but not on the granulosa cells of small follicles. Erickson et al. (1974) have also demonstrated that granulosa cells from immature follicles of the rabbit are unable to secrete progesterone when incubated in medium containing hCG whereas granulosa cells from large follicles secrete large amounts of progesterone after hCG stimulation.

Although more equivocal, several recent studies suggest that gap junctions and LH receptors may be lost from granulosa cells in preovulatory follicles just prior to ovulation. As noted previously,

Bjersing and Cajander (1974) have reported the interiorization of gap junctions in the granulosa cells of mature ovarian follicles of the rabbit after ovulating injections of hCG, and interiorized gap junctions have been observed in granulosa cells in mature follicles by several other investigators (Espey and Stutts, 1972; Merk *et al.*, 1972; Merk *et al.*, 1973; Albertini and Andersen, 1974b; Zamboni, 1974; Albertini *et al.*, 1975). Interestingly, Marsh *et al.* (1973) found that LH-stimulated adenylate cyclase activity disappears during the same 9–11 hour period following hCG injection, when gap junctions may be interiorized. One would like to know whether the decrease in LH-stimulated cyclase activity results from a loss of LH receptors. This question has not yet been resolved. Indeed it has been reported that radioactive hCG binding to rabbit follicles does not change significantly during the first 12 hours after mating (Mills and McPherson, 1974). (Some caution should be exercised in interpreting these experiments since 50–55% of the hCG binding in these follicles has been reported to be nonspecific.) Further questions regarding this issue have also been introduced by the finding that as much as 97% of the total cyclic-AMP production in isolated sheep follicles may occur in cells of the theca interna rather than in granulosa cells (Weiss *et al.*, 1976). These results might be explained by species' differences or by the variability observed in hCG binding to ovarian follicles during development. This problem may be resolved with careful hCG binding experiments and quantitative studies of gap junctional frequency on the granulosa and thecal cells in mature follicles of the rabbit after injections of ovulating doses of hCG.

Luteal cells in the rat appear to acquire gap junctions and LH receptors progressively during the first two weeks of pregnancy. Albertini and Andersen (1974a) demonstrated that gap junctions of rat and mouse luteal cells enlarge slowly during the first two weeks of pregnancy. Han *et al.* (1974) have directly demonstrated the binding of radioactive hCG to rat luteal cells in corpora lutea induced two days earlier by serum gonadotropin from pregnant mares. Especially interesting are the findings of Richards and Midgley (1976) who have demonstrated increasing numbers of LH receptors per μg of DNA in rat corpora lutea during the first two weeks of pregnancy.

1. Conclusion

There is thus some support for the idea that the frequency and size of gap junctions and the number of LH receptors in granulosa cells fluctuate in tandem during follicular maturation, ovulation and luteinization. However further studies correlating hormone binding and quantitative changes in the gap junctions in the granulosa cells of the same species are required. It will also be necessary to carefully define the involvement of other hormones in the maturation and luteinization process as suggested by the data of Richards and Midgley (1976). It must also be added that even if these correlations are correct, they do not provide the evidence necessary for identity of gap junctions and LH receptors in the granulosa cell.

B. Gap Junctions in other Peptide Hormone Target Tissues

Extensive intercellular gap junctions and annular gap junctions are prominent in other target tissues of peptide or glycoprotein hormones, such as the Leydig cells of the dog (Connell and Christensen, 1975), which are also targets of LH. Annular gap junctions have not yet been observed in the adrenal cortex, but as discussed previously, there is some reason to believe that gap junctions in this tissue are responsive to the peptide hormone ACTH.

Melanocyte stimulating hormone (MSH) is implicated in the synthesis of melanin in the skin through the activation of adenylate cyclase (Lee and Lee, 1973; McQuire, 1975; Johnson and Pastan, 1972; Wong et al., 1974; Geschwind et al., 1972) and it is for this reason that the appearance of extensive, invaginated gap junctions between pigmented cells in the ciliary epithelium of the rabbit and the occurrence of annular gap junctions in embryonic human pigmented retina may be relevant to this discussion. The annular junctions in embryonic pigmented retina and those occurring between pigment cells and neural retinal cells at this time may be of particular interest with regard to recent studies on the development of visual pathways (Sanderson et al., 1974; Guillery et al., 1975). These studies have shown that the development of abnormal visual pathways resulting from abnormalities in the growth of axons from the neural retina through the optic chiasm and to the geniculate bodies is always correlated with the condition of ocular albinism in several diverse mammalian species. This correlation is apparently only true for organisms in which the pigment retina develops immediately adjacent to the neural retina and not in organisms such as the axolotl (Guillery and Updyke, 1976) where these cell layers are not in direct contact. It will certainly be of interest to compare the apposed membranes of presumptive pigmented retina and neural retina in mammalian ocular albinos and normal individuals at the critical stages of development.

C. The Effect of other Hormones on Gap Junction Turnover

Estrogen appears to stimulate the formation and possible interiorization of gap junctions in several tissues. However, estrogen apparently interacts with cells at the level of cytoplasmic receptors, and it seems unlikely that its effects on gap junctions occur through the same mechanism as that hypothesized for the action of peptide hormones. Estrogen may, however, potentiate the effects of the glycoprotein hormone LH in granulosa cells, since it appears to be required for the synthesis of LH receptors through the stimulation of new mRNA and protein synthesis (Richards and Midgley, 1976). The action of vitamin A is also just beginning to be understood, and it appears that it may in some instances bind to specific cytoplasmic receptors (Bashor et al., 1973; Ong and Chytil, 1975a, 1975b; Ong et al., 1975).

D. Cyclase Localization Techniques; Calcium Deposits on Gap Junction Membrane?

The peptide and glycoprotein hormones discussed above all apparently produce their effects through the stimulation of adenylate cyclase by interacting with receptors located at the outer aspect of the plasma membrane. It is thought that alterations in receptor conformation may then lead to direct interactions with adenylate cyclase molecules located at the cytoplasmic aspect of the membrane where cyclic-AMP and inorganic pyrophosphate are formed by the cleavage of ATP (Robison et al., 1971).

Several attempts have been made to localize adenylate cyclase activity with techniques similar to that first introduced by Gomori (1941) for phosphatases. The basis for this procedure involves the precipitation of an enzymatically produced anion (usually phosphate) with a cation introduced in the reaction mixture. Specificity of the reaction is determined by utilizing specific substrates at specific pH, in the presence of specific inhibitors or stimulators of the enzyme in question. In the case of adenylate cyclase, lead is commonly included in a reaction mixture containing

5′-adenylyl-imidodiphosphate (AMP-PNP) based on the assumption that PNP will be precipitated as an electron dense deposit. AMP-PNP imparts additional specificity to the adenylate cyclase reaction since this is a substrate phosphatases are not able to utilize (Rodbell *et al.*, 1971; Rosenthal *et al.*, 1966). When this technique has been used for the ultrastructural localization of adenylate cyclase, however, the results have been generally disappointing. In experiments with rat liver cells (Reik *et al.*, 1970) and with rat pancreatic islets (Howell and Whitfield, 1972) reaction product appears on the outside rather than on the cytoplasmic side of the membrane where it is expected to occur on the basis of biochemical experiments. In some cases, little or no specificity for the adenylate cyclase reaction can be demonstrated, and biochemical controls have shown that both the fixative and lead used in these procedures inhibit a major proportion of the adenylate cyclase activity (Lemay and Jarett, 1975). In one study, however, reaction product did appear on the cytoplasmic aspect of the plasma membrane of *Dictyostelium discoideum* after incubation of intact cells or isolated membrane vesicles with AMP-PNP and lead (Cutler and Rossomando, 1975) and in controls omitting substrate or in boiled preparations little or no reaction product was formed.

It has recently been reported that the fixation of a tissue culture line in calcium-glutaraldehyde results in the formation of electron opaque deposits on the cytoplasmic faces of gap junction membrane (Larsen, 1975). These deposits are often precisely paired from cell to cell across the gap junctional membranes and paired deposits are frequently equivalent in size (Fig. 10). Among several possible explanations it was suggested that these deposits may have formed as calcium entered the cells with the fixative, trapping pyro-phosphate produced from endogenous ATP by adenylate cyclase located at the gap junction. Similar deposits have now been observed after glutaraldehyde-calcium fixation on the cytoplasmic faces of gap junction membrane in Leydig cells (Connell, 1975), granulosa cells (Larsen, 1975) and cultured Lesh-Nyhan cells (Larsen, 1975). Ultimate support for this or any other interpretation of these results awaits elemental analysis of the deposits and a specification of conditions required for this reaction.

The possibility that adenylate cyclase may be localized in regions of cell contact, however, is supported by other pieces of circumstantial evidence. It has now been demonstrated that in many cases the contact of normal cells in tissue culture is correlated with a significant rise in intracellular cyclic-AMP concentrations (Andersen

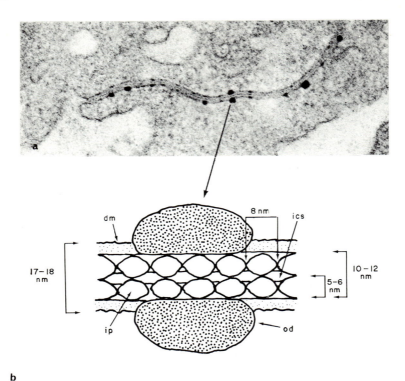

Fig. 10. a) Stained section of gap junction fixed in glutaraldehyde-calcium shows some pairing of deposits of equal size across the gap junctional membranes. x94,100. b) Diagramatic enlargement of area of gap junctional profile with paired deposits. This diagram illustrates a probable arrangement of major junctional elements. It is not known whether deposit centers are strictly aligned with the central axis of one intramembranous particle pair. Reproduced with permission of the author and Rockefeller University Press.

et al., 1973; Bannai and Shepard, 1974; D'Armiento et al., 1973; Heidrick and Ryan, 1971; Otten et al., 1971; Andersen and Pastan, 1975). If adenylate cyclase activity were dependent in part upon the aggregation of receptor-cyclase complexes in the membrane and their cross-linkage to other cells, these results might be readily explained if gap junctions were in part, aggregates of adenylate cyclase molecules. Needless to say, other equally attractive explanations for calcium deposition on the gap junction membrane are possible, and these will be discussed in a later section.

E. Biochemical Evidence for Protein Hormone Receptor
Aggregation and Loss

If we are to consider seriously the hypothesis that gap junctions
represent aggregates of receptor-cyclase complexes, we may ask
whether or not there is any evidence in the biochemical literature
supporting the idea that peptide or glycoprotein hormone receptors
do aggregate in the plasma membrane and turn over in response to
the presence of the hormones in question.

It has recently been suggested that many membrane receptors
patch or cap in the membrane after interaction with their specific
ligand (Edelman, 1976), but most of the evidence discussed in
support of this idea is derived from studies of receptors for H-2
antigens, concanavalin A, cholera toxin and insulin. Recent
biochemical studies on the interaction of the insulin molecule with
its receptor for example suggest that insulin receptors may interact in
such a way that increased occupancy of receptor sites leads to more
rapid dissociation rates of hormone and receptor. It has been
suggested that these results can be explained if insulin receptors exist
as oligomeric structures or clusters in the plasma membrane or are
induced to aggregate upon addition of the hormone (DeMeyts et al.,
1976). This phenomenon has also been demonstrated for receptors
of thyroid stimulating hormone (Li, 1973) and other hormone
receptor sites (DeMeyts et al., 1976). It has also been suggested that
the LH stimulated desensitization of LH-mediated adenylate cyclase
activity could result from an aggregation of LH receptors (Ryan and
Lee, 1976), and this receptor mediated, hormone specific
desensitization appears even to occur in cell-free particulate fractions
of porcine Graafian follicles (Bockaert et al., 1976).

Several electron microscopic studies have provided direct evidence
that insulin receptors are aggregated in the membrane rather than in
random array after the interaction of the receptor with
ferritin-labeled insulin. These results have been observed in isolated
plasma membranes of adipocytes (Jarett and Smith, 1974) and in
intact adipocytes (Jarett and Smith, 1975) and liver cells (Orci et al.,
1975).

Direct evidence for the aggregation of cholera toxin receptors on
rat lymphocyte membranes has been obtained with the use of
fluorescein labeled cholera toxin (Craig and Cuatrecasas, 1975), and
it has been proposed that aggregation is a basic component of the
mechanism leading to the stimulation of adenylate cyclase in these
cells.

In addition to the possible effects on receptor aggregation discussed above, recent reports suggest the possibility that incubation of intact cells with hormone for longer periods of time or at relatively high concentration may result in an actual loss of receptors from the limiting membrane of the cell. It has already been mentioned that high doses of hCG result in the loss of LH-mediated cyclase activity in mature ovarian follicles over a period of several hours (Marsh *et al.*, 1973). The number of insulin receptors per cell appears to decrease in lymphocyte membranes *in vitro* during chronic exposure to relatively high concentrations of insulin (Gavin *et al.*, 1974; Kahn *et al.*, 1973).

F. Gap Junctions as Aggregates of Cell-to-cell Conduits

Several alternative hypotheses related to the functional significance of the gap junction in inexcitable cells are based upon their apparent role in the cell-to-cell transfer of ions and small dye molecules in a variety of tissues. It has recently been suggested that molecules as large as approximately 1,000 daltons may be transferred from one cell interior to another (Loewenstein, 1975). I shall not discuss these experiments in any further detail since such a large number of excellent reviews already exist; but I will point out the one hypothesis assuming gap junctions as simple conduits between cells and elaborated in specific terms has been proposed by Sheridan (1976). Sheridan has suggested that cell division and differentiation may be regulated through the intercellular movement of cyclic nucleotides, and he has suggested that cyclic AMP and cyclic GMP themselves regulate the size of gap junctions (Sheridan and Johnson, 1975; Sheridan, 1976).

Other authors have proposed that the gap junction may provide permeable pathways between cells so that localized ionic imbalances may be buffered by the entire tissue volume (Loewenstein, 1966; Socolar, 1973) and to allow for the cell-to-cell diffusion of small metabolites (Gilula *et al.*, 1972; Gilula, 1974). It has also been suggested that these structures may provide pathways for the intercellular movement of "growth controlling signal molecules" (Loewenstein, 1975), but evidence involving the cell-to-cell movement of specific molecules in the control of specific biological phenomena in inexcitable tissues is not yet available. In addition, the recent finding of gap junctions between processes arising from the same cell suggest that these structures may not always be involved in

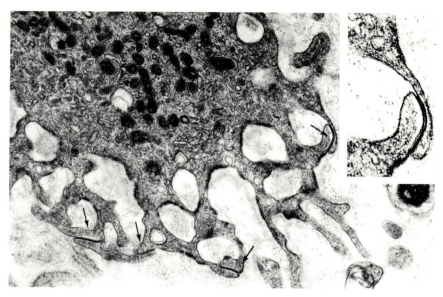

Fig. 11. Distal tips of pseudopodia arising from the same ovarian decidual cell interact to form gap junctions (arrows). x25,800. Inset x61,100 from Herr, 1976. Reproduced with permission of the author and Rockefeller University Press.

the cell-to-cell movement of small molecules (Herr, 1976). Gap junctions occurring between processes arising from the same ovarian decidual cell have been termed "reflexive gap junctions" (Herr, 1976) (Fig. 11). Similar junctions have been observed in vascular smooth muscle cells (Iwayama, 1971), mesangial and lacis cells of the kidney (Pricam *et al.*, 1974), in luteal cells (Enders, 1973), in Leydig cells of the dog and in a Dipteran salivary gland (J. L. Oschman, personal communication). Herr (1976) has suggested that reflexive gap junctions in ovarian decidual cells may be involved in the process of merocrine secretion, and it may therefore be significant that extensive, rapidly turning over gap junctions are so often associated with specific cell synthetic and secretory processes, some of which are mentioned below.

IV. CONCLUSIONS

The evidence supporting the functional hypotheses discussed in this review is indirect and circumstantial, and it is certainly possible that the apparent convergence of several lines of investigation is purely coincidental. For example, it is possible that hormone receptors do aggregate and turn over through endocytotic or exocytotic processes but that these phenomena have nothing to do with the behaviour of the intramembranous particles characterized as gap junctions. It does not seem likely, for example, that the aggregation of concanavalin A receptors is associated with concomitant aggregation of intramembranous particles. Although one recent report suggests that the aggregation of ferritin-labeled insulin on rat liver membranes may be associated with the aggregation of intramembranous particles, it seems likely that these aggregates occur on non-junctional membrane (Orci *et al.*, 1975). There is no unequivocal evidence for the association of hormones with gap junctions or for the localization of adenylate cyclase activity at the gap junction. Similarly, alternative explanations for calcium deposition on gap junction membranes include the precipitation of calcium with phosphate produced by phosphatases localized at the gap junction, and the binding of calcium to pre-existing binding sites which may regulate ionic movements through these membranes (Larsen, 1975).

Whether these structures are identical with hormone receptors or simply respond more indirectly to hormones, it does seem likely that the gap junction is more intimately related to hormonal action than has been previously suggested. In addition, these new insights into the response of gap junctions to various hormones also provide additional reason to believe that these structures may be involved in the control of cell growth and differentiation. The effects of various peptide hormones on the development and behavior of their respective targets, the effects of estrogen on specific cell synthesis (O'Malley and Means, 1974), concomitant effects on growth rate and gap junction frequency (Letourneau *et al.*, 1975), and the recently observed effects of vitamin A on the stimulation of cell synthesis and secretion (Marchok *et al.*, 1975; Clamon *et al.*, 1974; Prutkin, 1975; Levinson and Wolf, 1972) and upon the growth of certain tumor cells (Felix *et al.*, 1975; Smith *et al.*, 1975; Lasnitzki *et al.*, 1974; Prutkin, 1974) suggest that further studies of the gap junction will contribute to our understanding of the role of the cell membrane in these fundamental biological phenomena.

ACKNOWLEDGEMENTS

I gratefully acknowledge the many helpful discussions I have had with Drs David Garrison, R. Ho, John Marsh, David Smith, R. Azarnia, James Oschman, Betty Wall, Caryolyn Connell, David

Albertini, Paul Heidger, Asa Black, Kent Hermsmeyer and Mr John C. Herr. I thank Dr David Smith for use of the Balzers apparatus and Evelyn Jew for her skillfull drawing in Fig. 9. I am very grateful for the mature ovarian rabbit follicles given to me by Dr John Marsh. I also wish to thank Dr Ramsay for introducing me to the comparative method in Biological Science as an undergraduate through his book "Physiological Approach to the Lower Animals" (Cambridge University Press, 1962). Some of this work was supported by U.S. Public Health Service research grant CA14464, National Science Foundation grant BG36763X1 awarded to W. R. Loewenstein, and National Institutes of Health Postdoctoral Fellowship 1 F02 GM 51417-01 awarded to W. J. Larsen.

REFERENCES

Albertini, D. F. and Andersen, E. (1974a). *Anat. Record* **181**, 171—194.

Albertini, D. F. and Andersen, E. (1974b). *J. Cell Biol.* **63**, 234—250.

Albertini, D. F., Fawcett, D. W. and Olds, D. J. (1975). *Tissue and Cell* **7**, 389—405.

Amsterdam, A., Koch, Y., Lieberman, N. E. and Lindner, H. R. (1975). *J. Cell Biol.* **67**, 894—900.

Anderson, W. B. and Pastan, I. (1975). *In* "Advances in Cyclic Nucleotide Research" (Eds G. I. Drummond, P. Greengard and G. A. Robinson), Vol. V, pp. 681—698.

Anderson, W. B., Russell, T. R., Carchnan, R. A. and Pastan, I. (1973). *Proc. Natl Acad. Sci. U.S.A.* **70**, 3802—3805.

Asada, Y. and Bennett, M. V. L. (1971). *J. Cell Biol.* **49**, 159—172.

Azarnia, R. and Larsen, W. J. (1976). *In* "Intercellular Communication" (Ed. DeMello), Plenum, New York.

Azarnia, R. and Loewenstein, W. R. (1971). *J. Memb. Biol.* **6**, 368—385.

Azarnia, R., Larsen, W. J. and Loewenstein, W. R. (1974). *Proc. Natl Acad. Sci. U.S.A.* **71**, 880—884.

Bannai, S. and Shepard, J. R. (1974). *Nature, Lond.* **250**, 62—64.

Bashor, M. M., Toft, D. O. and Chytil, F. (1973). *Proc. Natl Acad. Sci. U.S.A.* **70**, 3483—3487.

Benedetti, E. L. and Emmelot, P. (1968). *J. Cell Biol.* **38**, 15—24.

Bennett, M. V. L. (1966). *Ann. N.Y. Acad. Sci.* **137**, 509—539.

Bennett, M. V. L. (1973). *Fed. Proc.* **32**, 65—75.

Bennett, M. V. L. (1974). *In* "Intercellular staining in Neurobiology" (Eds S. Kater and Nicholson), pp. 115—134. Springer-Verlag, New York.

Bennett, M. V. L. and Spira, M. E. (1971). *Biol. Bull.* **141**, 378.

Bennett, M. V. L. and Trinkaus, J. P. (1970). *J. Cell Biol.* **44**, 592—610.

Bennett, M. V. L., Spira, M. and Pappas, G. D. (1972). *Dev. Biol.* **29**, 419—435.

Bennett, M. V. L., Nakajima, Y. and Pappas, G. D. (1967a). *J. Neurophysiol.* **30**, 209—235.

Bennett, M. V. L., Pappas, G. D., Giminez, M. and Nakajima, Y. (1967b). *J. Neurophysiol.* **30**, 236—301.

Bergman, R. A. (1968). *J. Cell Biol.* **36**, 639—648.

Bjersing, L. and Cajander, S. (1974). *Cell Tiss. Res.* **153**, 1—14.

Bockaert, S., Hunzicker-Dunn, M. and Birnbaumer, L. (1976). *J. biol. Chem.* **251**, 2653—2663.

Brightman, M. W. (1965). *J. Cell Biol.* **26**, 99—123.

Chalcroft, J. P. and Bullivant, S. (1970). *J. Cell Biol.* **47**, 49—60.

Channing, C. P. and Kammerman, S. (1973). *Endocrin.* **92**, 531—540.

Clamon, G. H., Sporn, M. B., Smith, J. M. and Saffioti, U. (1974). *Nature* **250**, 64—66.

Connell, C. J. and Christensen, A. K. (1975). *Biol. Reprod.* **12**, 368—382.

Connell, C. J. and Connell, G. M. (1976). *In* "The Testis IV: 1970—1975" (Eds A. D. Johnson and W. R. Gomes), Academic Press, New York.

Craig, S. W. and Cuatrecasas, P. (1975). *P.N.A.S.* **72**, 3844—3848.

Cutler, L. S. and Rossomando, E. F. (1975). *Exp. Cell Res.* **95**, 79—87.

D'Armiento, M., Johnson, G. S. and Pastan, I. (1973). *Nat. New Biol.* **242**, 78—80.

David-Ferreira, J. F. and David-Ferreira, K. L. (1973). *J. Cell Biol.* **58**, 226—230.

De Meyts, P., Bianco, A. R. and Roth, J. (1976). *J. biol. Chem.* **251**, 1877—1888.

Decker, R. S. (1976a). *J. Cell Biol.* **69**, 669—685.

Decker, R. S. (1976b). *J. Cell Biol.* **70**, 412a (Abstr).

Decker, R. S. and Friend, D. (1974). *J. Cell Biol.* **62**, 32—47.

Edelman, G. M. (1976). *Science* **192**, 218—226.

Elias, P. M. and Friend, D. S. (1976). *J. Cell Biol.* **68**, 173—188.

Enders, A. C. (1973). *Biol. Reprod.* **8**, 158—182.

Epsey, L. L. and Stutts, R. H. (1972). *Biol. Reprod.* **6**, 168—175.

Erickson, G. F., Challis, J. R. G. and Ryan, K. J. (1974). *Dev. Biol.* **40**, 208—222.

Felix, E. L., Loyd, B. and Cohen, M. H. (1975). *Science* **189**, 886—888.

Fisher, S. K. and Linberg, K. A. (1975). *J. Ultrastruc. Res.* **51**, 69—78.

Fletcher, W. H. and Robertson, J. D. (1975). *J. Cell Biol.* **67**, 116 (Abstract).

Friend, D. S. and Gilula, N. B. (1972). *J. Cell Biol.* **53**, 148—163.

Furshpan, E. J. and Potter, D. D. (1957). *Nature* **180**, 342—343.

Furshpan, E. J. and Potter, D. D. (1968). *In* "Current Topics in Developmental Biology" (Eds A. A. Moscona and A. Monroy), Vol. III, pp. 95—127. Academic Press, New York.

Garant, P. R. (1972). *J. Ultrastruc. Res.* **40**, 333—348.

Gavin, J. R., Roth, J., Neville, D. M., DeMeyts, P. and Buell, D. N. (1974). *Proc. natl Acad. Sci. U.S.A.* **71**, 84—88.

Geschwind, I. I., Huseby, A. and Nishioka, R. (1972). *Progress in Hormone Res.* **28**, 91—130.

Gilula, N. B. (1974). *In* "Cell Communication" (Ed. R. P. Cox), pp. 1—29. John Wiley and Sons, New York.

Gilula, N. B., Reeves, O. R. and Steinbach, A. (1972). *Nature* 235, 262—265.
Gomori, G. (1941). *J. Cell. comp. Physiol.* 17, 71.
Goodenough, D. A. and Gilula, N. B. (1974). *J. Cell Biol.* 61, 575—590.
Goodenough, D. A. and Revel, J. P. (1970). *J. Cell Biol.* 45, 272—290.
Goodenough, D. A. and Stoeckenius, W. (1972). *J. Cell Biol.* 54, 646—656.
Guillery, R. W. and Updyke, B. V. (1976). *Brain Res.* 109, 235—244.
Guillery, R. W., Okoro, A. N. and Witkop, C. J. Jr. (1975). *Brain Res.* 96, 373—376.
Hagiwara, S. and Morita, H. (1962). *J. Neurophysiol.* 25, 721—731.
Han, S. S., Rajaniemi, H. J., Cho, M. I., Gershfield, A. N. and Midgley, A. R. Jr. (1974). *Endocrin.* 95, 589—598.
Heidrick, M. L. and Ryan, W. L. (1971). *Can. Res.* 31, 1313—1315.
Herr, J. (1976). *J. Cell Biol.* 69 (2), 495—501.
Howell, S. L. and Whitfield, M. (1972). *J. Histochem. Cytochem.* 20, 873—879.
Iwayama, T. (1971). *J. Cell Biol.* 49, 521—525.
Jarett, L. and Smith, R. M. (1974). *J. biol. Chem.* 249, 7024—7031.
Jarett, L. and Smith, R. M. (1975). *Proc. natl Acad. Sci. U.S.A.* 72, 3526—3530.
Johnson, G. S. and Pastan, I. (1972). *Nature New Biol.* 237, 267—268.
Johnson, R. and Sheridan, J. D. (1971). *Science* 174, 717—719.
Johnson, R. G., Hammer, M., Sheridan, J. D. and Revel, J. P. (1974). *Proc. natl Acad. Sci.* 71, 4536—4540.
Kahn, C. R., Neville, D. M. and Roth, J. (1973). *J. biol. Chem.* 248, 244—250.
Kogon, M. and Pappas, G. D. (1975). *J. Cell Biol.* 66, 671—676.
Larsen, W. J. (1975). *J. Cell Biol.* 67, 801—813.
Larsen, W. J., Azarnia, R. and Loewenstein, W. R. (1976). *E.M.S.A.* In press.
Lasnitzki, I. and Goodman, D. S. (1974). *Can. Res.* 34, 1564—1571.
Lee, T. H. and Lee, M. S. (1973). *Yale J. Biol. Med.* 46, 493—499.
Lemay, A. and Jarett, L. (1975). *J. Cell Biol.* 65, 39—50.
Letourneau, R. J., Li, J. J., Rosen, S. and Villee, C. (1975). *Can. Res.* 35, 6—10.
Levinson, S. S. and Wolf, G. (1972). *Can. Res.* 32, 2248—2252.
Li, C. H. (1973). "Advances in Human Growth Hormone Research" (Ed. S. Raite), pp. 321—341. Depart. of Health, Education and Welfare Publication No. (NIH 74-612, National Institutes of Health, Bethesda, Maryland).
Locke, M. (1969). *Tiss. and Cell* 1, 103—154.
Loewenstein, W. R. (1966). *Ann. N.Y. Acad. Sci.* 137, 441—472.
Loewenstein, W. R. (1974). *In* "Cell Membranes" (Eds G. Weissmann and R. Claiborne), pp. 113—122. H. P. Publishing Co., Inc., New York.
Loewenstein, W. R. (1975). *In* "The Nervous System" (Ed. D. B. Tower), Vol. I, pp. 419—436. The Basic Neuro-sciences. Raven Press, New York.
Loewenstein, W. R. and Kanno, Y. (1964). *J. Cell Biol.* 22, 565—586.
Loewenstein, W. R., Socolar, S. J., Higashire, Y., Kanno, Y. and Davidson, N. (1965). *Science* 149, 295—298.
Marchok, A. C., Cone, M. V. and Nettesheim, P. (1975). *Lab. Invest.* 33, 451—460.
Marsh, J. B., Mills, T. M. and LeMaire, W. J. (1973). *Biochim Biophys Acta* 304, 197—202.

McNutt, N. S. and Weinstein, R. S. (1973). *In* "Progress in Biophysics and Molecular Biology" (Eds J. A. V. Butler and D. Noble), pp. 45—101. Pergamon Press, Oxford and New York.

McQuire, J. (1975). *Clin. pharm. Therap.* **16**, 954—958.

Merk, F. B., Botticelli, C. R. and Albright, J. T. (1972). *Endocr.* **90**, 992—1007.

Merk, F. B., Albright, J. T. and Botticelli, C. R. (1973). *Anat. Rec.* **175**, 107—126.

Michalke, W. and Loewenstein, W. R. (1971). *Nature* **232**, 121—122.

Mills, T. M. and McPherson, J. C. (1974). *Proc. Soc. exp. Biol. Med.* **145**, 446—449.

Nicholls, J. G. and Purves, D. (1970). *J. Physiol. Lond.* **209**, 647—668.

O'Lague, P., Dalen, H., Rubin, H. and Tobias, C. (1970). *Science* **170**, 464—466.

O'Malley, B. W. and Means, A. R. (1974). *Science* **183**, 610—620.

Ong, D. E. and Chytil, F. (1975a). *J. biol. Chem.* **250**, 6113—6117.

Ong, D. E. and Chytil, F. (1975b). *Nature* **255**, 74—75.

Ong, D. E., Page, D. L. and Chytil, F. (1975). *Science* **190**, 60—61.

Orci, L., Rufener, C., Mallaisse-Lagae, F., Blondel, B., Amherdt, M., Bataille, D., Frecet, P. and Perrelet, A. (1975). *Israel J. Med. Sci.* **11**, 639—655.

Orwin, D. F. G., Thomson, R. W. and Flower, N. E. (1973). *J. Ultrastruc. Res.* **45**, 1—14.

Otten, J., Johnson, G. S. and Pastan, I. (1971). *Biochem. Biophys. Res. Commun.* **44**, 1192—1198.

Peracchia, C. (1973a). *J. Cell Biol.* **57**, 54—65.

Peracchia, C. (1973b). *J. Cell Biol.* **57**, 66—76.

Phillips, M. I. and Hoffman, W. E. (1976). *In* "International Symposium on the Central Actions of Angiotensin and Related Hormones." (Eds J. P. Buckley and C. Ferrario) In press. Pergamon Press.

Potter, D. D., Furshpan, E. J. and Lennox, E. S. (1966). *Proc. natl Acad. Sci. U.S.A.* **55**, 328—336.

Pricam, C., Humbert, F., Perrelet, A. and Orci, L. (1974). *J. Cell Biol.* **63**, 349—354.

Prutkin, L. (1974). *Experientia* **31**, 494.

Prutkin, L. (1975). *Can. Res.* **35**, 364—369.

Ramachandran, J. and Suyama, A. T. (1975). *Proc. natl Acad. Sci. U.S.A.* **72**, 113—117.

Reik, L., Petzold, G. L., Higgins, J. A., Greengard, P. and Barnett, R. J. (1970). *Science* **168**, 382—384.

Richards, J. S. and Midgely, A. R. (1976). *Biol. Reprod.* **14**, 82—94.

Robison, G. A., Butcher, R. W. and Sutherland, E. W. (1971). "Cyclic AMP." Academic Press, New York and London.

Rodbell, J., Birnbaumer, L., Pohl, S. L. and Krans, M. J. (1971). *J. biol. Chem.* **246**, 1877—1882.

Rosenthal, A. S., Moses, H. L., Beaver, D. L. and Schuffman, S. S. (1966). *J. Histochem. Cytochem.* **14**, 698—701.

Ryan, R. J. and Lee, C. Y. (1976). *Biol. Reprod.* **14**, 16—29.

Sanderson, K. J., Guillery, R. W. and Shackelford, R. M. (1974). *J. comp. Neurol.* **154**, 225—240.

Seifter, E., Zisblatt, M., Levine, N. and Rettura, G. (1973). *Life Sciences* **13**, 945—952.

Sheridan, J. D. (1966). *J. Cell Biol.* **31**, C1—C5.

Sheridan, J. D. (1970). *J. Cell Biol.* **45**, 91—99.

Sheridan, J. D. (1971a). *J. Cell Biol.* **50**, 795—893.

Sheridan, J. D. (1971b). *Dev. Biol.* **26**, 627—636.

Sheridan, J. D. (1976). *In* "B.B.A. (Membranes)" (Eds Nicolson and Post), Elsevier, Amsterdam.

Sheridan, J. D., Hammer, M. G. and Johnson, R. C. (1975). *J. Cell Biol.* **67**, 395 (Abstract).

Slack, C. and Palmer, J. F. (1969). *Expl. Cell Res.* **55**, 416—419.

Smith, D. M., Rodgers, A. E. and Newberne, P. M. (1975). *Can. Res.* **35**, 1485—1494.

Socolar, S. J. (1973). *Exp. Eye Res.* **15**, 693—698.

Tupper, J. T. and Saunders, J. W. (1972). *Dev. Biol.* **27**, 546—554.

Weidman, S. (1952). *J. of Physiol. Lond.* **118**, 348—360.

Weiss, T. J., Seamark, R. F., McIntosh, J. E. A. and Moore, R. M. (1976). *J. Reprod. Fertility* **46**, 347—353.

Wong, G., Pawelek, J., Sansone, M. and Morowitz, J. (1974). *Nature* **248**, 351—354.

Wood, R. L. (1959). *J. Biophys. Biochem. Cytol.* **6**, 343—351.

Yee, A. G. (1972). *J. Cell Biol.* **55**, 294 (Abstract).

Zamboni, L. (1974). *Biol. Reprod.* **10**, 125—149.

14. Protozoan Osmotic and Ionic Regulation

R. D. Prusch

Division of Biological and Medical Sciences, Brown University, Providence, Rhode Island, U.S.A.

Both freshwater and marine protozoa are capable of regulating their internal ionic environment. In addition, freshwater protozoa maintain themselves hyperosmotic with respect to their external environment. Elucidation of the mechanisms of these phenomena has depended in large part upon the development of several specialized techniques in the last 15—20 years. These include mass culturing of protozoa so that unidirectional and net ion movements can be followed in a population of cells, and techniques pioneered by Professor Ramsay for the manipulation of very small samples in order to investigate the freezing point depression and ionic contents of cytoplasmic and organelle microsamples. From different studies with protozoa, it has been established that maintenance of intracellular ionic and osmotic gradients in freshwater, and to a certain extent in marine protozoa is dependent upon various mechanisms for transfer of ions across the cell surface, including pinocytosis and active ion and water extrusion mechanisms involving the contractile vacuole.

I. IONIC REGULATION

The first comprehensive investigation of ionic regulatory processes in the protozoa was carried out with *Paramecium caudatum* by Akita (1941). It was established that the surface of the animal was permeable to cations, but suggested that it was impermeable to anions. The internal chloride concentration was found to be very low, and intracellular chloride was thought to be restricted to food vacuoles. In all probability the chloride determinations in this study were too low, due to inadequate quantitative techniques available at the time. Nevertheless, the relationship between internal and external cation concentrations was non-linear indicating a certain degree of cation regulation by *Paramecium*. The mechanism of this regulation presumably involved a cation exchange across the cell surface.

In the large ciliate *Spirostomum ambiguum* Carter (1957) attempted to explain the maintenance of a high internal K^+ concentration on the basis of a Donnan equilibrium: $K_i^+/K_o^+ = Cl_o^-/Cl_i^-$. The reported value of 7 mM K_i^+ in *Spirostomum*, obtained by tracer equilibration technique, is probably low and only represents the freely exchangeable fraction of internal K^+: a considerable amount of internal K^+ could be bound. Internal Na^+ was too low to be explained on the basis of a Donnan equilibrium in *Spirostomum* and consequently a Na^+-pump mechanism was proposed.

Klein (1959, 1961, 1964) showed in a series of studies with the soil amoeba *Acanthamoeba* sp. that Na^+ and K^+ were both actively transported across the cell membrane as well as being bound intracellularly. The methods used included tracer techniques, and measurement of net ion fluxes and effects of metabolic inhibitors on large numbers of cells. Changing K_o^+ or Na_o^+ did not appreciably effect the intracellular content of the other cation. This would indicate that if Na^+ and K^+ are both actively transported, their movements are not coupled.

A considerable amount of work on protozoan osmotic and ionic regulation has been done with the small ciliate *Tetrahymena pyriformis*. Large numbers of these animals can be grown and maintained simply in 2% proteose peptone. Dunham and Child (1961) established that *Tetrahymena* is capable of actively transporting both Na^+ and K^+, with Cl^- being distributed passively. Fractions of internal Na^+ and K^+ were restricted to an unexchangeable intracellular compartment. Andrus and Giese (1963), using low temperatures and several specific metabolic inhibitors, were also able to demonstrate active K^+ and Na^+ transport in *Tetrahymena*. Hoffmann and Kramhøft (1974) observed that ethacrynic acid inhibits Na^+ and K^+ transport processes. Since the Na^+ influx is decreased with ethacrynic acid, active Na^+ movements both into and out of the cell are indicated.

Chapman-Andresen and Dick (1962) postulated the active extrusion of Na^+ and passive distribution of Cl^- ions in the giant amoeba *Chaos chaos*. They felt that internal ion binding, especially Na^+, played little part in the ionic regulation of this animal. However, there was an unexchangeable Na_i^+ fraction whose presence could not be due to a proposed pumping mechanism. It was also speculated that the movement of K^+ across the plasmalemma of *Chaos* was linked to Na^+ movements. Riddle (1962) suggested that K^+ accumulation by *Chaos* also involved an active process.

Bruce and Marshall (1965) discounted the active transport of any ion species across the plasmalemma of *C. chaos*, partly because no change in internal ion concentrations was observed when the cells were exposed to $3°C$ for up to 20 minutes. The plasmalemma of *Chaos* was assumed to be permeable to Na^+ and K^+, but impermeable to Cl^-. They concluded that internal concentrations of Na^+ and K^+ were determined by a Donnan equilibrium.

In the smaller mononucleate freshwater *Amoeba proteus* both Na^+ and K^+ are actively transported across the plasmalemma (Prusch and Dunham, 1972). K^+ is actively accumulated from the medium and Na^+ is actively eliminated from the cells; Cl^- distribution is most likely passive. It was also observed in *A. proteus* that the ability of the cell to regulate internal Na^+ levels was dependent upon the level of Ca^{++} in the external medium. At very low external levels of Ca^{++}, increasing the external Na^+ concentration resulted in a linear increase of intracellular Na^+ (Fig. 1). As the external level of Ca^{++} is increased the intracellular concentration of Na^+ falls below the concentration in the external medium (Fig. 2), the maximum effect of external Ca^{++} on Na_i^+ being between 0.3 and 0.7 mM. This effect of Ca^{++} on internal Na^+ levels in *A. proteus* was explained on the basis of a decrease in Na^+ permeability of the plasmalemma with increasing external Ca^{++} and the existence of a Na^+-pump mechanism constantly moving Na^+ out of the cell.

The ciliate *Miamiensis avidus* is one of the very few marine protozoa in which osmotic and ionic regulatory mechanisms have been studied (Kaneshiro *et al.*, 1969). It was observed in this animal that both internal Na^+ and Cl^- were maintained at levels considerably below that of the sea water environment while internal K^+ was higher than that of the surrounding medium. Of the divalent ions which were analysed, internal Mg^{++} and Ca^{++} were maintained below environmental levels.

Ionic regulatory patterns in the protozoa are by no means

N

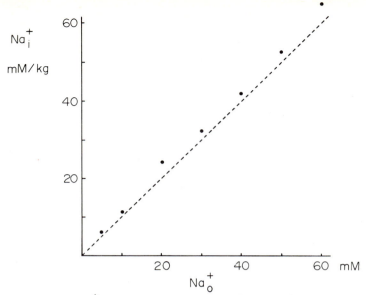

Fig. 1. Intracellular Na^+ in *Amoeba proteus* as a funtion of increasing external Na^+. Cells were equilibrated for three hours in Prescott-James medium (0.1 mM Ca^{++}) with the indicated external Na^+. The broken line represents the Na^+ "iso-concentration" line (from Prusch and Dunham, 1972).

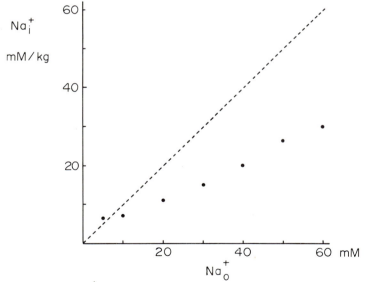

Fig. 2. Intracellular Na^+ in *Amoeba proteus* as a function of increasing external Na^+. Cells were equilibrated for three hours in Prescott-James medium with elevated Ca^{++} (1.0 mM) with the indicated external Na^+. The broken line represents the Na^+ "iso-concentration" curve (R. D. Prusch, unpublished).

consistent or clear and the mechanisms for maintenance of the various ionic gradients are even more obscure. Table I lists the measured intracellular and extracellular ion concentrations in different protozoa. There are variations in concentrations and ratios, but the most consistent observation is the similarity in internal K^+ concentrations. In two quite diverse protozoan classes, the Ciliata and Sarcodina (amoebae), internal K^+ is essentially the same — approximately 28 mM/kg wet weight. Internal Na^+ and Cl^- concentrations are not nearly as consistent between the species although internal Na^+ is apparently generally higher in the Ciliata than in the Sarcodina. The significance of these ionic patterns in

TABLE I

Internal protozoan ion concentrations

	Cells mM/kg	Medium mM/l	Reference
Paramecium caudatum	K 30 Na 20 Cl 0.36	K 10 Na 10 Cl 20	Akita (1941)
Tetrahymena pyriformis	K 31.7 Na 12.7 Cl 6.4	K 4.75 Na 36.5 Cl 28.7	Dunham and Child (1961)
Tetrahymena pyriformis	K 25 Na 5	K 2.5 Na 50	Andrus and Giese (1963)
Acanthamoeba	K 26.9 Na 14.3	dH_2O	Klein (1959)
Chaos chaos	K 33 Na 0.47	K 0.35 Na 0.22	Chapman-Andresen and Dick (1962)
Chaos chaos	K 34.5	K 0.23	Riddle (1962)
Chaos chaos	K 28.3 Na 0.35 Cl 16.5	K 0.43 Na 0.1 Cl 1.0	Bruce and Marshall (1965)
Chaos chaos	K 31 Na 0.57	K 0.08 Na 0.17	Riddick (1968)
Amoeba proteus	K 19 Na 3	K 0.1 Na 0.1	Batueva and Lev (1967)
Miamiensis avidus	K 73.7 Na 87.9 Cl 60.8	K 10 Na 450 Cl 550	Kaneshiro, Dunham and Holz (1969)
Tetrahymena pyriformis	K 28.8 Na 6.5 Cl 9.2	K 5 Na 31 Cl 36	Stoner and Dunham (1970)
Amoeba proteus	K 24.8 Na 1.08 Cl 9.73	K 0.14 Na 0.02 Cl 0.16	Prusch and Dunham (1972)

protozoa is not at all clear; the maintenance of an identical internal K^+ level in these diverse animals may reflect an optimal K^+ concentration in the cytoplasm or a compromise between a minimum requisite cytoplasmic K^+ level (Dunham and Child, 1961) and energy available to maintain this intracellular K^+ level in freshwater animals having such a relatively high surface to volume ratio.

Whatever the significance of the ion levels in protozoa, it is quite clear that they are established and maintained by a variety of active and passive transport processes. Uptake of ions across the cell surface involves both active transport and passive distribution due to electrical potential gradients and internal binding. In addition, mechanisms such as pinocytosis, which involves both active and passive processes, can account in part for the uptake of solutes.

II. PINOCYTOSIS

Pinocytosis, a term originated by Lewis (1931) occurs in a wide variety of cells although amoebae have been used mainly to investigate this process (Chapman-Andresen, 1962). The process of pinocytosis in amoeba was originally described by Mast and Doyle (1934). Its most obvious characteristics are the immediate cessation of movement upon the addition of an inducer to the external medium, assumption by the cell of a spherical form with "rosette" formation, and finally the formation of channels from the cell surface into the cytoplasm with pinching off of vesicles from their ends. Chapman-Andresen (1962) divides the substances which will induce pinocytosis into three groups;

    1. inorganic salts and amino acids,
    2. proteins and
    3. basic stains.

The common characteristic of this diverse group of substances is that they are only effective as pinocytotic inducers when they possess a net positive charge.

Apparently, the first step in pinocytosis is the binding of the inducer to the plasmalemma of the amoeba (Brandt, 1958; Schumaker, 1958). The plasmalemma of amoebae is a complex structure consisting of a unit-membrane 80 A thick, a clear space of 200 A devoid of any structural detail followed by a layer of "fibrous" material up to 2,000 A thick in contact with the outside medium (Pappas, 1960). The outside layer is a mucopolysaccharide, probably with fixed negative sites. De Terra and Rustad (1959) demonstrated quite convincingly that the binding of the inducer to the plasmalemma is not energy dependent, whereas the subsequent events in pinocytosis are.

Structural changes in the plasmalemma of amoeba associated with pinocytosis were characterized by Brandt and Freeman (1967). Elevation of external Na^+, an inducer of pinocytosis in amoeba, elicited a decrease in membrane resistance of *Amoeba proteus* and a concomitant increase in thickness of the plasma membrane. These changes in membrane structure and resistance could be blocked by increasing the external Ca^{++} concentration. Increased membrane thickness was thought to be brought about by hydration of the lipid layers of the plasma membrane, which would decrease its resistance by making it more ion permeable. Hydration of the plasma membrane during pinocytosis is also indicated by its increased permeability to polar nonelectrolytes in general (Brandt and Hendril, 1972).

Calcium ions may play a vital role in the underlying mechanism of pinocytosis. First of all, it has been noted that the presence of elevated Ca^{++} in the external medium increases the threshold concentration of inducer needed to elicit pinocytosis, suggesting that the sensitivity of the cell surface to the inducer is decreased by Ca^{++} (Brandt and Freeman, 1967; Josefsson, 1968). Secondly, external calcium apparently participates in controlling the overall permeability of the cell surface in protozoa (Prusch and Dunham, 1972). Elevated levels of external Ca^{++} decrease Na^+ permeability, even at Na^+ concentrations below that required to induce pinocytosis. Cooper (1968) has suggested that one step in pinocytosis may be displacement of Ca^{++} ions from the cell surface by the inducer. In the light of the preceding information, the sequential events in pinocytosis could thus be:

1. surface binding of the inducer to the plasmalemma or cell surface,
2. displacement of Ca^{++} ions from the cell surface by the inducer,
3. consequent general increase in cell surface permeability (with a concomittant non-selective increase in solute flux),
4. channel formation and
5. vesiculation and uptake into the cytoplasm of the specific inducer.

It is possible that channel formation and vesiculation may also be associated with localized Ca^{++} movements.

Although the primary function of pinocytosis has not been established in amoebae and has been the source of speculation in the past (Chapman-Andresen and Prescott, 1956), it must in part be associated with the uptake of specific, generally non-permeable, nutrient molecules. As the animal moves through its environment, it may periodically come in contact with localized relatively high concentrations of nutrient material, associated with the breakdown of animal or plant material. The process of pinocytosis would allow the animal to accumulate this material into its own cytoplasm, including material which would normally be too large to simply diffuse across the cell surface into the cytoplasm. Induction of pinocytosis by inorganic, monovalent cations such as Na^+ is most likely not primarily for the accumulation of this material, but simply because they fulfil a non-selective charge-density requirement which could trigger off the process of pinocytosis.

III. OSMOREGULATION: THE CONTRACTILE VACUOLE

Active uptake of material from the environment of protozoa, including the processes of active transport and pinocytosis, help maintain the osmotic and ionic integrity of these animals. In addition, the elimination of excess solute and water, which is carried out by the contractile vacuole, also contributes to overall internal osmotic and ionic homeostasis. Freshwater protozoa are distinctly hyperosmotic to their environment as has been demonstrated by a variety of techniques (Table II). In order to maintain themselves in this hyperosmotic condition, freshwater protozoa must possess a means of active solute uptake, which has been discussed previously, and a mechanism for the extrusion of excess water which enters the cell because of its hyperosmotic condition. The latter function is performed by the contractile vacuole. Contractile vacuoles are found primarily in free-living freshwater protozoa and sponges, and in marine ciliates. Structurally, the vacuole ranges from a transient organelle of relatively simple structure as is seen in the amoebae to a permanent cellular inclusion with complex structural relationships such as in the ciliates (Schneider, 1960; Elliot and Bak, 1964).

Indirect evidence obtained some time ago initially indicated that the contractile vacuole of freshwater protozoa was osmoregulatory in function. In general, it was found that the rate of contraction of the vacuole varied inversely with the osmolality of the external environment (Herfs, 1922). Thus, as the osmolality of the external

TABLE II
Intracellular osmolality of various protozoa

	mOsm/l	Method	Reference
Stentor	70	Electrical conductivity	Gelfan (1928)
Paramecium	70	Electrical conductivity	Gelfan (1928)
Spirostomum	77	Electrical conductivity	Gelfan (1928)
Euplotes	75	Electrical conductivity	Gelfan (1928)
Chaos	100	Vacuolar activity	Adolph (1926)
Chaos diffluens	5	Volume changes	Mast and Fowler (1935)
Paramecium	50	Vacuolar activity	Kamada (1935)
Spirostomum	52	Vapor pressure	Picken (1936)
Chaos chaos	94	Isotopic water	Løvtrup and Pigon (1951)
Discophyra piriformis	45	Vacuolar activity	Kitching (1956)
Carchesium aselli	45	Vacuolar activity	Kitching (1956)
Amoeba proteus	101	Freezing point depression	Schmidt-Nielsen and Schrauger (1963)
Chaos chaos	117	Freezing point depression	Riddick (1968)
Tetrahymena pyriformis	111	Freezing point depression (cellular extract)	Stoner and Dunham (1970)

medium is increased, the rate of fluid output by the vacuole declines. Presumably, increasing the external osmolality decreases the difference in osmo-concentration between the external medium and cell interior, which in turn decreases the amount of water entering the cell per unit time. Application of metabolic inhibitors, which also inhibits vacuolar function, causes freshwater protozoa to swell (Kitching, 1938).

The first direct evidence for the osmoregulatory role of the contractile vacuole in freshwater protozoa was obtained by Schmidt-Nielsen and Schrauger (1963) by determining the freezing point depression of samples obtained by direct micropuncture of the cytoplasm and contractile vacuole contents of *Amoeba proteus*. It was found that the cytoplasm had an osmolality of 101 mOsm, while the contents of the vacuole were 32 mOsm. The contents of the contractile vacuole are thus distinctly hyposmotic in respect to the cytoplasm, a situation consistent with the osmoregulatory function of the vacuole in freshwater protozoa. An essentially similar relationship between cytoplasmic and vacuolar osmolalities was found in the freshwater amoeba *Chaos chaos* (Riddick, 1968).

The question arises as to how this hyposmotic vacuolar fluid is being generated in freshwater protozoa. Various mechanisms have been suggested, including active water transport and filtration, which are not presently generally accepted. In general terms, the simplest mechanism for the formation of a hyposmotic fluid in the vacuole of amoebae would be a two stage process, separated spatially and/or temporally, in which an isosmotic fluid is first segregated from the cytoplasm and then solute is removed from this medium leaving a hyposmotic fluid. In amoebae, the contractile vacuole is a transitory structure surrounded by a layer of vesicles. It increases in size in a step-wise manner by a fusion of these vesicles, both their membranes and contents, with the main vacuole (Pappas and Brandt, 1958). If, as suggested by Schmidt-Nielsen and Schrauger (1963), the vesicles contained an isosmotic fluid which went into the generation of the contractile vacuole and its contents, then a hyposmotic fluid in the contractile vacuole could be formed by the removal of a particular solute(s), assuming a relatively low water permeability of the structure. This possibility was ruled out by evidence obtained by Riddick (1968) in which no correlation was found between vacuolar size and osmolality of the vacuolar contents. If the vacuole is initially formed by the fusion of vesicles with isosmotic fluid, and then preferentially removes solute, then the osmolality of the vacuolar contents should decrease with time as the size of the vacuole increases; this was not the case.

Riddick (1968) consequently suggested that the vesicles themselves contained a hyposmotic fluid and the vacuole itself formed from the coalescence of these vesicles would always contain a hyposmotic fluid. If this is the case, then an isosmotic fluid could be segregated into the vesicles which would then remove solute from this vesicular fluid. It has been known for some time, on the basis of indirect evidence, that Na^+ extrusion from freshwater protozoa and contractile vacuole activity are related (Dunham and Child, 1961; Chapman-Andresen and Dick, 1962). For example, if the external osmolality is increased by the addition of sucrose, the net flux of Na^+ from previously Na^+-loaded *A. proteus* is reduced (Fig. 3). Since vacuolar output is also reduced by increasing the external osmolality, then the reduced Na^+ flux from the cell may also be associated with the vacuole. By analysis of fluid obtained from the contractile vacuole of *C. chaos*, Riddick (1968) demonstrated that the contractile vacuole is removing Na^+ as well as water; the fluid being hyposmotic, low in K^+ and high in Na^+ in comparison with the cytoplasm. The possible sequence of events in the filling of the contractile vacuole could thus begin with the segregation of an isosmotic fluid into the vesicles. An active pump mechanism located in the membranes of the vesicles could move Na^+ into, and K^+ out of, the vesicles. If these ion movements were coupled in such a manner that several K^+ ions were moved out of the vesicle for each Na^+ moved in, and if the vesicular membrane possessed a low water

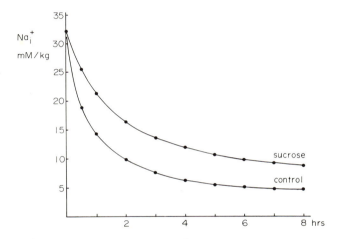

Fig. 3. Net Na^+ efflux from previously Na^+-loaded *Amoeba proteus*. Cells were loaded in Prescott-James medium with 30 mM Na^+ for four hours. The external Na^+ was reduced at zero time to approximately 0.2 mM and net Na^+ efflux was followed with time (control). Another group of cells was treated identically except the Na^+ efflux was followed in medium containing 100 mM sucrose (sucrose) (R. D. Prusch, unpublished).

permeability, then a high-Na^+, hyposmotic fluid could be generated. The vesicles would be very appropriate for this process because of their overall large surface to volume ratio, much larger than the surface/volume ratio of the contractile vacuole itself. The primary function of the vacuole then would be to act as a reservoir for the fluid generated by the vesicles. In freshwater ciliates, the generation of a hyposmotic fluid could occur in the tubular systems associated with the contractile vacuole (Schneider, 1960).

Since contractile vacuoles are also found in marine protozoa, the interesting question arises as to what is the function of this organelle in these marine animals. In freshwater protozoa the contractile vacuole has a primarily osmoregulatory role, but does it serve this function in marine protozoa as well? Indeed, are marine protozoans with contractile vacuoles capable of osmoregulation? It has already been established that marine ciliates regulate their ionic content (Kaneshiro *et al.*, 1969). Preliminary evidence was also obtained by Kaneshiro suggesting that the ciliate *Miamiensis avidus* is slightly hyperosmotic with respect to its sea water environment. This would suggest that the contractile vacuole in at least this marine protozoan may eliminate excess water, just as it does in freshwater protozoa. The possible osmoregulatory role of the contractile vacuole in marine protozoa may be secondary to the solute excretory functions associated with vacuolar activity. The excretion of Na^+, as well as water, is at least partially associated with the activity of the contractile vacuole in freshwater protozoa. If in marine ciliates, the contractile vacuole retains its Na^+ elimination mechanism, then its primary function may be for the regulation of cytoplasmic Na^+ concentration, which is considerably below that of sea water (Kaneshiro *et al.*, 1969). In freshwater protozoa the primary function of the contractile vacuole is elimination of excess water; Na^+ excretion being secondary and associated with the mechanism by which a hyposmotic vacuolar fluid is generated and perhaps for K^+ conservation. In marine protozoa, the necessity for maintaining a hyperosmotic condition is not clear, but in those with low cytoplasmic levels of Na^+ the primary role of the contractile vacuole could be for Na^+ excretion.

The mechanism by which the contractile vacuole empties its contents (systole), has also not been established. In ciliates a system of fibrils appears to be associated with the contractile vacuole, but whether or not these fibrils are themselves contractile has not been demonstrated. In amoeba, no structural contractile elements have been found to be associated with the vacuole, but the isolated contractile vacuole of *A. proteus* "contracts" upon the addition of ATP in the proper ionic environment (Prusch and Dunham, 1970). This would indicate that the contractile elements of the vacuole in amoebae may be associated with the vacuolar wall itself.

Electrical events associated with the systolic phase of the vacuolar cycle have been recorded in both *Paramecium caudatum* (Yamaguchi, 1960) and *A. proteus* (Josefsson, 1966; Prusch and Dunham, 1967). In both animals, a hyperpolarization of the membrane potential is recorded in the cytoplasm when the vacuole contracts (Fig. 4B). It is not known what causes this hyperpolar-

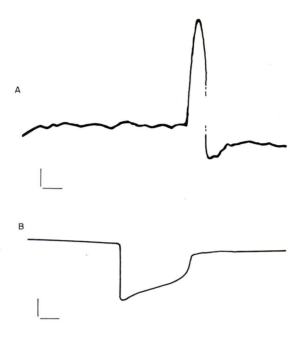

Fig. 4. Electrical potentials recorded from two different amoebae during the systolic phase of the contractile vacuole cycle. A. Microelectrode located in the contractile vacuole just prior and during to systole; vertical bar 10 mV, horizontal bar 0.1 sec, initial resting potential −50 mV. B. Microelectrode located in the cytoplasm adjacent to the contractile vacuole (within 15 microns) during systole; vertical bar 5 mV, horizontal bar 0.1 sec, initial resting potential −73 mV (from Prusch and Dunham, 1967).

ization, but it is seen in both amoebae and ciliates. If a microelectrode is positioned directly in an amoeba contractile vacuole, a depolarization is observed during systole (Fig. 4A). This is what would be expected, because initially the electrode is located intracellularly but during systole, the vacuole attaches to the plasmalemma and opens to the exterior and the potential would consequently fall towards zero.

The complete cycle of events of the vacuolar cycle in amoebae would then include:

1. generation of a hyposmotic fluid in the vesicles,
2. fusion of the vesicles to form the contractile vacuole,
3. contraction of the contractile vacuole with the elimination of the vacuolar contents, and

4. regeneration of the vesicles from the contractile bounding membrane (Mc Kanna, 1973), probably trapping and sealing off an isosmotic portion of the cytoplasm.

A considerable amount of information has been collected in the study of osmotic and ionic regulatory mechanisms in protozoa, including active membrane transport processes, pinocytosis and vacuolar activity. More work has to be done with this group of animals in order to fully understand and appreciate how unicellular (acellular) animals can achieve control of cytoplasmic solute concentrations. Some interesting problems that remain to be resolved in protozoa are possible anion excretion by the contractile vacuole, vacuolar function in marine ciliates and ionic regulation in marine protozoa without contractile vacuoles. (Also see Chapter 12.)

REFERENCES

Adolph, E. F. (1926). *J. exp. Zool.* **44**, 355—381.

Akita, Y. K. (1941). *Mem. Fac. Sci. Agric., Taihoku Imp. Univ.* **23**, 99—120.

Andrus, W. D. and Giese, A. C. (1963). *J. Cell. Comp. Physiol.* **61**, 17—30.

Batueva, I. V. and Lev, E. (1967). *Tsitologiya* **9**, 681—691.

Brandt, P. W. (1958). *Exp. Cell Res.* **15**, 300—313.

Brandt, P. W. and Freeman, A. R. (1967). *Science* **155**, 582—585.

Brandt, P. W. and Hendril, K. B. (1972). *C. r. Trav. Lab. Carlsberg* **38**, 423—443.

Bruce, D. L. and Marshall, J. M. (1965). *J. gen. Physiol.* **49**, 151—178.

Carter, L. (1957). *J. exp. Biol.* **34**, 71—84.

Chapman-Andresen, C. (1962). *C. r. Trav. Lab. Carlsberg* **33**, 73—264.

Chapman-Andresen, C. and Dick, D. A. T. (1962). *C. r. Trav. Lab. Carlsberg* **32**, 445—460.

Chapman-Andresen, C. and Prescott, D. M. (1956). *C. r. Trav. Lab. Carlsberg* **30**, 57—78.

Cooper, B. A. (1968). *C. r. Trav. Lab. Carlsberg* **36**, 385—403.

De Terra, N. and Rustad, R. C. (1959). *Exp. Cell Res.* **17**, 191—195.

Dunham, P. B. and Child, F. M. (1961). *Biol. Bull.* **121**, 129—140.

Elliot, A. M. and Bak, I. J. (1964). *J. Protozool.* **11**, 250—261.

Gelfan, S. (1928). *Protoplasma* **4**, 192—200.

Herfs, A. (1922). *Arch. Protistenk.* **44**, 227—260.

Hoffmann, E. K. and Kramhøft, B. (1974). *C. r. Trav. Lab. Carlsberg* **39**, 433—442.

Josefsson, J. O. (1966). *Acta Physiol. Scand.* **66**, 395—405.

Josefsson, J. O. (1968). *Acta Physiol. Scand.* **73**, 481—490.

Kamada, T. (1935). *J. Fac. Sci. Univ. Tokyo* **4**, 49—62.

Kaneshiro, E. S., Dunham, P. B. and Holz, G. G. (1969). *Biol. Bull.* **136**, 63—75.

Kitching, J. A. (1956). *Protoplasmatologia III*, D, 3a, 1—45.

Kitching, J. A. (1956). *Protoplasmatologia III*, D, 3a, 1—45.

Klein, R. L. (1959). *J. Cell. Comp. Physiol.* 53, 241—258.

Klein, R. L. (1961). *Exp. Cell Res.* 25, 571—584.

Klein, R. L. (1964). *Exp. Cell Res.* 34, 231—238.

Lewis, W. H. (1931). *John's Hopkins Hosp. Bull.* 49, 17—27.

Løvtrup, S. and Pigon, A. (1951). *C. r. Trav. Lab. Carlsberg* 28, 1—26.

Mast, S. O. and Doyle, W. L. (1934). *Protoplasma* 20, 555—560.

Mast, S. O. and Fowler, C. (1935). *J. Cell. Comp. Physiol.* 6, 151—167.

McKanna, J. A. (1973). *Science* 179, 88—90.

Pappas, G. D. (1960). *Xth Int. Congr. Soc. Cell Biol.*

Pappas, G. D. and Brandt, P. W. (1958). *J. biophys. biochem. Cytol.* 4, 485—488.

Picken, L. E. (1936). *J. exp. Biol.* 13, 387—392.

Prusch, R. D. and Dunham, P. B. (1967). *J. gen. Physiol.* 50, 1083.

Prusch, R. D. and Dunham, P. B. (1970). *J. Cell Biol.* 46, 431—434.

Prusch, R. D. and Dunham, P. B. (1972). *J. exp. Biol.* 56, 551—563.

Riddick, D. H. (1968). *Am. J. Physiol.* 215, 736—740.

Riddle, J. (1962). *Exp. Cell Res.* 26, 158—167.

Schmidt-Nielsen, B. and Schrauger, C. (1963). *Science* 139, 606—607.

Schneider, L. (1960). *J. Protozool.* 7, 75—90.

Schumaker, V. N. (1958). *Exp. Cell Res.* 15, 314—331.

Stoner, L. C. and Dunham, P. B. (1970). *J. exp. Biol.* 53, 391—399.

Yamaguchi, T. (1960). *J. Fac. Sci., Tokyo* 8, 573—591.

IV. Ion Transport at Tissue Level

15. Potassium Transport in Insect Midgut

K. Zerahn

Institute of Biological Chemistry A, August Krogh Institute, Copenhagen, Denmark

I. INTRODUCTION

Insects can be divided into groups according to the ionic composition of their haemolymph: group (a) has a high content of sodium in the haemolymph and usually a limited concentration of potassium, and group (b) has a high concentration of potassium and very often a small and varying concentration of sodium.

The active transport of Na has been known for a long time, at the beginning of the century Galeotti (1904) found that the isolated frog skin will maintain a potential difference of 50 to 100 mV when bathed in either sodium or lithium salt solution; other cations cannot substitute for Na or Li. Later Huf (1935) found that isolated pieces of frog skin transported Na, and Krogh (1937) found that frogs take up NaCl through the skin.

Since most insects have a high Na content in their haemolymph, most studies have been performed on animals of this group. The transport system seems sometimes to be similar to the transport system in vertebrates, with the concept of an ion exchange pump, where Na is exchanged for K. In vertebrates, the transport system is usually inhibited by ouabain, a specific inhibitor for the Na pump.

This is also seen in insects; O'Riordan (1969) found that the active transport of sodium and potassium in the midgut of the cockroach *Periplaneta americana* was inhibited by ouabain. But the K pump in the midgut of the american silkworm was not effected by ouabain (Haskell *et al.*, 1965) neither are K pumps in other insect tissue (see Maddrell this volume).

The active transport of K was first shown in an insect tissue in the Malpighian tubules by Ramsay in 1953. Later, another K pump was found in the midgut of the larva of the silkworm, *Hyalophora cecropia* by Harvey and Nedergaard (1964). The midgut of this larva can transport K independently of any other cation or any special anion. The midgut epithelium is one cell-layer thick, and when in potassium Ringer solution only K will be transported actively. Thus, it is no surprise that the tissue has been used extensively for studying K transport. The cecropia midgut is easy to handle because it is large, it has a high potential difference between the lumen and bloodside, and it has large fluxes of K. Hopefully, studies on the midgut can give new information on how a K pump can work (reviewed by Harvey and Zerahn, 1972). Two disadvantages are that the epithelium is highly folded and is comprised of two types of cells, the columnar cells and the goblet cells.

The K transported by the Malpighian tubules is linked to a transport of a fluid almost isotonic with the haemolymph. But the K transported by the midgut does not cause any significant movement of fluid; but the two K-pumps may still be closely related in other respects. The midgut is made up of different cells, namely, muscle cells, tracheal cells and the two types of epithelial cells. A schematic drawing is given in Fig. 6.

The muscle cells and trachea cells do not cover the blood-side as a continuous layer and the basement membrane is rather permeable to small ions. The lumen side is covered with the brushborder of microvilli from the columnar cells which also covers the goblet cells.

II. PROPERTIES OF THE POTASSIUM PUMP

The first indication of an ion pump may very well be the observation of a potential difference (pd) between the two sides of a membrane. But a pd is not always present, because a particular cation, e.g. K, may not be the only ion transported. There may be co-transport with an anion, as is often the case in the Malpighian tubules, or there may be a transport of another cation simultaneously, e.g. Na in the gut of *Periplaneta americana* (O'Riordan, 1969).

To test if K is transported, one can change the composition of the solution, and see if the transport occurs only in K solution, and if the potential difference (pd) depends on the K concentration. However, this is not always possible since some tissues may not survive the change of solution. However, the isolated midgut of cecropia larva works well and shows a steady pd with K as the only cation in the bathing solution. The anion can be varied widely, from Cl^- to $Fe(CN)_6^{4-}$ but it is necessary to have sucrose or a similar sugar in the solution. Although the Na pump in frog skin is generally considered to be an exchange pump, with K as the other ion, the K pump in cecropia midgut is certainly electrogenic because there is no other ion available for an exchange (Harvey *et al.*, 1968). Furthermore, as found by Harvey and Zerahn (1972, see also Fig. 1), very little change was found in pd with changing K concentration. Furthermore, the pd change is reversible with return to normal K concentration.

This behavior is different from that of the Na pump in frog skin where one finds that the pd depends on the Na concentration in the outside solution. Koefoed-Johnsen and Ussing (1958) found that when the outside Na concentration was decreased by a factor of ten, the potential difference decreased by about 60 mV.

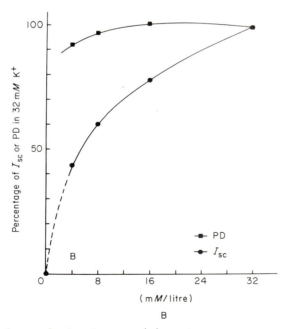

Fig. 1. Dependence of pd and I_{sc} on (K) in short term measurements in the absence of other cations. The K was changed by replacing an appropriate amount of KCl by an osmotically equivalent amount of sucrose. From Harvey and Zerahn (1972).

A practical way to detect and measure the active K transport is to short-circuit the potential difference and compare the short circuit current with the flux as measured by radioactive potassium. This method was used by Harvey and Nedergaard (1964). But if the K pump has to be verified, one often will use the flux ratio (Ussing, 1949), where the K flux in both directions is measured. The flux ratio has been measured in cecropia midgut (Harvey and Nedergaard, 1964), and also in some other insect tissues like the midgut of *Antheraea pernyi* (Wood, 1972) and the midgut of the cockroach (O'Riordan, 1969). The epithelium has to be isolated and set-up *in vitro* so that both sides can be labelled and the fluxes measured. This experimental treatment is not easy *in vivo*.

The results can be used with much confidence. The existing phenomena which will disturb the flux relation are known: e.g. single file diffusion (Hodgkin and Keynes, 1955), or exchange diffusion (Ussing, 1947). As an example, the results from Table I by Harvey and Nedergaard (1964) are given. The ratio between influx and efflux is quite different from what would be expected with a potential of zero, so an active transport of K is certain. To get more knowledge of the pump it is possible to calculate the voltage of the pump during short circuit; when this is done, the potential is often found to be too low. Is this a failure of the flux ratio method or is there another explanation? It is easy to find some explanation for

TABLE I

Potassium-42 fluxes measured concurrently with the short-circuit current[a]

Expt no.	Blood-directed flux 1	Lumen-directed flux	Blood-directed flux 2	Net flux	Current
63	5.0	8.5	0.7	5.6	9.1
67	1.2	18.7	—	17.5	18.7
79	1.8	18.8	—	17.0	19.8
80	1.9	18.5	2.7	16.2	19.5
82	7.0	14.0	9.4	5.8	9.9
83	1.5	14.6	12.9	7.4	7.3
115	2.7	12.8	7.1	10.8	11.4
135	3.2	24.0	2.2	21.3	26.9
136	3.2	19.1	14.0	10.5	19.1
88†	1.1	24.8	5.4	21.5	26.6
89†	2.2	28.6	2.9	26.0	25.8
137†	2.6	24.6	4.7	20.9	16.6
138†	3.3	14.9	4.6	11.0	17.5

[a] All values expressed as μ eq/hour.
The first nine rows are from experiments in which the midguts were maintained in the standard sodium-free solution. The experiments indicated by the daggers are from midguts with 2 mM sodium in the bathing solution. From Harvey and Nedergaard (1964).

the calculated low potential difference in the midgut because in the early investigations the guts may have been occasionally leaky. A leaky gut would produce too high an efflux, i.e. from the lumen to blood-side. There is also another explanation which is valid. My former student Claus Oerum found that under certain experimental conditions (when the midgut is distended), there is a high rate of exchange diffusion. This will decrease the flux ratio and make it possible to have a much higher potential in the open-circuited gut as can be found for the E_K during short-circuiting (see Ussing and Zerahn, 1951).

$$E_K = \frac{RT}{zF} \ln \frac{M_{in}}{M_{out}}$$

E_K is the transporting force, R is the gas constant, T is the absolute temperature, M_{in}/M_{out} flux ratio at pd = 0, z is the charge of the ion, and F is Faradays constant.

The study of Harvey and Nedergaard (1964) showed that the potassium ions are transported actively from blood-side to lumen. The larva of *Hyalophora cecropia* was chilled and the midgut dissected out and placed as a tube or sphere in an apparatus with circulation (see Fig. 2) of both inside and outside solutions of the gut. Both solutions are stirred with a stream of air. Solutions of different compositions have been used (Table II).

TABLE II

Composition of solutions used to study the midgut

Abbreviation[a]	Alkali metal ion	Cl	HCO$_3$	Ca	Mg	Sucrose
		(Concentration in millimoles per litre)				
32-K-S	32 mM K	50	2	5	5	166
32-Na-S	32 mM Na	50	2	5	5	166
32-Cs-S	32 mM Cs	50	2	5	5	166
100-K	100 mM K	98	2	0	0	166
32-K	32 mM K	30	3	0	0	166
4-K	4 mM K	2	2	0	0	166
32-Na	32 mM Na	30	2	0	0	166
32-Cs	32 mM Cs	30	2	0	0	166
16-K, 16-Na	16 mM K 16 mM Na	30	2	0	0	166
16-Na, 16-Cs	16 mM Na 16 mM Cs	30	2	0	0	166

a The first number of the abbreviation is the concentration in millimoles per litre of the alkali metal ion and the letter that follows is its symbol. Solutions which contain 5 mM Mg^{2+} and Ca^{2+} are identified by an "S" in the abbreviation. From Harvey and Zerahn (1972).

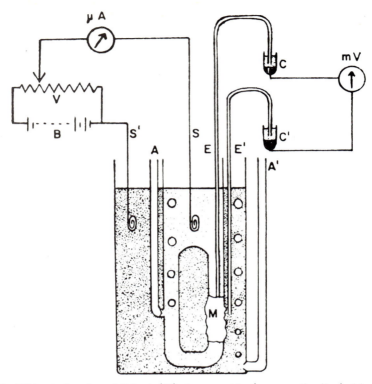

Fig. 2. Midgut chamber. Midgut (*M*); lumen side (sparse stippling); blood-side (close stippling); inlets for aerating and stirring gas (*A* and *A'*); calomel electrodes (*C* and *C'*); bridges between the bathing solutions and the calomel electrodes (*E* and *E'*); potentiometer (*mV*); dry-cell battery (*B*); voltage divider (*V*); microammeter (*μA*); silver-silver chloride electrodes (*S* and *S'*). From Harvey and Nedergaard (1964).

The midgut has a potential difference of 100 to 150 mV, lumen positive. When the gut is short-circuited, a current of several milliamperes can be drawn from a tissue weighing 100 mg.

The test for an active transport is that the flux from blood-side to lumen is often more than 10 times larger than the flux from lumen to blood-side when the potential difference is held at zero mV. It was further shown that the short-circuit current is almost equal to the net K flux within about 10%. Na^+ ions were not needed for the potential or generation of transport and did not influence the K flux as measured by the short-circuit current. Addition of Ca^{2+} or Mg^{2+} had no significant effect. The K flux needed a supply of O_2 to maintain the pd, aeration with N_2 decreased the potential to zero but return of O_2 at once restored the potential. Dinitrophenol will also inhibit the pd and K flux (see Haskell *et al.*, 1965, where other poisons of the K pump were also tested). Ouabain, which is considered a specific inhibitor for the Na pump in many vertebrates and also in a few insects, has no effect on the K pump in cecropia.

III. OXYGEN CONSUMPTION AND K TRANSPORT

In some epithelia like frog skin and toad bladder, the active Na transport is coupled to the energy supply from metabolism. If the Na transport is stopped, energy consumption is at a minimum, and with increasing Na transport, energy consumption increases.

The relationship between K transport and energy consumption in the midgut has been investigated by Harvey *et al.* (1967). The K pump does not increase the energy consumption, but there is no apparent reason why this should be so.

IV. KINETICS AND LAG TIME

When a cecropia midgut is placed in a solution and distended into a sphere, both sides of the gut can be bathed in stirred solutions which are saturated with oxygen. The gut can be either short-circuited or maintained with its natural potential difference. If the gut is bathed with the first solution used by Harvey and Nedergaard (1964, see Table II), it will be exposed to 32 mM K on both the lumen and the blood side. When the blood-side is then supplied with ^{42}K, the radioisotope will appear in the lumen solution after some time. The flux attains steady-state and is almost constant, only the decay by the tissue will show. The delay in reaching steady-state was plotted by Harvey and Zerahn (1969) by using the graphical method described by Andersen and Zerahn (1963). The amount of labelled K appearing in the lumen is plotted (see Fig. 3) on the ordinate and the time on the abscissa. The curve which can be drawn through the points obtained will have a linear part and when this is drawn to the abscissa, one can see the lag time. Of course this can only be done with a flux which is stable. No difference could be found in the lag time whether the flux was large or small, or whether the pd was maintained or short-circuited. More recently it has been found that the lag time measured in this way depends on and is almost

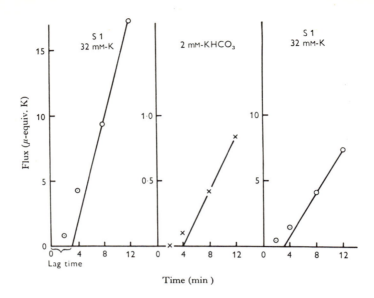

Fig. 3. The time-course of [42]K-movement from blood-side to lumen of an isolated midgut is plotted for a representative experiment in which the gut is equilibrated with S-1 (32 mM-K), with 2 mM-KHCO₃ (ordinate expanded 10X), and again with S-1. The lag time is estimated by extrapolating the steady-state line to the abscissa. Although the flux varied from 119 to 6.2 and back to 49 μ-equiv.K/hr the lag time was virtually unchanged. From Harvey and Zerahn (1969).

proportional to the thickness of the gut (Zerahn, 1973). Before we can interpret these results we need to know the K concentration in the gut cells and how this concentration varies with the K concentration in the bathing solution.

V. K CONCENTRATION IN THE MIDGUT

The procedure usually employed for determining cellular content of any compound is to measure the concentration in the whole tissue and the fraction in the extracellular space. The alternative procedure of washing the tissue with an inert solution before determining how much is left is less satisfactory because the cells may lose the compound during washing. When working with the gut *in vitro* the bathing solution contains a high concentration of sucrose (166 mM). The easiest way to determine the extracellular space is then to determine the sucrose content and assume that the sucrose is found only in the extracellular space. From these measurements it is easy to compute the percentage of extracellular space in the tissue and the amount of K in the cells.

From the content of K in the cells one can calculate the concentration, but only on the assumption that K is uniformly distributed in the cell water, but we know that this will hardly be the case. K concentration may be different in the different types of cells, e.g. columnar, goblet, trachea muscle, and regenerative cells. Table III gives the concentration of K in the midgut cells determined with different concentrations of K in the bathing solution. Harvey and Zerahn (1969) found that the K concentration in the midgut was from 50 to 65 mequiv. kg^{-1} weight of the gut, but they did not provide a value for cells. They considered the K concentration in the cells to be constant during the experiment. However, Harvey and Wood (1972) found that the K content in the gut tissue decreased significantly with time. This was tested by Zerahn (1975a) in a bathing solution of 32 mM K but it was not possible to find a

TABLE III

Potassium concentration in midgut cells with varying K concentration in bathing solution. Corrections for extracellular space is made by determining the sucrose content in the midgut

(K^+) in bathing solution	(K^+) in cells
32 mM	140 mM
8 mM	105 mM
2 mM	90 mM

significant change in cell concentration of K. Neither did the extracellular K vary with time. Furthermore, the K concentration of guts bathed in low K solution (2 mM $KHCO_3$ + 166 or 260 mM sucrose) was tested with time and no significant change could be found either (Zerahn, unpublished). The experiments by Harvey and Wood were done with pieces of midgut bathed in solution. Their experiments differ from all the other experiments where the guts were treated as a tube with solutions stirred on both sides by bubbling oxygen, and the potential differences controlled. The difference in handling of the tissue may be the reason for the difference in results.

VI. K EXCHANGE

A. 32 mM K Solution

Another factor which may influence the time of appearance of labelled K in the lumen is the rate of exchange of tissue K with K in the solution bathing the blood-side. Harvey and Zerahn (1969) found that the rate of exchange between the tissue and the bathing solution (32-K-S) was high: in 12 minutes the tissue was 71% labelled with blood-side K. Later Harvey and Wood (1972) found only 50% exchange in an hour. More recently Zerahn (1975a) has found that the K exchange is 50% in 2.5 min in 32 mM K (see Fig. 6).

One explanation for the differences in the rate of exchange may be that the results obtained by Harvey and Wood were made on pieces of midgut bathed in solution. In the solution with high K we thus have a half time for exchange of less than 3 min and a lag time for K transport of two to four min. As the lag time was found to be independent of external K concentration it was of interest to determine the exchange with a low external K concentration.

B. Exchange in 2 mM K Solution

Harvey and Zerahn (1969) needed a solution in which transport of K could occur without the fast exchange between blood-side K and tissue. A solution of 2 mM K was used containing only sucrose and K-bicarbonate.

They found that with this solution only 18% of tissue K was exchanged in 10 min, on the average it will be 2% min^{-1}. However, Harvey and Wood later (1972) found the exchange to be less than 10% in 30 min with a levelling off after that time. Again the experiments were done on pieces of midgut. So far these results do not interfere with the interpretation of the experiments if only the flux goes on; but the exchange experiments were repeated and the results are given in Fig. 4. The rate of exchange in Fig. 4 is slow at the start (4% per min) but increases to 30% after 10 min, in general agreement with the 18% found by Harvey and Zerahn.

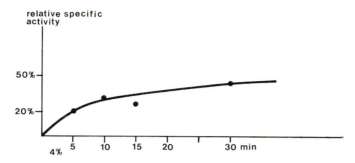

Fig. 4. Exchange of gut cell K with K in the bathing solution of 2 mM K. It is seen that the exchange is slow only about 4% per min in the beginning of the curve and the exchange after 30 min is below 50%. Procedure on distended guts as in (Zerahn, 1975) each point mean value of two or three guts.

1. Interpretation of Kinetics

Harvey and Zerahn (1969) observed that the lag time (Fig. 3) did not change with the size of the flux and that changing the external potassium concentration from 32 mM to 2 mM changed the K concentration of the tissue only from 65 to 50 mequiv. kg^{-1} wet weight. In the 32 mM K solution we observed a fast exchange of the cellular K with the blood-side solution. This will make it difficult to say whether the delay is short, because the ^{42}K from the blood-side enters the cells and rapidly exchanges with cell-K or whether some other more direct routes with little K are used.

Therefore, we had to consider the transport in low K (2 mM). The content of K in the tissue is about 30% lower under these conditions but still much K is left to delay a small K flux (5 mequiv. hr^{-1}). The exchange is slow between cell and blood-side (only a few % per min). If the slow ^{42}K flux of 5 mequiv. hr^{-1} should pass through the cells with a content of about 50 mequiv. K per kg of tissue sample, one should expect a lag time much longer than the measured two—four min.

We took this as evidence that the active K flux passes through the gut wall via a route with little K and does not mix with the major part of the cell-K. It is possible that the determined lag time has been too short due to inaccuracies in producing the curve, but a change of 10% or 20% would make no difference to the argument. The flux does often decay with time but the lag time is the same for different flux rates even if they are very slow.

Harvey and Wood claim (1972) that for obtaining an accurate lag time in the short-circuited guts it is necessary to use a correction for the effect of the flux decay on the lag time. The correction can only be made when the change in short-circuit current is measured accurately. However, we have another measure for the delay of ^{42}K through the gut epithelium, as we can determine the half time of the flux: the time it takes from adding the isotope on the blood-side, until half the value of the steady-state flux has been obtained. Such measurements have been made, and in agreement with the short lag time, the half times of the flux are also found to be short (Fig. 5). At the time when the half steady-state value has been obtained, the specific activity of the K appearing in the lumen will be half the steady-state specific activity and K in the transport route through the gut must at this time be much closer to the steady-state value. The lag time in Fig. 5 was found to be 3.6 minutes and the half-time for the flux 3.0 minutes. We could just as well choose the half-time to represent the delay of ^{42}K through the gut as the lag time. So if we doubt whether the lag time is correct, we can compare it with the

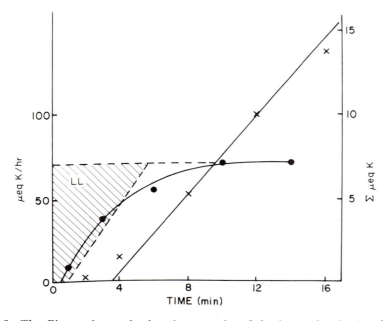

Fig. 5. The Figure shows the lag time equal to 3.6 min on the abscissa (curve marked with crosses). The time for obtaining half time flux value is 3 min (curve marked with dots). From Harvey and Zerahn (1972).

half-time for the flux and thus have a control. In experiments with 2 mM K solution the guts cannot be short-circuited, but we can observe if the flux is reasonably stable with time. As in high K we can determine the half-time for the flux which in low K will be correspondingly short like the lag time. As mentioned earlier, the objection raised by Harvey and Wood (1972) that the K content depends on the K concentration and varies with time, has not been confirmed. It was found that some K in the gut tissue would not exchange with the K in the solution (see later). This K can hardly be responsible for interfering with the flux of labelled K through the gut, because it will just stay unlabelled. So we will reconsider the conclusions by Harvey and Zerahn (1969). They left only two possible routes for the potassium transport across the midgut:

(a) either through only the goblet cells or

(b) through a special route which is not defined. The suggestions for (b) are either through the cells in specific channels or between the cells along the lateral surfaces.

Route (a) has been assumed to be reasonable, but has the difficulty that some phytophagous larvae such as the sawfly larva (*Simbex femorata*) studied by Turbeck (1976) have no goblet cells, but they are likely to transport K from blood-side to lumen of the midgut. In the common 32 mM K solution (Table II) they show a positive potential of more than 50 mV.

2. K Pools

There are unstirred layers (Dainty and House, 1966) on both sides of the gut, so when the blood-side is labelled with ^{42}K, the unstirred layer on that side will be labelled fast and then the labelled K will start to penetrate the gut. If the delay for steady-state of the flux is represented by the lag time, we can determine the amount of labelled K needed for filling the transport route before we obtain steady-state, which is the flux multiplied by the lag time, and this is the total pool for the transport. This pool will depend upon the flux. With a constant lag time it will be proportional to the flux. Where is this pool localized? If the ^{42}K does not mix with the cell K, the pool could be outside the cells. The pool could be mainly in the extracellular spaces on the blood-side. Some K could be present at the site of the K pump as a transport pool. Some K could also occur as a transported pool in the spaces on the lumen side. We have no way of determining the distribution of K between these various pools from the flux measurements, we only obtain the sum of the pools. When the flux is small there is no difficulty in concluding that the total K pool is in the extracellular spaces. But when the K flux is large, the pool will also be very large. When the flux is very large, as in short-circuited guts (120 μeq/hr in a 100 mg tissue and lag time three min), the total transport pool should be 6 μeq K. A piece of 100 mg gut (50% extracellular space) contains only 6 μeq of K from chemical determinations. Most of this K is in the cells. So where is the total K pool? The answer is straight forward: it is not measured when we make a chemical determination.

The high flux and the very large total transport pools are found in short-circuited guts in high K (32 mM). When we determine K in the gut, we stop the shorting. The flux in unshorted guts is only one-half to one-third, so the pool will decrease proportionally. We take the gut and remove it from the tubes to which it was tied and let the solution drain off. This removes the adhering solution which was almost in equilibrium with the solution in the extracellular spaces, so most of the extracellular transport pool will have diffused out at this time, and thus is removed. This explains why there is not a higher K concentration in shorted than in unshorted guts (Zerahn, unpublished observations).

3. Na Transport

When the cecropia midgut is bathed in solutions containing Ca^{2+} and Mg^{2+} they were found to transport Na poorly. But if Ca^{2+} and Mg^{2+} are omitted, Na can be substituted for K and a potential will be found. If the gut is shorted, a short-circuit current is kept for a half to one hour (Harvey and Zerahn, 1971).

The size of the short-circuit current is only about 50% of the value with K. But it was found that active transport of Na occurs and that all alkali metals can be transported under these conditions. It is exceptional to find a tissue where so many different ions can be actively transported – and the possibilities for new information on the pump were instantly clear.

4. Na-K Competition

Harvey and Zerahn observed that Na and K will compete. It was clear that K is the ion preferred by the pump. If 50% of the cations in the bathing solution is Na and 50% K, the pump will transport almost only K. It seems that it is the same pump which is used for all alkali metals because the ions do compete. Therefore, is the pump which will transport Na a common Na pump which will be inhibited by ouabain? The answer is no! Ouabain is just as inefficient an inhibitor for the midgut Na pump as it is for the midgut K pump.

The competition between the different ions was tested and the competition sequence for the pump determined. The procedure was the following: The solution was made up with the two competing

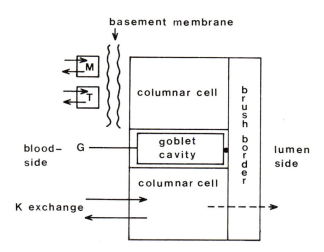

Fig. 6. Diagrammatic representation of the exchange of K in different cell types of the midgut. The largest compartment is represented by the columnar cells, although there will be some exchange with goblet cells (G) with their goblet cavities, and with muscle (M) and tracheolar (T) cells. The dashed arrow represents the hypothetical passive loss of K or of other alkali metals from the main K source, the columnar cells.

ions in a concentration ratio such that the flux of both competing ions could be conveniently measured. This was: Na-Li 1/1, Na-K 9/1, Cs-K 4/1, Rb-K 1/1. The blood-side solution could then be labelled with the appropriate isotopes and the fluxes measured. This could not be done for Na-Li competition, because there is no suitable radio-lithium. Instead, the gut was short-circuited and the difference between short-circuit current and Na-transport was taken as the Li flux. In this way the sequence for the competition was found to be: K > Rb ≫ Cs > Na > Li (Zerahn, 1971). If the ions are used alone, they will all be actively transported by the pump, but if there is more than one type of cation, there are some irregularities. Na in competition with K will sometimes be transported poorly. This happens when the gut is not shorted, or when the gut is distended, but it can also happen with shorted and undistended guts. The explanation may be that when the guts are highly folded the K in the fluid in the folds will be transported first and only when no K is left will Na be transported. New solution containing Na + K will diffuse into the folds. However, this result is seen only with competition between Na and K and not between K and Cs or Na and Cs; so the explanation applies only to Na and K and thus does not disturb the sequence.

5. Uptake of Alkali Metals by Gut

When the Cecropia midgut is bathed in Na-32, K will leave the gut. After five min the cellular concentration of Na is 100 mM, with time even more K will leave the gut cells but there is always some K left, just as there is some K which does not exchange with ^{42}K (about 10 to 20% of the gut K). Experiments were done to compare the ability of the different cations to exchange with cell K. In the guts used in the above-mentioned studies of competition for flux, the ion contents of the gut cells were measured and the sequence for uptake was the following: K > Rb > Na > Li > Cs (see Harvey and Zerahn, 1972). Differences between the two sequences are at once apparent; Cs has changed place with Na and Li and the difference is quite significant. Furthermore, in solutions with 50% K and 50% Na, Cs or Li where there is mainly a transport of K, the amount of Na and Cs in the cells is appreciable, Na is the highest. These sequences indicate that the transported ions do not follow a route mixing with most of the ions in the epithelial cells, but follow a route containing only a minor fraction of the alkali metal ions in the gut.

6. Determination of the Transport Pool and of Delay in 32-K

The exchange between the midgut K and the K in the blood-side solution was found to be fast, with a half-time of 2—3 min. This influences the interpretation of the delay of labelled K through the midgut wall. However another approach has been made for investigating this problem.

    When the midgut is bathed in 32-K and we suddenly replace the blood-side solution with 6.4-K and 25.6-Cs, containing both ^{42}K and

[137]Cs, at the start, we will have a cell K of 140 mM and neither Cs nor radioactivity will be present in the cells. Due to exchange between blood-side solution and the cells, the cells will soon contain significant amounts of both ions. If we assume that the ions in the cells are in the transport route there will be a significant longer lag time for the appearance of ^{137}Cs in the lumen solution than for the appearance of [42]K. The reason for this is that Cs has to increase in the cells to be transported. With 16-K, 16-Cs as the blood-side solution, we have found a cell content of 30 mM Cs and still no Cs was transported. When Cs is 100% transported, there is 80 mM Cs in the cells. In Fig. 7 we can see that it takes several minutes for cell Cs to reach an internal concentration sufficient for Cs transport, and several more minutes before the competition with K is finally obtained. If transport is dependent upon cell concentration, we expect that K is preferred at the beginning of the experiment. However, we find that this is not the case (Table IV). The ratio between transported K and transported Cs shows that there is no extra delay for the appearance of Cs in the lumen. This finding

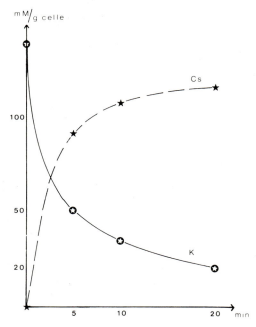

Fig. 7. Change in Cs and K concentration in midgut cells with time, after changing to a bathing solution of 32-Cs at zero time. From Zerahn (1973).

o

TABLE IV

Ratio of Cs flux to K flux at successive time periods after a sudden change of the solution on the blood-side from 32-K to 6.4-K, 25.6-Cs.

min	ratio μEqCs/hr-μEqK/hr
4	2.3
8	2.4
12	2.3
16	1.9
24	1.9

Each value is the mean of five experiments. From Zerahn (1973).

further supports the idea that the transported ions do not pass through the cell interior. Both K and Cs have the same mobility in solution and the same lag time in guts which are equilibrated. It is not the case when K and Na are compared, the mobility of Na in solution is 1.5x slower than that of K, and the lag time for the appearance of Na in the lumen is correspondingly longer. This could indicate that it is the mobility of the ions in free solution which determine the length of the lag times and not membrane permeabilities.

We must conclude that there is no delay. Our assumption of complete mixing within the cells is not justified since the cell content of K and Cs is not directly involved in transepithelial transport of these two ions. The transport pool, that is; the pool next to the pump must be small and rather quickly labelled with radio-Cs and radio-K.

7. Ratio of Ions in Cells

We have confidently determined the amount of the alkali metals and the volume of cells and calculated the concentrations. We have found that these ions are competing in the sequence K > Rb > Na > Li > Cs. We will now take a closer look and reconsider this.

(a) the alkali metal ions are in solution as ions with one positive charge. When we have a cell bathed in solution of two competing ions, both 16 mM, we should expect that when no active transport is present, the concentration inside and outside will be the same.

(b) we find that all the ions easily penetrate the blood-side membrane from 32-alkali metal ion. In 16-K, 16 Na we should expect the same concentration of K and Na inside the cells, but this is not the case. However there is some K in the tissue which will not be exchanged, so we deduct this as 20 mM K. We still have left about 75 mM K and 39 mM Na. For Cs we find an even more unequal distribution.

If we assume a K pump is present at the lumen (dashed arrow in Fig. 6), it should transport K out of the cells and only make the cellular K content smaller, so this would not explain the findings.

We have to assume either (a) an active uptake into the cells for the different ions or (b) that the concentration of ions is the same. Thus the concentration of the free ions would be similar while the amount of bound ions would be different since the different ions would bind with different affinities to the structures or compounds in the cells.

Wood *et al.* (1969), from their micropuncture studies consider the first suggestion (a) unlikely, so the suggestion (b) is left. In that case it is not correct to express the content of the cellular ions as concentrations even if this is practical for many purposes. The gut does not exchange or take up the ions from lumen (Harvey and Zerahn, 1969; also confirmed by Wood *et al.*, 1969) so the exchange of ions is on the blood-side. The exchange seems to be independent of the active transport rate. It can be seen easily in experiments when the gut is bathed with 16-Cs, 16-Na solution, the cellular Cs was only 38 mM and the cellular Na 87 mM (see Table V). This means that Na

TABLE V

Representative table for cell content of Na and Cs and left K in midgut of *Hyalophora cecropia* bathed in 16-Cs, 16-Na.

Date	mM Cs	mM Na	mM K	sum mM
2/6 A 1975	43	90	14	147
2/6 B 1975	32	88	19	139
3/6 A 1975	37	88	27	152
3/6 B 1975	39	79	33	151
3/6 C 1975	38	92	26	156
3/6 D 1975	36	82	24	142
mean value	38	87	24	149

Values are corrected for extracellular space by determining the sucrose content of the gut.

has a higher affinity for the cells than Cs. One cannot assume an active transport of Cs because the pump (dashed arrow, Fig. 6) should also be used by the other alkali metal ions which are competing for the pump. Potassium is not using this pump (see above), so this would mean that Cs and K are treated differently by the cells, e.g. Cs would be transported where K is not, and that is unlikely.

This result supports the earlier findings of Harvey and Zerahn (1969), that the cell content of ions is not directly related to the active transport through the midgut tissue.

The possibilities for harvesting information from the cecropia midgut (or similar tissue) is far from finished, as only few people have tried their efforts and so few approaches have been used. More studies will most likely bring many more details about how the K pump works in the midgut. (Also see Chapters 4, 9, 21 and 25.)

ACKNOWLEDGEMENTS

I thank Professor Svend Olav Andersen for his support and criticism in working out the manuscript, and Dr Signe Nedergaard for reading it. To Drs James Oschman and Betty Wall the best thanks for the inducement to write the paper and the help in reading and criticizing the manuscript.

REFERENCES

Andersen, B. and Zerahn, K. (1963). *Acta Physiol. Scand.* **59**, 319—329.

Dainty, J. and House, C. R. (1966). *J. Physiol. Lond.* **182**, 66—78.

Galeotti, G. (1904). *Hoppe-Seylers, Z.* **49**, 542.

Harvey, W. R. and Nedergaard, S. (1964). *Proc. Nat. Acad. Sci. U.S.A.* **51**, 757—765.

Harvey, W. R. and Wood, J. L. (1972). "Role of Membranes in secretory processes." (Eds L. Bolis, R. D. Keynes and W. Wilbrandt), North-Holland Publ. Amsterdam.

Harvey, W. R. and Zerahn, K. (1969). *J. exp. Biol.* **50**, 297—306.

Harvey, W. R. and Zerahn, K. (1971). *J. exp. Biol.* **54**, 269—274.

Harvey, W. R. and Zerahn, K. (1972). "Current Topics in Membranes and Transport 3." (Eds F. Bronner and A. Kleinzeller), Academic Press, New York and London.

Harvey, W. R., Haskell, J. A. and Zerahn, K. (1967). *J. exp. Biol.* **46**, 235—248.

Harvey, W. R., Haskell, J. A. and Nedergaard, S. (1968). *J. exp. Biol.* **48**, 1—12.

Haskell, J. A., Clemons, R. D. and Harvey, W. R. (1965). *J. Cell. Comp. Physiol.* **65**, 45—56.

Hodgkin, A. L. and Keynes, R. D. (1955). *J. Physiol. Lond.* **128**, 28—60.

Huf, E. (1935). *Pflf. Arch. ges. Physiol.* **235**, 655—673.

Koefoed-Johnsen, V. and Ussing, H. H. (1958). *Acta Physiol. Scand.* **42**, 298—308.

Krogh, A. (1937). *Skand. Arch. Physiol.* **76**, 60—73.

O'Riordan, A. M. (1969). *J. exp. Biol.* **51**, 699—714.

Ramsay, J. A. (1953). *J. exp. Biol.* **30**, 358—369.

Ussing, H. H. (1947). *Nature* **160**, 262—263.

Ussing, H. H. (1949). *Acta Physiol. Scand.* **19**, 45—56.

Ussing, H. H. and Zerahn, K. (1951). *Acta Physiol. Scand.* **23**, 110—127.

Wood, J. L. (1972). Some aspects of active potassium transport by the midgut of the silkworm, *Antheraea pernyi.* Ph.D. Thesis, University of Cambridge, England.

Wood, J. L. and Harvey, W. R. (1975). *J. exp. Biol.* **63**, 301—311.

Wood, J. L., Farrand, P. S. and Harvey, W. R. (1969). *J. exp. Biol.* **50**, 169—178.

Zerahn, K. (1971). *Phil. Trans. Roy. Soc. Lond.* **B 262**, 315—321.

Zerahn, K. (1973). "Transport Mechanisms in Epithelium" (Eds H. H. Ussing and N. A. Thorn), Munksgaard, Copenhagen.

Zerahn, K. (1975). *J. exp. Biol.* **63**, 295—300.

16. Cockroach Salivary Gland: a Secretory Epithelium with a Dopaminergic Innervation

C. R. House

Department of Physiology, Royal (Dick) School of Veterinary Studies, University of Edinburgh, Edinburgh, Scotland

I. INTRODUCTION

In some epithelia there is compelling evidence for hormonal control of ion and water transport. Others also possess an innervation and thus offer the opportunity to study neuroglandular transmission. With the advent of intracellular recording it became possible to observe an electrical response, called a secretory potential (Lundberg, 1955), that could be evoked by nerve stimulation or artificially applied neurotransmitters. As we see in Table I this change in potential may be a depolarization, hyperpolarization or biphasic response, each probably arising from the opening of specific ion channels. However, the ionic basis of the secretory potential elicited by nerve stimulation has not been described fully for any epithelium. Nor is it clear what relationship the secretory potential bears to secretion. Another false impression that might be gained from the table is that biphasic responses always consist of a depolarization followed by a hyperpolarization. In fact, the converse is observed in the cockroach salivary gland, which is the subject of this paper, and also in the cat submaxillary gland (Kagayama and Nishiyama, 1974). Thus, it can be inferred that secretion follows a variety of electrical signals and that probably no unique set of ionic movements constitutes the trigger for excitation-secretion coupling.

403

TABLE I

Some tissues and their secretory potentials

Cell	Resting potential (mV)	Stimulus[a]	Membrane response		Reference
			Secretory potential	Conductance increase	
Mouse pancreas	−40	{ P.n.	Depolarization	—	Dean and Matthews (1972)
		{ ACh	Depolarization	Na	Nishiyama and Petersen (1974a)
Rat brown adipocytes	−47	S.n.	Depolarization followed by hyperpolarization	—	Horwitz et al. (1969)
		NA	Depolarization	—	
		NA	Depolarization	Na (?)	Horowitz et al. (1971)
Guinea-pig liver	−36	NA	Hyperpolarization	K	Haylett and Jenkinson (1972a)
Cat sublingual gland	−33	P.n.; S.n.	Hyperpolarization	K	Lundberg (1957a,b)[b]
Dog submaxillary gland	−42	P.n.; ACh	Hyperpolarization	K	Imai (1965)
Mouse submaxillary gland	−60	ACh	Depolarization followed by hyperpolarization	Na	Nishiyama and Petersen (1974b)
				K	

[a] Parasympathetic nerve denoted by P.n., sympathetic by S.n., acetylcholine by ACh and noradrenaline by NA.

[b] Lundberg's results, in retrospect, suggest an increase in membrane potassium conductance whereas he proposed that ACh activated active Cl transport.

II. STRUCTURE

The structure of the cockroach salivary system has been described by several authors (see e.g. Whitehead, 1971). Paired glands lie on either side of the foregut and are accompanied by salivary reservoirs. The glands consist of acini and their associated ducts comprizing a branched system that ends with the main salivary ducts. The saliva issues from a single large duct formed by the union of the main salivary and reservoir ducts.

A. Acinar Cells

Different cell types are found in the acini (see e.g. Kessel and Beams, 1963; Bland and House, 1971). Peripheral cells possess an intracellular ductule with large microvilli and numerous mitochondria, and are generally pyramidal in shape (Fig. 1A). The other cells, called central, have extensive endoplasmic reticulum and large granules ($c.\ 2\ \mu$m diameter). Acinar cells are joined by septate desmosomes in their apical regions.

Bland and House (1971) showed that the salivary glands of the cockroach *Nauphoeta cinerea* contain amylase, invertase, maltase and protease and suggested that these enzymes are secreted by the central cells while the peripheral cells transport ions and water. This proposal was inferred from the structure of the cells, however, and direct evidence about function is lacking. Other suggestions about their secretory roles have been made (see Table I, Bland and House, 1971), but the majority appear inadmissible in the light of current evidence.

B. Innervation

The most comprehensive study of the innervation of the cockroach salivary gland is that of Whitehead (1971) on *Periplaneta americana*. The gland receives axons from the suboesophageal ganglion and the stomatogastric nerve; both sets of fibres innervate the reservoirs, the salivary ducts and acini.

The pair of salivary duct nerves from the suboesophageal ganglion pass along their respective reservoir and salivary ducts and branch where the latter bifurcate. Figure 1B shows an electron micrograph of a section through the reservoir duct and the salivary duct nerve in

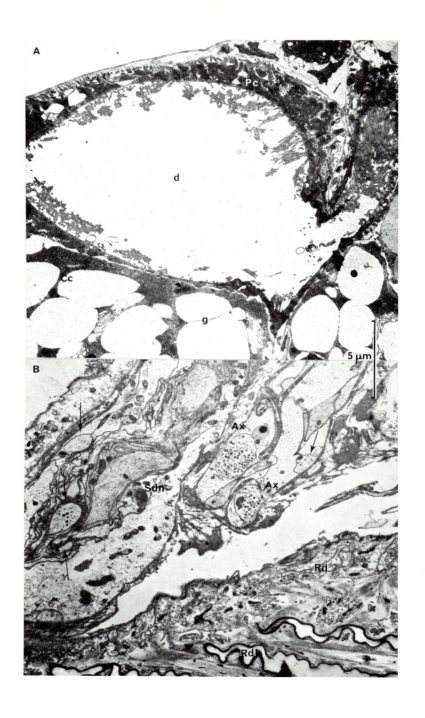

A

Pc

d

Cc

g

5 μm

B

Ax

Ax

Sdn

Rd

Rd

Nauphoeta. The nerve is composed of two large axons (*c.* 2 μm diameter) and several smaller ones, as Whitehead has reported for *Periplaneta*.

The branch of the stomatogastric nerve supplying the gland contains about a dozen axons, each less than 2 μm in diameter (Whitehead, 1971).

Electron micrographs often reveal the presence of several axons on the acinar surface. It has also been shown in preparations stained with methylene blue that the salivary nerves branch profusely in the neighbourhood of an acinus and form a nerve plexus on its surface. Moreover, the terminal axons have swellings of a similar size to the fluorescent varicosities observed in glands treated by the Falck-Hillarp method for monoamines (Bland *et al.*, 1973). Spectrofluorimetric analysis of the granules in the nerve endings indicated that they contain a catecholamine, probably dopamine or noradrenaline. A similar study of locust salivary gland nerves demonstrated the presence of dopamine (Klemm, 1972). This amine has also been detected in *Nauphoeta* glands by a radiochemical assay (Fry, *et al.*, 1974). It has been suggested that dopamine is probably the transmitter in tick (Megaw and Robertson, 1974) and moth salivary glands (Robertson, 1975).

C. Junction

The structure of the neuroglandular junction is not clear and requires further examination. According to Whitehead (1971) the nerve terminals may enter the spaces between central cells or "may merge with the basement membrane of the acinus where a more superficial penetration of the extracellular spaces" of peripheral and central cells occurs. There are also intra-acinar sympathetic nerve endings in mammalian salivary glands (Garrett, 1974). Whitehead found little evidence for a distinctive junction, perhaps because the nerve endings are small (<1 μm) and sparse. Terminals were found, however, that

---

Fig. 1. Electron micrographs of acinar cells and salivary duct nerve. A. Peripheral cell (Pc) with ductule (d) and adjacent central cells (Cc) containing large granules (g). B. Section through salivary duct nerve (Sdn) and adjacent wall of reservoir duct (Rd). Two relatively large axons (Ax) are accompanied by several smaller ones indicated by arrows. Lumen (Rdl) of reservoir duct is also evident. Both micrographs prepared by methods described previously (Bland and House, 1971) and photographed at same magnification.

had shed their glial coats and contained electron dense and lucent vesicles.

It may be concluded from Whitehead's study that both central and peripheral cells are innervated and that axons terminate on only a few cells in every acinus.

III. ELECTROPHYSIOLOGY

From the preceding anatomical description it is evident that the cockroach salivary gland offers some distinct advantages for electrophysiological work on neuroglandular transmission. The gland is not surrounded by a capsule, so that acini are accessible for microelectrode recording. It has no blood vessels, smooth muscle or myoepithelial cells; the reservoirs contain a few muscle fibres (Sutherland and Chillseyzn, 1968) but their movement is small and generally not troublesome since the reservoirs can be stretched and used to anchor the preparation.

A. Method

For the convenience of the reader the methods employed (Ginsborg *et al.*, 1974) are described briefly here. Paired salivary glands with ducts and reservoirs are dissected from *Nauphoeta* in a solution containing (mM) NaCl, 160; $CaCl_2$, 5; KCl, 1; $NaHCO_3$, 1; NaH_2PO_4, 1. The preparation is spread across a transparent pedestal in a perspex chamber (capacity 4 ml) which is perfused continuously by a peristaltic pump at 2 ml min^{-1} except when drug solutions are applied at 20 ml min^{-1}. Laminar perfusion is achieved by interposing nylon meshes between inflow and outflow pipes.

Microelectrodes (10–30 MΩ) filled with 3 M KCl are used for intracellular recording, the methods being conventional and requiring no description. Dye electrodes, filled with 5% aqueous solution of Procion 4 MRS, can be used for labelling cells, the Procion being ejected iontophoretically (House, 1975).

The reservoir and salivary ducts are drawn into a suction electrode for electrical stimulation of the embedded nerves. Bursts (1–100) of stimuli (0.5 msec, 30 V) are usually delivered at 100 Hz from a square pulse stimulator at intervals greater than two min. The acinar nerve endings can also be excited by electrical "field stimulation" (House, 1973).

B. Secretory Potential

When a microelectrode is inserted into an acinus an apparent resting potential of about −30 mV is observed. The potential remains stable often for periods exceeding 30 min and occasionally up to several hours. In these instances we can record reproducible secretory potentials (Fig. 2) from both peripheral and central cells (Fig. 2A) when the salivary duct nerve is stimulated. The secretory potential occurs after a latency of about 1 sec (Fig. 2B) and consists of a hyperpolarization sometimes followed by a smaller depolarization (cf. cat submaxillary gland, Kagayama and Nishiyama, 1974).

1. Amplitude

The strength of the nerve stimulus affects the size of the response in a manner suggesting that acini have a multiple innervation (see also House, 1973). In the example in Fig. 3 there is evidence for an axon with a threshold of 3.8 V and another between 6 and 8 V while between 80 and 100 V there might be several more making small contributions to the response. As judged by the stimulus-response relations of different cells the number of axons is rather variable (Ginsborg and House, 1976): a few acini appear to be innervated by one axon whereas others have at least two or three. The innervation of the cat submaxillary gland is possibly denser with about 5−10 axons per acinus (Lundberg, 1955), while in the possum (Creed and Wilson, 1969) the number may lie between two and five, somewhat akin to that in the cockroach.

Other electrophysiological experiments indicate that some acini are innervated not only by the ipsilateral salivary duct nerve, as invariably found, but also by the contralateral (Ginsborg and House, 1976).

In addition to the salivary duct nerve comprised of about 10 axons, the gland is also innervated by the stomatogastric nerve containing a similar number. However, responses evoked either by stimulation of the salivary duct nerve or by direct "field stimulation" are similar. Thus, the contribution made by the stomatogastric nerve is evidently inconspicuous and other kinds of experiment are required to establish its transmission properties.

It should be noted that the foregoing results inform us about the innervation of acini but not of individual cells because the responses recorded in any cell probably resemble those in its neighbours as acinar cells are coupled electrically (see Fig. 6).

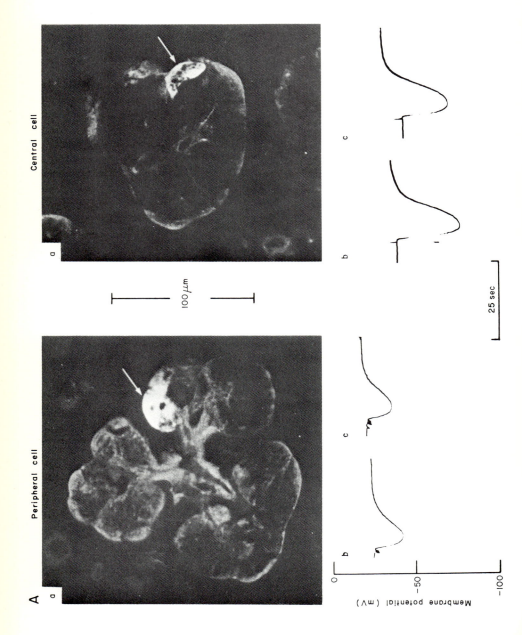

Central cell

Peripheral cell

100 μm

25 sec

Membrane potential (mV)

0

−50

−100

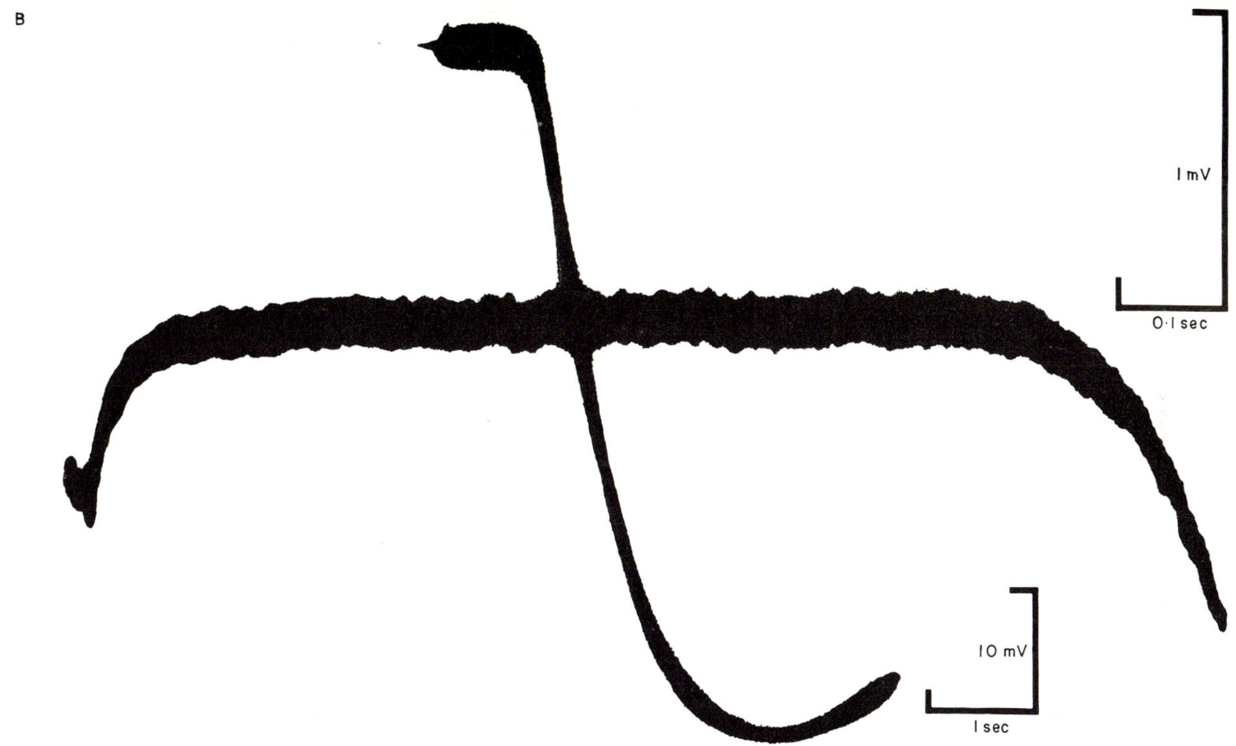

Fig. 2. Intracellular recording of secretory potentials from acinar cells. A. (a) are fluorescent microscope pictures of sections showing that each cell type has been stained with Procion injected from the recording electrodes (see House, 1975). (b) and (c) show the secretory potentials recorded from these cells before and 10 min after Procion injection. The peripheral cell response was evoked by 10 nerve stimuli at 100 Hz and that of the central cell by a single stimulus. B. A secretory potential recorded simultaneously at different gains and oscilloscope sweep speeds. Upper beam shows hyperpolarization attaining its peak (low gain, slow sweep) while lower beam shows latency between stimulus artefacts and start of hyperpolarization (high gain, fast sweep).

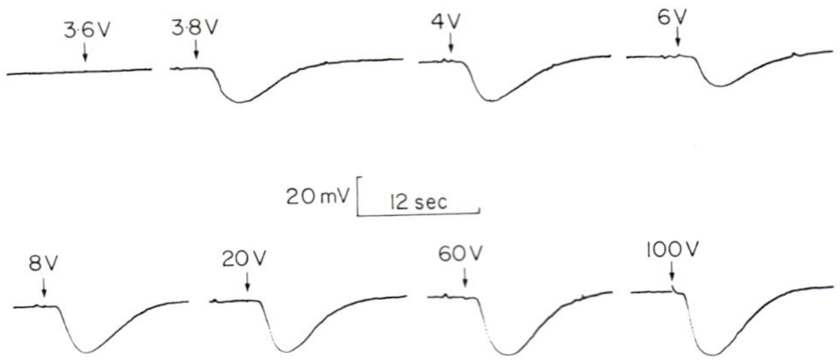

Fig. 3. Influence of strength of nerve stimulus on the amplitude of the secretory potential recorded from a salivary gland cell. In each case a single stimulus (0.5 msec) was delivered at the voltage shown.

The amplitude of the secretory potential also increases with the number of nerve stimuli (Fig. 4A). In this cell the response to a single stimulus was 14 mV and increased to a maximum of 78 mV. Presumably the gradation is due to sequential increments in the local concentration of transmitter at the neuroglandular junction. Maxima were obtained usually with about 30 stimuli and additional stimuli progressively increased the duration of the response (Fig. 4B). With prolonged stimulation the secretory potential is lengthened to an extent where it resembles responses to certain agonists, such as noradrenaline (Fig. 4C), and 5-hydroxytryptamine (5-HT) (House, 1973).

2. Ionic Basis

It seems likely that the peak value of the secretory potential is close to the potassium equilibrium potential for the basal membrane and therefore its dependence on potassium concentration, $[K]_o$, has been examined. Typical responses recorded at different $[K]_o$ are shown in the upper part of Fig. 5 while in the lower part measurements of resting and secretory potentials made in a separate series of experiments are plotted against log $[K]_o$. These results suggest that the transmitter increases potassium permeability, although other explanations are possible. Another approach to this question is membrane conductance measurements. To achieve this Ginsborg *et al.* (1974) inserted two electrodes into an acinus (Fig. 6), one for current injection and the other for recording. Current pulses produced electrotonic potentials at the recording site (Fig. 6A) and when the current electrode was converted to a

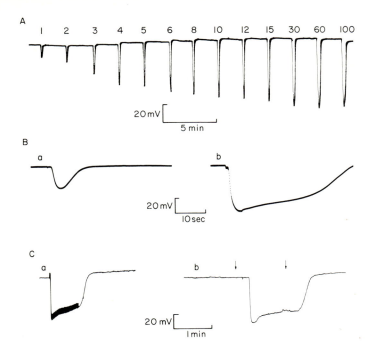

Fig. 4. Influence of the number of nerve stimuli on the amplitude and duration of secretory potentials. A. Secretory potentials evoked by different numbers of stimuli (shown above record) of the same strength. B. (a) shows secretory potential evoked by a single stimulus while (b) shows a larger and longer response from the same cell stimulated repetitively until the response had reached a maximum amplitude. C. (a) shows response to a train of nerve stimuli maintained for the period indicated by thickening of trace by stimulus artefacts. (b) shows response in the same cell to 5×10^{-6} M noradrenaline flowing through the bath for the period between the arrows.

recording one, similar secretory potentials were recorded at the different sites (Fig. 6B). Thus, there is coupling between acinar cells and it is feasible to attempt to measure conductance changes with two electrodes in the same acinus.

In Fig. 7A electrotonic potentials generated by current pulses are superimposed on the secretory potential evoked by nerve stimulation. Clearly the response is accompanied by a rise in membrane conductance and, as the longer records from a different cell illustrate (Fig. 7B), the conductance remains high for a period exceeding the hyperpolarizing phase of the response. In both cells

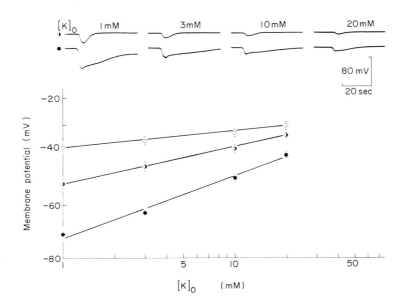

Fig. 5. Effect of external potassium concentration on the membrane potential before and after nerve stimulation. Upper part of figure shows some typical recordings at different $[K]_o$. Lower part shows resting potentials (O) and the peak values of responses to single shocks (◑) and stimulus trains (●) plotted against log $[K]_o$. Each point is the mean of 100 measurements and the bars signify ± S.E. values.

the results indicate that the secretory potential can be "inverted". For example in Fig. 7A the upper envelope of the potential trace gives the response at the resting potential whereas the lower envelope represents the response that would have been observed if a steady current had been passed equal to the pulsed current. From these experiments it was possible to estimate the "transmitter equilibrium potential" at the peak of the hyperpolarization and to deduce that the transmitter increases the membrane potassium conductance. A similar conclusion was reached for the hyperpolarizing response caused by bath applications of dopamine.

Inspection of the records in Fig. 7B strongly suggest that the transmitter also opens other ion channels since after the hyperpolarization there is a prolonged depolarization during which the membrane conductance declines slowly to its resting value.

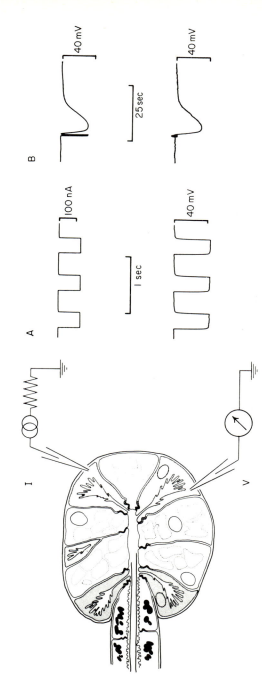

Fig. 6. Electrical coupling between gland cells. A. Upper trace shows injected current pulses (I) while lower is a record of membrane potential (V) illustrating electronic potentials. B. Secretory potentials recorded simultaneously by both intracellular electrodes.

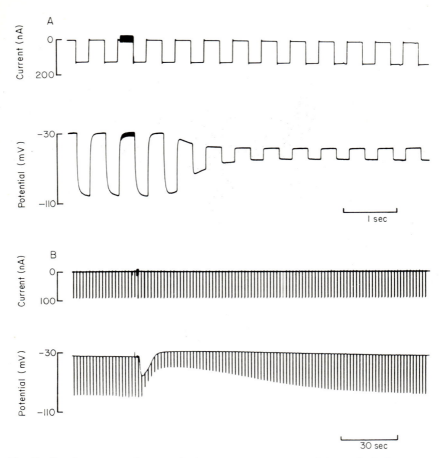

Fig. 7. Conductance change during secretory potential evoked by nerve stimulation. A. Upper trace shows injected current pulses while lower illustrates membrane potential. Period of nerve stimulation indicated by thickening of traces due to stimulus artefacts. B. Current and voltage traces, as in A, from a similar experiment on another cell.

IV. PHARMACOLOGY

As mentioned above, noradrenaline, dopamine and 5-HT cause a hyperpolarization of gland cells like that evoked by nerve stimulation. It was, therefore, of interest to examine the dose-response relationships of these agonists and other possible transmitter candidates (see Bowser-Riley and House, 1976).

A. Agonists

1. Biogenic Amines

Some typical responses to certain amines are shown in Fig. 8; these records were obtained by F. Bowser-Riley. Evidently the size of the response is graded with agonist concentration and the maxima equal the maxima evoked by nerve stimulation except for the partial agonist 5-HT. Of the catecholamines dopamine is the most potent (see Table II).

Another biogenic amine of some possible significance is octopamine. However, this substance is a very weak agonist, producing a maximal response considerably smaller than that caused by nerve stimulation.

It should also be pointed out that N-acetyldopamine, an important dopamine metabolite in insects (Sekeris and Karlson, 1966), fails to elicit an electrical response (Ginsborg *et al.*, 1976b).

2. α- and β-agonists

In view of the presence of catecholamines in salivary nerve endings it was considered worthwhile to investigate the actions of some established adrenergic agonists (House *et al.*, 1973). Neither amidephrine nor methoxamine, both α-agonists (Dungan *et al.*, 1965; Haylett and Jenkinson, 1972b), produced a hyperpolarization. The β-agonists, isoprenaline and salbutamol, also failed to elicit responses at low concentrations (<1 mM). Thus, the gland cell receptors cannot be regarded as classically adrenergic.

3. Acetylcholine

Even in the presence of an anticholinesterase ACh in high concentrations (1 mM) failed to mimic nerve stimulation. Carbachol behaved similarly to ACh. These substances, however, can influence transmission in this preparation (see Section IV, B).

4. Amino Acids

γ-aminobutyric acid, glutamic acid, glycine, alanine and aspartic acid, all of which are considered to be neurotransmitters at some

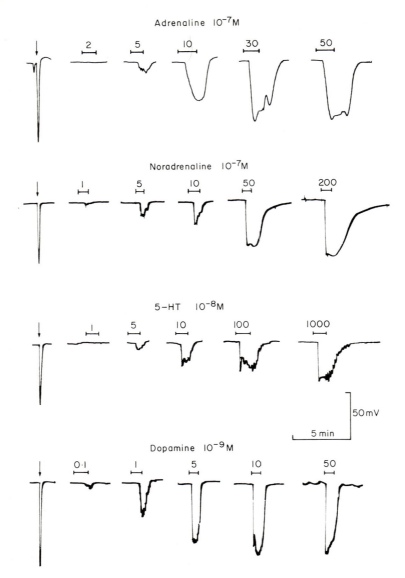

Fig. 8. Electrical responses produced by certain agonists. Each set of records for a given agonist was obtained in the same gland cell and is accompanied (on the left) by the maximal response (arrow) caused by nerve stimulation. Period of bath application is indicated by the horizontal bars and the concentration by the numbers above, viz. Adrenaline at 2×10^{-7}, 5×10^{-7}, 10^{-6}, 3×10^{-6} and 5×10^{-6} M.

TABLE II

Some pharmacological properties of cockroach salivary gland cells

Agonist	Concentration (μM) required to cause half-maximal agonist response	Antagonist	Affinity constant (10^6 M^{-1}) of antagonist for dopamine receptors
Dopamine	0.04	Ergometrine	250
5-HT	0.37	Phentolamine	2
Noradrenaline	1.3	Atropine	~1
Adrenaline	1.5	Methysergide	~0.2

Bowser-Riley and House (1976); Bowser-Riley (unpublished).

neuromuscular junctions and certain synapses (Gerschenfeld, 1973; Krnjević, 1974), have been tested. None produced an electrical response (Bowser-Riley and House, 1976).

B. Indirect Actions at Junctions

A number of agonists, as noted above, act directly on the acinar cell membrane and produce an electrical response. It is also possible that these substances cause the release of transmitter from nerve endings and this could be responsible for the brief hyperpolarizations superimposed on some of the agonist responses (see e.g. 5-HT, Fig. 8). The extent of this possible indirect effect might be established in experiments where iontophoresis is used rather than bath application of drugs.

House (1973) has reported that spontaneous "miniature secretory potentials" may be observed occasionally in unstimulated glands. Recently it has been found that several compounds can enhance the frequency of these "miniatures" and also their size. An example is illustrated in Fig. 9 where ACh in the presence of physostigmine gives rise to random potentials, some of large size, with a similar time course to evoked secretory potentials (marked by arrows). These actions are not confined to cholinergic agonists since bretylium (Silinsky, 1974) and some methylated derivatives of dopamine (Ginsborg et al., 1976b) are effective too. It is of interest also that ACh increases submaximal responses produced by nerve stimulation yet does not increase the dopamine sensitivity of the cells (Bowser-Riley and House, 1976). Carbachol acts similarly. Our tentative conclusion is that there are cholinergic receptors in the salivary nerve terminals that promote transmitter release.

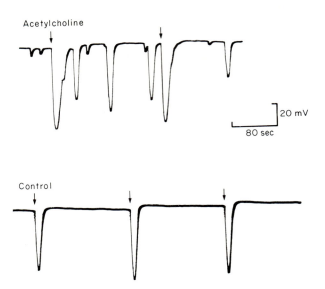

Fig. 9. Effect of acetylcholine (ACh) on the membrane potential of a gland cell. During the upper record ACh $(10^{-4}\,\text{M})$ was applied in the presence of physostigmine $(1.3 \times 10^{-5}\,\text{M})$. The lower trace was obtained 5—15 min after washout of these drugs with normal bathing solution. The secretory potentials produced by a single nerve stimulus are indicated by arrows (see Bowser-Riley and House (1976) for further details).

Also of interest is the finding that the sympathomimetic amine, tyramine, (see e.g. Trendelenburg, 1972) causes "miniature secretory potentials" (Ginsborg et al., 1976a).

A number of substances apparently reduce the amount of transmitter released per impulse as judged by their depression of the secretory potential without reduction of receptor sensitivity. In this category are guanethidine and bretylium (Silinsky, 1974). It seems likely that a similar decrement in low Ca and high Mg solutions arises in the same manner.

C. Antagonists

In view of the evidence for agonist activity it was conceivable that electrical responses to nerve stimulation could be blocked by adrenergic antagonists. Propranolol, a β-antagonist, had little or no effect in concentrations up to $2 \times 10^{-5}\,\text{M}$ but began to block above this level. On the other hand phentolamine, an α-antagonist, markedly reduced responses at about 10^{-5} M (Fig. 10A). It also effectively blocks responses to dopamine but not those to 5-HT (10B). An extended study of the antagonism exerted by phentolamine indicates that it is competitive (Bowser-Riley and

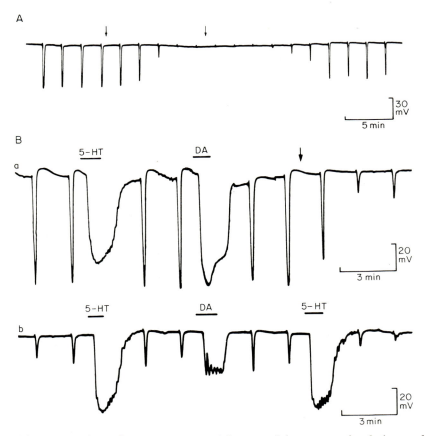

Fig. 10. Antagonism of secretory potentials, caused by nerve stimulation and dopamine, by phentolamine. A. Intracellular record showing that phentolamine (1.3×10^{-5} M) reversibly blocks responses evoked by single nerve stimuli. Phentolamine was present between the arrows. B. (a) shows secretory potentials evoked by nerve stimulation (25 stimuli at 100 Hz), 5-HT (10^{-6} M) and dopamine (DA, 3×10^{-7} M). Bars indicate periods of application. Phentolamine (5×10^{-5} M) was present after the arrow. (b) shows corresponding responses to the same stimuli as in (a) during the continual application of phentolamine. (a) and (b) are continuous records.

House, unpublished). Evidently this antagonist cannot distinguish between the receptors for dopamine and noradrenaline but can distinguish between these and the 5-HT receptors.

It is tempting to identify the dopamine receptors with those for the transmitter but the present evidence is circumstantial. Further indirect support is that ergometrine, which blocks dopamine receptors in other invertebrate systems (Ascher, 1972; Walker *et al.*, 1968; Berry and Cottrell, 1975), also blocks the responses to nerve stimulation and dopamine. Moreover, methysergide, an inhibitor of dopamine responses in ganglion cells of *Helix* (Woodruff *et al.*, 1971) and guinea-pig (Hirst and Silinsky, 1975), is also a blocking agent in the cockroach. Yet another antagonist is, surprisingly, the muscarinic blocker atropine, which acts apparently in a competitive manner (Bowser-Riley and House, 1976).

Table II gives estimates of the affinity constants of these antagonists for the acinar dopamine receptors.

V. CONCLUSION

The innervated cockroach salivary gland is a suitable preparation for the study of neuroglandular transmission. Membrane potential and conductance can be measured and estimates of the "transmitter equilibrium potential" obtained. Besides these postjunctional properties it should also be possible to investigate transmitter release under appropriate conditions.

Another interesting facet of the physiology of this tissue is its secretory behaviour. In this respect it is certain that the techniques originally developed by Ramsay (1954) for the study of ion and fluid transport in Malpighian tubules will again prove invaluable as they have for other epithelia. Indeed, Whitehead (1973) has used these methods to measure fluid secretion from *Periplaneta* glands stimulated by 5-HT. Recently R. K. Smith (unpublished) has observed salivary secretion from isolated glands of *Nauphoeta*. Secretory responses were elicited by applications of dopamine and electrical "field stimulation" of the salivary nerves and are illustrated in Fig. 11. The rate of secretion in the absence of stimulation is low, about $5 \ nl \ min^{-1}$ or less, while the maximal rate compares favourably with measurements in other epithelia, such as *Calliphora* salivary glands (Berridge and Patel, 1968) and *Rhodnius* Malpighian tubules (Maddrell, 1969). Thus, the results in Fig. 11 demonstrate the potential value of this tissue as a subject for the investigation of ion and fluid secretion. (Also see Chapters 9, 11, 21 and 25.)

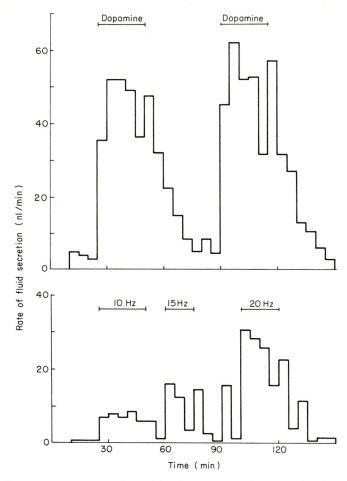

Fig. 11. Secretory responses elicited by dopamine and nerve stimulation. Fluid secretion was measured by the method of Ramsay (1954); acinar nerves were excited by electrical "field stimulation" (cf. House, 1973). The bathing fluid was identical to that given in Section IIIA with the addition of 2 mM glucose. Upper graph shows rate of fluid secretion produced by dopamine (10^{-7} M) while lower shows the results of nerve stimulation at different frequencies in another gland. Periods of stimulation by dopamine and nerve are indicated by bars.

Probably the main conclusion of the work described in this article is that the cockroach salivary gland's innervation is almost certainly dopaminergic. A catecholamine is present in the nerve terminals and a radiochemical assay confirmed the presence of dopamine in the gland. The postsynaptic action of nerve stimulation is imitated by dopamine. Finally, the pharmacological properties of dopamine receptors in the acinar cells are indistinguishable from those for the actual neurotransmitter.

In view of the possible importance of dopamine as a transmitter in the central nervous system of vertebrates (see e.g. Hornykiewicz, 1973; Iversen, 1975) it is valuable to have a peripheral dopaminergic system to examine.

ACKNOWLEDGEMENTS

I am greatly indebted to Bernard Ginsborg who shares responsibility for much of the work reported above and to Frank Bowser-Riley and Roy Smith for permission to quote unpublished observations. My thanks are also due to Mr C. M. Warwick for preparing the diagrams. Financial support was received from the University of Edinburgh and the Science Research Council and is gratefully acknowledged.

REFERENCES

Ascher, P. (1972). *J. Physiol.* **225**, 173—209.
Berridge, M. J. and Patel, N. G. (1968). *Science, New York* **162**, 462—463.
Berry, M. S. and Cottrell, G. A. (1975). *J. Physiol.* **244**, 589—612.
Bland, K. P. and House, C. R. (1971). *J. Insect Physiol.* **17**, 2069—2084.
Bland, K. P., House, C. R., Ginsborg, B. L. and Laszlo, I. (1973). *Nature, New Biol.* **244**, 26—27.
Bowser-Riley, F. and House, C. R. (1976). *J. exp. Biol.* **64**, 665—676.
Creed, K. E. and Wilson, J. A. F. (1969). *Aust. J. exp. Biol. med. Sci.* **47**, 135—144.
Dean, P. M. and Matthews, E. K. (1972). *J. Physiol.* **225**, 1—13.
Dungan, K. W., Stanton, H. C. and Lish, P. M. (1965). *Int. J. Neuropharmac.* **4**, 219—234.
Fry, J. P., House, C. R. and Sharman, D. F. (1974). *Br. J. Pharmac. Chemother.* **51**, 116—117P.
Garrett, J. R. (1974). *In* "Secretory Mechanisms of Exocrine Glands" (Eds N. A. Thorn and O. H. Petersen), pp. 17—28. Academic Press, New York.
Gerschenfeld, H. M. (1973). *Physiol. Rev.* **53**, 1—119.
Ginsborg, B. L. and House, C. R. (1976). *J. Physiol.* **262**, 477—487.
Ginsborg, B. L., House, C. R. and Silinsky, E. M. (1974). *J. Physiol.* **236**, 723—731.
Ginsborg, B. L., House, C. R. and Silinsky, E. M. (1976a). *J. Physiol.* **262**, 489—500.

Ginsborg, B. L., House, C. R. and Turnbull, K. W. (1976b). *Br. J. Pharmac. Chemother.* **57**, 133—140.

Haylett, D. G. and Jenkinson, D. H. (1972a). *J. Physiol.* **225**, 721—750.

Haylett, D. G. and Jenkinson, D. H. (1972b). *J. Physiol.* **225**, 751—772.

Hirst, G. D. S. and Silinsky, E. M. (1975). *J. Physiol.* **251**, 817—832.

Hornykiewicz, O. (1973). *Br. med. Bull.* **29**, 172—178.

Horowitz, J. M., Horwitz, B. A. and Smith, R. E. (1971). *Experientia* **27**, 1419—1421.

Horwitz, B. A., Horowitz, J. M. and Smith, R. E. (1969). *Proc. Natl Acad. Sci. U.S.A.* **64**, 113—120.

House, C. R. (1973). *J. exp. Biol.* **58**, 29—43.

House, C. R. (1975). *Experientia* **31**, 904—906.

House, C. R., Ginsborg, B. L. and Silinsky, E. M. (1973). *Nature, New Biol.* **245**, 63.

Imai, Y. (1965). *J. Physiol. Soc. Japan* **27**, 304—312.

Iversen, L. L. (1975). *Science, New York* **188**, 1084—1089.

Kagayama, M. and Nishiyama, A. (1974). *J. Physiol.* **242**, 157—172.

Kessel, R. G. and Beams, H. W. (1963). *Z. Zellforsch. mikrosk. Anat.* **59**, 857—877.

Klemm, N. (1972). *Comp. Biochem. Physiol.* **43A**, 207—211.

Krnjević, K. (1974). *Physiol. Rev.* **54**, 418—540.

Lundberg, A. (1955). *Acta physiol. Scand.* **35**, 1—25.

Lundberg, A. (1957a). *Acta physiol. Scand.* **40**, 21—34.

Lundberg, A. (1957b). *Acta physiol. Scand.* **40**, 35—58.

Maddrell, S. H. P. (1969). *J. exp. Biol.* **51**, 71—97.

Megaw, M. W. J. and Robertson, H. A. (1974). *Experientia* **30**, 1261—1262.

Nishiyama, A. and Petersen, O. H. (1974a). *J. Physiol.* **238**, 145—158.

Nishiyama, A. and Petersen, O. H. (1974b). *J. Physiol.* **242**, 173—188.

Ramsay, J. A. (1954). *J. exp. Biol.* **31**, 104—113.

Robertson, H. A. (1975). *J. exp. Biol.* **63**, 413—419.

Sekeris, C. E. and Karlson, P. (1966). *Pharmac. Rev.* **18**, 89—94.

Silinsky, E. M. (1974). *Br. J. Pharmac. Chemother.* **51**, 367—371.

Sutherland, D. J. and Chillseyzn, J. M. (1968). *J. Insect Physiol.* **14**, 21—31.

Trendelenburg, U. (1972). *Handb. exp. Pharmak.* **33**, 336—362.

Walker, R. J., Woodruff, G. M., Glaizner, B., Sedden, C. B. and Kerkut, G. A. (1968). *Comp. Biochem. Physiol.* **24**, 455—470.

Whitehead, A. T. (1971). *J. Morph.* **135**, 483—506.

Whitehead, A. T. (1973). *J. Insect Physiol.* **19**, 1961—1970.

Woodruff, G. N., Walker, R. J. and Kerkut, G. A. (1971). *Eur. J. Pharmac.* **14**, 77—80.

17. The Sodium Chloride Excreting Cells in Marine Vertebrates

L. B. Kirschner

Department of Zoology, Washington State University, Pullman, Washington, U.S.A.

I. INTRODUCTION

Every vertebrate class includes many species for which desiccation is a major environmental problem. Evaporation from the body surface is an obvious route of water loss for terrestrial forms, and marine teleosts and reptiles lose water by osmosis, because the body fluids are less concentrated than sea water. Excretion of urine increases the total water lost. The consequences are the same whether the mode of water loss is evaporative or osmotic, and adaptive responses have been developed that minimize their severity. Thus, mammals are normally antidiuretic as is shown by the onset of *diabetes insipidus* after hypophysectomy, and birds use both a counter-current system in the kidney and cloacal water reabsorption to concentrate the urine and minimize its volume. Fishes and reptiles are unable to produce urine more concentrated than blood, but the volume excreted is much smaller in sea water than in fresh water. In addition, these animals often develop mechanisms to lower the water permeability of the body surface. Such strategies reduce the magnitude of the problem, but however impressive they are, there remains a residual which can be solved only by water intake. For terrestrial animals this means drinking fresh water or eating dilute food, but such an avenue is not available in the ocean. Marine vertebrates must ingest sea water, or obtain free water from food often more concentrated than their body fluids. Details of the solution in teleost fishes were worked out nearly half a century ago (Smith, 1932), and they have become a paradigm for other groups as well. These fish drink sea water and absorb most of the water, NaCl and some divalent ions from the gut. The divalents are excreted by the kidney, NaCl is extruded across the gills into the more concentrated environment, and the free water generated balances osmotic and renal losses.

The story was extended by Schmidt-Nielsen and his collaborators who noted that marine birds and reptiles also face desiccation and showed that they cope in a similar fashion. Birds drink sea water and/or eat marine animals, and they absorb a concentrated solution from the gut. Free water is produced by excreting the salt, but in this case it is eliminated in the form of a hyperosmotic solution by a cranial organ called the nasal salt gland (Schmidt-Nielsen and Fänge, 1958a). A cranial salt gland was described at about the same time in a marine turtle (Schmidt-Nielsen and Fänge, 1958b). Subsequent work has extended the initial observations to many species of birds and reptiles, and research on these groups has been summarized recently (Peaker and Linzell, 1975; Dunson, 1976). The past decade has also witnessed intensive research on mechanisms of salt transfer by the teleost gill, and this has been described in several reviews (Motais, 1967; Maetz, 1971; Maetz and Bornancin, 1975; Evans, 1976).

Shortly after the first description of avian and reptilian salt glands, there appeared a report that elasmobranch fish also possess a salt-extruding organ capable of producing a hyperosmotic solution (Burger and Hess, 1960). This organ, unlike the others, is an appendage of the posterior intestine and is called the rectal gland. It is worth separate note, not because it is located in a different position, but because the nature of the stress facing these animals differs from that in marine teleosts, reptiles and birds (and from that in mammals as well). Elasmobranchs are well known to be isosmotic or even slightly hyperosmotic to sea water by virtue of their high blood urea concentration, hence they face no tendency to lose water across the body surface. But their plasma NaCl is less than half that of sea water, and salt loading by diffusion apparently exceeds the excretory capacity of the kidney; it requires the salt gland to maintain an ionic steady-state. The other groups must cope with the same diffusion gradient, but salt loading is markedly increased by the need to drink and absorb sea water. The reason, discussed in more detail later, is that water absorption from a hyperosmotic solution in the gut is obligatorily coupled to salt transport from lumen to blood (House and Green, 1965; Skadauge, 1969). In these animals water loss is the primary problem, and active ion transport pays the cost of volume regulation.

So much is part of the orthodox lore of the physiology of vertebrate osmoregulation. Recent research, much of it summarized in the reviews mentioned above, has been concerned with the cellular mechanisms of salt transfer. There is no *a priori* reason for supposing that the mechanisms of salt excretion by gills, cranial and rectal salt glands should be similar. Indeed, in addition to their different anatomical locations, data from physiological studies suggests differences in function. For example, the avian salt glands secrete intermittently, apparently under nervous control, while the teleost gill extrudes salt continuously with no evidence of nervous input. On the other hand, certain cellular features appear to be similar in all of these organs, and since they are rarely considered together, these similarities and their possible significance comprise one theme of this review. (For fish gills, also see Chapter 18.)

II. THE SALT EXCRETING CELLS

The marine teleost gill contains a distinct group of cells occupying the regions between respiratory lamellae and sometimes extending into the base of the lamellae (Keys and Willmer, 1932). These were called "chloride cells" by Keys and Willmer who suggested that they were responsible for net NaCl extrusion. Although this proposition had little experimental support for many years, recent studies have made the case fairly convincing. One of the striking findings is that some ultra-structural and enzymatic characteristics are similar in salt excreting cells from gills and glands.

Perhaps the most characteristic structural feature is an enormous infolding of basolateral membranes. In teleost fishes (Philpott and Copeland, 1963) and birds (Doyle, 1960; Komnick, 1963) the infoldings are mostly from the basal membrane, while in reptilian and elasmobranch salt glands it is primarily the lateral cell surfaces that are folded (Ellis and Abel, 1964; Doyle, 1962). The infoldings result in an extensive system of closely-spaced intracellular tubules in gills and in the avian salt gland. Reptile and elasmobranch epithelia have, instead, elaborately convoluted intercellular channels. From a functional point of view, the difference, illustrated schematically in Fig. 1, may be more apparent than real (cf. also Berridge and Oschman, 1972, Figs. 3b and 3f). Both sets of tubules communicate with the extracellular fluids (ECF) and extend far into the interior of the cell, nearly to the apical surface. By comparison, the mammalian eccrine sweat gland, another NaCl extruding organ, shows extensive lateral infoldings primarily near the basal region; these are much less

P

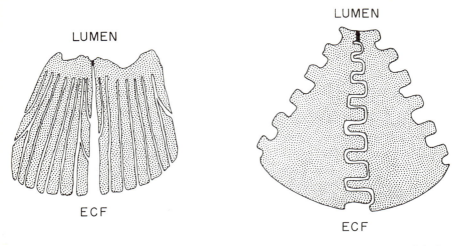

Fig. 1. Schematic representation of membrane infolding in cells responsible for hyperosmotic regulation in vertebrates. The infoldings are basal in teleost fish and in birds (left) and form an extensive network of intracellular tubules that open to the ECF. Infoldings are lateral in elasmobranch fish and in reptiles (right) and the cytoplasmic extensions of adjacent cells interdigitate.

impressive toward the apical end (Ellis and Montagna, 1961; Ellis, 1967).

The other common character is a high level of ATPase that requires both Na and K for activity (Na/K ATPase). This enzyme is known to be a reflection of the membrane transport system that extrudes Na from animal cells and maintains intracellular K high (Schwartz *et al.*, 1972). It has been found in most epithelia showing active Na transport (Bonting, 1970), but there seems, where comparison is possible, to be more in salt extruding organs that form hyperosmotic fluids. This is best illustrated in euryhaline species capable of regulating in both hyposmotic and hyperosmotic environments. For example, when ducklings were maintained on a high salt intake, enzyme levels in the salt gland increased for more than nine days by which time they were five times higher than in birds given fresh water to drink (Ernst *et al.*, 1967). Enzyme levels then declined when the birds were permitted to drink fresh water. Similarly, the killifish, adapted to sea water, had substantially higher gill Na/K ATPase levels than when adapted to fresh water (Epstein *et al.*, 1967), and two laboratories showed that many sea water-adapted fish have more enzyme in the gill than those in fresh water (Jampol and Epstein, 1970; Kamiya and Utida, 1969). In fish, as in birds, the increase in ATPase parallels the time course of adaptation to sea water, and its level declines when a euryhaline animal is returned to fresh water (Kamiya and Utida, 1968; Zaugg and McLain, 1970). The enzyme is also conspicuous in the salt glands in marine reptiles (Dunson, 1976) and elasmobranch fish (Bonting, 1966). Kamiya (1972) was able to separate chloride cells from other types in eel gill and showed that they contain the highest ATPase levels, a fact that had previously only been surmised.

A noteworthy characteristic of the ATPase is its intracellular localization. Until recently, there was little reason to have confidence in reports purporting to localize this enzyme histochemically, in part because of artifacts associated with the methods used (Ernst, 1972a), and also because the studies seldom showed that activity required alkali metals or that it could be abolished by cardiac glycosides like strophanthin and ouabain. Both characteristics are diagnostic for this enzyme. In a carefully controlled study, Ernst (1972a,b) developed a new technique which satisfied both critical requirements in the avian salt gland. His work (Fig. 2) showed that the enzyme is associated with the highly infolded basal membranes forming the intracellular tubular system. The technique has been extended to reptilian (Ellis and Goertemiller, 1974) and elasmobranch glands (Ellis, personal communication) and in both it is on the infolded (lateral) membranes as shown in Figs 3 and 4. The same distribution was noted in eel gill (Shirai, 1972), although older methods were used in this work, and it was not possible to show alkali metal activation or glycoside inhibition.

Such observations suggest that some of the cellular mechanisms of salt extrusion and/or water conservation in these organs may be similar in spite of known differences in their details. The rest of this paper will be devoted to a brief summary of current speculation about these mechanisms with emphasis on the possible roles of the tubular system and ATPase.

III. MODELS FOR SALT EXTRUSION

A. The Marine Teleost Gill

Since NaCl is extruded from blood (150–200 mM) into sea water (450–500 mM) active transport is clearly involved. What is less clear, and has been the subject of some controversy, is whether only one or

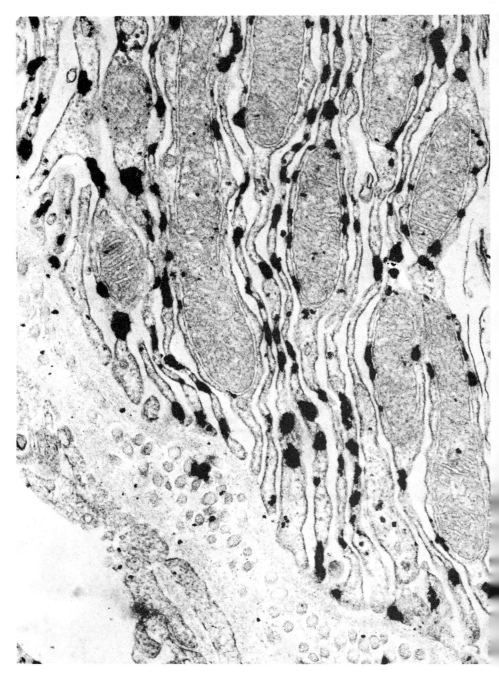

Fig. 2. Electron micrographs showing the membrane localization of the ouabain-sensitive ATP in avian (duck) salt gland. A. (Left) Basal region of the cell with dense deposits on the infolded membranes (×50,700). Note openings of

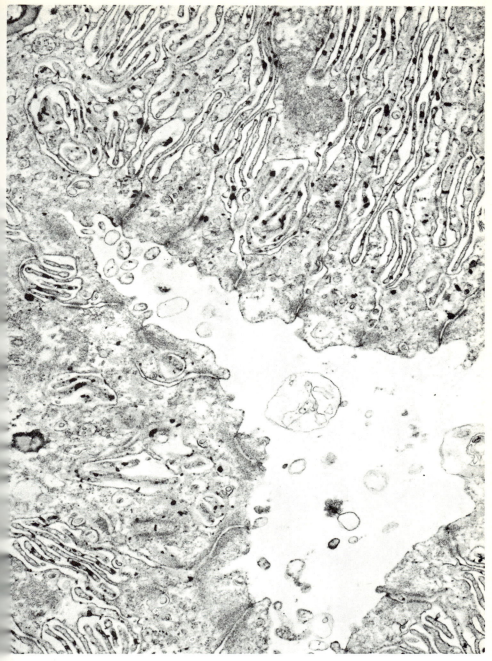

the intracellular tubules into the ECF. B. (Right) Apical region showing deposits on the internal tubular membranes, little or none on the luminal membrane (x23,400). The photomicrographs were provided by Dr Stephen Ernst.

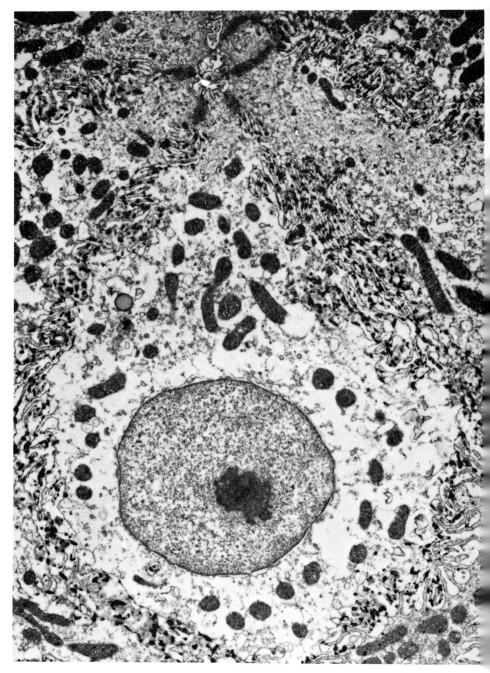

Fig. 3. Electron micrograph showing the ouabain-sensitive ATP on the lateral membranes in the green turtle (*Chelonia mydas*). Several cells abut the tubular lumen which can be seen near the top of the figure directly above the large nucleus. Little or no activity appears on the luminal membrane (x12,500). The photomicrograph was provided by Dr Richard Ellis.

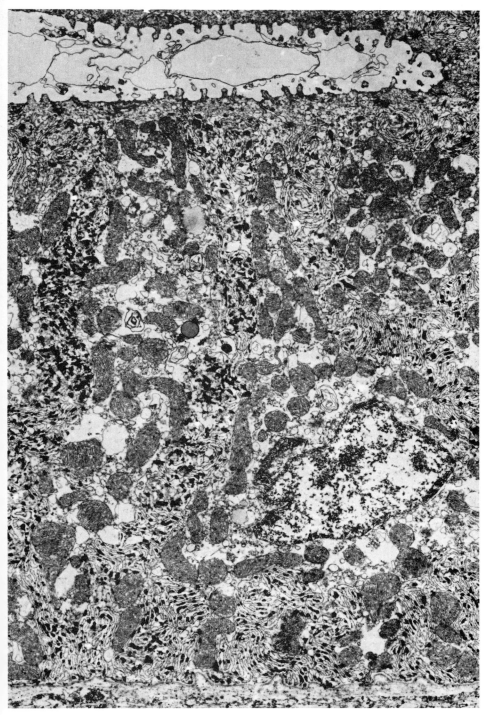

Fig. 4. ATPase activity in the dogfish rectal gland shows a distribution like that in Fig. 3 (×10,000). The photomicrograph was provided by Dr Richard Ellis.

both of the ions are extruded actively. I do not propose to review all the evidence bearing on the question; this has been summarized recently by Maetz and Bornancin (1975) and by Evans (1976). But a brief introduction will indicate the nature of the controversy and will also provide some background for comparing the mechanisms in different organs.

Among the criteria for delimiting active transport, the most frequently used is a demonstration that some solution of the diffusion equation is unable to describe the distribution or fluxes of an ion across an interface. In many, fish drinking contributes little to the total influx (J_{in}) and renal salt loss little to the efflux (J_{out}); most of the unidirectional movement occurs across the gills and $J_{in} \approx J_{out}$ across the body surface (e.g. Motais, 1967). Under these conditions an ion moving only under the influence of an electrochemical gradient (i.e. passively) will be distributed as predicted by the Nernst equation:

$$E = \frac{RT}{ZF} \ln \frac{a_2}{a_1} \tag{1}$$

Where E is the potential difference (PD) across the gills and a_1 and a_2 are the chemical activities in plasma and sea water. The test is to calculate, from chemical activities (usually approximated by concentrations) in blood and sea water, the magnitude of the voltage necessary to account for the steady-state of each ion. These values, denoted E_{Na} and E_{Cl}, are compared with the potential difference (PD) actually measured across the body surface (E). If the two values agree, diffusion suffices to explain the ion's behaviour. If not, it must be actively transported. In some fish (the rainbow trout is an example) drinking (Shehadeh and Gordon, 1969) can produce more than half the total influx of salt (Greenwald et al., 1974); hence unidirectional fluxes across the body surface are not equal. However, if the animals are in a steady-state and the ion in question is moving passively, then:

$$E = \frac{RT}{ZF} \ln \frac{J_{12}a_2}{J_{21}a_1} \tag{2}$$

Where J_{12} and J_{21} are unidirectional fluxes from compartment 1 (plasma) to 2 (sea water) and from compartment 2 to 1 respectively. In this case we compare the PD calculated for each ion from its fluxes and chemical activities with the PD measured by a pair of electrodes. Table I shows a group of representative values for marine teleosts. In no case is the value for E_{Cl} close to the measured voltage, and in most it has the wrong polarity. Hence, Cl must be extruded by active transport across the gill. For Na the situation is more equivocal. In many fish, E_{Na} is within a few millivolts of the measured values, and given the technical problems of implanting a bridge through the body wall with no electrical leak between sea water and body fluids, the agreement is quite good. One might conclude from these data that diffusion accounts for the behaviour

TABLE I

Gill TEP, E_{Na} and E_{Cl} in Marine Teleosts

Species	E (mV)	E_{Cl} (mV)	E_{Na} (mV)	References
Mugil capito	+22	−34	+25	Maetz and Bornancin (1975)
Serranus scriba	+25	−33	+23	Maetz and Bornancin (1975)
Platichthys flesus	+19	−34	+19	Potts and Eddy (1973)
Anguilla anguilla	+23	−36	+28	Maetz and Bornancin (1975)
Gillichthys mirabilis	+21	—	+24	R. Thompson (1973)
Blennius pholis	+23	−29	+23	House (1963)
Pholis gunnelus	+18	−35	+19	Evans (1969)
Salmo gairdneri	+11	−48	+45	Kirschner et al. (1974)
Hippocampus erectus	− 7	—	+29	Evans and Cooper (1976); Edwards and Condorelli (1928)
Achirus lineatus	− 4	—	—	Evans and Cooper (1976)

The potential difference (E) measured between body fluids and sea water is compared with the potential difference required for maintaining measured blood concentrations of Na (E_{Na}) and Cl (E_{Cl}) by diffusion. Values of E_{Na} and E_{Cl} for S. gairdneri were calculated from text equation (2); all others from equation (1). Values for E_{Na} are 1—2 mV lower than in the original publications because activities were used here, concentrations in the originals.

of Na. In the three cases at the bottom of the table the gill PD is much higher than E_{Na}, and if these fish can be shown to be in a steady-state during measurements, and without electrical leak around the bridge, then active sodium extrusion will have been demonstrated. Although such demonstrations do not yet exist, most investigators are convinced that Na is actively extruded (Maetz and Bornancin, 1975; Evans, 1976).

The discovery that sea water-adapted teleosts have a high Na-K-ATPase in the gill was accompanied by speculation that it functioned in active Na extrusion. For example, it was suggested that "this enzyme plays an important role in the active transport of Na^+ across the gill". (Epstein *et al.*, 1967). It was easy to see that the simplest model would find the ATPase located on the apical membrane where it could extrude Na from cytoplasm into sea water, at the same time transferring K from the latter into the cell. The apical membrane in such a system would have to be preferentially K-permeable to permit K to diffuse back into sea water. Such a model (Kirschner, 1969), emphasizing events at the apical membrane, is shown in Fig. 5. It received powerful support in experiments showing that Na efflux (J_{out}^{Na}) is dependent on K in the external medium, and that in some fish it is decreased by externally applied ouabain. The most compelling data were developed as follows. When a euryhaline fish (eel or flounder) is transferred from sea water to fresh water there is an immediate fall in J_{out}^{Na} to about 25% of the original value (Motais *et al.*, 1966). This is far in excess of

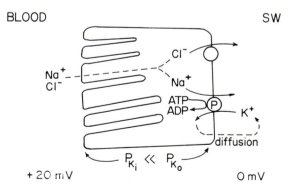

Fig. 5. Model of a teleost chloride-cell with the Na/K transport system located on the apical membrane. The apical membrane must be more K-permeable than the basal to permit K entering through the pump to diffuse back to the medium. Neither renal nor intestinal excretion could balance influx if K entered the ECF at rates approximating those of Na efflux.

the active transport component, which comprises only 2—5% of the unidirectional efflux in these fish, but what is germane to our story is that a Na-K pump on the apical membrane must be blocked under these conditions due to the absence of K in the external medium. The model predicts that if K is added to the dilute outside solution, J_{out}^{Na} should be stimulated. This is precisely what was found in flounder (Maetz, 1969). The J_{out}^{Na} increased with increasing external K, reaching 60% of the sea water value at 10 mM. Subsequent work in other laboratories has confirmed and extended these results to several species of fish. A second prediction of the model is that an apical transport system should bind and be inhibited by cardiac glycosides which are known to inhibit this system. In fact, binding of externally applied ouabain was reported in eel gills, and more binding was noted in the absence of external K than when it was present (Maetz and Bornancin, 1975). Motais and Isaia (1972) observed that externally applied ouabain reduced J_{out}^{Na} in the eel by 16% which is more than enough to account for complete inhibition of active extrusion in sea water. A similar result was obtained in another fish, the fat sleeper (Evans *et al.*, 1973).

Practically no attention has been paid to the means by which Na enters the cell across the basal membrane, and no measurements bearing on this step have been made. If intracellular [Na] is low and cytoplasm is negative to ECF as in most animal cells, Na might simply diffuse from blood into the cell at a rate equal to its transport across the apical membrane.

The handful of studies on Cl transport by the fish gill are summarized by Maetz and Bornancin (1975) and by Potts in Chapter 18.

B. The Avian Salt Gland

The literature up to 1974 was ably summarized by Peaker and Linzell (1975). We will consider only those aspects already discussed for fish, namely ion fluxes and electrochemical gradient in secreting glands, with the goals of describing which ions are actively transported, and briefly considering possible molecular mechanisms. In contrast to fish gill, both vascular and luminal surfaces of the secretory epithelium have been inaccessible, a fact that limits both experimental manipulations and the kinds of measurements that can be made. For example, it has not been possible to modify the composition of the luminal fluid, a maneuver used extensively in fish studies, or to measure unidirectional Na and Cl fluxes. Interpretation of some data may also be complicated by the fact that both chemical and electrical measurements on the luminal side are made in the main duct of the gland, and the latter is located far from the secretory membranes.

Although these limitations preclude a rigorous flux-force analysis, a few critical experiments have been used to characterize the magnitude of the electrochemical gradient and to draw inferences about active or passive ionic transfer. Plasma Na in marine birds is within the range for other terrestrial or marine vertebrates (150–200 mM). Plasma Cl is a little lower. Concentrations in the ductal fluid vary with experimental conditions but are usually in the range of 400–600 mM; Cl is usually slightly higher than Na. Thus, the concentration gradient across this epithelium is about the same as that across the fish gill. A crucial datum was provided by the observation that there was no PD between blood and lumen in the non-secreting gland, but when secretion was induced, the gland lumen became 40–60 mV positive to the blood (Thesleff and Schmidt-Nielsen, 1962). If this measurement, still the only one reported, represents the electrical situation at the secretory surface then Na moves from blood to lumen against a steep electrical as well as concentration gradient and must be actively transported. Further, the voltage is probably adequate to account for the high luminal Cl concentration, and active transport of this ion need not be invoked. This conclusion is obviously very different from that for fish. However, the electrical measurement has been repeated in no other bird, and the experimental conditions employed suggest the need for caution. The PD was determined between bridges inserted into blood and main duct of the gland, and the duct is far removed from the secretory epithelium. The electrode pair will measure voltage across the latter only if the electrical space constant of the ductal epithelium is very large. Expressed otherwise, the ductal epithelium must be absolutely impermeable to ions. This has, in fact, been argued on grounds of its histological appearance and the need to prevent osmotic equilibration with blood (Peaker and Linzell, 1975), but the datum is so critical that the proposition badly needs verification. At least one observation can be interpreted to suggest a different possibility. The salt concentration of the glandular fluid was shown to vary inversely with the rate of secretion (Hanwell *et al.*, 1971; Peaker and Linzell, 1975). Such behaviour might suggest that a primary secretion is followed by reabsorption of a hypotonic fluid distal to the secretory site and, if so, the electrical measurement may reflect properties of the distal (ductal) epithelium. I have seen no data to rule out this possibility which would be adaptive in a bird for which water loss engenders the salt loading. It is necessary to use this value for the PD since it is the only one extant, but with the reservation that not only its magnitude but its polarity may not really reflect the voltage across the secretory cells.

Another electrical measurement, the intracellular potential, was also measured by Thesleff and Schmidt-Nielsen, (1962) and has been used in reconstructing ion movements during secretion (Peaker, 1971). When a microelectrode was inserted into a cell, the PD between cell and blood, i.e. across the basal membrane, fell rapidly from an initial value of 60–80 mV to 20 mV (blood positive to cytoplasm), and when secretion was induced the value remained at 20 mV. This observation is also suspect for reasons discussed later, but again, it is the only one that exists.

In reconstructing molecular mechanisms in salt glands much more attention has been paid to transfer across the basal membrane than in fish gills. This is probably due in part to the fact that the avian gland is under cholinergic nervous control (Fänge et al., 1958) which raises the possibility that the transmitter modifies membrane permeability as at other neuroeffector junctions. Also, the rate of salt transfer can be exceptionally rapid; as much as 57% of the Na and 80% of the Cl were extracted from blood passing through the secreting gland in domestic geese (Hanwell et al., 1971). The electrical data, described above, suggest that membrane potentials across basal and apical surfaces are both 20 mV in the resting gland, since there is no potential difference across the epithelium; the cell is negative to both blood and lumen as shown in Fig. 6A. Intracellular ion concentrations have been estimated by several authors (Peaker, 1971a; B. Schmidt-Nielsen, 1976) and are also shown in the figure. On stimulation a transepithelial voltage of about 60 mV develops with no change in the PD across the basal membrane and little change in intracellular ion concentrations (Peaker, 1971a; Schmidt-Nielsen, 1976). One possible model accommodating these data is shown in Fig. 6B. Acetylcholine (Ach), released by efferent nerve impulses, makes the basal membrane of the secretory cell more ion-permeable, thus facilitating the entry of NaCl. Such a permeability change is supported by the observation that ^{24}Na exchange was increased in salt gland slices when an Ach analog was added to the medium (Van Rossum, 1966). Since intracellular concentrations remained constant the salt entering was extruded into the lumen at the same rate. The electrical data require a 60 mV change at the apical surface, and this was attributed to the action of

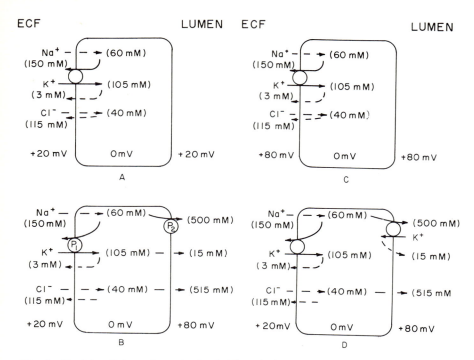

Fig. 6. Models of the avian salt gland. Electrical and concentration data are used to reconstruct the situation in a resting cell (A) and in the same cell after stimulation. (B) An alternative model (C and D) is based on the possibility that measurements of the basal membrane potential are grossly in error as described in the text.

an electrogenic Na pump (Peaker, 1971a). Although not characterized further by Peaker, the apical pump might be the Na/K ATPase acting in an electrogenic fashion (i.e. transferring more Na from cell to lumen than K from lumen to cell), because ouabain, applied via the gland lumen, abolished secretion (Thesleff and Schmidt-Nielsen, 1962). The model was modified (Peaker, 1961b) for two reasons. First, a change in ion permeability of the basal membrane should be reflected by a change in basal PD, but none was noted. And second, carbonic anhydrase inhibitors blocked secretion when injected intravenously (Fänge et al., 1958). This led to the suggestion that entry of Na and Cl was coupled to extrusion of H and HCO_3 produced by cellular metabolism. Such ion exchanges could be electrically neutral and would require no change in basal membrane potential on activation. Events at the luminal membrane in the revised model are the same as in the original: Na is extruded from cell to lumen by an electrogenic pump, while Cl transfer is passive. The precise nature of the events at the basal membrane is left unspecified in Fig. 6B which can therefore represent either variant.

It can be seen that the single measurement of intracellular potential in the herring gull salt gland has been critical in modeling events at both membranes. But the electrical measurement was made before the ultrastructure of the cell was understood, and it is clear today that insertion of a microelectrode is virtually impossible without destroying the intracellular tubular system which is in free communication with ECF. This can be appreciated by examining Fig. 2A (or Plate 2.2 in Peaker and Linzell, 1975). In fact, the PD across the basal membrane is likely to be 60—80 mV, the value observed immediately after impalement and characteristic of many postjunctional neuroeffector membranes. Its decline to 20 mV, and the failure of this value to change when secretion was initiated, can be attributed to massive damage to the tubular system with concomittant electrical leakage between cytoplasm and ECF. An alternative model consistent with this picture is illustrated in Figs 6C and 6D. The PD across the basal membrane of the resting cell is assumed to be 60—80 mV, and since there is no voltage between blood and duct, the same PD must exist across the luminal membrane. If neural stimulation depolarized the basal membrane by making it cation permeable the observed transepithelial PD could result in part or even completely from this event rather than solely from operation of an electrogenic pump on the luminal membrane. Such depolarization of the postjunctional membrane is consistent with the action of acetylcholine in other neuroeffector systems. An apical pump would be required to remove this salt as rapidly as it entered the cell, but the precise nature of this system cannot be specified. In Figure 6D the entire change in PD is assigned to the basal membrane, and the apical pump is electrically silent. On the basis of these data the latter could be the Na/K ATPase transferring one Na for each K.

There is no reason to insist on this model for salt transfer. It is presented merely as a plausible alternative to the one shown in Figs. 6A and B, especially to emphasize that the latter is based on two highly questionable electrical measurements.

C. Reptilian and Elasmobranch Salt Glands

Much less information is available in these animals which have received substantial attention only during the last few years. The

NaCl concentration in fluids secreted by a number of marine reptiles differs little from that in birds (Dunson, 1976), and the same appears to be true for rectal gland fluid from elasmobranchs (Burger and Hess, 1960; Burger, 1972). The ratio of salt concentrations in secretory fluid to plasma encompasses a range from about 2.5 in elasmobranchs which have a high plasma NaCl, to about six in some reptiles.

As with the bird salt gland, only one paper describes electrical measurements across secreting glands in this group. In an isolated, perfused rectal gland of dogfish the perfusion fluid (i.e. basal side) was positive by 3.6 ± 0.10 mV (Hayslett et al., 1973). This situation resembles that in teleost fish where the blood side is also positive. The PD is somewhat lower than E_{Na} (9.4 mV), and hence Na may be actively secreted in this preparation. Chloride is even further out of equilibrium, since it would require that the blood be negative by 10.3 mV to be concentrated passively in the secretory fluid. Needless to say, the reservations expressed above about the secretory potential in avian salt glands apply here as well.

IV. THE ROLE OF THE NA/K-ATPASE

The discovery that birds and teleost fish adapted to sea water have a high Na/K-ATPase in the salt excreting organs was accompanied by speculation that it functioned in active Na extrusion. The model developed for teleost fish was described earlier (Fig. 5) together with some of the experimental evidence supporting it. That such a system might operate in the avian salt gland as well could be inferred from the fact that infusion of ouabain into the collecting duct abolished both the secretory PD and fluid formation.

While these experiments were provocative, further investigation indicated that the model is probably not correct, and that they must have another explanation. In the first place, the gills of marine fish are preferentially cation permeable (Potts and Eddy, 1973; Kirschner et al., 1974), and the dependence of J_{out}^{Na} on external K is predicted for such an epithelium by the diffusion equation. There remains some uncertainty about whether diffusion is able to account for all of the stimulation observed (Maetz and Bornancin, 1975; Evans, 1976) but this bears more on the question of whether there is any Na transport than on the precise mechanism. The experiments with cardiac glycosides are even more equivocal. While some inhibition of J_{out}^{Na} was noted when ouabain was added to sea water bathing an eel, it had no effect on the flounder or sea perch (Motais and Isaia, 1972). This was attributed to a lower affinity of the pump for glycosides in the presence of external K in flounders and sea perch, but the explanation is unlikely because the concentration of inhibitor used inhibits the system (measured as ATPase activity) even in the presence of 20 mM K, twice the sea water concentration. Two cardiac glycosides, strophanthin K and ouabain, do not effect Na efflux across the gills of anesthetized rainbow trout. As shown in

TABLE II

Effect of ouabain on J_{out}^{Na} in K-free sea water

N	Sea water control	K-free sea water + Ouabain	
		1st hour	2nd hour
5	305 ± 79	252 ± 65	195 ± 66

J_{out}^{Na} was measured for one hour in sea water. The fish was then transferred to K-free sea water containing ouabain ($1-5 \times 10^{-5}$ M). Efflux was measured for another two hours in the presence of the glycoside. N = number of animals. Fluxes (mean ± S.E.) in μmole $(100 \text{ g})^{-1} \text{ hr}^{-1}$. Fluxes in these anesthetized fish decrease about 10% per hour even without experimental perturbation. The decrease after adding ouabain is not appreciably greater than would be seen in control animals.

Table II, they are inactive even when the gill is exposed to K-free sea water. Table III shows that they do not block the stimulation of J_{out}^{Na} when K is added to an ion-free medium bathing the gill. On the other hand, when ouabain is injected into the ventral aorta, where it has access to the basal rather than apical membrane, the ATPase is almost completely inhibited (Zaugg and McLain, 1971), and when applied internally it also blocks Na extrusion from isolated gills (Kamiya, 1967; Kamiya and Utida, 1968) and abolishes the potential

TABLE III

Effect of ouabain on K-stimulated J_{out}^{Na}

Inhibitor conc.	N	Sea water Control	DW	+K (10 mM)
0	2	111	19	81
$0.5-5 \times 10^{-4}$ M	8	146 ± 16	23 ± 1	115 ± 13
1×10^{-3} M	4	157 ± 14	23 ± 3	104 ± 15

J_{out}^{Na} was measured for one hour in sea water. The fish were then transferred to deionized water (DW) containing ouabain at the concentration shown, and the efflux was measured for another hour. K (10 mM) was then added and the efflux measured for another hour. Ouabain was present during the last two periods. Fluxes (mean ± S.E.) in μmole $(100 \text{ g})^{-1} \text{ hr}^{-1}$.

difference across isolated gill arches (Shuttleworth *et al.*, 1974). Externally applied ouabain had no effect on the isolated gills. The observations that ouabain, applied externally, is bound to the gill is apparently misleading, since it was subsequently shown to have entered the cells where it was probably bound to the intracellular tubules, not to the apical membrane (Karnacky *et al.*, 1976). Finally, the histochemical technique developed by Ernst (1972a,d) has shown that all, or nearly all of the enzyme is located on the basolateral membrane system in avian, reptilian and elasmobranch salt glands (Figs 2, 3, 4), and little or none is present on the apical membranes. Similar results were obtained for the eel gill (Shirai, 1972).

These data suggest that the model is untenable in its original form. The Na/K transport system is not located on the apical membrane (see, especially, Fig. 2B). As a corollary, stimulation of J_{out}^{Na} by K added to the apical surface and its inhibition, where this occurs, by ouabain on the apical side do not support the model as originally believed. But there is the stubborn fact that the transport system is present in high concentration in these organs. In an effort to rationalize this and take account of its intracellular localization, it was recently proposed that Na may be extruded into the tubular system of fish chloride cells to reach very high concentrations where they terminated near the apical surface. Since the region between the tubular endings and apical membrane has been reported to contain small vesicles, it was proposed that the latter are formed from the tubular endings and extrude the salt by exocytosis into sea water (Maetz and Bornancin, 1975). Such a suggestion was also made by Shirai (1972) and was suggested even earlier by Philpott and Copeland (1963) for the fish gill, and it could as easily be extended to the avian salt gland. It remains to be seen whether further experimental work supports such a mechanism, which would require some modification to work in the reptile-elasmobranch organs.

We have been preoccupied during the past two decades with active ion transport and the enzymatic correlates thereof. Identification of the Na/K-ATPase with active Na extrusion from cells has been a striking part of the picture, and the attempt to relate its presence in salt extruding cells to the same phenomenon has provided the basis for much of the modelling described above. I would like to suggest that it may play a different role only marginally related to ionic homeostasis. There is no necessary connection between salt excretion and the elaborately infolded membrane-plus-ATPase systems seen in gills and salt glands. Sweat glands are capable of respectable rates of secretion but have extensive lateral infoldings primarily in the basal region of the cells (cf. Fig. 7.7 in Ellis, 1967). The most striking difference between the sweat gland and the secretory organs we have been considering is that the primary secretion of the former is isosmotic or only slightly hyperosmotic to blood. In contrast, the apical surfaces of the latter are exposed to solutions 3—10 times more concentrated. Their potential for losing water osmotically is apparent, and the enzyme's major role may be to reduce water loss.

For a marine hyporegulator the costliest part of maintaining hydromineral balance is replacing water lost to the environment. The reason is that obtaining solute-free water is coupled to active ion transport. This proposition can be illustrated by comparing net water movements in fresh water and sea water-adapted fish. The former must cope with an osmotic gradient of about 250 mOsm favouring water entry. The rate of water uptake is determined by the permeability of the body surface, and the fish maintains its volume constant by excreting a dilute urine at the same rate. Formation of 1 ml of urine requires that about 100 μmole of NaCl be reabsorbed by the renal tubules from a glomerular filtrate. An additional 1—10 μmoles must be absorbed across the gills to balance salt lost in the urine. If we neglect the energy expended by the heart in maintaining filtration pressure, the cost of volume regulation is that engendered in reabsorbing this salt from the tubules and gills. In sea water the osmotic gradient is usually 3—4 times larger, about 600—800 mOsm, and if water permeability were the same as in fresh water the fish would lose about 3 ml per unit time. This must be replaced by drinking an equal volume of sea water containing nearly 1500 μmole NaCl. The ions must be absorbed by active transport from the gut as part of the mechanism transporting water from the hyperosmotic lumen, and then they must be extruded, again by active transport in the gill, from blood to sea water. We have no information about the energetics of active ion transport in these organs, but if it even approaches that of tubular transport in the fresh water kidney the cost of volume regulation in sea water would be greater by an order of magnitude. It would clearly be adaptive to reduce the osmotic permeability of salt excreting cells, and this is exactly what is found; osmotic permeability is much lower in sea water-adapted than in fresh water-adapted animals (Evans, 1969; Motais et al., 1969). The general argument will obviously apply to reptiles and birds, and water conservation may even be of value to elasmobranchs where limited osmotic inflow must operate both the kidney and rectal gland.

It is usually taken for granted that cells are uniformly isosmotic with body fluids, and if this were true in such exchange epithelia as we have been considering the gradient would develop entirely across the apical membrane. This raises the question of what changes occur that reduce osmotic permeability and still permit both rapid ion movement and a diffusional water permeability that is essentially the same as in fresh water (Evans, 1969; Motais *et al.*, 1969). The answer may be that the tubules and ATPase play this role. In all of the salt excreting cells the membrane infoldings occur in the basal or lateral membranes which are open to the ECF. It is reasonable to suppose that they have the usual passive characteristics of cell membranes, i.e. relatively high water permeability, but low permeability to ions, especially to Na. When the membrane transport system pumps Na into an intracellular tubule (or lateral intercellular space in reptiles and elasmobranchs) the latter becomes hyperosmotic to cytoplasm causing water to flow in osmotically. This would increase the hydrostatic pressure in the tubule, and since the latter is open at the base, the solution would flow back into the ECF, a situation depicted in Fig. 7A. The operation of such a system, with ions and water entering the tubule all along its length, could stratify the cell, as shown in Fig. 7B, so that the osmotic pressure of the apical cytoplasm approaches that of sea water or of the luminal fluid in a

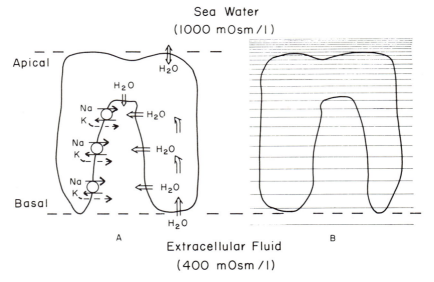

Fig. 7. Osmotic stratification of a cell by a tubular system containing the Na/K transport system.

gland. Under these conditions, no osmotic pressure gradient would exist at the exchange surface, hence no tendency for water to be lost to the medium. Instead, water flowing from the base of the cell toward its apex simply enters the tubular system and is returned to the ECF. The operation of such a system would have little effect on tracer water movement, but would result in a marked reduction in net water loss to the environment, and since the latter is used to calculate osmotic permeability, this value would be found to decrease. These predictions correspond to at least some extant data; diffusional water permeability differs little between fresh water and sea water fish, but osmotic permeability is much lower in sea water. In such a system the major role of the Na/K-ATPase is to minimize water loss rather than to extrude Na into the environment.

It should be pointed out that such a model is not novel. Conceptually, it is a variant of the standing gradient system developed to explain solute-linked water flow in other epithelia (Diamond and Bossert, 1967; but see Chapter 7) with solute recycling across basolateral membranes (Wall, 1971 and Chapter 23). Its application to reducing water loss through the teleost gill was suggested several years ago (Motais and Garcia Romeu, 1972). The model has the virtue of being able to account for both the large amount of ATPase and its distribution in these cells, and for their reduced osmotic permeability. It is otherwise unsupported by experimental evidence, a doubtful distinction that it shares with models relating the ATPase to Na extrusion from the cell. However, it should be amenable to test, although technical obstacles may be formidable. If the transport system operates to reduce osmotic permeability its inhibition, for example by ouabain injected into the ventral aorta, should increase the rate of water loss. The proposition that these cells are osmotically stratified may be demonstrable by observing their melting points, since the freezing point depression should be greatest at the apical pole of the cell and least at its base. If frozen sections can be prepared without abolishing the gradient, use of a freezing stage and polarization microscope ought to show this. In addition, if the stratified solutes are inorganic ions it may be possible to use electron probe X-ray spectrometry to display their asymmetric distribution from base to apex (see Gupta et al., Chapter 4).

In concluding, I would note that vesicular extrusion of Na is easily incorporated into this picture. Thus, if active Na transport is finally established, the intracellular tubular system with its associated Na/K pump could play a role both in water conservation and active Na excretion by these cells.

ACKNOWLEDGEMENTS

Research in this laboratory has been supported by grants from the National Institute of General Medical Sciences (GM 04254 and GM 01276) and the National Science Foundation (BMS75-00476). Technical assistance and the many ideas provided by a number of my colleagues is gratefully acknowledged. Contributions by Alan Koch, Lewis Greenwald, Klaus Beyenbach and Theodore Kerstetter were particularly important. I am especially indebted to Dr Stephen Ernst and Dr Richard Ellis for Figs 2, 3, and 4. Whatever finally proves to be the role of the Na/K transport system, Ernst's development of a convincing method for localizing it, was crucial in establishing a basis for further study.

REFERENCES

Berridge, M. J. and Oschman, J. L. (1972). "Transporting Epithelia." Academic Press, New York and London.

Bonting, S. L. (1966). *Comp. Biochem. Physiol.* 17, 953–966.

Bonting, S. L. (1970). *In* "Membranes and ion transport." (Ed. E. E. Bittar), Vol. I. Wiley-Interscience, New York.

Burger, J. W. (1972). *Comp. Biochem. Physiol.* 42A, 31–32.

Burger, J. W. and Hess, W. N. (1960). *Science* 131, 670–671.

Diamond, J. M. and Bossert, W. H. (1967). *J. gen. Physiol.* 50, 2061–2083.

Doyle, W. L. (1960). *Exp. Cell Res.* 21, 386–393.

Doyle, W. L. (1962). *Amer. J. Anat.* 111, 223–237.

Dunson, W. A. (1976). *In* "Biology of the reptilia." (Eds C. Gans, W. R. Dawson), Vol. 5. Academic Press, New York.

Edwards, J. S. and Condrelli, L. (1928). *Am. J. Physiol.* 86, 383–398.

Ellis, R. A. (1967). *In* "Ultrastructure of normal and abnormal skin." (Ed. A. S. Zelickson), Lea and Febiger, Philadelphia.

Ellis, R. A. and Abel, J. H. (1964). *Sience* 144, 1340–1342.

Ellis, R. A. and Montagna, W. (1961). *J. biophys. biochem. Cytol.* 9, 238–242.

Ellis, R. A. and Goertemiller, C. C. (1974). *Anat. Rec.* 180, 285–298.

Epstein, F. H., Katz, A. I. and Pickford, G. E. (1967). *Science* 156, 1245–1247.

Ernst, S. A. (1972a). *J. Histochem. Cytochem.* 20, 13–22.

Ernst, S. A. (1972b). *J. Histochem. Cytochem.* 20, 23–38.

Ernst, S. A., Goertemiller, C. C. and Ellis, R. A. (1967). *Biochim. biophys. Acta* 135, 682–692.

Evans, D. H. (1969). *J. exp. Biol.* 50, 689–703.

Evans, D. H. (1976). *In* "Comparative physiology of osmoregulation in animals." (Ed. G. M. O. Maloiy), Academic Press, New York.

Evans, D. H. and Cooper, K. (1976). *Nature Lond.* 259, 241–242.

Evans, D. H., Mallery, C. H. and Kravitz, L. (1973). *J. exp. Biol.* **58**, 627–636.
Fänge, R., Schmidt-Nielsen, K. and Robinson, M. (1958). *Am. J. Physiol.* **195**, 321–326.
Greenwald, L., Kirschner, L. B. and Sanders, M. (1974). *J. gen. Physiol.* **64**, 135–147.
Hanwell, A., Linzell, J. L. and Peaker, M. (1971). *J. Physiol.* **213**, 373–387.
Hayslett, J. P., Schon, D. A., Epstein, M. and Hogben, C. A. M. (1974). *Am. J. Physiol.* **226**, 1188–1192.
House, C. R. (1963). *J. exp. Biol.* **40**, 87–104.
House, C. R. and Green, K. (1965). *J. exp. Biol.* **42**, 177–189.
Jampol, L. M. and Epstein, F. H. (1970). *Am. J. Physiol.* **218**, 607–611.
Kamiya, M. (1967). *Annot. Zool. Japan.* **40**, 123–129.
Kamiya, M. (1972). *Comp. Biochem. Physiol.* **43B**, 611–617.
Kamiya, M. and Utida, S. (1968). *Comp. Biochem. Physiol.* **26**, 675–685.
Kamiya, M. and Utida, S. (1969). *Comp. Biochem. Physiol.* **31**, 671–674.
Karnacky, K. J., Kinter, L. B., Kinter, W. B. and Stirling, C. E. (1976). *J. Cell Biol.* (In press).
Keys, A. B. and Willmer, E. N. (1932). *J. Physiol. Lond.* **76**, 368–378.
Kirschner, L. B. (1969). *Comp. Biochem. Physiol.* **29**, 871–874.
Kirschner, L. B., Greenwald, L. and Sanders, M. (1974). *J. gen. Physiol.* **64**, 148–165.
Komnick, H. (1963). *Protoplasma* **56**, 605–636.
Maetz, J. (1969). *Science* **166**, 613–615.
Maetz, J. (1971). *Phil. Trans. R. Soc. Lond. B* **262**, 209–251.
Maetz, J. and Bornancin, M. (1975). *Fortsch. der Zool.* **23**, 322–362.
Motais, R. (1967). *Ann. Inst. Oceanog. Monaco* **45**, 1–84.
Motais, R. and Garcia Romeu, F. (1972). *Ann. Rev. Physiol.* **34**, 141–176.
Motais, R. and Isaia, J. (1972). *J. exp. Biol.* **57**, 367–373.
Motais, R., Garcia Romeu, F. and Maetz, J. (1966). *J. gen. Physiol.* **50**, 391–422.
Motais, R., Isaia, J., Rankin, J. C. and Maetz, J. (1969). *J. exp. Biol.* **51**, 529–546.
Peaker, M. (1971a). *J. Physiol. Lond.* **213**, 399–410.
Peaker, M. (1971b). *Phil. Trans. R. Soc. Lond. B* **262**, 289–300.
Peaker, M. and Linzell, J. L. (1975). "Salt glands in birds and reptiles." Cambridge University Press, London.
Philpott, C. W. and Copeland, D. E. (1963). *J. Cell Biol.* **18**, 389–404.
Potts, W. T. W. and Eddy, F. B. (1973). *J. Comp. Physiol.* **87**, 29–48.
Schmidt-Nielsen, B. (1976). *Am. J. Physiol.* **230**, 514–521.
Schmidt-Nielsen, K. and Fänge, R. (1958a). *Auk* **75**, 282–289.
Schmidt-Nielsen, K. and Fänge, R. (1958b). *Nature Lond.* **182**, 783–785.
Schwartz, A., Lindenmayer, G. E. and Allen, J. C. (1972). *In* "Current Topics in Membranes and Transport." (Eds Bronner, F. and Kleinzeller, A.), ????.
Shehadeh, Z. H. and Gordon, M. S. (1969). *Comp. Biochem. Physiol.* **30**, 397–418.
Shirai, N. (1972). *J. Fac. Sci. Univ. Tokyo IV.* **12**, 385–403.

Shuttleworth, T. J., Potts, W. T. W. and Harris, J. N. (1974). *J. Comp. Physiol.* **94**, 321–329.

Skadauge, E. (1969). *J. Physiol. Lond.* **204**, 135–158.

Smith, H. W. (1932). *Quant. Rev. Biol.* **7**, 1–26.

Thesleff, S. and Schmidt-Nielsen, K. (1962). *Am. J. Physiol.* **202**, 597–600.

Van Rossum, G. D. V. (1966). *Biochim. biophys. Acta* **126**, 338–349.

Wall, B. J. (1971). *Fed. Proc.* **30**, 42–48.

Zaugg, W. S. and McLain, L. R. (1970). *Comp. Biochem. Physiol.* **35**, 587–596.

Zaugg, W. S. and McLain, L. F. (1971). *Comp. Biochem. Physiol.* **38B**, 501–506.

18. Fish Gills

W. T. W. Potts

Department of Biology, University of Lancaster, Lancaster, England

I. INTRODUCTION

The epithelium of the teleost gill is remarkable in several respects. In marine teleosts the gills are the site of ion movements, both active and passive, which are amongst the most intense found anywhere in the living world. While in most secretory epithelia the process of active transport occurs in a preferred direction, the polarity of transport across the gills of euryhaline teleosts is reversible, the gills taking up ions in fresh water and excreting ions in sea water. The epithelium is simultaneously the site of several other vital processes including the elimination of carbon dioxide or bicarbonate ions and ammonium ions as well as the uptake of oxygen. These multifarious activities are performed by a mosaic of epithelial cells of several kinds. Most of the surface is composed of flattened epithelial cells with crenellated surfaces but these are interspersed with what are termed "chloride cells", particularly rich in mitochondria, and with mucus cells. As the "chloride cells" are more abundant and better developed in fish adapted to sea water and there is a correlation between the ability of fish to excrete ions and the number of chloride cells (Olivereau, 1970; Shirai and Utida, 1970) it is likely that they are the site of outward transport. The more common crenellated epithelial cell may be presumed to be the main site of oxygen diffusion and possibly also of ion uptake in fresh water and of ammonia excretion at all times, but direct evidence on this point is still lacking.

Perhaps the most remarkable feature of teleost gills is that they are the most active transporters of ions known. The sodium flux per unit area across the gill of the mullet (*Mugil*) is 2,000 times as

TABLE I

Sodium flux in various tissues
μeq cm^{-2} hr^{-1}

Mullet gill (SW)	20	Maetz and Bornancin (1975)
Crab muscle fibre	0.8	Bittar *et al.* (1972)
Barnacle muscle fibre	0.4	Bittar *et al.* (1972)
Frog skin (30 mM NaCl/l)	0.1	Kirschner (1955)
Frog skin (FW)	0.01	Kirschner (1955)

great as that across frog skin in its natural environment (Table I). The flux is even greater than that across the surface of a resting nerve or muscle cell in a marine invertebrate, where the ambient sodium levels are much higher than those around vertebrate cells. In the squid giant fibre the sodium flux rises momentarily to 240 μeq cm^{-2} hr^{-1} (670 μeq m^{-2} s^{-1}) during the passage of an action potential (Hodgkin and Huxley, 1952), about ten times that across the mullet gill. However, if the fluxes in the teleost gill are confined largely to the chloride cells, which constitute only a small fraction of the gill area, the continuous flux across these cells must be at least as large as that during the action potential in the squid giant axon and is probably a good deal higher. On adaptation to hypersaline conditions the fluxes across fish gills increase still further but this increase may be associated with the increase in the number of "chloride cells". Choride fluxes across teleost gills are similar to or slightly smaller than the sodium fluxes.

Much of the ionic flux across the teleost gill, as across any cell, takes place by passive diffusion but the active fluxes are proportionately large. The fish gill contains more active sites, as measured by ouabain binding, than any other tissue (Table II). In

TABLE II

Ouabain binding in various tissues
sites/mg wet weight

Fundulus gill (sea water)	1.5×10^{14}	Karnaky *et al.* (1976)
Guinea-pig brain	1.4×10^{12}	Baker and Willis (1972)
Guinea-pig kidney	1.0×10^{12}	Baker and Willis (1972)
Squid nerve	3.5×10^{11}	Baker and Willis (1972)
Electrophorus electroplaque	3.0×10^{11}	Rivas *et al.* (1972); La Torre *et al.* (1970)
Guinea-pig taenia coli	1.1×10^{11}	Brading and Widdicombe (1974)
Guinea-pig heart	8×10^{9}	Baker and Willis (1972)
Human erythrocyte	2.9×10^{9}	Baker and Willis (1972)

nerve and muscle cells the active transport system is equivalent to a trickle charger feeding an energy store which is discharged intermittently, but in the fish gill there is a continuous large actively maintained flux of salt.

In spite of their unique properties, teleost gills have attracted much less attention than such relatively inactive tissues as frog skin or the red blood cell. It has been known, at least since the time of Botazzi (1897), that the marine teleosts are peculiar in maintaining a blood concentration much lower than that of the surrounding sea water. The osmotic problem that this causes was investigated by Homer Smith (1930) who showed that marine teleosts drink sea water and absorb water from the gut to replace that lost by osmosis. The drinking rates of the fishes examined by Smith averaged about ½% of the body weight per hour. More detailed studies showed that water is taken up from the gut as a consequence of the uptake of sodium and chloride ions into the blood, while divalent ions and some water remain behind in the gut, to be voided through the rectum. The salt taken into the blood must be excreted back into the sea against a concentration gradient, and thus the original water problem is replaced by an ionic problem. The rate of urine production is low in marine teleosts and their urine contains little sodium or chloride. Smith postulated that in order to maintain salt balance the secretion of salt, or as he supposed of a hyperosmotic salt solution, must take place at the body surface, most probably where the blood and external medium are in most intimate contact at the gills (Fig. 1). Keys (1931) using perfused fish heads, confirmed that the active transport of chloride from blood to medium took

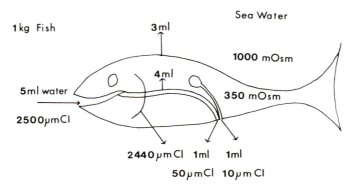

Fig. 1. Salt balance in marine teleosts as deduced from Homer Smith's experiments. The accepted model c. 1930–1965.

place in the head region. Keys and Willmer (1932) observed numerous acidophilic cells, showing some resemblance to the oxyntic cells of the vertebrate stomach, in the gills of eels adapted to sea water. As Keys had demonstrated the active transport of chloride through the eel head by silver nitrate titration and the oxyntic cells were believed to transport hydrochloric acid, these cells were termed "chloride cells" although later work has shown that the sodium fluxes across teleost gills are equal to or larger than the chloride fluxes.

Radioactive isotopes were first applied to the problems of teleost physiology by Mullins (1950) who measured the turnover of sodium and several other ions in the stickleback *Gasterosteus*. In the light of Smith's work he assumed that all the influx of salt took place through the gut. By chance Mullins' experiments were carried out in sea water of very low salinity and as a result the salt fluxes observed were small compared with those of marine teleosts in full strength sea water. The spurious "drinking rates" inferred from these experiments were high, *c.* 4% body weight per hour, but did not seem to be improbably large for a small fish. Smith's experiments had been carried out with fish weighing several hundred grams.

The inadequacy of Smith's model only became apparent when simultaneous measurements were made of both drinking rates and salt fluxes in the same fish, when it was found that the total influx of salt was five or ten times greater than that due to drinking alone (Motais and Maetz, 1965; Potts and Evans, 1967). The gills were evidently the site of both a massive, previously unsuspected, influx of salt and of an efflux which was much greater than that postulated in the previous model (Fig. 2). This discovery renewed interest in the character of these fluxes and the structure and properties of the

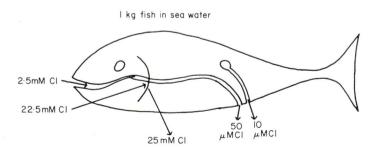

I kg fish in sea water

2·5mM Cl

22·5mM Cl

25 mM Cl

50
μMCl

10
μMCl

Fig. 2. Salt balance in marine teleost as disclosed by experiments with radioactive tracers.

chloride cell. In spite of a considerable amount of research during the last decade, the nature of these intense fluxes, found in one of the most common groups of animals, is still obscure. (Also see Chapter 17.)

II. THE MOVEMENT OF MONOVALENT IONS ACROSS THE GILLS OF MARINE TELEOSTS

The experimental methods used in the study of salt transport in the teleosts have, for practical reasons, been rather limited. For technical reasons effluxes are more easily measured with the aid of radioactives than are influxes, and measurements of sodium fluxes are less difficult than those of chloride fluxes. It is also a simple matter to change the external medium of a fish but very difficult to produce substantial changes in the composition of the internal medium. The most popular kind of investigation has therefore been a study of the effect of changes in the composition of the external medium on the rate of efflux of sodium. The study of the chloride fluxes has been comparatively neglected. More recently techniques have been developed to study the potential between the blood and external medium and studies of the influence of the composition of the external medium on this potential have thrown new light on the nature of the salt fluxes (Potts and Eddy, 1973a; House and Maetz, 1974). The gross manipulation of the composition of the internal medium can only be carried out in preparations of isolated perfused gills. These preparations have so far proved very intractable and no preparation is yet available which will maintain *in vitro* fluxes of the order of magnitude found *in vivo* (Shuttleworth, 1972; Rankin and Maetz, 1973) but a study of the potential across the isolated perfused gill has provided a third line of attack (Shuttleworth *et al.*, 1974).

Although the approach is rather indirect, investigations of the influence of the composition of the external medium on the salt efflux from marine teleosts have proved surprisingly fruitful. Early experiments by House (1963) demonstrated that the sodium efflux in the blenny (*Blennius pholis*) was dependent on the concentration of the external medium and dropped immediately to a low level on transfer to more dilute sea water. More detailed investigations by Motais *et al.* (1966) showed that in the flounder (*Platichthys flesus*) the sodium efflux, following immediate transfer from sea water to other media, was a function of the sodium content of the new medium, while the chloride efflux, though investigated in less detail, was similarly a function of the chloride concentration. The relationship between the external sodium concentration and the efflux appeared to follow Michaelis-Menten kinetics (Fig. 3). Wider investigations showed that in species such as the eel *Anguilla anguilla*, and the flounder, the rates of efflux were largely dependent on the composition of the external medium but in others, such as *Serranus scriba* and *Fundulus kansae*, they showed little change immediately after transfer, (Motais *et al.*, 1966; Potts and Fleming, 1970). Other experiments showed that in those species where the efflux was dependent on the composition of the medium the influx was similar to the efflux.

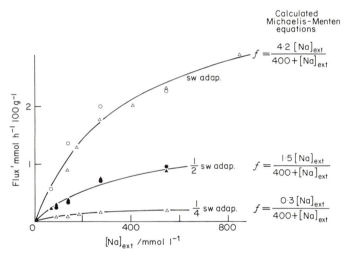

Fig. 3. The relationship between the sodium efflux from the flounder, when adapted to sea water and to various dilutions of sea water, and the sodium content of the external medium. (Motais *et al.*, 1966.)

It should be noted that following the transfer of euryhaline fishes such as *Platichthys flesus* to fresh water, changes in the fluxes take place after a period of an hour or so, which are not immediately reversible. These changes are associated with changes in the hormonal levels in the fishes and in the cell populations in the gills (Conte and Lin, 1967).

Although the sodium efflux in the flounder is almost independent of the external concentration of chloride, experiments with artificial solutions showed that it is dependent on the external concentration of potassium. At low concentrations potassium is even more effective in stimulating efflux than equivalent concentrations of sodium (Fig. 4). As American and Japanese workers (Epstein *et al.*, 1967; Kamiya and Utida, 1969) had shown that fish gills contained sodium-potassium dependent ATPases and the specific activity of these enzymes was found to be higher in sea water adapted fishes than in fish adapted to fresh water, Maetz (1969) suggested that the stimulation of sodium efflux by external potassium was brought about by sodium-potassium exchange processes mediated by the ATPase. According to this hypothesis the potassium pumped into the cells of the gill epithelium, in exchange for sodium, might diffuse out again down the concentration gradient. The concentration of sodium ions in sea water is almost 50 times as great as that of potassium. The very large sodium fluxes found in marine teleosts were postulated to be in the nature of a one for one exchange of sodium ions across the epithelium, mediated by the same transport system, an inadvertent consequence of the high concentration of sodium outside. Such an "exchange diffusion" might consume no energy and would cease immediately on transfer to a sodium free solution.

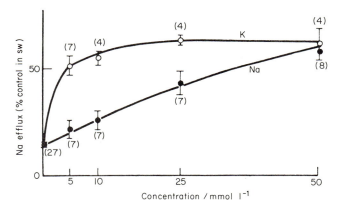

Fig. 4. The relationship between the sodium efflux from the flounder, when adapted to sea water, and the concentrations of sodium and potassium in the external medium. (Maetz, 1969.)

Several other pieces of evidence fitted well with this elegant model. In the flounder that proportion of the efflux in sea water which was attributable to the potassium in sea water, i.e. the difference between the efflux rates in normal sea water and in potassium-free sea water, was only about one third of the stimulation of efflux produced by adding the same concentration of potassium (10 mM) to fresh water. The lower effectiveness of potassium in sea water could be attributed to the competition of sodium ions for the enzyme. Again the potassium-linked efflux in sea water was approximately equal to the drinking rate in the flounder. In potassium free sea water the rate of turnover of sodium continues at a high level in the flounder but the sodium content of the fish gradually increases, as would be expected if the active outflux had been turned off. If a similar system were postulated for the transport of chloride, possibly in exchange for the bicarbonate ions in sea water, the major features of the salt fluxes in euryhaline fishes could be satisfactorily explained.

The first hint of an alternative explanation came from work on a crustacean, the brine shrimp *Artemia salina*. *Artemia* resembles the marine teleosts in maintaining the blood hypo-osmotic to the medium when in saline lakes, in drinking the medium to maintain water balance and in excreting the excess salt at the body surface. Even before any detailed work had been carried out on teleosts Croghan had shown that in *Artemia* the sodium efflux across the body surface is several times larger than that required to remove the salt ingested and that the efflux is a function of the sodium concentration in the medium, though it differs from that in the teleosts in showing no sign of saturation, even in sea water five times more concentrated than normal (Croghan, 1958a,b,c,d). Croghan attributed this large sodium flux to "exchange diffusion". An analysis of a flux into its active and passive components requires a knowledge of both the electrical and chemical gradients driving the flux. Smith (1969a,b) reinvestigated the fluxes in *Artemia*, measuring both the electrical and chemical gradients and showed that the dependence of the sodium efflux on the external sodium concentration was electrically mediated. *Artemia* proved to be similar to a sodium-specific electrode possessing a high permeability to sodium and a much lower permeability to chloride. When the external concentrations of sodium and chloride ions were high, the preferential permeability to sodium induced a positive charge in the body fluids, thus facilitating sodium efflux, while in dilute media the negative charge retarded the efflux of sodium. The potential E(mV) between the body fluids and external media in a wide variety of solutions could be predicted with reasonable accuracy from the equation:

$$E = 58 \log_{10} \left[\frac{[Na_o] + \alpha[K_o] + \beta[Cl_i]}{[Na_i] + \alpha[K_i] + \beta[Cl_o]} \right] \tag{1}$$

when $[Na_o]$, $[Na_i]$ etc. are the external and internal sodium concentrations etc, α and β the relative permeabilities to potassium and chloride ions in terms of the permeability to sodium. For *Artemia*, α has a value of about 0.6 and β has a value of about 0.03. The relationship between $[Na_o]$ and the passive sodium efflux in these circumstances is superficially similar to Michaelis-Menten kinetics (Fig. 5). While this hypothesis accounts more adequately than the "exchange diffusion" theory for the relationship between sodium flux and the composition of the external solution, because it takes account also of the potentials, it predicts that the chloride efflux should increase as the concentration of the external medium falls whereas in fact it decreases.

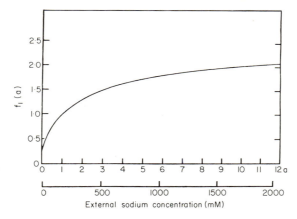

Fig. 5. The theoretical relationship between passive sodium efflux from an animal permeable only to sodium and chloride ions, in which the permeability to sodium is ten times greater than that to chloride, and the external sodium concentration. (Smith, 1969b.)

The relationship between the transepithelial potential and the composition of the external medium has recently been investigated in several teleosts and similar principles are found to apply: *Platichthys* (Potts and Eddy, 1973a), *Anguilla* (House and Maetz, 1974) and *Salmo* (Greenwald *et al.*, 1974; Kirschner *et al.*, 1974). In sea water and other sodium rich solutions the body fluids are strongly electropositive with respect to the external medium and sodium ions are close to electrochemical equilibrium across the gills. In contrast chloride ions are far removed from equilibrium and the greater part of the chloride outflux must be actively maintained.

The effect of potassium ions in enhancing sodium outflux could at first sight be accounted for by assuming that the gills were more permeable to potassium ions than to sodium ions, introducing a large term for potassium permeability into equation 1, but more detailed studies have shown that this explanation is inadequate. In the "fat sleeper" *Dormitator maculatus* the change of potential on transfer from seawater to potassium free seawater is sufficient to reduce the sodium efflux by only 1% whereas the observed reduction is 20% (Evans *et al.*, 1974). Similarly the enhancement of sodium efflux by potassium, immediately following transfer to freshwater, is much greater than the diffusion potential hypothesis would suggest. These discrepancies can of course be accounted for on the sodium/potassium exchange hypothesis. On the other hand the changes in sodium efflux for *Dormitator*, following changes in external sodium, were wholly accounted for by the changes in potential (Evans *et al.*, 1974). Similar discrepancies between the changes in sodium efflux and the changes in potential have been found in *Serranus* (Maetz and Dharmamba, quoted by Maetz and Bornancin, 1975) and in the mullet *Mugil capito* (Maetz and Pic, 1975). In both fishes the changes in potential induced by changes in external potassium were inadequate to account for the changes in sodium efflux. After transfer to freshwater it was possible in both cases to restore sodium efflux to the levels prevalent in seawater, by the addition of potassium, while the blood was still negative with respect to the medium. The evidence therefore suggests that the sodium fluxes are mediated partly by changes in potential but also by a sodium/potassium exchange mechanism.

While the better understanding of the electrochemical gradients obtained from the potential measurements has clarified the nature of the cation movements the problem of chloride movement has become further complicated. The relative permeability of the gills to chloride, as deduced from the potential measurements, is low and the active component of the chloride efflux required to balance both inward diffusion and ingestion is only a small part of the total efflux in seawater. This active component must be transported against a steep electrochemical gradient while the active sodium efflux, if any, occurs close to electrochemical equilibrium (Potts *et al.*, 1973). As the sodium influx and efflux are both close to electrochemical equilibrium it may be assumed that the greater part of these fluxes are passive. The passive chloride flux in any circumstances may be calculated from the sodium fluxes and the relative permeability to sodium and chloride as deduced from the potential (Equation 1; Fig. 6). In freshwater the negative potential generated by escaping

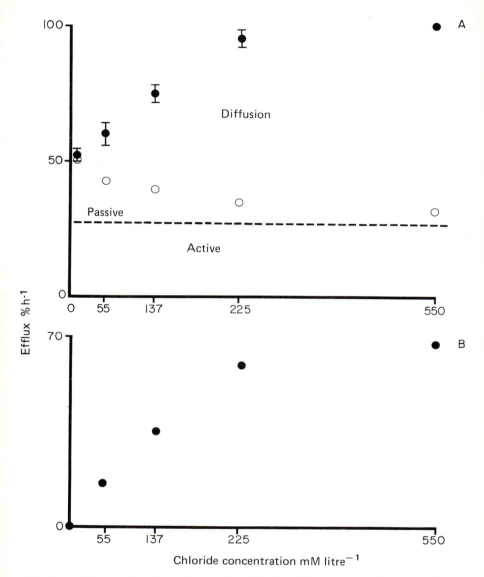

Fig. 6. A. The relationship between the chloride efflux from the flounder, when adapted to SW, and the chloride content of the external medium. Chloride efflux expressed as percentage of efflux in seawater. The passive component estimated from the transepithelial potential, the sodium efflux, and the relative permeabilities to sodium and chloride obtained from the best fit with equation 1. (Mashiter, unpublished.) B. The component of the chloride efflux dependent on the external concentration of chloride.

sodium facilitates the outward diffusion of chloride and the chloride efflux in the mullet, immediately following transfer to fresh water, can be wholly accounted for by simple diffusion; but this is not the case in the flounder (Appendix I and Fig. 6).

The non-passive part of the chloride efflux (Fig. 6) has many of the characteristics of a carrier mediated flux. A model analogous to that previously proposed for the sodium fluxes, consisting of three components, an active outflux in exchange for a counter ion, in this case possibly the bicarbonate ion in sea water, a large one to one chloride exchange, and a very small passive efflux, would account for the major features of the chloride fluxes. As the chloride/bicarbonate ratio in sea water is 230 to one, even a highly selective pump might show some chloride/chloride cycling. Some support for this hypothesis has been given by the demonstration of the presence of an anion stimulated ATPase in the gills of the goldfish *Carassius auratus*, the eel and *Fundulus heteroclitus* (Maetz and Bornancin, 1975). This ATPase activity is greater in sea water adapted fish and is found in both the mitochondrial and microsomal fractions of gill homogenates. Further support for a chloride/bicarbonate exchange is provided by the presence of a high activity of carbonic anhydrase in fish gills. In the eel the activity is again highest in the sea water adapted form. Unfortunately there is no detectable reduction in the rate of chloride efflux from marine teleosts when placed in bicarbonate free sea water (Fletcher, personal communication). Metabolic carbon dioxide makes it impossible to eliminate bicarbonate ions from the immediate neighbourhood of the gills of a living fish but experiments with isolated perfused gills, where the bicarbonate levels can be kept very low, again show no interaction between bicarbonate, either inside or outside the gill, and the chloride efflux (Potts and Harris, unpublished).

Thiocyanate is known to be an inhibitor of chloride transport in many tissues and following intraperitoneal injection of thiocyanate the chloride flux in the eel is reduced by over 70% (Epstein *et al.*, 1973). The magnitude of the reduction and the fact that the efflux in fresh water is unchanged indicate that both the active efflux and the exchange diffusion components are inhibited (Appendix 1). The potential across the mullet gill is not significantly affected by thiocyanate treatment (Maetz and Pic, 1975) indicating that the relative passive permeabilities to sodium and chloride ions are unchanged. As the permeability to chloride is relatively low any further reduction would have little effect on the potential in sea water, because the chloride term in Equation 1 is small, but in fresh water neither the potential nor chloride efflux, both of which are direct functions of chloride permeability, are consistently affected by thiocyanate (Maetz and Pic, 1975).

The active components of the sodium and chloride pumps are linked in some way. Activation of the sodium pump by the addition of potassium ions to the external medium stimulates the chloride efflux (Maetz and Bornancin, 1975) and is reciprocally blocked by thiocyanate (Maetz and Pic, 1975). These effects are most conveniently demonstrated immediately following transfer to fresh water. No doubt the same effects occur in sea water but the gross fluxes are then so large that the changes in active effluxes would be very difficult to detect. It should be noted that the quantities of sodium and chloride ions actively excreted in sea water may be approximately equivalent in spite of the large difference in permeability to the two ions.

In considering passive ion fluxes, we need to know how the trans-gill potential is related to the diffusion potential, at which the net fluxes of cations and anions are equivalent. With sea water outside and a vertebrate extracellular fluid inside, in both of which sodium and chloride ions predominate, the net influxes of these two ions will be almost equivalent at the diffusion potential. The water ingested by the fish will also contain similar quantities of the two ions. If the potential across the gills were maintained slightly above the diffusion potential the passive efflux of sodium would be increased while the influx would be decreased. In these circumstances sodium balance could be maintained in the absence of any active transport of sodium but the load on the chloride pump would be disproportionately increased as thermodynamic considerations show that the energy required for ionic regulation is minimal at the diffusion potential (Potts et al., 1973). It is not possible to discriminate in practice between the observed potential and the diffusion potential as the latter is a function of the ionic activities in sea water and body fluids which cannot be determined accurately enough. The quantities of ions passing through the chloride cell are very large and if they take the same route some coupling between the fluxes is likely. If they were not linked the quantities of counter ions required would place a considerable strain on the fishes' metabolism.

The chloride pump operates against a considerable electrochemical potential. If a sodium pump is also present it operates near electrochemical equilibrium. Most of the work is therefore carried out on the chloride ion and it may be significant that there is evidence that the combined pump is electrogenic, generating a positive potential inside. Symmetry requires that this potential must be of the same sign as the diffusion potential, which is the consequence of the influx which the pump is countering. In sea water the pump is operating in parallel with much larger passive fluxes and it is largely masked but the isolated perfused gill, with Ringer solution on both sides, can generate a potential of up to 10 mV which is largely abolished by 5 mM/l of thiocyanate inside or outside the gill. The potential is also abolished by as little as 10^{-5} M ouabain, when applied internally, though much larger concentrations applied externally have little effect (Shuttleworth *et al.*, 1974). The potential in the perfused gill is also sensitive to external potassium and this effect too is reduced by ouabain. Even after treatment with ouabain the transepithelial potential still shows some response to changes in external potassium, the consequence no doubt of the passive permeability of the gill to potassium, which must be greater than that to sodium (Potts, 1976), Fig. 7. *In vivo* experiments with the eel also suggest that the chloride pump may contribute to the potential in sea water but when the active pump is abolished by thiocyanate the diffusion potential remains and the reduction of potential which occurs is small (Maetz and Bornancin, 1975).

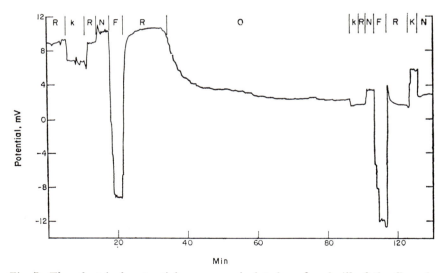

Fig. 7. The electrical potential across an isolated perfused gill of the flounder (*Platichthys flesus*) in various media before and after perfusion with 10^{-4} M ouabain. R: flounder Ringer solution, k: potassium-free Ringer solution, N 500 mM NaCl/l, F: fresh water, O: Ringer containing 10^{-4} M ouabain; K500 mM KCl.

III. DIVALENT IONS

Most studies with fish gills have been concerned with the interaction of the gills with monovalent ions but there is evidence that the gills may transport calcium to some extent and divalent ions in the medium have pronounced effects on both the permeability to monovalent ions and on the active transport of the monovalent ions.

A careful study of calcium balance in the southern flounder, *Paralichthys lethostigma*, has shown that the rectum and kidney together only account for 40% of the calcium ingested by drinking sea water. The remaining 60% must be removed by extra-renal pathways (Hickman, 1967). The most likely site for this activity would be the gills, although direct evidence is still lacking.

Calcium also plays an important indirect role by regulating the permeability of the gills to sodium ions and to water and to a lesser extent to chloride ions. Similar but smaller effects are also produced by magnesium ions.

The permeability of *Fundulus kansae* both to water and sodium ions is dramatically increased by the removal of calcium ions from sea water. Magnesium ions are less effective but they prolong survival in calcium free sea water. A similar effect on sodium permeability has been observed in the eel (Bornancin *et al.*, 1972), the mullet (Pic and Maetz, 1975) and in the flounder *P. flesus* (Potts, unpublished). The effects on the transepithelial potential in the mullet and the flounder suggest that calcium acts mainly on the permeability to sodium. In both sea water and fresh water the addition of calcium always reduces the potential, evidently by reducing permeability to sodium relative to the permeability to chloride. Direct measurements of the sodium and chloride effluxes from the mullet have shown that the addition of 10 mM Ca to deionized water reduces the sodium permeability by 85% and the chloride permeability by only 20% (Pic and Maetz, 1975; Maetz and Bornancin, 1975). A detailed analysis of the effects of calcium ions in the goldfish (Eddy, 1975) shows that calcium affects both ions, and again the effect is greater on sodium fluxes than on chloride fluxes. The fluxes of calcium across the gill are so small, compared with those of monovalent ions, that they make little direct contribution to the potential. Although the details of the interaction are uncertain, positively charged calcium ions are more likely to interfere with the movement of the similarly charged sodium ions than with the chloride ions of opposite charge. They might, for example, block negatively charged pores, through which the sodium ions may pass. Calcium ions also have a more general effect on permeability by reducing the hydration of mucopoly-saccharides in intercellular cements or on the surface of the cells, changes which could account for the reductions in permeability to water and to chloride.

The concentration of calcium ions in sea water is about 20 meq/l compared with only about 0.1 meq/l in soft fresh water and one to two meq/l in hard waters. In consequence, the transfer of a euryhaline fish from sea water to fresh water may be accompanied by calcium dependent changes in permeability which may complicate the interpretation of the experiments. In some cases, such as *F. kansae* or the eel, the calcium store in the gills is so firmly bound that the increase in permeability is only apparent after some hours or even days but in the mullet the effects are apparent immediately. Interaction between ions reduces the activity of calcium ions in sea water. A concentration of 20 meq of calcium/l in more dilute solutions would have a higher activity than the same concentration in sea water but the exact equivalent concentrations of calcium in sea water and fresh water cannot easily be computed. The effects of calcium ions are of considerable ecological interest. The ability of certain species of marine fish to survive in fresh water in coral limestone ponds in the Bahamas was noted as long ago as 1934 and reproduced artificially in the New York aquarium in fresh water with added calcium (Breder, 1934). More generally hard waters are an easier environment for fishes than soft.

There are indications that calcium is also involved in maintainance of the active transport of sodium and chloride ions. The increased uptake of sodium observed by Cuthbert and Maetz (1972) in goldfish which had been maintained in calcium free solutions, may have been no more than a response to the increased loss of sodium but Bornancin *et al.* (1972) found that eels which had been adapted for 15 hours to calcium free sea waters had a greatly depressed response to added potassium while the addition of calcium restored the ability of potassium to stimulate active sodium efflux. It is difficult to interpret this effect in terms of changes in passive permeability. Both sodium influx and efflux were considerably increased in the eels in calcium free sea water but the fish somehow maintained sodium balance. When adapted to low calcium sea water (2 meq/l Ca) *Fundulus kansae* showed a greatly increased rate of sodium exchange but managed to maintain the levels of total body sodium and of plasma sodium below their normal levels. The rate of incorporation of radio-phosphorus into gill RNA was greatly increased in the low calcium sea water but in view of the low levels of sodium in the body this cannot have been a direct response to a salt load (Fleming *et al.*, 1974). The results indicate rather a direct effect of calcium on the biochemistry of the gills.

IV. WATER PERMEABILITY

Most of the ionic problems facing teleosts are an indirect consequence of their permeability to water, which obliges them to drink to replace water lost by exosmosis. Comparative measurements of drinking and the rates of urine production of teleosts in sea water and fresh water respectively and of the rates of turnover of tritiated water in both media (Potts *et al.*, 1967; Evans, 1969) showed that in general marine fishes are less permeable to water than are fresh water fishes. The osmotic permeabilities of fresh water fishes, estimated from the urine flow, with allowance for any drinking which occurs, are larger than the diffusional permeabilities calculated from the rate of tritiated water exchanges but in marine fish the diffusional permeabilities and the osmotic permeabilities are fairly similar. However, there is a wide range of permeabilities amongst different fishes and Motais and Isaia (1972) reported that the diffusional permeability may apparently exceed osmotic permeability in the eel when in sea water. It is difficult to assess the osmotic permeability accurately from the drinking rates and urine flow rates as the handling and cannulation required may upset water balance for several days. Using incubated isolated gills, which do not suffer these disabilities, Shuttleworth and Freeman (1974) have found that the osmotic permeability of the eel *A. dieffenbachi* is only slightly greater in fresh water than in sea water. Higher osmotic permeability in fresh water may imply the presence of pores in the gill membrane. A wider variety of fishes needs to be examined but it is clear that osmotic water permeability is minimal in marine teleosts and generally higher in fresh water. The change in permeability is mediated in part by the higher levels of prolactin in circulation in fresh water-adapted fishes and in part by the higher levels of calcium present in sea water (Potts and Fleming, 1970, 1971). It is difficult to suggest any advantage accruing from the higher permeability found in fresh water teleosts but the increase may be related to the swelling of the surface crenellations of the epithelial cells noted recently by Harris (see below).

   Steen *et al.* (1975) estimate that in fresh water eels 98.5% of water diffusion takes place through the epithelial cells, the remaining 1.5%

diffusing through pores in the tight junctions between the cells. Ionic loss takes place mainly through these pores which are about 20 Å in diameter but have a total cross section area of only one part in 200,000 of the gill surface. Sea water adapted eels have not yet been examined in this respect but it may be presumed that once again water diffusion will take place mainly through the epithelial cells but a small part of the passive ion fluxes may occur through the intercellular pores between the epithelial cells.

V. THE STRUCTURE OF THE CHLORIDE CELL

The relationship between the physiological properties of the gill and its histological structure, first studied by Keys and Willmer in 1932, has recently been the subject of renewed interest. The greater part of the surface of the gill is composed of flattened epithelial cells with roughly concentric surface crenellations, rather like finger prints in appearance. Opening at intervals into depressions between these epithelial cells are the chloride cells and occasional mucus cells. The epithelial cells contain a moderate number of mitochondria and their surface ridges in section show some similarities in structure to microvilli, including a thickening of the surface membrane towards the apex which in microvilli have been associated with arrays of enzymes involved in active transport (Harris, in press). The role of the typical epithelial cell, apart from allowing the passive diffusion of oxygen, has been overshadowed in discussion by that of the chloride cell but its structure suggests that it is the site of some activity. One likely role for the epithelial cell is the exchange of sodium for ammonium ions. This transport system is probably of universal occurrence in fresh water fishes, where it helps to maintain body sodium and to eliminate ammonia but it probably occurs also in sea water fish, although here it must adversely increase the salt load (Evans et al., 1973). In the flounder the ridges are larger in fresh water than they are in sea water and they exhibit irregular swellings which further increases their surface area in fresh water, where salt uptake is essential but difficult.

The chloride cells are considerably deeper than the epithelial cells and are largely overlain by the latter. They are characterized by a dense population of mitochondria and glycogen granules which are concentrated mainly in the central and lower regions of the cell. A very well developed system of tubules of agranular reticulum is closely associated with the mitochondria and high resolution electron micrographs show that the tubules open to the extracellular fluid at the base of the cell (Shirai and Utida, 1970), a continuity confirmed by the rapid penetration of horseradish peroxidase into the reticulum (Phillpot, 1966). Towards the apex of the cell the tubular system breaks down into numerous vesicles; although a few tubules may reach to the apical surface of the cell, vesicles and tubules are rarely found opening to the exterior. The apical membrane is usually much folded and recessed below the general level of the gill surface to form an apical pit. The intercellular spaces between the chloride cell and its neighbours are closed by terminal bars. The extremely involved structure of the reticulum makes it impossible to establish any continuity between the apical portions of the reticulum and basal surface of the cell but the whole structure is consistent with the movement of ions from the extracellular fluid through the endoplasmic reticulum to the apex of the cell, the final stages of the journey possibly being completed by way of the vesicles, as suggested by Threadgold and Houston (1964). Various organic molecules such as inulin and methylene blue can penetrate readily through the chloride cell to the outside: methylene blue can be seen to flow out of the apical pit under suitable circumstances (Lam, 1968; Masoni and Payan, 1974). Selective staining of chloride ions by silver and of sodium ions by pyroantimonate has demonstrated the presence of both ions in the tubular system, as well as in the apical pit and the extracellular spaces (Petrik, 1968; Shirai, 1972). Autoradiographic techniques, while less precise in localizing the ions, have demonstrated that ^{22}Na and ^{36}Cl are accumulated in chloride cells within minutes of injection into the fish, and are concentrated towards the apical region of the cell. Masoni and Garcia Romeu (1973) have suggested that the ions are concentrated in the cytoplasm but Shirai (1972) has demonstrated, by selective staining, that the ions are localized in the vesicles in the upper regions of the cell. The tubules of the endoplasmic reticulum are sufficiently large in diameter that the normal laws of diffusion may be presumed and the total diffusive capacity of the system is similar to the flux of ions across the gills (Appendix II). If the ions were actively removed from the upper regions of the tubular system diffusion alone would not be sufficient to maintain the concentration of ions towards the apex of the cell and the concentration of ions at the apex would become depleted (Appendix III). The presence of large numbers of mitochondria around the tubules of the reticulum, together with evidence that the high concentration of Na/K ATPase in the chloride cells is probably located in the walls of the reticulum (Maetz and

Bornancin, 1975) should indicate that ions are being actively transported, and by inference, concentrated in the system. It may be that, in spite of the evidence that horseradish peroxidase and other large molecules rapidly permeate the cell, the reticular system is not at any time continuous but consists of several discrete portions through which sodium and chloride are progressively concentrated by transport through the cytoplasm from one section to the next. Such a system would reduce internal losses by diffusion and could sum the effects of a succession of ion pumps by placing them in series. The penetration of large molecules through the entire system implies that the disjunct portions, if they exist, fuse and break at intervals, a process which would reduce the overall level of efficiency but it should be remembered that in life cell contents are in continuous rapid, almost violent, movement as the result of Brownian motion and bulk movements of cytoplasm. A localized concentration of ions within the tubules might be achieved by the secretion into the tubules of some organic substance which by binding ions would lower their ionic activities. An amorphous material is visible in the lumen of the tubules, where it becomes concentrated towards the apex of the cell, and a similar material is also visible in the apical pit. If this were a polyanionic mucopolysaccharide it might bind ions to some extent (Maetz and Bornancin, 1975) but it is difficult to envisage that organic ions could be synthesized which would bind more than one inorganic ion for each carbon atom, in which case the loss of organic carbon would be insupportably large if the carrier were excreted along with the ions.

A model in which the sodium and chloride ions penetrate deep into the chloride cell by the reticular system and are then transported by vesicles to the apical surface and liberated in solution is superficially consistent with what is known of the structure and composition of the cell but more detailed considerations show that such a model is either inconsistent with what is known of the water movement across the gill, or is thermodynamically extremely inefficient.

The water balance of the fish requires that the ions are finally exported either in an extremely concentrated solution, or simply as ions, with no loss of water except possibly that of ionic hydration. Most teleosts in sea water export the equivalent of at least 20% of their total body sodium each hour; *Fundulus kansae* and the mullet export hourly the equivalent of the whole sodium content of the body. As the sodium space of the fish is equivalent to about 30% of the total body water, it follows that if the ions were exported in a solution of the same concentration as the blood, water losses would range between 6% and 30% of the total body water each hour. The observed drinking rates in fishes are only between 0.5% and 1% of the total body water each hour. Even if there were no loss of water by exosmosis or in the urine and all the water drunk were available for sodium chloride excretion, the degree of concentration in the vesicles required would be between six and 30 times the blood concentration or several times that of sea water. However measurements of the diffusional permeability of marine teleosts to tritiated water show that the minimal osmotic losses are similar in magnitude to the drinking rates. The simplest hypothesis is that marine fishes drink to compensate primarily for the osmotic loss rather than to replace excretory losses through the gills. It follows that little water will be available for excretion of the sodium chloride and if the ions are exported in solution the concentrations required must be extremely high and the energy consumed must be several times the thermodynamic minimum necessary. A possible escape from the water problem might be effected by supposing that water is also imported at the apical surface by pinocytosis and that the vesicles are then loaded with more sodium chloride before being returned to the cell surface. Calculation (Appendix IV) indicates that if the permeability of the vesicles were as low as that of some cell membranes it would be possible to maintain gross differences between the concentrations of the cytoplasm and the subcellular spaces without any significant water loss.

However the energy required to remove the salt will be a function of the maximal concentration generated, which in this model would necessarily be greater than that of sea water. Such a system would still be inefficient to operate. The most efficient model, consistent with water balance, requires that sodium and chloride be transported to the immediate neighbourhood of the apical membrane at a concentration not greater than that of sea water and then be released across the membrane without water loss. Water balance is most easily maintained if the vesicles operate a shuttle service between the reticular system and the apical membrane but do not fuse with the latter. The problem of salt transfer between the vesicle and the sea water, in the absence of an immediate energy source, leads to further speculation. If the concentration in the vesicles and possibly in the terminal regions of the reticulum, were raised to a level slightly greater than that of sea water and if the membranes at the point of contact were permeable to both sodium and chloride both an exchange of ions and an efflux of ions would take place. Electroneutrality requires that the net efflux of cations and anions would be equivalent but potassium could enter in exchange for sodium. The ratio of the net efflux to the total exchange would be in the ratio of the excess concentration above sea water to the concentration of the sea water. In a marine teleost this might be only 5 or 10%. Such a system would not be far removed from the thermodynamically most efficient condition where the vesicles would be isotonic with sea water. If the vesicles were slightly hyperosmotic to the medium they might even abstract some water from the sea and return it to the fish, in which event the diffusional permeability to water might exceed the osmotic permeability as Motais and Isaia (1972) observed in the eel although this is not an essential feature of the model.

A number of secondary problems arise from this model. The gross fluxes of sodium are generally greater than the gross fluxes of chloride in most teleosts and, more pertinently, the sodium fluxes are mediated by the transepithelial potential. This linkage between the potential and the sodium fluxes requires that much of the sodium passes across the gill by diffusion rather than in discrete packets. A substantial part of the sodium flux may therefore diffuse through the cells. A high permeability to sodium would cause no osmotic problem provided that the permeability to chloride remained low and the chloride content could be regulated. The concentration of sodium at any point would be dependent on the availability of anions. In most tissues the cell volume is regulated by sodium extrusion, but in the chloride cell the volume may be regulated by chloride extrusion.

Some inequality between the sodium and chloride flux might be achieved even in the absence of sodium movement through the cytoplasm, if the membrane of the vesicle were preferentially permeable to sodium and the period during which the vesicle was in contact with the apical membrane were too short for complete equilibration between vesicles and medium to be attained. However it is difficult to reconcile such a selectively permeable membrane with the observations that methylene blue and even inulin can penetrate the cell; it seems more likely that the excess sodium flux over the carrier mediated chloride flux takes place through the cytoplasm or through the tight junctions between the cells.

The question of whether there are sufficient vesicles to maintain the chloride fluxes must be considered briefly. If it is assumed that the proportion of the chloride flux which is not due to passive diffusion is carried in the vesicles an estimate can be made of the rate of turnover. The chloride flux in the mullet is about 0.05 mole kg^{-1} hr^{-1} (Maetz and Pic, 1975). Perhaps 20% of this could be accounted for by passive diffusion through the cytoplasm.

If the vesicles contain 0.6 M Cl and have a total volume of 3×10^{-4} cm^{-3}/kg (Appendix 4) the quantity of chloride in transit at any time would be $3 \times 0.6 \times 10^{-7}$ or 2×10^{-7} mole/kg fish. To export 4×10^{-2} mole Cl/hr would require a turnover rate of 2×10^5/hr. Every vesicle would have to make the round trip 50 times a second! This is improbably fast but it might just be possible to save the hypothesis by revising the estimate. The chloride flux used above is that of one of the most physiologically active fishes, the mullet, the proportion of chloride cells in the gill is based on one of the most inactive, the flounder. The chloride flux in the flounder is only one-fifth that of the mullet; if the estimated proportion of the cell occupied by vesicles were also increased the rate could be reduced further. However the vesicles must be exceedingly active if they are the vehicles of the carrier mediated chloride flux.

When hypothesis is piled on hypothesis the whole structure must at some point collapse like a house of cards. Perhaps this point is now close. However such speculations serve to clarify the problems that arise in thinking quantitatively about this extremely active tissue. If they evoke either better hypotheses or critical experiments they will have served their purpose. What is the point of building a house of cards if not to enjoy its collapse? (see Addendum.)

ACKNOWLEDGEMENTS

I am grateful to Drs Fletcher and Maetz for their helpful comments and suggestions.

REFERENCES

Baker, P. F. and Willis, J. S. (1972). *J. Physiol. Lond.* **224**, 441–462.

Bittar, E. E., Chen, S., Davidson, Bo. G., Hartmann, H. A. and Tong, E. Y. (1972). *J. Physiol. Lond.* **221**, 389–414.

Bornancin, M., Cuthbert, A. W. and Maetz, J. (1972). *J. Physiol. Lond.* **222**, 487–496.

Botazzi, F. (1897). *Arch. ital. Biol.* **28**, 61–76.

Brading, A. F. and Widdicombe, J. H. (1974). *J. Physiol. Lond.* **238**, 235–249.

Bryan, G. W. (1960). *J. exp. Biol.* **37**, 83–128.

Breder, C. M. (1934). *Zoologica,* **18**, 57–91.

Conte, F. P. and Lin, D. H. Y. (1967). *Comp. Biochem. Physiol.* **23**, 945–957.

Croghan, P. C. (1958a). *J. exp. Biol.* **35**, 219–233.

Croghan, P. C. (1958b). *J. exp. Biol.* **35**, 234–241.

Croghan, P. C. (1958c). *J. exp. Biol.* **35**, 243–249.

Croghan, P. C. (1958d). *J. exp. Biol.* **35**, 425–436.

Cuthbert, A. W. and Maetz, J. (1972). *J. Physiol. Lond.* **221**, 633–643.

De Silva, P., Saloman, R., Spokes, K. and Epstein, F. H. (1977). *J. exp. Zool.* **199**, 419–26.

Eddy, F. B. (1975). *J. comp. Physiol.* **96**, 131—142.

Epstein, F. H., Katz, A. I. and Pickford, G. E. (1967). *Science* **156**, 1245—1247.

Epstein, F. H., Maetz, J. and De Renzis, G. (1973). *Am. J. Physiol.* **224**, 1295—1299.

Evans, D. H. (1969). *J. exp. Biol.* **50**, 689—704.

Evans, D. H., Mallery, C. H. and Kravitz, L. (1973). *J. exp. Biol.* **58**, 627—636.

Evans, D. H., Carrier, J. C. and Boyan, M. B. (1974). *J. exp. Biol.* **61**, 277—283.

Fleming, W. R., Nichols, J. and Potts, W. T. W. (1974). *J. exp. Biol.* **60**, 267—273.

Fletcher, C. R. (1977). *J. comp. Physiol.* in press.

Greenwald, L., Kirschner, L. B. and Sanders, M. (1974). *J. gen. Physiol.* **64**, 135—147.

Hickman, C. P. (1967). *Canad. J. Zool.* **46**, 457—466.

Hodgkin, A. L. and Huxley, A. F. (1952). *J. Physiol. Lond.* **116**, 449—472.

House, C. R. (1963). *J. exp. Biol.* **40**, 87—104.

House, C. R. and Maetz, J. (1974). *Comp. Biochem. Physiol.* **47A**, 912—924.

Hughes, G. M. and Morgan, M. (1973). *Biol. Rev.* **48**, 419—475.

Kamiya, M. and Utida, S. (1969). *Comp. Biochem. Physiol.* **31**, 671—674.

Karnaky, K. J. and Kinter, W. B. (1977). *J. exp. Zool.* **199**, 355—364.

Karnaky, K. J., Degnan, K. J. and Zadunaisky, J. A. (1977). *Science* **195**, 203—5.

Karnaky, K. J., Kinter, L. B., Kinter, W. B. and Stirling, C. E. (1976). *J. Cell. Biol.* **70**, 157—177.

Keys, A. B. (1931). *Z. vergl. Physiol.* **15**, 364—388.

Keys, A. B. and Willmer, E. N. (1932). *J. Physiol. Lond.* **76**, 368—381.

Kirschner, L. B. (1955). *J. cell. comp. Physiol.* **45**, 61—87.

Kirschner, L. B., Greenwald, L. and Sanders, M. (1974). *J. gen. Physiol.* **64**, 148—165.

Lam, T. J. (1968). *Comp. Biochem. Physiol.* **28**, 259—465.

La Torre, J. L., Lunt, G. S. and De Robertis, E. (1970). *Proc. U.S. Nat. Acad. Sci.* **65**, 716—720.

Maetz, J. (1969). *Science* **166**, 613—615.

Maetz, J. and Bornancin, M. (1975). *Fortschritte der Zoologie* **23**, 322—362.

Maetz, J. and Pic, P. (1975). *J. comp. Physiol.* **102**, 85—101.

Maetz, J. and Pic, P. (1977). *J. exp. Zool.* **197**, 325—37.

Marshall, W. S. (1977). *J. comp. Physiol.* **114**, 157—65.

Masoni, A. and Garcia Romeu, F. (1973). *Z. Zellforsch.* **141**, 575—578.

Masoni, A. and Payan, P. (1974). *Comp. Biochem. Physiol.* **47A**, 1241—1244.

Motais, R. and Isaia, J. (1972). *J. exp. Biol.* **56**, 587—600.

Motais, R. and Maetz, J. (1965). *C. r. hebd. Seanc. Acad. Sci. Paris* **261**, 532—535.

Motais, R., Garcia Romeu, F. and Maetz, J. (1966). *J. gen. Physiol.* **50**, 391—422.

Mullins, J. (1950). *Acta physiol. Scand.* **21**, 303—314.

Olivereau, M. (1970). *C. r. Soc. Biol.* **164**, 1951—1955.

Petrik, P. (1968). *Z. Zellforsch.* **92**, 422—427.

Phillpott, C. W. (1966). *J. cell Biol.* **31**, 86A.

Pic, P. and Maetz, J. (1975). *C. r. Acad. Sci. Paris.* **280**, 983—6.

Potts, W. T. W. (1976). "Perspectives in Experimental Biology." (Ed. P. Spencer Davies) Vol. I, pp. 65—75. Pergamon Press, Oxford.

Potts, W. T. W. and Eddy, F. B. (1973a). *J. comp. Physiol.* **87**, 29—48.
Potts, W. T. W. and Eddy, F. B. (1973b). *J. comp. Physiol.* **87**, 305—315.
Potts, W. T. W. and Evans, D. H. (1967). *Biol. Bull. mar. biol. Lab. Woods Hole* **133**, 411—430.
Potts, W. T. W. and Fleming, W. R. (1970). *J. exp. Biol.* **53**, 317—327.
Potts, W. T. W. and Fleming, W. R. (1971). *J. exp. Biol.* **55**, 63—76.
Potts, W. T. W., Foster, M. A., Rudy, P. P. and Parry Howells, G. (1967). *J. exp. Biol.* **47**, 461—470.
Potts, W. T. W., Fletcher, C. R. and Eddy, B. (1973). *J. comp. Physiol.* **87**, 21—28.
Rankin, J. C. and Maetz, J. (1973). *J. Endocr.* **51**, 621—635.
Rivas, E., Lew, V. and De Robertis, E. (1972). *Biochim. biophys. Acta* **290**, 419—423.
Shirai, N. (1972). *J. Fac. Sci. Univ. Tokyo VI* **12**, 385—403.
Shirai, N. and Utida, S. (1970). *Z. Zellforsch.* **103**, 247—264.
Shuttleworth, T. J. (1972). *Comp. Biochem. Physiol.* **43A**, 59—64.
Shuttleworth, T. J. and Freeman, R. F. H. (1974). *J. exp. Biol.* **60**, 769—82.
Shuttleworth, T. J., Potts, W. T. W. and Harris, J. N. (1974). *J. comp. Physiol.* **94**, 321—329.
Smith, H. W. (1930). *Am. J. Physiol.* **93**, 480—505.
Smith, P. G. (1969a). *J. exp. Biol.* **51**, 727—738.
Smith, P. G. (1969b). *J. exp. Biol.* **51**, 739—758.
Steen, J. B. and Stray-Pedersen, S. (1975). *Acta physiol. Scand.* **95**, 6—20.
Threadgold, L. T. and Houston, A. H. (1964). *Exp. Cell. Res.* **34**, 1—23.

APPENDIX I. PASSIVE CHLORIDE LOSS IN FRESH WATER

The chloride efflux that remains after transfer to fresh water may be attributed to the passive efflux of chloride. Maetz and Pic (1975) provide data on the sodium and chloride fluxes and the simultaneous potentials in the mullet in a variety of conditions.

	Potl. mV	Na flux μeq/100 g/h	Cl flux μeq/100 g/h	Plasma Na meq/l	Cl meq/l
SW					
Normal animals	+22.3	8255	4165	170	153
		7386	5168		
SW					
SCN treated animals	+20.6	7809	2550	199	182
FW					
Normal animals	−43.4	1746	805	—	—
FW					
SCN treated animals	−43.0	835	572	—	—
		1966	1686		

If it is assumed that the sea water contained 470 mE Na and 550 meq Cl/L then if pCl/pNa (Eq. 1) = 0.05 the predicted potential would be +23 mV. After SCN treatment the elevation of the blood concentration would reduce this to 19 mV. After transfer to fresh water a pCl/pNa ratio of 0.05 would produce a diffusion potential of −78 mV, about the same as that obtained on several occasions (see Maetz and Pic) but the mean observed potential of −43 mV is consistent with a pCl/pNa ratio of 0.20. The decrease may be attributed, at least in part, to the lower calcium content of the fresh water.

In sea water the passive sodium efflux is enhanced by the positive potential while in fresh water the passive chloride efflux will be enhanced by a negative potential. The passive ion effluxes can be obtained from the Constant Field Theory as:

$$M_{Na} = \frac{E_1 F}{RT} \times pNa \times \frac{[Na_i]}{1 - e^{-E_1 F/RT}}$$

$$M_{Cl} = \frac{-E_2 F}{RT} \times pCl \times \frac{[Cl_i]}{1 - e^{E_2 F/RT}}$$

Using M_{Na} = 7800, E_1 = 22.3 mV, E_2 = −43.4 mV, blood sodium and chloride levels of 170 and 150 mM, and a relative permeability pCl/pNa of 0.05, this gives a calculated chloride efflux M_{Cl} of 479 μeq/100 g/h. Increasing pCl/pNa to 0.2 increases the efflux to 1916 μeq/100 g/h. If the blood chloride declines when in fresh water the chloride efflux will be reduced in proportion. The value of the passive sodium efflux will be smaller than the value of the total efflux, 7800 μeq/100 g/h, used in these calculations. This may contain an active component which might amount to 10 or 20% of the total. The computed chloride loss should then be reduced in proportion. The observed values of the chloride loss in fresh water lie between the values computed on the assumption that pCl = 0.05 and 0.2 pNa.

APPENDIX II. MAXIMUM PASSIVE SODIUM AND CHLORIDE FLUX THROUGH THE ENDOPLASMIC RETICULUM OF CHLORIDE CELLS IN FLOUNDER GILL

Assumptions: Area of gill = 3 $\times$ 10^3 cm^2/kg fish (Hughes and Morgan, 1973). Sea water = 0.5 M Na Cl/l. Proportion of chloride cell/unit area gill surface 0.01. (estimate from electron-micrographs.)

Proportion of endoplasmic reticulum in cross section of chloride cell (estimated from electron micrographs) 0.1. Depth of chloride cell (estimate from electron micrographs) 15 μ. Path length of diffusion 1.5 cell depth. Diffusion constant of NaCl 1.484×10^{-5} cm^2/sec. Area available for diffusion $= 3 \times 10^3 \times 0.1 \times 0.01$ cm^2/kg fish $= 3$ cm^2. Diffusion distance $15 \times 10^{-4} \times 1.5 = 2.25 \times 10^{-3}$ cm. Concentration difference driving ions inwards $= 0.5 \times 10^{-3}$ M/cm^3.

$$\text{Flux} = D\frac{dc}{dx}A$$

$$= 1\cdot484 \times 10^{-5} \times \frac{0\cdot5 \times 10^{-3}}{2\cdot25 \times 10^{-3}}$$

$$= 1 \times 10^{-5} \text{ M/kg fish/sec}$$

$$= 3\cdot6 \times 10^{-2} \text{ M/kg fish/hr}$$

Observed flux through flounder gill c. 1×10^{-2} mole/kg/hr.

APPENDIX III. CONCENTRATION DIFFERENCE BETWEEN ENDS OF TUBULES REQUIRED TO MAINTAIN EFFLUX

Assumptions: Efflux $= 1 \times 10^{-2}$ mole/kg/hr: $A = 3$ cm^2/kg. $D = 1.484 \times 10^{-5}$ cm^2/sec. Effective path length 2.25×10^{-3} cm. Concentration gradient required to maintain flux:

$$\frac{dc}{dx} = \frac{1 \times 10^{-2}}{3600 \times 3 \times 1\cdot484 \times 10^{-5}}$$

$$\Delta c = \frac{1 \times 10^{-2} \times 2\cdot25 \times 10^{-3}}{3600 \times 3 \times 1\cdot484 \times 10^{-5}}$$

$$= 140 \text{ mM}$$

i.e. The concentration difference across the cell required to maintain a flux of 1×10^{-2} mole/kg/hr would be 140 mM.

APPENDIX IV. WATER LOSS FROM HYPEROSMOTIC VESICLES IN CHLORIDE CELLS

Assumptions: Upper 5 μ of cells consists of 10% by volume of cubic vesicles 0.1 μ in breadth. Total cross section of chloride cells = 3 cm^2/kg (See appendix II). Water permeability of vesicular membrane 0.002 μm^{-1} sec (Potts and Eddy, 1973b). Concentration of vesicles contents 5 M (5 x SW). Total volume of vesicles 3 x 0.1 x 5 x 10^{-4} cm^3 = 1.5 x 10^{-4} cm^3. Volume of 1 vesicle = 1.0 x 10^{-15} cm^3. Number of vesicles = 1.5 x 10^{-4}/1 x 10^{-15} = 1.5 x 10^{11}. Surface area of vesicles = 6 x 10^{-10} x 1.5 x 10^{11} cm^2. Gross water flux = 90 x 3600 x 0.002 x 10^{-4} = 6.5 x 10^{-2} cm^2/hr.

$$\text{Net water flux} = \frac{6.5 \times 10^{-2} \times 5}{55.4} = 0.006 \ cm^3/hr.$$

ADDENDUM

A number of studies which significantly further our knowledge of the chloride cell have been published since this chapter was written in 1975. The isolated gill has proved to be a somewhat intractable preparation. Epithelia, consisting largely of chloride cells, have recently been described which will greatly facilitate the study of the transport properties of these cells. Preliminary results have been published from two of these preparations (Karnaky and Kinter, 1977; Karnaky et al., 1977; Marshall, 1977). These preparations maintain potentials similar to those found in the intact fish. With Ringer on both sides the average chloride efflux, across the *Fundulus* epithelium, was six times as large as the influx and the net chloride efflux was equal to the short circuit current (S.C.C.), an unambiguous demonstration of active chloride transport. The potential and the S.C.C. in both are dependent on bicarbonate and are greatly reduced in the presence of the carbonic anhydrase inhibitor furosemide.

Colchicine, with disrupts microtubules, abolishes the potassium stimulation of the chloride efflux in whole fish and reduces the potassium stimulation of the sodium efflux to the level to be expected from the change of potential alone (Maetz and Pic, 1977). This confirms that the stimulation of the sodium efflux by potassium is due in part to the effect on potential and in part to active transport, the latter being linked to the potassium stimulated chloride efflux.

De Silva et al. (1977) have confirmed that ouabain only blocks gill ATPase when it has access to the serosal surface of the cell. When the ATPase is 90% inhibited the sodium and chloride effluxes are reduced by a similar amount, but the tritiated water flux is also greatly reduced suggesting that the whole circulation is disturbed. These authors propose a possible model of the chloride cell, which accounts for the position of the Na, K, ATPase at the base of the cell. In this model the cell base contains, in addition to the usual sodium potassium exchange pump, a sodium chloride linked carrier so that sodium diffusing into the cell down the electrochemical gradient also drives chloride into the cell. This system would extrude chloride if the intracellular electrochemical potential of chloride were sufficiently high but it is difficult to see why such a pump should be dependent on bicarbonate ions.

An entirely different approach has been suggested recently by Fletcher (1977) who points out that any ion pump driven by a biochemical reaction is analogous to any other biochemical reaction and both forward and back reactions will be in progress simultaneously, the balance depending on the activation energy of the reaction and the concentrations on the two sides. In this model any forward flux should be accompanied by a back flux superficially resembling an exchange diffusion, or a leaking pump such as Bryan described in *Astacus* (Bryan, 1960). From the balance of the two fluxes the activation energy can be calculated. This elegant theory would seem to account for many anomolous fluxes which have been vaguely attributed in the past to exchange diffusion.

19. Homeostasis of The Brain Micro-environment: A Comparative Account

N. J. Abbott and J. E. Treherne*

Department of Physiology, King's College, University of London, England and
*Department of Zoology, University of Cambridge, Cambridge, England

I. INTRODUCTION

Many nervous systems possess sophisticated mechanisms for the regulation of the fluid environment of the neurones. The need for such regulatory devices results from the particular vulnerability of neurones, and especially synapses, to changes in the chemical composition of the bathing medium.

There appear to be four main reasons for the development of systems that regulate the composition of the immediate fluid environment of central neurones:

(1) to boost extracellular ion levels in the brain, especially when the blood is dilute and, particularly, deficient in sodium ions;

(2) to maintain ionic ratios in extracellular fluid which are significantly different from those of the blood;

(3) for internal stability, when the blood composition fluctuates or cerebral metabolism causes extracellular accumulation of substances beyond the tolerance of central neurones;

(4) to exclude extraneous pharmacologically active or toxic molecules, and retain others (e.g. central transmitters).

Where no regulatory mechanisms are present, it is usually found that the blood plasma is an adequate medium for neuronal function, and is either itself homeostatically controlled, or fluctuates in composition within limits which can be tolerated by neurones.

Two main types of regulatory mechanism can be distinguished:
  (a) "Barrier" systems: passive and active mechanisms located at
  the interfaces between blood and brain, which to some extent
  control the degree and nature of exchange with the bulk
  extracellular fluid of the CNS.
  (b) Glial-neuronal regulation: homeostatic control over the local
  microenvironment in the immediate vicinity of the neurones,
  by glia or neurones themselves. This local regulation will
  obviously be more effective if the composition of the bulk
  extracellular fluid is already regulated by barrier mechanisms.
As the glia may contribute both to barrier mechanisms at the
blood-brain interface, and to local regulation, there is obviously some
overlap between the two categories.

II. THE BLOOD-BRAIN INTERFACE

Regulation of the chemical composition of the extra-neuronal fluid
will clearly be facilitated by any barrier mechanisms that tend to
isolate the brain microenvironment from other fluid compartments
of the body.

A. Leaky Systems

Among most invertebrates, however, there is little evidence for
barrier structures, and the blood-brain interface appears relatively
leaky. Intercellular access of water-soluble ions and molecules from
blood to the neuronal surfaces is not appreciably restricted in the
relatively solid, and frequently avascular, central nervous structures
(Fig. 1) of most of the higher invertebrate groups. Ultrastructural
studies, for example, have revealed apparent uninterrupted
communication between the blood and neuronal surfaces in the
ganglia and connectives of an annelid, the leech (Fig. 1a) (Coggeshall
and Fawcett, 1964; Nicholls and Kuffler, 1964) of various gastropod
(Fig. 1c) (Sattelle and Lane, 1972; Mirolli and Crayton, 1968; Mirolli
and Gorman, 1973; Pentreath and Cottrell, 1970) and lamellibranch
species (Fig. 1b) (Gupta *et al.*, 1969; Lane and Treherne, 1972a;
Sattelle and Howes, 1975) and in cephalopod brain (Stephens and
Young, 1969). Ultrastructural studies have also shown that the
intercellular clefts of the cerebral capillaries in a crustacean ganglion
(Fig. 1d) are relatively leaky and admit particles as large as those of

ferritin (Abbott, 1970; 1972), although various junctional complexes, which could restrict intercellular diffusion, are observed in the perineurial and perivascular glial clefts (Lane and Abbott, 1975; Abbott *et al.*, 1975).

Electrophysiological and radioisotopic investigations have also indicated that the organization of the central nervous tissues in annelids and molluscs imposes little restriction on the diffusion of small, water-soluble, ions and molecules between the blood, or bathing medium, and the neuronal surfaces. In the leech ganglion, for example, the electrical responses of the neurones and glial cells were shown to be consistent with theoretical half-times calculated for the linear diffusion of cations and sucrose molecules along the morphologically determined intercellular clefts leading to the neuronal surfaces (Nicholls and Kuffler, 1964). Essentially similar conclusions have been drawn from quantitative electrophysiological analyses in molluscan ganglia (Sattelle and Lane, 1972; Sattelle, 1973b; Mirolli and Gorman, 1973) and central nervous connectives (Sattelle, 1973a; Sattelle and Howes, 1975) and from the kinetics of radiosodium fluxes in the central nervous tissues of a lamellibranch species (Mellon and Treherne, 1969).

As might be expected in species with a leaky blood-brain interface, in most of these groups there is no evidence that the neuronal environment can be maintained with a different ionic composition from that of blood. However, in decapod crustacea and certain molluscs, a limited "buffering" capacity appears to be present, which can minimize fluctuations in important ions such as Na and K (see below). This buffering is necessarily only as a short-term measure, and cannot maintain long-term differences between blood and neuronal environment; this implies that the normal ionic composition of blood must be an adequate neuronal medium.

It appears, therefore, that neuronal function in annelids, molluscs and crustaceans depends upon the ability of these invertebrates to regulate the osmotic and ionic concentrations of their body fluids or, in the case of euryhaline osmoconformers, on the tolerance of the neurones. In a stenohaline osmoconformer, the spider crab (*Maia squinado*), for example, the axonal spike-generating system suffers irreversible damage at the same osmotic concentration of the blood as is found to be lethal in the whole animal (Pichon, Y. and Treherne, J. E., unpublished observations). In freshwater molluscs, the ability of neurones to function in the extremely dilute body fluids results from the maintenance of relatively low intracellular sodium levels, so that an adequate electrochemical gradient for

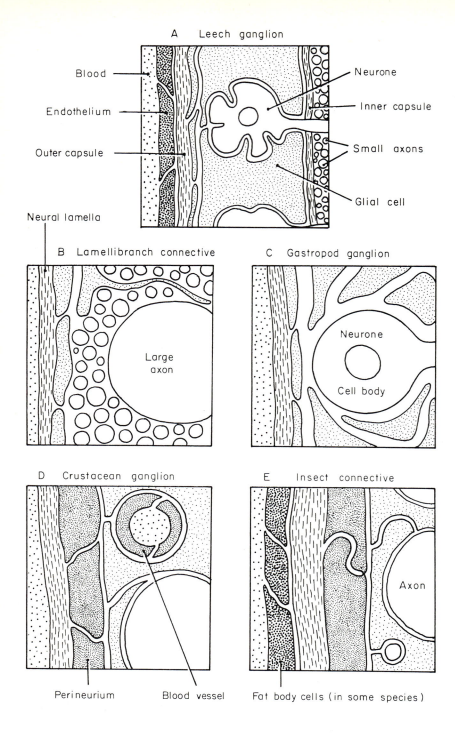

A Leech ganglion

Blood — Neurone

Endothelium — Inner capsule

Outer capsule — Small axons

Glial cell

Neural lamella

B Lamellibranch connective

Large axon

C Gastropod ganglion

Neurone

Cell body

D Crustacean ganglion

E Insect connective

Axon

Perineurium   Blood vessel   Fat body cells (in some species) sodium ions exists across the neuronal membranes. In the central axons of the lamellibranch, *Anodonta cygnea*, for example, sodium-mediated action potentials are maintained (Treherne *et al.*, 1969b; Carlson and Treherne, 1969) despite exceptionally dilute blood ($Na^+ = 15.6$ mM) (Potts, 1954), by the inward sodium gradient resulting from an estimated intracellular concentration of 8.6 mM Na^+ (Mellon and Treherne, 1969).

B. Non-leaky Systems

It is only in insects and vertebrates that there is evidence for significant regulation of the chemical composition of the neuronal environment and, in both groups, control at the blood-brain interface and at the level of glia and neurones within the brain parenchyma can be demonstrated.

1. Insects

In the avascular central nervous systems of insects (Fig. 1e) the central neurones are protected by well developed blood-brain barrier systems (cf. Treherne and Pichon, 1972; Treherne, 1974). Such protection is particularly necessary in herbivorous insects which frequently possess blood of exceedingly low sodium concentration, often lower than that of potassium (cf. Florkin and Jeuniaux, 1964;

Fig. 1. Diagrammatic representations of the organization of the blood-brain interfaces and the structure of the underlying glial and neuronal elements in the central nervous systems of some higher invertebrates. The superficial connective tissue sheaths are directly bathed by the blood except in the leech, in which the nervous system is overlaid by a leaky endothelium (1A), and some herbivorous insects, in which a permeable fat body sheath may be present (1E). In arthropods (1D and E) the peripheral glia form a specialized cellular layer, the perineurium. An intracerebral blood supply is present in decapod crustaceans (1D). The glial coverage of the neurones varies considerably, being minimal in lamellibranch molluscs (1B), slightly greater in gastropods (1C), and extensive in annelid (1A) and arthropod nervous systems (1D and E). Based on ultrastructural studies of the leech, *Hirudo medicinalis* (Coggeshall and Fawcett, 1964) (1A); the bivalve, *Anodonta cygnea* (Gupta *et al.*, 1969; Sattelle and Howes, 1975) (1B); the water snail, *Limnaea stagnalis* (Sattelle and Lane, 1972) (1C); the shore crab, *Carcinus maenas* (Abbott, 1971a,b) (1D) and the insects, *Carausius morosus* and *Periplaneta americana* (Smith and Treherne, 1963; Maddrell and Treherne, 1966, 1967; Lane and Treherne, 1972b) (1E).

Treherne, 1966) and yet show "conventional", sodium-dependent, action potentials which require relatively high concentrations of sodium at the axon surfaces (Treherne and Maddrell, 1967; Weidler and Diecke, 1969; Pichon *et al.*, 1972). Even in the cockroach in which the activity of blood sodium is relatively high (a_{Na} = 0.088 M) (Treherne *et al.*, 1975), the amplitude of the action potentials is reduced, in surgically isolated axons, as a result of the relatively high activity (a_K = 0.010 M) of potassium ions in the blood (Thomas and Treherne, 1975).

Intercellular access of water-soluble ions and molecules to the axon surfaces in insect central nervous systems appears to be severely restricted by junctional complexes which occlude the perineurial clefts (Fig. 2) (cf. Lane, 1974). In cockroach central connectives, for example, the inward movement of macroperoxidase, microperoxidase and ionic lanthanum (Lane and Treherne, 1970, 1972b), into the extracellular system appears to be prevented by *zonulae occludentes* in the perineurial clefts (Skaer and Lane, 1974; Lane *et al.*, 1975). The potassium-induced extraneuronal potentials, which occur in the absence of significant axonal depolarization, may result from the effects of potassium on the outwardly-directed perineurial membranes, access to the inwardly-directed ones being restricted by the intercellular occlusion in the perineurial clefts (Treherne *et al.*, 1970; Pichon and Treherne, 1970; Pichon *et al.*, 1971). The ionic permeability of the outwardly-directed perineurial membrane is largely dominated by that for potassium, only small electrical responses resulting from changes in external sodium concentration (Pichon *et al.*, 1971). In connectives of a lepidopteran, *Manduca sexta*, on the other hand, a more pronounced permeability to sodium has been observed (Pichon *et al.*, 1972), while in the phasmid, *Carausius morosus*, no significant responses were recorded to changes in external sodium or potassium concentrations (Schofield, 1975). In the latter species, however, a significant permeability of the outer perineurial membrane to acetate and nitrate is indicated (Schofield, 1975). It appears, therefore, that insect perineurial membranes show considerable variation in passive ionic permeabilities.

Despite the relatively low passive sodium permeability of the perineurium, rapid recovery of action potentials can be recorded when Na-depleted connectives are exposed to normal external sodium concentrations (Fig. 3). Recovery of the action potentials did not occur if lithium ions were substituted for sodium in the bathing medium, although lithium can sustain axonal function in desheathed preparations (Schofield and Treherne, 1975). The specificity for

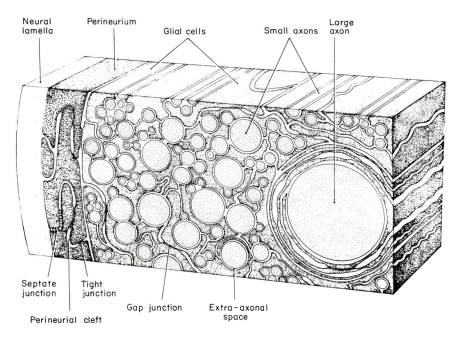

Fig. 2. The periphery of a cockroach central nervous connective showing the superficial connective tissue sheath, the neural lamella, which overlies the cellular perineurium. The intercellular clefts which traverse the perineurium are occluded by junctional complexes. The axons are enveloped by glial processes which delimit a complex system of narrow intercellular channels. The glial elements are linked to one another and to adjacent perineurial cells by gap junctions. (From Schofield and Treherne, 1975.)

sodium ions is unlikely to result from passive properties of the perineurium and suggests that there is an inwardly-directed transport of sodium ions to the axon surfaces in intact preparations. The inability of lithium ions to gain access to the axon surfaces does not appear to result from an exclusion by the outer perineurial membranes: lithium has been shown to accumulate to levels equivalent to those of sodium ions within the connectives, and to be retained even when the electrochemical gradient is reversed by exposure to normal saline (Bennett *et al.*, 1975). It has been suggested that the net inward movement of sodium ions to the axon surfaces may involve a unidirectional transport across the outer perineurial membrane by a pump which will also accept lithium ions (Schofield and Treherne, 1975).

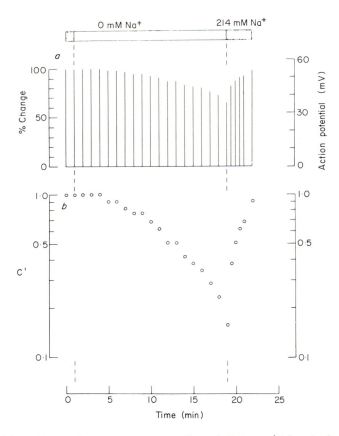

Fig. 3. (a) Effect of exposure to sodium-deficient (Tris-substituted) and subsequent return to normal Ringer on the amplitude of the action potentials recorded in an intact central nervous connective from a cockroach (*Periplaneta americana*). (b) Change in relative extra-axonal sodium concentration (C′) estimated from the percentage change in the action potential amplitude shown in (a). C′ was calculated from the Nernst relationship between sodium concentration and the percentage change in action potential amplitude measured in surgically exposed axons. (From Schofield and Treherne, 1975.)

2. Vertebrates

In vertebrate brain, potential sites of exchange between blood and brain are the choroid plexuses, the arachnoid membrane, and the walls of parenchymal capillaries. At all these interfaces, structural and functional "barrier" specializations are found. *Zonulae occludentes* are present, restricting free diffusion via an extracellular pathway between blood and extracellular fluid (Brightman and Reese, 1969; Shabo and Maxwell, 1971). The location of the barrier sites and the interrelationships between the blood and brain compartments are shown in Fig. 4.

In addition to the usual interstitial fluid compartment found in invertebrates, a second extracellular fluid compartment, containing cerebrospinal fluid (CSF) is present in vertebrate brain. CSF fills the internal cavities of the brain, the ventricles, and flows out through cisterns to bathe the outer surface, in the subarachnoid space. There is little restriction to diffusion from CSF to brain either at the ependymal or pial surface (Fig. 4) (Brightman, 1965a,b). Studies on the movement of H^+ ions between CSF and brain (Fencl *et al.*, 1966) and on K^+ fluxes (van Harreveld, 1966), provide strong evidence that the interstitial fluid (ISF) of the brain is in dynamic equilibrium with the CSF. Direct determinations of the intracortical K activity (Lux *et al.*, 1972; Vyskočil *et al.*, 1972) have confirmed this finding. It follows that any homeostasis of the CSF composition will be reflected in homeostasis of the interstitial fluid. This is also an important point experimentally because CSF is readily available for sampling, whereas ISF is not. Under appropriate conditions, the composition of the ISF can be inferred from that of CSF, although the latter can only reflect the average ISF composition, and cannot give information about local variation or fluctuations.

Evidence that CSF (and hence ISF) composition is regulated more efficiently than that of plasma is available for H^+, K^+, Mg^{++} and Ca^{++} (see Katzman and Pappius, 1973). Where the experimental technique has allowed calculation or measurement of ISF composition, it is found that this is similarly regulated, and in some cases it is possible to show that ISF regulation is not merely a passive consequence of CSF regulation, but can occur independently (see below).

The mechanism for CSF homeostasis may involve several components. CSF is produced by the choroid plexuses (and possibly by other, extrachoroidal sites) as a secretion, not a simple ultrafiltrate, and its composition is unlike that of an ultrafiltrate of plasma (Rougemont *et al.*, 1960). Passive and active mechanisms at the choroid plexus interface between blood and brain can, therefore, contribute to homeostasis of the neuronal environment. However, that substances as large as inulin can penetrate the choroid plexus epithelium (Welch, 1965; Welch and Sadler, 1966), that only a small potential difference can be measured across the secretory layer

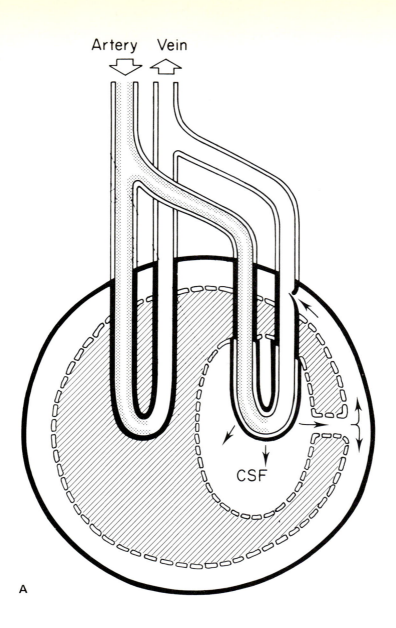

A

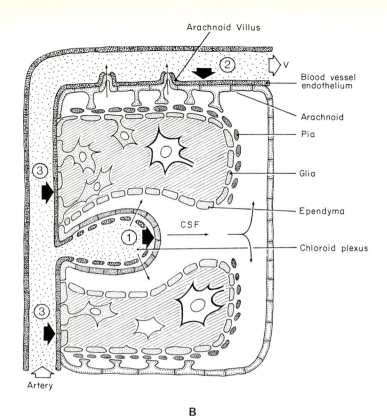

Fig. 4. (A) Schematic representation of the compartmentation and blood supply of vertebrate brain, showing a capillary in brain parenchyma (left) and choroid plexus (right). Direction of blood circulation is indicated by broad arrows; CSF secretion, circulation and outflow by thin arrows. Heavy black line indicates site of barrier mechanisms. CSF, cerebrospinal fluid. (B) Diagram to illustrate the relationships of brain compartments, and cell layers participating in barrier mechanisms (broad black arrows) at the blood-brain interface. (1) choroid plexus epithelium (2) arachnoid membrane (3) parenchymal capillary wall. Broad open arrows, direction of blood flow (V, vein); thin arrows, route of CSF secretion, circulation and outflow. (Modified from Davson, 1972.)

(Patlak *et al.*, 1966; Welch and Sadler, 1965), and other evidence, suggest that the choroid plexus is a rather leaky epithelium (Frömter and Diamond, 1972), which is unlikely to be able to produce a freshly-secreted CSF grossly different from a plasma ultrafiltrate. Nevertheless, the K concentration of newly-secreted CSF is remarkably constant with alterations in plasma K (Ames *et al.*, 1965), suggesting that a degree of homeostasis is possible in spite of the leakiness of the epithelium. By contrast, the blood-brain barrier (see below) is more impermeable, and the permeability of this latter interface might exert a dominating influence over the composition of the ISF, and hence indirectly, the bulk CSF. It should however be noted, that as vertebrate plasma contains the important ions in approximately the correct concentrations and proportions for neuronal function, and demonstrates a degree of homeostasis, gross differences between plasma and CSF/ISF composition are probably not required.

The mechanism of CSF production has been extensively studied *in vivo* and *in vitro* (Wright, 1970, 1972). A ouabain-sensitive, Na-K-ATPase is located on the apical (CSF-facing) choroid plexus epithelium membrane, and secretion is thought to be explicable according to Na transport-induced water movement (Diamond and Bossert, 1967, 1968). In addition, active processes are present for the removal of particular substances from the CSF into blood, including anions such as I, Br, SCN and SO_4 (see Davson, 1972), organic acids (Pappenheimer *et al.*, 1961; Bass and Lundborg, 1973), and organic bases, including quaternary ammonium compounds and metabolites of central nervous transmitters (see Davson, 1972). The control of CSF (and hence ISF) composition by the choroid plexus thus contributes to the homeostasis of the neuronal microenvironment.

CSF drains back into the venous system via bulk transport across the arachnoid membrane, in specialized regions where the arachnoid projects into venous sinuses. The mechanism of drainage is not entirely clear, but appears to be sensitive to the hydrostatic pressure difference between CSF and venous blood (Davson *et al.*, 1970; Welch and Pollay, 1961); flow via labile openings between endothelial cells (Gomez *et al.*, 1973), or via transient vesicles or vacuoles (Tripathi, 1974) has been suggested. There is little evidence that this outflow route shows any specificity or selectivity, and it is therefore unlikely to be involved in homeostasis except as a bulk excretory pathway. However, the arachnoid membrane in other regions may be the site of carrier-mediated active and passive exchange of substances between CSF and blood (Wright, 1972,

1974), and therefore contribute to homeostasis of the interstitial fluid, particularly in regions bathed by CSF in the subarachnoid space. The arachnoid is thus a possible second interface exerting control over the brain microenvironment.

If CSF is collected at different points in the ventricular system, its composition is seen to vary (Table I) (Ames *et al.*, 1964); in particular, the concentrations of K and Ca become less like those of a

TABLE I

Concentrations of ions (mEq/kg H_2O) in plasma, plasma ultrafiltrate, cerebrospinal fluid from the choroid plexus (freshly secreted) and from the cisterna magna (after equilibrating with the brain) of the cat

	Cl	Na	K	Ca	Mg
Plasma	132	163	4.4	2.62	1.35
Plasma ultrafiltrate	136	151	3.3	1.83	0.95
Choroid plexus fluid	138	158	3.28	1.67	1.47
Cisterna magna fluid	144	158	2.69	1.50	1.33

From Ames *et al.*, 1964. The value for K in plasma is probably an overestimate, due to contamination from cells.

plasma ultrafiltrate, the further from its site of secretion it is sampled. The most reasonable explanation of this alteration is that the freshly-secreted fluid is modified as it flows over the brain surface, by the metabolic activities of brain cells, or by mixing with fluid of different origin (e.g. produced at the parenchymal capillary surface). There is considerable evidence that the blood-brain barrier interface is less permeable than the choroid plexus, (e.g. it is impermeable to inulin (Crone, 1965; Katzman *et al.*, 1968)), and restricts entry of substances as small as ions (see Katzman and Pappius, 1973). Moreover, the capillary-brain interface has been implicated as the site of passive and active carrier-mediated transfer processes which can transport particular substances from or into the brain ISF (e.g. Davson and Pollay, 1963; Pollay and Curl, 1967; Bradbury and Stulcová, 1970) and recent studies justify this localization (Davson and Hollingsworth, 1973; Pardridge and Oldendorf, 1975). The larger surface area of the parenchymal capillaries compared with the choroid plexuses, and the intimate relationship between the blood vessels and brain ISF, means that exchange at the capillary interface dominates the composition of the ISF, and probably also the bulk CSF. Thus a third and key site for the control of the brain microenvironment is the blood-brain barrier.

R

The studies outlined above implied that the CSF is in free diffusional communication with the brain ISF, and that the latter forms the microenvironment of neurones. Evidence from electron-dense tracer studies (Brightman, 1965a,b; Brightman and Reese, 1969), diffusion of polysaccharide extracellular markers (Katzman *et al.*, 1968; Levin *et al.*, 1971) and ion movements monitored by electrophysiological techniques (Cohen *et al.*, 1968) show that the brain extracellular space is indeed patent, and provides little diffusional restriction even to substances the size of inulin. There is now general agreement that the interstitial space of the brain accounts for approximately 20—22% of the wet weight of the tissue (see Katzman and Pappius, 1973). Earlier estimates based on electron microscopy or extracellular markers perfused via the blood were much smaller, suggesting that the extracellular space in brain was very restricted or even absent, but these studies probably introduced artefacts due to asphxia during fixation (van Harreveld *et al.*, 1965) or lack of equilibration of the extracellular markers due to the low permeability of the blood-brain barrier and a "sink" action of the CSF (Oldendorf and Davson, 1967). Thus the ISF can be regarded as the true microenvironment of the neurones. As the ISF is generally present in a rather narrow space between cells, rapid alterations in the concentration of this space can occur as a result of neuronal (or glial) activity. It is therefore reasonable to expect that mechanisms should be present to limit the extent of these acute fluctuations in ISF composition, and that glia and/or neurones might show specializations for controlling their own microenvironment.

III. GLIAL-NEURONAL REGULATION

A. Techniques

Evidence that the extracellular concentration is controlled in local regions of the CNS, in addition to the gross and longer-term regulation at the "barrier-sites" described above, is more difficult to obtain, largely because of the problem of sampling from the ISF. However, two particular techniques have been used successfully *in vivo*. The first relies on the sensitivity of the potentials of cell membranes within the brain to alterations in their bathing ionic medium. If the cell response can first be "calibrated" under equilibrium conditions, then the membrane potential can be used as an indicator of the concentration of the relevant ion in the neuronal medium in subsequent experimental studies. This technique was first used by Frankenhaeuser and Hodgkin (1956) in the squid axon, then by Kuffler and co-workers in the leech ganglion (see Kuffler and Nicholls, 1966), and later extended to insects (Treherne et al., 1970) and vertebrates (Kuffler et al., 1966; Orkand et al., 1966; Cohen et al., 1968). The second technique is to use ion- or molecular-selective microelectrodes to measure directly the composition of the ISF. When used in conjunction with potential recording electrodes, this method is particularly valuable.

B. Species showing No regulation

The first technique has provided evidence that the extracellular space of the central nervous system in molluscs and annelids is freely accessible to ions from the bathing medium (see above). Since their ganglia and connectives seem leaky, it is not surprising that in most molluscs and annelids there is no evidence of any significant regulation of the ionic concentration of the neuronal interstitial fluid (cf. Treherne and Moreton, 1970). This is most convincingly seen in the leech, in which identical action and resting potentials were recorded in isolated neurones as compared with those in intact ganglia (Nicholls and Kuffler, 1964). Stimulation of neurones caused depolarization of the glia (Baylor and Nicholls, 1969), probably due to local accumulation of K produced by the active neurones. The subsequent time course of recovery was consistent with K diffusing away via the extracellular cleft, and there was no evidence of passive or active regulation by the glia.

C. Passive Regulation Mechanisms

Although most molluscs studied show no evidence of regulation, an exception would appear to be the freshwater lamellibranch, Anodonta cygnea (Fig. 1b), in which the large axons are able to conduct impulses in sodium-deficient media (Treherne et al., 1969b). This has been postulated to result from a small sodium store in the vicinity of the axon surfaces, but whether glia or neurones are responsible for the store is not known (Treherne et al., 1969a; Carlson and Treherne, 1969).

In decapod crustacea, application of an electrophysiological technique showed that K entry to the axon surface is much slower than expected for free diffusion via the extracellular cleft, while K efflux and Na movements are fairly rapid (Abbott *et al.*, 1975). Active mechanisms for impeding the build-up of K at the axon seemed ruled out by inhibitor studies (Fig. 5), but passive mechanisms controlling K and Na distribution seemed likely (Abbott and Pichon, 1976, and in preparation). Extracellular or intracellular ion binding, or a passive glial contribution (see below) appeared compatible with the observations.

In insects, there is no experimental evidence that long-term extracellular cation regulation is achieved by passive mechanisms. It has been proposed, however, that the anion groups associated with the extracellular hyaluronic acid (Ashhurst, 1961; Ashhurst and Costin, 1971) could maintain a cation-reservoir in the immediate vicinity of the neuronal surfaces (Treherne, 1962, 1967). In the absence of active processes the thermodynamic activity of cations associated with such an anion matrix could not exceed that of the external medium (Treherne, 1967). Thus an extracellular anion matrix could not, for example, lead to significant elevation of the effective extracellular sodium concentration. It is conceivable, however, that such an extracellular sodium reservoir could serve a homeostatic function by contributing to the maintenance of a stable ionic environment for the neurones. Depletion of free sodium ions in the region of the neuronal surfaces, as a result of sustained nervous activity, could be compensated by sodium ions released from the matrix. Similarly, potassium ions would undergo a reduction in activity coefficient when released into the matrix during the terminal phase of the action potential. This mechanism could contribute to the relatively rapid decay of the negative after-potential ($t_{0.5}$ = 9.2 msec) recorded in cockroach axons (Narahashi and Yamasaki, 1960), as compared with that observed in the squid giant axon ($t_{0.5}$ = 30–100 msec) (Frankenhaeuser and Hodgkin, 1956), and could be reinforced by the geometry of the adjacent extracellular system, as suggested by Smith and Treherne (1963).

When the electrophysiological technique was extended to lower vertebrates, the amphibians *Rana* and *Necturus* (Kuffler *et al.*, 1966; Orkand *et al.*, 1966), it was found that the glial membrane potential responded to the bathing K concentration in a similar manner to invertebrate glia, and recent studies have shown that the response is as expected for a more or less perfect K electrode (Bracho *et al.*, 1975).

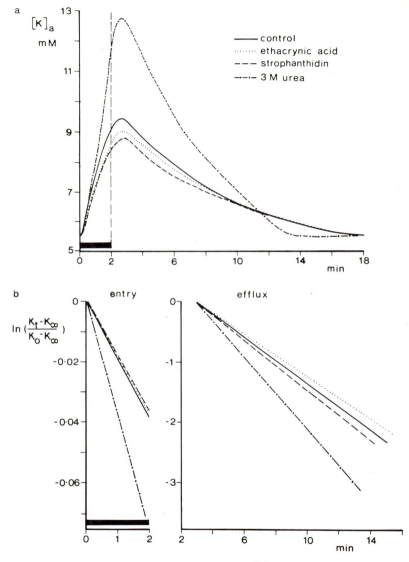

Fig. 5. Crayfish central nervous connective. (a) Effect of 2 min 100 mM K-Ringer pulse (horizontal bar) on K concentration at the axon surface $[K]_a$, in control and after treatment with inhibitors or urea. SO_4 replaced Cl in the K-Ringer. Results with KCl-Ringer were comparable. (b) Curves from Fig. 6a replotted semilogarithmically, to show that Na transport inhibitors have little effect on rate (slope of each line) of K entry and efflux. (From Abbott and Pichon, 1976.)

As in the leech, axon stimulation in amphibia causes measurable depolarization of the adjacent glial cells, presumably due to K accumulation (Kuffler *et al.*, 1966; Orkand *et al.*, 1966). It was pointed out that this could be the basis for a mechanism for passive redistribution of K ions within the glial compartment, preventing excessive rise of K around active neurones. Glia are known to be electrotonically coupled by gap junctions (see Kuffler and Nicholls, 1966), and so can be envisaged as a functional syncytium. Depolarization of glial membranes adjacent to active neurones will create a potential difference within the glial network, as glial membranes adjacent to inactive neurones will not be depolarized. This PD will cause K to enter in the depolarized region, and leave in the relatively undepolarized terminals; current circulating back through the extracellular space will largely be carried by Na ions. Thus as long as the glial membranes are depolarized asymmetrically, the glial compartment will act as a "spatial buffer" for K, preventing excessive rise in K concentration in the extracellular space around active neurones.

Evidence for such a passive spatial buffering action is surprisingly difficult to produce, even in lower forms. There is some evidence that mammalian glial cells may not behave as perfect K electrodes, suggesting that other ions have appreciable permeability through the glial membrane, or that an electrogenic component may be present (Pape and Katzman, 1971; Ransom and Goldring, 1973; Dennis and Gerschenfeld, 1969). In either case, the ability of the glial membrane to buffer extracellular K levels might be limited. However, deviation of the membrane potential response from the 58 mV slope predicted by the Nernst equation, may be partly explicable as non-specific damage caused by electrode penetration in the small mammalian glia, so this factor itself is not sufficient to rule out passive K buffering by mammalian glial cells.

Indirect support for a passive glial buffer comes from the study of Trachtenburg and Pollen (1970), who measured membrane properties of mammalian astrocytic glia *in vivo*, and pointed out that the short time constant and low membrane specific resistance, and hence large length constant, made these glia particularly suitable for removing K passively and conveying it intracellularly to a distance. Extracellular current flow recorded during activity is also consistent with passive K redistribution in amphibian optic nerve (Orkand *et al.*, 1966), amphibian retina (Miller and Dowling, 1970), and mammalian spinal cord (Somjen, 1973). However, an active component to K redistribution needs also to be considered. This factor would seem to have been eliminated in amphibian optic nerve by studies of the temperature dependence of recovery (Bracho and Orkand, 1972), but information is not available in the other systems. Thus while passive K regulation by glial "buffering" may make a significant contribution to short term K homeostasis within the brain microenvironment in lower vertebrates, its importance in mammalian brain is difficult to assess.

It should be noted that all the passive "buffering" systems described, either in invertebrates or vertebrates, can only act to reduce the effect of short-term fluctuations in the blood or neuronal medium, and cannot alone maintain activities of ions in the neuronal environment chronically different from those in blood.

D. Active Regulation Mechanisms

An alternative to passive redistribution, is active regulation of the neuronal milieu by energy-dependent mechanisms, especially Na-K transport systems. It is well known that neurones can take up K and extrude Na against a concentration gradient, due to the presence of a specific ouabain-sensitive Na-K-ATPase present in the membrane (Baker, 1965; Skou, 1957); moreover, K lost during activity must eventually be regained by the neurone. The neurones themselves might therefore be the most obvious site for regulation of the ion content of their own microenvironment. However, it is found that glial cells also possess membrane-bound Na-K-ATPase, possibly at a higher activity than in neurones (Haljamae and Hamburger, 1971; Henn *et al.*, 1972), and stimulated by the kind of K concentrations that occur during axonal activity (Hertz, 1966). Neurones do not appear to show the same dependence of ATPase activity on K concentration (Henn *et al.*, 1972). It is thus possible that either neurones or glia, or both, might be involved in active homeostasis of the neuronal microenvironment in vertebrate brain.

Involvement of glial elements in long-term extracellular cation regulation is indicated by electrophysiological studies on insect nervous systems. It has been shown, for example, that the net inward and outward movements between the extra-axonal fluid and the bathing medium exhibit a marked asymmetry. Exposure of intact cockroach connectives to sodium-deficient-saline results in a slow decline in the amplitude of the recorded action potentials, a rapid return being observed on restoration of the normal external sodium concentrations (Schofield and Treherne, 1975). The recovery was reduced in the presence of dilute dinitrophenol and ethacrynic acid (Fig. 6) and was abolished when lithium ions were substituted for those of sodium in the bathing medium. The inhibitory effect of the ethacrynic acid is of interest, for electrical responses of the axons (measured in intact and desheathed preparations) indicate that a sodium pump associated with the glial and/or perineurial elements is ethacrynic-sensitive, the axonal one being ouabain-sensitive (Pichon

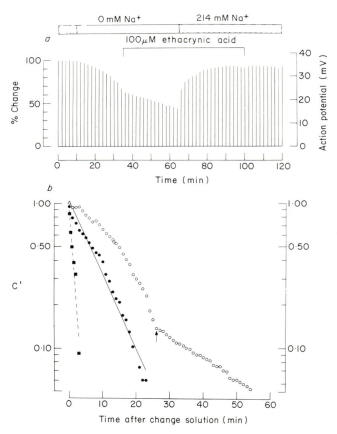

Fig. 6. (a) Effects of dilute ethacrynic acid on the amplitude of action potentials recorded in intact cockroach connectives during exposure to sodium-deficient and normal Ringer. (b) Change in relative extra-axonal sodium concentration (C') estimated from (a), showing reduction in the rate of decline (○) following addition of ethacrynic acid (arrow). The rate of recovery (plotted as $1-C'$) in the presence of ethacrynic acid (●) was slower than that observed in an untreated preparation (■). (From Schofield and Treherne, 1975.)

and Treherne, 1974). The above evidence can be interpreted in terms of a perineurial-glial mediated transport of sodium ions from the bathing medium to the extra-axonal fluid (Schofield and Treherne, 1975; Treherne, 1975). The inability of lithium ions to gain access to the extra-axonal fluid does not appear to result from exclusion by the outer perineurial membranes (p. 487) and it is suggested that the demonstrated accumulation of this cation within the connectives (Bennett *et al.*, 1975) results from the inability of the inwardly-directed perineurial and glial membranes to transport lithium ions (Schofield and Treherne, 1975). This inability contrasts with the rapid net movement of sodium ions to the axon surfaces and suggests that sodium is transported across the glial membranes by a conventional pump which, as in crustacean neurones (Baker, 1965), will not accept lithium ions. The presence of such a pump on the glial, as well as the axonal membranes, could also serve a regulatory function by transporting potassium ions away from the axon surfaces. Electrophysiological evidence for such a potassium regulation has been advanced by Thomas and Treherne (1975) and is also indicated by the ouabain-induced axonal depolarization of around 8 mV, observed in desheathed cockroach connectives (Wilson, 1973). This observation cannot be wholly explained in terms of the effect on an electrogenic sodium pump. It is suggested that the depolarization resulted from a changed equilibrium potential which was caused by an accumulation of potassium ions in the immediate vicinity of the axon surfaces in the presence of the cardiac glycoside (Wilson, 1973).

Direct evidence for active regulation of the K concentration by local mechanisms in mammalian brain has come from studies using ion-selective microelectrodes. Use of the membrane potential of glia (or the negative after-potential of neurones) as a measure of extracellular K is complicated by two features; it is necessarily indirect and it depends on homogeneity of the glial or neuronal population, for K concentration is derived by extrapolation from the behaviour of the cells used to construct the calibration curve. Small differences in the condition of the cells, or damage caused by microelectrode penetration, may introduce large errors into the K estimation. A more direct method is to use K-selective microelectrodes which can be introduced into the extracellular space. When combined in a double-barrelled system with an electrode for monitoring intracellular or extracellular potential, this can give a technique with high resolution and reliability (Walker, 1971; Vyskočil and Kříž, 1972).

Such a technique has been used by Kříž et al. (1975). They introduced double-barrelled K-sensitive microelectrodes into the lumbar spinal cord of cats, and recorded K activity $[K]_e$ and focal extracellular potentials. During peripheral nerve stimulation, the extracellular $[K]_e$ increased from the 3 mM resting level up to a maximum of 9 mM, with an approximately exponential time course. The use of double-barrelled electrodes allowed recording of $[K]_e$ very close to the discharging neurones, and the non-anaesthetized (decerebrate) preparation meant that approximately normal activity was probably recorded. Of particular interest was the observation that during the recovery phase, $[K]_e$ decreased to a value as much as 0.5 mM lower than the resting level, before finally returning to normal (Fig. 7). This subnormal phase of $[K]_e$ was dependent on the duration and frequency of stimulation and was abolished by anoxia (which had less effect on the rising phase of K accumulation), from which it was concluded that the subnormal phase reflects active processes responsible for taking up K from the extracellular space. A similar conclusion was reached by Krnjević and Morris (1975), who showed that the subnormal phase of $[K]_e$ was abolished, and the peak amplitude of $[K]_e$ increased, by application of strophanthidin, a specific inhibitor of Na-K-ATPase, in cat brain.

The location of mechanisms for K removal is of particular interest, especially in view of the evidence from isolated glial and neuronal fractions, which suggests that the glial Na-K-ATPase is more active in the physiological range of $[K]_e$ than that of neurones. Kříž et al. (1975) observed that the subnormal phase of $[K]_e$ is only seen when the microelectrode is located very close to the discharging neurones, although elevated levels of $[K]_e$ may be measured more peripherally. They claim that this suggests the neurones themselves, and not glia, are responsible for the active K reabsorption, on the basis that while the active neurones are located in the depths of the grey matter, glia are distributed throughout the cross section. However, many other factors could be responsible. Thus the geometry of the extracellular space could favour resolution of K changes near the neurones rather than more peripherally, or the Na-K-ATPase could be activated only at the high levels of K immediately adjacent to the neurones, or peripheral zones might show a higher concentration of oligodendrocytes with little regulatory activity, while high activity astrocytes were concentrated around neurones.

Further information on the site of regulation has come from developmental studies. It is well known that the rat brain shows considerable immaturity at birth compared with other experimental

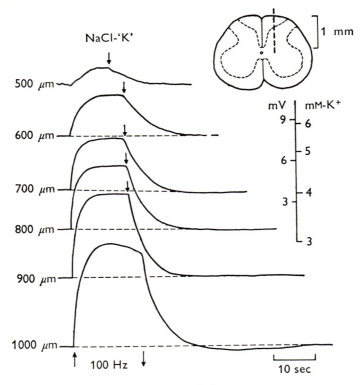

Fig. 7. Changes in extracellular potassium $[K]_e$ in cat spinal cord produced by peripheral stimulation of the common peroneal nerve at 100 Hz. No reduction in $[K]_e$ was observed down to a depth of 900 μm, although the $[K]_e$ increased by 2 mM. Subnormal phase of $[K]_e$ appeared at a depth of 1000 μm. The direction of the microelectrode track is shown schematically in the drawing. (From Kříž *et al.*, 1975.)

species such as guinea pig or rabbit (Dobbing, 1968; Ford, 1973). In particular, although neurogenesis is reasonably complete by 15–20 days postnatally, glia at this stage, and until approximately 30 days, are relatively undeveloped. Rat brain of 20 days thus represents a virtually glial-free preparation. In adult rat, cerebral cortex slices incubated *in vitro* in elevated K media (20 mM cf. normal 7 mM) are known to show increased intracellular K and decreased intracellular Na, and it is suggested that this "cationic exchange" reflects the activity of a Na-K-ATPase on the glia (Franck *et al.*, 1968; Franck and Schoffeniels, 1972). By contrast, 20 day rats show little change in intracellular Na and K when incubated in 20 mM K (Grisar and

Franck, 1975). This may be correlated with the absence of glia, particularly astrocytes, at this stage. However, evidence is needed that the increased K in the adult cortex slices is not simply a passive phenomenon.

The relative importance of the activity of glia and neurones in controlling extracellular potassium has frequently been inferred from enzymic studies on isolated cell fractions. Methods are now available for separation of reasonably pure glial and neuronal fractions from whole brain and spinal cord (Blomstrand and Hamberger, 1969). It is found that the Na-K-ATPase from glial-enriched fractions in adult rabbits is more active (8.16 ± 2.26 μmoles Pi/mg protein/h) than the neuronal ATPase (4.91 ± 1.21 μmoles) (Grisar and Franck, 1975). It is, however, dangerous to argue from these results, because they could easily reflect differences in the geometry of the cell (surface/volume ratio), and especially the nature of the isolated material. Thus the isolated neuronal fraction is rich in perikarya, but largely shorn of processes, where membrane ATPase might be expected to be concentrated. On the other hand, the astrocytic glial fraction is generally more intact, with numerous processes. Thus apparent relative activities of the enzymes may be artefactual.

Differences in optimal K concentrations for ATPase activity are, however, probably valid. The optimal activity of the glial ATPase occurred around 20 mM of potassium, while that of neurones was approximately 5 mM (Grisar and Franck, 1975). This would indicate that the glial ATPase is more likely to be stimulated by the kinds of K concentration that occur during neuronal activity. It is interesting that in young rabbits (14 days old) there was no difference between glial and neuronal enzyme stimulation, both being optimal at 5 mM K; this may suggest that the specialized glial ATPase is not yet fully developed at this age (Grisar and Franck, 1975).

Attempts to further differentiate oligodendrocytic and astrocytic glial contributions to the regulation of the extracellular microenvironment have been made in preparations in which one or other class of glia is selectively damaged or absent. Interestingly, absence or damage to olidogendrocytes (e.g. in Jimpy mouse, Sidman et al., 1964; Keen et al., 1976) is associated with reduction in the characteristic K stimulated increase in oxygen consumption which is found in normal brain, and thought to be related to activation of glial Na-K-ATPase. It would be interesting to see whether a similar reduction is found when astrocytic function is selectively impaired (e.g. by porto-caval anastomosis, Zamora et al., 1973). Meanwhile, it is premature to conclude that the major Na-K-ATPase activity of vertebrate brain is associated with any one cell type (oligodendrocyte, astrocyte or neurone), or that such activity is responsible for control of the ionic composition of the extracellular space in the immediate vicinity of neurones.

IV. CONCLUSIONS

1. Regulation and homeostasis of the fluid environment of neurones can be achieved by a combination of "barrier" and local glial-neuronal mechanisms within the central nervous system, superimposed on mechanisms for the background homeostasis of body fluids such as blood. Where a species shows no evidence of regulatory mechanisms within the brain, it usually implies that the composition and homeostasis of blood is adequate for normal neuronal function, or that the tolerance of the neurones is appreciable.

2. Most invertebrates have a "leaky" blood-brain interface, and no evidence of local mechanisms for regulation of the brain interstitial fluid composition.

3. "Non-leaky" systems are found only in insects and vertebrates; in insects the barrier at the blood-brain interface is due to the perineurial glial layer, while in the more complex vertebrate brain, equivalent barriers can be identified at the choroid plexus, arachnoid membrane and parenchymal capillary wall.

4. These barriers are characterized by passive impermeability to ions and large molecules, and active and passive transfer processes. Thus barrier phenomena at the blood-brain interface are generally due to a combination of passive and active mechanisms.

5. Glial-neuronal regulation, controlling the extracellular fluid composition in the immediate vicinity of the neurones, is found among insects and vertebrates. However, it is also found in rudimentary form in certain species which have a relatively leaky blood-brain interface, such as decapod crustacea and certain molluscs. Regulation in such leaky systems is necessarily fairly inefficient, and seems confined to passive "buffering" mechanisms which minimize the effect of acute fluctuations in extracellular fluid concentration, but cannot control long-term composition.

6. Local regulation can be much more efficient in nervous systems protected by a relatively impermeable blood-brain interface, i.e. in insects and vertebrates. Although passive buffering mechanisms possibly contribute to this regulation, active processes, especially those dependent on ion pumps such as Na-K-ATPase, appear especially important.

7. In insects, local regulation can be seen as an extension of the activity of the perineurial (glial) barrier, and seems specialized for long term ionic regulation of the extracellular fluid, probably reflecting the unsuitability of insect blood as a neuronal medium in many species.

8. In vertebrate central nervous systems, although long-term regulation can occur, and can be shown to depend on activity of the barrier mechanisms at the blood-brain interface under conditions of chronically altered blood composition, this appears less crucial under normal conditions than in insects. Certainly the normal ionic composition of vertebrate plasma is not greatly different from that of the brain extracellular fluid. Instead, regulation within the brain seems designed to cope with short-term fluctuations, especially those due to activity of the central nervous system itself. The relative contributions of glia and neurones to this local regulation are not yet clear.

It will be interesting to see whether further evidence of brain microenvironment regulation can be found among groups which appear to have a leaky blood-brain interface, especially in those species which are exposed to extremes or fluctuations in composition in the external environment. And the striking structural similarity between the well-developed barriers of insects and vertebrates needs to be examined in more detail, to establish whether the parallel extends to the level of molecular mechanisms for passive and active transfer. (For ion-binding in extracellular materials see Chapter 4.)

REFERENCES

Abbott, N. J. (1970). *Nature Lond.* **225**, 291–293.

Abbott, N. J. (1971a). *Z. Zellforsch.* **120**, 386–400.

Abbott, N. J. (1971b). *Z. Zellforsch.* **120**, 401–419.

Abbott, N. J. (1972). *Tissue and Cell* **4**, 99–104.

Abbott, N. J. and Pichon, Y. (1976). *In* "Transport Phenomena in the Nervous System." (Eds G. Levi, L. Battistin and A. Lajtha), Plenum, New York.

Abbott, N. J., Moreton, R. B. and Pichon, Y. (1975). *J. exp. Biol.* **63**, 85–115.

Ames, A., Sakanoue, M. and Endo, S. (1964). *J. Neurophysiol.* **27**, 672—681.

Ames, A., Higashi, K. and Nesbett, F. B. (1965). *J. Physiol. Lond.* **181**, 506—515.

Ashhurst, D. E. (1961). *Nature Lond.* **191**, 1224—1225.

Ashhurst, D. E. and Costin, N. M. (1971). *Histochem. J.* **3**, 297—310.

Baker, P. F. (1965). *J. Physiol. Lond.* **180**, 383—423.

Bass, N. H. and Lundborg, P. (1973). *Brain Res.* **56**, 285—298.

Baylor, D. A. and Nicholls, J. G. (1969). *J. Physiol. Lond.* **203**, 555—569.

Bennett, R. R., Buchan, P. B. and Treherne, J. E. (1975). *J. exp. Biol.* **62**, 231—241.

Blomstrand, C. and Hamberger, A. (1969). *J. Neurochem.* **16**, 1401—1407.

Bracho, H. and Orkand, R. K. (1972). *Brain Res.* **36**, 416—419.

Bracho, H., Orkand, P. M. and Orkand, R. K. (1975). *J. Neurobiol.* **6**, 395—410.

Bradbury, M. W. B. and Stulcová, B. (1970). *J. Physiol. Lond.* **208**, 415—430.

Brightman, M. W. (1965a). *J. Cell Biol.* **26**, 99—123.

Brightman, M. W. (1965b). *Am. J. Anat.* **117**, 193—219.

Brightman, M. W. and Reese, T. S. (1969). *J. Cell Biol.* **40**, 648—677.

Carlson, A. D. and Treherne, J. E. (1969). *J. exp. Biol.* **51**, 297—318.

Coggeshall, R. E. and Fawcett, D. W. (1964). *J. Neurophysiol.* **27**, 229—289.

Cohen, M. W., Gerschenfeld, H. M. and Kuffler, S. W. (1968). *J. Physiol. Lond.* **197**, 363—380.

Crone, C. (1965). *J. Physiol. Lond.* **181**, 103—113.

Davson, H. (1972). *In* "The Structure and Function of the Nervous System." (Ed. G. H. Bourne), Vol. IV, pp. 321—445. Academic Press, New York and London.

Davson, H. and Hollingsworth, J. R. (1973). *J. Physiol. Lond.* **233**, 327—347.

Davson, H. and Pollay, M. (1963). *J. Physiol. Lond.* **167**, 247—255.

Davson, H., Hollingsworth, G. and Segal, M. B. (1970). *Brain* **93**, 665—678.

Dennis, M. J. and Gerschenfeld, H. M. (1969). *J. Physiol. Lond.* **203**, 211—222.

Diamond, J. M. and Bossert, W. H. (1967). *J. gen. Physiol.* **50**, 2061—2083.

Diamond, J. M. and Bossert, W. H. (1968). *J. Cell Biol.* **37**, 694—702.

Dobbing, J. (1968). *In* "Applied Neurochemistry." (Eds A. N. Davison and J. Dobbing), pp. 287—316. Blackwells, Oxford.

Fencl, V., Miller, T. B. and Pappenheimer, J. R. (1966). *Am. J. Physiol.* **210**, 459—472.

Florkin, M. and Jeuniaux, C. (1964). *In* "The Physiology of Insecta." (Ed. M. Rockstein), 3, pp. 109—152. Academic Press, New York and London.

Ford, D. H. (1973). *In* "Development and Aging in the Nervous System." (Ed. M. Rockstein), pp. 63—88. Academic Press, New York and London.

Franck, G. and Schoffeniels, E. (1972). *J. Neurochem.* **19**, 395—402.

Franck, G., Cornette, M. and Schoffeniels, E. (1968). *J. Neurochem.* **15**, 843—857.

Frankenhaeuser, B. and Hodgkin, A. L. (1956). *J. Physiol. Lond.* **131**, 341—376.

Frömter, E. and Diamond, J. (1972). *Nature New Biol.* **235**, 9—13.

Gomez, D. G., Potts, D. G., Deonarine, V. and Reilly, K. F. (1973). *Lab. Invest.* **28**, 648—657.

Grisar, T. and Franck, G. (1975). Fifth International Meeting of the Int. Soc. Neurochem. (Abstr. p. 228).

Gupta, B. L., Mellon, D. and Treherne, J. E. (1969). *Tissue and Cell* 1, 1–30.

Haljamäe, H. and Hamberger, A. (1971). *J. Neurochem.* 18, 1903–1912.

Henn, F. A., Haljamäe, H. and Hamberger, A. (1972). *Brain Res.* 43, 437–443.

Hertz, L. (1966). *J. Neurochem.* 13, 1373–1387.

Katzman, R. and Pappius, H. M. (1973). "Brain Electrolytes and Fluid Metabolism." Williams and Wilkins, Baltimore.

Katzman, R., Schimmel, H. and Wilson, C. E. (1968). *Proc. Virchow Med. Soc. N.Y.* 26, 254–280.

Keen, P., Osborne, R. H. and Pehrson, U. M. M. (1976). *J. Physiol. Lond.* (In press.)

Kříž, N., Syková, E. and Vyklický, L. (1975). *J. Physiol. Lond.* 249, 167–182.

Krnjević, K. and Morris, M. E. (1975). *J. Physiol. Lond.* 250, 36–37P.

Kuffler, S. W. and Nicholls, J. G. (1966). *Ergebn. Physiol.* 57, 1–90.

Kuffler, S. W., Nicholls, J. G. and Orkand, R. K. (1966). *J. Neurophysiol.* 29, 768–787.

Lane, N. J. (1974). *In* "Insect Neurobiology." (Ed. J. E. Treherne), North Holland, Amsterdam.

Lane, N. J. and Abbott, N. J. (1975). *Cell and Tissue Res.* 156, 173–187.

Lane, N. J. and Treherne, J. E. (1970). *Tissue and Cell* 2, 413–425.

Lane, N. J. and Treherne, J. E. (1972a). *J. exp. Biol.* 56, 493–499.

Lane, N. J. and Treherne, J. E. (1972b). *Tissue and Cell* 4, 427–436.

Lane, N. J., Skaer, H. leB. and Swales, L. S. (1975). *J. Cell Biol.* (In press).

Levin, E., Arieff, A. and Kleeman, C. R. (1971). *Am. J. Physiol.* 221, 1319–1326.

Lux, H. D., Neher, E. and Prince, D. A. (1972). *Pflügers Arch. ges. Physiol.* 332, R 89.

Maddrell, S. H. P. and Treherne, J. E. (1966). *Nature* 211, 215–216.

Maddrell, S. H. P. and Treherne, J. E. (1967). *J. Cell Sci.* 2, 119–128.

Mellon, D. and Treherne, J. E. (1969). *J. exp. Biol.* 51, 287–296.

Miller, R. F. and Dowling, J. E. (1970). *J. Neurophysiol.* 33, 323–341.

Mirolli, M. and Crayton, J. W. (1968). *J. Cell Biol.* 39, 92–93a.

Mirolli, M. and Gorman, A. L. F. (1973). *J. exp. Biol.* 58, 423–435.

Narahashi, T. and Yamasaki, T. (1960). *J. Physiol.* 151, 75–88.

Nicholls, J. G. and Kuffler, S. W. (1964). *J. Neurophysiol.* 27, 645–676.

Oldendorf, W. H. and Davson, H. (1967). *Arch. Neurol.* 17, 196–205.

Orkand, R. K., Nicholls, J. G. and Kuffler, S. W. (1966). *J. Neurophysiol.* 29, 788–806.

Pape, L. G. and Katzman, R. (1972). *Brain Res.* 38, 71–92.

Pappenheimer, J. R., Heisey, S. R. and Jordon, E. F. (1961). *Am. J. Physiol.* 200, 1–10.

Pardridge, W. M. and Oldendorf, W. H. (1975). *Biochim. biophys. Acta* 382, 377–392.

Patlak, C. S., Adamson, R. H., Oppelt, W. W. and Rall, D. P. (1966). *Life Sci.* 5, 2011–2015.

Pentreath, V. W. and Cottrell, G. A. (1970). *Z. Zellforsch.* 111, 160—178.

Pichon, Y. and Treherne, J. E. (1970). *J. exp. Biol.* 53, 485—493.

Pichon, Y. and Treherne, J. E. (1974). *J. exp. Biol.* 61, 203—218.

Pichon, Y., Moreton, R. B. and Treherne, J. E. (1971). *J. exp. Biol.* 54, 757—777.

Pichon, Y., Sattelle, D. B. and Lane, N. J. (1972). *J. exp. Biol.* 56, 717—734.

Pollay, M. and Curl, F. (1967). *Am. J. Physiol.* 213, 1031—1038.

Potts, W. T. W. (1954). *J. exp. Biol.* 31, 376—385.

Ransom, B. R. and Goldring, S. (1973). *J. Neurophysiol.* 36, 855—868.

Rougemont, J. de, Ames, A. Nesbett, F. B. and Hoffmann, H. F. (1960). *J. Neurophysiol.* 23, 485—495.

Sattelle, D. B. (1973a). *J. exp. Biol.* 58, 1—14.

Sattelle, D. B. (1973b). *J. exp. Biol.* 58, 15—28.

Sattelle, D. B. and Howes, E. A. (1975). *J. exp. Biol.* 63, 421—431.

Sattelle, D. B. and Lane, N. J. (1972). *Tissue and Cell* 4, 253—270.

Schofield, P. K. (1975). Extra-axonal cation regulation in insects. Ph.D. Thesis, Cambridge University.

Schofield, P. K. and Treherne, J. E. (1975). *Nature Lond.* 255, 723—725.

Shabo, A. L. and Maxwell, D. S. (1971). *J. Neuropath. exp. Neurol.* 30, 506—524.

Sidman, R. L., Dickie, M. M. and Appel, S. M. (1964). *Science N.Y.* 144, 309—311.

Skaer, H. le B. and Lane, N. J. (1974). *Tissue and Cell* 6, 695—718.

Skou, J. C. (1957). *Biochim. biophys. Acta* 23, 394—401.

Smith, D. S. and Treherne, J. E. (1963). *In* "Advances in Insect Physiology." (Eds J. W. L. Beament, J. E. Treherne and V. B. Wigglesworth), 1, pp. 401—484. Academic Press, London and New York.

Somjen, G. G. (1973). *In* "Progress in Neurobiology." (Eds Kerkut, G. A. and Phillis, J. W.), Ch 6, pp. 201—232. Pergamon, Oxford and New York.

Stephens, R. R. and Young, J. Z. (1969). *Phil. Trans. R. Soc. Lond. B.* 255, 1—12.

Thomas, M. V. and Treherne, J. E. (1975). *J. exp. Biol.* 63, 801—810.

Trachtenberg, M. C. and Pollen, D. A. (1970). *Science N.Y.* 167, 1248—1251.

Treherne, J. E. (1962). *J. exp. Biol.* 39, 193—217.

Treherne, J. E. (1966). "The Neurochemistry of Arthropods." Cambridge University Press, Cambridge.

Treherne, J. E. (1967). *In* "Insects and Physiology." (Eds J. W. L. Beament and J. E. Treherne), Oliver and Boyd, Edinburgh and London.

Treherne, J. E. (1974). *In* "Insect Neurobiology." (Ed. J. E. Treherne), pp. 187—244. North-Holland, Amsterdam.

Treherne, J. E. (1975). *In* "Fluid Environment of the Brain." (Eds H. F. Cserr, J. D. Fenstermacher and V. Fencl), pp. 105—122. Academic Press, New York and London.

Treherne, J. E. and Maddrell, S. H. P. (1967). *J. exp. Biol.* 47, 235—247.

Treherne, J. E. and Moreton, R. B. (1970). *Int. Rev. Cytol.* 28, 45—88.

Treherne, J. E. and Pichon, Y. (1972). *In* "Advances in Insect Physiology." (Eds J. E. Treherne, M. J. Berridge and V. B. Wigglesworth), 9, pp. 257—308. Academic Press, London and New York.

Treherne, J. E., Carlson, A. D. and Gupta, B. L. (1969a). *Nature Lond.* **223**, 377—380.

Treherne, J. E., Mellon, D. and Carlson, A. D. (1969b). *J. exp. Biol.* **50**, 711—722.

Treherne, J. E., Lane, N. J., Moreton, R. B. and Pichon, Y. (1970). *J. exp. Biol.* **53**, 109—136.

Treherne, J. E., Buchan, P. B. and Bennett, R. R. (1975). *J. exp. Biol.* **62**, 721—732.

Tripathi, R. (1974). *Brain Res.* **80**, 503—506.

Van Harreveld, A. (1966). "Brain Tissue Electrolytes." Butterworth, Washington.

Van Harreveld, A., Crowell, J. and Malhotra, S. K. (1965). *J. Cell Biol.* **25**, 117—137.

Vyskočil, F. and Kříž, N. (1972). *Pflügers Arch. ges. Physiol.* **337**, 265—276.

Vyskočil, F., Kříž, N. and Bureš, J. (1972). *Brain Res.* **39**, 255—259.

Walker, J. L. (1971). *Anal. Chem.* **43**, 89—92A.

Weidler, D. J. and Diecke, F. P. J. (1969). *Z. Vergl. Physiol.* **64**, 372—399.

Welch, K. (1965). *In* "Cerebrospinal Fluid and the Regulation of Ventilation." (Eds Brooks, C. Mc C., Kao, F. F. and Lloyd, B. B.), pp. 413—421. Blackwell, Oxford.

Welch, K. and Pollay, M. (1961). *Am. J. Physiol.* **201**, 651—654.

Welch, K. and Sadler, K. (1965). *J. Neurosurg.* **22**, 344—349.

Welch, K. and Sadler, K. (1966). *Am. J. Physiol.* **210**, 652—660.

Wilson, M. C. L. (1973). "Cold acclimation in insect nerves." Ph.D. Thesis, Cambridge University, Cambridge.

Wright, E. M. (1970). *Brain Res.* **23**, 302—304.

Wright, E. M. (1972). *J. Physiol. Lond.* **226**, 545—571.

Wright, E. M. (1974). *Brain Res.* **76**, 354—358.

Zamora, A. J., Cavanagh, J. B. and Kyu, M. H. (1973). *J. Neurol. Sci.* **18**, 25—45.

20. The Vertebrate Gall-Bladder – The Routes of Ion Transport

T. Zeuthen

Laboratory of Physiology, University of Cambridge, Cambridge, England

The prominence of gall-bladder in the study of epithelial ion and water transport in the vertebrates is due to its relative simplicity. Compared with any segment of the intestine the epithelial cell layer of the gall-bladder contains only one type of cell with apparently only one function; to absorb salt and water. Unlike the intestine, there are no mechanisms for transport of sugars, amino acids or fats. The bile produced in the liver is passed into the gall-bladder. Here the epithelial cell layer which lines the internal surface absorbs mainly NaCl, NaHCO$_3$, and water, leaving behind a concentrated solution of sodium salts of the bile-acids and the bile pigments. This concentrated solution is then periodically passed on into the intestine. The absorbed salt and water enters the capillary system which is situated just below the epithelial cell layer.

The epithelial cells of the gall-bladder are 20 to 40 μm long and have a diameter varying between 5 μm in the rabbit to about 40 μm in *Necturus*. They are closely packed in a columnar fashion with their mucosal ends facing the lumen of the gall-bladder and the serosal ends abutting the basement membrane and facing the underlying capillaries and connective tissue. The narrow spaces between the cells, lined by the serosal membrane are called the lateral intercellular spaces. These spaces terminate towards the mucosal surface where the cells make contact via the *Zonulae occludentes* and *Zonulae adherentes*. These *Zonulae* as a whole are usually named "the tight junction". In this study they will be named "the leaky junctions" as they have been found to offer relatively little resistance to ion movements (Frömter and Diamond,

1972; Frömter, 1972). Studies with electron microscopy and freeze-fracturing suggest that such resistance to transepithelial ion movement as exists, is mainly in the *Zonulae occludentes* and that these consist of strands or "hoops" around the cells, (Staehelin, 1975); via these the cells make frequent but discontinuous contact.

The macroanatomy of the gall-bladder is relatively simple. There' are no extensive muscle layers as in the intestine, only a few intertwined smooth muscle fibres in the layer of connective tissue which, lined by the peritonial membrane, faces the abdominal cavity. The excised gall-bladder from cold-blooded animals can be mounted so that the epithelial cell layer becomes a flat layer without *villi*.

Up to the mid-sixties ion and water transport in the gall-bladder was studied mostly by the "black box" approach (see review by Diamond, 1968). The transmural properties were investigated: The chemical and electrical changes in the mucosal and serosal solution were determined in different transport situations. The main finding was that salt transport into the tissue was the prime mover in water absorption. From then on the research continued in three, obviously interconnected, lines. (i) The coupling mechanism between salt and water. Here the discussion has centred around two possible models of coupling; the three compartment model (Curran, 1960) as applied to the gall-bladder by Kaye *et al.* (1966) and the standing gradient model (Diamond and Bossert, 1967). Coupling mechanisms between solute and solvent are discussed in Chapters 5 and 7. (ii) The properties of the leaky junctions. After it was confirmed by Frömter and Diamond, (1972) and Frömter, (1972) that the junctions were in fact leaky and constituted the major route for transmurally induced current flows, the permselective properties of the leaky junctions were determined (see review by Moreno and Diamond, 1974). At physiological pH the junction is largely cation selective in the period immediately after discussion. With time the anion selectivity increases. At low pH the leaky junctions become anion selective. (iii) The route of ion flows. As the movement of ions into and across the tissue from the lumen is the primary event in the overall transport process, the central question is: which forces move the ions? A clue to the answer must be to determine the route taken by the different ions. The determination of the forces acting upon the ions, and the routes they take, are the subject of this review.

The transport of ions can be studied in two ways. One can either induce an ion or current flow in the tissue and determine tissue parameters from the pattern of flow; or one can study the spontaneously transporting tissue by making deductions from tissue parameters measured during this transport. The first alternative is the subject of the first two parts. Section I deals with the effects of an induced transmural electrical potential-difference, and the tissue parameters that can be deduced from this. Section II deals with the effects of an induced intraepithelial current flow and the significance of the tissue parameters so derived. The second alternative is the subject of Section III: from measurements of intra and extracellular ion activities and electrical potentials one can discuss the possible routes of ion transport.

When discussing routes of ion transport it is necessary to define a model of the tissue. The behaviour of the chosen model will obviously be such as to simulate the behaviour of the tissue; however, the model will always be simplified as compared to the tissue. Misconceptions will arise if one attempts to derive statements about the tissue behaviour from a model which is too simplified in respect to the phenomena under study. A discussion of present models of the tissue will therefore constitute a large part of the text which may sometimes seem "technical" but the reward for constructing a good model of the tissue will be to avoid excessive "kissing in the black box" (using a slightly polished phrase coined by Diamond, 1974). The difficulties in constructing an equivalent electrical model for a tissue is exemplified in Fig. 1 A and B, which shows electron-micrographs of the gall-bladder epithelium of the *Necturus*. The most striking features are the complexity of the lateral intercellular spaces and the density of micro-organelles in the cytoplasm. Some of the consequences of this are discussed in Sections I and II.

I. THE ROUTES OF EXTERNALLY INDUCED TRANSEPITHELIAL ION TRANSPORT

A. Summary

When an external osmotic gradient is imposed across the epithelial cell layer of the gall-bladder the lateral intercellular spaces distend when waterflow is in the direction from mucosa to serosa and collapse when the flow is reversed. When the spaces distend the transmural direct-current impedance decreases in the *Necturus* (Frömter, 1972) and frog gall-bladder (Bindslev et al., 1974) but no change is found for the rabbit gall-bladder (Smulders et al., 1972). If

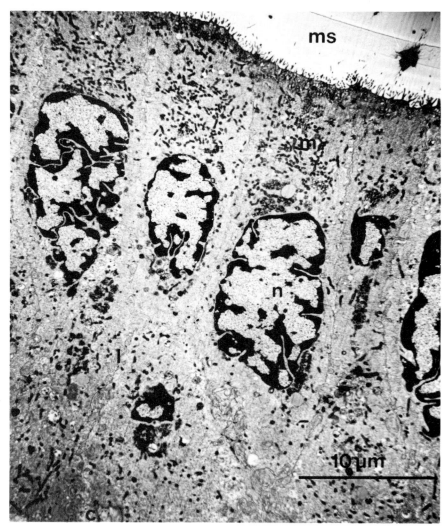

Fig. 1. Electron micrographs of *Necturus* gall-bladder epithelium; courtesy of Dr B. S. Hill. The tissue was fixed unstretched. A. (Left) The most prominent features are the high density of mucous-filled vesicles (v) and mitochondria (m) in the mucosal end of the cell, the large nucleus (n) and the extensive tortuosity of the lateral intercellular spaces (1). The plane of the section is not perpendicular to the mucosal surface wherefore there seems to be two cell layers; in fact there is only one layer between the mucosal solution (ms) and the underlying connective tissue (c). B. (Right) At higher magnification an occasionally planar or tubular matrix (mc) is seen throughout the cytoplasm. Note also the complexity of the lateral spaces.

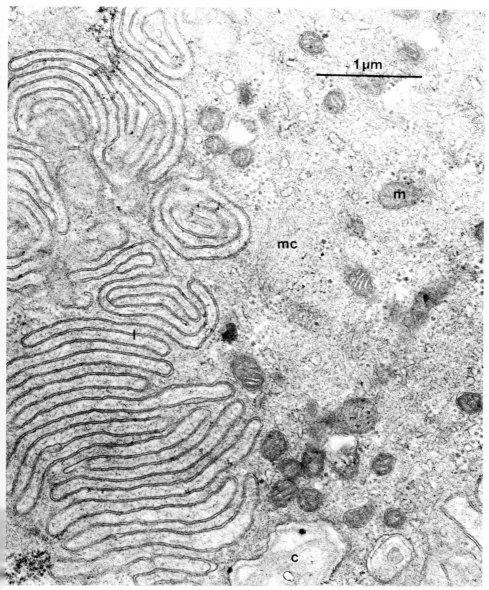

This review deals with the difficulties in constructing electrical models of these tissues; i.e. the justification of considering the serosal membrane as being flat and parallel to the mucosal membrane, and the justification of considering the cytoplasm to have the same properties as saline with—respect to diffusion of ions.

this decrease in impedance is associated solely with the change in the geometry of the lateral spaces (Frömter, 1972) the calculated length-constant of the lateral spaces is of the same order as the length of the spaces. In this case a neglect of the route of current flow across the leaky junction into the cell via the lateral serosal membrane leads to an overestimate of the impedance of the serosal membrane. An alternative explanation, that the decrease in transmural impedance is partly due to a change in the impedance of the leaky junction has also been proposed (Bindslev *et al.*, 1974).

Before it was confirmed that the "tight junctions" were leaky (Frömter and Diamond, 1972; Frömter, 1972), it was apparent that the lateral intercellular spaces constituted part of the route for ion and water flux across the epithelial cell layer of the gall-bladder. In connection with the standing gradient model (Diamond and Bossert, 1967) for coupling between the movements of water and NaCl, Tormey and Diamond (1967) suggested that NaCl was moved from the luminal solution across the mucosal membrane and cell cytoplasm, and was transported across the serosal membrane into the lateral intercellular spaces. From here NaCl was moved with the water into the blood stream *in vivo*, or into the serosal solution *in vitro*. The water was found to enter the lateral intercellular spaces as these spaces were observed, in light or electron micrographs, to distend during transport from the mucosal to the serosal solution and to collapse when transport was stopped, either by metabolic inhibitors, or by substituting Cl^- with $SO_4^=$ in the mucosal solution. From the pattern of potentials evoked by intra or extracellular applied currents (see Section II) it was later deduced (Frömter and Diamond, 1972; Frömter, 1972) that in the *Necturus* gall-bladder 96% of the transepithelial current bypassed the cells and was conducted via the leaky junctions and the lateral intercellular spaces. Measurements of unidirectional fluxes across the gall-bladder of the rabbit (Frizzell *et al.*, 1975) led to a similar conclusion for this tissue, although the uncertainty in determining the magnitude of the paracellular shunt was larger. From these measurements it was clear that the paracellular route constituted a main pathway for transmurally induced current flows.

This pathway has been studied by correlating the change in dimensions of the lateral spaces with changes in transmural electrical impedances. The geometry of the lateral intercellular space changes markedly when external osmotic or electrical gradients are imposed between the mucosal and serosal solutions. When 600 mM sucrose is added to the saline solution on the mucosal side, water flows from

TABLE I

Transmural electrical impedance as a function of transmural osmotic gr₍ ₎nts

	Transmural impedance [Ω cm^2]			
	Serosa +600 mM Sucrose	Serosa and mucosa isotonic	Mucosa +600 mM sucrose	
Necturus	110—180	200—500		Frömter (1972)
Rabbit	28	28[a]	84	Smulders et al. (1972)
Frog	89	114	154	Bindslev et al. (1974)

[a] Wright and Diamond (1968); Wright et al. (1971).

the serosal side into the mucosal solution. From the microscopical study of the anatomy of tissues fixed during this transport it is seen to be associated with a virtual disappearance of the lateral intercellular spaces as the serosal membranes of two adjacent cells now are separated by less than 200 A (Smulders et al., 1972; Bindslev et al., 1974). Conversely, when water flow is induced in the opposite direction the spaces are observed to distend, e.g. when the Necturus gall-bladder is bathed in saline with 600 mM sucrose added to the serosal solution, the dilation of the spaces in response to the mucosa to serosa induced waterflow can even be observed under a dissection microscope (Frömter, 1972).

Quantitative information about the dimensions and electrical properties of the lateral intercellular spaces can be deduced from the changes in transepithelial electrical resistance resulting from osmotically induced transmural flow of water (Table I) but in doing so three questions arise. (i) Is the impedance of the tissue during maximal dilation of the lateral spaces equal to the impedance of the leaky junction alone? (ii) Does the impedance of the leaky junction change when different external gradients are imposed? (iii) In case the impedance of the leaky junction remains constant, can the change in transmural impedance be attributed to changes in the dimensions of the lateral intercellular spaces alone, or is the route of current flow during the impedance measurement not only along the space but also across the serosal cell membranes (Fig. 2)? In other words, is the length-constant of the lateral intercellular spaces comparable to their length?

In answer to the first question it is not clear whether the lateral intercellular spaces dilate in the region immediately below the leaky junction during maximal dilatation of the spaces. The impedance

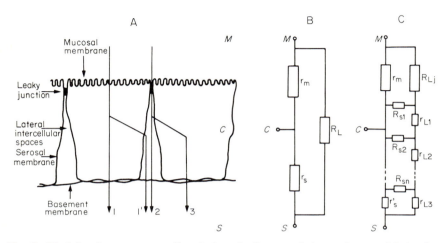

Fig. 2. Models for transmurally induced flows of ions. In model A the nomenclature as used in this study. M designates the mucosal or luminal solution, C the cell interior and S the serosal side of the tissue. The effect of the connective tissue is neglected and S thus constitutes the final compartment for transport. The various possible transport routes are indicated by numbered arrows. No. 1 shows the transcellular route in which the ions leave the cells across the extreme basal end of the cell. No. 1 is also a transcellular route but here the ions cross the serosal membrane and enter the lateral intercellular space. No. 2 is the paracellular route. Depending on the length-constant of the lateral extracellular spaces, some of the ions which enter the tissue via the leaky junctions will proceed via a transcellular route, No. 3.

The electrical model B is valid in case the length-constant of the lateral intercellular space is large compared to the actual length of the space. In this case the route 3 for current flow is unimportant, and the impedance of leaky junctions and the lateral intercellular space can be represented by one resistor only (R_L). The mucosal membrane is represented by r_m and the serosal by r_s. When the length-constant of the spaces is of the same order as the length of the cell, the impedance of the serosal membrane and the lateral spaces must be distributed as a leaky cable. The serosal membrane is described by the resistors R_{s1} to R_{sn} terminating with r_s', which represents that part of the serosal membrane which is not attached to the lateral spaces. The lateral space is described by the resistors r_{L1} to r_{Ln}, and the leaky junction itself by the Resistor R_{Lj}. If for a particular tissue the length-constant of the lateral intercellular space is comparable to the length of the space, the impedance of the serosal membrane will be overestimated if the simple model B is applied. This is discussed in detail in the text.

must therefore be the sum of the junctional impedance plus the impedance of the undilated space. Based on geometrical considera-tion alone, however, the impedance of this undilated space should be small as compared with the leaky junction. In answer to the second question Bindslev et al. (1974) calculated from electron micrographs the impedance of the lateral intercellular space alone to be 5Ω cm^2 for the frog gall-bladder when no external osmotic gradient was imposed. Therefore the drop in impedance of 25Ω cm^2 when the spaces were dilated by imposing mucosa-to-serosa water flow could not be accounted for by the dilatation of the spaces alone. These authors therefore raised the possibility that the impedance of the leaky junction also decreased during mucosa to serosa induced water flow; but they emphasized the uncertainties in calculating the impedance of the lateral intercellular spaces from electron micrographs due to lack of knowledge of the tortuosity of these spaces. In connection with the third question Frömter (1972) attributed all of the change (200Ω cm^2) in transmural impedance of the *Necturus* gall-bladder to the change in dimensions of the lateral intercellular spaces during mucosa to serosa induced waterflow. In this interpretation the spaces therefore represented an impedance of about 200Ω cm^2 when the tissue was transporting normally without externally imposed osmotic gradients. With the numerical values for other tissue parameters derived by Frömter (see Section II) we can make an estimate of the length-constant of the lateral intercellular spaces of the epithelial cell layer in *Necturus* gall-bladder.

With a cell height of 30 μm and a cell-junction length of 1150 cm per cm^2 of tissue (apparent mucosal surface area), the width of the lateral intercellular spaces during isotonic transport can be calculated from Ohm's law to be 100–300 A if there is no tortuosity of the spaces and if the specific resistivity in the spaces (R) is similar to that of the external solutions (about 100Ω cm). In order to give the same transmural impedance, a larger tortuosity (τ) allows for a larger width (a) of the lateral spaces, therefore $a = a_0 \tau$, where $a_0 = 100$–300 A. With a tortuosity of 10 the width would be 1000–3000 A. The impedance of the serosal membrane (r_s) was estimated as 2880Ω cm^2 (cm^2 of apparent area). If we further assume the tortuosity to be of a kind that linearly increases the length (l) of the lateral intercellular spaces this length can be expressed as τh where h is the height of the cell. With a cubic cell the area of the serosal membrane would be about five times larger than the apparent mucosal area if there was no tortuosity, therefore the true impedance per cm^2 of serosal membrane is $5\tau r_s$ where r_s is

$2880 \ \Omega \ cm^2$. The ratio between the length constant λ and the actual length of the lateral space (l) will then be calculated as:

$$\frac{\lambda}{l} = \frac{\lambda}{h\tau} = \frac{1}{h\tau}\sqrt{\frac{5\tau r_s a_0 \tau}{2R}} = \sqrt{\frac{5 \ 2880(100 \ \text{to} \ 300)10^{-8}}{2(30 \ 10^{-4})^2 100}} \tag{1}$$

$$= 2 \ \text{to} \ 5$$

where the formulae for λ is taken from Katz (1966).

This calculation is based on the assumption that the specific resistivity of the medium external to the spaces, which in this case is the cytoplasm is equal to that of the lateral spaces. As discussed in Section II the cytoplasmic specific resistivity could be ten times higher, in which case the factor determined by formula 1 would be larger. If the length constant is only 2—5 times larger than the actual length of the lateral intercellular spaces, it will mean that from 15 to 50% of the current that enters the tissue via the leaky junctions will cross the lateral serosal membrane into the cell and further into the serosal solution (Fig. 2a, route no. 3). Now consider this possibility of a significant length-constant in relation to two electrical models for an epithelial tissue (Fig. 2b and c). If there is current dissipation via the lateral serosal membrane the impedance of the serosal membrane and the lateral spaces must be distributed as a cable and the extended model in Fig. 2c applies. It is seen that during transmural current flow, e.g. from mucosa to serosa, more current will leave the cell across the basal serosal membrane than enters the cell across the mucosal membrane. Furthermore, due to a more positive potential in the mucosal end of the lateral space, current flow out of the cell will be confined to this basal end of the serosal membrane. The simple model in Fig. 2b is based on the assumption that the current that enters via the leaky junction leaves the tissue via the lateral intercellular spaces (Fig. 2a, route 2), so that the tissue can be modelled by three resistors only: that of the mucosal membrane (r_m), in series with that of the serosal membrane (r_s), both shunted by that of the leaky junction and the lateral space (R_L). The difference between the potential of the intracellular compartment (C) recorded by a microelectrode, and the potential of the serosal solution is a measure of the impedance of the serosal membrane. If, however, the length-constant of the lateral intercellular spaces is short, but the measured values are assigned to the simple model of Fig. 2b, the impedance of the serosal membrane will therefore be overestimated. Values reported for the *Necturus* gall-bladder suggest that the length-constant of the lateral intercellular spaces is of the same order as the length of the spaces (Frömter, 1972), but the value for the impedance of the serosal membrane ($2880 \, \Omega \, cm^2$) was derived from the simple model in Fig. 2b, and is therefore probably too large. This in turn would mean that the length constant would be even smaller than that calculated from formula 1 above.

A lower impedance of the serosal membrane could also account for the fact that when the lateral intercellular spaces are collapsed as with osmotically induced waterflow into the mucosa (Table I), surprisingly low transmural impedances are recorded. The possibility exists that, when the lateral spaces are collapsed, some of the transmural current passes via the leaky junction and the serosal membrane, thus bypassing the mucosal cell membrane and the collapsed lateral spaces.

II. THE ROUTE OF INTRAEPITHELIAL INDUCED CURRENT FLOW

A. Summary

The potential evoked in a flat epithelium by intracellularly injected currents can be used for a determination of the impedances of the cytoplasm, the cellular membranes and the paracellular shunt. The application of this method is discussed for two different situations; one in which the cytoplasmic specific resistivity is low, similar to the external physiological salt solutions, and one in which the resistivity is 10 times higher. The error involved in assuming too low a cytoplasmic resistivity is illustrated by a numerical example.

When current is injected by means of an intracellular electrode into one cell of an epithelial cell layer the resulting evoked potential will be a function of the electrical parameters of the tissue. A theoretical formulation of the connection between the tissue parameters and the evoked potential profile has been given by Eisenberg and Johnson (1970) and in a simplified situation by Shiba (1971) in the case of a two dimensional flat epithelium. Combined with the measurement of tissue parameters during transmurally induced current flows (see preceding section), this method has been used for assigning impedance values to the mucosal and serosal cellular membrane, the paracellular shunt, and the impedance of the cellular coupling in *Necturus* gall-bladder epithelium (Frömter, 1972; Reuss and Finn, 1975), toad urinary bladder (Reuss and Finn, 1974) and *Necturus* gastric mucosa (Spenney *et al.*, 1974). In this chapter I shall discuss the applicability of the method to the gall-bladder and point out some difficulties in using it with too simplified a model for the epithelium, illustrating this with a numerical example.

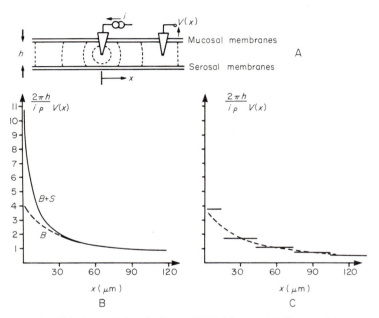

Fig. 3. A simplified model of the gall-bladder epithelium with respect to intracellularly injected-currents. In model A the tissue is equated to a continuous sheet of height h equal to the cell height, and a conductivity which is a conglomerate of the cytoplasmic resistivity and the resistance of the cellular junctions. The mucosal membranes are equated to a continuous membrane (MM) with an impedance r_m (Ω cm^2) and the serosal membranes to a flat membrane (SM) of impedance r_s (Ω cm^2). As discussed in the text, this is only strictly valid in the case where the length-constant of the lateral intercellular spaces is large.

When current i is injected between an intracellular microelectrode and a large external electrode (not shown in the Figure) the current will spread spherically in the vicinity of the tip, and cylindrically further along the plane of the epithelium eventually leaving the tissue via the mucosal and serosal membrane. The surfaces of equal potential are indicated by broken lines. This evoked potential $V(x)$ can be recorded by a second microelectrode as a function of distance x, from the current-injecting electrode.

Two different cases are discussed. In model B the evoked potential is shown when the cytoplasmic conductivity is low compared to the conductivity of the cellular junctions. Here the potential will drop sharply in the vicinity of the

Consider a current injected by means of an intracellular microelectrode into one cell of a two dimensional flat preparation, e.g. the gall-bladder epithelium bathed in saline (Fig. 3a). The current will first spread spherically in the vicinity of the tip, and then horizontally through the epithelial cell layer via the cellular junctions and out of the tissue across the mucosal and the serosal membranes and the lateral intercellular spaces. The current will evoke a potential along the epithelium. The magnitude of this evoked potential will be a decreasing function of distance from the current injecting electrode, and the shape of the potential profile will depend on the resistivities of the membranes, the cytoplasm and the external solutions together with the geometry of the cells. One can measure this evoked potential $V(x)$ (Fig. 3a) by means of a second microelectrode inserted intracellularly into the tissue at different distances (x) from the current-injecting electrode. The measured potential profile can then be compared to the profile predicted from the behaviour of a simplified model of the tissue. Those model parameters that cause the measured values to coincide with those

---

current-injecting electrode. In mathematical terms, the potential will be described by the sum of a spherical function (S) and a cylindrical function (B). For large distances away from the current-injecting electrode S vanishes (see formula 2 in the text). The curves are drawn to scale for the following numerical values: conductivity of the sheet $\rho = 3000 \ \Omega$ cm, $r_m = 3000 \ \Omega$ cm^2, $r_s = 1000 \ \Omega$ cm^2, $h = 30 \ \mu$m. The ordinate is given in dimensionless units $V(x) \ 2\pi h/i\rho$. Model C shows the case where the cytoplasmic conductivity is high compared to the conductivity of the cellular junctions. The width of the cells is assumed to be 30 μm. The centre of the current-injected cell is assumed to be at $x = 0$, and in this case the evoked potential will be constant inside each cell (solid line) and the potential drop will be at the cell junctions. The potential drop in the vicinity of the current-injecting tip will be negligible (see text), and the potential drop as a function of distance x can be fitted to the function B (broken line).

From the shape of the cylindrical part of the voltage attenuation, (model B), the values of the tissue parameters can be determined. Thus, in the case illustrated in B it will be necessary to correct for the spherical potential attenuation (S) in the vicinity of the current-injecting tip. If this is omitted errors will be involved, as the tissue parameters are now determined from that cylindrical function that happens to fit the sum of the spherical and cylindrical functions $(B + S)$. The correction will depend on how close the potential is recorded to the current-injecting electrode. If for example the shortest distance is 30 μm the argument of the function B would be underestimated by half and the length constant overestimated by x2. If the shortest distance is 6 μm, the argument of the function B would be overestimated by a factor of 2 and the length constant underestimated by a half. By comparison with formula 2, the error involved in the determination of the tissue parameters can be assessed.

predicted from the model can now be assigned to the cellular elements. Thus one of the main problems in this method is to choose a necessarily simplified model of the tissue.

One simplified representation of the tissue consists of a two-dimensional infinite sheet of thickness h (equal to the cell-height) covered by membranes, on the two sides (Fig. 3a). The analytical expression for the potential evoked in this model by the injected current has been derived by Eisenberg and Johnson (1970). In the vicinity of the current injecting tip the evoked potential can be described by a sum of two parts (Fig. 3b) a spherically-symmetric function (S) which is independent of the membrane properties and which is due to the spherical symmetry of the current flow near the tip, and a cylindrically-symmetrical function (B) which is due to the current flow horizontally through the tissue:

$$V(x) = \frac{i\rho}{2\pi h} (B + S) = \frac{i\rho}{2\pi h} \left[K_0 \left(x \sqrt{\frac{\rho (r_m + r_s)}{r_m r_s h}} \right) + S \left(\frac{x}{h} \right) \right] \quad (2)$$

$V(x)$ is the evoked potential as a function of the horizontal distance (x) of the potential recording electrode from the current injecting electrode, ρ is the specific conductivity of the sheet, h its height and r_m and r_s the impedances of the two membranes covering the two-dimensional slab; i is the injected current and K_0 is the modified Bessel function of the second kind (Abramowitz and Stegun, 1967). S is the spherical function, the numerical value of which is shown for one numerical example (see below) in Fig. 3b. Its analytical expression is given by Eisenberg and Johnson (1970). With increasing distance from the current-injecting tip the spherical function vanishes and the potential is described by the cylindrical function alone.

In order to use this membrane bound sheet and the derived expression for the evoked potential, as a model for the epithelium, some simplifying assumptions must be made. First, the serosal membranes must be equated to a planar membrane. In other words, the length-constant of the cell interior and the lateral intercellular spaces must be ignored. This is probably justifiable in respect to the cell interior, but as discussed in the preceding part, is probably not true for the lateral intercellular spaces, at least in case of *Necturus* gall-bladder. As a result relatively more of the intracellularly injected current will traverse the basal part of the serosal membrane than the apical part, and thus the measured value of the impedance of this membrane will be higher than the actual value. Secondly, the alternating sequence of cytoplasm and cellular junctions along the epithelial cell layer, which forms the horizontal path of current, must be equated to a continuous slab of constant conductivity. This can be done to a good approximation in two cases: (i) When the specific resistivity of the cytoplasm is either so low that the potential attenuation in the cytoplasm is small compared to the attenuation across the cellular junctions. (ii) When the cytoplasmic resistivity is so high that the attenuation of the potential is entirely in the cytoplasm.

The first case can be illustrated by a numerical example: When currents of the order 5×10^8 A were injected into one cell of a *Necturus* gall-bladder epithelium (Frömter, 1972), an evoked potential of 50 mV was recorded from the same cell. If the cytoplasmic specific resistivity is close to that of saline, about 100 Ω cm, this current would only cause a small potential difference in the cytoplasm itself, less than 4 mV for distances more than 1 μm away from the tip.

From this example, it is seen that when the cytoplasmic resistivity is low, the potential changes mainly at the cellular junctions. The evoked potentials within the cells are nearly constant and are the values to which the function B should be fitted (Fig. 3c). Furthermore as the potential in the current-injected cell varies less than 10% for $x > 1 \mu$m the spheric part S of the potential attenuation can be neglected, as in practice the potential sensing electrode will always be more than 1 μm away from the tip of the current electrode. Using this measured function and formula 2 it is possible to assign impedance values to the tissue elements. Values found for the *Necturus* gall-bladder were (Frömter, 1972).

$$\rho = 3000 \ \Omega \, \text{cm}, \ \frac{r_m r_s}{r_m + r_s} = 1750 \ \Omega \, \text{cm}^2.$$

By additional measurements of transepithelial current (see Part I) r_m was determined as 4470 Ω cm^2, r_s as 2880 Ω cm^2 and the paracellular shunt R_L (Fig. 2b) as 300 Ω cm^2.

It has been suggested recently (Zeuthen, 1977) that the specific resistivity of the cytoplasm in the epithelial cells of the *Necturus* gall-bladder is higher than that of the saline by a factor of up to 10. This was based on measurements with three microelectrodes inside one cell. Current was injected between one electrode and a large external electrode. The potential attenuation in the cytoplasm was assessed by the potential difference recorded between the two other electrodes, which were placed at different distances from the current injecting electrode, but close to it as compared with the dimensions of the cell. This potential difference was compared to the potential difference obtained in the saline, distances and currents being equal, to determine the resistivity of the cytoplasm. From this it seems that the potential attenuation in the current-injected cell is not negligible, and the potential as a function of horizontal distance x from the current-injecting electrode would be of the kind depicted in Fig. 3b; where the spherical part (S in formula 2) cannot be ignored. In order to determine the parameters of the tissue in this case it would be necessary to obtain the evoked potentials as a function of x and to fit them to the sum of the cylindrical function B and the spherical function S.

The error involved in assuming that the intracellular resistivity is low can be illustrated by a numerical example: assuming that one of the two membranes that covers the sheet (of height 30 μm) has an impedance of 3000 Ω cm^2, the other of 1000 Ω cm^2, and the resistivity of the sheet is 3000 Ω cm, the potential profile in the horizontal direction will then have the numerical values used in Fig. 3b, as can be calculated from the curves given by Eisenberg and Johnson (1970). From inspection of these curves and comparison with formula 2, it is seen that if the potential of the current-injected cell is recorded 30 μm away from the current-injecting electrode, and this value is taken to represent the potential of this cell cytoplasm, ρ will be assessed at half its real value (1500 Ω cm) and r_m and r_s will be assessed twice as large (6000 Ω cm^2 and 2000 Ω cm^2). Similarly if the potential recording is systematically done 6 μm from the current injecting electrode ρ will be over-estimated by a factor of two and the impedances r_m and r_s estimated about eight times too small. The error involved in each case will depend on how much weight the potentials recorded at small x are given relative to the points recorded for large x.

These considerations involving a higher intracellular resistivity than expected, could account partly for the fact that the reported values of tissue parameters derived by this method have large standard variations. Thus the individual values in the studies on *Necturus* gall-bladder (Frömter, 1972; Reuss and Finn, 1975), *Necturus* gastric mucosa (Spenney *et al.*, 1974), vary up to ten-fold. In some of the experiments on *Necturus* gastric mucosa the method yielded negative values for the paracellular shunt conductance.

III. ROUTES OF METABOLICALLY DEPENDENT
ION TRANSPORT

A. Summary

A model for the transporting gall-bladder has to explain the isotonic transport of NaCl concurrently with a low transmural electrical potential difference. A simplified model in which the lateral intercellular spaces and the serosal solution have the same activities and potentials is discussed. This model excludes the possibility of an electrical coupling between an actively transported ion (e.g. Na^+) and a passively transported ion (e.g. Cl^-). However, in a detailed model where the lateral spaces have a cable-like structure such a coupling is possible. Some of the consequences of recirculation of the actively transported ion via the leaky junctions, back into the mucosal solution, are discussed. Finally some criteria for simultaneous passive solute and solvent flow across the mucosal membrane are discussed.

It is well established that gall-bladder transports NaCl from the luminal solution into the serosal solution, and that H_2O is transported concurrently as a result of the salt transport (see review by Diamond, 1968). As most early work was done on the rabbit or fish gall-bladder, the transmural potential difference was thought to be practically zero when transport was between identical physiological salines. However, recent investigations with other species have shown that the serosal potential is often a few mV positive (Table II). The coupling of water to the ion transport is dealt with in other places in this volume, and I shall only discuss the possible

TABLE II

Transmural electrical potential differences across vertebrate gall-bladders (Serosa positive)

Fish (roach)	<1 mV	Diamond (1962b)
Rabbit	<2 mV	Wheeler (1963)
Necturus	2 mV	Frömter (1972)
Frog	3 mV	Bindslev et al. (1974)
Goose	4 mV	Gelarden and Rose (1974)
Monkey	3 mV ⎫	
Dog	0 mV ⎬	Rose et al. (1973)
Man	8 mV ⎭	

routes of Na^+ and Cl^- transport, their interactions, and the interconnection between the transport routes and the transmural potential.

B. Transport across the Serosal Membrane

The site of metabolically-mediated transport is generally believed to be at the serosal membrane, as ATPase is found mainly to be located there (Kaye *et al.*, 1966; Van Os and Slegers, 1971). The first idea put forward to explain the transport was a neutral NaCl-pump (Diamond, 1962b) which was located at the serosal membrane and explained the absence of a significant transmural potential difference. This explanation is still a possibility, although the recent finding of a significant transmural potential in some species has revived the discussion whether only one of the ions Na^+ or Cl^- is actively transported and that the other follows by an electrical coupling.

Thus one of the questions is: if say, only Na^+ is actively transported from the cell interior across the serosal membrane into the lateral intercellular space, how is Cl^- coupled to this transport? Does Cl^- move via the transcellular route or does it move through the leaky junctions in response to a positive potential, set-up in the lateral intercellular spaces by the active Na^+-transport (Fig. 4a)? A similar question can be posed if Cl^- is considered to be the only ion transported actively. This question has been discussed in connection with a simplified electrical model (Fig. 4b) for the rabbit gall-bladder by Frizzell *et al.* (1975). In this model the battery E_m at the mucosal membrane represents the membrane potential resulting from a diffusion potential together with an electrical pump. This battery is in series with a resistance r_{m1} due to ion movements in this channel. The ability of ions to bypass the channel is represented by a resistor r_{m2} in parallel with the battery. The serosal membrane and the paracellular shunt are likewise represented. The battery in the paracellular shunt represents a diffusion potential caused by a gradient of ions in this space. As 95% of the total transmural conductance is constituted by the paracellular shunt in *Necturus* and rabbit, the transmural potential difference can be expressed as:

$$V_{MS} = 0 \cdot 05 \left(E_M \frac{r_{m2}}{r_{m1} + r_{m2}} - E_S \frac{r_{s2}}{r_{s1} + r_{s2}} \right) \qquad (3)$$

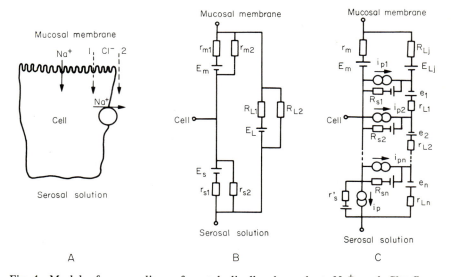

Fig. 4. Models for coupling of metabolically dependent Na^+ and Cl^--flows across the gall-bladder epithelium. A. In some models Na^+ alone is considered to be actively transported from the mucosal solution (M) into the cell (C) and across the serosal membrane into the serosal solution (S). In this case which route does Cl^- follow, the transcellular route, indicated by 1, or the paracellular route, indicated by 2? This can be discussed in terms of a simple model B, where the mucosal membrane is represented by the resistors r_{m1} and r_{m2}, and the mucosal membrane potential by the battery E_m. Similarly the serosal membrane is given by r_{s1}, r_{s2} and E_s and the paracellular pathway by R_{L1}, R_{L2} and E_L. As shown in the text (formula 3) the transmural potential in this model is not large enough to allow the observed Cl^--fluxes to be solely paracellular, (a, route 2). When discussing whether Cl^- can move by the transcellular route $(A, 1)$ it is necessary to consider the more detailed model C. This model accounts for the distributed nature of the serosal membrane and lateral intercellular spaces. The mucosal membrane is represented by the resistor r_m and the battery E_m. The serosal membrane is given by the resistors R_{si} and the batteries. The active transport capability of this membrane is represented by the generators i_{pi}. These can represent an Na^+-pump, a Cl^--pump or a NaCl-pump. The leaky junction is given by the resistor R_{LJ} and the battery E_{LJ}. The lateral space is represented by the chain of resistors r_{Li} and any diffusion potential in this space is given by the batteries e_i. When i_{pi} represents Na^+-pumps, a gradient of electrochemical potential will exist in the lateral intercellular spaces. Cl^- will move across the serosal membrane in response to this potential gradient. Thus the potential measured between M and S is different from the intraepithelial potential that moves Cl^-.

This equation expresses the potential as a function of the tissue parameters. The expression in brackets represents that potential which would exist across the tissue in the case of no paracellular shunt. In other words the tissue is self short-circuited and produces a current flow in the lateral intercellular spaces. The implications of such a recirculation in terms of the energy consumption will be discussed briefly at the end of this section. In the simplified model of Fig. 4b it is thus the attentuated potential V_{ms} which is available to move Cl^- into the serosal solution. As the paracellular conductance is about 20 times larger than the transcellular conductance the Cl^- ions should move through the paracellular route. Now with a transmural potential of max 2 mV (serosa positive) and a paracellular shunt conductance of about 40 mho/cm^2 for the rabbit (Frizzell *et al.*, 1975), and 2 mV and 3 mho/cm^2 for *Necturus* (Frömter, 1972), these transmural potentials can only drive the observed Cl^- flux of 14 μmol/hour cm^2 (equivalent to a current of 0.4 mA/cm^2) in the rabbit, and 2 μmol/hour cm^2 (or 0.06 mA/cm^2) in *Necturus*, provided the paracellular conductance is highly Cl^--selective. It is, however, established that the leaky junctions are cation-selective (see review by Moreno and Diamond, 1974) with a $Na^+ : Cl^-$ permeability ratio of up to 10 : 1. Therefore the observed Cl^--fluxes probably cannot be accounted for by passive electrical coupling via the paracellular pathway. If Cl^- is the actively transported ion, the observed transmural potential difference has the wrong sign (serosa positive) for a passively induced flux of Na^+. The above type of argument has been put forward by Frizzell *et al.* (1975) and Cremaschi and Henin (1975) against a simple passive electrical coupling between the Na^+- and Cl^--fluxes and by Diamond (1962b) in support of the notion of a neutral active transport of NaCl.

Yet another discouraging aspect of the model where the cation and anion follow different routes across the tissue, is the potentials that would be generated by the transcellular current flows. For example the flux of Na^+ (14 μmol/cm^2 hour) would be equivalent to a current of 0.4 mA/cm^2. This would cause a potential drop of 400 mV across a membrane with a resistance of 1000 Ω cm^2, in disagreement with the observed potentials across the mucosal membrane.

With our present knowledge it thus seems that both Na^+ and Cl^- must mainly follow the transcellular route. In order to discuss whether one of these ions, Na^+ or Cl^-, is transported passively by this route in response to the active transport of the other, it is necessary to consider a more detailed model than that in Fig. 4b.

This simple model implies that the serosal compartment into which transport occurs across the serosal membrane is of uniform concentration, that is to say that it is well-stirred. A more detailed model is depicted in Fig. 4c where the paracellular shunt is represented by an impedance for the leaky junction R_{Lj} and a distributed impedance for the lateral intercellular spaces (r_{Li}). These impedances are in series with batteries which describe the diffusion potentials arising from the unequal mobilities and distribution of ions. The mucosal membrane is represented by a membrane impedance r_m in series with a battery E_m describing the potential difference across this membrane. The serosal membrane connects with the lateral intercellular spaces via a distributed impedance (R_{Si}) combined with a battery. The serosal membrane's ability to actively pump Na^+ or Cl^- is symbolized by the generator i_{pi}.

Without deriving an analytical expression for the behaviour of this model in terms of coupling between ion-flows it is possible to make some qualitative statements. In the case where Na^+ is the only ion which is actively transported into the lateral intercellular space, an electrochemical potential gradient will exist in the lateral spaces. This gradient will depend on whether the pumping rate (i_{pi}) is uniform over the serosal membrane, and on the geometry of the lateral intercellular spaces. Thus the impedance (r_{Li}) will decrease towards the serosal side where the spaces are wider. Finally, the potential gradient will depend on the diffusional potentials (e_i). It is in response to this potential gradient that Cl^- moves across the serosal membrane. Thus in this picture, Cl^- is coupled to the active Na^+ transport in a manner reminiscent of the stading gradient theory for coupling of solvent to solute flow. In the case where the flux of Cl^- across the serosal membrane is smaller than the flux of Na^+ the resulting positive current will tend to create a positive potential in the lateral intercellular spaces. This potential would in turn cause a flux of Na^+ back into the lumen via the leaky junctions, and this in turn would ensure equal transmural fluxes of Na^+ and Cl^-. Thus in this extended model (Fig. 4c) the positive potential causes passive electrical coupling inside the epithelium (Keynes, 1969) and is different to that measured by external electrodes. The small positive electrical potential found in the serosal solution of some gall-bladders (Table II) is no longer responsible for moving Cl^- as in the simple model (Fig. 4a). Similar considerations can be put forward if one considers the i_{pi}'s in Fig. 4c to be metabolically dependent Cl-pumps and Na^+ to be passively and electrically coupled to this transport. Finally if i_{pi} is a neutral pump, causing a metabolically dependent neutral flux of NaCl, one arrives at the concept originally suggested by Diamond (1962b).

The models discussed above utilizing an intraepithelial electrical coupling between an actively transported Na^+ and a passively transported Cl^- has recently been questioned by Frizzell *et al.* (1975) on the basis of measurements of unidirectional fluxes. When Cl^- in the mucosal solution of the rabbit gall-bladder was replaced by the impermeable ion $SO_4^=$ transmural net-transport of Na^+ ceased. This was found to be due to a decrease in the unidirectional flux of Na^+ from the mucosal to serosal solution only, the unidirectional flux of Na^+ from serosa to mucosa remaining unchanged. Thus when $SO_4^=$ replaced Cl^-, the two unidirectional fluxes became equal. Now if the Na^+ transport across the mucosal and serosal membrane remained constant after the anion-substitution, one would expect the intraepithelial positive potential to be larger, as it is now uncompensated by any anion flux. This again would cause a back flux of Na^+ into the lumen. As a result the unidirectional flux from serosa to mucosa should increase when Cl^- was replaced by $SO_4^=$ in disagreement with the observed decrease in the flux from mucosa to serosa only. It was therefore concluded that intraepithelial electrical coupling was an unlikely explanation for the coupling between Na^+ and Cl^-. These considerations, however, are based on a three compartment model of the tissue consisting of the mucosal compartment, the serosal compartment and the cellular compartment; the lateral intercellular space is considered to be part of the serosal compartment. This model is therefore too simple to be used for arguing against intraepithelial coupling in an unstirred lateral intercellular space which ought to be treated as a fourth compartment, for example, if recirculation of Na^+ occurs (see below); Na^+ from the lumen which enters the cells via the mucosal membrane, would return via the serosal membrane, the mucosal end of the lateral spaces and the leaky junctions. This would necessitate an interpretation of the measured unidirectional fluxes in the framework of a four-compartment model.

C. Recirculation

The possibility exists that some of the actively transported ions return via the lateral intercellular spaces and the leaky junctions into the lumen. At least in the gall-bladders of the frog and *Necturus* where the impedances of the lateral intercellular spaces are similar to the impedance of the leaky junctions between one-third and a half of the ions transported into the mucosal end of the lateral intercellular spaces could return to the mucosal solution. With recirculation the flux of actively transported ions into the tissue would be larger than the net flux across the tissue. This would allow for coupling-mechanisms to be operative in which the ratio of passively transported ions to actively transported ions is smaller than with no recirculation. For example the active flux of Na^+ into the tissue needed to bring about the observed net Cl^--flux via an interepithelial potential could be larger than the observed net flux of Na^+. Recirculation would also play a part in solute-solvent coupling: the flux of NaCl into the tissue needed to bring about the flux of H_2O could be larger than the observed net flux of NaCl.*

It is a usually neglected fact that the shunting of a transepithelial potential difference, or recirculation, involves an additional energy dissipation. As discussed above, it might be useful energy but nevertheless its contribution to the total oxygen consumption has to be considered. With recirculation the ratio of total pumped Na^+ to consumed O_2 would be larger than the ratio of the net transported Na^+ to consumed O_2. This again would have bearing upon whether or not ATP was responsible for Na^+ transport in these tissues. The only study on the oxygen consumption of the gall-bladder (Martin and Diamond, 1967), does not exclude the possibility of recirculation. The oxygen consumption of the rabbit gall-bladder fell by about 9% when net transport was abolished by substituting mucosal Cl^- by $SO_4^=$ but by 45% when the total transport was stopped by poisoning the tissue with ouabain. Thus these data do not exclude a recirculation ratio of five, i.e. a five times higher flux of Na^+ into the tissue than flux of Na^+ across the tissue. The data on ATPase activity in this tissue (Van Os and Slegers, 1971) are compatible with a recirculation ratio of about two.

---

* In the original standing gradient model for isotonic water-transport the junctions were considered to be tight (Diamond and Bossert, 1967). The NaCl which was transported into the lateral intercellular spaces had therefore, two functions: (i). As it was present in these spaces as a hypertonic solution it had to cause the water to flow into these spaces. (ii). It also constituted the NaCl which was to be transported into the serosal solution. Within given limits of the geometry of the spaces, the distribution of pumps on the serosal membrane and the diffusion constant in the lateral spaces the absorption should be isotonic with the mucosal solution. It has recently been calculated (Hill, 1975a) that with physiologically occurring dimensions for the spaces and magnitudes of the water permeabilities, absorption should be hypertonic in case of the junction being tight and the pump rate uniform over the serosal surface. However, if NaCl recirculates it would contribute to the osmotic forces without being transported into the serosal solution. This may be important in isotonic transport within the limits of existing tissue geometries and water permeabilities.

D. Transport across the Mucosal Membrane

As discussed above, it seems that even if the junctions are leaky, the major route for NaCl into the tissue is across the mucosal membrane and the water transport is probably also via the transcellular route as the water permeabilities of the leaky junctions are too low to account for the observed flows (Wright *et al.*, 1972 and Hill, 1975b). However, the observed water permeabilities are subject to some uncertainty as they are influenced by unstirred layers (see Wright *et al.*, 1972).

The transport across mucosal membrane differs from that of the serosal membrane in that no metabolically dependent pump has yet been demonstrated in this membrane, e.g. ouabain is only active in stopping transport when applied to the serosal membrane. Thus transport across the mucosal membrane must be a result of a difference in ionic environment on its two sides. In order for energy to be available for the simultaneous transport of Na^+ and Cl^- across this membrane the following inequality must be fulfilled.

$$RT \ln \frac{[Na_o^+]}{[Na_i^+]} + RT \ln \frac{[Cl_o^-]}{[Cl_i^-]} > 0, \quad RT = 616 \text{ cal/M} \qquad (4)$$

or

$$[Na_o^+][Cl_o^-] > [Na_i^+][Cl_i^-] \qquad (5)$$

where the subscript *o* refers to the outside or mucosal activities and *i* to the intracellular activities. This describes the necessary condition for the simultaneous movement of Na^+ and Cl^- into the cell, where energy is invested in the transfer of the two ions. The intracellular ion concentrations have been determined for the rabbit gall-bladder as $Na^+ = 66$, $Cl^- = 84$ and $K^+ = 85$ expressed as mM per litre of intracellular water (Frizzell *et al.*, 1975). Na^+ and K^+ were determined from flame-photometry and Cl^- from uptake of radioactive Cl_{36}^-. Similar values were reported by Cremaschi *et al.* (1974). The values are close to those determined in the fish gall-bladder (roach) by Diamond (1962a) as $Na^+ = 57$, $K^+ = 81$ and $Cl = 91$. Thus with 140 mM NaCl in the mucosal solution relation (5) is fulfilled and a passive influx of NaCl is energetically possible. With an intracellular electrical potentials of -45 mV (Frizzell *et al.*, 1975) and -50 mV (Henin and Cremaschi, 1974) it can be seen that Cl^- moves against its electrochemical gradient and Na^+ moves down a steep electrochemical gradient. Based on such considerations Henin and Cremaschi (1974) and Frizzell *et al.* (1975) have proposed the existence of a carrier mediated neutral flux of NaCl across the mucosal membrane. It is, however, well established that the gall-bladder continues to transport NaCl and H_2O at roughly the same rate even if mucosal NaCl is partly replaced isotonically by sucrose (Diamond, 1962a; Dietchy, 1964): the transport rate is: reduced by less than 10% when 50% of the NaCl is replaced by sucrose and by 50% when 90% of the NaCl is replaced. As no studies are available on the change in intracellular activities when sucrose replaces NaCl in the lumen, it remains to be seen whether the passive influx mechanism described above alone can account for a passive movement of NaCl across the mucosal membrane at low external NaCl concentrations.

It has been reported recently (Zeuthen, 1976) that the cells of *Necturus* gall-bladder bathed in Na-saline have intracellular gradients of ion activities and electrical potentials (Table III). By means of

TABLE III

Intracellular gradients of activities and electrical potentials in *Necturus* gall-bladder bathed in Na-saline.[a]

	Activity (mM)			Electrical potential
	Na^+	K^+	Cl	(mV)
Mucosal end of cell	46	64	100	−26
Serosal end of cell	13	182	23	−56

[a] (115 Na^+, 3 K^+, 3 Ca^{++}, 122 Cl^-, 2 HCO_3^-, mM). Zeuthen (1976).

ion-selective microelectrodes it was found that the K^+-activity increased and that of Cl^- and Na^+ decreased towards the serosal end of the cell. This means that the values reported above for the intracellular environment, based on measurements of the whole tissue, must be considered as average values; if the existence of intracellular gradients is confirmed by other methods* the possibility remains that the intracellular environment in the region of the microvilli is different to that given by the average activities of the whole cytoplasm.

* Gradients of similar magnitude have been found in the salivary glands of adult blow-fly (Gupta *et al.*, Chapter 4) and in rabbit ileum (Gupta, Naftalin and Hall, in preparation) by means of the electron-probe X-ray analysis.

ACKNOWLEDGEMENTS

I would like to thank Miss J. Hannant for typing the manuscript; Dr
A. E. Hill, Professor R. D. Keynes and Professor S. G. Schultz for
discussions on the contents. The study was supported by the Medical
Research Council, England.

REFERENCES

Abramowitz, M. and Stegun, J. A. (1967). "Handbook of Mathematical
    functions." National Bureau of Standards, 6th Printing, Washington.
Bindslev, N., Tormey, J. McD. and Wright, E. M. (1974). *J. Membr. Biol.* **19**,
    357—380.
Cremaschi, D. and Henin, S. (1975). *Pflügers Arch.* **361**, 33—41.
Cremaschi, D., Henin, S. and Ferroni, A. (1974). *In* "Bioelectrochem. Bioener-
    getic." Vol. 1. 208—216.
Curran, P. F. (1960). *J. gen. Physiol.* **43**, 1137—1148.
Diamond, J. M. (1962a). *J. Physiol. Lond.* **161**, 442—473.
Diamond, J. M. (1962b). *J. Physiol. Lond.* **161**, 474—502.
Diamond, J. M. (1968). *In* "Handbook of physiology, Alimentary Canal."
    Vol. III, pp. 2451—2482. American Physiological Society, Washington D.C.
Diamond, J. M. (1974). *Fedn Proc. Fedn. Am. Socs exp. Biol.* **33**, 2220—2224.
Diamond, J. M. and Bossert, W. H. (1967). *J. gen. Physiol.* **50**, 2061—2083.
Dietschy, J. M. (1964). *Gastroenterology* **47**, 395—408.
Eisenberg, R. S. and Johnson, E. A. (1970). *Prog. Biophys. mol. Biol.* **20**, 1—65.
Frizzell, R. A., Dugas, M. C. and Schultz, S. G. (1975). *J. gen. Physiol.* **65**,
    769—795.
Frömter, E. (1972). *J. Membr. Biol.* **8**, 259—301.
Frömter, E. and Diamond, J. M. (1972). *Nature, New Biology* **235**, 9.
Gelarden, R. T. and Rose, R. C. (1974). *J. Membr. Biol.* **19**, 37—54.
Henin, S. and Cremaschi, D. (1974). *Pflügers Arch.* **355**, 125—139.
Hill, A. E. (1975a). *Proc. R. Soc. Lond. B.* **190**, 99—114.
Hill, A. E. (1975b). *Proc. R. Soc. B.* **191**, 537—547.
Katz, B. (1966). "Nerve, muscle and synapse." McGraw-Hill Book Co., New
    York.
Kaye, G. I., Wheeler, H. O., Whitlock, R. T. and Lane, N. (1966). *J. Cell. Biol.*
    30, 237—268.
Keynes, R. D. (1969). *Quart. Rev. Biophys.* **2**, 177—290.
Martin, W. M. and Diamond, J. M. (1967). *J. gen. Physiol.* **50**, 295—315.
Moreno, J. H. and Diamond, J. M. (1974). *In* "Membranes — A Series of
    Advances." (Ed. G. Eisenmann), Vol. 3. M. Dekker Inc, New York.
Reuss, L. and Finn, A. L. (1974). *J. gen. Physiol.* **64**, 1—25.
Reuss, L. and Finn, A. L. (1975). *J. Membr. Biol.* **25**, 115—139.

Rose, R. C., Gelarden, R. T. and Nahrwald, D. L. (1973). *Am. J. Physiol.* **224**, 1320—1326.

Shiba, H. (1971). *J. theoret. Biol.* **30**, 59—64.

Smulders, A. P., Tormey, J. M. and Wright, E. M. (1972). *J. Membr. Biol.* **7**, 164—197.

Spenney, J. G., Shoemaker, R. L. and Sachs, G. (1974). *J. Membr. Biol.* **19**, 105—128.

Staehelin, L. A. (1975). *Int. Rev. Cytol.* **39**, 191—283.

Tormey, J. McD. and Diamond, J. M. (1967). *J. Physiol. Lond.* **50**, 2031—2059.

Van Os, C. H. and Slegers, J. F. G. (1971). *Biochim. biophys. Acta* **241**, 89—96.

Wheeler, H. O. (1963). *Am. J. Physiol.* **205**, 427—438.

Wright, E. M. and Diamond, J. M. (1968). *Biochim. biophys. Acta* **163**, 57—?

Wright, E. M., Barry, P. H. and Diamond, J. M. (1971). *J. Membr. Biol.* **4**, 331—?

Wright, E. M., Smulders, A. P. and Tormey, J. McD. (1972). *J. Membr. Biol.* **7**, 198—219.

Zeuthen, T. (1976). *J. Physiol. Lond.* **256**, 32P.

Zeuthen, T. (1977). *J. Membr. Biol.* **33**, 281—309.

V. Fluid Transport in Epithelia

21. Insect Malpighian Tubules

S. H. P. Maddrell

Department of Zoology, University of Cambridge, Cambridge, England

I. INTRODUCTION

The Malpighian tubules of insects are elongate tubular structures which lie in the haemocoel (blood space) of the abdomen. One end of each tubule is closed and the other end leads into the alimentary canal at the junction of the midgut with the anterior part of the hindgut (Fig. 1). The main role of the Malpighian tubules in excretion is to deliver to the hindgut a flow of fluid containing many of the haemolymph constituents at concentrations in proportion to their concentrations in the haemolymph. The hindgut reabsorbs those constituents required by the insect and rejects the others; in this way the composition and volume of the haemolymph is kept relatively

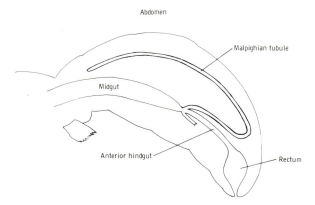

Fig. 1. Diagram to show the abdominal organs in insects that are of major importance in excretion.

constant or is adjusted to meet the needs of the insect. Different species of insects have as few as two or as many as 150 tubules; however, in insects with many tubules, they are usually short, so that the total surface area of the tubules in relation to the volume of the haemolymph is probably roughly constant. It was after the appearance of Ramsay's papers on Malpighian tubules (Ramsay 1953, 1954, 1955, 1956, 1958) that there first emerged a clear picture of how the tubules operate. Broadly this was as follows. The tubules secrete fluid iso-osmotic to haemolymph by a process dependent on and perhaps driven by active transport of potassium into the tubule lumen. Because the tubule wall is permeable to compounds of relatively low mol. wt such as amino acids and sugars, diffusion leads to the appearance in the secreted fluid of substances from the haemolymph at concentrations more or less inversely dependent on their molecular size. Phospate ions appear in the secreted fluid at concentrations above that of the bathing medium.

It will be the main aim of this article to describe the more important ways in which recent research has added to and amplified this picture.

II. MECHANISM OF FLUID SECRETION BY MALPIGHIAN TUBULES

A. Ion Transport

As discussed on p. 543, water movements across the walls of Malpighian tubules seem to be consequent upon ion movements. We shall therefore first consider the proposed mechanism for ion transport.

The lumen of a tubule is typically at a potential about 30 mV positive to the bathing solution, so the movement of potassium ions into it is clearly thermodynamically uphill. The rate of fluid secretion in nearly all the cases examined depends on the potassium concentration in the bathing saline and, in the absence of potassium, the rate is usually less than 10% of the maximum. These facts suggest that potassium transport is active and is important to fluid transport. Recent intracellular measurements of potential (unpublished results of Prince and Maddrell) show that the cell interior is negative to the bathing solution by up to 50 mV with a positive step of up to 80 mV on the luminal side. It seems probable therefore that a potassium pump exists on the apical cell membrane. From the rapid depolarization of the basal membrane that follows an increase in the potassium concentration of the bathing solution, it seems that the basal membrane is essentially potassium selective.

Given these facts it is possible, with some assumptions, to produce a relatively simple model to account for the ionic movements which underlie fluid secretion by insect Malpighian tubules. Evidence is accumulating to suggest that the "potassium pump" on the apical membrane may in fact have a higher affinity for sodium ions than for potassium ions (p. 550). Given the higher potassium permeability of the basal membrane the action of such a pump will be to deplete the cell of sodium ions, after which it will mainly pump the potassium ions which will enter the cell to replace those which are pumped out across the luminal face. To understand why ions should enter the cell across the basal membrane, water movements have to be taken into account. As will be discussed, ion transport out across the apical membrane carries water molecules in the same direction. The cell thus tends to shrink. Provided it has some elastic resistance to this process and/or contains osmotically active material, such as protein, not able to leave the cell, water will enter across the basal membrane, This leads to a reduction in the concentration of ions in the cell so that they tend also to enter passively. The extensive area of the basal membrane acts to facilitate these processes. The model is shown in diagrammatic form in Fig. 2. Potassium ions are pumped across the apical membrane and cross the basal membrane passively. Chloride ions cross the tubule wall passively, entering the cell from the bathing solution where their activity is sufficiently greater than that in the cell to overcome the adverse potential gradient across the basal membrane. They then follow potassium ions out of the cell on the luminal side where the potential gradient is more steeply downhill than the activity gradient is uphill.

B. Fluid Transport

The following evidence suggests that water movements are a secondary consequence of ion movements. The fluid secreted by nearly all Malpighian tubules is marginally but consistently hyperosmotic to the bathing fluid over a wide range of osmotic concentrations of the bathing solution (Maddrell, 1971). The rates of fluid flow produced by Malpighian tubules are in a fairly close inverse relationship to the osmotic concentration of the bathing solution (Maddrell, 1971). In other words, the rate of solute movement is approximately constant

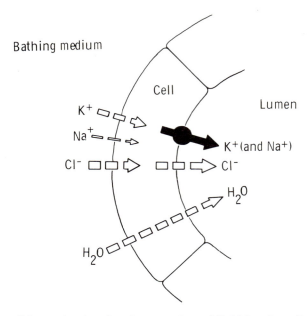

Fig. 2. A possible mechanism for the secretion of fluid by the cells of a typical Malpighian tubule. Broken lines indicate passive processes and the solid line the active transport of cations.

but water movements change so that the fluid produced is slightly hyperosmotic. Exactly how solute movements give rise to water movements is not yet clear. Ion transport across the apical microvilli on the luminal side is likely to lead to locally raised osmotic concentrations in the spaces between the microvilli. It has been suggested that such osmotic concentration differences could act to draw water through the cell membrane concerned (Diamond and Bossert, 1967; Maddrell, 1971). This interpretation faces difficulties in Malpighian tubules because the infoldings and microvilli are shorter than in such fluid transporting tissues as the gall-bladder and the proximal tubules of vertebrate kidneys for which the model was originally developed. Recently the theoretical basis for this model has come under criticism (Hill, 1975a and Chapter 7); it seems that the osmotic permeabilities of the cell membranes required to ensure that osmotic concentration of the fluid produced is close to that of the bathing solution are so high as to be virtually impossible. Instead Hill (1975b) suggests that ion movements might entrain water movements by electro-osmosis. How this would work say for apical microvilli of Malpighian tubules would be that the action of the electrogenic cation pump would produce an electrical potential difference across the membrane. This gradient would draw chloride ions out from the cell through the membrane. In crossing the membrane the chloride ions would frictionally interact with water molecules and cause them also to move out of the cell. This mechanism relies on the maintenance of a potential gradient across this cell membrane so that it would be important that the apical wall should be so arranged that it is not bathed by fluids other than its own secretion. Folding of the membranes as in the microvilli would serve such a purpose and would also allow an effectively higher density of pump sites because of the increase in membrane area. It is important to point out that the two models are not exclusive, as water movements might partly result from a osmotic gradient and partly from electro-osmotic coupling with passive ion movements. As far as Malpighian tubules go the situation is somewhat more complex than has yet been analysed in that it is not only the extracellular spaces which are long and narrow; the cytoplasm close to the apical and basal membranes is also arranged as long thin channels or sheets. It has been argued that this could act to promote osmotic coupling between ion and water flows (Maddrell, 1971), though its effectiveness in this needs numerical analysis. A further feature whose importance needs evaluation is that there is in the rather narrow extracellular spaces of Malpighian tubules, surface material adhering to the cell membrane. This substance might act to slow ion movements relative to water movements which would again allow a closer coupling of ion and water flow. It is also worth mentioning that the cell membranes at the innermost ends of the basal infoldings and the apical microvilli are sharply curved. It is known that membranes with a small radius of curvature are more permeable to water (for a discussion of this point see Oschman *et al.*, 1974), so that the osmotic permeability of the steeply curved parts of the basal and apical membrane of Malpighian tubules may be higher than might otherwise be predicted. These mechanisms do not exhaust the possibilities; such processes as ion-recycling may well play a significant role. Although rather unsatisfactory, until the picture becomes clearer, it will be prudent to say no more than that in Malpighian tubules ion transport is so well coupled to water movements that the fluid they produce is virtually iso-osmotic. From the design point of view this makes sense for the main function of the tubules is to produce a flow of fluid into which materials from the haemolymph can diffuse or be transported.

III. FAST SODIUM TRANSPORT BY TUBULES OF BLOOD-SUCKING INSECTS

As discussed above the Malpighian tubules of most insects secrete potassium-rich fluid at a rate which is sensitive to the potassium concentration of the medium and they will only produce fluid very slowly in K-poor solutions. Such behaviour would be of little use to blood-sucking insects after a meal when they need to rid themselves rapidly of much of the sodium rich plasma from the meal so as to concentrate the protein-rich blood cells prior to their digestion. In such insects the concentration of potassium in the haemolymph is only about 5—10 mM, so that the fast formation of K-rich fluid by the tubules would make it necessary to reabsorb K at high speed if the haemolymph is not soon to lose most of its content of potassium. Further, if the rate of fluid secretion were as sensitive to K levels in these insects as in others, they would not easily be able to achieve the high rates of fluid elimination observed. The hemipteran blood sucker, *Rhodnius prolixus*, has Malpighian tubules which, under hormonal stimulation, will rapidly secrete fluid in media of low or even zero potassium concentration (Maddrell, 1969). Oddly enough, however, the upper, secretory lengths of the tubules secrete a fluid many times richer in potassium than the bathing fluid when this contains less than 20 mM K (Fig. 3). Most of the potassium

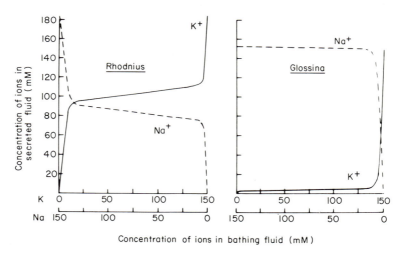

Fig. 3. Secretion of Na and K by Malpighian tubules of *Rhodnius* and *Glossina* in salines of differing composition (data from Maddrell, 1969; Gee, 1976).

(chloride) has to be retrieved during the passage of fluid through the lower lengths of the tubules (Maddrell and Phillips, 1975b). Malpighian tubules from the tsetse fly, *Glossina morsitans*, can secrete fluid at a high rate in solutions containing as little as 3 mM K (Gee, 1976), but what sets them apart from other tubules is that, under most conditions, sodium almost entirely replaces potassium as the transported cation (Fig. 3). In addition, the rate of fluid secretion markedly depends on the bathing sodium concentration rather as fluid secretion by tubules of most other insects depends on the bathing potassium concentration. In some recent experiments the behaviour of Malpighian tubules from adult *Aedes taeniorhynchus*, mosquitoes whose females suck blood, has been compared with those from larvae of the same insect. The results (Fig. 4) show that the fluid produced by the adult tubules is markedly richer in sodium than is fluid secreted by tubules from the larvae. Fluid secreted by tubules from adult females is significantly richer in sodium than is the fluid from tubules of males and taking into account the fact that the larger tubules of females secrete fluid about three times faster

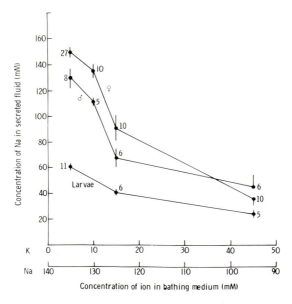

Fig. 4. Secretion of Na by Malpighian tubules from larvae and adults of *Aedes taeniorhynchus* in salines of differing composition. The vertical lines attached to the points represent ± the standard error of the mean; the adjacent figures indicate the number of measurements on which the mean value is based.

than those of males it is clear that they can excrete sodium very
much more rapidly than males. These results show that *in vivo* the
Malpighian tubules of adult mosquitoes probably secrete fluid which
is predominantly sodium based whereas those of the larvae are more
like the tubules of most other insects in secreting a potassium-rich
fluid.

A seemingly inappropriate feature of the operation of Malpighian
tubules of female mosquitoes is that the rate of fluid secretion is
potassium sensitive (Fig. 5) so that, under the conditions found in
the haemolymph (5—10 mM potassium), fluid secretion would be
considerably slower than its maximum rate. The very large size of the
tubules in female mosquitoes compensates for this to some extent.

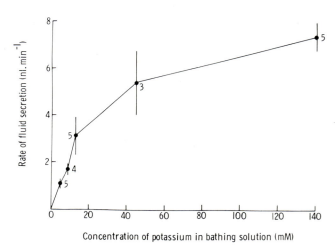

Fig. 5. The dependence of the rate of fluid secretion by Malpighian tubules
from adult female *Aedes taeniorhynchus* on the potassium concentration of the
bathing medium. The vertical lines attached to the points represent ± the
standard error of the mean; the adjacent figures indicate the number of
measurements on which the mean value is based.

IV. A POSSIBLE COMMON CATION PUMP

That the Malpighian tubules of some insects can secrete sodium at
high rates, while those of most others cannot, at first suggests that no
one model could embrace both types of behaviour. However an
ingenious and simple way of explaining the results has been suggested

to me by Dr J. L. Wood. The proposal is that Malpighian tubules have an electrogenic cation pump on the membrane facing the lumen and that this pump has a higher affinity for Na than for K. The pump would act to maintain the intracellular level of Na lower than that of K. However, the actual rate at which cations are pumped across the tubules from the bathing solution into the lumen by this pump depends not only on the affinity of the pump for the two ions but also on how fast the ions enter the cell. This latter depends partly on the electrochemical gradients across the cell membrane facing the bathing solution but also on the permeability of this membrane to these ions. There is indirect evidence that the basal cell membranes of Malpighian tubules of *Carausius* and *Rhodnius* are more permeable to K than Na (Pilcher, 1970; Maddrell, 1971). Direct measurements of the effects of changes in potassium concentration on the potential difference across the basal cell membranes of *Rhodnius* Malpighian tubules have confirmed that this side of the cell is more permeable to K than Na (unpublished results of Prince and Maddrell). These findings suggest that although the electrochemical gradient favouring Na movements into Malpighian tubules cells may be steeper than those favouring K entry, K movements may occur at comparable rates because of the higher K permeability. Possibly, then, comparatively small changes in the relative permeability of the basal membrane to Na and K may cause large changes in the ionic composition of the fluid secreted by Malpighian tubules. The ability of tubules from *Glossina* to secrete a sodium rich fluid at a high rate, for example, might be simply explained by their having a higher permeability to Na than have other tubules; as a result Na^+ ions enter faster than do K^+ ions and as the pump has a higher affinity for Na in any case, it is these ions which are transported. In K-secreting tubules by contrast, K ions enter faster, sodium ions being virtually excluded, and so it is potassium ions which are transported. On this view the very slow secretion by tubules in K-free media follows from the very slow entry of Na into the cells. It has been found that Malpighian tubules of *Calliphora* will rapidly secrete Na-rich fluid if bathed in a saline lacking Ca^{2+} ions as well as K (Maddrell, unpublished results). These tubules ordinarily secrete only very slowly in K-free, Na-rich solutions (Berridge, 1968). A somewhat similar effect of omitting Ca from the bathing medium occurs in experiments on preparations of midgut epithelium from larvae of silk moths. This epithelium which normally transports no sodium can be induced to do so in a Ca-free medium (Harvey and Zerahn, 1971). It seems likely that the explanation of both these results is that Ca omission increases the permeability of the cell membranes so as to allow Na$^+$ ions access to the cell interior and so to the cation pump which rapidly transports them into the lumen.

One piece of information which would throw light on the possibilities so far discussed is some knowledge of the intracellular levels of Na and K during fluid transport. Some recent work on *Rhodnius* tubules has supplied this (Maddrell, unpublished work). *Rhodnius* Malpighian tubules are particularly suitable for this as they will transport Na or K at equally high rates and the proportion of Na transport to K transport can be changed in a predictable way by changing the Na/K ratio of the bathing medium (Maddrell, 1969). By using ^{22}Na and ^{42}K as tracers it was possible to measure the intracellular levels of both Na and K and at the same time measure the Na and K levels in the secreted fluid and bathing fluid. The results are shown in Fig. 6 from which it is clear that there is a correlation between the intracellular level of an ion and the rate at which it is transported. In addition it is also clear that Na transport is much faster than K transport at low intracellular levels — as would be expected if transport depends on a cation pump having a higher affinity for Na than K.

Obviously a good deal more work is needed before one can conclude that Malpighian tubules are driven by a cation pump able to pump either Na or K — for example one needs evidence that the intracellular pool is that from which ions are drawn for transport — but the evidence so far is at least compatible with the idea.

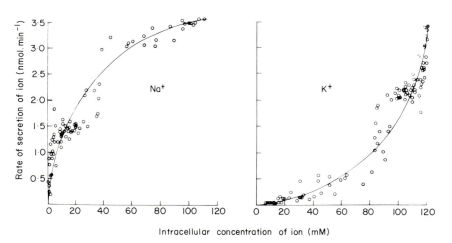

Fig. 6. Secretion of Na$^+$ and K$^+$ by Malpighian tubules of *Rhodnius* as a function of the intracellular concentrations of these ions.

V. TRANSPORT OF CHLORIDE IONS

The lumina of the Malpighian tubules of most insects are at a potential positive with respect to that of the bathing solution. The Malpighian tubules of *Rhodnius* are unusual in that during fast fluid secretion the lumen is at a potential about 30 mV *negative* to the surrounding medium (Maddrell, 1971). As the secreted fluid contains about 180 mM Cl when the bathing fluid contains 155 mM, movements of Cl$^-$ ions seem to be thermodynamically uphill. In fluids of lower Cl content, chloride transport has been shown to occur against both a six-fold concentration gradient and a 20 mV potential difference (Maddrell, 1971). It seems very likely that these tubules have a chloride pump. In the earlier account it was suggested that this pump might be situated on the apical side of the cells. From measurements of the potential difference across the basal and apical cell membranes (unpublished results of Prince and Maddrell) the basal side of the cell now seems a more likely site. Typically the cell interior is at a potential some 50 mV negative to the bathing solution while the lumen is about 20 mV positive to the cell interior so that chloride ions may well follow cation transport passively from inside the cell into the lumen but may have to be pumped into the cell against the electrical gradient. This interpretation rests on the assumption that the intracellular Cl level is lower than in the bathing and luminal fluids; direct evidence that this is the case has recently been obtained (Gupta *et al.*, 1976 and Chapter 4). Other evidence also suggests a basal site for the Cl pump. When the bathing solution Cl concentration is lowered from 155 to 9 mM (SO$_4$ replacing Cl) the potential difference across the basal side of the cell decreases by only 3–4 mV; by contrast a similar change in potassium concentration produces potential changes of 50–60 mV. The permeability of the basal side of the cell to Cl would seem to be much lower than to K, which is in line with the suggestion that the Cl pump is sited on this side of the cell.

VI. TRANSPORT OF PHOSPHATE

Ramsay (1956) found that Malpighian tubules of *Carausius morosus* secreted fluid rich in phosphate and that replacing chloride ions in the bathing medium with phosphate caused an acceleration of fluid secretion. Berridge (1969) found that the ability of saline solutions to support fluid secretion by *Calliphora* tubules depended critically on the size of the anions present with the notable exception of phosphate ions; in solutions containing phosphate, fluid secretion was faster than with any other anion. However, phosphate ions do not support fluid secretion by Malpighian tubules of *Rhodnius* (Maddrell, 1969). The position has become clearer since it was discovered that Malpighian tubules possess an active transport system for organic anions such as p-aminohippuric acid and indigo carmine (Maddrell *et al.*, 1974).

In *Rhodnius*, this organic anion transport goes on at a rate dependent on the physiological state of the insect; it only occurs at a high rate in the period following a meal (Maddrell and Gardiner, 1975). Interestingly, phosphate transport by *Rhodnius* tubules also only occurs at a high rate in the period after a meal (unpublished results of Maddrell and Gardiner). This fast transport is much slowed by the presence of p-aminohippurate ions. It seems very likely that phosphate ions are transported by the system responsible for removal of organic anions from the haemolymph. It is worth recalling that arsenate ions, which Berridge (1969) found to slow phosphate transport by *Calliphora* tubules, also slows transport of indigo carmine by *Rhodnius* tubules (Maddrell *et al.*, 1974). Also in line with his new interpretation is the finding that *Calliphora* tubules can transport organic anions a good deal faster than can *Rhodnius* tubules (Maddrell *et al.*, 1974) and yet secrete fluid more slowly (Maddrell, 1969); not surprisingly, therefore, phosphate ions reach a much higher concentration in the fluid secreted by *Calliphora* tubules (cf. Chapter 4.) Phosphate ions are found at quite high concentrations in the lumen of unstimulated tubules of *Rhodnius* (Gupta *et al.*, 1976). In the unstimulated condition these tubules secrete fluid only very slowly so that even a low rate of phosphate transport could account for the high concentration of phosphate found in the lumen.

It seems probable that the ability of phosphate ions to support secretion by some Malpighian tubules may come from the fact that they can be rapidly transported across the tubule wall by the organic anion transport system.

VII. TRANSPORT OF MG AND SO$_4$ BY MALPIGHIAN TUBULES

Mosquito larvae which live in hyperosmotic waters (Scudder, 1969) cope with osmotic water loss by drinking and absorbing the water in which they live and then excreting the excess salts (Kiceniuk and Phillips 1975), behaviour which parallels that of marine fish. Unlike marine fish, however, divalent ions from the ingested fluid are rapidly absorbed into the haemolymph. The question arises as to how they are subsequently excreted. In larvae of the mosquito *Aedes campestris*, the Malpighian tubules carry out active transport of magnesium ions (Phillips and Maddrell, 1974) and of sulphate ions (Maddrell and Phillips, 1975a). The characteristics of these processes are shown in Fig. 7.

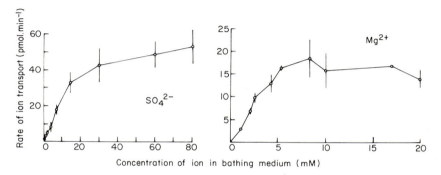

Fig. 7. Secretion of Mg^{2+} and SO_4^{2-} by Malpighian tubules of *Aedes campestris* (data from Maddrell and Phillips, 1975a; Phillips and Maddrell, 1974).

It has been calculated that magnesium ions are transported by isolated Malpighian tubules fast enough to account for the observed excretion of magnesium in living larvae (Phillips and Maddrell, 1974). However, similar calculations suggested that the Malpighian tubules would only be able to excrete all of the sulphate ingested by larvae living in waters containing less than 100 mM SO_4 and might not be able to account for the larger amount ingested in waters richer in sulphate than this (Maddrell and Phillips, 1975a). It has now been discovered that in larvae of *A. taeniorhynchus* the rate at which sulphate ions are transported by the Malpighian tubules is much higher in tubules from larvae raised in sulphate-rich water than in tubules from larvae raised in water containing only low levels of sulphate (Maddrell and Phillips, 1977). Increased rates of sulphate transport can be induced in a few hours by transferring larvae to water containing higher sulphate concentrations. This being the case, it seems very probable that sulphate transport by the Malpighian tubules of *A. campestris* from sulphate rich waters would be faster than shown in Fig. 7, and could after all account for the removal of all the ingested sulphate ions. It remains to be seen whether increased rates of Mg transport can also be induced by rearing larvae in Mg-rich media.

The concentrations of Mg and SO$_4$ in fluid secreted by tubules of these mosquito larvae can reach high levels. For example, tubules of *A. campestris* bathed in a solution containing 8 mM Mg were found to secrete fluid containing 55 mM Mg and tubules from *A. taeniorhynchus* secreted fluid containing 135 mM SO$_4$ from a medium containing 30 mM sulphate. (Also see Chapter 28.)

VIII. ACTIVE TRANSPORT OF ORGANIC COMPOUNDS BY MALPIGHIAN TUBULES

A. Organic Anions

For a long time it has been known that insect Malpighian tubules can concentrate acidic dyes from dilute solutions (Lison, 1937; Palm, 1952). It now seems that this ability is attributable to an organic anion transport system somewhat similar to that known for vertebrate kidney tubules. Two types of compounds are handled by this system: acylamides, such as p-aminohippuric acid (PAH) and sulphonates, such as indigo carmine and amaranth (Maddrell *et al.*, 1974). The transport system has a high affinity for these substances, being half saturated at concentrations of the order of 0.1 mM. Organic anion transport by at least some insect tubules seems to differ from that of vertebrate kidney tubules in that acylamides and sulphonates appear not to compete with one another and so presumably are handled by separate mechanisms. On the other hand, Nijhout (1975) has been able to show a depression of acidic dye excretion by Malpighian tubules of larvae of *Manduca sexta* bathed in fluid containing 100 mM sodium hippurate.

The question arises as to the function of this ability of Malpighian tubules to transport organic anions. From studies on the metabolism of insecticides it is known that many toxic compounds are detoxified by alterations in their structure (see, for example, Smith, 1962). The products are such compounds as hippurates, ethereal sulphates, β-glucosides, β-glucuronides and acetamido derivatives, just the sorts of substances carried by the organic anion transport system of Malpighian tubules. Thus the ability of Malpighian tubules to transport organic anions probably serves to eliminate some of the products of metabolism of potentially harmful substances.

Just as with sulphate transport, the activity of the organic anion transport system seems to vary with the load put upon it. Tubules taken from unfed *Rhodnius* secrete such compounds as PAH at a slow rate; after a protein-containing meal, however, the capacity for PAH transport increases steeply for 2–3 days before falling slowly away again (Maddrell and Gardiner, 1975). Similarly, Malpighian tubules isolated from larvae of *Aedes taeniorhynchus* that have been starved for two days, concentrate phenol red from the bathing solution into the secreted fluid to a much lesser extent than do tubules from fed larvae (Maddrell, unpublished results).

B. Organic Cations

Various workers have investigated the ability of insect Malpighian tubules to secrete basic dyes. Tubules of larvae of *Chironomus* can rapidly concentrate neutral red (Salkind, 1930) but tubules of *Blatta orientalis*, *Forficula auricularia* and *Carausius morosus* cannot (Lison, 1938). Tubules of larvae of *Manduca sexta* secrete such basic dyes as methyl green, methylene blue and neutral red (Nijhout, 1975). The possible significance of this has become apparent from recent work on the excretion of alkaloids, which of course are also organic bases.

Alkaloids occur in many plants, particularly of the families *Papaveraceae*, *Papilionaceae*, *Ranunculaceae* and *Solanaceae* and are believed to have been evolved as a protective measure against herbivores. Some insects can nevertheless thrive on a diet of such alkaloid containing plants. It turns out that the Malpighian tubules of several insects can remove such alkaloids as nicotine, atropine and morphine from the haemolymph at high rates and against steep concentration gradients (Maddrell and Gardiner, 1976). This must be an important element in the ability of insects to exploit a wide range of food plants.

Larvae of *M. sexta* feed largely on plants of the family *Solanaceae* which characteristically contain alkaloids. Tubules from these insects can transport nicotine unchanged and at a high rate but tubules from adult *Manduca* which are nectar-feeders, appropriately show no evidence of being able to excrete nicotine other than by passive means.

Unexpectedly, tubules of the blood-sucking insect, *Rhodnius prolixus* will also transport nicotine unchanged and against concentration gradients as steep as 40 : 1. This ability is not coupled to the ion and water transport processes of the tubule and unlike dye transport seems not to be affected by the physiological state of the animal. For the upper, fluid secreting tubules from 5th stage larvae of *Rhodnius*, the rate of nicotine transport has a maximum value of 700 pmol min^{-1} per tubule and is half saturated at a concentration of 2–3 mM in the bathing medium (Fig. 8). The high rate of transport leads to very high concentrations of nicotine in fluid secreted by the tubules; with 20 mM nicotine in the bathing solution, nicotine may appear at concentrations as high as 150 mM in the secreted fluid.

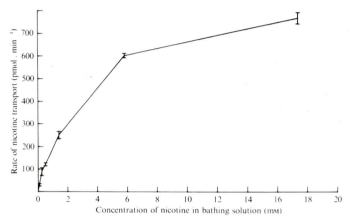

Fig. 8. The dependence of the rate at which nicotine is transported by the upper Malpighian tubules of *Rhodnius* on the concentration of nicotine in the bathing medium (from Maddrell and Gardiner, 1976).

Rhodnius tubules can also transport morphine and atropine at similarly high rates. Since these alkaloids and nicotine compete with one another they are presumably carried by the same transport system.

Tubules from larvae of *Pieris brassicae* and from adult *Calliphora erythrocephala* and adult *Musca domestica* all remove nicotine from solutions bathing them. *Pieris* tubules transport nicotine unchanged but an unknown metabolite appears in the fluid secreted by the tubules of the other two species (Maddrell and Gardiner, 1976).

The ability of some insect Malpighian tubules to concentrate basic dyes may thus be attributable to the ability of the organic cation transport system which acts to excrete toxic alkaloids. In line with this suggestion are some recent experiments in which we have found that *Rhodnius* tubules that rapidly transport alkaloids also rapidly transport the basic dye methylene blue, whereas Malpighian tubules of the wasp *Vespula germanica*, secreted neither atropine nor methylene blue (Maddrell and Gardiner, unpublished results).

C. Ouabain Excretion

Many transport processes in vertebrates are known to be inhibited by the glycoside, ouabain. Until recently it seemed that this poison had no effect on fluid secretion by Malpighian tubules, and indeed it has been found that the Malpighian tubules of *Oncopeltus fasciatus* can transport ouabain against a concentration gradient (Bernstein, unpublished results). Fluid secretion by the Malpighian tubules of *Locusta migratoria*, however, is slowed by concentrations of ouabain greater than about $10^{-6} - 10^{-7}$ M and, at 10^{-3} M fluid secretion is cut to only 5% of the control rate (Anstee and Bell, 1975). It seems that in this insect the ouabain sensitive $Na^+ - K^+$-activated ATPase found in the tubules may play a role in fluid transport. Just how widespread is this ouabain sensitivity remains to be seen; we have found no effect of ouabain, even at 10^{-3} M, on fluid secretion by isolated Malpighian tubules of *Schistocerca gregaria*.

IX. RESORPTIVE PROCESSES IN MALPIGHIAN TUBULES

A. KCl Resorption

Rhodnius freshly fed on blood rapidly excrete a hypo-osmotic fluid that is rich in sodium but poor in potassium (Maddrell, 1964). Fluid secreted by the upper lengths of the Malpighian tubules is iso-osmotic with the haemolymph and is relatively rich in potassium. The lower lengths of the Malpighian tubules are largely responsible for the necessary alteration of the composition of the fluid that leaves the upper tubule (Maddrell and Phillips, 1975b). By working with whole isolated tubules it was shown that the osmotic concentration and the K content of the secreted fluid varied in direct relation to one another (Fig. 9). This suggested that the change in K concentration caused the osmotic pressure change, possibly by a reabsorption of KCl without osmotically compensating amounts of water. It was possible however that the reduction of osmotic and K

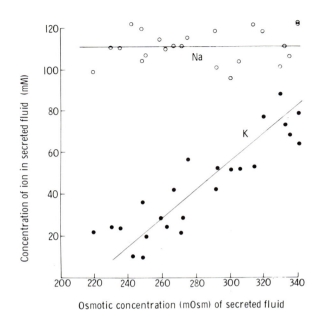

Fig. 9. The relation between the osmotic concentration and the concentrations of sodium and potassium in fluid secreted by whole Malpighian tubules of *Rhodnius* (from Maddrell and Phillips, 1975b).

concentration could also have been achieved by inward movements of water.

That this is not the case was shown in experiments where initially iso-osmotic K-rich fluid was perfused through isolated cannulated lengths of the lower Malpighian tubule. From a knowledge of the rate at which fluid left the cannula and from the rate at which it emerged from the lower tubule it could be determined whether fluid had left or entered the tubule during passage along it. The results showed that a reduction in osmotic concentration of the perfused fluid was always accompanied by a slow movement of water *from* the lumen into the bathing solution. Clearly the fall in osmotic concentration is caused by KCl absorption and not by inward movements of water.

From the rate of KCl absorbtion and water loss from the lumen it was calculated that each lower tubule absorbs about 6 nl min^{-1} of 900 mM KCl solution. This can result in very rapid changes in the composition of the fluid flowing through the lumen — its osmotic concentration may fall by 8 mOsm sec^{-1} and its potassium (and chloride) content decline at a rate of 4.5 mM sec^{-1}.

Rubidium ions are partially effective substitutes for K in this system but Na$^+$ ions are ineffective.

The operation of this system is under hormonal control (Maddrell and Phillips, 1976) and as with several hormonally controlled processes in insects, hormone action can be mimicked by treatment with 5 hydroxytryptamine (5-HT) — in this case by concentrations greater than about 10^{-8}M. In addition to hormonal control, this resorptive system is also much affected by the concentration of K in the fluid bathing the lower tubule. The lower the K concentration the more KCl is resorbed and so the lower is the osmotic pressure of the fluid leaving the lower tubule. It follows that, *in vivo*, the concentration of K will automatically tend to be held constant, as an increase in the level of K in the haemolymph would lead both to its more rapid removal by the upper tubule and a slower return to the haemolymph by the lower tubule, and *vice versa*. This is the first evidence in insects of control at such a local level but there are some similar, less well substantiated and more preliminary findings in one or two other cases, notably that of the locust rectum. It may be that insects possess a form of autonomous control of excretion in which the system reacts directly to changes in the fluid bathing it and responds appropriately.

It is at the moment open to question whether KCl resorption by the lower tubule of *Rhodnius* is active or passive. Because resorption only goes on from K-rich fluids in the lumen and in bathing solutions containing less than 20—30 mM K, K resorption always occurs down a concentration gradient. The lumen of the tubule is at a potential of the order of 20—30 mV negative to the bathing fluid when the lumen contains 100 mM K and 155 mM Cl and the bathing solution contains 5 mM K and Cl (Prince and Maddrell unpublished results), so that for both K and Cl the electrochemical potential difference is such as to favour passive movement out from the lumen. Finally it has not been possible to obtain any evidence that KCl absorption is saturable. Tubules perfused with fluid containing twice the level of K normally found in the lumen and at a rate twice as high as the natural rate, respond by absorbing KCl from the lumen at a proportionally higher rate. At the moment, therefore, it is possible that the large potassium concentration difference set up by the activity of the upper Malpighian tubule is simply passively exchanged in the lower tubule for an osmotic concentration difference. Whatever the mechanism, potassium and chloride ions which enter the upper tubule each accompanied by about 150 water molecules leave the lower tubule with only about 30 water molecules following each ion.

B. Reabsorption of Sugars

Insect Malpighian tubules are permeable structures and such useful substances as sugars and amino acids rapidly diffuse into the luminal fluid from the bathing solution (Ramsay, 1958; Maddrell and Gardiner, 1974). There is evidence that the rectum of *Schistocerca* returns amino acids to the haemolymph (Balshin and Phillips, 1971) and it has been supposed that sugars might also be reabsorbed in the anterior hindgut or in the rectum (Maddrell, 1971). Unexpectedly, it now appears that the Malpighian tubules themselves may act to reabsorb physiologically important sugars and so limit their loss from the haemolymph. Knowles (1975) investigated the relation between the concentration of sugars in solutions bathing isolated Malpighian tubules of *Calliphora vomitoria* and their concentration in the secreted fluid. He found that glucose and trehalose appeared at much lower concentration than did other sugars of similar size (Fig. 10). This difference was less marked at higher concentrations of glucose in the bathing medium and was also suppressed by treatment with phlorizin, an inhibitor of glucose transport in other tissues. These findings suggest that glucose is actively reabsorbed from the tubule lumen by a saturable system inhibited by phlorizin.

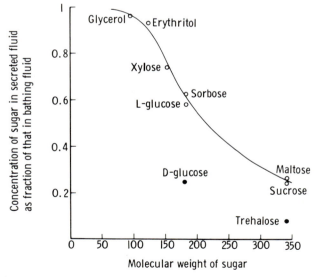

Fig. 10. The concentrations at which sugars of different molecular sizes appear in the fluid secreted by isolated Malpighian tubules of *Calliphora vomitoria* (data from Knowles, 1975).

X. THE PASSIVE PERMEABILITY OF MALPIGHIAN TUBULES

So far we have considered the way in which active transport of ions underlies fluid secretion and how various materials are actively pumped into and out of the tubule lumen. However, as was pointed out at the outset, Malpighian tubules are permeable structures and many substances passively find their way into the secreted fluid by diffusion across the tubule wall. Ramsay (1958) was the first to demonstrate this permeable nature of the tubule wall and to point out that this is an essential feature in that it provides for the automatic removal of toxic material from the haemolymph. Another consequence is that such useful compounds as amino acids and sugars enter the primary excretory fluid and have to be reabsorbed.

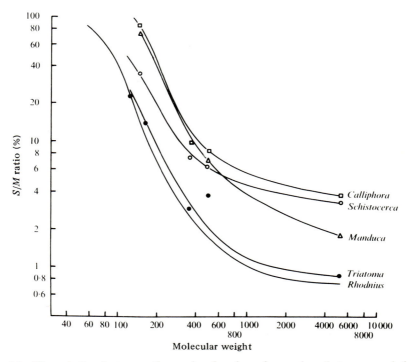

Fig. 11. The relation between the molecular size of organic substances and the concentration (as % of that in the bathing solution (S/M ratio)) that they reach in the fluid secreted by Malpighian tubules of various insects. For example, inulin (mol. wt 5200) appears in fluid secreted by isolated tubules of *Calliphora* at about 4% of the concentration at which it is present in the bathing solution (from Maddrell and Gardiner, 1974).

Maddrell and Gardiner (1974) have investigated the permeability of a range of insect Malpighian tubules. Their results are summarized in Fig. 11. In essence, compounds of mol. wt below about 400 freely cross the tubule wall while larger compounds enter more slowly; inulin (MW 5200) for example, is found in the secreted fluid at concentrations only about 1—5% of that in the bathing fluid.

What is surprising is that compared with the high permeability of the glomerulus of the vertebrate kidney or even of the Malpighian tubules of the millipede, *Glomeris marginata* (Farquharson, 1974), insect Malpighian tubules are very much less permeable. For example, toxic materials of mol. wt of the order of 5—10,000 would be cleared rapidly by glomeruli but would be excreted only slowly in insects. The significance of this difference is not clear. One suggestion is that insects as small terrestrial animals are unlikely in any case to be able to regulate their haemolymph composition closely at all times. Most of their tissues must have evolved a tolerance to changes in the composition of the haemolymph, while the relatively intolerant and conservative nervous system shelters behind the remarkable regulative ability of the perineurium, which forms a well developed blood-brain barrier system (Treherne, 1974). Given such tolerance or avoidance, the speed with which unwelcome materials are removed from the haemolymph may not be as crucial as the certainty of their eventual removal. A less permeable excretory system would both cater for this and have the advantage that energy consuming reabsorption of useful substances could also be relatively slow.

XI. DENSE BODIES (MINERALIZED DEPOSITS)

Most structural investigations of Malpighian tubules have noted the existence of granular structures, often termed dense bodies or mineralized deposits or concretions in the cytoplasm and the appearance of somewhat similar structures in the lumen (see Sohal (1974) for references). These bodies have been shown to contain various minerals, proteins and mucopolysaccharides. Recent work by Dr R. Sohal on these structures in the Malpighian tubules of adult *Musca domestica* has thrown light on their ontogeny and on their composition and I am most grateful to him for permission to quote from his unpublished work in the account which follows.

Dense bodies first appear in the cytoplasm of the tubules of newly emerged adult *Musca* as clear vacuoles (Fig. 12) which gradually accumulate electron opaque contents. In old flies relatively large areas of the cytoplasm are occupied by dense bodies (Fig. 13). Adult

Musca tubules have as many as four distinct cell types in them (Sohal, 1974). Dense bodies are found in both type I and type III cells (Fig. 14), which together are much the most abundant cells.

Electron probe X-ray microanalysis of the dense bodies shows that they are rich in phosphorus (presumably as phosphate) and sulphur, and also contain chloride, calcium, zinc, iron and copper. The contents of dense bodies in the lumen of the tubules are very similar. The obvious suggestion that these intraluminal bodies are derived from those in the cytoplasm is weakened by the fact that no ultrastructural evidence of such movement has been seen; they may, therefore, arise in the lumen.

A further point worth making is that in aged flies tertiary lysosomes, containing high levels of acid phosphatase activity, are also found to contain phosphate, sulphur, chloride, calcium and copper suggesting that the cytoplasmic dense bodies may become incorporated into the developing secondary lysosomes.

The occurrence of such metals as calcium, zinc, iron and copper in the dense bodies suggests that they may play a role in achieving an effective excretion of these substances. Insects concentrate their excretory material to a greater extent than do other organisms. It can be argued that this ability depends on the cuticular lining of the hindgut, which at least in the rectum (Phillips and Dockrill, 1968) has a permeability low enough to prevent substances larger than disaccharides affecting the operation of the rectal cells which so effectively absorb water from the rectal lumen. Very high concentrations of potentially injurious substances can thus accumulate with no ill effects. Metal ions unlike organic toxicants, whose molecules are larger, would not so easily be retained in the rectum and so they would tend to diffuse back into the haemolymph through the rectal wall. Particularly for insects living in dry environments, which are forced to remove as much water as possible from the excreted material, it may thus be very difficult to excrete excess metal ions by the normal route. Organic material in dense bodies rich in such strongly negative groups as phosphates and sulphides would strongly bind divalent cations and it is possible that this binding, which increases with age as one would expect, represents the insects' answer to this problem. In order to establish that the dense bodies are important sites of excretion of metal ions as suggested, requires measurements of the fate of such ions in the food. In particular, what fraction of ingested metal ions ends up in the dense bodies? The answer to this problem awaits further research. (Also see Gupta *et al.*, Chapter 4.)

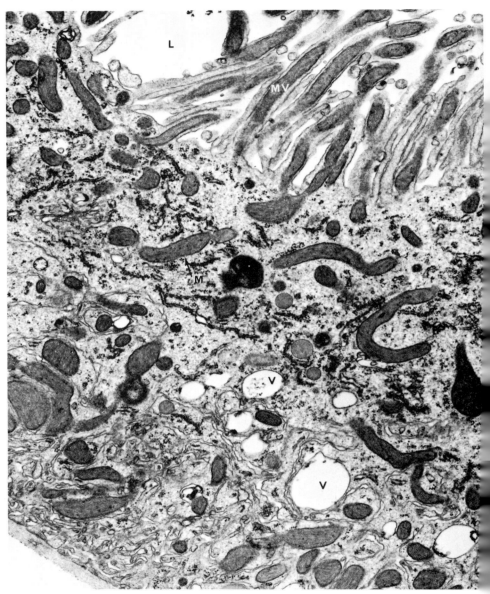

Fig. 12. Electronmicrograph of a cross section through the Malpighian tubule of a 5-day old female housefly showing a Type I cell. L, lumen of the tubule; M, mitochondria; MV, microvilli; V, vacuoles beginning to accumulate electron opaque contents. ×21,000. Micrograph by courtesy of Dr R. Sohal.

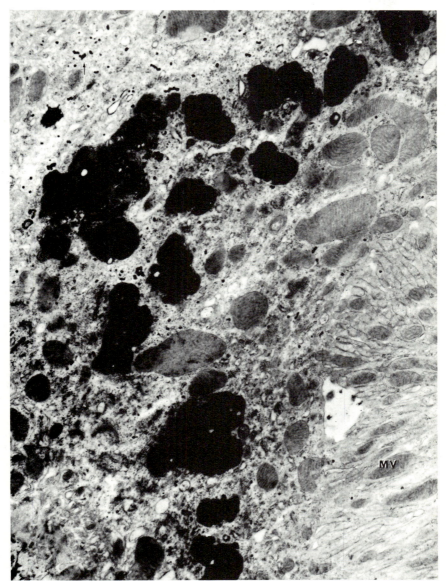

Fig. 13. Type I cell from a 35-day old female fly showing several dense mineralized structures. MV, microvilli. ×15,000. Micrograph by courtesy of Dr R. Sohal.

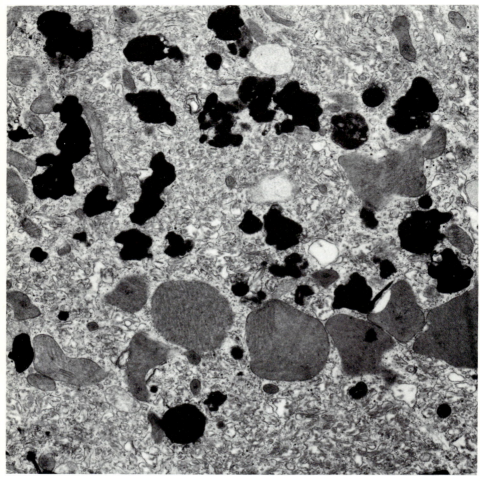

Fig. 14. Type III cell from a 30-day old fly showing numerous mineralized bodies. ×15,000. Micrograph by courtesy of Dr R. Sohal.

XII. CONCLUDING REMARKS

It is the main feature of recent studies that they have made much more complex the picture that one has of the operation and functions of insect Malpighian tubules. In particular, transport systems for a wide range of substances have been uncovered. This should not surprise us, as vertebrate nephrons which fulfil some of

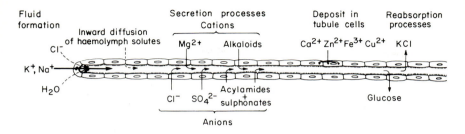

Fig. 15. Summary of the excretory processes known to occur in insect Malpighian tubules. Not all these processes occur in tubules from a single species. The apparent regionalization of processes in the diagram is of no significance; fluid formation, for example, is often found to occur along the whole length of a tubule, though in other cases it is restricted to the regions of tubule furthest from its junction with the alimentary canal.

the same functions have an almost bewildering array of such systems.

Figure 15 attempts to summarize in diagrammatic form the processes now known to be involved in the excretory functions of Malpighian tubules. In addition to the secretion of fluid driven by transport of K+ (and/or Na^+) ions, Malpighian tubules often have the ability to transport other inorganic ions (Cl^-, Mg^{2+} and SO_4^{2-}), organic anions (acylamides and sulphonates), organic cations (nicotine and other alkaloids) and glycosides (ouabain), usually when these substances accumulate in the haemolymph at rates too high for them to be eliminated simply by diffusion through the tubule wall. Because of the passive permeability of the tubule, most compounds present in the haemolymph will steadily diffuse into the tubule lumen. The rate of this process is less than in the analogous parts of most other excretory systems, a luxury which insects may enjoy because of the tolerance of their tissues and because of their effective blood-brain barrier. Useful compounds such as ions and sugars are resorbed by the walls of some Malpighian tubules. The effective excretion of metal ions may to some extent be achieved by binding to organic materials in dense bodies in the tubule walls.

It is certain that there are many other aspects of Malpighian tubule function yet to be discovered. The most obvious area of ignorance is the function of the different parts of those Malpighian tubules that have several regions differing markedly in their appearance and structure. That such morphological differences are likely to be correlated with functional differences is clear from Irvine's pioneer work on the regionalized tubules of *Calpodes* (Irvine, 1969) and from recent work on the lower lengths of tubules in *Rhodnius* (Maddrell and Phillips, 1975b). (For related topics see Chapters 9, 11, 15, 22, 23 and 25.)

REFERENCES

Anstee, J. H. and Bell, D. M. (1975). *J. Insect Physiol.* **21**, 1779—1784.
Balshin, M. and Phillips, J. E. (1971). *Nature New Biol.* **233**, 53—55.
Berridge, M. J. (1968). *J. exp. Biol.* **48**, 159—174.
Berridge, M. J. (1969). *J. exp. Biol.* **50**, 15—28.
Diamond, J. M. and Bossert, W. H. (1967). *J. gen. Physiol.* **50**, 2061—2083.
Farquharson, P. A. (1974). *J. exp. Biol.* **60**, 41—51.
Gee, J. D. (1976). *J. exp. Biol.* **65**, 323—332.
Gupta, B. L., Hall, T. A., Maddrell, S. H. P. and Moreton, R. B. (1976). *Nature, Lond.* **264**, 284—287.
Harvey, W. R. and Zerahn, K. (1971). *J. exp. Biol.* **54**, 269-274.
Hill, A. E. (1975a). *Proc. R. Soc. Lond. B.* **190**, 99—114.
Hill, A. E. (1975b). *Proc. R. Soc. Lond. B.* **190**, 115—134.
Irvine, H. B. (1969). *Am. J. Physiol.* **217**, 1520—1527.
Kiceniuk, J. W. and Phillips, J. E. (1974). *J. exp. Biol.* **61**, 749—760.
Knowles, G. (1975). *J. exp. Biol.* **62**, 327—340.
Lison, L. (1937). *Archs. Biol. Paris* **48**, 321—360.
Lison, L. (1938). *Z. Zellforsch.* **28**, 179—209.
Maddrell, S. H. P. (1964). *J. exp. Biol.* **41**, 163—176.
Maddrell, S. H. P. (1969). *J. exp. Biol.* **51**, 71—97.
Maddrell, S. H. P. (1971). *Adv. Insect Physiol.* **8**, 199—331.
Maddrell, S. H. P. and Gardiner, B. O. C. (1974). *J. exp. Bio.* **60**, 641—652.
Maddrell, S. H. P. and Gardiner, B. O. C. (1975). *J. exp. Biol.* **63**, 755—761.
Maddrell, S. H. P. and Gardiner, B. O. C. (1976). *J. exp. Biol.* **64**, 267—281.
Maddrell, S. H. P. and Phillips, J. E. (1975a). *J. exp. Biol.* **62**, 367—378.
Maddrell, S. H. P. and Phillips, J. E. (1975b). *J. exp. Biol.* **62**, 671—683.
Maddrell, S. H. P. and Phillips, J. E. (1976). In "Perspectives in Experimental Biology." Vol. I Zoology. pp. 179—185. Pergamon Press, Oxford.
Maddrell, S. H. P. and Phillips, J. E. (1977). *J. exp. Biol.* in press.
Maddrell, S. H. P., Gardiner, B. O. C., Pilcher, D. E. M. and Reynolds, S. E. (1974). *J. exp. Biol.* **61**, 357—377.
Nijhout, H. F. (1975). *J. exp. Biol.* **62**, 221—230.
Oschman, J. L., Wall, B. J. and Gupta B. L. (1974). *Symp. Soc. exp. Biol.* **28**, 305—350.
Palm, N-B, (1952). *Ark. Zool.* **3**, 195—272.
Phillips, J. E. and Dockrill, A. A. (1968). *J. exp. Biol.* **48**, 521—532.
Phillips, J. E. and Maddrell, S. H. P. (1974). *J. exp. Biol.* **61**, 761—771.
Pilcher, D. E. M. (1970). *J. exp. Biol.* **53**, 465—484.
Ramsay, J. A. (1953). *J. exp. Biol.* **30**, 358—369.

Ramsay, J. A. (1954). *J. exp. Biol.* **31**, 104—113.
Ramsay, J. A. (1955). *J. exp. Biol.* **32**, 200—216.
Ramsay, J. A. (1956). *J. exp. Biol.* **33**, 697—709.
Ramsay, J. A. (1958). *J. exp. Biol.* **35**, 871—891.
Salkind, S. J. (1930). *Z. Zellforsch.* **10**, 53—72.
Scudder, G. G. E. (1969). *Vehr. internal. Verein. Limnol.* **17**, 430—439.
Smith, J. N. (1962). *Ann. Rev. Ent.* **7**, 465—480.
Sohal, R. S. (1974). *Tissue and Cell* **6**, 719—728.
Treherne, J. E. (1974). In "Insect Neurobiology." (Ed. J. E. Treherne), pp. 187—244. North Holland Publishing Co., Amsterdam.

22. Active Transport of Water Vapour

J. Noble-Nesbitt

School of Biological Sciences, University of East Anglia, Norwich, England

I. INTRODUCTION

At the time when Professor Ramsay was drawing attention to the need to consider critically the physical parameters associated with the cuticular membrane in membrane-limited evaporation from insects, the first stirrings in the new world of reversed transpiration, or the uptake of water from atmospheres with humidities well below saturation, were occurring with the publication of the work of Buxton (1930) and Mellanby (1932) on the larva of the mealworm, *Tenebrio molitor*. In the four decades since Mellanby correctly attributed the increased water content of previously desiccated mealworms following their exposure to atmospheres with 90% or higher Relative Humidity (RH), to uptake of water vapour from the atmosphere, the phenomenon has been described in a number of different insects, mites and ticks (see Table I for summary of published observations). In conjunction with some of these reports, a number of suggestions as to the site and mechanism of uptake have been made.

In recent years, some notable strides have been made towards a more precise localization of the anatomical region of the body where uptake occurs, and it is not uncharacteristic that the painstakingly careful work of Professor Ramsay (1964; Grimstone *et al.*, 1968) on *Tenebrio*'s cryptonephric excretory system helped to pave the way for the present author's studies on both *Tenebrio* and the firebrat *Thermobia domestica* (a junior synonym for *Lepismodes inquilinus* — Watson, 1967) specifically directed to this end (Noble-Nesbitt, 1970a,b, 1973, 1975). In turn, this has regalvanized interest in the search for the site (or sites) of uptake generally. With better

Arachnida					
Acarina — Ticks					
Ixodes ricinus	92.0	0.92	4.8	110	Lees (1946)
Ixodes canisuga	92.0	0.92	4.8	110	Lees (1946)
Ixodes hexagonus	94.0	0.94	3.6	80	Lees (1946)
Ornithodorus moubata	82.0	0.82	12.2	270	Lees (1946)
Rhipicephalus sanguineus	84.0	0.84	10.8	240	Lees (1946)
Amblyomma maculatum	88.0	0.88	7.8	180	Lees (1946)
Amblyomma cajennense					
adult	90.0	0.90	6.2	140	Lees (1946)
larva	80.0	0.80	13.9	300	Knülle (1965)
Dermacentor andersoni					
adult	86.0	0.86	9.2	210	Lees (1946)
larva	80.0	0.80	13.9	300	Knülle (1965)
Dermacentor reticulatus	86.0	0.86	9.2	210	Lees (1946)
Dermacentor variabilis	80.0	0.80	13.9	300	Knülle (1965)
Acarina — Mites					
Acarus siro	71.0	0.71	22.6	460	Knülle (1962), Solomon (1962)
Echinolaelaps echidninus					
(= *Laelaps echidnina*)	90.0	0.90	6.2	140	Wharton and Kanungo (1962)
Dermatophagiodes farinae	70.0	0.70	23.8	480	Larson (1969)

TABLE I

Arthropods which can absorb water vapour from subsaturated atmospheres

	CEH (%)	CEA	Osmotic pressure (osmoles)	(Atm)	References
Insecta					
Thysanura					
Ctenolepisma longicaudata	60.0	0.60	37.0	700	Heeg (1967)
Thermobia domestica	45.0	0.45	67.0	1100	Beament et al. (1964)
Orthoptera					
Chortophaga viridifasciata	82.0	0.82	12.2	270	Ludwig (1937)
Arenivaga investigata	82.5	0.83	11.8	260	Edney (1966)
Siphonaptera					
Xenopsylla brasiliensis	50.0	0.50	55.5	940	Edney (1947)
Xenopsylla cheopsis	65.0	0.65	30.0	580	Knülle (1967)
Ceratophyllus gallinae	82.0	0.82	12.2	270	Humphries (1967)
Psocoptera					
Liposcelis rufus, L. knullei, and L. bostryohophilus	58.0	0.58	40.2	740	Knülle and Spadafora (1969)
Coleoptera					
Tenebrio molitor	88.0	0.88	7.8	180	Mellanby (1932)
Lasioderma serricorne	43.0	0.43	73.6	1170	Knülle and Spadafora (1970)

localization of the site(s), investigation of possible mechanisms of uptake can be better founded and, we hope, more successful.

In any terms, the ability shown by these arthropods is truly impressive. The equivalent concentrations of solutions in equilibrium with the atmospheric water vapour from which they gain their water (Table I) show that uptake occurs against gradients of water chemical potential of up to 1200 Atmospheres. In whatever terms the gradient is expressed, the enormity of the task performed by these small creatures is obvious, and it is this which has attracted the attention and comment of many observers. Any hypothesis advanced to explain the uptake of water from the atmosphere must recognize and account for it.

The topic has been reviewed a number of times, notably by Edney (1957), Beament (1954, 1964, 1965), Schmidt-Nielsen (1969), Berridge (1970), Maddrell (1971), Noble-Nesbitt (1973), House (1974), Ebeling (1974), Stobbart and Shaw (1974) and Professor Ramsay himself (1971). Though some reappraisal is now called for, these reviews remain valuable sources of information and ideas on the subject. It is my intention here to review the results of past investigations in the light of recent discoveries, including some of my own current investigations, and to attempt to indicate the direction in which the study may advance in the future.

II. UPTAKE OF WATER VAPOUR

The gain in weight shown by larvae of the mealworm, *Tenebrio molitor*, when kept in subsaturated atmospheres of high humidity (88% RH or greater) without access to food or liquid water was first attributed to production and retention of metabolic water by Buxton (1930). In his careful and detailed extension of this work Mellanby (1932) showed that this explanation did not account for the observed weight gains, and correctly attributed the gain in weight to absorption of water vapour from the atmosphere. He also showed that these gains were most marked when the water content of the mealworm had been previously depleted during a period of desiccation, and that they tended to cease when the water lost during desiccation had been fully recovered.

From his work and that carried out subsequently on the various other arthropods which show a capacity for extracting water from subsaturated atmospheres, a number of fairly general points can be made.

A. Uptake as an Active Physiological Process

Uptake from subsaturated atmospheres occurs only in living specimens, and ceases on death. In a number of cases (e.g. argasid tick — Browning, 1954; desert cockroach *Arenivaga* — Edney, 1966; flea — Knülle, 1967; firebrat — Noble-Nesbitt, 1969; ixodid tick — Knülle and Devine, 1972) carbon dioxide anaesthesia has been shown to prevent uptake which recommences after recovery. Uptake ceases in *Ixodes* if the ticks are asphyxiated, poisoned with cyanide or unduly desiccated (Lees, 1946). Prolonged starvation can also lead to impairment of uptake (ticks — Lees, 1946, 1948, 1964; firebrat — Okasha, 1972). To this extent, uptake can be said to be an active physiological process, dependent upon continuing metabolic activity.

The energy required to transport water at the rates observed does not amount to more than a small proportion of the total metabolic activity of the body. This is supported by the calculations of Lees (1948), Kanungo (1965), Edney (1966), Maddrell (1971) and House (1974), and by direct observations of the loss of dry weight in firebrats over a period of three weeks during which they were either subjected to five cycles of desiccation and rehydration, or to continuous hydration. The dry weight losses were similar in both regimes (Okasha, 1971). If uptake is confined to a small area of the body, however, the tissue concerned may be metabolically very active indeed (Maddrell, 1971).

B. Relationship to Atmospheric Relative Humidity

As pointed out by Mellanby (1932) in his studies on *Tenebrio* larva, there is a specific RH below which no gain in weight occurs, but above which gain in weight is possible. At this "equilibrium RH" any loss of water to the air is just balanced by uptake. This has since been termed the critical equilibrium humidity (CEH) by Knülle and Wharton (1964), later modified to critical equilibrium activity (CEA) by Wharton and Devine (1968), and it varies according to the species, but is constant within each. In Table I these values are given for each species mentioned. The lowest recorded values (for flea prepupa, firebrat and cigarette beetle) lie in the range 40—50% RH. The value does not change with temperature, though there is some evidence that it may be related to physiological state, such as state of nutrition (ticks — Lees, 1948, 1964; Browning, 1954) or degree of water stress (firebrat — Noble-Nesbitt, 1975). It is to this value that the air inside a small sealed container is brought by one of these arthropods, irrespective of whether it starts well above or below this level, as shown by Mellanby (1932) for the mealworm, Edney (1947) for the flea prepupa, Rudolph and Knülle (1974) for ixodid ticks, and Noble-Nesbitt (1975) for the firebrat, for instance. On death, the humidity within the container rises nearly to saturation point. This relationship of uptake to RH rather than to vapour pressure difference resembles that of hygroscopic substances (Mellanby, 1932) and is a feature which must be accounted for in any proposed scheme purporting to explain the phenomenon of uptake of water vapour (Schmidt-Nielsen, 1969, 1975).

C. Regulation of Uptake

Uptake is usually demonstrated by exposure to high humidity following a period of desiccation to deplete the water reserves of the test specimen. In most cases, uptake under such conditions continues until the water lost during desiccation is regained (making due allowance for dry weight losses during the additional period of starvation, which is known to result in an increase in the proportion of water in the body, a feature which has been ascribed to volume regulation — firebrat, Noble-Nesbitt, 1969, 1973; Okasha, 1971, 1972). Subsequently, small fluctuations in weight, indicative of slight net gains and losses of water, may ensue, indicating that uptake is fairly precisely regulated so that any losses of water are just compensated by gains (Noble-Nesbitt, 1969, 1973). Considerable exchange of water between the animal and the atmosphere occurs however, as shown by radio-isotope investigations using tritiated water, for instance in mites (Wharton and Devine, 1968; Devine and Wharton, 1973; Arlian and Wharton, 1974) and ticks (Knülle and Devine, 1972). In these the "active" component of water uptake is said to cease below CEH (Arlian and Wharton, 1974; Knülle and Devine, 1972) confirming earlier conclusions reached with the firebrat that the uptake mechanism does not operate below CEH (Noble-Nesbitt, 1969) and so resolving (in these species at least) the question as to whether or not at humidities below CEH the uptake mechanism continues in operation, restricting transpiration but being swamped by it: it apparently does not.

Regulation of uptake possibly provides an explanation for the lack of uptake found in engorged ticks (Lees, 1946). Engorged ticks are likely to be above the volume threshold needed for net uptake to occur, though initiation of the moult may also affect uptake.

D. Effects of Injury

If the cuticle of the tick *Ixodes* is subjected to minute abrasion, sufficient only to interrupt the superficial wax layer over a limited area, then uptake following desiccation is arrested whilst the cuticle is being repaired (Lees, 1947). Complete repair is not required, and uptake recommences before the cuticle regains its normal degree of impermeability. Light abrasion of the cuticle interrupts uptake for about one day in *Arenivaga* and its effects can be prevented by painting over the abraded area (Edney, 1966). Removal of scales in *Thermobia* has little if any effect on uptake (Beament *et al.*, 1964). Application of small amounts of hot wax to the cuticle can temporarily arrest uptake in *Thermobia*, recovery being complete in a few hours (Noble-Nesbitt, 1970a,b), whilst local burning with a hot needle and other forms of reversible stress do not noticeably reduce uptake as measured over one day (Okasha, 1971). Short exposures to high temperatures (<1 minute at 55°C) do not impair uptake in *Thermobia* though longer exposures result in irreversible damage and death of the insect (Noble-Nesbitt, unpublished observations). It seems clear that where no permanent injury is sustained, uptake suffers only a temporary halt whilst some repair occurs.

E. Effects of Moulting

Moulting arrests uptake in ticks (Browning, 1954), *Tenebrio* (Locke, quoted in Beament, 1954; Machin, 1975), *Arenivaga* (Edney, 1966) and *Thermobia* (Okasha, 1971; Noble-Nesbitt, 1973). In *Thermobia*, uptake is arrested for at least one day prior to ecdysis, but is fully restored within four hours after ecdysis and weight loss during the moult is then compensated for (Noble-Nesbitt, unpublished observations). During prolonged starvation, in 93% RH, moulting of *Tenebrio* is often followed by a large, overcompensating period of uptake (Noble-Nesbitt, unpublished observations).

F. Nutritional Effects

In order to eliminate weight changes due to feeding, experiments on water vapour are usually carried out with specimens denied access to food. Weight losses due to starvation are usually small, and as they act in the sense opposite to weight gains due to uptake of water vapour, are usually discounted. This technique has been noted by Locke (1964) who suggested a mechanism for uptake based on resorption of cuticular materials similar to that occurring during starvation and moulting. However, there is some evidence that uptake of water vapour will occur also in feeding individuals, for example in *Thermobia* (Okasha, 1971), and *Arenivaga* (Noble-Nesbitt, unpublished results). In culture conditions (83% RH) *Thermobia* is in water balance, and is above its CEH. When desiccated and rehydrated in the presence of food, rapid weight gain occurs in the rehydration period, making good the weight loss during desiccation. It is probable that this is uptake of water vapour. At the same time feeding apparently goes on, since the dry matter at the end of five cycles of desiccation and rehydration is much greater than in insects starved over the same period (Okasha, 1971). In *Arenivaga* a similar deduction follows from a slightly different experimental approach. During feeding at 93% RH the weight of the food and the insect were individually monitored with food blanks acting as a control. The food absorbed some water vapour hygroscopically. After making allowance for this effect, a large increment of weight gain was still apparent in the combined weight of food plus insect. The insect's weight gain whilst feeding in high humidity could therefore be related to the two components of weight gain from food ingested and by uptake of water vapour (Noble-Nesbitt, unpublished observations).

This suggests that the state of nutrition is not an important factor in determining whether uptake will occur or not, except as food reserves become seriously depleted, when uptake begins to fail. Lees (1946, 1948, 1964) has shown that uptake is progressively reduced and the CEH rises, in starving ticks, and this also seems to explain the eventual failure of uptake during prolonged starvation in *Thermobia* (Okasha, 1972), since subsequent feeding restores the capacity of surviving specimens to absorb water vapour (Noble-Nesbitt, unpublished observations).

During prolonged starvation at high humidities, the mealworm may become "dropsical" (Buxton, 1930; Mellanby, 1932), and this was suggested to result from a failure to excrete or transpire water rapidly enough. The reason for this is not immediately obvious; reduced excretory loss may result from the cessation of feeding, since faecal pellet formation slows down, which is known to prevent water excretion in locusts (Phillips, 1964b).

Some specimens also over-compensate for water losses during desiccation, when returned to a high humidity. Regulation is not nearly so precise as in some other insects, as has been noted by a number of observers (Edney, 1957; Browning, quoted in Edney, 1957; Dunbar and Winston, 1975; Machin, 1975) and confirmed by the present author.

In other cases of prolonged starvation the situation is usually much more clear cut, with uptake regulated in a manner which allows for a good deal of volume regulation, and in the firebrat, for instance, prolonged starvation results in an increasing proportion of water in the body, (Noble-Nesbitt, 1969, 1973; Okasha, 1971).

When such insects are refed, the proportion of water in the body returns to the value normally found in culture insects, (*Thermobia* — Okasha, 1971) or lower (*Tenebrio* — Machin, 1975), the process taking about a week. Machin has noted that this period is one of rapid weight gain (which he equates with growth). He concludes that water vapour uptake thus releases the insect from the need to produce water metabolically. This interesting observation agrees with the more rapid growth and maturation found in moist cultures fed regularly with foodstuffs of high water content. Machin goes on to suggest that water lack may in fact be responsible for the retardation of subsequent growth under dry culture conditions.

Scrutiny of the data he presents in fact indicates that in dry bran cultures (RH 70% or 77%) there is a slight negative balance between the water gained from the bran consumed and the combined loss of water in the faeces and by evapotranspiration, which presumably must be offset by metabolic water production, or lead to a permanent negative water balance. Such insects may therefore be regarded as being under water stress. This in turn could account for water gains observed in dry-culture insects transferred to high humidities without prior desiccation, and in part for overcompensation following prior desiccation.

III. ANATOMICAL LOCATION OF SITE OF UPTAKE

From the first, attempts to associate the uptake of water vapour with some specific anatomical part of the animal have been made, but only in recent years has clear experimental evidence indicating the precise location been obtained, and then only in two or three cases. The small size of some of the animals (e.g. mites) makes experimental investigation of the site of uptake difficult. Some of the suggestions that have been made are considered below.

A. The Tracheal System

Mellanby (1932) produced the first suggestion as to the site of uptake, indicating that the tracheal system of *Tenebrio* may be involved. He suggested that water vapour was drawn in through the spiracles and absorbed in the tracheoles. Beament (1964) demonstrated that tracheal air has a humidity of close to 99% RH in the cockroach, and suggested that uptake via the tracheal system could therefore be discounted. Certainly spiracles appear to be devices to restrict water loss from the tracheal system. Water loss usually increases markedly when the spiracles are caused to remain open longer, either by exposure to high carbon dioxide concentrations (5% or more), or because of increased metabolism (Mellanby, 1934; Edney, 1957; Wigglesworth (1972). Ramsay (1935) himself demonstrated that water loss in stimulated *Blatta* increased from 3.9 to 6.0 mg.hr^{-1} for instance. Blocking of the spiracles is not a reliable means of testing whether the tracheal system is involved in water vapour uptake, because of the possible effects of anoxia (Lees, 1946; Browning, 1954; Beament, 1964). Where cutaneous respiration is possible, however, as in some ixodid ticks, blocking of the spiracles does not prevent uptake (Lees, 1946), and uptake is also possible in other ticks (Knülle and Devine, 1972) and some mites (Knülle, 1965), which do not possess a tracheal system. It is not likely, therefore, that the tracheal system is the site of uptake.

B. The Integument of the General Body Surface

Abrasion and repair of the cuticle and moulting arrest uptake. These events have been taken to indicate that the epidermis, as a whole, is involved in uptake, in the absence of any suggestion or evidence indicating that any specialized region is specifically responsible for water vapour uptake (Beament, 1954, 1964). However, as Beament pointed out, the effect of abrasion may be so greatly to increase water loss that any uptake occurring elsewhere is masked, and when abraded areas are painted over uptake is not prevented in *Arenivaga* (Edney, 1966). Furthermore, if a specialized region is involved it could equally be the case that uptake ceases in favour of repair processes until the immediate damage is repaired. There is some evidence for short-term arrest of uptake in the firebrat following experimental treatment, though there is little if any residual effect seen 24 hrs after superficial abrasion (Beament *et al.*, 1964), treatment with hot wax (Noble-Nesbitt, 1970a,b) or other kinds of stress (Okasha, 1971); yet a specialized region involved with uptake has been demonstrated in this insect (Noble-Nesbitt, 1970a,b, 1973, 1975 — see also p. 581). Moulting affects specialized regions and cannot therefore be taken as a definite indicator that the whole of the body surface is involved in uptake. Where such a specialized region is shown to be involved in uptake, the general body surface may be eliminated as a site for uptake, but in the absence of any such clear demonstration in other specific instances it may be as well not to rule out completely the possible involvement of the general body surface.

C. The Insect Rectum

Since Wigglesworth (1932) suggested that an important function of the rectum is the removal of water from the faeces, this idea has been confirmed in a number of insects. Even in insects living on moist food, such as the locust, blowfly and cockroach, the ability of the rectum to remove water from its contents can be quite impressive, with the osmolarity of the rectal contents being increased to about 1.0 osmol.litre^{-1} (Phillips 1964a, 1969; Wall and Oschman, 1970). In the mealworm, which lives in stored products without access to liquid water, the drying process is even more impressive, producing faecal pellets enclosed in an air space with an RH of as low as 75%. A solution in equilibrium with this humidity would have a concentration in excess of 13 osmol.litre^{-1} and a freezing point depression of about 25°C (Ramsay, 1964, 1971; Grimstone et al., 1968). These findings suggested a reappraisal of the location of the site of uptake of water vapour from the atmosphere and the possible role of the rectum in this process. The involvement of the rectum in uptake has now been demonstrated in the firebrat, *Thermobia domestica* (Noble-Nesbitt, 1970a,b, 1973; 1975) and mealworm, *Tenebrio molitor* (Noble-Nesbitt, 1970a,b, 1973; Dunbar and Winston, 1975; Machin, 1975).

In the firebrat, uptake is arrested when the anus is experimentally blocked with wax, whereas uptake is scarcely affected by the application of wax to other parts of the body, including blocking the mouth (Noble-Nesbitt, 1970a,b). By carefully sealing the insect into a double-chamber array so that its head-end and its tail-end were contained in separate chambers which could be opened and closed independently, the effects of exposing the head-end and tail-end independently to RH's of 83—85% were followed (Noble-Nesbitt, 1973, 1975). Uptake occurs with the tail-end exposed, but not with the head-end exposed and this reversible arrest of uptake can be demonstrated repeatedly in sequence. This applies even if only the most posterior extremity of the insect, bearing the anal valves, is

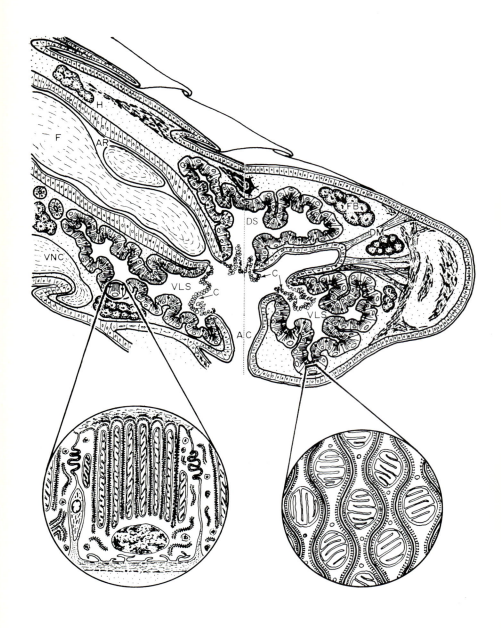

allowed to project into the tail-end chambers. The humidity within the closed tail-end chamber is reduced to about 50% RH, as shown by the gain or loss of water by small crystals of suitable salts enclosed within the chamber. This does not occur in the closed head-end chamber, where the humidity slowly rises. Movements of the anal valves indicate that the rectal system is operating. The involvement of the rectum in uptake in this insect is clear from these results, and the system will be further considered below (see p. 590).

Blockage of the anus also prevents uptake in the mealworm larva (Noble-Nesbitt, 1970a,b, 1973; Machin, 1975). Removal of the wax-resin cap used to block the anus is possible in this case, and when done early enough uptake can be restored fully. Delayed removal can lead to secondary blockage or even necrosis, affecting subsequent uptake (Noble-Nesbitt, 1973). Ligatures placed round the posterior abdomen prevent uptake, whereas more anteriorly placed ligatures do not (Dunbar and Winston, 1975). When the mealworm is carefully sealed in a double-chambered array uptake can take place if the tail-end chamber is open to an atmosphere of 93% RH at 20°C, but not if it is closed (Noble-Nesbitt, unpublished observations). Since the air spaces within the rectum are known to have a RH of 88% on average from Ramsay's own work, it seems justifiable to conclude that this is the site of uptake in this insect too. I shall deal with the rectal system of this insect in relation to uptake in more detail below (see p. 589).

Fig. 1. The posterior rectal sac complex of *Thermobia*. The main diagram shows to the left, vertical longitudinal and to the right, transverse profiles of the posterior region of the abdomen and indicates the arrangement of the sacs in relation to surrounding structures. Note the convoluted nature of the sac epithelium, forming smaller saccules within the three major sacs. Left inset: Enlarged view of a single sac epithelial cell, showing the hypertrophied apical plasma membrane associated with elongated mitochondria. Right inset: A still further enlarged view of a section parallel to the luminal surface of a sac epithelial cell, showing the hexagonal packing of the mitochondria surrounded by pleated infolds of the apical plasma membranes, in the normal condition with the adjacent plasma membranes closely opposed. A particulate coating covers the cytoplasmic surface of the plasma membrane. Key: AC = anal canal, AR = anterior rectum, BM = basement membrane, C = cuticle, lifted away from sac apithelial surface, DM = dilator muscle of posterior rectum, DS = dorsal sac of posterior rectum, F = faeces/faecal pellet, FB = fatbody, H = haemococele/haemolymph, VLS = ventro-lateral sacs of posterior rectum, VNC = ventral nerve cord.

However, it is unlikely that rectal uptake of water vapour will occur in all arthropods possessing this ability, or even in all insects. It is a peculiarity that in insects the rectum has the combined task of removing water from the products of digestion, and from the urine produced by the Malpighian tubules which empty into the gut. In cases where the two systems are separate we should be prepared to look for alternative sites of uptake, possibly associated with one system or the other. Such a case exists in ixodid ticks, where anal blocking does not arrest uptake (McEnroe, 1973; Knülle and Devine, 1972), and is considered in the next section. In the insects, some new observations also indicate the likelihood of alternative sites. Edney (personal communication) has reported that anal blockage arrests uptake in the desert cockroach, *Arenivaga investigata*, but so does blockage of the mouth, and it is possible that in this case the insect is reacting against the experimental procedure. Professor Edney and his colleagues kindly supplied me with some living specimens whilst I was on sabbatical leave at the University of Colorado at Boulder, and I was able to seal a few specimens successfully into a double-chambered array. These specimens were able to take up water from 93% RH. This uptake continued when the tail-end chamber (containing the posterior tip of the abdomen) was closed, but was arrested when the head-end was closed. Another site is apparently involved in this case (see also next section).

D. The Salivary Gland-buccal Cavity Complex

In ixodid ticks uptake can be arrested by complete blocking of the mouth with wax and restored by removing the wax according to Rudolph and Knülle (1974), despite earlier suggestions that the mouth was probably not involved in uptake in ticks (c.f. Beament, 1964). Using a technique which allowed the isolation of the mouthparts in a separate chamber, Rudolph and Knülle demonstrated that the ambient humidity surrounding the mouth approached 93% RH and they suggest that the saliva may be involved in this. They concluded that active uptake of water vapour takes place through the mouth parts in *Amblyomma variegatum, Rhipicephalus bursa, Hyalomma anatolieum excavatum* and *Hyalomma schulzei* (all of which were included in their experiments) and suggested that this may apply to all ixodid ticks. It is interesting that in ixodid ticks the salivary glands are excretory organs and this may account for their apparent involvement in water vapour uptake.

In the desert cockroach, *Arenivaga*, salivary secretions are frequently extruded from the mouthparts, it being common for a large droplet to appear and then to be withdrawn. Although this response has not been rigorously tested in relation to ambient humidity or to the state of the insect's water reserves, in view of the results outlined in the previous section, and the situation in ixodid ticks, the possibility that it is related to uptake of water vapour from the atmosphere must not be dismissed at this stage.

IV. THE MECHANISM OF UPTAKE OF WATER VAPOUR

The transport of water vapour from the atmosphere to the liquid phase of the arthropod's internal tissues clearly depends upon the life processes of the animal, and is disrupted during anaesthesia. To this extent it is an active physiological process. Whether it is brought about by the primary transport of an inorganic or organic solute not part of the actual pumping mechanism (and therefore classified as a case of secondary transport) or is the result of primary transport of water (and can therefore be termed active in the strict sense) has been the subject of considerable controversy, especially since many other examples of apparent primary transport of water can be explained on the basis of local recycling of solute within the transporting epithelium. In reviewing the function of the insect rectum in the transport of water, Ramsay (1971) called for the specific enunciation of the criteria required to satisfy the claim that a primary transport of water occurs in any system. This call has been answered by House (1974) and in essence his basic criterion is that water must be transported in the absence of any other flow of material save that involved in the metabolic reactions driving the water pumping mechanism itself. Whilst conceding that the uptake of water vapour, often against extremely steep gradients of chemical potential, is possibly the most compelling case for active transport of water, House remains unconvinced and as a verdict prefers one of "not-proven".

It is against this exacting standard that the transport of water vapour must be judged, but before considering a number of hypotheses which have been proposed to account for water transport and judging how far they are applicable to the uptake of water vapour, it is perhaps worth mentioning a few general points. First of all, many of the hypotheses derive from systems where the gradients are relatively small. Secondly, many are based on systems operating in insects which are not able to absorb water vapour. Thirdly, some systems would be incapable of producing the large gradient required. Fourthly, although some of them are reasonably well based on evidence accruing from varied sources, others are little more than speculation. Fifthly, whilst some of the less extreme cases of uptake may be satisfactorily explained by one mechanism or another, this in no way diminishes the requirement to demonstrate a suitable mechanism to account for the most extreme cases, wherein lie the greatest difficulties and the greatest challenge.

A. Temperature and Thermo-osmosis

As House (1974) points out, little experimental data on thermo-osmosis exists, and as a device for moving water it remains entirely speculative. In relation to water moving from the vapour phase to the liquid phase, a small drop in temperature can bring about condenstation from near-saturated air. This property formed the basis of Beament's (1964) demonstration of the high humidity within the insect tracheal system. Conceivably, greater temperature differences, or gradients, could cause condensation from lower humidities, and this has been advanced as a possible mechanism for making water vapour available to the tissues. For the water so condensed would be "pure" (distilled) water and freely able to diffuse into the tissues. In fact, this offers a device for reversing the gradient of chemical potential between tissue and atmosphere. Schmidt-Nielsen (1969) has considered just such a possiblity, estimating the thermal-gradient required between tissue and atmosphere to condense water at the interface between integument and atmosphere. For condensation from 45% RH (required in the case of *Thermobia*) the temperature difference would have to be about $14°C$ which could hardly be expected in any case and certainly not under constant temperature conditions where ambient temperature is maintained at $±1°C$ or better. As a general mechanism he rejected it, but House (1974) suggests that the integument-atmosphere interface may behave in such a way that quite small temperature gradients could exert marked effects on water uptake, though he stresses that this is mere speculation. Such behaviour would be quite different from the known behaviour of aqueous solution-air interfaces (where relative lowering of the vapour pressure is well understood) or solid-air interfaces (where dew point measurements provide extremely accurate estimates of humidity levels). The cuticle does have some interesting properties, but it seems unlikely that its behaviour in this respect will be sufficiently anomalous to account for the uptake of water vapour. One further point militates against this possibility too. Condensation requires a colder surface, yet the metabolizing insect is more likely to have a hotter surface because of internal heat production, and release of latent heat during water uptake would only exacerbate this situation (c.f. Lees, 1948).

B. Pinocytosis

It is possible that a secretion may be returned to the epidermal cells in the form of small vesicles of concentrated solution, taking with it water absorbed from the vapour phase, by characteristic pinocytotic activity of the apical membrane of the cells (Locke, 1964; Berridge, 1970). Once within the cell, metabolic activity may release the entrained water. There would nevertheless remain the problem of producing a sufficiently concentrated secretion capable of absorbing water hygroscopically from the air in the first place (c.f. Noble-Nesbitt, 1973, 1975).

C. Double-membrane and Local Osmosis Models

These depend upon producing high enough local concentrations of solute to draw water by osmosis across the membrane. Based on inorganic ions, these models have been applied successfully to many transporting epithelia, including the insect rectum (e.g. blowfly — Berridge and Gupta, 1967; cockroach — Oschman and Wall, 1969; Wall and Oschman, 1970; Wall et al., 1970), and evidently account for water transport up appreciable gradients. These insect rectal systems involve mitochondria-membrane complexes on the lateral margins of the rectal epithelial cells, in association with lateral sinuses, and contrast with the arrangement of the rectal tissue in *Tenebrio* and *Thermobia*. It is perhaps as well to keep the extent of the problem in perspective. At $20-25°C$, saturated KCl solution is in equilibrium with 85% RH, 7 M NH_4Cl solution with 80% RH, saturated NaCl solution with 75% RH, saturated K_2CO_3 solution with 43% RH and saturated $CaCl_2$ solution with 32% RH. Obviously, exceptional conditions would need to prevail if any of these salts are involved in the mechanism of uptake. The very high concentrations required should be readily detectable with modern techniques.

D. Recycling of Organic Molecules

The secretion of an organic molecule infinitely soluble in water (e.g. glycerol) has been suggested as a basis for uptake (Weis-Fogh, personal communication; Diamond, quoted in House, 1974). In theory, this could account for uptake from all RH's, but this does not fit in well with the observed species specific CEH's, which would have to be explained on some other basis. Once more we are faced with the problem of the release of the entrained water. In this case, it could possibly be achieved by metabolic degradation of the organic molecule within the epithelial cell. Conformational changes (possibly under the influence of ionic or pH changes within the cells) of hydrophilic macromolecules having low melting points and not easily crystallized have been suggested as a possible basis for uptake by Maddrell (1971), who refers to the compound with a mol. wt 10,000–12,000 found in the perirectal space of *Tenebrio* by Ramsay (Grimstone *et al.*, 1968) as a possible agent. If the macromolecules were located in the apical plasma membrane, then an energy-consuming system like this could constitute a primary water pump. Clearly, such a possibility deserves further investigation.

V. ORAL UPTAKE OF WATER VAPOUR IN IXODID TICKS

As we have seen before (p. 584) the recent work of Rudolph and Knülle (1974) has demonstrated that blockage of the mouth prevents uptake of water vapour in ixodid ticks. During uptake, a salivary discharge rich in K^+ and Na^+ forms on the mouthparts and the fluid is, presumably, taken into the mouth and so into the gut, where it may be assumed the water is made available to the internal tissues and the ions recycled to the salivary gland for further discharge. The solids of the salivary discharge are hygroscopic and absorb water from RH's of 80% or greater, but crystallize out in RH's of 75% or less. Thus the production of a highly concentrated solution within the salivary glands and its discharge onto the mouthparts would allow for absorption of water vapour from the surrounding atmosphere, with consequent dilution of the discharged fluid which could then be sucked into the alimentary tract, carrying with it entrained atmospheric water. The CEH for these ixodid ticks (80% RH) is within the range, for instance, of NaCl solutions (CEH = 75% RH), though the anions in the salivary discharge as yet remain to be identified. Just how the salivary glands are able to produce concentrated secretions of this degree is an intriguing question, an answer to which would be of great importance in our understanding of other systems of atmospheric water vapour uptake. Salivary glands clearly are able to handle water to a considerable degree, but this represents altogether a different level to that observed in usual salivary secretion.

One possibility is that active solute secretion by the salivary gland epithelium occurs through a membrane (perhaps the apical plasma membrane) which is highly impermeable to water and that the secretion is discharged before water can be withdrawn from the tissues. The alternative of withdrawal of water from a more dilute secretion before discharge is attended with much the same difficulties as uptake of water vapour and would merely serve to locate the problem at the site of water withdrawal.

Further information on this system would clearly be welcome.

VI.   RECTAL UPTAKE OF WATER VAPOUR IN THE MEALWORM

The cryptonephric system of *Tenebrio* has been described in relation to its physiology by Ramsay (Ramsay, 1964, 1971; Grimstone *et al.*, 1968). In this system the distal (blind) ends of the Malpighian tubules are hidden within a membranous bag surrounding the rectum, hence the term cryptonephric. The rectal epithelium is relatively unspecialized, without folded elaborations of the lateral, basal or apical plasma membranes closely associated with mitochondria, or particulate decorations on the cytoplasmic side of the apical plasma membrane. In these respects it is unlike either the rectal gland epithelium of the cockroach, locust or blowfly, or the rectal sac epithelium of the firebrat. The main elaborations of this system lie in the arrangement of the distal ends of the Malpighian tubules, and the perinephric membrane which separates the whole of the system from the haemolymph. This gives a series of compartments across the complex. The rectal lumen is separated from the perirectal space by the rectal epithelium. The perirectal space in turn is separated from the lumen of the Malpighian tubules by the tubule epithelium, and from the haemolymph by the perinephric membrane, which also separates the tubule lumen from the haemolymph. The perinephric membrane is a multilayered structure. At certain points, the leptophragmata, it is thinner and at these points the tubule wall is also thinner, associated with the occurrence of a special cell. It is these areas which are thought to be

U

responsible for active potassium transport without isotonic transfer of water, forming an important stage in the concentrating mechanism of the system leading to the drawing of water into the tubule lumen from the rectal side. Active transport of potassium by the anterior rectal epithelium results in the transport of water into the perirectal space from the rectal lumen. However, there is a steep gradient of water potential maintained over the posterior rectal epithelium between rectal lumen (5.5 osmol.litre^{-1}) and perirectal space (2.5 osmol.litre^{-1}), and this reflects a transport of water by the rectal epithelium by some method at present unknown. The posterior region of the rectal lumen contains dehydrated faecal pellets (which equilibrate with 88% RH on average, though some do so with as low as 75% RH.) in air spaces, presumably with RH also of these values (Ramsay, 1964). Since gut fluid entering the anterior part of the rectal complex is isosmolar to the haemolymph (0.55 osmol.litre^{-1}), the dehydrating mechanism undoubtedly resides within the complex. Since air spaces with RH as low as 75% occur within the rectal lumen, and since these air spaces can be brought into direct contact with the outside atmosphere by opening the anus, an avenue for the uptake of water from the atmosphere exists. It is this avenue which anal blocking (see p. 583) interrupts and so prevents uptake.

The lower limit of RH (75%) found to exist in the rectum represents an impressive ability for the transport of water vapour. At this RH, common salt (NaCl) would be crystallized out of solution. Any mechanism proposed to account for this uptake must take account of this.

Maddrell (1971) has developed the idea that the large organic molecules found in the perirectal space (Grimstone et al., 1968) may be involved in this water transport, but there remains the problem of presenting these molecules alternately to the rectal lumen and to the haemocoele, tissues or perirectal space and of releasing the water entrained by them and making it available to the internal tissues (Noble-Nesbitt, 1973). A change about the isoelectric point caused by ionic or pH changes is one possibility, but some valve-like arrangement may also be needed.

VII. RECTAL UPTAKE OF WATER VAPOUR IN THE FIREBRAT

In *Thermobia*, the anal sacs of the posterior rectum open to the exterior through a roughly triangular opening closed by three valves, one dorsal and two ventro-lateral. Of the three main lobes of the sacs, one is dorsal and two are ventro-lateral, corresponding to the anal valves. The sac epithelium in each lobe is thrown into further smaller folds, increasing the surface area. The maximum volume of these sacs in a 30 mg insect is approximately 250 nl. The epithelium itself is a simple layer of specialized cells with no clear subepithelial (perirectal) space or elaborations of the lateral or basal plasma membranes but with a deeply infolded apical (luminal) plasma membrane associated with a dense and regularly arranged concentration of elongated rod-like mitochondria, and having a 15 nm wide particulate coating on its inner cytoplasmic surface in very close association with the outer membrane of the adjoining mitochondrion. The cuticle overlying the epithelium is usually lifted away and the extensive subcuticular region contains a faintly granular material, which includes mucopolysaccharide which penetrates into the infolds of the plasma membrane (Noble-Nesbitt, 1970a,b, 1973; Noirot and Noirot-Timothée, 1971).

The sacs are approximately 1.0 mm long, 0.6 mm across and 0.7 mm deep in a 30 mg insect, but the nature of the external opening means that the distance from the exterior to the most remote parts of the sacs is not greater than approximately 0.7 mm. These dimensions are of importance in considering the rate at which water vapour from the ambient air can be presented to the sac epithelium. Maddrell (1971) concluded that diffusion over distances of not more than 1 mm would account for the observed rates of uptake, basing his calculations on a cylindrical rectum 1.5 mm long and 0.4 mm in diameter. A reconsideration of these calculations in relation to the measured dimensions given above indicates that with the anal valves fully open diffusion alone is amply sufficient. However, the valves during uptake show two types of behaviour. Periodic wide-opening of the valves occurs in some cases (Noble-Nesbitt, 1973), though it is more usual for incipient opening and closing at rates of up to 4—5 cycles sec^{-1} to occur (Noble-Nesbitt, unpublished observations). This latter behaviour suggests rapid tidal ventilatory cycles. When the dimensions of the very narrow Y-shaped opening under these conditions are taken into account, diffusion alone is insufficient, and requires augmentation by ventilation at approximately 2—4 cycles sec^{-1}. In desiccating conditions, the valves close tightly and thereby reduce water loss.

During the moult, when uptake is arrested, the sac epithelium undergoes profound changes with apparently complete disruption of the specialized apical region of the cells. This is reformed towards the end of the moult and reaches its full development at about the time of ecdysis (Noble-Nesbitt, 1973). Uptake recommences within four hours of ecdysis.

Changes in the ultrastructure of the cells also occur during cycles of desiccation and rehydration (Noble-Nesbitt, 1973). Hydrated insects show no gap between neighbouring outer plasma membrane surfaces in the apical infolds, except occasionally at the base of the infolds. In desiccated insects, distinct extracellular spaces can be seen in the infolds, unless the osmolarity of the fixative is increased, suggesting that these areas have a high osmotic activity. This may be of significance in the mechanism of uptake.

Some possible mechanisms of uptake will now be considered.

A. Pressure Changes

In considering compression of the rectum as a possible factor in uptake in the firebrat, Maddrell (1971) estimated the number of cycles of filling and emptying of the rectum (which he estimated can contain about 100 nl of air) required per sec. to achieve the maximal rates of uptake achieved by this insect ($5 \ \mu g.min^{-1}$ — Noble-Nesbitt, 1969), concluding that the rectum must fill and empty more than 80 times per sec. In fact, his estimate of the rectal volume is on the low side. Measurements of specimens of known weight and dimensions obtained from serial sections, indicate that the total volume of the rectal sacs is approximately 250 nl. Even so, the need to fill and empty some 30 times per sec would seem to be too high a frequency, and is much higher than the rate of pulsations observed, which occur at 2–5 times per sec (Noble-Nesbitt, 1973 and unpublished observations). Pressure changes probably do not play a part in uptake in *Thermobia*.

A similar conclusion is reached when volume changes are measured. Using an enclosed system containing a firebrat, it has been possible to monitor volume changes of down to 0.25 nl, well below those expected if compression of the rectal air spaces plays an important, or even only a contributory, part in uptake of water vapour. In fact only minute volume changes were recorded, even though the behaviour of the anal valves indicated that ventilation of the rectal sacs was occurring.

B. Inorganic Ion Transport

The CEH of the firebrat (45% RH) means that solutions of KCl (CEH = 83%) and NaCl (CEH = 75%) can be discounted as the vehicles for uptake. However, K_2CO_3 (44%), KNO_2 (43%) and $CaCl_2$ (32%) remain possibilities. If any of these salts or ions are involved in the transport of water across the rectal epithelium, they should occur at least locally in very high concentrations, well above the normal physiological concentrations found in cells or haemolymph. It would certainly be interesting to have some information about the presence or absence of these ions or salts, and their concentrations, within the rectal sac epithelium. To this end, I have been studying ionic profiles across the sac epithelium and comparing them with those of the haemolymph and other tissues, using electron-probe X-ray micro-analysis. Preliminary results indicate that no unusual distribution or concentration of the inorganic ions K, Na, Ca or Cl occurs, strongly suggesting that inorganic ion transport does not provide the key to water absorption in this case.

This conclusion is supported by the ultrastructure of the sac epithelium, which exhibits none of the lateral plasma membrane-mitochondria complexes or lateral sinuses found in other cases of ion-driven water transport in insect recta. (Also see Chapters 4 and 23.)

C. Organic Carrier Molecules

We have little evidence concerning organic carrier molecules as yet. It is conceivable that the muco-polysaccharide of the subcuticular region forms an essential part of the uptake mechanism, especially if it is continually being produced and reabsorbed (Locke model) or otherwise altered cyclically (Beament model). Moreover the apical membrane of the sac cell is decorated on its cytoplasmic side with 15 nm particles thought to be the site of ATPase activity (Gupta and Berridge, 1966; Berridge and Gupta, 1968) and these may be involved in transport of macromolecules across the membrane. Once inside, a conformational change (induced by ionic or pH changes) might occur, releasing the entrained water. It is possible that the elongated mitochondria within the apical folds may be involved in producing these pH or ionic changes, perhaps even maintaining a gradient along their length (Noble-Nesbitt, 1973), and that carrier molecules occur within the apical plasma membrane, with conformational changes occurring there. Such a system would be energy consuming, and would constitute a primary water pump (cf. Maddrell, 1971). It would also explain the very dense and regular array of elongated mitochondria in close association with a hypertrophied and infolded apical plasma membrane, though it does not preclude other explanations.

VIII. CONCLUSIONS

Improved location of the site of uptake is leading towards the answering of at least some of the outstanding questions surrounding the uptake of water vapour. In ixodid ticks a solution rich in Na and K presumably produced by the salivary glands is involved, and this suggests that transport of water vapour in these arthropods is secondary. Indeed many of the lesser cases of uptake, within the range of subsaturated NaCl or KCl solutions for instance, may prove to be driven by transport of inorganic ions. In the mealworm, however, the ionic composition of the tubule and perirectal fluids does not seem to account for uptake, which depends upon some as yet unresolved concentrating mechanism across the rectal epithelium. This may involve a macromolecule of 10,000–12,000 mol. wt. Transport of inorganic ions does not seem to provide the answer in the case of the firebrat, either, and a number of other possible mechanisms have been ruled out. Again the most promising line for future advance seems to lie in the involvement of a macromolecular mechanism, possibly involving a primary water pump. To this extent, the most exacting criteria required for the demonstration of active transport of water, are being met. The next stage lies in the further elucidation of the precise mechanism of uptake of water vapour in these specific cases where transport of inorganic ions apparently does not provide an answer. Whilst we may still have to concur with House's verdict of "not-proven", the case for active transport of water, so cogently argued by Ramsay, is further strengthened.

ACKNOWLEDGEMENTS

Some of the results quoted for *Tenebrio* and *Arenivaga* derive from work carried out whilst I was on sabbatical leave as a Senior Fulbright-Hayes Scholar at the Department of Environmental, Population and Organismic Biology, University of Colorado, Boulder under N.S.F. grant GB 14167 awarded to Professor P. W. Winston, to whom I am indebted for hospitality and facilities. The living *Arenivaga* were supplied by Professor E. B. Edney's laboratory and I am grateful to him and to Mr P. Franco who collected and despatched them for me. The work on electron-probe X-ray microanalysis of *Thermobia* rectum was carried out in the Biological Microprobe Laboratory, Department of Zoology, University of

Cambridge, supported by a grant from Science Research Council to (Late) Professor T. Weis-Fogh and Drs P. Echlin, B. L. Gupta, T. A. Hall and R. B. Moreton. I am also grateful to Dr D. A. Parry for facilities and Dr A. Saubermann for advice.

REFERENCES

Arlian, L. G. and Wharton, G. W. (1974). *J. Insect Physiol.* **20**, 1063—1077.

Beament, J. W. L. (1954). *Symp. Soc. exp. Biol.* **8**, 94—117.

Beament, J. W. L. (1964). *Adv. Insect Physiol.* **2**, 67—129.

Beament, J. W. L. (1965). *Symp. Soc. exp. Biol.* **19**, 273—298.

Beament, J. W. L., Noble-Nesbitt, J. and Watson, J. A. L. (1964). *J. exp. Biol.* **41**, 323—330.

Berridge, M. J. (1970). *In* "Chemical Zoology." (Ed. M. Florkin and B. T. Scheer), Vol. 5 pp. 287—319. Academic Press, New York and London

Berridge, M. J. and Gupta, B. L. (1967). *J. Cell Sci.* **2**, 89—112.

Berridge, M. J. and Gupta, B. L. (1968). *J. Cell Sci.* **3**, 17—32.

Browning, T. O. (1954). *J. exp. Biol.* **31**, 331—340.

Buxton, P. A. (1930). *Proc. R. Soc. Lond. B.* **106**, 560—577.

Devine, T. L. and Wharton, G. W. (1973). *J. Insect. Physiol.* **19**, 243—254.

Dunbar, B. S. and Winston, P. W. (1975). *J. Insect Physiol.* **21**, 495—500.

Ebeling, W. (1974). *In* "The Physiology of Insecta." (Ed. M. Rockstein), 2nd Ed. Vol. VI, pp. 271—343. Academic Press, New York and London.

Edney, E. B. (1947). *Bull. ent. Res.* **38**, 263—280.

Edney, E. B. (1957). "The Water Relations of Terrestrial Arthropods." Cambridge University Press, Cambridge.

Edney, E. B. (1966). *Comp. Biochem. Physiol.* **19**, 387—408.

Grimstone, A. V., Mullinger, A. M. and Ramsay, J. A. (1968). *Phil. Trans. R. Soc. Lond. B* **253**, 343—382.

Gupta, B. L. and Berridge, M. J. (1966). *J. Morph.* **120**, 23—81.

Heeg, J. (1967). *Zoologica Africana* **3**, 21—41.

House, C. R. (1974). "Water Transport in Cells and Tissues." Edward Arnold, London.

Humphries, D. A. (1967). *Nature Lond.* **214**, 426.

Kanungo, K. (1965). *J. Insect Physiol.* **11**, 557—568.

Knülle, W. (1962). *Z. vergl. Physiol.* **45**, 233—246.

Knülle, W. (1965). *Z. vergl. Physiol.* **49**, 586—604.

Knülle, W. (1967). *J. Insect Physiol.* **13**, 333—357.

Knülle, W. and Devine, T. L. (1972). *J. Insect. Physiol.* **18**, 1653—1664.

Knülle, W. and Spadafora, R. R. (1969). *J. stored Prod. Res.* **5**, 49—55.

Knülle, W. and Spadafora, R. R. (1970). *J. econ. Entomol.* **63**, 1069—1070.

Knülle, W. and Wharton, G. W. (1964). *Proc. 1st int. Congr. Acarology, Acarologia* **6**, 299—306.

Larson, D. (1969). "The critical equilibrium activity of adult females of the house dust mite, *Dermatophagoides farinae* Hughes." Ph.D. Dissertation, Ohio State University, Columbus, Ohio.

Lees, A. D. (1946). Parasitology 37, 1—20.

Lees, A. D. (1947). *J. exp. Biol.* 23, 379—410.

Lees, A. D. (1948). *Discuss. Faraday Soc.* 3, 187—192.

Lees, A. D. (1964). *Proc. 1st int. Congr. Acarology, Acarologia* 6, 315—323.

Locke, M. (1964). *In* "The Physiology of Insecta." (Ed. M. Rockstein), Vol. III, pp. 379—470. Academic Press, New York, and London.

Ludwig, D. (1937). *Physiol. Zool.* 10, 342—351.

Machin, J. (1975). *J. Comp. Physiol. B* 101, 121—132.

Maddrell, S. H. P. (1971). *In* "Advances in Insect Physiology." (Ed. J. W. L. Beament, J. E. Treherne and V. B. Wigglesworth), Vol 8, pp. 199—331. Academic Press, London and New York.

McEnroe, W. D. (1973). *Acarologia* 14, 542—543.

Mellanby, K. (1932). *Proc. R. Soc. Lond. B.* 111, 376—390.

Mellanby, K. (1934). *Proc. R. Soc. Lond. B.* 116, 139—149.

Noble-Nesbitt, J. (1969). *J. exp. Biol.* 50, 745—769.

Noble-Nesbitt, J. (1970a). *Nature, Lond.* 225, 753—754.

Noble-Nesbitt, J. (1970b). *J. exp. Biol.* 52, 193—200.

Noble-Nesbitt, J. (1973). *In* "Comparative Physiology." (Eds L. Bolis, K. Schmidt-Nielsen and S. H. P. Maddrell), pp. 333—351. North-Holland Publishing Co., Amsterdam; Elsevier, New York.

Noble-Nesbitt, J. (1975). *J. exp. Biol.* 62, 657—669.

Noirot, Ch. and Noirot-Timothée, C. (1971). *J. Ultrastruct. Res.* 37, 335—350.

Okasha, A. Y. K. (1971). *J. exp. Biol.* 55, 435—448.

Okasha, A. Y. K. (1972). *J. exp. Biol.* 57, 285—296.

Oschman, J. L. and Wall, B. J. (1969). *J. Morph.* 127, 475—510.

Phillips, J. E. (1964a). *J. exp. Biol.* 41, 15—38.

Phillips, J. E. (1964b). *J. exp. Biol.* 41, 68—80.

Phillips, J. E. (1969). *Canad. J. Zool.* 47, 851—863.

Ramsay, J. A. (1935). *J. exp. Biol.* 12, 373—383.

Ramsay, J. A. (1964). *Phil. Trans. R. Soc. Lond. B.* 248, 279—315.

Ramsay, J. A. (1971). *Phil. Trans. R. Soc. Lond. B.* 262, 251—260.

Rudolph, D. and Knülle, W. (1974). *Nature, Lond.* 249, 84—85.

Schmidt-Nielsen, K. (1969). *Q. Rev. Biophys.* 2, 283—304.

Schmidt-Nielsen, K. (1975). "Animal Physiology: Adaptation and Environment." Cambridge University Press, Cambridge.

Solomon, M. E. (1962). *Ann. appl. Biol.* 50, 178—184.

Stobbart, R. H. and Shaw, J. (1974). *In* "The Physiology of Insecta." (Ed. M. Rockstein), 2nd Ed. Vol. V, pp. 361—446. Academic Press, New York and London.

Wall, B. J. and Oschman, J. L. (1970). *Am. J. Physiol.* 218, 1208—1215.

Wall, B. J., Oschman, J. L. and Schmidt-Nielsen, B. (1970). *Science* 167, 1497—1498.

Watson, J. A. L. (1967). *J. Insect Physiol.* 13, 1689—1698.

Wharton, G. W. and Devine, T. L. (1968). *J. Insect. Physiol.* 14, 1303—1318.
Wharton, G. W. and Kanungo, K. (1962). *Ann. ent. Soc. Am.* 55, 483—492.
Wigglesworth, V. B. (1932). *Q. J. microsc. Sci.* 75, 131—150.
Wigglesworth, V. B. (1972). "The Principles of Insect Physiology." 7th Edition.
    Chapman and Hall, London.

23. Fluid Transport in The Cockroach Rectum

B. J. Wall

Marine Biological Laboratory, Woods Hole, Massachusetts, U.S.A.

I. INTRODUCTION

One of the highlights of our frequent visits to Cambridge since 1967 has been our discussions with Professor Ramsay. I had enjoyed his description of the device he used to collect and weigh fecal pellets of *Tenebrio* (Ramsay, 1964), and it was therefore most exciting to be shown the original model, as well as the other apparatus he had constructed and used during his various studies. He always had a lively interest in our work, and we are indebted to him for many timely (and sometimes humorous) suggestions.

During the last few years we have witnessed portions of Ramsay's struggle with the cryptonephric system of various Lepidoptera. Each year a few more pieces were added to the puzzle, but he always expressed doubt about whether the outcome would be worth publishing. Fortunately he succeeded in the end, and the result (Ramsay, 1976) is another inspiring classic.

In this paper I shall discuss an aspect of rectal function in which Ramsay has long been interested in: the possibility that organic solutes are involved in the absorption of water by the rectum. In his studies on the cryptonephric complex of *Tenebrio*, Ramsay (1964) found that the fluid in the space surrounding the rectum (perirectal space) has a high osmotic pressure — as high as 4500 milliosmoles (mOs). The high osmotic pressure is largely due to a non-electrolyte that renders the fluid viscous and makes it difficult to determine its freezing point depression. Faecal pellets in the posterior portion of *Tenebrio* rectum are not in direct contact with the rectal epithelium but are separated from it by an air space. Hence the final step in water absorption is from a vapour phase. The precise mechanism of water absorption from the vapour phase is unknown, but the peculiar non-electrolytes that accumulate in the posterior portion of the perirectal space may be involved.

The firebrat, *Thermobia*, is capable of absorbing water from atmospheres down to 45% relative humidity. The posterior part of the hindgut, the anal sac, is thought to be responsible for the uptake of water vapour (Noble-Nesbitt, 1969a,b, 1970, 1975 and in this volume). Ultrastructural studies on the hindgut of *Thermobia* have revealed that there is an electron-dense material present in the subcuticular space of the anal sac (see Noirot and Noirot-Timothée, 1971). The dense material may be functionally analogous to the non-electrolyte in the perirectal spaces of *Tenebrio* or the electron-dense material secreted into the intercellular spaces of *Calliphora* rectal papillae following the injection of distilled water into the lumen (Berridge and Gupta, 1967). In the recta from dehydrated cockroaches there is also an electron-dense material in the dilated intercellular spaces (Oschman and Wall, 1969). From the results of studies I shall describe in this paper, it seems that electrolytes make up only 50—60% of the osmotically active solutes in the rectal tissue. This would be consistent with an involvement of non-electrolytes in water absorption by the cockroach rectum.

II. ANALYSIS OF RECTAL TISSUE

These studies were done to determine how the osmolality and Na and K concentrations of the rectal tissue compared with those of haemolymph and rectal lumen contents. Both dehydrated and hydrated cockroaches, *Periplaneta americana*, were used. Rectal tissue and colon tissue were dissected from the animals and placed in preweighed Beckman tubes containing 10 μl H_2O. The tubes were sealed, weighed, frozen, boiled and then stored in the refrigerator for 24 hours. The fluid samples were analysed for osmotic pressure, using the method of Ramsay and Brown (1955), and for Na and K concentrations using a Baird-Atomic model KY-2 flame photometer. Samples from the haemolymph, rectal lumen and colon lumen were collected and stored under liquid paraffin and then analysed.

Figure 1 shows that in dehydrated animals the osmolality of the rectal tissue is lower than that of the fluid in the rectal lumen. The Na concentration in the tissue is similar to that of the fluid in the rectal lumen. The K concentration of the tissue is less than half that

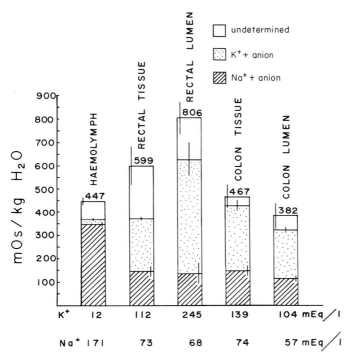

Fig. 1. Composition of samples obtained from dehydrated animals (provided with food but not water). Values are means ± 2SE, N = 11. Values for colon lumen are from Wall and Oschman (1970).

of the lumen. Of the total osmolality in rectal tissue, 38% is accounted for by solutes other than Na + K + accompanying anions. In the rectal lumen, 22% of the total osmolality is unaccounted for.

Figure 2 shows that in hydrated animals the osmolality of the rectal tissue is higher than that of fluid in the rectal lumen. The Na concentration in the tissue is higher than that of the fluid in the rectal lumen. The K concentration of the tissue is also higher than that of the lumen. Of the total osmolality of rectal tissue, 36% is accounted for by solutes other than Na + K + accompanying anions. In the rectal lumen, 50% of the total osmolality is unaccounted for.

The sum of Na + K in rectal tissue of dehydrated cockroaches is higher compared to Na + K in hydrated cockroaches. However, the proportion of Na + K + accompanying anions of the total osmotic pressure in rectal tissue of both dehydrated and hydrated cockroaches are 62% and 64%, respectively, even though the rectal

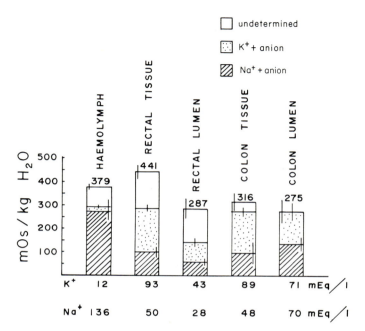

Fig. 2. Composition of samples obtained from hydrated animals (provided with water but not food). Values are means ± 2SE, N = 4. Values for haemolymph and colon lumen are from Wall and Oschman (1970).

tissue of hydrated animals is less concentrated than that of dehydrated animals. Furthermore, the contribution of Na (11 or 12%) or K (19 or 21%) to the total osmotic pressure is similar, whether from dehydrated or hydrated animals. Thus the composition of the tissue seems to be regulated so that the proportion of various solutes remains the same even though the total osmotic concentration changes.

Comparisons of the osmotic pressures and Na and K concentrations of the rectal tissue with that of the rectal lumen shows that in hydrated animals ions are concentrated in the tissue. This and results of other studies in which sinus fluid was collected and analysed (see Wall and Oschman, 1970) suggest that hydrated animals can absorb Na and K from the rectal lumen and return these ions to the blood (see Fig. 2). The fluid which enters the rectal lumen from the colon is rich in Na and K and the osmotic pressure is totally accounted for by Na + K + accompanying anion (Wall and Oschman, 1970).

In dehydrated animals there is also evidence that Na is reabsorbed from the rectal lumen. The fluid from the colon which enters the rectal lumen has an osmolality of 382 mOs, contains 57 mEquivalent l^{-1} Na and 104 mEquivalent l^{-1} K (Wall and Oschman, 1970). This fluid becomes concentrated in the rectal lumen, but when the fluid in the rectal lumen is 806 mOs, the Na is only 68 mEquivalent l^{-1}, about half that expected from the increased osmotic pressure if all the Na remained in the lumen (see Fig. 1). The K concentration in the rectal lumen is what one would expect if K remained in the lumen. There is also evidence that Na may be actively absorbed from the rectal lumen in the locust (Phillips, 1964b).

The rectum is able to return to the blood a fluid which is less concentrated than the fluid in the rectal lumen since the sinus fluid is hypo-osmotic to rectal lumen fluid. Since the whole rectal tissue is less concentrated than the lumen, either water moves against its gradient (which seems unlikely) or the tissue is not homogeneous in its osmotic concentration. The latter possibility is being confirmed in studies now being conducted on frozen hydrated sections of *Calliphora* rectal papillae using electron microprobe x-ray analysis (Gupta *et al.*, unpublished observations; see Chapter 4).

III.   ANALYSIS OF FLUID IN INTERCELLULAR SPACES

Curran (1960) and Diamond (1964) have suggested models in which there are compartments (e.g. intercellular spaces) within epithelia which could have a higher osmotic pressure than the cells and the fluid bathing the cells. The movement of water is thought to be the consequence of an osmotic gradient established by active transport of some solute into a confined space or infolding of the cell surface.

I had observed that the intercellular spaces in the rectal pad of the cockroach opened up when dehydrated animals were given water to drink (see Fig. 3 for a diagram of the structure of the rectal pad). It was possible to obtain samples of fluid from the intercellular spaces of the rectal pads and to determine the osmolality, and Na and K concentration. First the animals were dehydrated, then the rectum was injected with fluid which contained a dye to make the lumen dark and aid in viewing the intercellular spaces. The spaces could be seen opening up within a minute after injection. A pipette filled with oil could be introduced into the intercellular space so that the tip could be seen in the space. When the intercellular fluid was sucked up into the pipette, the intercellular space collapsed. The sample was then removed and analysed for osmolality and Na and K

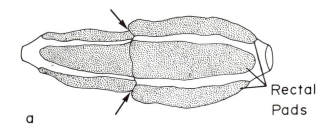

Rectal
Pads

a

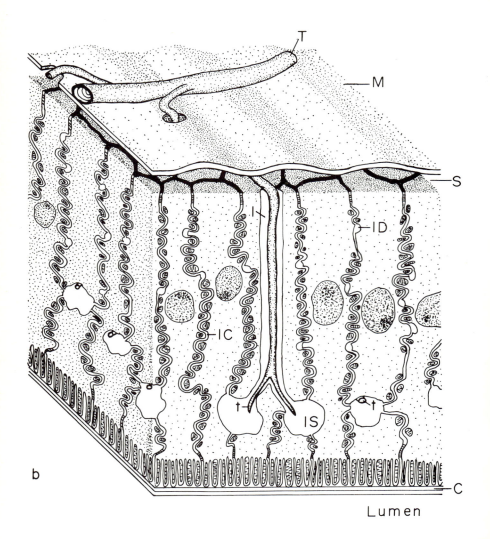

b

Lumen

concentrations using a micro-flame photometer (Hampel Scientific Instruments, Frankfurt).

Table I shows the results of the measurements. Fluid in the intercellular spaces is more concentrated than the fluid in the rectal lumen (see also Wall *et al.*, 1970), except for one determination (animal no. 3, sample 1). When the injected fluid contained 100 mOs NaCl, the Na : K ratio was 2.8 : 1 for the fluid in the intercellular space. When the injected fluid contained 100 mOs KCl, the Na : K ratio was 1 : 1.2 for the fluid in the intercellular space.

The composition of the fluid of the intercellular space may be accounted for partly by what was already present when the rectal lumen was injected. However, since the Na : K ratios are not exactly opposite depending on whether Na or K was injected, this may mean that when Na is present in the lumen this ion is quickly pumped into the intercellular spaces, but that the same does not happen with K.

When the injected fluid contained trehalose and either NaCl or KCl, the total Na + K + accompanying anion was about 54% of the total osmolality of the fluid of the intercellular space. Since nearly half of the total osmolality is unaccounted for, it seems likely that non-electrolytes may be involved in creating a high osmotic concentration in the intercellular space.

IV. NON-ELECTROLYTES

A. Amino Acids

What are the non-electrolytes which may be present in the rectum and involved in water absorption? Amino acids seem to be absorbed by the rectum of the cockroach (Wall and Oschman, 1970) and

Fig. 3. Organization of rectal pads in *Periplaneta americana*, based on Oschman and Wall (1969). a. Rectum as seen *in vivo*, showing rectal pads and radial fold (between arrows) in muscle. b. Diagrammatic representation of portion of rectal pad, based on electron microscopy. A layer of cuticle C faces lumen and a muscle sheath (M) faces haemolymph. Between pads and muscle layer is a narrow subepithelial sinus (S). Columnar pad cells are characterized by presence of numerous mitochondria located in close contact to highly folded lateral and apical plasma membranes. Between cells are narrow 200 Å wide intercellular channels (IC) with occasional dilations (ID). Septate desmosomes seal intercellular channels from rectal lumen and from sinus. Tracheae (T) penetrate through muscle layer and extend along basement membrane-lined indentations (I) of basal cell surface. In apical region tracheae branch into fine tracheoles (t) that enter large intercellular spaces (IS). Narrow intercellular channels between cells open into larger intercellular spaces. (From Wall and Oschman, 1970).

TABLE I

Analysis of fluid in intercellular spaces

	Osmolality mOs/kg H_2O	Na mEq/kg H_2O	K mEq/kg H_2O
1 Animal dehydrated 11 days, defecated with CO_2 anesthesia, injected 500 mOs solution into rectal lumen (400 mOs trehalose, 100 mOs NaCl). I.S. = fluid from intercellular space.			
Sample			
haemolymph	394		
rectal lumen after sampling I.S.	543		
fluid from intercellular space 1	585	102	36
fluid from intercellular space 2	563		
2 Animal dehydrated 5 days, other conditions same as above			
excreted pellet	1024		
haemolymph	399		
rectal lumen after sampling I.S.	631		
fluid from intercellular space 1	671	126	70
fluid from intercellular space 2	1275	323	93
3 Animal dehydrated 7 days, injection fluid (400 mOs trehalose, 100 mOs KCl), other conditions same as above			
haemolymph	392		
rectal lumen after sampling I.S.	511		
fluid from intercellular space 1	505	69	63
fluid from intercellular space 2	521	36	59
fluid from intercellular space 3	558	39	78
fluid from intercellular space 4	558	93	81

locust (Balshin and Phillips, 1971). Balshin and Phillips also found that the total amino acid content in the rectal tissue was 125 mM l^{-1} (60% proline). The physiological state of the locusts was not reported and may influence the total amino acid content of the tissue. However, amino acids could account for a significant portion of the osmolality of the tissue and of the fluid in the intercellular spaces.

The amino acid content in the blood of another locust varies inversely with the protein content during dehydration and rehydration of the animal (Djajakusumah and Miles, 1966). When the blood volume decreases during dehydration, the amino acid content decreases, and when the blood volume increases, the amino acids increase and contribute one-third of the osmotically active solutes added to the blood. These studies suggest that the amino acid content of the blood can be closely controlled and that amino acids can contribute to the osmotic activity of the blood.

B. Sugars

In the cockroach glucose absorbed across the midgut is rapidly converted to trehalose which is the main blood sugar (Treherne, 1957). Cockroach blood contains 2.2 mM l^{-1} glucose and 36.9 mM l^{-1} trehalose (Treherne, 1960). In the present experiments most of the remainder of the non-ionic haemolymph osmolality is probably accounted for by trehalose. Perhaps this is also true for the sinus fluid.

Colon fluid is derived from fluid secreted by the Malpighian tubules and/or fluid from the midgut. Malpighian tubules of other insects (locusts, blowflies, stick insects, blood-sucking bugs) are permeable to a whole range of organic solutes (Ramsay, 1958; Maddrell and Gardiner, 1974). Although cockroaches have not been studied, it seems likely that their Malpighian tubules also filter amino acids and sugars from the blood. The fluid may be modified in the colon, both by absorption across the colon wall and by symbiotic organisms residing in the colon lumen. When the colon fluid enters the rectal lumen it still contains amino acids, but we do not know if it contains trehalose or glucose.

The rectal cuticle of the locust is relatively impermeable to disaccharides (Phillips and Dockrill, 1968). Phillips (1964a) injected a hyperosmotic trehalose solution into the rectal lumen and found that water but not trehalose was absorbed. Similar studies have not been done on the cockroach rectum and we do not know if trehalose can cross the cuticle or if it is absorbed. If trehalose does enter the locust (or cockroach) rectal lumen from the colon, then it seems likely it is excreted. It is conceivable that the symbionts in the colon hydrolyse trehalose into glucose, which could be absorbed by the colon and/or rectum. Glucose can cross the rectal cuticle (Phillips and Dockrill, 1968).

C. Glycerol

Both H. H. Ussing and the late T. Weis-Fogh (1970, personal communication) have suggested that an organic molecule that is both highly water soluble and readily cycled through cellular metabolic pathways could be utilized repeatedly to generate osmotic forces to cause water movement. Both Ussing and Weis-Fogh mentioned glycerol as a likely candidate for this role. Glycerol could be synthesized from glycogen that is found in the rectal epithelial cells (see Grimstone et al., 1968; Berridge and Gupta, 1967).

Interestingly, glycerol is produced in very high concentrations in insects that over-winter, and brings about a considerable depression in the temperature at which these insects freeze (see Wyatt, 1967; Wyatt and Meyer, 1959; Chino, 1957, 1958). Glycerol is synthesized in these insects from glycogen as the temperature drops, and is readily reconverted to glycogen (Baust, 1972). The possibility thus arises that comparable metabolic pathways and physico-chemical tricks might be utilized by insects to carry out the functions of protecting against freezing and absorbing water, since both physiological problems are solved by producing highly concentrated aqueous solutions of non-electrolytes.

D. Glycoproteins

One can now look to other mechanisms used by animals to protect themselves from freezing to see if the molecules involved might have potential as water absorbing agents. For example, antarctic fish utilize "antifreeze" glycoproteins that have remarkable colligative properties. These molecules depress the freezing point several hundred times further than expected on the basis of their mol. wt (see Feeney, 1974; DeVries *et al.*, 1970). Much work has been done on the physical chemistry of these molecules by the authors cited above, and it is apparent that a greatly expanded "domain" of the glycoproteins is responsible for their unusual properties in aqueous solutions. Interestingly, glycoproteins are present in high concentration in insect haemolymph. For example, about half of the blood protein in *Periplaneta* is glycoprotein (Wigglesworth, 1972).

V. CONCLUSIONS

The lateral membranes of the blowfly rectal papillae and cockroach rectal pads are closely associated with mitochondria and form narrow intercellular channels which open into large intercellular spaces (see Gupta and Berridge, 1966; Oschman and Wall, 1969). The lateral membranes are likely sites for active solute transport. An ATPase has been localized at the lateral membranes of the blowfly rectal papillae (Berridge and Gupta, 1968). Gupta and Berridge (1966) also suggested that ions may be recycled within the rectal epithelium or recruited from the blood.

Both Na and amino acids may be actively absorbed by the rectal pad cells in the cockroach, and may be recycled within the rectal pad along with other solutes. In a variety of organisms (see Potts, 1968), amino acids and other organic solutes are used to increase the concentration of blood and tissues. It is reasonable that insects might use these molecules to reabsorb water. As discussed earlier, in the cockroach, organic solutes may account for a large part of the total osmolality in the tissue as well as in the fluid of the intercellular spaces.

None of the studies done to date have clearly established the mechanism of fluid absorption by the insect rectum. However, from the information available one can suggest a scheme that will provide a framework for further investigations (see Fig. 4).

Sodium and organic solutes, e.g. amino acids, are probably transported into the cell across the apical folds. In dehydrated animals water may be absorbed from the lumen into the cell, although the coupling mechanism is unclear. The permeability of the apical membrane may be under hormonal control, and under antidiuretic conditions may become more permeable (see Chapter 11). Solutes within the cell are probably transported into the intercellular channels. The membranes of the intercellular channels may be relatively impermeable to water, so a high concentration of solutes accumulates and then flows into the larger intercellular spaces which are lined by more permeable membranes. Water could then flow into the intercellular spaces from the lumen either via the cell and/or the junction between the cells. The solutes to be recycled could then be taken back into the cell along the basal membrane as the absorbed fluid leaves the rectum and enters the haemolymph. The basal membranes may also be relatively impermeable to water. The solutes would then be recycled back into the intercellular channels to maintain water absorption.

One would expect to find a higher concentration of recycling organic solutes within the rectal epithelium when the cockroach is dehydrated and absorbing water against a very high osmotic gradient. One would predict that the fluid in the intercellular spaces would show the highest concentration of the organic solutes compared to that in the cells, lumen and sinus fluids. As mentioned above, micro-probe analysis is providing fresh information on the distribution of electrolytes within the rectal tissue. It is perhaps somewhat ironical that this long-awaited information is becoming available at a time when we are beginning to suspect that non-electrolytes may play a key role in rectal absorption. (Also see Chapter 22.)

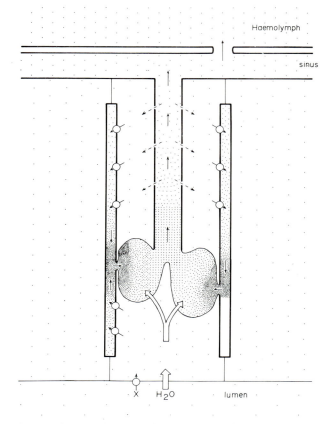

Fig. 4. Proposed explanation of water uptake by rectal pads of cockroach, *Periplaneta americana*. Some solute, designated as X, is absorbed from rectal lumen and transported into narrow intercellular channels. Concentrated fluid flows into larger intercellular spaces, generating local osmotic gradients that bring about osmotic water uptake from lumen into spaces. Absorbed fluid flows along indentations of basal plasma membrane (solid arrows), into sinus, and through gaps in muscle layer into haemolymph. When solute X is not present in lumen, system continues to function by recycling (dashed arrows). Solute X could also be an organic molecule that is recycled through cellular metabolic pathways. Solute concentration gradient between indentations and sinus on one hand and cell interior on the other (indicated by density of dots) may be such that movement of solute back into the cell is downhill. Thicker lines along intercellular channels and basal surface are to indicate that the cell membranes in these regions may be relatively permeable to solute but not to water. (From Wall and Oschman, 1975).

ACKNOWLEDGEMENTS

I thank J. L. Oschman, B. L. Gupta and R. B. Moreton for their comments on the manuscript. Part of this work was done in the laboratory of B. Schmidt-Nielsen and I thank her for her support. This work has been supported by National Institutes of Health grants FR-7028 and AM-14993 to J. L. Oschman, AM 09975-04 to B. Schmidt-Nielsen.

REFERENCES

Balshin, M. and Phillips, J. E. (1971). *Nature New Biol. Lond.* **233**, 53—55.

Baust, J. G. (1972). *Nature Lond.* **236**, 219—220.

Berridge, M. J. and Gupta, B. L. (1967). *J. Cell Sci.* **2**, 89—112.

Berridge, M. J. and Gupta, B. L. (1968). *J. Cell Sci.* **3**, 17—32.

Chino, H. (1957). *Nature Lond.* **180**, 606—607.

Chino, H. (1958). *J. Insect Physiol.* **2**, 1—12.

Curran, P. F. (1960). *J. gen. Physiol.* **43**, 1137—1148.

DeVries, A. L., Kamatsu, S. K. and Feeney, R. E. (1970). *J. biol. Chem.* **245**, 2901—2908.

Diamond, J. M. (1964). *J. gen. Physiol.* **48**, 15—42.

Djajakusumah, T. and Miles, P. W. (1966). *Aust. J. biol. Sci.* **19**, 1081—1094.

Feeney, R. E. (1974). *Am. Scient.* **62**, 712—719.

Grimstone, A. V., Mullinger, A. M. and Ramsay, J. A. (1968). *Phil. Trans. R. Soc. Lond. B* **253**, 343—382.

Gupta, B. L. and Berridge, M. J. (1966). *J. Morph.* **120**, 23—81.

Maddrell, S. H. P. and Gardiner, B. O. C. (1974). *J. exp. Biol.* **60**, 641—652.

Noble-Nesbitt, J. (1969a). *Nature Lond.* **225**, 753—754.

Noble-Nesbitt, J. (1969b). *J. exp. Biol.* **50**, 745—769.

Noble-Nesbitt, J. (1970). *J. exp. Biol.* **52**, 193—200.

Noble-Nesbitt, J. (1975). *J. exp. Biol.* **62**, 657—669.

Noirot, C. and Noirot-Timothee, C. (1971). *J. Ultrastruct. Res.* **37**, 335—350.

Oschman, J. L. and Wall, B. J. (1969). *J. Morph.* **127**, 475—510.

Phillips, J. E. (1964a). *J. exp. Biol.* **41**, 15—38.

Phillips, J. E. (1964b). *J. exp. Biol.* **41**, 39—67.

Phillips, J. E. and Dockrill, A. A. (1968). *J. exp. Biol.* **48**, 521—532.

Potts, W. T. W. (1968). *A. Rev. Physiol.* **30**, 73—104.

Ramsay, J. A. (1958). *J. exp. Biol.* **35**, 871—891.

Ramsay, J. A. (1964). *Phil. Trans. R. Soc. London B* **248**, 279—314.

Ramsay, J. A. (1976). *Phil. Trans. R. Soc. London B* **274**, 203—226.

Ramsay, J. A. and Brown, R. H. J. (1955). *J. sci. Instrum.* **32**, 372—375.

Treherne, J. E. (1957). *J. exp. Biol.* **34**, 478—485.

Treherne, J. E. (1960). *J. exp. Biol.* **37**, 513—533.

Wall, B. J. and Oschman, J. L. (1970). *Am. J. Physiol.* **218**, 1208—1215.

Wall, B. J. and Oschman, J. L. (1975). *In* "Excretion." (Ed A. R. E. Wessing), Fortschritte der Zoologie Vol. 23, pp. 193—222. Gustav Fischer Verlag, Stuttgart.

Wall, B. J., Oschman, J. L. and Schmidt-Nielsen, B. (1970). *Science* **167**, 1497—1498.

Wigglesworth, V. B. (1972). "The Principles of Insect Physiology." 7th edition. Chapman and Hall, London.

Wyatt, G. R. (1967). *In* "Advances in Insect Physiology." Vol. 4, pp. 287—360. Academic Press, London and New York.

Wyatt, G. R. and Meyer, W. L. (1959). *J. gen. Physiol.* **42**, 1005—1011.

24. Fluid Movement Through The Crayfish Antennal Gland

J. A. Riegel

Department of Zoology, Westfield College, University of London, London, England

This contribution contains ideas which are the product of my own research and that of others. It also contains many ideas which owe their origins to discussions held with Arthur Ramsay. Probably he will not recognize them in the form in which they are presented here. Usually they arose out of my reaction to advice which was always sound and criticism which was unfailingly constructive.

I. INTRODUCTION

The antennal glands ("kidneys") of the crayfish are paired organs lying in the second antennal segment of the head. As shown in Fig. 1, each gland consists of a coelomosac, labyrinth, tubule and bladder. The coelomosac is a remnant of the coelom which, in arthropods in general, is replaced by extensive blood spaces (haemocoels). The labyrinth consists of a thin layer of spongy tissue which encloses the rest of the gland parts except the bladder. The tubule is quite long (two centimetres or so); its tubular shape is purely superficial; the interior is divided into a myriad of spaces which, like those of the labyrinth, are formed by anastomoses of the walls of an original simple tube. The bladder is large and overlies the rest of the antennal gland.

The antennal gland of the crayfish is well served by an arterial blood supply. The antenno-renal artery has a branch which serves the coelomosac exclusively. This same artery sends branches to the

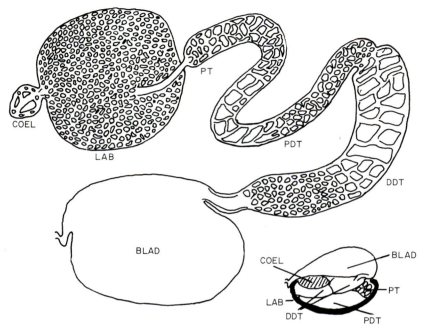

Fig. 1. (Left). View of a teased-out antennal gland, illustrating the relative shapes and sizes of the parts and the spaces within. (Right). Sagittal section of the antennal gland as it is *in situ*. Abbreviations: COEL = coelomosac, LAB = labyrinth, PT = proximal tubule, PDT = proximal part of the distal tubule, DDT = distal part of the distal tubule, BLAD = bladder.

anterior portion of the labyrinth, tubule and bladder. The posterior portion of the labyrinth and tubule are served by the sternal artery. There is no equivalent of a venous blood supply as found in animals with closed circulatory systems. The foregoing description is based on the work of Maluf (1939), Peters (1935) and personal observations.

In recent years the ultrastructure of several parts of the antennal gland has been studied. The coelomosac and labyrinth have significance for the present discussion; their fine structure will be reviewed. Figure 2 is a drawing of a few epithelial cells of the coelomosac. The cells are very tall, and they are unconnected at their lateral borders. They are podocytic: that is, their bases form into interdigitating processes which, in transverse section, make it appear as if the part of the cell abutting on the basement membrane is columnar. Between the interdigitating basal processes are mem-

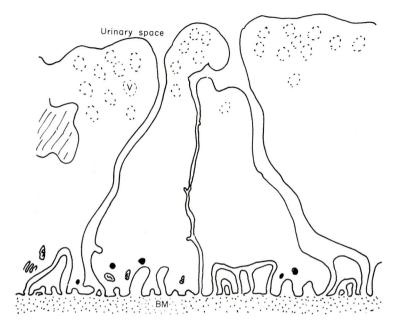

Fig. 2. Diagrammatic representation of the ultrastructure of coelomosac epithelial cells. Abbreviations: BM = basement membrane, v = apical vacuole.

branes. The ultrastructure of the epithelial cells of the coelomosac is very similar to the ultrastructure of the cells of the visceral epithelium of Bowman's capsule in the vertebrate glomerulus. Probably the same function is subserved by both. The coelomosac cells have numerous inclusions and organelles which suggests that they carry on active metabolism. Characteristically the apical part of the cell contains structures which have the appearance of vacuoles (Kümmel, 1964). At the bases of the cells are mitochondria and Golgi configurations (Kümmel, 1964; Cook, 1973).

The ultrastructure of cells of the labyrinth is characteristic of metabolically-active cells; numerous mitochondria and Golgi structures are in evidence. Furthermore, the apical borders of the cells have microvilli, and the basal borders are greatly infolded. In the basal region may be seen numerous lysosomes. Possibly these are responsible for the green or yellow colour taken on by the labyrinth at various times of the year. The colour can be attributed to granules identifiable in light-microscope studies, which lie in the same position as the lysosomes seen in electron-microscope studies.

II. FLUID MOVEMENT IN THE ANTENNAL GLAND OF THE CRAYFISH

Direct studies of fluid movement in the antennal gland began when Schlieper and Herrmann (1930) analysed urine from the nephropore. They speculated that the low concentration of the urine was due to reabsorption of solute from an "ultrafiltrate" of the blood. The results of the first micropuncture study of the antennal gland (Peters, 1935) also supported the concept of the antennal gland as a filtration : reabsorption kidney. Peters found that the concentration of chloride in the presumptive urine fell sharply in the tubule. This observation lent support to the contention that the long tubules observed by Grobben (1881) and others in the excretory organs of freshwater and brackish water animals are responsible for diluting the urine.

Maluf (1941a,b,c) carried out what may be termed the first systematic study of excretion in the crayfish. He noted that the histological anatomy of the antennal gland changes during various physiological states. He also studied the excretion of various vital dyes. Maluf was obviously impressed with the secretory appearance and abilities of the antennal gland. Perhaps it was this that led him to conclude that the antennal gland forms urine by a secretory process. Amongst the compounds Maluf believed to be "secreted" was the renal test substance, inulin. Of course, inulin was and still is the mainstay of the "filtration : reabsorption" theory of renal function.

Maluf's assertion that inulin is secreted by the antennal gland rested primarily upon two observations. Firstly, the concentration of inulin excreted in the final urine seemed to bear no relationship to the haemolymph inulin concentration. Secondly, the inulin concentration in the final urine was always well in excess of the haemolymph inulin concentration. If inulin were filtered, the latter observation would suggest that water was reabsorbed from the presumptive urine. At the time, the reabsorption of water from the urine of a freshwater animal did not seem a reasonable proposition.

Had Maluf asserted that inulin is secreted by a vertebrate kidney, the assertion would not have stood unchallenged for several years (as witness the recent controversy over the reabsorption of inulin in vertebrates (Thurau et al., 1968)). Nevertheless, it was almost 20 years (1960) before Riegel and Kirschner restudied the problem of inulin excretion by the crayfish antennal gland. They found that when crayfishes are made antidiuretic by osmotic stress or too-frequent handling the inulin concentration in the final urine rises remarkably, from a usual value of two to three times the concentration in the haemolymph, to over 30 times the haemolymph inulin concentration (Table I). Riegel and Kirschner concluded that both their results and those of Maluf could be rationalized within the framework of a filtration : reabsorption mechanism.

Between 1963 and 1968 the writer carried out a series of micropuncture studies of the crayfish antennal gland. These studies revealed that the antennal gland varies in its activity during the year. During the spring and early summer the animals become very diuretic. The diuresis is reflected by a greatly swollen antennal gland. At this time, the total solute concentration in all parts of the antennal gland (Fig. 3) is similar to the total solute concentration of the haemolymph. During the balance of the year: late summer to winter, the total solute concentration in the distal part of the tubule falls to about three-quarters of the value of the haemolymph. Whilst the total solute concentration falls in the tubule, the concentration of inulin rises in all parts of the gland. This indicates that from one-half to two-thirds of the water entering the antennal gland in the coelomosac is reabsorbed prior to entry of the presumptive urine into the bladder. Since the total solute concentration falls only slightly, the reabsorbed fluid is probably slightly hyperosmotic.

The actual reduction in the concentration of solutes to levels found in final urine appears to be the responsibility of the bladder.

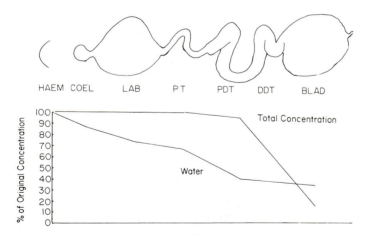

Fig. 3. Changes in the water content (reciprocal of the inulin : haemolymph concentration ratio) and total solute concentration (osmotic pressure) in the urine as it passes through the antennal gland.

TABLE I

Urine : haemolymph concentration ratios (U/H) of inulin in crayfishes during various experimental conditions

Species	Experimental conditions	Inulin U/H	Reference
Procambarus clarkii *Orconectes virilis*	Freshwater Freshwater (handling stress)	2–3 up to 28 }	Riegel and Kirschner (1960)
Austropotamobius pallipes	Freshwater 50% seawater (osmotic stress)	2–3 up to 31	Riegel (1965) Bryan and Ward (1962)
Pacifistacus leniusculus	Freshwater Freshwater (handling stress) 20, 40, 70% seawater	2–3 up to 28 } c. 2	Riegel and Kirschner (1960) Pritchard and Kerley (1970)

Two lines of evidence support this. Firstly, Kamemoto *et al.* (1962) found that the transport of sodium out of the bladder was active; it could be halted by eserine, an inhibitor of cholinesterase, an enzyme implicated in active sodium transport in a variety of tissues. Secondly, studies done in the author's laboratory indicate that isolated bladders have a marked ability to transport sodium chloride. The sodium chloride concentration of haemolymph-isotonic saline placed in the isolated bladders was reduced to less than one-half in a few hours. Further, some work done by Cook (1973) using intact crayfishes also suggests that the bladder has remarkable solute-transporting capabilities. As shown in Table II, when crayfishes are handled excessively, the sodium and potassium concentrations of the final urine may rise to very high levels; a few hours later, the solute concentration may be quite low.

The results discussed thus far are all compatible with the contention that the crayfish antennal gland is a renal organ operating on the "filtration : reabsorption" principle. Still more support for this conclusion derives from the studies of Kirschner and Wagner (1965) who studied the permeability of the antennal gland to large molecules. They found that dextran whose mol. wt ranged from 15,000 to 20,000, was excreted by the antennal gland as readily as inulin. A dextran of greater mol. wt (60,000–90,000) was excreted only about 70% as effectively as inulin, however. This result suggests that the higher mol. wt compound was restricted in its entry into the presumptive urine.

The crayfish antennal gland also exhibits abundant secretory activity. Studies of the secretory abilities of the antennal gland have concerned the accumulation of dyes or the concentration of dyes in the final urine. As shown by numerous investigators (e.g. Lison,

TABLE II

Concentrations of sodium and potassium (mM) in urine samples taken from three crayfishes suffering from handling stress

Sampling Time (hr)	Crayfish B		Crayfish C		Crayfish D	
	Na	K	Na	K	Na	K
3	22.0	—[b]	118	—	8.8	0.17
6	17.7	0.44	136	150	206	89.1
9	414	96.9	329	121	115	114
12	nso	nso[a]	158	nso	7.1	2.22
15	49.0	29.4	14.0	26.4	12.6	2.00

[a] nso = no sample obtainable.
[b] – – = no attempt made to obtain sample.

1942; Maluf, 1941c) a variety of dyes are accumulated by the crayfish antennal gland. Many of these are also secreted, and often, these dyes become highly concentrated in the process of secretion.

Processes of cellular accumulation and secretion may have been clarified by a recent study by Cook (1973). She followed the sequence of secretion of the organic acids phenol red (phenol-sulfonephthalein or PSP) and para-amino hippuric acid (PAH) both by histological and chemical analyses. Both organic acids become highly concentrated in the urine. The clearance of PAH averaged nine % of the haemolymph per hour. The clearance of PSP did not reach a steady value but fell from about 50% to about 17% of the haemolymph per hour. The two organic acids compete for secretion (Fig. 4). Phenol red is cleared about three times more rapidly than

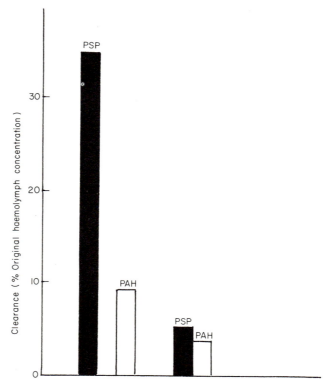

Fig. 4. Clearance of phenol red (PSP) and para-amino hippuric acid (PAH) from the haemolymph of the crayfish. Left-hand histograms refer to clearance when the organic acids are excreted individually. Right-hand histograms refer to simultaneous clearance of the two organic acids.

PAH. However, when the two acids are being excreted simultaneously, PSP clearance falls to about one-seventh its maximum value, whilst PAH clearance is reduced only by about one-half. The clearances of both PSP and PAH when excreted simultaneously is approximately equal to the clearance of the more slowly-secreted acid (PAH) secreted individually. An interesting feature of Cook's results was the finding that up to one-half the PSP excreted by the crayfish may be in the form of a pH-insensitive yellow compound. A similar compound secreted into the bile of rats proved to be a conjugate of glucuronic acid and PSP (Hart and Schanker, 1966).

Cook was able to localize the sites of secretion of organic acids by micropuncture. She found that the cells of the coelomosac and labyrinth were involved. She attempted to visualize the passage of organic acid across the cells by light and electron microscopy. The cells of the coelomosac exhibited no profound change in structure during acid secretion, but there was an apparent increase in the number of lysosomes in the basal region. Cells of the labyrinth exhibited striking structural changes during organic acid secretion. As observed by light microscopy, PSP was accumulated in discrete granules at the bases of the cells. Cook was unable to determine further events in the secretion sequence of PSP. However, in size and position, the PSP-containing granules were similar to green or yellow granules seen in numerous light-microscopical studies of the crayfish antennal gland.

Cook made ultrastructural studies of labyrinth cells during secretion of PAH. The cytoplasm of the cells became filled with multimembranous bodies which had an electron-dense core (Fig. 5). The bodies appeared to expand and bud off membrane-bound vesicles which migrated toward the apical region of the cell and were released into the lumen at the apico-lateral border. Malaczyńska-Suchcitz and Uciński (1962) have observed that the green granules also migrate to the apical border of labyrinth cells. Organic acids are secreted, at least in part, within vesicles (Cook, 1973), so it is quite possible that the process that Cook has observed is the sequence of events pertaining during organic acid secretion. This process can be summarized as follows: organic acids move into the labyrinth cells by an unknown sequence of events. In the cells, the acids become incorporated, probably with lysosomes, into membranous structures within which they may be subjected to chemical alteration by hydrolysis. The products of hydrolysis may increase the osmotic pressure of the membrane-bound structures, attracting water into the

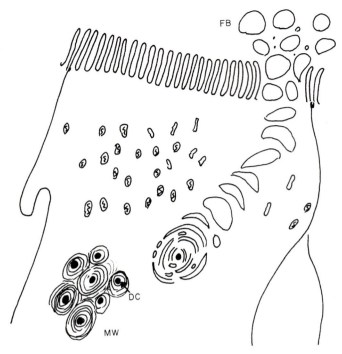

Fig. 5. Diagrammatic representation of the ultrastructure of the apical region of a labyrinth cell which is secreting PAH. Abbreviations: MW = membrane whorl, DC = dense core, FB = formed bodies.

cell. The increased turgor due to the hydration of the cell then forces the vesicles along the path of least resistance out of the cell. Presumably this path is along the lateral and apical portions of the cell.

The foregoing probably describes a general mechanism rather than one confined to organic acid secretion. For example, the crayfish antennal gland excretes Blue Dextran 2000 in vesicles also (Riegel, 1972), and the dyes Congo red and cyanol are secreted in vesicles (personal observation of the writer).

III. FORMED BODIES IN THE CRAYFISH

Secretion vesicles or "blebs" have been described repeatedly in histological studies of the crayfish antennal gland and a variety of other tissues. It has been customary for researchers to accord them scant attention other then to assume that they are either vehicles of cellular secretion or that they are artifacts. During micropuncture studies it was discovered that the vesicular elements of the antennal gland are present in fluid removed from all parts of the antennal gland, except, usually, the bladder. The vesicular elements, which the writer has termed "formed bodies" appear to be of two basic kinds. The first, called spheroids, have a smooth, optically-dense limiting membrane, and they are regularly spherical. Spheroids appear to be produced by the cells of the coelomosac. They vary in size from less than the resolving power of the light microscope (c. 0.3 μm) to over 50 μm. Vesicles are produced by the cells of the labyrinth. Their size is generally 20 μm or larger. When viewed under phase-contrast illumination, the limiting membrane of vesicles appears to be varigated. This give them a coarse-coated appearance. The limiting membrane of spheroids is stained readily with Sudan dyes, especially dyes which preferentially stain neutral lipid (e.g. Sudan III). The limiting membrane of vesicles does not stain significantly with Sudan dyes, although the contents usually are lightly stained.

One of the rather frustrating aspects of formed bodies is that they seem to be observable only in the "living" state. Possibly because of their lipid nature, spheroids, especially, disappear in sections prepared for light- or electron-microscopical studies. Vesicles may have protein in their limiting membranes. They are seen in some light and electron microscopical studies so it is possible that protein is less likely to be disrupted in solvents used in microscopy.

Observation of "living" formed bodies in micropuncture samples indicate that they undergo changes in density and size as they pass through the antennal gland. In the coelomosac, spheroids are distributed at all planes of observation suggesting that their density is close to that of the fluid that surrounds them. In the proximal tubule and in the proximal distal tubule the spheroids appear to be more dense than their medium. In the distal portion of the distal tubule the spheroids are found in the uppermost plane of observation, so they have become less dense than their medium. The spheroids increase enormously in size as they pass through the antennal gland; by the time they reach the distal tubule most are quite large (50 μm). Vesicles (normally), are found only in the labyrinth where they appear to be more dense than their medium.

Three characteristics are known to parallel the above-described density and size changes. Firstly, as shown in Table III, in parts of the tubule where the density of the formed bodies equals or exceeds that of their medium, the pH is normally alkaline. Where the density

TABLE III

Hydrogen ion concentration (pH) in the haemolymph and in various parts of the antennal gland of control crayfishes and crayfishes which were excreting injected organic acids

Haemolymph	Coelomosac	Labyrinth	Proximal Tubule	Proximal distal tubule	Distal distal tubule	Bladder
			Control			
7.88 ± .03	7.53 ± .02	7.12 ± .05	7.24 ± .10	7.12 ± .08	6.83 ± .11	6.61 ± .15
			PSP excreting			
7.28 ± .06	7.09 ± .11	6.63 ± .18	6.67 ± .13	6.83 ± .15	6.41 ± .10	6.87 ± .16
			PAH excreting			
7.26 ± .06	6.82 ± .09	6.43 ± .12	6.39 ± 1	6.42 ± .12	6.34 ± .1	6.44 ± .08

of the formed bodies is less than that of their medium, the pH is normally mildly acid. Secondly, in the distal tubule, the concentration of amino acids rises markedly (Riegel, 1966b). Finally, in an acid medium, the formed bodies become susceptible to osmotic disruption.

One final observation is worth mentioning since it may be connected with acidification of the urine. During dialysis experiments which will be described in detail later, small volumes of micropuncture fluid were dialysed against larger volumes of Ringer or isosmotic sucrose made slightly alkaline. Quite often during these dialyses the entire volume became acidified as indicated by the fact that the pH-indicator dye (phenol red) which the dialysing solution contained turned yellow. Similar volumes of Ringer left for similar periods did not become acidified, so it seems unlikely that the result was due to diffusion of carbon dioxide into the samples. Acidification due to bacterial action could probably be ruled out because of the brief time involved (generally four hours or less). It seems possible that the formed bodies are the source of the hydrogen ion. Formed bodies contain proteolytic enzymes which are active in mildly alkaline and mildly acidic media (Riegel, 1966a). It is therefore possible that the increase in acidity of fluid in the distal tubule of the antennal gland and the change in density that formed bodies undergo there are related. That is, the increased acidity may bring about hydrolytic activity within the formed bodies decreasing their density. In what way the hydrogen ions fit into this scheme is not clear, especially since they appear to arise from within the formed bodies.

IV. FLUID MOVEMENT DUE TO FORMED BODIES

In an attempt to separate the formed elements of micropuncture samples from their surrounding fluid, a procedure called "micro-dialysis" was developed. Droplets of micropuncture fluid were dialysed against droplets of haemolymph-isosmotic sucrose. Millipore filters having an average pore diameter of 0.1 μm were used as dialysis membranes. The whole procedure was carried out under liquid paraffin.

Two droplets of an aqueous solution appose each other across a Millipore filter wetted so that fluid can pass through; the whole is kept under liquid paraffin which invades the unwetted portion of the Millipore filter preventing water passing through. A pressure will be exerted upon the droplets due to the surface tension at the liquid-paraffin: water interface. This pressure is proportional to the surface tension (T) and inversely proportional to the radius (r) of the droplets. The amount of the pressure developed depends upon the shape of the droplets: For a spherical droplet,

$$P = \frac{4T}{r}. \text{ For a hemispherical droplet, } P = \frac{2T}{r}.$$

Where two droplets of exactly equal shape and volume appose each other all parameters will be equal and neither droplet should change volume. However, where one droplet is smaller than the other, the pressure exerted on it will be greater and its volume will diminish by transfer of fluid to the larger droplet.

Droplets of micropuncture fluid of about 65 nl were dialysed against larger droplets (0.5 to 5 μl) of isosmotic sucrose. It was hoped that before the small droplet collapsed into the larger droplet, solutes in the micropuncture droplet which were free to do so would "dialyse" leaving behind the formed bodies. It would then be a simple matter to chemically analyse the Millipore filter to establish the composition of the formed bodies. In only a few instances did the events predicted above occur. Generally, the volume of the micropuncture droplets shrank to some fraction of original and remained at that volume or increased slightly. In some cases the increase in volume was such that all of the fluid in the larger droplet was transferred to the smaller droplet. These results are explicable only if there is some component of the micropuncture-sample droplets which is capable of generating a very large osmotic pressure. The only obvious component of micropuncture samples which qualifies from the viewpoint of size is the formed bodies (Riegel, 1970a).

Evidence from the following experiments suggest that formed bodies in micropuncture samples are responsible for fluid movement. Droplets of fluid taken from the coelomosac and droplets of Ringer which contained colloid were apposed across small discs of Millipore "Pellicon" ultrafiltration membrane. The ultrafiltration membrane prevents passage of solutes whose mol. wt exceeds about 40,000. The small droplets were put under pressure until their volume had diminished to a few nanolitres. The unwetted portion of the dialysing membrane was then punctured, equalizing pressure on both sides. The growth in volume of the small droplet was then measured and timed. As shown in Fig. 6, fluid movement into micropuncture samples was both more rapid and more sustained than fluid movement into droplets not containing formed bodies.

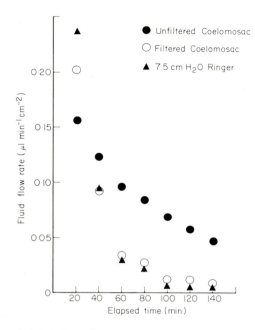

Fig. 6. The rate of fluid flow from Ringer containing colloid equivalent to 10 cm H_2O into droplets of filtered coelomosac fluid, unfiltered coelomosac fluid and Ringer containing 7.5 cm H_2O colloid.

In early experiments it was found that fluid movement into micropuncture-sample droplets could be halted by application of a very small opposing hydrostatic pressure (c. 2 cm H_2O). However, fluid can be made to flow against very large gradients of colloid osmotic pressure (up to 80 cm H_2O). All these observations fit in well with the characteristics required of the force effecting filtration in the antennal gland and other fluid-transporting epithelia. It has been possible to utilize the known properties of formed bodies to construct a model of the function of the antennal gland (Riegel, 1970b). Aspects of that model will be reviewed below.

V. A MODEL OF FLUID MOVEMENT IN THE CRAYFISH ANTENNAL GLAND

There seems little doubt that the process underlying formation of the primary urine in the antennal gland is filtration across the podocytic epithelium of the coelomosac. Use of the term "filtration" implies only that some solutes enter the urine by a selective process which does not depend upon the metabolism of the epithelial cells. The impetus for fluid movement is assumed to be a hydrodynamic force. That is, either hydrostatic pressure or osmotic pressure (or both). It is therefore assumed that water flows rather than diffuses, simply because the coelomosac epithelium appears to be freely permeable to molecules which are many times larger than water.

In the crayfish, conditions seem to be conducive for hydrostatic-pressure filtration. The general haemocoelar pressure, although highly variable, averages about 15 cm H_2O (Picken, 1936). The colloid osmotic pressure of the haemolymph averages 6.8 cm H_2O. The colloid osmotic pressure of the primary urine (coelomosac fluid) averages 8.5 cm H_2O (Riegel and Cook, 1975). It is problematical whether or not the haemolymph circulation is the sole source of hydrostatic pressure, bearing in mind the water-attracting properties of formed bodies. The following model may illustrate the influence that formed bodies might have on the processes of urine formation and elaboration. It is believed that two known attributes of formed bodies are of major significance to formation and elaboration of the urine:

1. Formed bodies attract water and in so doing swell and burst.
2. As formed bodies swell, they exclude solutes, thereby raising the osmotic potential of their environment.

As outlined above, in *in vitro* studies, formed bodies have been shown to have a remarkable water-attracting ability. It is possible that this ability is associated with their proteolytic activity. The potential osmotic pressure of protein is enormous. For example, the complete hydrolysis of an amount of protein of 100,000 mol. wt to amino acids (*c.* 100 mol. wt) raises the osmotic pressure 1000-fold. It is not possible to make an empirical calculation of the influence that formed bodies may have on primary urine formation in the coelomosac. However, *in vitro* studies can provide at least an approximation of their influence. In Fig. 6 it can be seen that the maximum rate of fluid flow into coelomosac fluid droplets was about 0.15 μl per minute per square centimeter. This occurred under conditions where the droplets were reduced in size to a few nanolitres; a condition which may approximate that prevailing at the site of primary urine formation. The rate of primary urine flow is about 2.5 μl per minute. Were this due entirely to formed-body swelling, a surface of about 16 cm^2 would be required to effect filtration. It is not known whether or not the filtering surface of two coelomosacs approaches this figure. However, even if it seems doubtful that the filtration rate observed in the crayfish is due exclusively to the activities of formed bodies, that doubt doesn't invalidate the basic proposition. Namely, formed bodies play a role in primary urine formation in the crayfish, and that role probably varies in significance with other physiological parameters such as haemolymph colloid concentration and hydrostatic pressure. In addition to causing fluid movement, formed bodies are capable of generating considerable solute movement in a direction opposite to water movement. This has been shown by *in vitro* studies (Riegel, 1970a).

The sequence of events depicted in Fig. 7 may be those which bring about fluid and solute movement due to the swelling of formed bodies. Formed bodies may be ejected from cells into an adjacent enclosed space. The formed bodies are permeable to water, but not to solutes. Therefore, as fluid moves into the space, water enters the formed bodies whilst solute remains outside the formed bodies. This tends to concentrate solute outside the formed bodies. If the volume of the space is small relative to the volume of the formed bodies, the osmotic pressure of the space will increase. This will attract more water into the space and cause solute to leave the space.

Swelling of formed bodies within enclosed spaces would tend to create a volume restriction in those spaces; this, in turn, would give rise to an hydrostatic pressure sufficient to cause fluid to flow out of the spaces. The direction of fluid flow would depend upon the hydraulic resistances of the pathways giving access to the spaces. In the diagram, the hydraulic resistance of the basement-membrane-like structure (lower part of figure) is greater than the resistance of the passages leaving the space; therefore, fluid flows in the direction indicated by the stipled arrows. At one and the same time, formed bodies could be responsible for both primary urine formation and fluid reabsorption. In the former case, Fig. 7 depicts the events occurring at the bases of the coelomosac cells. In the latter case, Fig. 7 depicts events occurring at the bases of cells in other parts of the antennal gland.

In the antennal gland, formed bodies in the lumen also could contribute to fluid and solute movements. As shown in Fig. 1, many parts of the antennal gland are characterized by very small lumenal spaces. This is true of the labyrinth and distal portion of the distal tubule. It is quite possible that formed bodies swelling in these enclosed spaces aid fluid reabsorption.

The foregoing represents a brief summary of the application of the formed-body model of fluid flow (Riegel, 1970b, 1972) to the

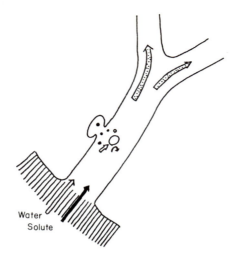

Fig. 7. Diagrammatic representation of events which may occur when formed bodies are released from a cell into an adjacent confined space. The swelling of the formed bodies causes water and solutes to move into the space. The water enters the formed bodies; the solutes remain in the space attracting more water. The volume restriction of the space causes fluid (stippled arrows) to move out of the space by passageways offering the least resistance.

crayfish antennal gland. There is still no clear understanding of the activities that occur in formed bodies or, indeed, how they are formed in cells. It appears that sodium and potassium are important in their function because those ions are bound or localized by formed bodies. It is possible that formed bodies are the site of adenosine triphosphatase or that that enzyme is important in their formation, which would explain the importance of sodium and potassium ions to formed-body function. (Also see Chapters 21 and 27.)

REFERENCES

Bryan, G. W. and Ward, E. (1962). *J. mar. biol. ass.* **42**, 199—241.

Cook, M. A. (1973). "Organic acid secretion by the antennal gland of the crayfish, *Austropotamobius pallipes pallipes* (Lereboullet)." Doctoral Dissertation, University of London.

Grobben, C. (1881). *Arb. Zool. Inst. Univ. Wien u. Stat. in Triest.* **3**, 93—110.

Hart, L. G. and Schanker, L. (1966). *Proc. Soc. Exp. Biol. Med.* **123**, 433—435.

Kamemoto, F. I., Keister, S. M. and Spalding, A. E. (1962). *Comp. biochem. Physiol.* **7**, 81—87.

Kirschner, L. B. and Wagner, S. (1965). *J. exp. Biol.* **43**, 385—395.

Kümmel, G. (1964). *Zool. Beitr.* **10**, 227—252.

Lison, L. (1942). *Mem. Acad. Belg. 2 cl. Sci.* **19**, 1—107.

Malaczyńska-Suchcitz, Z. and Uciński, B. (1962). *Folia Biol.* **10**, 251—292.

Maluf, N. S. R. (1939). *Zool. Jb. (Abt 3)* **59**, 515—534.

Maluf, N. S. R. (1941a). *Biol. Bull. Woods Hole* **81**, 127—133.

Maluf, N. S. R. (1941b). *Biol. Bull. Woods Hole* **81**, 134—148.

Maluf, N. S. R. (1941c). *Biol. Bull. Woods Hole* **81**, 235—260.

Peters, H. (1935). *Z. Morph. Ökol. Tiere* **30**, 355—381.

Picken, L. E. R. (1936). *J. exp. Biol.* **13**, 309—328.

Pritchard, A. W. and Kerley, D. E. (1970). *Comp. biochem. Physiol.* **35**, 427—437.

Riegel, J. A. (1963). *J. exp. Biol.* **40**, 487—492.

Riegel, J. A. (1965). *J. exp. Biol.* **42**, 379—384.

Riegel, J. A. (1966a). *J. exp. Biol.* **44**, 379—385.

Riegel, J. A. (1966b). *J. exp. Biol.* **44**, 387—395.

Riegel, J. A. (1968). *J. exp. Biol.* **48**, 587—596.

Riegel, J. A. (1970a). *Comp. biochem. Physiol.* **35**, 843—856.

Riegel, J. A. (1970b). *Comp. biochem. Physiol.* **36**, 403—410.

Riegel, J. A. (1972). "Comparative Physiology of Renal Excretion." Oliver and Boyd, Edinburgh.

Riegel, J. A. and Cook, M. A. (1975). "Recent studies of excretion in Crustacea." Fortschritte der Zoologie Bd. 23, Heft 2/3, pp. 48—75.

Riegel, J. A. and Kirschner, L. B. (1960). *Biol. Bull. Woods Hole* **118**, 296—307.

Schlieper, C. and Herrmann, F. (1930). *Zool. Jb. (Abt. f. Anat. u. Ont. der Tiere)* **52**, 624—630.

Thurau, K., Valtin, H. and Schnermann, J. (1968). *Ann. rev. Physiol.* **30**, 441—524.

25.   Fluid Secretion in Exocrine Glands

W. T. Prince

Department of Zoology, University of Cambridge, Cambridge, England

I. INTRODUCTION

In 1954, Thaysen, Thorn and Schwartz suggested a two-stage hypothesis for secretion by the mammalian salivary gland. They proposed that a plasma-like primary secretion is formed in the acini and is modified as it passes through the ducts. Subsequent work has confirmed this hypothesis and indicated that it applies equally to other exocrine tissues such as pancreas, sweat glands, and lacrimal glands. Moreover, studies employing microelectrodes, micropuncture, and microperfusion have provided details of the cellular mechanisms involved in generating and controlling secretion. The precise mechanism by which ion secretion is coupled to water movement is unknown, and some of the hypotheses are described in Chapters 5 and 7 in this volume (see also Gupta, 1976). The present paper summarizes what is known about the properties of exocrine gland cells, beginning with a detailed description of the mammalian salivary gland.

II. MAMMALIAN SALIVARY GLANDS

A. Introduction

The primary secretion of mammalian salivary glands is formed in the acini (Fig. 1a). Secretion can occur against an applied pressure gradient, in which case the saliva formed is hypertonic (Imai *et al.*, 1973). The acini are innervated by both parasympathetic

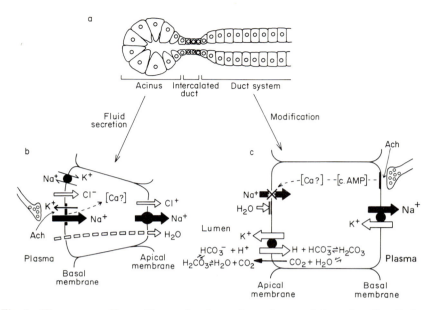

Fig. 1. The mammalian salivary gland consists of several functionally distinct regions (a). The acinus consists of fluid secreting cells (b). In the complex duct system the primary secretion is modified by potassium and hydrogen exchange and sodium reabsorption (c). These two regions are connected by the intercalated ducts. Acetylcholine increases the permeability of the basal membrane of the acinar cells to sodium and potassium (b). Fluid secretion is driven by a sodium pump on the apical membrane, but the means of stimulus-secretion coupling is not known. Possibilities include stimulation by an increased intracellular concentration of sodium or calcium. Acetylcholine also decreases reabsorption of sodium in the duct system. The intracellular mediator here is probably cyclic AMP.

(cholinergic) and sympathetic (noradrenergic) neurons. Cholinergic agents induce a flow ten times that induced by adrenergic agents (Young et al., 1971). In both cases the electrolyte composition of the primary secretion is similar to that of the saliva produced by the unstimulated gland and to plasma.

B. The Basal Cell Surface

The primary event in the secretory response appears to be an increase in the passive permeability of the basal plasma membrane of the acinar cells (Fig. 1b). This permeability change causes an influx of sodium and efflux of potassium down their electrochemical gradients. Three types of studies have demonstrated that this permeability increase occurs in response to acetylcholine or to parasympathetic stimulation: a) studies of the fluxes of ions between the blood and the acinar cells; b) measurements of the resting and secretory potentials; and c) measurements of the resistance of the basal membrane. Some details of these studies follow.

1. Ionic Fluxes

Poulsen (1974a) has studied the movements of ions between the blood and the cells by following changes in the concentration of Na and K in the venous effluent from isolated perfused salivary glands. Figure 2a, taken from his work, shows that a short pulse of acetylcholine causes a transient entry of Na into the cells followed by an Na efflux. The Na entry occurs simultaneously with a somewhat larger K efflux (Fig. 2a). Poulsen (1974a) has suggested that the difference occurs because some of the Na taken up from the blood may be secreted into the lumen and flows into the ducts, where it is reabsorbed back into the blood. Ouabain inhibits secretion and thereby prevents Na reabsorption in the ducts. In the presence of ouabain the Na influx is equivalent to the K efflux (Fig. 2b), indicating that both fluxes are passive. Moreover, ouabain also abolishes the subsequent reaccumulation of K and extrusion of Na by the cells (Fig. 2b). These K and Na fluxes are thus thought to be due to a ouabain sensitive Na/K exchange pump on the basal surface.

2. Membrane Potentials

The changes in ionic fluxes noted above are reflected in changes in membrane potential recorded across the basal membrane upon stimulation (Poulsen, 1974a; Nishiyama and Petersen, 1974). The membrane potentials of exocrine gland cells are generally lower (closer to the Na equilibrium potential) than in excitable cells such as nerve and muscle. This is probably because the ratio of the K to Na permeability is lower (Williams, 1970). Chloride is assumed to be distributed passively. In the salivary gland, potentials of -20 to -70 mV have been measured (Petersen, 1972, 1973). Recently, Nishiyama and Petersen (1974) have obtained a mean value close to

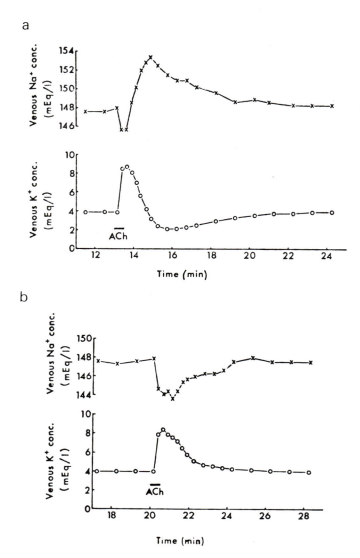

Fig. 2. Effects of pulses of acetylcholine (10^{-5} M) on Na (X) and K (O) concentrations in the venous effluent from isolated salivary glands perfused with Locke solution, from Poulsen (1974a). (a) Acetylcholine causes a transient entry of Na and a larger loss of K from the cells, followed by reaccumulation of K and loss of Na. (b) In the presence of ouabain (10^{-4} M) the initial Na influx is equivalent to the K efflux, and the subsequent reaccumulation phase is abolished. Secretion rates were (a) 380 μl/2 min. and (b) 30 μl/2 min.

−70 mV in the mouse parotid gland and −55 mV in mouse submaxillary gland. They found a 49 mV change in membrane potential with a 10-fold change in potassium concentration. During acetylcholine stimulation the basal membrane hyperpolarizes. The hyperpolarization is not affected by ouabain, but is greater if the external potassium concentration is lowered (Petersen, 1970a). These findings confirm the conclusion from flux measurements that acetylcholine increases the passive permeability of the basal cell surface. In the absence of sodium the secretory potentials (hyperpolarizations) are larger, indicating that the outward movement of potassium is normally partly short-circuited by an inward flow of sodium (Petersen, 1970a).

3. Resistance Measurements

Stimulation of the parasympathetic nerve supply to the rabbit submaxillary gland produces either a monophasic or biphasic response accompanied by a fall in resistance, as shown in Fig. 3 (Nishiyama and Kagayama, 1973). Although the secretory potential is a hyperpolarization, the biphasic responses, obtained from cells with high resting potentials, have an initial depolarization Fig. 3). The decrease in resistance occurs within 100 msec. of stimulation, whereas hyperpolarization occurs within 500 msec. The initial depolarization seen in the biphasic responses have also been observed in the mouse (Petersen, 1973) and cat (Nishiyama and Kagayama, 1973).

   In the absence of chloride, the hyperpolarizations are significantly reduced (Lundberg, 1958; Nishiyama and Petersen, 1974).

C. The Apical Cell Surface

It has been more difficult to study the properties of the apical (luminal) membranes because of the small size of the acinar lumen. Imai (1974), studying the dog submaxillary gland has found that the apical surface hyperpolarizes at about the same time as the basal surface, although with a faster rise-time. This initial hyperpolarization is followed by a slow depolarization. Imai concludes that the initial hyperpolarization may be caused by stimulation of an electrogenic sodium pump. The presence of such a pump is

R

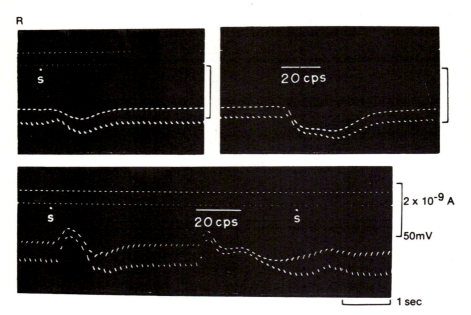

Fig. 3. The effect of nerve stimulation on membrane potential and resistance in the rabbit submaxillary gland. The lingual nerve was stimulated with single shocks (s) or at 20 c/s. The lower trace shows biphasic potential responses. There is an initial depolarization followed by hyperpolarization. (From Nishiyama and Kagayama, 1973).

supported by the finding that secretion is inhibited by ethacrynic acid (Petersen, 1971) and by ouabain injected into the acinar lumen (Poulsen, 1974b). The nature of this pump is unclear at present.

D. Control of Secretion

From the information summarized above (see Fig. 3b), it seems that acetylcholine increases the permeability of the basal membrane to electrolytes. Na and K and possibly Ca flow down their electrochemical gradients. The Na pump on the apical surface is activated, and secretion begins. The second messenger that triggers Na extrusion into the lumen has not been established. One possibility is that it is the increase in internal Na that stimulates the apical pump, as has been proposed for the avian salt gland (Peaker, 1971). However, Poulsen (1973) was unable to evoke secretion by artificially raising the intracellular Na concentration, by lowering the temperature or the external K. Alternatively, an influx of Ca may stimulate the Na pump (Petersen, 1970b) as in other secretory systems (Douglas and Poisner, 1963; Petersen et al., 1967).

E. Modification of the Primary Secretion

During passage through the duct system the primary secretion is modified by active reabsorption of sodium and secretion of ions such as potassium or bicarbonate (see Schneyer et al., 1972). These processes occur to differing extents in salivary glands of different species or even at different secretory rates in the same gland. Thus the ionic composition of the secreted saliva can vary widely. In many salivary glands the final secretion is hypotonic to the plasma. In these cases sodium chloride is reabsorbed and potassium and perhaps bicarbonate are secreted. There seems to be little or no net movement of water in this region since there is no concentration of radioactive inulin by isolated ducts (Young et al., 1967; Martin and Young, 1971). In the final saliva, sodium and chloride concentrations are generally below those of plasma and potassium and bicarbonate are higher.

Much new information has been obtained by micropuncture and microperfusion and from in vitro preparations of isolated excretory ducts. The latter technique has the advantage that it allows control of the fluid composition on both sides of the duct epithelium.

1. Sodium Reabsorption

Sodium reabsorption is similar to that in other transporting epithelia such as frog skin or toad bladder (Koefoed-Johnsen and Ussing, 1958). That is, sodium enters the cell across the apical membrane passively and is then pumped out of the cell across the basal membrane by active transport (Fig. 1c). Chloride is believed to follow passively to maintain electroneutrality.

The duct lumen is negative with respect to the bathing medium (Martin and Young, 1971; Martin et al., 1973). The apical membrane seems to be permeable to sodium rather than to potassium since the potential across the apical surface is affected by changes in lumen sodium concentration but not by changes in the potassium concentration (Young et al., 1967; Schneyer, 1969; Knauf and Frömter, 1970; Knauf et al., 1972; Field and Young, 1973).

Amiloride, which is thought to inhibit the passive influx of sodium into cells, reduces both sodium transport and the potential difference across the epithelium (Knauf, 1973; Schneyer, 1974a). Ouabain has no effect on potential or sodium flux when perfused through the secretory duct of the rat submaxillary gland (Young *et al.*, 1967; Schneyer, 1969). Measurements of sodium flux indicate that sodium is actively transported (Schneyer, 1969).

Sodium flux is affected by the calcium concentration in the lumen but not by that in the bathing medium (Schneyer, 1974a). Perfusion with a medium containing EGTA increases sodium flux from the lumen to the bathing medium and increases the potential difference. Increasing the calcium concentration has the opposite effect.

The basal membrane behaves like a potassium electrode even after treatment with metabolic poisons (Knauf *et al.*, 1971). Transport of sodium across this membrane probably utilizes a conventional sodium-potassium exchange pump since it is inhibited by ouabain and is dependent on the potassium concentration of the bathing medium (Knauf *et al.*, 1971; Knauf, 1972).

The rate of sodium transport can be altered by stimulation of either the parasympathetic or sympathetic nerve supply or by adding pharmacological agents such as isoproterenol, carbachol or pilocarpine (Knauf and Frömter, 1970; Young *et al.*, 1971) to the bathing medium. The effect of isoproterenol is blocked by propanalol, and that of carbachol by atropine (Martin *et al.*, 1973; Denniss and Young, 1974, 1975). Stimulation of either adrenergic or cholinergic receptors leads to a decrease in sodium reabsorption. Both types of stimulation are equally effective, in contrast to their actions on acinar cells where parasympathetic stimulation induces ten times faster secretion than sympathetic stimulation (Young *et al.*, 1971). In the rabbit submaxillary duct carbachol increases the resistance of the apical membrane to sodium and the slope for a 10-fold change in sodium concentration in the lumen is reduced from 53 mV at rest to 33 mV during stimulation (Knauf *et al.*, 1972). No change in shunt resistance could be detected.

The cellular mechanisms involved in control of reabsorption are poorly understood. Acetylcholine acts on the basal surface of the cell since it is ineffective when perfused through the lumen. Its effect can be mimicked by cyclic AMP or cyclic GMP. Cyclic AMP is only active when applied to the apical surface of the cell, as in the proximal tubule of the kidney (Denniss and Young, 1974, 1975). Theophylline, which can raise the level of cyclic AMP by preventing its degradation, also mimics the effect of acetylcholine (Denniss and

Young, 1974). By analogy with other systems it seems unlikely that the cyclic nucleotides affect membrane resistance directly. It is more likely that they act indirectly by altering the calcium concentration within the cell. For example, inhibition of active transport by treatment with cyanide, DNP or ouabain also increases membrane resistance (Knauf *et al.*, 1971).

2. Potassium Secretion

Potassium is actively transported into the lumen (Field and Young, 1973). Potassium secretion is increased when the transepithelial potential is increased (Knauf *et al.*, 1975). Thaysen and Tarding (1974) have suggested that there is a 1 : 1 exchange of Na for K in the sheep parotid gland. However, under certain conditions potassium and sodium transport can be separated. When sodium is not present in the fluid perfusing the ducts, potassium flux is reduced but not abolished (Young *et al.*, 1967; Knauf and Lübeka, 1975). Pharmacological agents which affect sodium transport affect potassium transport relatively little. If transport of sodium and the transepithelial potential difference are reduced with amiloride, potassium transport still occurs (Knauf *et al.*, 1975; Knauf and Lübeka, 1975). Potassium transport is not inhibitied by ouabain (Schneyer, 1969) and is not affected by aldosterone as sodium transport is. In adrenalectomized rats the sodium transporting capability of the excretory ducts is impaired whilst potassium transport is relatively unaffected. The effect on sodium transport can be overcome by the adminstration of aldosterone (Gruber *et al.*, 1973). Actinomycin D reduces sodium transport without affecting potassium transport. Conversely, at certain doses cytochalasin B has a greater inhibitory effect on potassium transport than on sodium transport (Schneyer, 1974b). In spite of these findings, potassium transport does decrease when the sodium concentration in the lumen is raised (Martin and Young, 1971; Knauf and Lübeka, 1975). The nature of this linkage is not understood.

3. Bicarbonate Secretion

The ducts of some salivary glands secrete bicarbonate. The coupling between bicarbonate transport and the fluxes of other solutes (Na, K, and H) is poorly understood. Under normal conditions and during alkalosis (increased bicarbonate in the bathing medium) potassium and bicarbonate transport can be maintained in the absence of sodium transport (Knauf *et al.*, 1975). One possibility is that hydrogen ions are transported from the lumen into the cell in exchange for potassium, with bicarbonate entering the lumen down the pH gradient so created (Fig. 3c). In support of this, Knauf (1973) has shown that when sodium transport is inhibited with amiloride the potassium transported is equivalent to bicarbonate transport. However, the potassium transported under these conditions is only about 20% of that transported in the absence of amiloride. Under normal conditions, potassium is secreted in excess of bicarbonate. Moreover, during acidosis, potassium and hydrogen are both secreted, i.e. the direction of hydrogen transport is reversed.

4. *Site of Modification of Primary Secretion*

The main excretory duct is the most thoroughly studied part of the system because it is more accessible. To what extent do the other parts of the duct system participate in active transport? Histological studies demonstrate that the cells forming the granular, striated and excretory ducts all have the appearances of active ion transporting cells with many mitochondria and elaborate basal infoldings. In contrast, the cells of the intercalated ducts are relatively unspecialized (Tamarin and Sreebny, 1965). Enzymes associated with the high metabolic activity of transporting cells (cytochrome oxidase and succinate dehydrogenase) have been localized histochemically in the granular and striated ducts but are absent from the intercalated ducts (Harrison, 1974).

The granular and excretory ducts are composed of more than one cell type, suggesting the possibility that the different cell types have different functions. The excretory ducts are composed of light, dark and basal cells. Tamarin and Sreebny (1965) say that the dark cells seem to be structured more for transport than the light cells. The latter have mitochondria localized at the basal membrane while the dark cells have more mitochondria, distributed throughout the cells. A functional distinction between these cell types has been indicated by experiments which demonstrated that the number of dark cells increased during metabolic acidosis (Knauf *et al.*, 1974). Under these conditions the transporting capabilities of the tissue change too. Hydrogen ions are secreted rather than reabsorbed across the apical membrane while bicarbonate is reabsorbed. It is also possible that potassium secretion and sodium reabsorption occur in different types of cells. There may be structural differences between ducts that modify the primary secretion differently, for example, between the submaxillary gland of the rat which transports bicarbonate, and that of the rabbit which does not. It is known that the rat possesses a bicarbonate-stimulated ATPase and that there is much less of this ATPase in the rabbit (Knauf *et al.*, 1974; Wais and Knauf, 1975).

5. *Variations in Saliva Composition*

That primary secretion and the different modifying processes can occur independently explains why the final saliva can vary widely in composition between different species and even in the same gland at different secretory rates. For example, stimulation of the sympathetic or parasympathetic nerve supply to the rat submaxillary gland yields saliva of different compositions (Young *et al.*, 1971). In the resting state, sodium and potassium in the primary secretion were 136 mM and 8.4 mM respectively. On stimulation with carbachol or isoproterenol these values changed only to 139 mM and 4.0 mM. However, the final saliva in the resting state contained 2 mM Na^+ and 100 mM K^+. With stimulation these values changed to 40 mM Na^+ and 30 mM K^+ with carbachol and 10 mM Na^+ and 150 mM K^+ with isoproterenol. These differences in composition of the saliva can be explained by the different effects of these stimulants on primary secretion and on secondary modifying processes. Carbachol and isoproterenol both effect potassium secretion and sodium reabsorption equally but the rate of primary fluid production induced by carbachol is ten times that for isoproterenol. In the resting state sodium is reabsorbed maximally and the primary secretion flows slowly through the ducts giving ample opportunity for sodium reabsorption and potassium secretion. With stimulation sodium reabsorption is reduced but not completely inhibited but the length of the duct still allows the sodium to be lowered significantly. However, sodium reabsorption is incomplete at high flow rates, and the sodium concentration rises with carbachol stimulation. Also, since potassium secretion is inhibited by high sodium in the ducts, carbachol stimulation yields a lower potassium concentration than isoproterenol.

F. Conclusions about Mammalian Salivary Glands

The structure and characteristics of mammalian salivary glands are summarized in Fig. 1. The gland consists of acini, where the primary secretion is formed, and the ducts, where the primary secretion is modified by secretion and reabsorption (Fig. 1a). In the acinus (Fig. 1b) secretion is initiated by acetylcholine released from the parasympathetic neurons. Acetylcholine appears to increase the passive permeability of the basal cell surface to Na and K, which diffuse down their electrochemical gradients. Secretion is driven by a sodium pump on the apical surface, with chloride and water following passively. The mechanism of stimulus-secretion coupling is unclear, but it is possible that it is an increase in intracellular Na or Ca that activates the apical pump.

In the ducts (Fig. 1c) sodium is reabsorbed while potassium and bicarbonate are secreted. The mechanism of bicarbonate secretion is unclear, but may involve an exchange of potassium for hydrogen across the apical cell border. The apical membrane is probably impermeable to water. Acetylcholine inhibits sodium reabsorption, with cyclic AMP and calcium as the probable second messengers.

The basic scheme of primary secretion followed by modification in the ducts applies generally to exocrine glands, and some other examples will be described next.

III. INSECT SALIVARY GLANDS

Because of their simple anatomy and viability *in vitro*, more is known about the mechanism and control of secretion in salivary glands of insects than in mammals. In both the blowfly, *Calliphora*, and in the moth, *Manduca sexta*, the primary secretion is a potassium-rich fluid and reabsorption of potassium occurs in the ducts so that the final saliva is hypotonic (Oschman and Berridge, 1970; Kafatos, 1968; Robertson, 1974). Studies of the blowfly salivary gland have been particularly valuable in evaluating the roles of calcium and cyclic nucleotides in control of secretion (Berridge *et al.*, 1975; Prince *et al.*, 1972; Prince and Berridge, 1973).

The present hypothesis is summarized in Fig. 4. The stimulus, 5-HT, increases the intracellular level of two second messengers, cyclic AMP (cAMP) and calcium. Cyclic AMP stimulates an electrogenic potassium pump on the apical membrane whereas calcium increases the permeability of both basal and apical

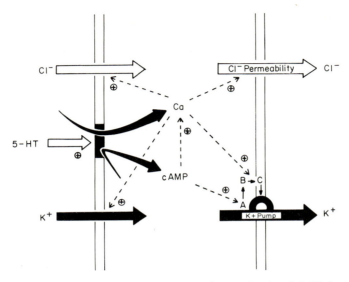

Fig. 4. A model for fluid secretion by the salivary glands of *Calliphora*. 5-HT increases the intracellular level of two intermediaries, cyclic AMP (cAMP) and calcium. Cyclic AMP stimulates a potassium pump on the apical membrane, whereas calcium increases the permeability of both apical and basal membranes to chloride. By analogy with other systems, cyclic AMP may activate a protein kinase (A) leading to phosphorylation (B) and consequent activation (C) of a potassium carrier. There is some evidence that calcium may be involved in this process.

membranes to chloride. Since the resistance to chloride flow is low, the cation pump does not normally produce much effect on the transepithelial potential. However, if chloride is replaced by a less permeable anion such as isethionate, stimulation causes the apical membrane to hyperpolarize. The same effect can be produced by preventing the normal increase in chloride permeability. This can be done by stimulating the apical potassium pump by itself with cyclic AMP, or by stimulating in the absence of external calcium (Berridge *et al.*, 1975).

As in the mammalian salivary gland, there is a secretory potential caused by an increase in the passive permeability of the basal surface, both in the blowfly and in the cockroach *Periplaneta* (Prince and Berridge, 1972; House, 1973). The size of the potential depends on the potassium concentration of the bathing medium (Ginsborg *et al.*, 1974; Berridge *et al.*, 1976). In the blowfly, stimulation by 5-hydroxytryptamine causes a 60-fold increase in fluid secretion that is accompanied by a fall in resistance of the basal cell surface (Berridge *et al.*, 1975). As in the mammalian gland, the secretory potential is reduced when chloride is absent from the bathing fluid. Study of these glands has indicated how changes in intracellular chloride and in chloride permeability may be an important factor in the control mechanism. Berridge *et al.* (1975, 1976) have suggested that stimulation increases the transport of potassium into the lumen and increases the chloride conductance of the apical membrane. Both potassium and chloride enter the lumen, lowering the intracellular chloride concentration. The resulting gradient for chloride entry into the cell hyperpolarizes the basal membrane, providing an electrical gradient for the entry of potassium. This hypothesis explains why the hyperpolarization develops more slowly than the resistance change, and how the transported ions can enter the cell passively.

The processes involved in modification of the primary secretion have not been thoroughly studied in insects. However, in another insect secretory system, the Malpighian tubules of *Rhodnius*, potassium appears to be reabsorbed passively down its electro-chemical gradient (Maddrell, personal communication). Secondary modification of the secretion has also been demonstrated in both *Calliphora* and *Manduca* (Oschman and Berridge, 1970; Robertson, 1974). (Also see Chapters 4 and 9 and Gupta *et al.*, 1977.)

IV. EXOCRINE PANCREAS

The exocrine pancreas differs from most salivary glands in that there is little reabsorption of sodium or potassium, and the final fluid secreted is isotonic. These ions are secreted in plasma-like concentrations by the centro-acinar cells and by the cells of the intra- and extra-lobular ducts (Mangos and McSherry, 1971; Swanson and Solomon, 1973). In the mouse pancreas the centro-acinar cells in the different regions function similarly electrophysiologically (Green-well, 1975). Secretin, the hormone which stimulates fluid secretion in the pancreas, stimulates the centro-acinar cells but does not produce a response in the enzyme-secreting acinar cells (Greenwell, 1975).

The current hypothesis for pancreatic secretion is summarized in Fig. 5. Active transport of sodium across the apical cell membrane is probably the driving force for secretion, although the evidence is not conclusive.

The main problem is that ouabain and other cardiac glycosides

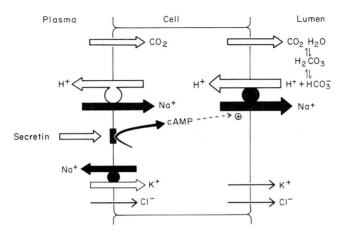

Fig. 5. A model for fluid secretion by pancreas. Secretin stimulates the secretion of a sodium rich fluid, with cyclic AMP acting as an intermediary. Active sodium extrusion across the apical membrane is coupled to hydrogen ion exchange.

inhibit secretion when applied to the basal surface of the cells while the pump is thought to be located on the apical surface (Ridderstap and Bonting, 1969; Case and Scratchard, 1974). Moreover, low doses of ouabain actually stimulate secretion and stimulate the Na-K ATPase (Ridderstap and Bonting, 1969).

Swanson and Solomon (1975) have suggested that unless ouabain can penetrate through the epithelium and act at the apical surface, it must be inhibiting a Na-K pump on the basal surface. Amiloride inhibits secretion and the Na-K ATPase (Wizemann and Schultz, 1973). Ethacrynic acid and metabolic poisons such as DNP and cyanide also inhibit secretion (Case and Scratcherd, 1974).

In the rabbit pancreas, active secretion of sodium appears to be coupled to reabsorption of hydrogen ions, so that the primary secretion is rich in bicarbonate which enters the acinar lumen because of the pH gradient (Mangos and McSherry, 1971; Swanson and Solomon, 1973). As yet, sodium and hydrogen transport have not been dissociated, and factors that affect one affect the other. Swanson and Solomon (1972, 1975) suggest that sodium exchanges for hydrogen across the basal membrane. This is thought to be a downhill process, driven by the gradient of sodium maintained by the sodium-potassium exchange pump. In addition they suggest uphill secretion of sodium across the apical membrane in exchange for hydrogen. This hypothesis can partly explain the effects of ouabain and low sodium, both of which inhibit secretion. Their effects differ in that ouabain does not affect cellular pH, whereas low sodium causes a marked pH increase. Low sodium decreases secretion by reducing the intracellular sodium concentration and therefore the sodium available for the mucosal pump. Consequently hydrogen ion transport is inhibited and the cellular pH rises. However, the effect of ouabain is still difficult to explain, unless the inhibition of a sodium-potassium exchange pump on the basal surface can inhibit sodium-hydrogen exchange at both cell surfaces, as suggested by Swanson and Solomon (1975).

Under normal circumstances bicarbonate is the counter-ion for sodium in the secreted fluid. Carbon dioxide equilibrates according to the pH gradient across the epithelium. A bicarbonate stimulated ATPase has been demonstrated in the pancreas. This ATPase is closely associated with the plasma membrane and with carbonic anhydrase, and is inhibited by thiocyanate, which also inhibits secretion (Simon and Thomas, 1972). The enzyme is present throughout the duct system of the cat pancreas while Na-K ATPase is found mainly in the acini, suggesting that the former is more important in secretion (Wizemann et al., 1974). However, the pancreas does not show an absolute requirement for bicarbonate, because when bicarbonate is replaced with acetate or other anions secretion still occurs (Swanson and Solomon, 1975). The rate is somewhat reduced and is equivalent to the rate seen with bicarbonate in the presence of Diamox, an inhibitor of carbonic anhydrase. Thus, it is possible that the bicarbonate-ATPase facilitates bicarbonate secretion rather than directly driving it. Bicarbonate and acetate transport are probably driven by a common mechanism involving active transport of hydrogen. The effects of serosal bicarbonate concentration and pH all seem consistent with this view (Swanson and Solomon, 1975).

Electrophysiological investigations of the fluid secreting cells of the pancreas are few. Greenwell (1975) has found that the cells hyperpolarize in response to secretin. The hyperpolarization is long-lasting, but takes 3—4 minutes to develop. This is slower than the onset of secretion. The hyperpolarization is probably not associated with the efflux of potassium that occurs at the onset of stimulation (Case et al., 1969), as has been postulated for salivary glands. The potassium efflux is short-lived and is over by the time the hyperpolarization is developed. Greenwell (1975) suggests that hyperpolarization may result from the movement of hydrogen ions out of the cell preceding the influx of sodium.

In the cat, dog and rabbit, the pancreas secretes an isotonic fluid rich in sodium and bicarbonate. At low rates of secretion the fluid is rich in chloride because of a chloride-bicarbonate exchange in the ducts (Schultz *et al.*, 1969; Mangos and McSherry, 1971). At higher rates of secretion the exchange mechanism becomes saturated and the bicarbonate concentration rises (Schultz *et al.*, 1969; Mangos and McSherry, 1971). The rat is different in that the chloride concentration rises as the flow increases (Mangos and McSherry, 1971).

Fluid secretion is stimulated by secretin which probably uses cyclic AMP as a mediator. Secretin stimulates cyclic AMP production and its effect is potentiated by theophylline (Case *et al.*, 1972; Deschodt-Lanckman *et al.*, 1974). Nicotinic acid and imidazole lower cyclic AMP levels by stimulating phosphodiesterase, the enzyme responsible for cAMP breakdown, and reduce secretion (Bonting *et al.*, 1974). However, the details of stimulus-secretion coupling are poorly understood.

V. SWEAT AND LACRIMAL GLANDS

Like salivary glands, sweat and lacrimal glands produce a primary secretion that is modified by its passage through the duct system. In both tissues the primary secretion is basically plasma-like, having high concentrations of sodium and chloride (Schultz *et al.*, 1965; Alexander *et al.*, 1972). During passage through the duct system there is reabsorption of sodium in sweat glands but not in lacrimal glands (Schultz *et al.*, 1965; Alexander *et al.*, 1972; Kaiser *et al.*, 1974). Potassium is secreted by both glands in the duct system so that the final secretion may be hyperosmotic (Alexander *et al.*, 1972; Slegers and Moons, 1973). Kaiser *et al.* (1974) postulate that hydrogen ions are secreted in exchange for sodium in the duct system of human sweat glands, and that potassium secretion may be passive.

VI. CONCLUSIONS

In general, exocrine glands produce primary secretions that are approximately isotonic to plasma. Water is presumed to follow the active movement of ions because of local osmotic gradients created by ion movement (see Gupta, 1976). This process depends on the activity of a cation pump located at the apical membrane. In some tissues, like insect salivary glands, chloride is able to follow the transported cation passively. This is not always the case because in some tissues (e.g. pancreas) sodium transport is probably in exchange for hydrogen ions, and the result is bicarbonate secretion. Ions pumped into the lumen must be replaced from the bathing medium. This replacement does not seem to require active uptake across the basal surface because the ionic gradients favour influx of the transported solutes. Active transport is thought to maintain these gradients. In the resting state this would be accomplished by the classical ouabain-sensitive Na-K exchange pump, while in the active state the pump at the apical membrane would be involved.

Although there is much information in the literature regarding movements of ions across fluid secreting epithelia, very little is known about regulation of cell volume during fluid secretion. Maddrell (1976) suggests that there is either some elastic resistance or some nondiffusible osmotically active material within the cell which helps to counter any volume change. Regulation of cell volume has been studied in nontransporting tissues (see Hoffmann, Chapter 12). Volume changes have been observed in transporting tissues (Gupta, 1976) but their control has yet to be elucidated.

REFERENCES

Alexander, J. H., von Lennep, E. W. and Young, J. A. (1972). *Pflügers Arch. ges. Physiol.* 337, 299—309.

Berridge, M. J., Lindley, B. D. and Prince, W. T. (1975). *J. Physiol. Lond.* 244, 549—567.

Berridge, M. J., Lindley, B. D. and Prince, W. T. (1976). *J. exp. Biol.* (In press).

Bonting, S. L., Case, R. M., de Pont, J. J. H. H. M., Kempen, H. J. M. and Scratcherd, T. (1974). *J. Physiol. Lond.* 240, 34—35P.

Case, R. M. and Scratcherd, T. (1974). *J. Physiol. Lond.* 242, 415—428.

Case, R. M., Harper, A. A. and Scratcherd, T. (1969). *In* "The Exocrine Glands." (Eds Botelho, S. Y., Brooks, F. P. and Shelley, W. B.), pp. 39—60. University of Pennsylvania Press, Philadelphia.

Case, R. M., Johnson, M., Scratcherd, T. and Sherratt, H. S. A. (1972). *J. Physiol. Lond.* 223, 669—684.

Denniss, A. R. and Young, J. A. (1974). *Proc. Aust. Physiol. and Pharmacol. Soc.* 5, 189.

Denniss, A. R. and Young, J. A. (1975). *Pflügers Arch. ges. Physiol.* 357, 77—89.

Deschodt-Lanckman, M., Robberecht, P., De Neef, Ph. and Christophe, J. (1974). *Arch. Int. Physiol. et de Biochemie* 82, 180.

Douglas, W. W. and Poisner, A. M. (1963). *J. Physiol. Lond.* 165, 528—541.

Field, M. J. and Young, J. A. (1973). *Pflügers Arch. ges. Physiol.* 345, 207—220.

Ginsborg, B. L., House, C. R. and Silinsky, E. M. (1974). *J. Physiol. Lond.* 236, 723—731.

Greenwell, J. R. (1975). *Pflügers Arch. ges. Physiol.* 353, 159—170.

Gruber, W. D., Knauf, H. and Frömter, E. (1973). *Pflügers Arch. ges. Physiol.* 344, 33—49.

Gupta, B. L. (1976). *In* "Perspectives in Experimental Biology." (Ed. P. Spencer-Davies), pp. 25—42. Pergamon Press, Oxford.

Gupta, B. L., Berridge, M. J., Hall, T. A. and Moreton, R. B. (1977). *J. exp. Biol.* in press.

Harrison, J. D. (1974). *J. Histochem.* 6, 649—664.

House, C. R. (1973). *J. exp. Biol.* 58, 29—43.

Imai, Y. (1974). *In* "Secretory Mechanisms of Exocrine Glands." (Ed. N. A. Thorn and O. H. Petersen), pp. 199—212. Munksgaard, Copenhagen.

Imai, Y., Nishikawa, H., Yoshikazo, K. and Watari, H. (1973). *Jap. J. Physiol.* 23, 635—644.

Kafatos, F. C. (1968). *J. exp. Biol.* 48, 435—453.

Kaiser, D., Songo-Williams, R. and Drack, E. (1974). *Pflügers Arch. ges. Physiol.* 349, 63—72.

Knauf, H. (1972). *Pflügers Arch. ges. Physiol.* 333, 326—336.

Knauf, H. (1973). *Pflügers Arch. ges. Physiol.* 343, R63.

Knauf, H. and Frömter, E. (1970). *Pflügers Arch. ges. Physiol.* 316, 238—258.

Knauf, H. and Lübcke, R. (1975). *Pflügers Arch. ges. Physiol.* 361, 55—59.

Knauf, H., Frömter, E. and Gebler, B. (1971). *In* "Proceedings of the International Union of Physiological Sciences." Vol. IX, pp. 310. German Physiological Society.

Knauf, H., Gebler, B., Martin, C. J. and Young, J. A. (1972). *Pflügers Arch. ges. Physiol.* 335, R60.

Knauf, H., Baumann, K., Röttger, P. and Wais, U. (1974). *Pflügers Arch. ges. Physiol.*

Knauf, H., Röttger, P., Wais, U. and Baumann, K. (1975). *In* "Excretion." (Ed. A. Wessing), Fortschritte der Zoologie, Band 23, Heft 2/3. pp. 307—321. Gustav Fischer Verlag, Stuttgart.

Koefoed-Johnsen, V. and Ussing, H. H. (1958). *Acta physiol. Scand.* 42, 298—308.

Lundberg, A. (1957). *Acta physiol. Scand.* 40, 35—58.

Lundberg, A. (1958). *Physiol. Rev.* 38, 21—40.

Maddrell, S. H. P. (1976). *In* "Handbook on Transport across Biological Membranes." Vol. 3. (Ed. G. Giebisch), (In press).

Mangos, J. A. and McSherry, N. R. (1971). *Am. J. Physiol.* 221, 496—503.

Martin, C. J. and Young, J. A. (1971). *Pflügers Arch. ges. Physiol.* 327, 303—323.

Martin, C. J., Frömter, E., Gebler, B., Knauf, H. and Young, J. A. (1973). *Pflügers Arch. ges. Physiol.* 341, 131—142.

Nishiyama, A. and Kagayama, M. (1973). *Experientia* 29, 161—163.

Nishiyama, A. and Petersen, O. H. (1974). *In* "Secretory Mechanisms of Exocrine Glands." Alfred Benzon Symposium VII (Eds Thorn, N. A. and Peterson, O. H.), pp. 216–224. Munksgaard, Copenhagen.

Oschman, J. L. and Berridge, M. J. (1970). *Tissue and Cell* 2, 281–310.

Peaker, M. (1971). *Phil. Trans. R. Soc. Lond. B* 262, 289–300.

Petersen, O. H. (1970a). *J. Physiol. Lond.* 210, 205–215.

Petersen, O. H. (1970b). *In* "Electrophysiology of Epithelial Cells." (Ed. G. Giebisch), pp. 208–221. Schattauer Verlag. Stuttgart, New York.

Petersen, O. H. (1971). *J. Physiol. Lond.* 216, 129–142.

Petersen, O. H. (1972). *Acta Physiol. Scand. Suppl.* 86/381, 1–57.

Petersen, O. H. (1973). *Experientia* 29, 160–161.

Petersen, O. H., Poulsen, J. H. and Thorn, N. A. (1967). *Acta Physiol. Scand.* 71, 203–210.

Poulsen, J. H. (1973). *Pflügers Arch. ges. Physiol.* 338, 201–206.

Poulsen, J. H. (1974a). *Pflügers Arch. ges. Physiol.* 349, 215–220.

Poulsen, J. H. (1974b). *In* Secretory Mechanisms of Exocrine Glands." (Eds Thorn, N. A. and Petersen, O. H.), pp. 570–581. Munksgaard, Copenhagen.

Prince, W. T. and Berridge, M. J. (1972). *J. exp. Biol.* 56, 323–333.

Prince, W. T. and Berridge, M. J. (1973). *J. exp. Biol.* 58, 367–384.

Prince, W. T., Berridge, M. J. and Rasmussen, H. (1972). *Proc. Nat. Acad. Sci. USA* 69, 553–557.

Ridderstap, A. S. and Bouting, S. L. (1969). *Am. J. Physiol.* 217, 1721–1727.

Roberstson, H. A. (1974). "Structure, function and innervation of the salivary gland of the moth *Manduca sexta.*" Ph.D. Thesis. University of Cambridge.

Schneyer, L. H. (1969). *Am. J. Physiol.* 217, 1324–1329.

Schneyer, L. H. (1974a). *Am. J. Physiol.* 226, 821–826.

Schneyer, L. H. (1974b). *Am. J. Physiol.* 227, 606–612.

Schneyer, L. H., Young, J. A. and Schneyer, C. A. (1972). *Physiol. Rev.* 52, 720–777.

Schultz, I., Ullrich, K. J., Fromter, E., Holgreve, H., Frick, A. and Hegel, U. (1965). *Pflügers Arch. ges. Physiol.* 284, 360–372.

Schultz, I., Yamagata, A. and Weske, M. (1969). *Pflügers Arch. ges. Physiol.* 308, 277–290.

Simon, B. and Thomas, L. (1972). *Biochim. biophys. Acta* 288, 434–442.

Slegers, J. F. G. and Moons, W. M. (1973). *Pflügers Arch. ges. Physiol.* 343, 49–63.

Swanson, C. H. and Solomon, A. K. (1972). *Nature Lond.* 236, 183–184.

Swanson, C. H. and Solomon, A. K. (1973). *J. gen. Physiol.* 62, 407–429.

Swanson, C. H. and Solomon, A. K. (1975). *J. gen. Physiol.* 65, 22–45.

Tamarin, A. and Sreebny, L. M. (1965). *J. Morph.* 117, 295–352.

Thaysen, J. H. and Tarding, F. (1974). *In* "Secretory Mechanisms of Exocrine Glands." (Eds N. A. Thorn and O. H. Petersen), pp. 464–472. Munksgaard, Copenhagen.

Thaysen, J. H., Thorn, N. A. and Schwartz, I. L. (1954). *Am. J. Physiol.* 178, 155–159.

Wais, U. and Knauf, H. (1975). *Pflügers Arch. ges. Physiol.* 361, 61–64.

Williams, J. A. (1970). *J. theor. Biol.* **28**, 287—296.

Wizemann, V. and Schultz, I. (1973). *Pflügers Arch. ges. Physiol.* **339**, 317—338.

Wizemann, V., Christian, A.-L., Wiechmann, J. and Schultz, I. (1974). *Pflügers Arch. ges. Physiol.* **347**, 39—47.

Young, J. A., Fromter, E., Schögel, E. and Hamann, K. F. (1967). *Pflügers Arch. ges. Physiol.* **295**, 157—172.

Young, J. A., Martin, C. J. and Weber, F. D. (1971). *Pflügers Arch. ges. Physiol.* **327**, 285—302.

x

VI.   Osmoregulation

26. Transpiration in Land Arthropods

E. B. Edney

Laboratory for Medicine and Radiation Biology, University of California, Los Angeles, California, U.S.A.

I. INTRODUCTION

Whether or not transpiration through the cuticle of an arthropod forms a major part of overall water flux depends of course on the size of the other components. But most arthropods have large surface/volume ratios, so that the significance of transpiration is often large, and their ability to live in rather dry habitats and to feed on rather dry food, depends on the presence of a relatively impermeable cuticle. The means by which this is achieved (and perhaps regulated) while still permitting the integument to carry out all its many other functions in various and variable environmental circumstances has been the object of research for many years.

The lines of approach in this research have been (i) to find the effects of climatic factors upon transpiration and to relate this information to ecological affairs, particularly with regard to distribution and survival of individuals and species, and (ii) to discover the morphological, chemical and physical structure of the cuticle, by direct observation and by experimental manipulation, taking advantage wherever possible of natural processes such as moulting, that provide good research leverage. These approaches have been mutually supportive in so far as transpiration rates have been used to throw light on structural details, and vice versa. Today we know a good deal more than we did 40 years ago when this search began, but the most important problems are still unresolved. This chapter will refer to the inter-relationships between transpiration and structure as these two fields of enquiry have developed, and will consider the relevance of transpiration to ecological affairs.

II. CUTICLE STRUCTURE

The biology of arthropod cuticles has recently been reviewed by Neville (1975). For present purposes the basic features may be described as follows, further details being considered later as occasion demands.

The cuticle is produced by a single cell layered epidermis and consists of a relatively thick inner procuticle and a thin (up to 1 μm) epicuticle. The procuticle is composed of protein and chitin (a polymerized, acetylated gluocosamine), the outer part of which may be hardened (and sometimes darkened) by quinone tanning. This outer portion is then known as the exocuticle, to distinguish it from the unhardened endocuticle. The thin epicuticle is a stable but very complex structure, consisting mainly of proteins and lipids whose details will be considered below. The procuticle and at least part of the epicuticle is traversed by numerous fine, often twisted ribbon-like, pore canals, through which materials may be transported from the epidermal cells to the exterior of the epicuticle.

III. THE MEASUREMENT OF TRANSPIRATION

The measurement of transpiration is complicated by the fact that in arthropods the outer surface is not homogenous but is perforated by spiracles that lead to a complex system of branching tracheae whose internal surfaces form respiratory membranes. Experimentally it is not simple to separate the respiratory and cuticular components of transpiration, but what evidence there is suggests that respiratory loss is relatively small in the resting state (references in Edney, 1977). Here we are concerned with cuticular loss alone, recognizing that for active animals in real life, the respiratory component may become very important.

If the cuticle were a homogeneous membrane (it is not), movement of water through it could be defined by a form of Fick's equation,

$$J_w = \frac{D\Delta C}{\Delta X} \tag{1}$$

where J_w is the flux of water in mg cm^{-2} sec^{-1}, D is the diffusion coefficient of the cuticle in cm^2 sec^{-1}, ΔC is the water concentration gradient across the cuticle in mg cm^{-3}, and ΔX is the thickness of the cuticle in cm.

Most recent work on cuticular water loss has been reported in mg cm^{-2} hr^{-1} (mm Hg)$^{-1}$, using vapour pressure gradient (ΔP) in mm Hg as a measure of the force tending to move water across the cuticle; and this is a useful measure of permeability. For some purposes (e.g. for comparison with data for plant cuticles, where different units are commonly used) it is convenient to express permeability in cm sec^{-1}, and its reciprocal, resistance, in sec cm^{-1}. In this case D/X in equation (1) may be replaced by R^{-1} (resistance) in sec cm^{-1}, and the equation (1) modified to give,

$$J_w = \frac{k(C_i - C_o)}{R_c + R_a} \tag{2}$$

where k is a constant concerned with units and geometry, C_i and C_o are concentrations of water in mg cm^{-3} inside the cuticle and in the air at a distance from it, and R_c and R_a are resistances of the cuticle and the air respectively, in sec cm^{-1}. (This implies the simplifying assumption that water passes through the cuticle as water vapor).

Unfortunately the real situation is by no means so simple, as stressed by Beament (1961) for the cuticle is multi-layered and it is entirely possible not only that the resistance of each layer may differ, but that the resistance of one or more layers may vary with their water content. In other words, D may be a function of C in equation 1. Futhermore, if R_c is very low, J_w will be affected greatly by R_a, and air movement, for example, becomes very important in determining transpiration. Different R_a's in such cases may in turn affect water concentrations in the cuticle and thus cause changes in R_c. Measurements of transpiration when R_c is low compared with R_a will therefore provide little information about R_c; however, where R_c is large (as it usually is), and unchanged by water absorption, measurement of cuticle resistance is more feasible.

Some indication of the size of this effect may be derived from published values for air boundary layer resistances next to plant leaves. For a leaf 5 cm long, in air at 20°C moving at 10 cm sec^{-1}, the unstirred boundary layer is about 0.25 cm thick and $R_a = 1$ sec cm^{-1}; for an unstirred layer 0.025 cm thick $R_a = 0.1$ sec cm^{-1} (Nobel, 1974). Since R_{total} for most arthropods is very much greater than this (see Table I) we need not be very concerned with errors caused by air resistance in field conditions. In laboratory experiments in nearly still air, R_a may be more important.

TABLE 1[a]

Taxon	Habitat	Permeability in $\mu g\ cm^{-2}\ h^{-1}\ mmHg^{-1}$	Resistance in sec cm^{-1}
Isopods			
Porcellio scaber	hygric	110	31
Venezillo arizonicus	xeric	32	106
Insects			
Hepialus larvae	hygric	190	18
Tenebrio molitor larvae	xeric	5	680
Periplaneta americana	mesic	55	62
Arenivaga investigata	xeric	12	275
Calliphora erythrocephala	mesic	51	67
Glossina morsitans	mesic-xeric	8	425
Calosis amabilis	xeric	1.09	3125
Trigonopus sp.	mesic	4.13	823
Arachnids			
Ixodes ricinus	mesic	60	57
Ornithodorus moubata	xeric	4.0	833
Pandinus imperator	mesic	76	45
Hadrurus arizonensis	xeric	1.22	2790

[a] References in Edney (1974, 1977).

IV. TRANSPIRATION AND TEMPERATURE

In 1933 Gunn reported that in living cockroaches water loss increases rapidly at temperatures above about 30°C, and ascribed this to the onset of active respiratory movements. He was probably correct in part, but Ramsay's (1935) percipient experiments with *Periplaneta* showed that the increase in transpiration occurs even in dead cockroaches with sealed spiracles, and is not all accounted for by increasing ΔP (vapour pressure gradient from insect blood to dry air) as temperature rises. In this way the significance of the structure and properties of insect cuticle in relation to water affairs was firmly established.

Another important step was taken when Wigglesworth (1945), Beament (1945) and Lees (1947) carried out experiments on a wide variety of insects and ticks, and the lipids extracted from their cuticles. These experiments showed that in dead animals with sealed spiracles, transpiration remains low as ambient temperature rises, until a point is reached (the "critical" or "transition" temperature) above which transpiration increases rapidly with any further rise in temperature. In general they found transition temperatures to be higher, and transpiration itself to be lower, in arthropods from hotter, drier habitats. Transition temperatures were often different in different developmental stages (e.g. between larvae and adults of the same species), and between different ages within one instar. Of great interest, too, was Wigglesworth's demonstration that very slight abrasion (visible only histochemically) of the insect's surface is sufficient to raise transpiration greatly and to abolish the transition effect.

Beament (1945) further found that if lipids extracted from cast skins (largely epicuticles) are deposited on inert membranes of tanned gelatin or cleaned butterfly wings, such membranes then show permeabilities and transition phenomena very similar to those of the intact cuticles from which the lipids came.

Wigglesworth and Beament then proposed that permeability is controlled not so much by the cuticle or the epicuticle as a whole, but rather by a monolayer of lipid molecules oriented on the epicuticular surface with their hydrophilic, polar ends inwards and their hydrophobic non-polar hydrocarbon chains outwards. Beament (1964) subsequently proposed that the hydrocarbon chains are oriented at an angle of about 25° from normal to the substrate, in which position they are in van der Waal contact, and form a crystalline layer highly impermeable to water. Transition to a more permeable state would occur at a particular temperature as a result of a physical change of the monolayer to a more mobile state — the hydrocarbon chains being free to move and permit the passage of water molecules. An alternative to this model (Davis, 1974a) is referred to below, but whatever the correct interpretation at the molecular level, the evidence strongly points to a superficial layer in the epicuticle as an important barrier to transpiration.

Since 1947 a bewildering variety of temperature/transpiration measurements have been published, but space permits only a brief reference to the major conclusions that have emerged and the problems that remain. For further consideration see Beament (1961), Neville (1975) and Edney (1977). In general, measurements confirm the presence of transition phenomena — at least in some circumstances and in some groups — although conflicts do remain. Transition phenomena are more firmly established in insects and arachnids than in other arthropod groups, and contrary to some early suggestions, increases in transpiration in the "transition" groups are not accounted for by increasing ΔP as temperature rises. Double transition regions have been reported (but not always confirmed) and ascribed to additional wax layers, a cement layer, or to lipids deeper in the cuticle. To account for conflicting results cuticle surface temperature has been emphasized in some cases, vulnerability of surface waxes to abrasion by handling in others. Conflicts are also ascribable to differences in techniques of measurement (these have been very great), and to differences in physiological condition of the animals concerned. Differences in the state of cuticle hydration from time to time may be very important. Thus, in many arthropods water is lost rapidly from the cuticle at first, leading to slower transpiration later, so that measurements in one animal over long periods of time can be very difficult to interpret.

The general pattern that emerges is that transition is a real phenomenon, but that its results vary from a very abrupt increase in transpiration above a sharply defined temperature, to a gentle increase over a wide temperature range. It may be important that in all cases permeability (*sensu stricto*) continues to increase with increasing temperature above the transition range. Differences in the temperature at which transition occurs are also very great (from around 28°C in *Thermobia* to over 50° in several insects and ticks). Furthermore these large differences in abruptness and in actual temperature levels are found not only between taxa, but also between different developmental stages, sexes, ages and states of nutrition in the same taxon, or even in one individual, as Davis (1974b) found for the tick *Haemaphysalis*.

All this, of course, points to the need for information about the nature and disposition of the lipids involved, and we shall return to this after considering two other matters concerning cuticular properties relevant to transpiration.

V. ASYMMETRICAL PERMEABILITY OF THE CUTICLE

Some time ago Hurst (1941) and Beament (1945) found that *in vitro* the permeability to water of arthropod cuticles is greater inwards than outwards. If the epicuticle surface is exposed to water and the procuticle to air, total resistance is less than it is in the reverse situation. Differences of 2 : 1 or more (Richards *et al.*, 1953) have been measured. The rapid absorption of water vapour where this occurs (see Chapter 22 for details) seems to provide evidence for the same effect *in vivo*, but recent work has shown that such absorption occurs not through the body cuticle, but through the rectal wall in *Thermobia* (Noble-Nesbitt, 1970) and *Tenebrio* larvae (Locke, 1964; Noble-Nesbitt, 1973), and even through the mouth in the tick

Amblyomma (Rudolph and Knulle, 1974). If these methods of entry prove to be general, there will be no evidence of asymmetry of cuticular permeability in living insects. However, there is still evidence of asymmetry *in vitro*, and this merits further consideration since it clearly has implications for cuticular structure and for transpiration (even though the procuticle *in vivo* is, of course, never exposed to air).

Several attempts to explain asymmetry, none of them entirely satisfactory, have been made. As Hartley (1948) pointed out, in any bilaminar membrane, if the resistance of one layer (R_i) varies with its water content while that of the other (R_{ii}) does not, then in an experimental situation where one surface is exposed to water, the other to air, R_{total} will differ according to the orientation of the membrane. In arthropod cuticles, if we assume that the resistance of the procuticle is greater when its water content is lower, and that the resistance of the epicuticle is unaffected (or affected to a lesser extent), then R_{total} will be lower in the system: water-procuticle-epicuticle-air, than in the reverse system. But this compounds rather than solves the problem, for *in vitro* (and perhaps *in vivo*), resistance inwards (epicuticle to procuticle) is less than that in the reverse direction.

Several other proposals have been made to account for asymmetry, all of which involve unproved assumptions, e.g. that there is a monolayer of lipid molecules at the surface whose state is affected by water activity outside the cuticle, and that fine capillaries pass through the epicuticle to the surface. This is the basis of Locke's (1965, 1974) suggestion and of Noble-Nesbitt's (1969) modification of a proposal by Beament (1965). Both offer attractive models but experimental verification is lacking, and asymmetrical permeability, insofar as it exists either *in vitro* or *in vivo*, deserves further investigation.

VI. EPIDERMAL CELLS AS WATER BARRIERS

There is accumulating evidence that water activity in the cuticle of an arthropod may not be in equilibrium with that in its blood. Bursell (1955) found this in isopods, Winston (1967) in locusts and cockroaches, and Loveridge (1968) in locusts. It has been used as evidence for the existence of a pump (of an unspecified nature), that moves water from the cuticle to the epidermal cells; but as Berridge (1970) pointed out, such a mechanism is unnecessary if there is a barrier to the movement of water from cell to cuticle. Such a barrier might involve the outer (apical) cell membrane, and if the resistance here were high, water activity in the cuticle would tend to come into equilibrium with that in the air outside rather than with that in the blood. Regulation of transpiration (such as that observed by Loveridge) would then be achieved by varying the resistance of the apical cell membrane.

There is no experimental evidence at all that this occurs. On the other hand there is evidence that the process of plasticization in *Rhodnius* cuticle (to permit stretching after a blood meal) does involve an increase in cuticular water content (Reynolds, 1974a,b; 1975), and there is a recent report (Treherne and Willmer 1975) that cuticle permeability is under hormonal control in *Rhodnius*. If, in addition, recent evidence that body water content, cuticle hydration and ambient humidity may all affect cuticle permeability in some arthropods turns out to be generally true, a tentative hypothesis would be to suppose that information about both internal and external parameters may be fed through sensory channels to the central nervous system, there to affect the production and release of a hormone which in turn controls water balance by its action on the apical membrane of epidermal cells. However, experimental evidence is again lacking. The nature and function of plasma membranes in relation to water transport is, of course, central to this problem and the subject has recently been reviewed by Oschman *et al.*, (1974). (Also see Chapter 6.)

VII. THE STRUCTURE OF THE EPICUTICLE

We come now to the second method of approach, namely, observation of the formation and structure of the cuticle, and the nature and function of the lipids found in the epicuticle. These matters have recently been reviewed by Neville (1975) and Edney (1977), and it is the present purpose only to draw attention to recent work and to its significance in relation to transpiration. Inevitably there has been a good deal of confusion with regard to terminology: here we follow Wigglesworth (1975) with cross references where necessary.

In *Rhodnius* (Wigglesworth, 1975) the first layers of the new cuticle to be laid down are the outer and the inner epicuticle, in that order. Both layers contain abundant lipid, the outer layer appears first as two electron dense lines separated by a space; the inner, much thicker layer, is composed largely of "cuticulin", a term used by

Wigglesworth to refer to a highly stabilized lipo-protein complex. (Cuticulin is now known to occur also in the cement layer and probably in parts of the procuticle). Both layers contain fine, lipophilic channels — the epicuticular channels, about 20—25 nm wide, through which at this stage the moulting fluid may pass. After the epicuticle has been formed, the epidermal cells proceed to lay down the procuticle, perforated by pore canals which are continuous with the epicuticular channels. Shortly before ecdysis the epicuticular channels contain silver-binding material, perhaps tyrosine as a forerunner of the polyphenols that appear later, and this seems to be discharged mainly at the junction of the outer and inner epicuticles, from where it permeates much of the inner epicuticle. But the material is also visible as a thin membrane over the outer surface of the outer epicuticle.

About this time, lipid material appears in the epicuticular channels, and is discharged onto the outer surface, at first covering, but later (just before ecdysis) permeating the silver binding material with which it combines, and perhaps becomes polymerized, to form a thin (< 10 nm) highly stabilized, unreactive, hydrophobic surface — the "wax layer". Finally, after ecdysis, a material, perhaps a mucopolysaccharide, from the dermal glands, is discharged via the gland ducts over the wax layer, with which it in turn becomes very closely associated.

The above account refers to *Rhodnius*, but it may well apply with modification to many other insects with hard waxes. It suggests that the layer previously referred to as the oriented lipid layer by Locke (1966) is in fact a multimolecular layer of stabilized lipid and phenolic materials. On the other hand, in insects where the surface is softer and grease-like (as in cockroaches), an oriented monolayer may be formed, for Ramsay (1935) observed that grease from a living cockroach cuticle spreads over a water drop sprayed onto the cuticle, and virtually prevents evaporation from it. The whole film is multilayered since it shows interference patterns, but the innermost layer may have consisted of oriented molecules.

VIII. THE NATURE OF EPICUTICULAR LIPIDS

There have been several recent reviews dealing with what is known of cuticular lipids (refs in Hackman, 1974; Neville, 1975), but there is unfortunately very little work correlating the presence of particular lipids with particular cuticular properties. Hydrocarbons are usually present in high proportions — up to 77% in *Periplaneta,* but even in insects with hard waxes hydrocarbons are usually abundant. These are nonpolar lipids, and as such would not be expected to form monolayers. Most other classes of lipids contain polar groups, and of these, alcohols, phospholipids and fatty acids are known to form stable crystalline monolayers which may undergo transition in physical state to a more liquid form at temperatures determined largely by the carbon chain length and the extent of branching in the nonpolar hydrocarbon moieties. Alcohols and phospholipids are conspicuously absent from arthropod cuticles (save apparently from *Tenebrio* larvae) but fatty acids are usually abundant enough and could be involved in monolayer formation.

One of the few papers that relate lipid composition to biological significance is that of Armold *et al.* (1969) on the stone-fly *Pteronarcys*, where the adult has twice as much lipid as the larva. The preponderant adult lipids are free fatty acids (49%), sterols (18%) and hydrocarbons (12%). The authors believe that the adult lipids would have higher transition ranges and be less permeable to water than those of the larvae, and that this would be adaptive since the larvae are aquatic while the adults are terrestrial (and probably do not eat or drink).

Another recent example of work relating composition to function is that of Davis (1974a,b) on the rabbit tick, *Haemaphysalis*, which shows quite striking changes in transpiration phenomena in different physiological states. Davis estimated the proportions of various solvent classes of epicuticular lipids by thin layer chromotography and found that as permeabilities change, so do the presence and proportions of different lipids. There are, for example, different kinds of sterols in ticks with or without abrupt transition points. Unfortunately the particular lipids were not identified, so that the results, suggestive as they are, need amplification.

As a result of this work, Davis proposes that the lipids responsible for permeability control are crystalline at room temperatures, but become more mobile at higher temperatures. Increased transpiration, she suggests, results not from the disruption of a monolayer, but from changes in the packing of hydrocarbon chains of many molecules throughout the epicuticle. The presence, abruptness and temperature levels of transition would then depend upon the degree of homogeneity of the lipid mixture — single lipids giving an abrupt transition, complex mixtures leading to very smooth or no transitions.

This is an attractive hypothesis, but of course virtually the same effects could be expected from monolayers according to their degree of homogeneity and the nature of the lipids involved. One can quite readily envisage a monolayer functioning as a barrier to the passage of water and exhibiting transition phenomena — and these properties might well persist even if the lipids are in some way polymerized. It is less clear how stabilized lipids distributed throughout a relatively thick membrane would associate to confer impermeability and transition phenomena. However, a lipid monolayer has not so far been satisfactorily identified in electron micrographs, and lipids are certainly widely distributed in the "wax layer" and elsewhere (Wigglesworth, 1975) of *Rhodnius*. Thus the evidence is somewhat favourable to Davis's hypothesis, but the monolayer theory has certainly not been disproved. If the "bulk" lipid hypothesis proves to be correct, it will be the task of future research to identify the nature and configuration of the lipids concerned.

Finally, identification of an impermeable layer at or near the surface does not preclude the possibility that other layers may be additionally or alternatively involved. The thick cuticle of the stag beetle *Lucanus* is not rendered permeable by even deep abrasion (Lafon, 1943; Beament, 1961), and the same may be said of *Sarcophaga* larvae and of isopods (Bursell, 1955; Edney, 1951) where impermeability is preserved until the whole epicuticle has been pierced. (Richards *et al.*, 1953.)

IX. TRANSPIRATION AND ECOLOGY

Despite the complexities referred to above and the consequent difficulty of interpreting measurements of transpiration in terms of cuticle structure, comparative measurements of transpiration rates are valid for ecological purposes. Such measurements have been made by different workers with animals in a variety of different physiological conditions, but it is often possible to arrive at an approximate figure for overall cuticular loss and to compare species from different habitats in this respect. When this is done, very large differences between species are found, and in general, as might be expected, low permeabilities go together with xeric or hot habitats. A small selection of the data available is shown in Table I, to illustrate the enormous range in cuticle permeabilities and the correspondence with habitat. Further examples are given by Bursell (1970) and Edney (1974, 1977).

However, it is important to recognize that integumental permeability is by no means a complete guide to ecological adaptation with respect to water balance, because it says nothing about size, and therefore about the proportion of water reserves lost in unit time for unit permeability. Animals that have both low permeability as well as large size (as do several desert scorpions whose weight may be several grams) are exceedingly well adapted to withstand desiccating conditions for long periods.

This point is further illustrated in Fig. 1 where rates of water loss are compared in several Namib desert insects of various sizes (Edney, 1971). If the different species are compared by using the rate of loss appropriate to the mean size for each species, there is a significant negative correlation between proportion of weight lost and mean size (both on a log scale), when $y = 3.65 - 0.26x$, and $r = 0.88$. If point (1) in Fig. 1 (which refers to a thysanuran) is omitted, the y intercept is 4.095, the slope is -0.33 and r is 0.775.

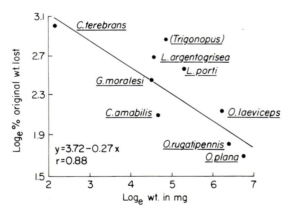

Fig. 1. The relation between weight and weight specific water loss (both on a log scale) during five days in air at $27°C$ in living tenebrionid beetles from the Namib desert. For discussion see the text. Redrawn from Edney (1971).

The data illustrate several points. Firstly, they show an overall effect of size: the largest beetles (*Onymacris plana*) lose a smaller proportion of their original weight per day than the smaller ones when all are tested in dry air at 27°C. Secondly some insects of similar size (e.g. *Lepidochora argentogrisea*, *Gyrosis moralesi*, *Calosis amabilis* all of which weigh about 90 mg) show different weight specific water loss rates, and such differences probably reflect differences in cuticle permeability. Thirdly, whatever the cause of differences in proportional water loss, such differences certainly correspond with ecological facts. Thus *Onymacris plana*, which is active by day in the summer and runs on open sand, has the lowest proportional water loss rate of all. Of the three intermediate sized species *Calosis ambilis* is most desertic, being active by day in exposed areas. It might therefore be expected to have a low permeability (area specific water loss), and indeed it is the least permeable of all the beetles studied. *Onymacris laeviceps* with evening activity has a higher proportional rate of loss, while the two species of *Lepidochora*, active largely by night, have the highest proportional rates of loss except for *Ctenolepisma*, and this insect absorbs water vapor. The beetle *Trigonopus sp.* is a mesic tenebrionid found in pine woods, and was included in the above measurements for comparative purposes. Its proportional water loss is about twice as high as the highest desert species and its permeability is higher than all the rest.

Finally, it is interesting to compare the permeabilities we have been discussing with those measured, albeit by very different means, in plants. Plant physiologists think of "leaf resistance" to water vapor moving from the mesophyll cell wall to the outside air as being offered by two pathways in parallel: one consisting of the intercellular air spaces and stomatal pores in series, the other consisting of the cuticle. When the stomata are open, the total resistance is usually low (by arthropod standards), in the region of $2-10$ sec cm^{-1} for mesophytes and $5-20$ sec cm^{-1} for xerophytes. When the stomata close, resistance increases, and the residual cuticular resistance is usually $20-80$ sec cm^{-1} for mesophytes and $100-200$ or more sec cm^{-1} for xerophytes (Nobel, 1974). These values compare with about 10^4 sec cm^{-1} for tsetse pupae, 3125 sec cm^{-1} for the tenebrionid *Calosis amabilis* to a minimum recorded resistance of 12.6 sec cm^{-1} for the centipede *Lithobius*.

X. THE SIGNIFICANCE OF THE TRANSITION TEMPERATURE

The biological relevance of different cuticular permeabilities is clear. It is possible to think of a particular permeability characteristic as being the result of an interplay of various selective advantages and disadvantages (even though the selective advantage of high cuticular permeability can at present only be guessed at). Is it also true that the temperature and abruptness of transition is independently selected? At present this seems unlikely. Firstly, we cannot equate the onset of rapid transpiration with evaporative cooling because transition often occurs at temperatures above those normally encountered, or even above the lethal limit (about 50°C for *Tenebrio* larvae and *Rhodnius*). On the other hand transition appears to occur in some species at temperatures below the insects' preferred range (about 28°C in *Thermobia*). There are indeed cases where lower permeability seems to correspond with higher transition temperature: for example Ahearn (1970) found transition temperatures of 40°, 47.5° and 50°C for the desert tenebrionids *Eleodes armata, Cryptoglossa verrucosa*, and *Centrioptera muricata* respectivley, and the beetles stand also in that order of decreasing permeability. *Periplaneta* has a low transition range and a high permeability. But there are many different shapes of temperature transpiration curves: sharp, as in *Schistocera* adult (Beament, 1959), intermediate, as in locusts (Loveridge, 1968); hardly discernible, as in *Schistocerca* nymphs (Beament, 1959), or highly variable within one individual in different physiological states, as in *Haemaphysalis* (Davis, 1974a). It has been suggested by Camin (quoted by Davis, 1974a) that in ticks, a low transition range leading to rapid loss of water immediately before feeding may stimulate rapid engorgement, but this remains to be proved. Transition points may be single, as in *Calliphora* (Beament, 1959) and *Arenivaga* (Edney and McFarlane, 1974), or double, as in *Oniscus* (Bursell, 1955), *Rhodnius, Tenebrio* and *Pieris* pupae (Beament, 1959) and the scorpion *Hadrurus arizonensis* (Hadley, 1974). Unfortunately we do not yet know what the molecular basis for these great differences in permeability and in permeability/temperature relationships is. Perhaps the position of the transition region, where it exists, is a by-product of particular molecular structures determining permeability characteristics. Such characteristics are selected for, while the accompanying transition phenomena that necessarily follow are selectively neutral, or at least of minor importance.

XI. CONCLUSIONS

It is fairly clear that if our interests lie in the relationship of cuticle structure to water balance, and through this to arthropod biology, we shall not get very much further by continuing to measure evaporation from whole arthropods, even dead ones, because it is almost impossible to interpret the results. Much more profitable for this field of enquiry will be measurements of the properties of isolated integuments, preferably alive, using autoradiography and labelled water for the accurate assessment of net movments. In this connection it is important to recognize that in many experimental systems water molecules move through the cuticle in both directions simultaneously, and that net flow (the parameter often measured) is simply the algebraic sum of possibly much greater gross fluxes in both directions. Work along these lines has already been started by Wharton, Knulle, Kanungo, and their school with interesting results (e.g. Arlian and Wharton, 1974).

On the other hand, if one's immediate interest has an ecological slant it may be important to know the extent of cuticular water loss (a) in comparison with other sources of loss, and (b) in relation to questions of distribution and biological strategies for living in difficult habitats. For these purposes measurements of total transpiration and of the effect of temperature on this, are not only justifiable but essential, and useful work along these lines will no doubt continue. Such work will have increased validity if it turns out that the use of radioactive isotopes for the measurement of water fluxes and metabolic rates in free running animals proves to be as suitable for arthropods as it has been shown to be for vertebrates (Nagy and Shoemaker, 1975). Work employing such techniques is being undertaken in several laboratories and results may be expected before long.

REFERENCES

Ahearn, G. A. (1970). *J. exp. Biol.* **53**, 573—595.

Arlian, L. G. and Wharton, G. W. (1974). *J. Insect Physiol.* **20**, 1063—1077.

Armold, M. T., Blomquist, G. J. and Jackson, L. L. (1969). *Comp. Biochem. Physiol.* **31**, 685—692.

Beament, J. W. L. (1945). *J. exp. Biol.* **21**, 115—131.

Beament, J. W. L. (1959). *J. exp. Biol.* **36**, 391—422.

Beament, J. W. L. (1961). *Biol. Rev. Cambridge Phil. Soc.* **36**, 281—320.

Beament, J. W. L. (1964). *In* "Advances in Insect Physiology." Vol. 2. pp. 67—129. Academic Press London and New York.

Beament, J. W. L. (1965). *In* "Symposia Society of Experimental Biology." (Ed. Fogg, G. E.), pp. 273—298. Cambridge.

Berridge, M. J. (1970). *In* "Chemical Zoology." Arthropoda Part A (Eds Florkin, M. and Scheer, B. T.), Vol. 5, pp. 287—320. Academic Press, New York and London.

Bursell, E. (1955). *J. exp. Biol.* **32**, 238—255.

Bursell, E. (1970). *In* "An Introduction to Insect Physiology." pp. 276 Academic Press, New York and London.

Davis, M. T. B. (1974a). *J. Insect Physiol.* **20**, 1087—1100.

Davis, M. T. B. (1974b). *J. exp. Biol.* **60**, 85—94.

Edney, E. B. (1951). *J. exp. Biol.* **28**, 91—115.

Edney, E. B. (1971). *Physiol. Zool.* **44**, 61—76.

Edney, E. B. (1974). *In* "Desert Biology." (Ed. Brown, G. W.), Vol. 2, pp. 311—384. Academic Press, New York and London.

Edney, E. B. (1977). "Water balance in land arthropods." pp. 282. Springer Verlag, Heidelberg.

Edney, E. B. and McFarlane, J. (1974). *Physiol. Zool.* **47**, 1—12.

Gunn, D. L. (1933). *J. exp. Biol.* **10**, 274—85.

Hackman, R. H. (1974). *In* "The Physiology of Insecta." (Ed. Rockstein, M.) Vol. VI, pp. 216—270.

Hadley, N. F. (1974). *J. Arachnol.* **2**, 11—23.

Hartley, G. S. (1948). *Disc. Faraday Soc.* **3**, 223.

Hurst, H. (1941). *Nature, Lond.* **147**, 388—389.

Lafon, M. (1943). *Ann. Sci. Natur.* **11**, 113—146.

Lees, A. D. (1947). *J. exp. Biol.* **23**, 379—410.

Locke, M. (1964). *In* "The Physiology of Insecta." (Ed. Rockstein, M.), Vol. III, pp. 379—470. Academic Press, New York and London.

Locke, M. (1965). *Science* **147**, 295—298.

Locke, M. (1966). *J. Morph.* **118**, 461—494.

Locke, M. (1974). *In* "The Physiology of Insecta." (Ed. Rockstein, M.), Vol. VI, pp. 124—213.

Loveridge, J. P. (1968). *J. exp. Biol.* **49**, 1—13.

Nagy, K. A. and Shoemaker, V. H. (1975). *Physiol. Zool.* **48**, 252—262.

Neville, A. C. (1975). "Biology of the arthropod cuticle." pp. 448. Springer Verlag, New York.

Nobel, P. (1974). "Plant cell physiology. A physicochemical approach." pp. 488. W. H. Freeman and Co., San Francisco.

Noble-Nesbitt, J. (1969). *J. exp. Biol.* **50**, 745—769.

Noble-Nesbitt, J. (1970). *J. exp. Biol.* **52**, 193—200.

Noble-Nesbitt, J. (1973). *In* "Comparative Physiology." (Ed. Bolis, L., Schmidt-Nielsen, K. and Maddrell, S. H. P.), pp. 333—351. North Holland.

Oschman, J. L., Wall, B. J. and Gupta, B. L. (1974). *In* "Symp. Soc. exp. Biol." (Eds Sleigh, M. H. and Jennings, D. H.), vol. 28, pp. 305—350. Cambridge.

Ramsay, J. A. (1935). *J. exp. Biol.* **12**, 373—383.

Reynolds, S. E. (1974a). *J. Insect Physiol.* **20**, 1957—1962.

Reynolds, S. E. (1974b). *J. exp. Biol.* **61**, 705—718.

Reynolds, S. E. (1975). *J. exp. Biol.* **62**, 69—80.

Richards, A. G., Clausen, M. B. and Smith, M. N. (1953). *J. cell. comp. Physiol.* **42**, 395—414.

Rudolph, D. and Knulle, W. (1974). *Nature Lond.* **249**, 84—85.

Treherne, J. E. and Willmer, P. C. (1975). *J. exp. Biol.* **63**, 143—159.

Wigglesworth, V. B. (1945). *J. exp. Biol.* **21**, 97—114.

Wigglesworth, V. B. (1975). *J. Cell Sci.* **19**, 459—485.

Winston, P. W. (1967). *Nature Lond.* **214**, 383—384.

27. Transport and Osmoregulation in Crustacea

A. P. M. Lockwood

Department of Oceanography, University of Southampton, Southampton, England

Twenty years ago when I first began to study the osmoregulation of crustacea a well-meaning friend advised me against the topic on the grounds that the subject was worked out. However, like Charles the Second, it has been an "unconscionable time dying" and, since my earnest colleague's dire prognostications have broadened into a variety of new facets.

At the present time a new dimension is being given to the field by the increasing emphasis on studies dealing with the responses of crustacea to fluctuating salinities and the physiological changes occurring during the nonsteady-state phase and it is this aspect which will be taken as a central theme in this review. Hopefully I shall be able to communicate the fascination of a living and rapidly advancing subject, a fascination which has kept me in the field since Arthur Ramsay first suggested to a tyro research student that he go away and study the Water Hog Louse!

I. BLOOD CONCENTRATION IN RELATION TO CONCENTRATION OF THE MEDIUM

Marine species of crustacea usually have body fluids which are close to being isotonic with their medium. Exceptions to this generalization occur in a number of species of isopod, prawn, mysid and crab, though the forms concerned are mostly neritic or estuarine in distribution.

In media differing in concentration from normal sea water (c. 35‰ salinity) various types of relationship occur (Fig. 1). These include:

(a) Osmoconformers: species unable to maintain the body fluid concentration markedly different from that of the medium but able to tolerate some measure of salinity change.

(b) Hypertonic-isotonic regulators: species which, though they may be isotonic with the medium in high salinities, maintain the body fluids hypertonic to more dilute media.

(c) Hypertonic-hypotonic regulators: species which are hypertonic in low salinities and hypotonic in high salinities.

(d) Hypertonic regulators: forms strongly hypertonic to the medium in both low and high salinities.

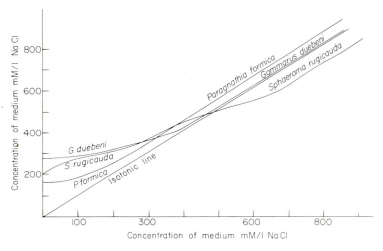

Fig. 1. Three of the fundamental types of relationships maintained between blood and medium. (a) Hypertonic − isotonic regulation, *Gammarus duebeni*; (b) Hypertonic regulation, *Paragnathia formica* (after Appelbee, 1975); hyper-hypotonic regulation *Sphaeroma rugicauda* (after Harris, 1967). Osmoconformers would have blood following the isotonic line to the limit of their tolerance.

A. Osmoconformers

Many species, normally regarded as fully marine, will tolerate some measure of dilution of their medium though unable to regulate body extra-cellular fluid concentration. Examples include sublittoral species such as *Maia squinado* and *Hyas araneus* (Schlieper, 1929). In addition some littoral species of crab, including *Cancer pagurus* (Schlieper, 1929), *Pisida longicornis* and *Porcellana platycheles*, show only a trivial degree of hypertonic regulation. The salinity range of such forms is limited by the degree to which the cells will either tolerate haemolymph dilution and the consequent tendency for water to shift from blood to cells, or alternatively the degree to which they can adjust cell internal osmotic concentration.

B. Hypertonic-isotonic Regulators

The majority of the estuarine and mid-littoral species so far examined fall into this category though the range over which hypertonicity is maintained, degree of hypertonicity and salinity and range over which isotonicity is tolerated all vary widely. In addition to variation between species it should also be noted that other factors such as sex (*Callinectes*, Tan and van Engel, 1966), temperature (*Asellus*, Lockwood, 1959), size (*Carcinus*, Gilbert, 1959), time of year (*Hemigrapsus*, Dehnel, 1962), starvation (*Corophium*, McLusky, 1970), stage in the moult cycle (*Carcinus*, Robertson, 1960) and previous salinity history of the animals, all influence the maintained gradient at the level of the individual. Even when such factors do not appear to be operating, wide individual variation is commonly found (e.g. *Australoplax tridentata*, Barnes, 1968).

C. Hypertonic-hypotonic Regulators

Certain species of Decapoda natantia (*Palaemon serratus*, Panikkar, 1950), Mysidacea (*Neomysis integer*, Ralph, 1965), Decapoda anomura (*Uca crenulata*, Jones, 1941), Decapoda brachyura (*Pachygrapsus crassipes*, Jones, 1941), Branchiopoda (*Artemia salina*, Croghan, 1958), and Isopoda (*Gnorimosphaeroma oregonensis*, Riegel, 1959) regulate hypertonically in dilute media and hypotonically in more saline media. The degree of hypo- or hypertonicity ranges widely (Fig. 2). Animals displaying hyper-hypotonic regulation are generally either littoral, estuarine or salt marsh forms and related but fully marine species may lack this regulatory capacity. For example, the marine prawns *Pandalus montagui* and *Pandalina brevirostris* differ from *P. serratus* in being isotonic with sea water (Panikkar, 1950).

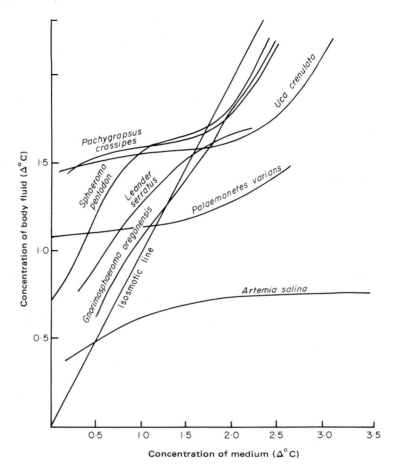

Fig. 2. Various levels of hypo-osmotic and hyperosmotic regulation, showing the range between almost homoiosmotic regulation and almost complete conformity with the medium. Data from the following authorities: *Artemia salina* from Croghan (1958), *Palaemonetes varians* and *Leander serratus* from Panikkar (1941), *Sphaeroma pentodon* and *Gnorimosphaeroma oregonensis* from Riegel (1959), *Uca crenulata* and *Pacchygrapsus crassipes* from Jones (1941). (After Lockwood, 1968).

D. Purely Hypertonic Regulators

Until very recently it would have been considered that the above mentioned forms of regulation, together with forms isosmotic within their tolerance range, constituted the only three naturally occurring types of relationship between blood and medium. However, Appelbee (1975) has demonstrated the presence of a fourth style of regulation in the praniza larva of the isopod *Paragnathia formica*. The larval stage of this species is hyperosmotic to the medium by some 300 m Osmoles when in very dilute media ($<1\%_0$) but, unlike the typical hyperosmotic regulator, also remains strongly hypertonic to more concentrated media, at least up to salinities of 180% sea water (Fig. 1). Consideration of the life style of the praniza larva offers a possible explanation for what would appear to be an energetically expensive means of regulation. The larva is ectoparasitic on the gills of *Pleuronectes limanda* and other estuarine species of fish and, in the particular region of salt marsh where the specimens were collected, the salinity changes between $2\%_0$ and $26\%_0$ over a tidal cycle (Springs). Since the larvae perforce go where the fish goes they can, potentially, be exposed to rapid and extensive increases in salinity of the medium. The pranizas are of small size and relatively permeable to water, consequently they would be in grave risk of suffering extensive dehydration on transfer to raised salinity unless protected in some way. The fact that they are always some 140 m Osmoles more concentrated than any medium to which they have acclimated presumably offers a measure of protection against osmotic dehydration should the fish carry them to more concentrated water since no water withdrawal can occur until the salinity has risen by at least 140 m Osmoles.

Another euryhaline species of small body size, the copepod *Acartia tonsa* also maintains a comparable degree of hypertonicity over a relatively wide salinity range (Lance, 1965) suggesting that the effect is not limited to parasitic forms.

No purely hypotonic regulators are known though *Artemia salina* tends towards this condition being hypertonic in only a small part of its salinity tolerance range (Croghan, 1958).

II. REGULATORY MECHANISMS

Regulation of the body fluids at a level more, or less, concentrated than the medium, involves a variety of physiological adaptations including restriction of the rate of passive movement of ions and water across the body surface, development (or hypertrophy) of processes responsible for the transport of ions at the body surface (including the gut) and, in some cases, the production of urine hypotonic to the blood. In addition, species which tolerate wide variations in blood concentration show the ability to vary cell osmotic pressure by means of adjustment to the concentration of intracellular ions and free amino acid levels. Although not regulating the blood concentration, isosmotic forms do require to slow changes in blood ion levels as much as possible in order to give adequate time for compensatory changes in cell osmotic pressure to occur.

The degree of reliance placed upon individual regulatory mechanisms differs from one species to another but it seems clear, in view of the multiplicity of forms which have successfully colonized non-marine waters, that the physiological and biochemical mechanisms necessary for enabling survival in either dilute or very concentrated media have evolved independently on many occasions in phylogenetically separated groups. The inference which may be drawn from such repeated development of comparable systems is that the fundamental apparatus for such processes as the regulation of free amino acids, active transport of ions at the body surface etc. are already present is isotonic marine species subserving other functions and that natural selection has merely acted to modify and hypertrophy such existing systems in the non-marine forms.

A. Volume Regulation

Stenohaline marine crustacea such as *Maia squinado, Galathea squamifera, Porcellana longicornis* and *Eupagurus bernhardus* all swell after sudden exposure to dilute sea water. In *Galathea, Eupagurus* and *Porcellana* the rate of increase in weight is some 12%/hr after transfer to 60% sea water and both *Galathea* and *Porcellana* generally die after a weight increment of some 12% (Davenport, 1972). *Acartia tonsa* may even swell so rapidly that it bursts when its medium is suddenly diluted from 100 to 10% sea water (Lance, 1975).

Euryhaline forms are more effective in matching osmotic intake of water by urine output. Thus the crab *Pachygrapsus* is estimated to produce some 15 times as much urine when in 50% sea water as it does in 100% sea water (Gross and Marshall, 1960). Comparable increases in urine flow occur on transference of *Gammarus oceanicus* from sea water to dilute media (Werntz, 1963). Rapidity of response to changed salinity is essential to euryhaline forms and study of the rate of loss of the inert marker ^{51}Cr-EDTA from *G. duebeni* in sea water and after transfer to dilute media suggest that most individuals can expedite urine flow rate to a level capable of matching water inflow within five minutes of dilution of the medium (Lockwood and Inman, unpublished).

Adjustment of urine flow rate in the crab *Carcinus* seems to involve at least three processes (1) variation in filtration rate due to hydrostatic pressure changes (2) change in pore dimensions at the site of filtration and (3) modification in the proportion of the primary urine which is subsequently reabsorbed (Norfolk, 1976).

Perfusion of 1% T 40 Dextran into the antennal artery results in a urine flow rate proportional to the applied pressure once a threshold value of 6.46 cm H_2O is exceeded (calculated colloid O.P. of the Dextran is 5.8 cm water). The urinary increment is 1.4% crab weight/day/cm H_2O applied arterial pressure for each excretory organ. The pressure required to effect the rate of urine production observed in 50% sea water would be higher than that observed in the intact crab if pressure alone were controlling urine production rate. However, it is found that extracts of the brain of *Carcinus* added to the perfusate substantially increase urine flow rate. Brain extracts from crabs in 100% sea water more than double the flow rate (at a concentration of 1 brain from 100 gm crab per 2 ml perfusate) whilst a similar brain concentration after the donor crabs have spent one hour in 25% sea water has a still more striking effect (Fig. 3) (Norfolk, 1976).

It remains to be established that the active extract from the brain is the agent responsible *in vivo* for changing urine flow rate, but Norfolk's technique clearly offers considerable promise for further experimentation with crustacean urinary control systems.

Carcinus and some other crabs and prawns (Riegel and Lockwood, 1961; Riegel *et al.*, 1974; Franklin, 1975) have a ^{U}/B ratio* for ^{51}Cr EDTA or inulin in excess of unity, implying that some of the primary urine is reabsorbed prior to release of the definitive urine. Harris (1975) also suggests that reabsorption of water is the most likely explanation of high inulin ^{U}/B in the crab *Potamon*. Not all crustacea show such a high ^{U}/B as the crabs and prawns; in *Homarus* the ^{U}/B for inulin is close to unity (Burger, 1957) and the same is true for the ^{U}/B ratio for ^{51}/Cr EDTA for *G. duebeni* in sea water. On dilution of the medium the ^{U}/B ratio declines towards unity as the flow of urine increases both in *Carcinus* (Norfolk, 1972) and *Palaemon serratus* (Franklin, 1975).

---

* Ratio of the concentration of material in the urine to that in the blood.

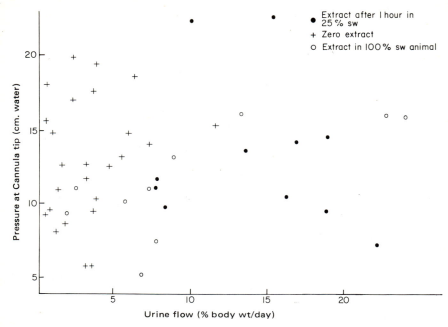

Fig. 3. The effect of perfused brain extracts from *Carcinus* in 100% sea water and from animals one hour after transfer to 25% sea water, on urine production rate per unit applied blood pressure. (Redrawn from Norfolk, 1976).

B. Cell Osmotic Regulation

Effective overall volume regulation does not diminish the potential problems of volume regulation at the cellular level when the blood is diluted. Cells appear to be in osmotic equilibrium with the blood and any sudden decline in concentration of the latter will consequently tend to result in a shift of water into the cells to restore the osmotic balance. Evidence of this effect has been found in *Pachygrapsus* (Gross, 1957) and *Sphaeroma* (Harris, 1969). In the absence of compensatory measures quite small changes in blood concentration would be expected to result in large variations in blood volume as a result of such water shifts (Croghan; in Lockwood, 1968). Since volume changes could result in circulatory embarrassment it is necessary that they be counteracted. The principal expedient adopted is the adjustment of intracellular concentration so as to prevent or lessen water shifts. Amino acids contribute substantially to changes in cell osmotic pressure.

C. Free Amino Acids and Volume Regulation

Some potential for adjusting cell free amino acid (FAA) levels seems to be present in all crustacea since the effect has been observed to a limited extent in stenohaline species such as the fresh water crayfish *Astacus* (Duchâteau-Bosson and Florkin, 1961) and hyper-hypotonic regulators, e.g. *Crangon crangon* (Weber and van Marrewijk, 1972) in addition to those forms which experience wider variations in blood concentration. In species such as the isopod *Sphaeroma rugicauda*, which tolerate wide changes in blood concentration (Fig. 1) adjustment to total FAA following transfer of the animal from 100% sea water to 2% sea water is both substantial and rapid, lagging but little behind the decline in haemolymph sodium concentration (Fig. 4) (Harris, 1969); thus ensuring that water shift from the blood to the cells is limited during the adjustment phase. The accuracy of the adjustment in individuals which have had time to acclimatize to a changed salinity is attested by the observation that in another euryhaline species, *Gammarus duebeni*, there is no detectable difference in the blood volume of animals in 100% and 2% sea water (Lockwood and Inman, 1973a). Similarly the water content of the muscles of *Carcinus maenas* is only 3.8% greater when the animals are in 40% sea water than when they are in 100% sea water (Shaw, 1958).

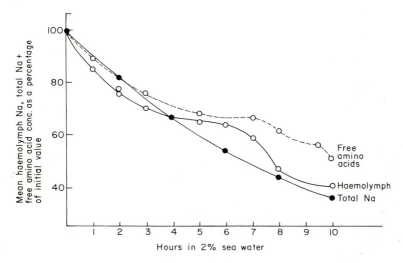

Fig. 4. The relationship between cell free amino acid concentration, blood sodium concentration and total sodium concentration in *Sphaeroma rugicauda* after transfer from sea water to 2% sea water. (Redrawn from Harris, 1967).

Since it takes a finite time to adjust cell osmotic concentration it behoves species inhabiting regions of fluctuating salinities to decrease, as far as possible, the rate of change of blood concentration. Decrease in surface permeability to water and salts, production of urine hypotonic to the blood and maintenance of blood concentration by active ion transport all contribute to this process.

D. Rate of Salt Loss

The rate of salt loss from species inhabiting brackish and fresh water is markedly lower than that of related species from more saline media (Table I). Not only does such a decreased loss rate assist in slowing blood concentration changes in animals inhabiting a region of fluctuating salinity, it also decreases the energy requirements for maintenance of blood concentration.

Reduced permeability of the general body surface to salts is also apparent in littoral and sub-littoral crabs (Gross, 1957) (Fig. 5).

In addition to general differences in loss rates observed between marine and brackish water species, variation is also apparent at the level of the individual. Transfer of the brackish water amphipod, *Gammarus duebeni*, from a high salinity (c. 35‰) to a low salinity (c. 0.7‰) results in an initially high rate of sodium loss due largely to increased urine flow. The rate of loss declines within about two hours however. Factors contributing to the decrease include decrease in the gradient between blood and medium, reduction in permeability to water (see p. 696) and production of urine hypotonic to the blood (see p. 691).

A further complicating feature is a tendency for an animal's permeability to bear a relationship to the length of time its medium has remained constant. Thus, when *G. duebeni* is acclimatized to low salinities for extended periods both the loss and uptake of sodium appear to decline (Lockwood and Inman, unpublished).

Permeability to water also varies according to whether the animal is in a constant or cycling salinity.

III. FACTORS INFLUENCING THE RATE OF ION TRANSPORT

Seven factors are known which can influence the rate of active transport of the major inorganic ions. These are:

(1) The concentration of the body fluids.

(2) The concentration in the external medium of the ion being transported.

TABLE I

Adapted from Sutcliffe (1968)

	Habitat	Avg. wt. (g)	Na loss rate mM/l blood/hr	Blood (Na) mM/l	Medium mM/l Na	T (°C)
(1) Carcinus maenas	SW – B-W	50	51	300	180	15–20
(2) Pachygrapsus crassipes	SW – B-W	24	43	280	190	15
(3) Eriocheir sinensis	B-W – FW	153	6.0	280	10	15
(4) Potamon niloticus	FW	15	2.3	259	0.5	24
(5) Palaemonetes varians	B-W	0.15	51.5	204	10	15
(6) Palaemonetes antennarius	FW	0.15	13.7	177	0.5	15
(7) Asellus aquaticus	FW	0.06	2.2	137	0.4	15–22
(8) Sphaeroma rugicauda	B-W		53.1			
(8) Sphaeroma serratum	SW		96.9			
(13) Mesidotea entomon (FW race)	FW	0.25	8.8	260	0.17	10
(9) Daphnia magna	FW	0.002	4.7	65	0.3	20
(10) Triops longicaudatus	FW	0.5	3.2	74	0.25	25
(9) Marinogammarus finmarchicus	SW	0.055	57.0	c300	60–120	
(9) M. obtusatus	SW	0.036	80.0	–	60–120	
(9) G. tigrinus	B-W	0.030	20.0	220	10	
(11) G. duebeni	B-W	0.071	10.9	273	10	
(9) G. zaddachi	B-W	0.048	13.7	252	10	
(11) G. pulex	FW	0.047	4.9	135	0.3–0.5	
(12) G. lacustris	FW	0.048	4.0	135	0.3–0.5	
(9) Crangonyx pseudogracilis	FW	0.005	7.1	c130	0.3	10

(1) Shaw (1961a) (2) Rudy (1966) (3) Shaw (1961b) (4) Shaw (1959) (5) Potts and Parry (1964) (6) Parry and Potts (1965) (7) Lockwood (1959) (8) Harris (1967) (9) Sutcliffe (1968) (10) Horne (1967) (11) Sutcliffe (1967) (12) Sutcliffe and Shaw (1967) (13) Croghan and Lockwood (unpublished). SW = sea water, B-W = brackish water, FW = fresh water.

(3) The concentration of other ion species which may interfere or compete with the ion being transported.
(4) The volume of the body fluids.
(5) The stage in the moult cycle.
(6) Sudden dilution of the medium.
(7) Temperature.

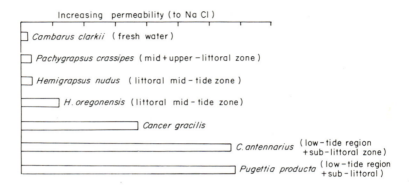

Fig. 5. Differences in the salt permeability of the cuticle of crustacea from different environments. (From Lockwood 1963, after Gross).

A. The Concentration of the Body Fluids

For animals in a steady-state, blood concentration is the major factor determining ion transport rate. In crayfish a blood concentration decrease of some 1—2% below the "normal" level results in some increment of sodium uptake and a drop of 6—8% produces the maximal transport of which the animal is capable (Shaw, 1959b). Comparable effects occur in estuarine forms though there is considerable variation in the level at which transport increase is effected. In the crab *Carcinus* decrease in the salinity of sea water produces little change in sodium uptake until the blood concentration has fallen to about 400 mM/l Na. A further drop of some 20 mM/l results in a 15-fold increase in transport rate (Shaw, 1961a).

Rate of intake of sodium is controlled by the level of sodium ion in the blood. The transport rate of other ions is not necessarily determined solely by the level of that ion in the blood. Thus, Shaw (1964) finds that chloride ion uptake, although not prevented by the absence of sodium in the medium, is ultimately determined by the blood sodium level.

For convenience when considering the effect of various factors on ion transport, the overall rate of uptake may be regarded as being the product of the transport rate per site and the number of sites operative. The animal is assumed to have control of the latter whilst the former may be influenced by physical and chemical factors in the medium such as ion concentration, temperature and pH.

B. The Concentration of Ions in the External Medium

When the concentration of the medium is below a given level the external concentration influences transport rate, influx declining hyperbolically in relation to a linear reduction of ion level (Fig. 6).

A curve of this form might be expected if the sodium were being transported by a carrier system whose saturation varies with the external concentration of the ion. The transport parameters fit the Michaelis-Menten equation in the form:

$$\text{Uptake} = K_2 \ \frac{C}{K_m + C} \quad \text{(Shaw, 1959b)}$$

where K_2 is the maximum rate of transport when the system is fully saturated, C the concentration of ion in the medium and K_m the concentration of the ion at which half maximal transport is achieved. A consequence of this relationship is that the rate of uptake will not be influenced by the external concentration when the latter is sufficiently high to result in the transport sites tending towards saturation but at low external concentrations the intake will decline.

Decline in the rate of transport per site will necessarily result in a drop in blood concentration if the loss of ions remains at a constant level, and such a drop secondarily initiates activation of more transport capacity by the animal as discussed above. Consequently, the rate of uptake will only decline rapidly once all reserve transport capacity has been mobilized.

Patently, if drop in blood concentration is to be avoided, it is desirable for species which live in, or experience, dilute media to have both a transport system which saturates at low concentration (low K_m) and a measure of transport capability in reserve. Appropriately, during the course of their evolution species which live in fresh water have indeed tended to develop transport systems with a lower K_m than forms inhabiting more saline waters (Table II). Species such as *Gammarus duebeni* and *Mesidotea entomon*, which

Y

TABLE II

The relationship between habitat and K_m value for various species of crustacea

	Usual medium	K_m mM/l Na	References
Carcinus maenas	Sea and B-W	20	Shaw (1961a)
Marinogammarus finmarchicus	Sea	6–10	Sutcliffe (1968)
Mesidotea entomon	Baltic (B-W)	9	Croghan and Lockwood (1968)
Mesidotea entomon	Fresh water	2–3	Croghan and Lockwood (1968)
Cyathura carinata	B-W	5.6	Roberts (1969)
Gammarus duebeni	B-W	1.5–2.0	Sutcliffe (1967)
G. zaddachi	B-W	1.0–1.5	Sutcliffe (1968, 1971c)
G. duebeni (celticus)	FW	0.4	Sutcliffe and Shaw (1967b)
G. pulex	FW	0.1–0.15	Sutcliffe (1971b)
G. lacustris	FW	0.1–0.15	Sutcliffe and Shaw (1967)

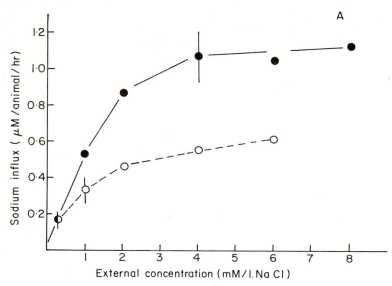

Fig. 6A. The relation between the external concentration and sodium influx in *Gammarus zaddachi* acclimatized to 10 mM/l NaCl (open circles) and 0.3 mM/l NaCl (solid circles) at 9°C. Vertical lines indicate the extent of the standard deviation from the mean. (Redrawn from Sutcliffe, 1968).

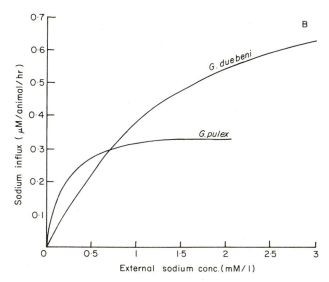

Fig. 6B. Comparison of the influx curves of *Gammarus duebeni* and *G. pulex*. (Redrawn from Shaw and Sutcliffe, 1961).

have races in both brackish water and fresh water, have different K_m values in the two stocks, that of the fresh water individuals being the lower. The K_m of these species, both of which are thought to have had a shorter history in fresh water than the majority of fresh water species, are higher than those of more fully adapted forms. Within limits, however, there is enough genetic lability, at least in the brackish-water stock of *G. duebeni*, to enable some measure of modification of the K_m. Thus, when individuals of the brackish-water race were maintained in Windermere fresh water, the progeny two years later had a K_m of 0.6 mM/l Na, which is akin to that of the Irish fresh water form *G. duebeni celticus* (Sutcliffe, 1971a).

The interaction of external concentration and the control of transport rate by the animal is shown in Fig. 6a. Comparison is made of the influx of sodium at different external concentrations by *Gammarus zaddachi* which had previously been acclimatized to 0.3 and 10 mM/l NaCl respectively. The animal from the lower concentration, which would be expected to have its transport system in a more activated state, has a higher rate of transport at any given salinity than the animal from 10 mM/l (Sutcliffe, 1968). In addition, the effect of sodium concentration on the rate of uptake is also apparent in the lower concentrations though the sites are tending towards saturation in the more concentrated media. Comparison of the influx curves of the fresh water species *G. pulex* and *G. duebeni* illustrates the advantage of a low K_m to a fresh water form. In brackish waters *G. duebeni* can take up sodium more rapidly than *G. pulex*, but in low salinities *G. pulex* can maintain its uptake at a lower concentration than can *G. duebeni*; the relative uptake rates are thus reversed (Fig. 6b). For a given surface permeability, animals with a low K_m can achieve ion balance at a lower external concentration than those with a high K_m (Fig. 7).

C. The Influence of the Concentration of Other Ions in the Medium

Inorganic ions are lost at different rates from the body, for example, loss of Cl from crayfish is only some 72% of the Na loss rate (Shaw, 1960a) and consequently independent uptake systems are necessary. Such independent systems have long been known, sodium being taken up from $NaHCO_3$ and Cl from $CaCl_2$, KCl, NaCl, and NH_4Cl (Krogh, 1939). The rate of uptake of sodium from NaCl, Na_2SO_4, $NaNO_3$ and $NaHCO_3$ is substantially the same and so too are the Michaelis-Menten curves for sodium uptake from NaCl and Na_2SO_4.

These anions do not, therefore, apparently exert a gross influence on sodium uptake in crayfish (Shaw, 1960a), though in *Cyathura carinata* influx from NaCl is substantially faster (0.08 μM/animal/hr) than from Na_2SO_4 (0.02 μM/animal/hr) or $NaNO_3$ (0.05 μM/animal/hr) (Roberts, 1969). Sodium uptake from Na_2SO_4 by the crayfish occurs in the absence of any anion and must therefore involve exchange with another cation. The ammonium ion, which is lost by the crayfish at a rate some 80% of the sodium uptake rate (Shaw, 1960a), is the most probable candidate and further confirmation of this comes from the observation that increasing the concentration of ammonium ion in the medium to circa 1 mM/l decreases sodium uptake by 70—80%. The rate of ammonium ion release does not, however, vary with the rate of sodium uptake and it seems likely that the difference between the ionic uptake of sodium and release of NH_4 is accounted for by hydrogen ions (at least when anion transport is unable to maintain charge balance). As would be expected if hydrogen ions play this role, low external pH (pH 4, 0.1 mM/l HCl) also decreases sodium uptake.

Divalent ions can also directly or indirectly affect transport rate but the effect is more variable.

D. The Volume of the Body Fluids

If the blood concentration of *Gammarus duebeni* is kept constant but the volume is decreased by removal of blood, the rate of active intake of sodium is increased. The same result occurs if fluid removal is effected by placing the animals in isosmotic non-electrolyte so that water loss accompanies ion loss without altering blood concentration (Lockwood, 1970). It thus appears that volume, as well as concentration, can influence the control system responsible for regulating transport rate. However, control of concentration and volume interact, since, if the animals are placed in hypertonic external media so that a rise in blood concentration accompanies osmotic withdrawal of fluid, no increment in transport rate occurs (Lockwood, 1970). Poisoning the uptake of ions by placing the basic dye thionine in the external medium results, when the animal is in sea water, in a marked decline in the rate of urine production. This suggests that isosmotic fluid uptake normally contributes to the fluid which is ultimately excreted as urine. Indeed it is interesting to speculate as to the possibility that a requirement for a polarized transport of ions to drive isosmotic fluid transfer in marine forms isotonic with their medium might have provided the basic mechanism on which selection has acted to produce transport systems in estuarine and fresh water forms capable of maintaining the body fluids hypertonic to the medium. Such a possibility would go some way towards explaining the frequency with which the active intake of ions and hypertonicity has evolved independently in brackish-water forms. Of course hypertrophy of transport processes originally involved in isosmotic water transfer could only result in hypertonicity of the body fluids if associated with a reduction in the rate of transfer of water relative to ions. There is, however, evidence that some crustacean species are capable of effecting variations in water permeability.

E. Moult Cycle Changes

Active intake of sodium is greatly increased at around the time of ecdysis and during post-ecdysis when expansion of the body volume is occurring. In *Gammarus duebeni* increment in transport rate seems to start a few hours before ecdysis, rises to a peak shortly after moult and then declines back to the intermoult level over three to four days. For animals moulting in sea water the peak rate of transport can be 20—40 times that shown by the same animals during the intermoult phase (Lockwood and Andrews, 1969) (Fig. 8). At its maximum rate the intake of sodium at moult by *Gammarus* actually exceeds the uptake rate of intermoult members of the same species adapted to fresh water or dilute brackish water.

Much of the fluid intake which occurs in *Carcinus* at moult can be attributed to water absorbed from the gut (Robertson, 1960) but the drinking rate in *G. duebeni* is inadequate to account for the amount of sodium taken up at moult. Dandrifosse (1966) concludes that a considerable proportion of the fluid which enters the spider crab *Maia* at moult does so across the body surface in association with an active uptake of ions. The observation that replacement of sea water by a non-electrolytic medium stops fluid uptake by *Maia* (Dandrifosse, 1966) offers further support for the assumption that the *raison d'etre* of the increment in ion uptake rate at moult by forms isosmotic with the medium may be in order to effect volume increase by taking up fluid by some form of isosmotic fluid transfer process. The stimulus initiating the increment in ion uptake is unknown but the coincidence that in *Gammarus* the increase in transport is virtually contemporaneous with the separation of the old cuticle suggests the possibility that a modification of the volume

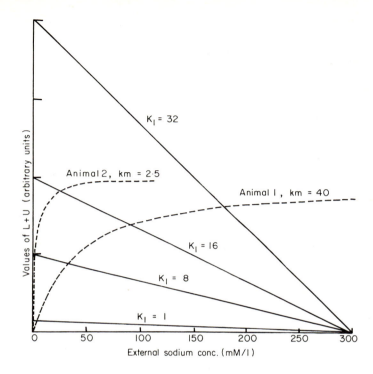

Fig. 7. The balance conditions for two hypothetical animals maintaining the same blood sodium concentration (300 mM/l). The straight lines represent the net passive loss rate (L) according to the relation $L = K(B - C)$ for different values of K_1. The dotted lines represent the active uptake rate (U) according to the relationship $U = K_2[C/(K_m + C)]$ for the two animals. For animal 1 $K_m = 40$ mM/l and for animal 2 $K_m = 2.5$ mM/l. Both animals have the same maximum rate of uptake (K_2). (B = blood concentration, C = concentration of the medium). (Redrawn from Shaw, 1961b).

regulation process could be involved. A system maintaining body volume might well be stimulated at moult if it involved stretch receptors located on the cuticle, since following ecdysis these would not be fully extended until the volume expansion was complete. Such a possibility is of course speculative and, on the basis of current evidence, it could equally be true that the increment in ion uptake is due to a process which is solely an adjunct of moult (Lockwood and Andrews, 1969).

F. Sudden Dilution of the Medium

Part of the accepted dogma of comparative physiological studies has been that animals respond to internal changes but not directly to variations in the external environment. Perhaps however, it is time to question the validity of this view. Various items of evidence suggest that, though the prime response is indeed directed towards the internal situation, in some species sudden changes in the concentration of the external medium may provoke a response more appropriate to the condition which will ultimately apply in that medium rather than to the internal concentration pertaining at the time. The physiological response may therefore be regarded as being an anticipatory one since, not only does it initiate the corrective measures which will ultimately be necessary when the animal reaches a steady-state in the new medium, but the rate of change of the body fluids towards the new steady-state is slowed thus permitting additional time for osmotic adjustments at the cellular level. Consider the evidence for such responses to external changes. When *Carcinus* immersed in 100% sea water have just their antennules perfused by a stream of 50% sea water the rate of urine production is increased. No such change occurs if the jet stream is 100% sea water (Norfolk, 1976). Transference of the isopod *Sphaeroma rugicauda* from a high salinity, in which active sodium uptake would normally be low, to a salinity where uptake would be faster, results in a rapid increase in uptake before the blood concentration has changed appreciably and certainly before it has fallen to the level appropriate to the raised rate of intake (Harris, 1970), (Fig. 9). *Carcinus*, transferred from 100% sea water to 50% sea water for five minutes before retransfer back to 100% sea water, increase the rate of urine production and subsequently lose more water in the urine than that taken up during the five minutes in the dilute medium (Norfolk, 1976). Finally, we may cite the change in urine concentration of *Gammarus duebeni* when transferred from a high concentration to a lower one. This brackish water amphipod, when in a steady-state with its medium, only produces urine hypotonic to the blood in concentrations lower than about 40—50% sea water (Lockwood, 1961). However, when transferred from 160% sea water to 2% sea water the animal begins to excrete urine hypotonic to the blood within about 1½ hours despite the fact that at this time its blood concentration is still in excess of 90% sea water (Fig. 10). This level is well above the concentration at which dilute urine would be

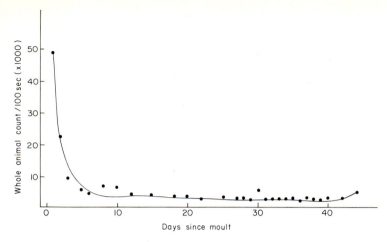

Fig.,8. Influx of sodium from 10 μM/l NaCl plus sucrose solution isotonic with 100% sea water by an individual *Gammarus duebeni* on the days following moult. Between the $\frac{1}{2}$ hour periods of uptake each day the animal was maintained in sea water. (Redrawn from Lockwood and Andrews, 1969).

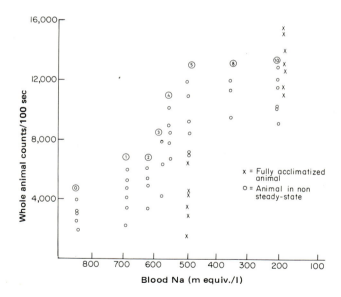

Fig. 9. After transfer from 100% sea water to 2% *Sphaeroma rugicauda* show a markedly higher rate of sodium uptake after five hours than steady-state controls with the same blood concentration. X, controls; O, experimental animals. Numbers are the hours since transfer from 100% sea water. (Redrawn from Harris, 1967).

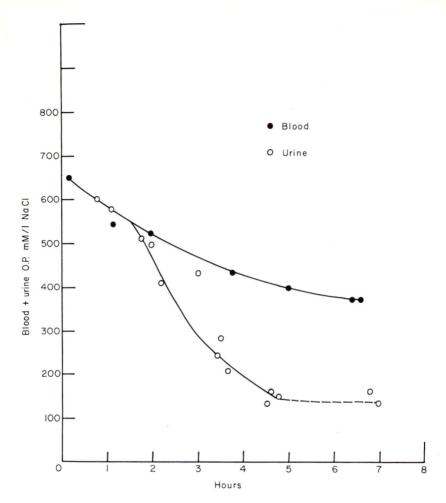

Fig. 10. Dilute urine production at high blood concentration in the amphipod *Gammarus duebeni* following transfer from hyper-concentrated sea water to fresh water.

produced by an individual in a steady-state. These effects doubtless contribute to the slowing in the rate of decline in blood concentration which occurs in both *Sphaeroma* and *Gammarus* about 1½—2 hours after transfer from a high to a low salinity. To confuse the issue further not all pericaridans show this effect. Transfer of *Cyathura carinata* from 100% sea water to dilute medium results in a fall, not a rise in active influx (Roberts, 1969) (Table III).

TABLE III

Active uptake of sodium by *Cyathura carinata* acclimatized to different salinities and then transferred to 4% sea water for measurement of ^{22}Na influx

Acclimatization medium	Uncorrected total influx (μE Na/animal/hr)	Active uptake (i.e. total flux corrected for effect of PD) μE Na/animal/hr
100% SW	0.20 ± 0.04	0.14
75% SW	0.17 ± 0.03	0.11
50% SW	0.06 ± 0.01	0.003
25% SW	0.08 ± 0.01	0.02
10% SW	0.13 ± 0.02	0.07
4% SW	0.16 ± 0.02	0.11

(From Roberts, 1969.)

G. Temperature

A change in temperature tends to result in an alteration in the rates of both loss and uptake of ions and, where there is a differential effect on loss and uptake, the steady-state blood concentration changes. Variation in blood concentration with temperature is however, usually of only limited degree since any change in concentration naturally tends to initiate the normal response to variation in blood concentration. Fig. 11 illustrates this point for a hypothetical fresh water form.

Any tendency for loss to increase relative to uptake will cause the blood concentration to decline. Decline in blood concentration results in activation of more transport capacity so that the overall blood concentration change is small. Conversely, if loss decreases relative to uptake, the blood concentration will rise. Again the change will be small since a rise in blood concentration results in a decrease in transport activation (Lockwood, 1968). In animals such as *Asellus aquaticus*, where the coefficient for uptake of sodium is greater than that for ion loss, a decrease in temperature results in a fall in blood concentration (Lockwood, 1960). *Potamon niloticus* behaves similarly (Shaw, 1959). Conversely, in *Gammarus duebeni* the temperature coefficient for loss is greater than that for uptake and consequently blood concentration rises as the temperature declines.

IV. THE REGIONS RESPONSIBLE FOR TRANSPORT

It seems probable that where species of crustacea have gills these subserve the function of ion transport at the body surface. For most species this assumption is speculative only but active transport by the gills has been demonstrated unequivocably on *Artemia* (Croghan, 1958), *Eriocheir* (Koch *et al.*, 1954) and *Astacus* (Bialawski, 1964). However, the gills are not necessarily the sole site of transport. An area which stains with silver in a manner analogous to the gills is found in the upper part of the branchial chamber in the isopod *Asellus aquaticus* (Hrabe, 1949) and the ventro-posterior region of the branchiostegite region similarly stains in the mysid *Neomysis integer* (Ralph, 1965). Areas which silver stain are presumed to include the regions where transport can occur. A more unusual transport region is the fenestra dorsalis of the anaspids *Allanaspides belonomus* and *A. hickmoni*. This structure, lying in the mid-dorsal region of the cephalothoracic tergite, has the typical ultramicroscopic appearance of transporting tissue (Lake *et al.*, 1974) and indeed resembles micrographs of crustacean gill tissues such as that of *Gammarus duebeni* (Lockwood *et al.*, 1973).

Ion uptake also occurs from the gut, at least in forms hypotonic to the medium (Croghan, 1958; Teinsongrusmee, 1976).

V. NA-K-ATPASE

Although Na-K-ATPase was first discovered in crustacean nerve (Skou, 1957, 1959) subsequent studies on the enzyme in crustacea have not been extensive. It has however been established that its role and properties in respect of sodium transport are comparable with those in other animal groups.

The amount of Na-K-ATPase in the gills of the grapsoid crab *Metapograpsus messor* increases as the concentration of the medium is decreased (Kato, 1968) and comparable results have been found for *Cardiosoma guanhami* (Quinn and Lane, 1966) and *Carcinus* (McWilliams, 1975). Conversely, in *Artemia salina*, an animal which generally has body fluids hypotonic to the medium, the ATPase level increases with a rise in concentration of the medium (Augenfeld, 1969).

Change in ATPase level in the gills of *Carcinus* can be approximately equated to variation in transport of sodium in so far as a major increase in ATPase level occurs when the haemolymph sodium level drops below about 400 mM/l (McWilliams, 1975) which is the approximate concentration at which sodium uptake is activated in this species (Shaw, 1961a). Transference of *Carcinus* from 25% to 100% sea water results in a three-fold decline in gill

ATPase within a period of 4.5 hours, a finding commensurate with the decline in sodium uptake which occurs on such transfer. Control of ATPase level appears to be linked to internal rather than external ion levels if one may judge from the observation that treatment of *Carcinus* in 75% sea water with thionine (which interferes with sodium uptake) results in a sharp rise in gill ATPase level concomitant with a decline in blood sodium concentration (McWilliams, 1975).

Both a reduction of potassium level in the medium and the presence of ouabain externally partially inhibit the efflux of sodium from *Palaemon serratus* in full strength sea water. This finding implies that the sodium transport process is of the typical potassium-linked, ouabain sensitive type associated with Na-K-ATPase (Teinsongrusmee, 1976).

Bilateral extirpation of the eyestalks decreases the Na-K-ATPase levels in the crab *Metopograpsus* but injections of brain extract tend to counteract the effect of eyestalk ablation (Kamemoto and Tullis, 1972).

The ability of the bladder wall to concentrate magnesium in the urine of both *Carcinus* (Norfolk, 1972) and *Palaemon serratus* (Franklin, 1975) is grossly interfered with by dosing the animals with the inhibitor ethacrynic acid. It is possible, therefore, that an ATPase other than the mundane Na-K-ATPase may also be involved in some forms of cation transport in crustacea.

VI. WATER PERMEABILITY

Maintenance of an osmotic gradient between blood and medium makes it desirable that the permeability to water be limited, and indeed most euryhaline and fresh water species tend to have markedly slower water flux constants than isotonic marine species, cf. Table IV.

Furthermore, many, though not all, euryhaline species show variations in the influx constants when the concentration of the medium is changed. Perhaps the most striking example of such modifications is that of *Artemia salina,* an animal in which the influx constant is 5.60 ± 0.95 hr in 38% sea water, 7.85 ± 2.25 in 100% sea water and 17.85 ± 5.50 in 566% sea water (Stewart, 1974). This species is however unusual in two respects. Firstly, as can be seen from Table IV, its apparent permeability is some 20 to 100-fold less than that of other crustacea of comparable size, and secondly the

TABLE IV

Half time of water exchange in crustacea from various environments

Species	Normal habitat	Temperature	Medium	$t\frac{1}{2}$ min	K (hourly water exchange)	
Species of large body size						
Macropipus	Marine	10°C	100% SW	17.4	2.39	Rudy (1967)
Carcinus maenas	Marine brackish	10°C	100% SW	52.6	0.79	Rudy (1967)
Hemigrapsus nudus	Marine brackish	10°C	95% SW	44.2	0.94	Smith and Rudy (1972)
Astacus fluviatilis	FW	10°C	FW	207.9	0.2	Rudy (1967)
Species of small body size						
Idotea linearis	Marine, estuarine	19°C	100% SW[a]	3.5		Lockwood and Inman (1973c)
I. linearis (moulted)		19°C	100% SW	1.7		Lockwood and Inman (1973c)
Cyathura carinata	Estuarine	18.5°C	100% SW	12.9		Hussain (1973)
C. carinata		18.5°C	2% SW	19.8		Hussain (1973)
Gammarus duebeni	Estuarine	19°C	2% SW	16.8		Lockwood and Inman (1973c)
G. duebeni		19°C	100% SW	7.3		Lockwood and Inman (1973c)
G. duebeni (moulted)		19°C	100% SW	3.8		Lockwood and Inman (1973c)
C. carinata		4.5°C	100% SW	21.2		Hussain (1973)
C. carinata		10.5°C	100% SW	15.2		Hussain (1973)
Chirocephalus diaphanus	Fresh water	15°C	FW	11.0		Stewart (1974)
Artemia salina	High saline waters	15°C	38% SW	315		Stewart (1974)
A. salina		15°C	57% SW	364		Stewart (1974)
A. salina		15°C	100% SW	474		Stewart (1974)
A. salina		15°C	293% SW	715		Stewart (1974)
A. salina		15°C	566% SW	1070		Stewart (1974)

[a] 100% SW is taken as being equivalent to approximately 34‰ salinity.

apparent permeability of *Artemia* decreases with increasing external concentration as does that of the teleosts, whilst those of other euryhaline crustacea such as *Rhithropanopaeus harrisi* (Smith, 1967), *Gammarus duebeni* (Lockwood *et al*., 1973) and *Cyathura carinata* (Hussain, 1973) all tend to increase.

The total change in apparent water permeability, at constant temperature, in *Gammarus duebeni* is about two and a half fold over the salinity range 2% sea water to 150% sea water and an appreciable part of this change occurs in the salinity range 50—75% sea water

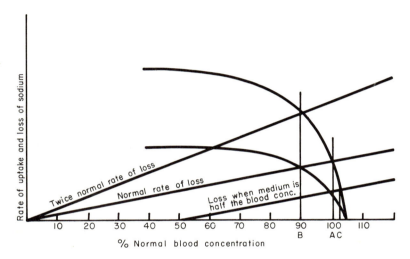

Fig. 11. Hypothetical interaction between blood concentration and uptake and loss of sodium. The two curves represent respectively the uptake rate with the transport sites fully saturated (upper) and with the sites half saturated (or at a lower temperature) (lower). (From Lockwood, 1963).

(Fig. 12). The results for the other species mentioned above suggest a more linear relationship between salinity and apparent permeability.

If the medium is diluted suddenly from 100% sea water to 2% sea water the corresponding change in apparent permeability to water is complete within five minutes (Lockwood *et al*., 1973), a time so short as to imply that physical or chemical processes rather than a biological response largely account for the change. Transfer of *G. duebeni* from sea water to mannitol isotonic with sea water results in a sharp decrease in the animal's apparent permeability suggesting that the response may be one to decreased ion level (or potential

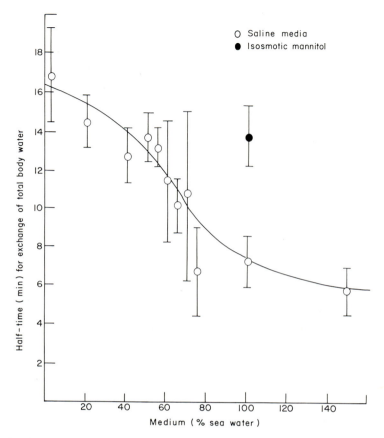

Fig. 12. The relationship between concentration of the medium and water flux in *Gammarus duebeni*. ●, Isosmotic mannitol; ○, saline media. Redrawn from Lockwood *et al.* (1973).

difference across the body surface) rather than to change in external osmotic pressure *per se*.

The reverse transfer, from a dilute to a more saline medium initially produces a further decrease in apparent permeability to water. This decline is however followed later by a more gradual increase back toward the level found in fully acclimatized animals in sea water (Lockwood *et al.*, 1973).

If the apparent permeability reflects the actual permeability to water and, in at least the case of *Gammarus duebeni*, the evidence suggests that there is such a relationship (Lockwood and Inman,

1973a,b), then one may infer that these various permeability changes confer several practical advantages on the animal. Decrease in permeability on exposure to dilute media will not only serve initially to slow the rate at which blood concentration declines and consequently assist in providing time for intracellular osmotic adjustment, but will also decrease the relative rate of osmotic influx of water once the steady-state condition has been reached. Conversely, the initial decline in permeability on transfer to a high salinity will decrease osmotic water loss during the period when the blood is hypotonic to the medium whilst the ultimate increase in water permeability in animals isosmotic with their medium may perhaps aid ion-driven isosmotic water transfer processes.

Not all crustacea display such changes. Rudy (1967), has demonstrated that the prawn *Palaemonetes varians* shows no variation in apparent permeability with changing salinity and this finding has been more recently confirmed on another prawn *Palaemon serratus* (Tin Tun, 1975).

Conformational changes are reported to occur in the subcuticular surface of the epithelial cells of *Gammarus* gills in response to salinity change though whether these modifications can be correlated with variation in transport rate of ions remains uncertain (Lockwood *et al.*, 1973; Milne and Ellis, 1973).

Other factors which can influence the apparent permeability to water include moult, temperature and divalent ion level in the medium. Immediately following ecdysis the apparent permeability of *Gammarus duebeni* and *Idotea linearis* is about twice that of intermoult individuals. The intermoult level is achieved about 3—5 days after moult (Lockwood and Inman, 1973c).

The Q_{10} for THO and DHO influx in the crab *Hemigrapsus nudus* is of the order of 1.6 in the temperature range 10—20°C (Smith and Rudy, 1972) and values have been found for *Palaemon serratus* and *Artemia salina* in the range 1.56—2.76 (Tin Tun, 1975; Stewart, 1974).

The effect of divalent ions is less uniform. A raised magnesium level in the medium results in a decrease in the apparent permeability in *Artemia salina* and decreasing the magnesium level increases the flux; by contrast lowering the Ca^{2+} level has little effect (Stewart, 1974). Similarly calcium has little effect on the water fluxes of *Chirocephalus diaphanus*, the exchange being identical in 1 mM/l NaCl and 1 mM/l NaCl + 2 mM/CaCl$_2$ (Stewart, 1974). By contrast the water flux in *Palaemon serratus* is substantially affected by the amount of calcium present (Table V).

TABLE V

The influence of calcium in the water flux of *Palaemon serratus* (T = 14°C)

Medium	$t\frac{1}{2}$ (min)[a]	N
100% SW (Ca^{++}-free)	15.3 ± 1.1	(7)
100% SW (0.1 mM/l Ca^{++})	18.2 ± 2.3	(6)
100% SW (1.0 mM/l Ca^{++})	20.0 ± 2.1	(7)
100% SW (10.0 mM/l Ca^{++})	22.0 ± 1.4	(7)
2% SW (Ca^{++}-free)	14.5 ± 1.1	(6)
2% SW (0.1 mM/l Ca^{++})	19.5 ± 1.8	(8)
2% SW (1.0 mM/l Ca^{++})	20.3 ± 1.8	(6)
2% SW (10.0 mM/l Ca^{++})	21.4 ± 2.0	(8)

[a] $t\frac{1}{2}$ is the time taken for exchange of half the water in the body to be effected. (From Tin Tun, 1975.)

VII. CONTROL SYSTEMS

By comparison with the level of knowledge of ion and water fluxes, blood concentration and rates of excretion, information about the neural and hormonal processes controlling these parameters is extremely limited. The neurohormonal areas of the eyestalks (in decapods), brain and thoracic ganglion have been implicated in various processes involving volume regulation, and ion or water transport but detailed knowledge of the nature or mode of action of the systems is lacking.

The involvement of the eyestalks in water balance control was first demonstrated by Abramowicz and Abramowicz (1940) who observed that *Uca pugilator* with the eyestalks ablated expanded further at the next moult than did controls. Scudamore (1942, 1947) extended the study to the crayfish *Cambarus immunis* and showed that, whilst animals from which the sinus gland or eyestalk had been removed had a greater postmoult water content than controls, the affect could be in part counteracted by the implantation of sinus glands in eyestalkless animals. Comparable counteraction of the effects of eyestalk ablation by sinus gland extracts also occurs in *Carcinus* (Carlisle, 1956). The increased weight of eyestalkless crabs (*Eriocheir*) at moult can be attributed to fluid increase only rather than to increased tissue formation (Koch, 1952).

Eyestalk removal in fresh water intermoult crayfish does not result in a net increase in water content (Kamemoto *et al.*, 1966) but, if the nephropores are blocked, the rate of weight increase in eyestalkless *Procambarus* and *Metapograpsus* is greater than that of intact but nephropore-blocked control animals (Kamemoto *et al.*, 1966; Kato and Kamemoto, 1969). This finding, coupled with the observation by Kamemoto and Ono (1969) and De Leersnyder (1967) that urine production is more rapid in eyestalkless than control *Metapograpsus* and *Eriocheir*, suggests that fluid uptake from the medium is more rapid after eyestalk removal. However, although the experiments with *Procambarus, Eriocheir* and *Metapograpsus* were conducted under conditions in which the animals' body fluids were hypertonic to the medium, it cannot necessarily be concluded that eyestalk removal influences fluid intake by merely causing an increase in surface permeability to water. Two other observations seem to militate against such an interpretation: (a) the land crab *Gecarcinus* shows an abnormal increase in size at moult after eyestalk removal, as do the aquatic forms (Bliss *et al.*, 1966), and (b) eyestalkless, nephopore-blocked prawns (*Palaemon serratus*) also increase in weight more rapidly than nephropore-blocked control prawns in 100% sea water despite the fact that in this medium the blood is markedly *hypotonic* to the medium (Teinsongrusmee, 1976). Such an effect implies that fluid intake is an active rather than a passive process and that removal of the eyestalks serves to amplify the fluid uptake. Presumably this indicates that eyestalkless prawns expedite fluid uptake from the gut and, since such fluid uptake may be expected to accompany ion uptake, one explanation of the effect might be that removal of the eyestalks expedites ion uptake. Possible confirmation of such a suggestion comes from (a) the observation that after transfer from 100% to 50% sea water, eyestalkless prawns maintain a markedly higher blood ion concentration than do intact prawns (Teinsongrusmee, 1976) and (b) that removal of the eyestalks in crayfish does not impede their ability to take up ions from the medium (Bryan, 1960). However, ligaturing the eyestalks of prawns markedly decreases their capability to secrete magnesium into the urine (Franklin, 1975). The overall picture in respect of ion balance is, however, further confused by indications that, despite the observation that net sodium transport remains normal in eyestalkless crayfish (*Austropotamobius*), the blood concentration ultimately falls (Bryan, 1960) and that a similar decline in ion level occurs in *Metapograpsus* in 25% sea water (Kato and Kamemoto, 1969) and in *Procambarus clarkii* (Kamemoto *et al.*, 1966). Possibly this latter series of observations reflects a decline in Na-K-ATPase levels since Kamemoto and Tullis (1972) find that eyestalk removal results in a decline in the level of this enzyme in the excretory tubule cells of *P. clarkii* which can be reversed by injections of eyestalk extract.

However, in view of the complex of systems (ion transport, water and ion permeability, blood volume and relative and absolute excretion rates of ion and water) which contribute to the concentration of the blood, it is probably unwise at the present state of knowledge to attempt detailed interpretation of cause and effect in relation to neurohormonal control of concentrations. A potentially more profitable approach would seem to be that involving isolated organ systems and, in this context experiments by Mantel (1967) suggesting that extracts of the ventral ganglion increase the water permeability of *Carcinus*; by Norfolk (1976) indicating that brain extracts of *Carcinus* increase the rate of filtration per unit of applied filtration pressure in the excretory system; by Dandrifosse (1966) showing that blood from a freshly moulted animal expedites ion driven water transport across the body surface of newly ecdysed *Maia* and by Kamemoto and Tullis (1972) suggesting that brain extracts of *P. clarkii* may increase the sodium influx at the surface, all merit further substantiation and extension.

A major problem still to be resolved is the manner by which cerebral control is exercized. The time delay between change in concentration of the medium and change in rate of ion transport in *Gammarus* (Lockwood, 1964) suggests that control is by hormonal rather than nervous means but it is not clear in the examples cited above (except that of Dandrifosse) whether the effect is normally produced by a general release of hormone into the blood or localized release at the target site. Furthermore, it is still far from certain that crude extracts of eyestalk, brain and thoracic ganglion necessarily owe their hormonal activity to accumulations of a natural hormone rather than to the presence of nonspecific surface-active materials. The techniques are now available, however, to permit separation and characterization of trace amounts of organic materials and the elucidation of a control system which seems, at least superficially, to rival that of the vertebrates in its complexities presents a fascinating challenge for the next decade.

ACKNOWLEDGEMENTS

I would like to express my appreciation to Dr Betty Wall and Mr Christopher Inman for their assistance in checking the draft version of this paper.

REFERENCES

Abramowicz, R. K. and Abramowicz, A. A. (1940). *Biol. Bull. mar. biol. Lab. Woods Hole* **78**, 179.

Appelbee, J. F. (1975). "An investigation into some aspects of the osmoregulation of *Paragnathia formica* (Hesse)." M.Sc. dissertation, University of Southampton.

Augenfield, J. M. (1969). *Life Sciences* **8**, 973—978.

Barnes, R. S. K. (1968). *Comp. Biochem. Physiol.* **27**, 447—450.

Bialawski, J. (1964). *Comp. Biochem. Physiol.* **13**, 423.

Bliss, D. E., Wang, S. M. E. and Martinez, E. A. (1966). *Am. Zool.* **6**, 197—212.

Bryan, G. W. (1960). *J. exp. Biol.* **37**, 100—112.

Burger, J. W. (1957). *Biol. Bull. mar. biol. Lab. Woods Hole* **113**, 207—223.

Carlisle, D. B. (1956). *Pubbl. staz. Zool. Napoli* **28**, 227—231.

Croghan, P. C. (1958). *J. exp. Biol.* **35**, 219—233.

Croghan, P. C. and Lockwood, A. P. M. (1968). *J. exp. Biol.* **48**, 141—158.

Dandrifosse, G. (1966). *Herbst. Arch. int. Physiol. Biochim.* **74**, 329—331.

Davenport, J. (1972). *J. mar. biol. Assoc. U.K.* **52**, 863—877.

Dehnel, P. A. (1962). *Biol. Bull. mar. biol. Lab. Woods Hole* **122**, 208—227.

De Leersnyder, M. (1967). *Cah. Biol. mar.* **8**, 295—321.

Dûchateau-Bosson, G. and Florkin, M. (1961). *Comp. Biochem. Physiol.* **3**, 245—249.

Franklin, S. (1975). "Investigation into some aspects of the secretion of magnesium by *Palaemon serratus*." M.Sc. dissertation, Southampton.

Gilbert, A. B. (1959). *J. exp. Biol.* **36**, 113—119.

Gross, W. J. (1957). *Biol. Bull. mar. biol. Lab. Woods Hole* **112**, 43—62.

Gross, W. J. and Marshall, L. A. (1960). *Biol. Bull. mar. biol. Lab. Woods Hole* **119**, 440—453.

Harris, R. R. (1967). "Aspects of ionic and osmotic regulation in two species of *Sphaeroma* (Isopoda)." Ph.D. Thesis, Southampton.

Harris, R. R. (1969). *J. exp. Biol.* **50**, 319—326.

Harris, R. R. (1970). *Comp. Biochem. Physiol.* **32**, 763—773.

Harris, R. R. (1975). Urine production and urinary loss in the freshwater crab, *Potamon edulis*.

Horne, F. R. (1967). *Comp. Biochem. Physiol.* **21**, 525—531.

Hrǎbe, S. (1949). *Zpravy Anthropol. Spol.* **11**, 1.

Hussain, N. A. (1973). "The influence of environmental salinity and temperature on the water fluxes of the isopod crustacean, *Cyathura carinata* (Krøyer)." M.Sc. dissertation, Southampton University.

Jones, L. L. (1941). *J. cell. comp. Physiol.* **18**, 79—91.

Kamemoto, F. I. and Ono, J. K. (1969). *Comp. Biochem. Physiol.* **29**, 393—401.

Kamemoto, F. I. and Tullis, R. E. (1972). *Gen. comp. Endocrinol.* **3**, 299—307.

Kamemoto, F., Kato, K. N. and Tucker, L. E. (1966). *Am. Zool.* **6**, 213—219.

Kato, K. N. (1968). "Neuroendocrine regulation of salt and water balance in the grapsoid crab, *Metopograpsus messor* (Forskal)." Ph.D. Thesis, University of Hawaii.

Kato, K. N. and Kamemoto, F. I. (1969). *Comp. Biochem. Physiol.* **28**, 665–674.

Koch, H. (1952). *Mededel. Kon. Vl. Acad. Wet.* **14**, 1–11.

Koch, H., Evans, J. and Schicks, E. (1954). *Mededel. Vl. Acd. Wet Kl Wet* **16**, 5.

Krogh, A. (1939). "Osmotic and Ionic Regulation in Aquatic Organisms." Cambridge University Press, Cambridge.

Lake, P. S., Swain, R. and Ong, J. E. (1974). *Z. zellforsch.* **147**, 335–351.

Lance, J. (1965). *Comp. Biochem. Physiol.* **14**, 155–565.

Lockwood, A. P. M. (1959). *J. exp. Biol.* **36**, 556–561.

Lockwood, A. P. M. (1960). *J. exp. Biol.* **37**, 614–630.

Lockwood, A. P. M. (1961). *J. exp. Biol.* **38**, 647–658.

Lockwood, A. P. M. (1963). "Animal Body Fluids and Their Regulation." Heinemann, London.

Lockwood, A. P. M. (1964). *J. exp. Biol.* **41**, 447–458.

Lockwood, A. P. M. (1968). "Aspects of the Physiology of Crustacea." Oliver and Boyd, Edinburgh.

Lockwood, A. P. M. (1970). *J. exp. Biol.* **53**, 737–751.

Lockwood, A. P. M. and Andrews, W. R. H. (1969). *J. exp. Biol.* **51**, 591–605.

Lockwood, A. P. M. and Inman, C. B. E. (1973a). *Comp. Biochem. Physiol.* **44A**, 935–941.

Lockwood, A. P. M. and Inman, C. B. E. (1973b). *J. exp. Biol.* **58**, 149–163.

Lockwood, A. P. M. and Inman, C. B. E. (1973c). *Comp. Biochem. Physiol.* **44A**, 943–952.

Lockwood, A. P. M., Inman, C. B. E. and Courtenay, T. H. (1973). *J. exp. Biol.* **58**, 137–148.

Lockwood, A. P. M., Croghan, P. C. and Sutcliffe, D. W (1976). *In* "Perspectives in Experimental Biology." (Ed. P. Spencer Davies), Vol. I. Zoology. Pergamon Press, Oxford and New York.

McLusky, D. S. (1970). *Comp. Biochem. Physiol.* **35**, 303–306.

McWilliams, P. G. (1975). "The role of a Na^+/K^+ stimulated, ouabain-sensitive adenosine triphosphate in the adaptation to salinity of the shore crab, *Carcinus maenas*." M.Sc. dissertation. University of Southampton.

Milne, D. J. and Ellis, R. A. (1973). *Z. Zellforsch* **139**, 311–318.

Norfolk, J. R. W. (1972). "A preliminary investigation of the effect of ethacrynic acid on magnesium excretion in *Carcinus maenas*." M.Sc. dissertation, Southampton University.

Norfolk, J. R. W. (1976). "The control of urine production in *Carcinus maenas*." Ph.D. thesis, University of Southampton.

Panikkar, N. J. (1941). *J. mar. biol. Assoc. U.K.* **25**, 317–359.

Panikkar, N. K. (1950). *Proc. Indo-Pacif. Fish Council* **2**, 168–175.

Parry, G. and Potts, W. T. W. (1965). *J. exp. Biol.* **42**, 415–422.

Potts, W. T. W. and Parry, G. (1964). *J. exp. Biol.* **41**, 591–601.

Quinn, D. J. and Lane, C. E. (1966). *Comp. Biochem. Physiol.* **19**, 533–543.

Ralph, R. (1965). "Some aspects of the ecology and osmotic regulation of *Neomysis integer*. Leach." Ph.D. Thesis, Southampton University.

Riegel, J. A. (1959). *Biol. Bull. mar. biol. Lab. Woods Hole* **116**, 272–284.

Riegel, J. A. and Lockwood, A. P. M. (1961). *J. exp. Biol.* **38**, 491—499.

Riegel, J. A., Lockwood, A. P. M., Norfolk, J. R. G., Bulleid, N. C. and Taylor, P. A. (1974). *J. exp. Biol.* **60**, 167—181.

Roberts, R. A. (1969). "An experimental study of certain aspects of the ionic and osmotic physiology of the Anthurid isopod, *Cyathura carinata* (Krøyer)." Ph.D. thesis, Southampton University.

Robertson, J. D. (1960). *Comp. Biochem. Physiol.* **1**, 183—212.

Rudy, P. P. (1966). *Comp. Biochem. Physiol.* **18**, 881—907.

Rudy, P. P. (1967). *Comp. Biochem. Physiol.* **22**, 581—589.

Schlieper, C. (1929). *Verh. deut. zool. Ges. Grag.* **33**, 214—218.

Scudamore, H. H. (1942). *Anat. Record.* **84**, 515.

Scudamore, H. H. (1947). *Physiol. Zool.* **20**, 187—208.

Shaw, J. (1958). *J. exp. Biol.* **35**, 920—929.

Shaw, J. (1959a). *J. exp. Biol.* **36**, 157—176.

Shaw, J. (1959b). *J. exp. Biol.* **36**, 126—144.

Shaw, J. (1960a). *J. exp. Biol.* **37**, 557—572.

Shaw, J. (1960b). *J. exp. Biol.* **37**, 548—56.

Shaw, J. (1960c). *J. exp. Biol.* **37**, 534—47.

Shaw, J. (1961a). *J. exp. Biol.* **38**, 135—152.

Shaw, J. (1961b). *J. exp. Biol.* **38**, 153—162.

Shaw, J. (1964). *Symp. Soc. exp. Biol.* **18**, 237—254.

Shaw, J. and Sutcliffe, D. W. (1961). *J. exp. Biol.* **38**, 1—15.

Skou, J. C. (1957). *Biochim. biophys Acta* **23**, 394.

Skou, J. C. (1959). *Biochem. J.* **85**, 495—507.

Smith, R. I. (1967). *Biol. Bull. mar. biol. Lab. Woods Hole* **133**, 643—658.

Smith, R. I. (1970). *Biol. Bull. mar. biol. Lab. Woods Hole* **139**, 351—362.

Smith, R. I. and Rudy, P. P. (1972). *Biol. Bull. mar. biol. Lab. Woods Hole* **143**, 234—246.

Stewart, A. J. (1974). "The influence of environmental salinity, temperature and ionic composition on the water fluxes of *Artemia salina* (L.)." M.Sc. Thesis, Southampton University.

Sutcliffe, D. W. (1967). *J. exp. Biol.* **46**, 529—550.

Sutcliffe, D. W. (1968). *J. exp. Biol.* **48**, 359—380.

Sutcliffe, D. W. (1971a). *J. exp. Biol.* **55**, 325—344.

Sutcliffe, D. W. (1971b). *J. exp. Biol.* **55**, 345—355.

Sutcliffe, D. W. (1971c). *J. exp. Biol.* **55**, 357—369.

Sutcliffe, D. W. and Shaw, J. (1967). *J. exp. Biol.* **46**, 519—528.

Tan, E. C. and Van Engel, W. A. (1966). *Chesapeake Sci.* **7**, 30—35.

Teinsongrusmee, B. (1976). "Aspects of Osmoregulation in the common prawn, *Palaemon serratus* (Pennant, 1777)." Ph.D. Thesis, Southampton University.

Tin Tun, M. (1975). "The influence of environmental calcium and temperature on the water fluxes of *Palaemonetes varians* (Leach)." M.Sc. dissertation, University of Southampton.

Weber, R. E. and Van Marrewijk, W. J. A. (1972). *Life Sciences* **11**, 589—595.

Werntz, H. O. (1963). *Biol. Bull. mar. biol. Lab. Woods Hole* **124**, 225—239.

28. Osmotic and Ionic Regulation in Saline – Water Mosquito Larvae

J. E. Phillips and T. J. Bradley

Department of Zoology, University of British Columbia, Vancouver, British Columbia, Canada

I. INTRODUCTION

A consideration of osmotic and ionic regulation in mosquito larvae is most appropriate to this volume. Ramsay's studies (1950–53) of the excretory system in these larvae were amongst his earliest and they led to studies which revealed the general nature of the excretory process in insects. The techniques and experimental approaches which he developed during these studies are largely responsible for the present wealth of knowledge concerning the Malpighian tubule – rectal system of insects.

While the mechanisms of osmotic regulation in freshwater species of mosquito larvae are relatively well understood (reviewed by Stobbart and Shaw, 1974), particularly the function of the anal papillae as extra-renal organs concerned with active absorption of NaCl (Wigglesworth, 1933–38; Koch, 1938; Stobbart, 1959–71), the same could not be said for euryhaline species until very recently. This paper will therefore summarize recent published and unpublished observations concerning the process of osmotic and ionic regulation in saline-water species of mosquito larvae.

II. HABITATS AND REGULATORY CAPACITY

Several species of mosquito larvae are known to thrive both in fresh water and in saline-waters considerably hyperosmotic (at least 2–4x) to the haemolymph (*Aedes detritus*, Beadle, 1939; *A. campestris*, Phillips and Meredith, 1969a,b; *Opifex fuscus*, Nicholson, 1972;

709

A. taeniorhynchus, Nayar and Sauerman, 1974, and Bradley and Phillips, 1975). Survival of *A. taeniorhynchus* in tide pools containing 300% saline water is the highest salinity tolerance reported to date. More remarkable perhaps is the ability of a single species, *A. campestris*, to thrive in a variety of alkaline salt-lakes which are common in semi-arid regions of North America and in which the dominant ions may be $NaHCO_3$ (pH > 10) or equal amounts of $MgSO_4$ and Na_2SO_4 (Scudder, 1969; Phillips and Meredith, 1969a,b; Kiceniuk and Phillips, 1974; Phillips and Maddrell, 1974; Maddrell and Phillips, 1975). Recently we have found that larvae from Ctenocladus pond (700 mOsM (Mg + Na)SO_4) have a very high rate of survival in 700 mOsM $NaHCO_3$, 700 mOsM NaCl and FW; indeed, better than in their natural habitat (Bradley and Phillips, in preparation). We must conclude that the occurrence of this species in waters of various chemical types reflects regulatory adaptability rather than the existence of distinct physiological races.

Beadle (1939), in his study of *A. detritus* in saline water, observed that haemolymph Cl^- and osmotic concentrations are similar to those of freshwater mosquito larvae, i.e. about one-third seawater levels. He concluded that the adaptations for survival included minimization of passive diffusion across the body wall, and in particular a reduction in the size of the anal papillae, which are the principal site of such exchange in freshwater species. Secondly, he suggested that regulation was achieved solely by the excretory system, which produced hyperosmotic urine, while the external anal papillae were thought to be rudimentary. Because the rate at which water and ions exchanged between these larvae and their environment was not precisely determined, the attainment of a steady-state following the transfer of larvae to other waters, and hence regulatory ability of the larvae, has been questioned (Stobbart and Shaw, 1964, 1974).

However, Phillips and Meredith (1969a,b) found that when *A. campestris* larvae were transferred to tap water, 800 mM NaCl or solutions of intermediate salinity, most of the changes in blood Na^+, Cl^-, K^+ and osmotic concentrations occurred during the first day, so that steady values were observed within two days. This is also the case for Mg^{++} and $SO_4^=$ (Kiceniuk and Phillips, 1974; Maddrell and Phillips, 1975). Following adaptation to waters varying in ionic concentration by 5000-fold, the steady-state haemolymph concentrations of Na^+, Mg^{++}, K^+, Cl^- and total solutes (osmolality) change at most by 2-fold, except in very fresh water ([NaCl] < 1−2 mM). Sulphate represents an exception. The haemolymph concentration of this anion is regulated at low levels (<10 mM) until the external concentration rises above 100 mM; thereafter, blood levels rise in parallel with external values to 100 mM without an obvious detrimental effect on the larvae over the next several days. Similar results have been reported for *A. taeniorhynchus* larvae reared in various media from fresh water to 300% saline water (Nayar and Sauerman, 1974; Bradley and Phillips, in preparation). These observations testify to the remarkable regulatory capacity of saline-water mosquito larvae.

III. SITES AND RATES OF EXCHANGE

To appreciate the regulatory load which is imposed upon the excretory system and anal papillae of euryhaline mosquito larvae under extreme conditions, the net exchange rate of water and major ions between larvae and their environment must be considered. Ingestion has been estimated from the initial linear rate at which whole larvae accumulate large labelled polymers (eg. ^{14}C-inulin, ^{131}I-PVP) from the media (Kiceniuk and Phillips, 1974; Bradley and Phillips, 1975). Extending the methods of Wigglesworth (1933—38), if larvae are ligated posterior to the 5th abdominal segment so as to prevent loss of Malpighian tubule secretion and the mouth is sealed shut, osmosis across the body can then be determined from changes in body weight. Likewise, if the portion of the larvae anterior or posterior to the ligature is removed, water movement via the body wall and anal papillae can be differentiated. By the same isolation procedures, the sites of influx and efflux of [22]Na, [36]Cl, and tritiated water (T_2O) can be determined. A number of unpublished experiments by one of the authors (J. E. Phillips) on *A. campestris* larvae adapted to either hyposmotic (5 mM NaCl) or natural hyperosmotic pond water containing mostly sodium bicarbonate, but with isotonic levels of chloride at pH 10, are summarized in Fig. 1. Since experiments in different years were not all conducted under strictly identical conditions, and since some parameters were estimated by more than one type of experimental procedure, the diagrams contain typical values for exchanges, and are therefore approximations. Loss via the anus and bicarbonate flux across the body wall were not measured but rather they were estimated from the balance of other exchanges. Each site of exchange will subsequently be considered in detail.

The rapid adjustment of haemolymph composition observed within 1—2 days of transferring *A. campestris* to waters of different salinity

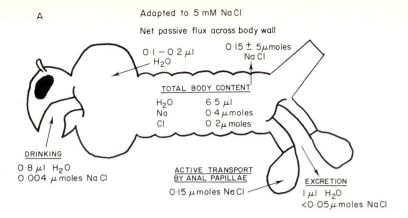

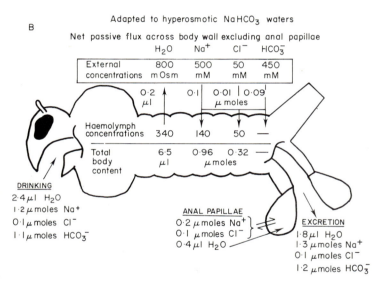

Fig. 1. Net exchanges per day via different body sites for 8 mg larvae of *A. campestris* in hyposmotic (A) or hyperosmotic (B) waters (Phillips, unpublished observations). A. Net passive fluxes across the body wall include the anal papillae. The two values for net water movement indicate simple diffusional and osmotic permabilities respectively. Excretion was calculated to balance measured rates of net exchange by all other sites. B. Net passive flux of water across the anal papillae and the remainder of the body wall are distinguished in this situation; values are for osmotic flow. Values for net movement of HCO_3^- by all routes except drinking and all of those for excretion were calculated to balance other measured exchanges. The exchanges of NaCl across the anal papillae, which are indicated by half-arrows ($\rightleftharpoons$), are unidirectional fluxes. Since influx of these ions through terminal segments equalled efflux, no net exchange or transport of salt could be shown experimentally across anal papillae in this situation (discussed in text, p. 729).

is consistent with the exchanges indicated in Fig. 1. The total body content of water and salts is turned over relatively slowly in fresh water. With increasing hyperosmolality of the external media, a dramatic reduction in turnover times is observed (Table I). This is largely due to increased ion concentration in the ingested fluid, which becomes almost the exclusive means of entry of water and most ions in hypertonic media, as first suggested by Beadle (1939).

TABLE I

Time in days for turnover of total body content due to net exchange with external media

External media	Body constituents				Reference
	H_2O	Na^+	Cl^-	Mg^{++}	
	A. taeniorhynchus				
Sea water (800 mOsM)	1.3	0.3	0.2	—	Bradley and Phillips (1975)
	0.4				Bradley and Phillips (in preparation)
	A. campestris				
Fresh water (5 mM NaCl)	6.0	3.0	1.3	—	Fig. 1a
NaHCO₃ pond (800 mOsM)	2.7	0.67	1.5	—	Fig. 1b
NaHCO₃ pond (1200 mOsM)	1.2	—	—	—	Phillips, unpublished data
(Na + Mg)SO₄ pond[a] (350 mM Na^+, 85 mM Mg^{++})	0.33	—	—	0.13	Kiceniuk and Phillips (1974)

[a] Larvae grossly shrunken; near upper tolerance limits for this type of water.

A. Ingestion and Assimilation

It is surprising that over 80% of the net uptake of water by *A. campestris* larvae adapted to fresh water occurs by drinking, since a common strategy amongst hyper-regulators is to minimize ingestion. Kiceniuk and Phillips (1974) offer an explanation. While there is abundant organic nutrient in the flocculent sediment on which larvae appear to feed, only a small percentage of the ingested carbon compounds are available in forms which can be assimilated and the water content of this material is high. To take advantage of an abundant but dilute energy source, high rates of fluid intake and the associated high cost of osmoregulation cannot be avoided. In support of this view that high rates of fluid intake are an unavoidable consequence of feeding activity, we have observed a two-fold increase in ingestion following the feeding of liver, as a dried powder, to saline water-adapted larvae (*A. taeniorhynchus*) which had been starved for two days. Moreover, larvae can be reared in sterile culture medium in the absence of particulate matter or micro-organisms (reviewed by Clements, 1963; Nayar, 1966). Since many natural saline-waters in which larvae thrive are undoubtedly very rich in dissolved nutrients and micro-organisms, larvae may normally indiscriminately ingest some external medium in an attempt to obtain nutrients.

A. campestris larvae respond to rising external salinities by increasing their drinking rate by up to six-fold (Table I; Phillips and Meredith, 1969a,b). However, most of this increase occurs only near the upper tolerance limits when a reduction in body volume, and hence internal hydrostatic pressure, threatens the mobility and therefore the survival of larvae (Kiceniuk and Phillips, 1974). However, Bradley and Phillips (in preparation) did not observe any difference in the very high drinking rate of *A. taeniorhynchus* reared in either fresh water, sea water or double-strength sea water (Table I). Another possible explanation for high drinking rates is suggested on page 726.

Kiceniuk and Phillips (1974) and Maddrell and Phillips (1975) have demonstrated that virtually all of the water, $SO_4^=$ and Mg^{++} ingested by *A. campestris* larvae living in hyperosmotic $(Na + Mg)SO_4$ waters is absorbed in the midgut. It seems probable that this is also the case for other major ions in various saline-waters. This may serve to concentrate a dilute food source in preparation for digestion. The almost complete absorption of Mg^{++} and $SO_4^=$ in the midgut contrasts with the relatively small amount of intestinal absorption of divalent ions in marine fish (reviewed by Conte, 1969). Considering the relatively large hydrated size of $SO_4^=$ and its rapid rate of absorption, Maddrell and Phillips (1975) have postulated uptake in the midgut by facilitated diffusion or active transport. Recent observations (Maddrell and Phillips, in preparation) on *A. taeniorhynchus* adapted to sea water containing 50 mM $SO_4^=$ failed to demonstrate uptake of this anion from the midgut against concentration gradients when water movement was minimized by sucrose. There is therefore no need to postulate active transport of $SO_4^=$ at this time. Moreover, larvae reared in normal sea water and others in sulphate-free sea water both absorbed over 87% of ingested $^{35}SO_4^=$ within one day following transfer to labelled normal sea water. There is therefore no evidence that adaptation to high sulphate waters involves the turning on or induction of carrier-mediated processes for $SO_4^=$ in the midgut, as occurs in Malpighian tubules (see p. 719).

In conclusion, because of the almost complete absorption of ingested ions and water in the midgut, a reasonably good estimation of the load imposed upon the regulatory organs in hyperosmotic media can be obtained simply from a knowledge of ingestion rate and composition of external media. Before dealing with regulation however, a consideration of the body wall permeability is in order.

B. Passive Exchange via the Body Wall

Beadle (1939) and Ramsay (1950) judged the cuticle of *A. detritus* larvae to be virtually impermeable to water and salts. Stobbart (1971c), however, observed considerable shrinkage when larvae of this species which had their neck and anal segment ligated were placed in sea water. Nicholson and Leader (1974) reported the first quantitative estimate of body wall permeability to water for *Opifex fuscus* larvae, which have a mean weight similar to *A. campestris* larvae. The diffusional permeability of these two species is rather similar (4.8×10^{-3} and 2×10^{-3} cm h^{-1} respectively; *A. campestris* value from Fig. 1a). This is 3 to 10 times lower than values reported for various freshwater insect larvae (reviewed by Nicholson and Leader, 1974). As commonly observed for biological membranes, osmotic permeability of the body wall exceeds diffusional permeability by some 5x in *O. fuscus* and also in *A. campestris*. This is possibly due to bulk or laminar flow rather than simple diffusion. Some unpublished measurements (Phillips) of T$_2$O influx into *A. campestris* support this view. Blocking the mouth and ligating anal papillae have no significant effect on the unidirectional influx rate of T$_2$O. This indicates that nearly all exchange by simple diffusion is across general body surfaces rather than the anal papillae. Of course net flux in this case represents a small difference between two very large unidirectional fluxes. However, weight loss studies indicate that 66% of osmotic loss occurs through anal papillae (Fig. 1b), which represents a very small fraction ($< 1\%$) of the total body surface. Obviously osmotic permeability of the latter organs is about two orders of magnitude higher than values given above. This bulk water movement may be associated with ion transport activities of anal papillae, discussed on page 727. Influx of T$_2$O was not markedly different under the hyposmotic and hyperosmotic conditions described in Fig. 1; clearly, permeability does not change during adaptation to these extreme conditions.

The body wall is the major site of net salt loss from *A. campestris* larvae in fresh water (Fig. 1a). The relative importance of anal papillae and general body surface in this regard has not been clearly differentiated. Under the hyperosmotic conditions of Fig. 1b, net influx of ions across the general body surface only accounts for about 8% of total net uptake. The ratio of ^{22}Na: ^{36}Cl influx by this route (10 : 1) reflects the relative concentrations of these ions in the external medium. This would seem to exclude chloride selectivity over bicarbonate at this site as a mechanism of regulation in unbalanced media (i.e. where $[Na^+] > [Cl^-]$). However, the 2 : 1 ratio for influx of these two ions by the anal papillae does suggest such a selectivity in this organ. As a result, unidirectional influx of ^{36}Cl$^-$ through anal papillae is 10x greater than through the body wall surface but this difference is only 2x for ^{22}Na. The net flux of ions across the anal papillae under these conditions, if any, is not known because of uncertainty concerning the site of efflux from terminal segments (see pp. 729—30).

IV. MALPIGHIAN TUBULE SECRETION

Ramsay (1950) showed that the fluid produced by the Malpighian tubules of *A. detritus* was slightly hyperosmotic or hyposmotic to the haemolymph in saline water or fresh water respectively and that adjustment of urine osmolality to achieve osmotic balance in these larvae occurred primarily in the rectum. Further information on Malpighian tubule secretion of saline species has only recently been provided by Phillips and Maddrell (1974), and Maddrell and Phillips (1975) using *A. campestris* larvae from (Mg + Na)SO$_4$ waters of moderate salinity (50% of haemolymph osmolality). Each larva has five tubules (3 mm long) of similar appearance and without obvious divisions into distinct segments at the light microscope level. These tubules are considerably smaller than those of other insect species which have been successfully studied using the *in vitro* preparation of Ramsay (1954). Nonetheless, by suitable miniaturization of the techniques, the same methods could be applied.

In Ringer having an ionic composition very close to that of natural haemolymph, complete set of tubules produces an isosmotic fluid at a rate of approximately 0.250 picolitres per minute. After corrections for the fact that only 50% of the tubules is in the bathing drop and also allowing for decline in activity *in vitro* (factor of 2), this corresponds to a total secretion of 1.4 μl larvae^{-1} day^{-1}, which is in reasonable agreement with estimates of total fluid intake by drinking and osmosis across the body wall in moderately hyposmotic waters (Fig. 1). Like tubules of the freshwater species, *A. aegypti* (Ramsay, 1952, 1953a), and many other insects (reviewed by Maddrell, 1971), those of *A. campestris* produce a fluid more concentrated in K^+ (eg. 8x) and less concentrated in Na^+ (eg. 0.3x) than the bathing fluid. Raising the level of external K^+ from 4 to 76 mM by replacing Na^+ increases fluid excretion rate by 3.3 times. Rather surprisingly and unlike the Malpighian tubules of most other insects (Maddrell, 1971), these tubules had a transwall potential difference in which the lumen was negative with respect to the bathing solution. Nevertheless, secretion still occurred at a reduced rate in a K-rich, Cl-poor medium when the average transwall potential difference was 16 mV, lumen positive. These results are consistent with the view that fluid secretion is coupled to a potassium transport process, in parallel with a chloride (or anion) pump, as is proposed for some other insects (Maddrell, 1971). Secretion stops if all chloride is replaced with sulphate, even though the latter anion, in the presence of some Cl^-, is actively secreted into the tubule lumen to create concentrations as high as 90 mM. Under such conditions it may be concluded from the osmolality of the fluid that sulphate accounts for most of the total anions which are secreted.

There are obviously occasions when fluid exchange by drinking and osmosis may change during larval development, eg. upon feeding and during changes in external osmolality (p. 714). Some experiments on the control of fluid secretion have been conducted on isolated tubules from *A. taeniorhynchus* larvae adapted to sea water (Maddrell, 1976 and in this volume). The initial rate of fluid secretion does appear to be higher (2x) in feeding versus starved larvae. The steady rate of secretion which occurs within 10—20 minutes of isolation can be increased ten-fold by adding 5HT, cAMP, or extracts of the head or the brain of the larvae. The secretion rate is presumably under the control of a diuretic hormone similar to those of other insects (reviewed by Maddrell, 1971). Interestingly, this hormone probably stimulates sulphate transport because the concentration of this ion in the secretion does not decline substantially after stimulation.

While Malpighian tubules of saline-water mosquito larvae may not be unusual in their mechanism of fluid secretion, this is not the case for secretion of divalent ions. *A. campestris* larvae survive in waters in which Mg^{++} concentration is 100 mM and in which sulphate, at 350 mM, accounts for 90% of the total anions. Since these ions are also relatively high in sea water (Hoar, 1975), the problem of maintaining low levels of Mg^{++} and $SO_4^=$ in the haemolymph is probably common to most insect larvae which live in this environment. Active transport of magnesium and sulphate has been shown to occur against large electrochemical gradients under appropriate experimental conditions in *A. campestris* and also, in the case of sulphate, in *A. taeniorhynchus* (Phillips and Maddrell, 1974; Maddrell and Phillips, 1975). This appears to be the first report of the transport of these two ions in insects. Tubules of *Rhodnius* and *Carausius* are unable to secrete these ions against a concentration gradient. In saline-water mosquito larvae however, concentration ratios (lumen : haemolymph) across the tubule wall of 12 : 1 for Mg^{++} and of 5 : 1 for $SO_4^=$ were observed and maximum concentrations in the secretion of 45 mM and 90 mM respectively were achieved. Magnesium ions are not required for fluid transport, which proceeds independently of magnesium concentration in the bathing media. As a result, fluid which is secreted slowly contains higher concentrations of magnesium than that which is secreted more rapidly. Both transport mechanisms exhibit saturation kinetics with K_m values of 2.5 mM and 10 mM and V_{max} of 15 and 50 p-mole min^{-1} $tubule^{-1}$ for Mg^{++} and $SO_4^=$ respectively.

Over a wide range of external concentrations, haemolymph levels of Mg^{++} and $SO_4^=$ are normally maintained at relatively constant values of 1.5–4 and 2–7 mM respectively. This is near or below the K_m values for the transport processes; i.e. in the region of the curve where the rate of ion transport is more or less directly proportional to Mg^{++} and $SO_4^=$ levels in the blood. Thus, rather small increases in haemolymph levels of these ions due to ingestion automatically result in their increased elimination as a direct consequence of the kinetic properties of the transport mechanisms. At low and moderate salinities, the calculated rates of Mg^{++} and $SO_4^=$ secretion by Malpighian tubules of *A. campestris* (Phillips and Maddrell, 1974; Maddrell and Phillips, 1975) are adequate to account for elimination of the total amounts of these ions which are ingested. This may be the first example amongst insects of ionic regulation which is a direct consequence of selective secretion by the Malpighian tubules, as opposed to selective reabsorption in the rectum.

At higher external concentrations of Mg^{++} and $SO_4^=$, the transport capacities of Malpighian tubules reported by the above authors appear to be inadequate to eliminate all the ingested ions. In the case of Mg^{++} a more concentrated secretion can be turned on in the rectum (discussed below) and the possibility of elimination by the anal papillae cannot be excluded (Kiceniuk and Phillips, 1974; Phillips and Maddrell, 1974). Another form of response has recently been observed by Maddrell (1976), similar to that reported for anal papillae by Phillips and Meredith (1969a). If *A. taeniorhynchus* larvae are reared in sulphate-free sea water, the tubular secretion contains only low levels of sulphate, well below those of the bathing Ringer. Other larvae reared in normal sea water or in sea water with the sulphate concentration doubled, secrete this ion against large concentration differences at several times the rate observed for larvae reared in the absence of sulphate. Indeed, under standard test conditions, the rate of sulphate secretion at any haemolymph concentration reflects the external level during adaptation. Since it requires about 16 hr to turn on sulphate transport in larvae reared in SO_4-free sea water, the synthesis of new protein carrier is presumably involved (i.e. induction). This time is much faster than that required to turn on Na^+ transport in anal papillae (see p. 728). This form of regulatory response also seems to be the rule for ion transport processes in the rectum of saline-water larvae.

V. THE RECTUM

Ramsay (1950) provided the first information on the function of the rectum in both freshwater and saline-water mosquito larvae. In both cases the fluid entering this organ is nearly isosmotic with the haemolymph, as expected of a fluid derived from the Malpighian tubules. The osmolality falls to low values in the rectum of both types of larvae when they are placed in fresh water, presumably due to ion reabsorption against large concentration gradients without accompanying water movement. When placed in single or double strength Ringer, the fresh water form dies, whereas the saline-water species are able to produce a hyperosmotic urine (slightly less than the saline water in total concentration) and survive. These observations were confirmed for *A. campestris* (Fig. 1b) which can produce urine 3—4x blood osmolality (Phillips and Meredith, 1969a,b). It has since been assumed that saline-water insects produce hyperosmotic urine in the same manner as terrestrial insects, by absorbing water without proportional amounts of solute (i.e. by transferring hyposmotic absorbate in the rectum). Ramsay associated the ability of the saline-water larvae to produce hyperosmotic urine with an additional segment absent in the freshwater species. This is also the case for *A. campestris* (Meredith and Phillips, 1973a).

However, Phillips and Meredith (1969b) and Meredith and Phillips (1973a) found that the ultrastructural features associated with the production of hyperosmotic urine in the rectum of most terrestrial insects, namely elaborate development of the lateral plasma membranes which are associated with most of the mitochondria of the cells, were absent in the rectal epithelium of *A. campestris* (Fig. 2). The anterior rectum of these larvae consists of a single layer of cells containing one cell type. These are similar to those observed in the rectum of *A. aegypti*, which is of uniform structure throughout its length (also Asakura, 1970), both apical and basal membranes, the latter in particular, are highly folded. Mitochondria are evenly distributed throughout the cell and their profiles constitute 22% of the cell area seen in micrographs of *A. campestris*, as compared to 13% in *A. aegypti*. In both species the apical infoldings extend 16—20% of the distance across the cells. This membrane is coated

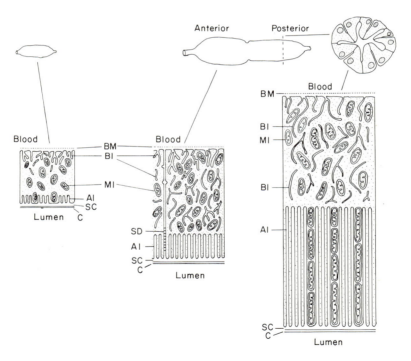

Fig. 2. Morphology of rectum of freshwater Mosquito (Left) (*A. aegypti*) and saline-water mosquito (Right) (*A. campestris*) larvae. Based on electron micrographs by Meredith and Phillips (1973a) and drawn by Wall and Oschman (1975). Drawn approximately to scale. BM: basement membrane, BI: basal infold, MI: mitochondrion, AI: apical infold, SD: septate junction, SC: subcuticular space, C: cuticle.

with particles similar to those described by Gupta and Berridge (1966) and Noirot and Noirot-Timothée (1966). The anterior rectum of *A. campestris* is separated from the posterior segment by a narrow band of unspecialized junctional cells lacking apical infoldings. The epithelia of the ileum and anal regions of the hindgut are much reduced in size and are without infolding or abundant mitochondria. These observations suggest that the rectum is the dominant site of transepithelial transport processes in the hindgut.

The posterior rectal epithelium of *A. campestris* is twice as thick (62 μm) as the anterior portion and again contains only one cell type, which is rather similar in general structure to the anterior cells. However, the apical membranes are much more numerous and regular, and extend 60% of the distance across the cell, (i.e. some 6x longer than those of anterior cells). Moreover, most of the mitochondrial profiles, which cover 22% of cell area of micrographs, are closely associated with the apical infoldings. Lateral membranes are relatively straight, with septate junctions. When larvae acclimated to fresh water and saline water (2x haemolymph) were compared, no qualitative or major quantitative differences in ultrastructure were observed.

Since saline-water larvae were found to drink hyperosmotic external media at high rates (Table I), Phillips and Meredith (1969b) and Meredith and Phillips (1973a) suggested that larvae did not have to conserve water in the rectum as must terrestrial insects. Rather they need only rid themselves of the excess ions so ingested. They proposed, therefore, that hyperosmotic urine is produced by active secretion of these excess ions across the elaborately developed apical membranes of the posterior rectum.

This has been confirmed by Bradley and Phillips (1975). Sea water-adapted larvae were ligated at the sixth and anal segments so as to prevent movement of fluid into or from the rectum except by net transfer across the rectal wall. These terminal segments were suspended from the surface of Ringer solutions by means of the respiratory siphon, thereby assuring a constant supply of air to the rectum by normal tracheal connections. The integument was torn to permit exposure of the rectum to the external Ringer, thus assuring constant conditions on the hemocoel side. Under such conditions, the lumen rapidly fills with a hyperosmotic secretion similar to sea water in both total osmotic pressure and ionic composition, with one exception: Potassium levels are about 18x higher than in sea water. The secretion of Na^+, K^+, Mg^{++}, and Cl^- all occur against large concentration differences of 2—18x. When similar experiments are conducted with a ligature now placed between the anterior and posterior rectum, the posterior rectum continues to produce a strongly hyperosmotic secretion (Bradley and Phillips, in preparation). The average transepithelial potential in the latter segment, as measured with glass micro-electrodes, decreases from 10 to 6 mV (lumen positive) during the course of secretion by ligated recta over the 2 hr study period. It is therefore necessary to postulate active transport of all four ions to the lumen side. Rectal secretion was not observed in larvae reared in fresh water, and the small amount of fluid which could be collected by micropuncture had an osmolality which was 40% of haemolymph level, in agreement with the observation of Ramsay (1950) on *A. detritus*. Clearly, secretion by the posterior rectum is only turned on in hyperosmotic media.

The concentration of the rectal secretion from whole larvae reared in 200% sea water was substantially higher than that from larvae living in 100% sea water. Since the mean ionic concentrations in such secretions were slightly below those of the external medium (although standard errors of the former overlapped sea water levels), it may be necessary to postulate some ion secretion by another site, such as the anal papillae, to explain completely osmotic regulation under these conditions (see p. 730).

These ion-secretory processes exhibit unusual kinetic properties, as the following experiment indicates (Bradley and Phillips, in preparation). Ligated recta of larvae adapted to sea water were exposed to Ringers in which the Na^+ or Cl^- concentrations were varied independently (by substituting choline and sulphate respectively, while keeping osmotic pressure constant), and the rate and concentration of rectal secretion were measured. The saturation kinetics of Michaelis-Menten, which characterize most transport processes of insects and other organisms, are not observed (Fig. 3). Rather, volume and ionic concentration of secretion appear to increase most rapidly at the higher haemolymph concentrations. This is suggestive of an allosteric enzyme which is only turned on at higher substrate concentrations. Possibly rectal transport processes of these larvae possess some intrinsic capacity to respond automatically to increasing NaCl concentration in the haemolymph since sharply increased secretion of this salt will occur only when normal haemolymph levels are exceeded. This matter requires further study.

The ion transport processes of the posterior rectum are adjusted during prolonged adaptation to new external conditions, presumably due either to the synthesis of more protein carriers (i.e. induction) or to the turning on of inactive sites by neural or hormonal mechanisms. This has been demonstrated by analysis of the secretion

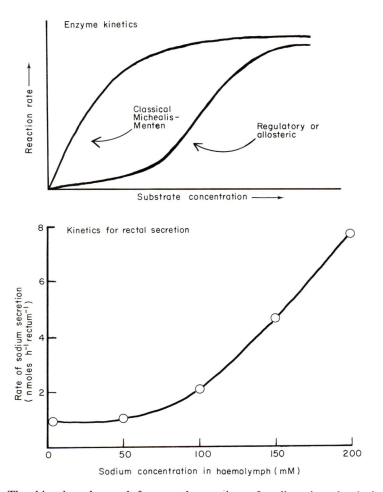

Fig. 3. The kinetics observed for rectal secretion of sodium ions by isolated recta of *A. taeniorhynchus*. Sodium concentrations in the Ringer solutions for bathing preparations were varied by substituting choline for sodium ions. The total rate of sodium secretion was calculated from the concentration of this cation in rectal fluid collected by micropuncture and from the volume of rectal secretion which could be collected after 1.5 hr. The sodium level in the haemolymph of normal larvae from a wide range of waters is 140—160 mM. The kinetics observed for rectal secretion of Na^+ and Cl^- ions resemble those for regulatory or allosteric enzymes rather than the classical kinetics of Michealis-Menten, which are exhibited by other insect transport processes which have been reported to date (Bradley and Phillips, in preparation).

from the ligated recta of larvae reared in three different media (all 700 mOsm; i.e. 2x haemolymph concentration) in which the dominant ions were either NaCl, or $NaHCO_3$, or $(Mg + Na)SO_4$. The recta from these larvae were all exposed to the same test Ringer solution by tearing the body wall. Bicarbonate was not measured. With the exception of sulphate, the ratios and concentrations of all other ions (Na^+, K^+, Mg^{++}, Cl^-) in the secretion tended to adjust toward those prevailing in the external medium to which the larvae were adapted (Bradley and Phillips, in preparation). For example, Mg^{++} levels in the three external media varied from 8 to 126 mM. When ligated recta from these three groups of animals were all tested in Ringer containing 5 mM Mg (normal haemolymph levels in all animals is 4—6 mM) the level of this ion in the secretion was almost identical to that in the corresponding external medium to which the larvae had been previously adapted. A similar proportional response was observed for K^+ secretion although here the ion concentration was consistently maintained at about 10x external levels. The long-term response of Na^+ secretion to different external levels of this ion have been referred to above. Long-term adaptation is not always as precise as observed for Mg^{++} and K^+. When tested in a Ringer containing 102 mM Cl^-, recta from sea water-adapted (368 mM Cl^-) larvae produced a rectal secretion containing 498 ± 53 mM Cl^-, whereas that from larvae reared in $NaHCO_3$ with 26 mM Cl^- or $(Mg + Na)SO_4$ with 18 mM Cl^- contained half as much Cl^- (225 ± 41 and 230 ± 15 mM Cl^- respectively). The latter values were unexpectedly high but were undoubtedly the consequence of using a Ringer with Cl^- levels much higher than those found in haemolymph of larvae reared in these two media (49—56 mM Cl^-). When the Cl^- concentration in the test Ringer was reduced from 102 to 20 mM, Cl^- levels in rectal secretion from larvae reared in the $(Mg + Na)SO_4$ medium dropped from 230 to 40 ± 6 mM. The high level of Cl^- in the first test Ringer was presumably interpreted as indicating a high external Cl^- concentration, since even in sea water, the haemolymph Cl^- level is only 78 mM.

Unlike the other ions, sulphate levels in secretion from ligated recta remain low (5—8 mM) even when the external media levels are 377 mM. However, substantial amounts of the anion (132 mM) are observed in fluid collected from the recta of whole larvae, similarly reared, but with their anus blocked for 1 hr to allow accumulation of fluid. Two possible explanations of this paradox are suggested. Secretion of sulphate by the rectum is initiated and maintained by a neural or hormonal control mechanism which is removed by the process of ligation. Alternately, all sulphate secretion may occur solely in the Malpighian tubules, with further concentration of MgSO₄ as a consequence of fluid and KCl removal in the anterior rectum (Fig. 4). The sulphate observed in rectal fluid from whole animals may simply have come from the Malpighian tubules since communication with the latter organs was not prevented in this preparation. However, the rectal secretion is presumed to flow anteriorly when the anus is so blocked, thus opposing introduction of Malpighian tubule fluid into the rectum. This would explain the lower level of $SO_4^=$ observed in rectal fluid collected from intact larvae with their anuses blocked as compared to concentrations in the external media.

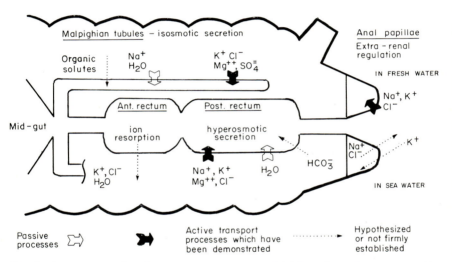

Fig. 4. A summary of active transport processes which have been demonstrated in various parts of the excretory system and the anal papillae of either *A campestris* or *A. taeniorhynchus*. Other net ionic transfers which are hypothesized or for which evidence is not conclusive are included and are distinguished by narrow dotted arrows. Other processes which have been studied and which appear to be passive movements on the basis of available data are indicated by open arrows. (See text for references).

This raises the question of the balance between anions and cations in the rectal secretion from larvae in these various external media. In larvae adapted to hyperosmotic NaHCO₃ and (Na + Mg)SO₄ media, the ratio of Na : Cl in the rectal secretion is about 8 : 1. Since $SO_4^=$ is not secreted by ligated recta, it is probable that HCO_3^- is actively secreted into the rectal lumen with the Na⁺ under these conditions. Natural waters in which $SO_4^=$ constitutes 90% of the anion are often strongly alkaline and contain substantial amounts of HCO_3^-. Moreover, since Malpighian tubules in high-sulphate, low-magnesium waters secrete largely K_2SO_4, some of the Na^+ secretion in the posterior rectum may occur in exchange for K^+ uptake in the anterior rectum to yield Na_2SO_4.

Rectal secretion must compensate for osmotic loss across the body wall. The final solute concentration of the urine might therefore be expected to be considerably higher than that of external media. It is therefore surprising that in some, but not all situations, the observed osmolality of rectal secretions only equals that of the external hyperosmotic medium (eg. *A. taeniorhynchus* in 100% saline water). As discussed by Bradley and Phillips (1975), this may be an experimental artifact, because there is evidence of some ion binding to fecal material so that effective osmolality of the total excreta from whole larvae is equal to 137% of sea water levels. Extra-renal regulation may account for additional salt removal from the animal (p. 730). However, these observations suggest an additional possible explanation for high rates of drinking in saline-water mosquito larvae (Table I). If the rate of fluid ingestion, rather than being only of sufficient magnitude simply to replace water loss by osmosis across the body wall, is several times larger, then the required concentration of secretion from the rectum to achieve osmotic balance is reduced and ultimately approaches that of the external medium. For example, if twice as much fluid is imbibed as is lost by osmosis, then the excreta must be twice as concentrated as the external medium. If the ratio of imbibition to osmotic loss is increased to 4 : 1, as indicated in Fig. 1b, then the required total effective concentration of excreta need only be 133% of external osmolality. Possibly the rectal epithelia can accommodate very large increases in total transport capacity of ions, resulting from increased drinking; however, this membrane may be more restricted in the osmotic concentration difference against which this secretion can occur. Increased drinking is an obvious solution to this problem. This might explain both the very high drinking rates of larvae which occur near the upper tolerance limits (Table I), and also the advantage of a lower osmotic permeability of the body wall in saline water as compared to freshwater species of mosquito larvae (p. 715).

VI. ANAL PAPILLAE

Beadle (1939) concluded that the external anal papillae of *A. detritus* were rudimentary and impermeable structures which did not function as organs of extra-renal regulation, as do those of strictly freshwater species. His conclusion was based on three observations. The anal papillae are much reduced in size compared to those of freshwater forms, a fact more recently confirmed in *A. campestris, A. taeniorhynchus* and *A. togoi* (Meredith and Phillips, 1973b,c). The anal papillae are not destroyed by concentrated NaCl solutions as are those of *A. aegypti*. This is perhaps not unexpected considering the severe environments in which saline-water species are often found. Finally, when *A. detritus* larvae were transferred from saline water to fresh water the anal papillae did not take up significant amounts of Cl^- ions from the external medium after one day. However, this adaptation period to fresh water was probably too short to demonstrate induction of NaCl uptake, as indicated below.

A regulatory function for anal papillae has been demonstrated in at least one saline-water species of mosquito larvae, *A. campestris* Phillips and Meredith, 1969a). Indeed, this is the first report of extra-renal regulation amongst saline-water insects in general. When these larvae are reared in fresh water and exposed to distilled water for 1–2 days prior to experiments, net uptake of NaCl through the anal papillae against large electrochemical gradients occurs from 0.5 but not 0.05 mM NaCl. The strictly freshwater larvae of *A. aegypti* can do the same from external solutions at least one order of magnitude more dilute (Stobbart, 1965). The active uptake by anal papillae of *A. campestris* exhibits Michaelis-Menten type kinetics with a K_m of 2 mM and V_{max} of 1.5 nanomoles h^{-1} mg body weight^{-1}. This K_m value is 4 times higher than that reported for *A. aegypti* (Stobbart, 1967). In terms of transport per mg wet weight of anal papillae, V_{max} was estimated to be similar for *A. campestris* and *A. aegypti,* but of course the total body weight of anal papillae in the latter species, relative to total body weight, is much greater. In all, these differences may account for the regulatory advantages which strictly freshwater species have over saline-water species in very dilute natural waters. When saline-water larvae are transferred from hyperosmotic media to fresh water, no significant influx of ^{22}Na via the anal papillae is observed initially. However, over a period of 5–15 days, a large influx of this isotope through these organs develops (Fig. 5). Clearly several days are required to turn on Na^+ transport in this species. Sodium and chloride transport can be turned on independently, since some net uptake of chloride occurs from 5 mM KCl, NH_4Cl, LiCl, $CaCl_2$ and choline Cl, while uptake of Na^+ will occur from the same concentration of $NaNO_3$, Na_2HPO_4 and Na_3 citrate. In this regard anal papillae of freshwater-adapted *A. campestris* appear similar to those of the freshwater species, *A. aegypti* (Stobbart, 1965, 1967). Similar unpublished experiments (Phillips)

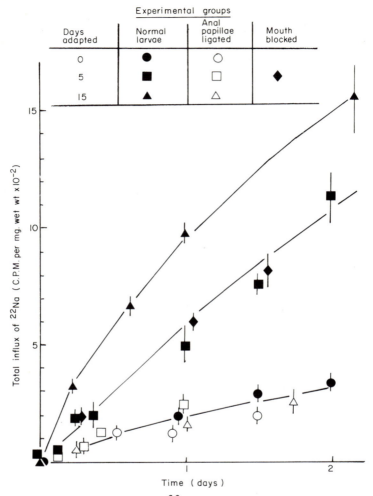

Fig. 5. The unidirectional influx of ^{22}Na into larvae of *A. campestris* at various times after transferring animals from hyperosmotic waters (1.2 mOsM NaHCO$_3$) to hyposmotic media (4 mM ^{22}NaCl: constant specific activity). Uptake of ^{22}Na by normal larvae (controls) is compared with that by animals which had their mouths blocked with beeswax-resin to prevent drinking or others in which the anal segment was ligated to prevent uptake by the anal papillae. Immediately after transfer (0 days) only a small influx of ^{22}Na is observed and this does not occur through the anal papillae. A large increase in influx appears within 5 to 15 days of transfer and all of this increment is due to uptake by the anal papillae because blocking the mouth has no effect while ligating the papillae totally abolishes the increased ^{22}Na influx. (Phillips, unpublished; net uptake of Na by anal papillae, Phillips and Meredith, 1969a).

using ^{86}Rb indicate that this ion is also actively absorbed by a saturable mechanism through anal papillae of fresh water-adapted *A. campestris* larvae. Since K$^+$ inhibits this ^{86}Rb influx, we assume that the uptake is by a K$^+$ pump. Competition between Rb$^+$ and K$^+$ for a common transport mechanism has been reported in other insect tissues (reviewed by Maddrell, 1971).

Ultrastructural studies support the physiological demonstration of active transport by these organs. Those of *A. campestris* and *A. togoi* (saline-water species: Meredith and Phillips, 1973b,c) and *A. aegypti* (freshwater species: Copeland, 1964; Sohal and Copeland, 1966) have similar structures, not unlike the anterior rectum of the former species. The papillae contain only one cell type arranged as a single layer covered with cuticle externally. Internally only a basement membrane separates the cell from the haemocoel. Both apical and basal borders are highly folded. Mitochondria are abundant, more so in the saline-water species where their profiles constitute 30% of the cell area seen in electron micrographs. Surprisingly, no major differences in structure were evident between papillae of *A. campestris* larvae adapted for two weeks to fresh water or strongly hyperosmotic NaHCO$_3$ water, nor in *A. togoi* which were living in rock pools containing 130% sea water. The ultrastructure suggests active transport activity under all conditions. This raises the question as to the function of these organs in hyperosmotic waters.

An obvious role for anal papillae in unbalanced hyperosmotic natural waters (eg. NaHCO$_3$ and (Mg + Na)SO$_4$) is to absorb actively those essential ions which are present in the external environment at only low concentrations; eg. Cl$^-$ and K$^+$ in the case of the waters just mentioned. Certainly in NaHCO$_3$ waters containing Cl$^-$ at a level isotonic with the haemolymph, half of the total influx and over 90% of the influx across the body surface occurs through the anal papillae (p. 716 and Fig. 1b). The ^{36}Cl influx via the anal papillae under these conditions retains the saturation kinetics and the same K$_m$ of 2 mM observed for freshwater-adapted larvae. However, the V$_{max}$ (i.e. the amount of carrier) is reduced by an order of magnitude to the value of 0.1—0.2 nanomoles h^{-1} mg body weight^{-1} (Fig. 1b), another example of longterm adjustment in transport capacity. Using isolated terminal segments of larvae, similar to those used to study rectal function but without the second ligation on the anal segment, an estimate of influx and efflux was obtained in an attempt to demonstrate net uptake of ^{36}Cl$^-$ under such conditions, i.e. in the absence of fluid release from the Malpighian tubules. While these preparations were left for an hour to permit voiding of rectal contents, we cannot exclude the possibility that part of the efflux occurred from the rectum, which was subsequently shown to secrete ions. Since total efflux by all routes from these terminal segments, including any loss by excretion, was found to equal total influx (i.e. no net flux: Fig. 1b), net uptake of Cl^- by anal papillae may indeed occur. Be this as it may, one can definitely conclude that less than half of the total net uptake of $^{36}Cl^-$ occurs through the anal papillae and the rest largely by drinking.

Using the same approach as for Cl^-, the influx and efflux of $^{22}Na^+$ between terminal segments and the external media of Fig. 1b were also found to be equal. But in this case there is a 10-fold concentration gradient favouring net diffusion of Na^+ into the larvae. This is presumably balanced either by rectal secretion or active transport of Na^+ to the exterior through the anal papillae. While the situation is again not clear, one can at least conclude that such active secretion by anal papillae could account for less than 20% of total net efflux; that is, the excretory system is clearly much more important in eliminating Na^+ under these particular conditions.

There is some other circumstantial evidence to suggest that anal papillae can, like the gills of marine teleosts, reverse direction of active transport and secrete NaCl to the external medium in saline water. The anal papillae of larvae reared in isosmotic saline were destroyed by $AgNO_3$ treatment (Phillips and Meredith, 1969a). When such larvae were subsequently transferred to saline water, the level of Cl^- in their haemolymph rose 20% above that of control larvae. However, some other effect of $AgNO_3$ treatment in addition to that on the anal papillae can not be excluded. As mentioned above, ultrastructural observations have indicated very active anal papillae in *A. togoi*, which were living in 130% sea water, where both Na^+ and Cl^- were in excess. Moreover, rectal secretion may not be able to account totally for ionic regulation in some hyperosmotic waters (see p. 726). In summary, the possibility of active secretion of ions to the exterior by anal papillae of larvae living in hyperosmotic waters is supported by a considerable body of circumstantial evidence but must remain an open question. The answer awaits a technical solution to the problem of plugging the anus without damaging the anal papillae.

VII. SUMMARY

Having discussed each of the regulatory organs of saline-water mosquito larvae individually, the nature of the overall process of osmotic regulation, as we presently envisage it, will now be considered. Fig. 4 summarizes the processes which have been demonstrated or hypothesized for the Malpighian tubules, rectum and anal papillae.

We propose the following events in saline-water larvae adapted to fresh water: The Malpighian tubules secrete an isosmotic fluid containing largely KCl; indeed, the transport of these two ions is responsible for the fluid movement. These two ions and possibly others (eg. Na^+), as well as essential metabolities, are reabsorbed in the anterior rectum without water. This creates a hyposmotic fluid, containing largely nitrogenous and other waste products; i.e. the system functions much like that of other freshwater insects (reviewed by Stobbart and Shaw, 1964, 1974). Under these conditions the posterior rectum is inactive and fluid passes through this segment to the exterior without major change. Externally net uptake of K^+, Na^+, and Cl^- occurs through the anal papillae in any natural fresh water containing more than 0.1 mM of these ions.

The pattern amongst larvae adapted to hyperosmotic waters is as follows: the Malpighian tubules continue to produce isosmotic fluid rich in KCl. However, if the natural water contains high levels of Mg^{++} or $SO_4^=$, active secretion of these ions accounts for a proportionally greater fraction of the total solute. The total transport capacity (i.e. amount of carrier) for $SO_4^=$ and possibly Mg^{++} is increased over a period of a day in response to prolonged exposure to high external levels of these ions. In the anterior rectum most of the KCl and water in proportional amounts is reabsorbed, leaving a smaller volume of isosmotic fluid enriched in $MgSO_4$ and other waste products. Secretion of a hyperosmotic fluid is turned on in the posterior rectum. The relative rates at which various ions (Na^+, K^+, Mg^{++}, Cl^- and possibly HCO_3^-) are transported depends on their concentration in the external medium to which larvae are adapted, as well as on haemolymph levels. In more concentrated waters, the total rate of ion transport and the volume of secretion increases. However, proportionally less water follows the ion movement, so that osmolality of the secretion also increases. This fluid is then eliminated via the anus. The net uptake of individual ions from the external medium may continue in unbalanced waters; eg. Cl^- uptake from $NaHCO_3$ waters low in the latter anion. In sea water where both Na^+ and Cl^- are hypertonic to the haemolymph, the uptake of both these ions by anal papillae is turned off. Indeed, NaCl is possibly transported in the opposite direction, and thus eliminated from the larvae. We suggest that some of the Na^+ secreted in this manner may be linked to a net influx of K^+, as occurs in gills of marine teleosts. This would yield the following advantages: the downhill gradient for K^+ entry from saline water could, by coupled exchange, provide energy for outward movement of Na^+ against an electrochemical gradient. Secondly, the intake of K^+ by anal papillae might replace the large quantities of this ion lost by secretion in the posterior rectum. Some preliminary observations in our laboratory by Leadem support this view of anal function in *A. taeniorhynchus*.

Finally, we believe that saline-water mosquito larvae may offer unusual opportunities to study the molecular nature and control of the ion transport processes of insects. Many of these processes in mosquito larvae, like those in bacteria, are possibly inducable. This opens the possibility, rare amongst metazoans, of applying techniques so successfully employed by microbiologists to isolate carrier proteins. Biochemical isolation should be aided by the fact that most of the regulatory organs are discrete and homogenous; i.e. they are simple epithelia of one cell type without connective tissue or muscle layers in some cases. The question as to which cell contains the physiologically demonstrated transport or associated biochemical activity is indisputable. These features are in contrast to the situation in euryhaline fish gills, for example, which can also turn on ion transport. The individual cells of saline water larvae are large enough to be visible under a highpower dissecting microscope and to permit intracellular recording with glass micro-electrodes. Thus transport activities of the apical and basal plasma membranes may be distinguishable. Finally, the short life history of some species, one week for *A. taeniorhynchus*, makes genetic studies of transport feasible, with the added feature that giant chromosomes are reported in these regulatory epithelia.

ACKNOWLEDGEMENTS

We wish to acknowledge the financial support of the North Atlantic Treaty Organization and the National Research Council of Canada during recent studies which were conducted by Dr Simon Maddrell and the two authors and which are described in this paper.

REFERENCES

Asakura, K. (1970). *Sci. Rep. Kanazawa Univ.* 1, 37—55.
Beadle, L. C. (1939). *J. exp. Biol.* 16, 346—362.
Bradley, T. J. and Phillips, J. E. (1975). *J. exp. Biol.* 63, 331—342.

Clements, A. M. (1963). "The Physiology of Mosquitos." pp. 33. The MacMillan Co., New York.

Conte, F. P. (1969). *In* "Fish Physiology." (Eds W. S. Hoar and D. J. Randall), Vol. I, pp. 241–292. Academic Press, New York and London.

Copeland, E. (1964). *J. Cell Biol.* **23**, 253–264.

Gupta, B. L. and Berridge, M. J. (1966). *J. Cell Biol.* **29**, 376–382.

Hoar, W. S. (1975). "General and Comparative Physiology." Prentice Hall, Toronto.

Kiceniuk, J. W. and Phillips, J. E. (1974). *J. exp. Biol.* **61**, 749–760.

Koch, H. J. (1938). *J. exp. Biol.* **15**, 152–160.

Maddrell, S. H. P. (1971). *In* "Advances in Insect Physiology." **8**, 199–331.

Maddrell, S. H. P. (1976). *In* "Transport Across Biological Membranes." (Eds H. Ussing, D. C. Tosteson and G. Geibisch), In press. Springer-Verlag.

Maddrell, S. H. P. and Phillips, J. E. (1975). *J. exp. Biol.* **62**, 367–378.

Meredith, J. and Phillips, J. E. (1973a). *J. Zellforsch.* **138**, 1–22.

Meredith, J. and Phillips, J. E. (1973b). *J. Insect Physiol.* **19**, 1157–1172.

Meredith, J. and Phillips, J. E. (1973c). *Canad. J. Zool.* **51**, 349–353.

Nayar, J. K. (1966). *Ann. Ent. Soc. Amer.* **50**, 1283–1285.

Nayar, J. K. and Sauerman Jr., D. M. (1974). *Entomologia exp. appl.* **17**, 367–380.

Nicholson, S. W. (1972). *J. Ent.* **47**, 101–108.

Nicholson, S. W. and Leader, J. P. (1974). *J. exp. Biol.* **60**, 593–604.

Noirot, C. and Noirot-Timothée, C. (1966). *C.r. Acad. Sci. Paris* **263**, 1099–1102.

Phillips, J. E. and Maddrell, S. H. P. (1974). *J. exp. Biol.* **61**, 761–771.

Phillips, J. E. and Meredith, J. (1969a). *Nature Lond.* **222**, 168–169.

Phillips, J. E. and Meredith, J. (1969b). *Am. Zool.* **9**, 588.

Ramsay, J. A. (1950). *J. exp. Biol.* **27**, 145–157.

Ramsay, J. A. (1951). *J. exp. Biol.* **28**, 62–73.

Ramsay, J. A. (1952). *J. exp. Biol.* **29**, 110–126.

Ramsay, J. A. (1953a). *J. exp. Biol.* **30**, 79–89.

Ramsay, J. A. (1953b). *J. exp. Biol.* **30**, 358–369.

Ramsay, J. A. (1954). *J. exp. Biol.* **32**, 183–199.

Scudder, G. G. E. (1969). *Verh. Int. Verein. Limnol.* **17**, 430–439.

Sohal, R. S. and Copeland, E. (1966). *J. Insect Physiol.* **12**, 429–439.

Stobbart, R. H. (1959). *J. exp. Biol.* **36**, 641–653.

Stobbart, R. H. (1960). *J. exp. Biol.* **37**, 594–605.

Stobbart, R. H. (1965). *J. exp. Biol.* **42**, 29–43.

Stobbart, R. H. (1967). *J. exp. Biol.* **47**, 35–57.

Stobbart, R. H. (1971a). *J. exp. Biol.* **54**, 19–27.

Stobbart, R. H. (1971b). *J. exp. Biol.* **54**, 29–66.

Stobbart, R. H. (1971c). *J. exp. Biol.* **54**, 67–82.

Stobbart, R. H. and Shaw, J. (1964). *In* "The Physiology of Insecta." (Ed. M. Rockstein), Vol. III, pp. 190–258. Academic Press, New York and London.

Stobbart, R. H. and Shaw, J. (1974). *In* "The Physiology of Insecta." (Ed. M. Rockstein), 2nd edition, Vol. V, pp. 362–446. Academic Press, New York and London.

Wall, B. J. and Oschman, J. L. (1975). *Fortschritte der Zoologie* **23**, Heft 2/3,
    pp. 192—222.
Wigglesworth, V. B. (1933a). *J. exp. Biol.* **10**, 1—15.
Wigglesworth, V. B. (1933b). *J. exp. Biol.* **10**, 16—26.
Wigglesworth, V. B. (1933c). *J. exp. Biol.* **10**, 27—37.
Wigglesworth, V. B. (1938). *J. exp. Biol.* **15**, 235—247.

29. Role of Integument in Molluscs

J. Machin

Department of Zoology, University of Toronto, Toronto, Canada

I. INTRODUCTION

To those who know Professor Ramsay for his transport studies on nephridia, Malpighian tubules and the insect rectum, it may come as a surprise that some of his early work was concerned with the laws governing evaporation of water from animals and the methods used in its measurement (Ramsay, 1935a,b). His division of evaporation into two distinct processes, "evasion" and "diffusion", was a helpful starting point in my own studies and led to a separate consideration of the nature of the evaporating surfaces in land snails (Machin, 1964a) and the atmospheric diffusion of water vapour from them (Machin, 1964b,c). It will be seen later in this chapter that the treatment of evaporation as a number of separate but related diffusive processes persists in my approach to the subject.

Ramsay has returned in the latter part of his career to evaporation and the equilibria which exist at the gas-liquid interfaces of a biological system. Nobody can help but be amused by his delightfully ingenious studies of water exchange in freshly eliminated mealworm faecal pellets (Ramsay, 1964). Perhaps it requires personal involvement in the problems of mealworm water balance (Machin, 1975b) to realize the full value of measurement of equilibrium humidity by that simple but elegant technique in providing a means of studying without interference the physiological state of an insect's rectal surfaces.

The epidermis of a land snail is another biological gas-liquid interface whose complex and sometimes unusual physiological properties are only just beginning to be understood (Schmidt-Nielsen, 1969). Views concerning the structure and functions of the molluscan epidermis along with those of other soft bodied invertebrates have changed in recent years. There is now a growing realization that the permeable epidermis is a dynamic structure, engaged in a wide variety of active as well as passive transport processes. At present physiological information from a wide variety of molluscs is lacking. However, the remarkable uniformity of structure of the epidermal cells throughout the Mollusca permits a broader interpretation of the existing data with some confidence. One aspect, having some personal fascination, is how a structurally unspecialized epidermal layer is able to function so close to the air-water boundary.

II. GENERAL STRUCTURAL FEATURES OF THE MOLLUSCAN EPIDERMIS

Ideas relating to the physiology of the epidermis have always been greatly influenced by knowledge gained from arthropods and vertebrates. We have come to think of the epidermis in all groups as a single layer of cells, protected and, to a certain extent, isolated from the external environment. Before the electron microscope came into use, light microscopists extended this concept, perhaps unintentionally, to the Mollusca and other soft bodied invertebrate phyla, by describing distinctive surface layers of the epidermis as "cuticles". The persistence of the idea of a protected, rather inactive epidermis is surprising in view of the fact that it has long been known that the integument of soft bodied invertebrates is comparatively permeable to water and solutes, and consequently less influenced by the internally regulated body fluids and more exposed to the exigencies of the environment. Despite the fact that active transport can only be sustained in metabolically active epidermal cells so long as their degree of hydration remains within viable limits, most permeable-skinned groups inhabit a wide range of osmotic environments, from the marine and fresh water to the fully terrestrial.

Perhaps the most significant discovery leading to a more dynamic view of the permeable epidermis, was that cells making up this layer had many ultrastructural features in common with well established transporting epithelia. The most obvious feature, a surface zone of microvilli, appears to be widespread in the invertebrates. Epidermal microvilli have been demonstrated in molluscs, turbellarians, including parasitic (Lumsden, 1975), free-living aquatic (Lyons, 1973) and even terrestrial species (Storch and Abraham, 1972), as well as in some cestodes and nemertines (Lumsden, 1975). Epidermal microvilli are also found in some groups bearing an external fibrous cuticle, such as annelids and Pogonophora (Gupta and Little, 1970). Among the annelids, terrestrial oligochaetes (Coggeshall, 1966), free-living polychaetes (Storch and Welsch, 1970; Chien *et al*., 1972) and even air breathing forms (Storch and Welsch, 1972) have external microvilli, whereas tube dwelling polychaetes do not. In the above examples, the open network of collagenous fibres of the cuticle are regularly traversed by the elongated microvilli, permitting rather than preventing exchanges with the outside.

The earliest ultrastructural studies of the molluscan epidermis indicated the presence of apical microvilli on the ciliated gill epithelia of lamellibranchs (Fawcett and Porter, 1954) and on the adhesive epithelium of lamellibranchs and gastropods (Hubendick, 1958). Because of the specialized nature of these epithelia, wider implications of the presence of surface microvilli were not considered. Slightly later work on the lamellibranch mantle (Kawaguti and Ikemoto, 1962a,b) showed structural evidence of transport, but this was to be expected in a tissue so clearly associated with the formation of shell. The first suggestion that the type of transport which requires active cellular participation was a general feature of the molluscan epidermis was obtained from studies of the general external surfaces of land pulmonates: tissues which seemed least likely to be engaged in this type of transport. Schwalback and Lickfield (1962) working with *Helix pomatia* and Lane (1963) with *H. aspersa* and *Arion hortensis* showed that the external surfaces of these animals were covered with a layer of microvilli. Since then further ultrastructural studies have appeared which confirm the presence of the microvilli on all regions of a land pulmonate's body. They are found on the general epidermis of *A. rufus* (Wondrak, 1968a), *Oxychilus helveticus* (Lloyd, 1969) *H. aspersa*, including the sole of the foot (Rogers, 1971), *H. pomatia* (Zylstra, 1972a), *A. hortensis* and *Agriolimax reticulatus* (Newell, 1977) and the mantle of *H. pomatia* (Saleuddin, 1970). It is now clear that the same ultrastructural features, including surface microvilli, are also characteristic of the general epidermal cells of the freshwater pulmonates *Lymnaea stagnalis* and *Biomphalaria pfeifferi* (Zylstra, 1972a), the marine prosobranch *Nassarius reticulatus* (Crisp, 1971), the nudibranch *Trinchesia granosa* (Schmekel and Wechsler, 1967) and the squid *Loligo opalescens* (Cloney and Florey, 1968). It also seems that cells in epidermal tissues having some special function, frequently retain some of their transport characteristics, particularly

the microvilli. Examples of this are the ciliated cells of the lamellibranch ctenidium (Fawcett and Porter, 1954) and the ciliated
supporting cells of the prosobranch osphradium (Welsch and Storch,
1969) and hypobranchial gland (Hunt, 1973). Although Welsch
(1968) has assigned a purely passive role to the microvilli as simply
aiding the adhesion of mucus to an epidermal surface, the presence
of internal organelles associated with transport amply confirm a
more active and dynamic role. Intracellular vacuoles, lysosomes and
abundant mitochondria are prominent in the majority of molluscan
epidermal cells.

The typical epidermal cell junction in the Mollusca consists of an
apical *Zonula adherens* (Farquhar and Palade, 1963) with a septate
zone (Locke, 1965) beneath. This leaves the major part of adjacent
lateral cell membranes unconnected by special structures except for
the occasional *macula adherens* (Wondrak, 1968a; Zylstra, 1972a)
towards the base. In terrestrial molluscs neighbouring epidermal cells
are markedly interdigitated on the lateral boders. In recent years cell
junctions and spaces between epithelial cells have been shown to play
an important part in the transport of water and solutes. In several
molluscan epithelia an intercellular route has been proposed for
passive ion transport (Satir and Gilula, 1970; Neff, 1972; Newell and
Skelding, 1973). Although intercellular spaces are frequently seen in
micrographs of molluscan integuments, their association with water
transport has yet to be demonstrated.

Permeable integuments are usually equipped with a second
transport system, the gland cells. These cells are specialized
principally for the synthesis of macromolecules and for their
mechanical transport to the epithelial surface. The simplest and
apparently most primitive type of glands are the single "goblet" cells
which are the same height as the surrounding epithelium. Cells of this
type are common in the epidermis of marine molluscs (Fretter and
Graham, 1962). Molluscan epidermal glands show great variability
both in structure and in the chemistry of their secretions (Fretter
and Graham, 1962; Campion, 1961; Wondrak, 1967, 1968a,b;
Zylstra, 1972a). Single celled glands, particularly on the exposed
surfaces of the mantle and foot margins, have become enormously
enlarged in fresh water, terrestrial and intertidal species, so that the
cell body extends well beneath the surrounding epidermis. In glands
which perforate the sole of the foot, the same subepidermal
arrangement occurs, with an elongated neck between the cell body
and the surface. In either case the basement membrane remains
intact by passing around the base of each gland cell. In some

specialized regions of the body, gland cells become aggregated into special mucus-producing tracts such as the hypobranchial gland in prosobranchs (Hunt, 1973). In others, multicellular mucus glands may communicate with the exterior only by a single opening, as in the various pedal glands of gastropods (Fretter and Graham, 1962).

Although there are variations, all gland cells have the ultrastructural characteristics of secretory cells: extensive rough or granular endoplasmic reticulum, numerous Golgi bodies, mitochondria and membrane-bound secretory granules or droplets. The mechanism by which stored secretory material is discharged from the many types of gland cell is still poorly understood. It would appear that exocytosis by membrane fusion (Palade, 1975) does not apply to some molluscan mucus glands. Hunt's (1973) micrographs indicate that secretory droplets in the hypobranchial gland lose their membranes before discharge, while in other glands (Zylstra, 1972a) secretory droplets are extruded with their membranes still intact. In the larger gland cells, forceful or rapid discharge of material would surely rupture the cell membrane. There is some structural (Campion, 1961) and experimental (Machin, 1964a) evidence that extrusion in large cutaneous mucus glands of pulmonates is muscular. Contraction of fibres associated with the gland or general muscular activity of the body wall can bring about a local squeezing effect. It is possible, however, that alternative mechanical mechanisms of glandular discharge exist, such as the contraction of intracellular fibrils suggested by Palade (1975), or osmotic swelling perhaps produced by local membrane changes. These mechanisms might apply particularly to the smaller type of gland cell located within the epidermal layer, where the forces for extrusion are more obscure.

In several species it has been possible on the basis of histochemistry and fine structure (Campion, 1961; Wondrak, 1967, 1968b; Zylstra, 1972a) to distinguish a bewildering variety of epidermal glands. Although, for example Zylstra listed 14 types in *Lymnaea stagnalis*, these could be reduced to four, depending on whether they contained histochemically identified protein, sulphated mucopolysaccharide or nonsulphated slightly acid to neutral mucopolysaccharide, or were eosinophilic goblet cells containing an unknown nonmucous secretion. The mixed distribution of epidermal glands makes it impossible to separate secretions by gland type. Indeed there is evidence that some essential characteristics of mucus are only developed after the different secretions have been extruded. Important viscoelastic properties of *Buccinum* hypobranchial mucus develop only after the mixing and interaction of the products of two or more cell types (Hunt, 1970). It is clear from chemical studies of bulk secretions that the principal constituents of epidermal mucus, throughout the phylum, are various sulphated polysaccharides and glycoproteins (Hunt, 1970). The hypobranchial mucus of marine prosobranch snails contains one or more linear polysaccharides with covalently linked estersulphate groups (e.g. glucan sulphate), together with glycoprotein bound to small amounts of hexosamine and neutral sugar. In some species hypobranchial mucus may contain other sulphated polysaccharides or acid mucopolysaccharides (Hunt, 1970). The slime secreted by *Otella lactea* (Pancake and Karnovsky, 1971), which by their description is almost certainly the land snail *Otala lactea* (Müll.), contains sulphated mucopolysaccharide in which galacturonic acid is the major component. Perhaps of greatest significance to transport mechanisms are the viscous and electrostatic properties common to all these molecules. In epidermal mucus the polysaccharides are the most highly charged due to the ester sulphate groups or to uronic acid residues. Electrostatic forces in macromolecules are responsible for a variety of complex interactions, including those with water and ions, especially divalent cations (Scott, 1968; Hunt, 1970).

III. MANTLE TRANSPORT ASSOCIATED WITH SHELL FORMATION

Regrettably few direct transport studies have been performed using the exposed molluscan integument. On the other hand the exchange of materials associated with shell formation and maintenance, an activity not restricted to the mantle edge but shared by the entire external mantle epidermis, has been extensively investigated. Freshwater and marine lamellibranchs have proved the most convenient for experimental study; however there is sufficient information from parallel studies with gastropods, including terrestrial species, to suggest that the following account applies generally to all shelled molluscs.

Even in terrestrial species, exchanges with the shell are mediated through the fluid filled extrapallial space lying between the shell and the mantle. Materials are transported in both directions, associated with shell formation, as well as erosion during anaerobiosis (Crenshaw and Neff, 1969; Greenaway, 1971b). The ionic constituents of extrapallial fluid in fresh water and marine lamellibranchs are similar but not identical to that of the haemolymph (Wilbur, 1972). In the extrapallial fluid of the marine

Mercenaria (Crenshaw, 1972) the concentrations of cations are higher and chloride lower, suggesting a Donnan equilibrium. However there are complicating factors since dialysis experiments have identified an extrapallial bound calcium fraction and a bound potassium component in the haemolymph. There is some evidence (Wilbur, 1972) that the distribution of trace elements such as Mn, V, Cu and Fe in the tissues and shells of several marine bivalves are also governed by processes other than simple diffusion.

The extrapallial fluid also contains a measurable nondialysable organic fraction containing acid mucopolysaccharide and protein (Kobayashi, 1964; Crenshaw, 1972). Although the precise origin of these macromolecules is in some doubt owing to a scarcity or complete lack of gland cells in the extrapallial epidermis (Istin and Kirschner, 1968; Zylstra, 1972a), it seems likely that macromolecules play an important role in the calcification process. Kobayashi (1964) obtained correlations between the number of different extrapallial proteins and mucopolysaccharide fractions and the type of crystal structure being deposited. Weiner and Hood (1975) have obtained evidence which suggests that some soluble proteins act as templates for shell formation. Sulphated mucopolysaccharides and glycoproteins have been identified in the organic matrix of the shell (Simkiss, 1963; Hunt, 1970).

Isotope (Istin and Maetz, 1964) and electrophysiological (Kirschner *et al.*, 1960; Kirschner, 1962; Istin and Kirschner, 1968) studies using freshwater lamellibranchs have shown the mantle to be highly permeable to calcium. Passive calcium efflux apparently generates a measurable electrical potential with the extrapallial side positive. From ultrastructural studies of the distribution of calcium in the mantle epithelium of *Mercenaria*, Neff (1972) has reinterpreted the electrophysiological data, emphasizing an intercellular route of permeation for calcium. This is consistent with the model proposed by Frömter and Diamond (1972) for permeable vertebrate epithelia. Neff (1972) concluded that much of the calcium moves through the mantle epithelium in a soluble ionized form, making its way between the epithelial cells and through selectively permeable *zonulae adherentes* and septate desmosomes as described by Barry and Diamond (1971). Neff (1972) also observed mineral-rich granules in the intercellular spaces of the epidermal cells. These, he concluded, were probably a complex of calcium bound to mucopolysaccharide or glycoprotein, which is engulfed by the epidermal cells at the level of the septate desmosome and extruded through the microvilli into the extrapallial cavity.

Mantle epidermal cells are also capable of absorbing particulate matter from the extrapallial space by pinocytosis (Bevelander and Nakahara, 1966; Nakahara and Bevelander, 1967). The fact that pinocytotic vesicles coalesce in dense bodies identified histochemically by the presence of acid phosphatases as lysosomes, suggests a digestive role. During anaerobiosis, measurable quantities of calcium are dissolved from the shell and removed to the tissues and body fluids (Crenshaw and Neff, 1969). It is possible that the inward movement of calcium is brought about by the absorption of mineralized organic granules which are later digested. A precedent for this type of inward transport, together with an outward calcium flux in a predominantly ionic form, is to be found in the amphipod *Orchestia* (Graf, 1971). Transport processes in the extrapallial epidermis are summarized in Fig. 1.

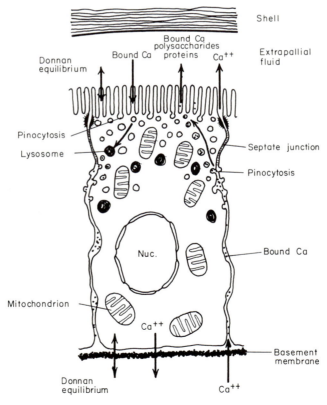

Fig. 1. Summary of transport activities in the extrapallial epidermis of molluscs. After Nakahara and Bevelander, 1967; Greenaway, 1971b; Neff, 1972.

IV. MARINE MOLLUSCS

The body fluids of marine molluscs are essentially iso-osmotic to sea water (Robertson, 1964; Schoffeniels and Gilles, 1972). In spite of this all species investigated (Robertson, 1953; Harrison, 1962; Harrison and Martin, 1965; Little, 1967) continuously produce significant amounts of urine at rates between 1.8 and 21 ml.kg^{-1}. h^{-1}. It is also possible that water is lost in the process of mucus formation and extrusion.

Unfortunately the mechanism by which water is replaced by marine molluscs is entirely unknown. Experiments with the gut, one of the most likely sites of water gain (Bethe, 1934; Lawrence and Lawrence, 1967; Little, 1967; Hanisch and Lawrence, 1972) have failed to demonstrate water transport sufficient to balance urinary losses. The integument of marine molluscs is readily permeable to water and hence is a potential site of water intake. This is shown in a variety of marine species, which rapidly gain weight (4 to 12% body weight h^{-1}) following transfer to 75% sea water (Bethe, 1934; Tucker, 1970; Pierce, 1971). Electronmicrographs of the molluscan epidermis frequently show lateral and basal intercellular spaces which may be associated with iso-osmotic water transport (Tormey and Diamond, 1967; Schmidt-Nielsen and Davis, 1968). Unfortunately, the only available general ultrastructural survey of the epidermis of a marine mollusc, is of the highly specialized eolid nudibranch *Trinchesia* (Schmekel and Wechsler, 1967). At the present time convincing experimental and ultrastructural information which might implicate the epidermis in water absorption in marine species is lacking.

Various measurements of radioactive calcium and phosphorus uptake in marine molluscs have been made as part of the study of shell formation (Ronkin, 1950; Bevelander, 1952; Pomeroy and Haskin, 1954). Pomeroy and Haskin showed that P^{32} is absorbed more rapidly by the gills than by the mantle or labial palps in the oyster *Crassostrea*. Evidence that other ionic constituents of the haemolymph may be regulated by exchange processes in the integument is less direct. Comparisons between normal haemolymph ion concentrations and those determined after experimental dialysis show some consistency between all marine molluscs, which appear actively to accumulate potassium and eliminate sulphate (Robertson, 1953, 1964; Little, 1967; Webber and Dehnel, 1968). In specific molluscs other ions do not distribute themselves according to Donnan equilibria, presumably because they are actively transported or bound to macromolecules. Haemolymph calcium is higher than predicted in *Pecten, Neptunea* and *Archidoris*, whereas cephalopod haemolymph is low in sodium and magnesium. Comparison between an observed transepidermal potential of -1.24 mV and calculated Nernst potentials for the individual ions have led Little (1967) to conclude that potassium and chloride are taken up actively by *Strombus* while calcium and sulphate are actively eliminated. Although these observations indicate nothing specific about the site of transport, parallel investigations of urine composition suggest that some ion regulation must be extrarenal, as for example potassium and sulphate in *Strombus* (Little, 1967) and sodium in *Sepia* (Robertson, 1964). A possibility suggested by Little, and deserving further investigation, is that mucus secretion might be used to eliminate specific ions such as sulphate. Although molluscan mucus is rich in sulphate covalently bound to polysaccharides (Hunt, 1970) it has yet to be demonstrated that this is of ionic or haemolymph origin. It is possible that electrostatic association between polar mucus macromolecules and calcium (Scott, 1968) may also contribute to the active elimination of this ion in *Strombus*.

It has been suggested (Potts, 1967; Stephens, 1968) that marine invertebrates including molluscs are probably able to absorb organic molecules, especially amino acids, in dissolved, colloidal or particulate form directly from the environment. Although the site and mechanism is unknown, the ability of the epidermis such as the ctenidia in *Mytilus* (Pasteels, 1961) to engulf particulate matter suggests that the body surface might be involved in the transport of organic molecules as well as inorganic material. Transport activities in the epidermis of marine molluscs are summarized in Fig. 2.

V. FRESHWATER MOLLUSCS

All molluscs living in fresh water have very low blood concentrations, *Anodonta cygnea* having the lowest on record for any metazoan (42 mOsm kg^{-1}) (Potts, 1953, 1954). Despite small osmotic gradients across the integument, urine flows are sizeable (19 to 110 ml. kg^{-1}. h^{-1} or 430 to 870 ml. kg^{-1}. h^{-1}. $Osmol^{-1}$:) (Potts, 1954; Little, 1965b; Little, 1968; Van Aardt, 1968), indicating a correspondingly large osmotic flow through a permeable integument. The principal extracellular ions of freshwater species, sodium and chloride, are never in equilibrium with the external medium and many of those in low concentration in the blood are only rarely so. This leads to the passive net loss of sodium chloride and of other unequally

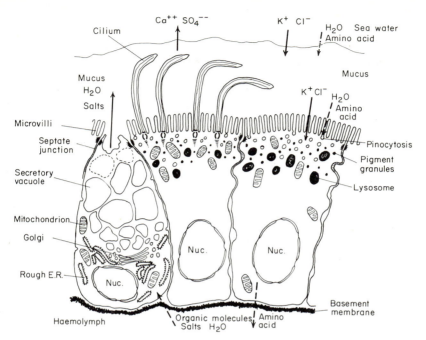

Fig. 2. Summary of transport activities in the epidermis of marine molluscs. Continuous arrows indicate established mechanisms, dotted arrows, hypothetical exchanges. After Welsch and Storch, 1969; Crisp, 1971 and Hunt, 1973.

distributed diffusable solutes. Nondiffusable calcium shown in *Anodonta* (Schoffeniels, 1951) to be 29% of the total haemolymph concentration, would not be subject to this exchange. Although Greenaway (1970, 1971a) was unable to separate efflux measurements into their integumentary and urinary components, ionic losses through the integument, particularly in gillbearing species, must have an important effect on ion balance. The work of Satir and Gilula (1970) suggests in the gill epithelium of the freshwater mussel *Elliptio*, that an intercellular route of passive loss through the septate junctions is probably important. Tracer studies indicate a significant exchange component of sodium and calcium fluxes in *Lymnaea* (Greenaway, 1970, 1971a).

It has been known since the classical work of Krogh (1939) that food-deprived freshwater animals are able to remain in ionic balance as long as certain minimum ion levels exist in their environments. These levels are 0.006 mM Na l^{-1} and 0.2 mM Ca l^{-1} in *Viviparus viviparus* (Little, 1965a) and 0.025 mM Na l^{-1} (Greenaway, 1970)

0.062 mM Ca l^{-1} (Greenaway, 1971a) or 0.05 mM Ca l^{-1} (Van der Borght and van Puymbroeck, 1964) in *Lymnaea stagnalis*. Since these values are well below corresponding concentrations in the final urine in *Viviparus* (Little, 1965b) there must be an extrarenal site of ion absorption. At present there is no evidence for lamellibranch and prosobranch gastropods, as there is for crustaceans and fish, to suggest that uptake is confined to gill epithelia. In pulmonate molluscs with an air-filled lung, uptake must be restricted to other parts of the integument, as the use of the gut for this purpose in nonfeeding animals would seem unlikely. Influx of sodium and calcium in *Lymnaea* (Greenaway, 1970, 1971a) and sodium and chloride in *Margaritana* (Chaisemartin *et al.*, 1968) shows high affinity, Michaelis-Menten saturation kinetics. The absorption mechanism in *Lymnaea* was half saturated at 0.25 mM Na l^{-1} and 0.3 mM Ca l^{-1}. In *Lymnaea* there is a potential difference between haemolymph and medium of up to 39 mV with the inside negative. Comparisons between observed values and those calculated from the Nernst equation suggest that sodium must be absorbed actively. The same applies to calcium in lower environmental concentrations (Greenaway, 1970, 1971a; Van der Borght and van Puymbroeck, 1964). Ion absorption mechanisms show evidence of homeostatic control, responding to ion depletion or reduced blood volume by increased absorption rates (Krogh, 1939; Hiscock, 1953; Greenaway, 1970, 1971a).

It is possible that mucus secretion by the epidermal glands has a significant effect on water and ion balance in freshwater molluscs. Wilson (1968) has shown that the epidermal mucus of immersed *Lymnaea truncatula* contains 91.8% water, measurable quantities of free sugars, amino acids and lipids as well as 60 mM Na.kg^{-1} and 17 mM K.kg^{-1}. Sodium concentrations are close to those of the haemolymph (Pullin, 1971) while those of potassium, which is probably of intercellular origin (Burton, 1965) are seven times greater. Zylstra (1972a) has shown that ciliated cells are scattered throughout the body surface of *Lymnaea*, probably serving to disperse and spread the mucus. However, the rate at which this is done is unlikely to prevent the mucus from behaving as an "unstirred layer" (Dainty and House, 1966) retarding diffusion of materials across the integument. Although in most instances this would be an advantage to a fresh water mollusc, it would hinder the dispersal of excretory material. Wilson (1968) has commented upon the presence of large amounts of ammonia in the mucus of *Lymnaea truncatula*.

Zylstra (1972b) has shown that general epidermal cells of *Lymnaea* are able to take up particulate material in the manner previously described for the external mantle epidermis. The mechanism is similarly associated with high acid phosphatase concentrations and digestive lysosomal activity. Zylstra suggested that this system is involved in mucus digestion. If so, such recycling would considerably reduce the loss of macromolecules and ions from the animal. The electrostatic attraction and adsorption of ions to charged polysaccharides might even bring about a net gain from the environment. Transport processes in the epidermis of freshwater molluscs are summarized in Fig. 3.

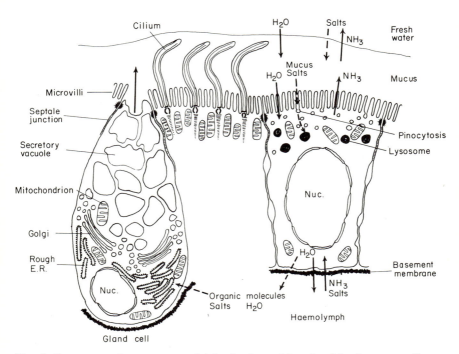

Fig. 3. Summary of transport activities in the epidermis of freshwater molluscs. Continuous arrows indicate established mechanisms, dotted arrows, hypothetical exchanges. After Zylstra, 1972a.

VI. TERRESTRIAL MOLLUSCS

A. Introduction

In aqueous media the haemolymph osmotic pressure is an important factor in determining the direction and rate of water exchange through the integument. In spite of the notorious variability of individual haemolymph concentrations in terrestrial, semiterrestrial and intertidal molluscs (Todd, 1964; Avens and Sleigh, 1965; Little, 1968; McAlister and Fisher, 1968; Rumsey, 1972; Machin, 1975a) osmotic pressure differences in vapour pressure terms are negligible compared with much larger humidity fluctuations in the atmosphere. It has been estimated that vapour pressure gradients between haemolymph and environment, driving water from land snails are commonly tens or even hundreds of times greater than the gradients promoting water uptake in their freshwater relatives (Machin, 1975a). The theoretical possibility of passive uptake of water vapour due to haemolymph solute vapour pressure lowering has been the subject of much speculation in the past. In carefully controlled temperatures, vapour uptake by direct condensation has been observed experimentally in snails (Machin, 1972). However, because of the very high haemolymph equilibrium humidities, the level of temperature control required to maintain a humidity from which uptake could occur, without forming liquid droplets, must be better than $\pm 0.1°C$. Temperature stability at this level would never be found in nature and environmental humidities high enough to absorb from, would condense as dewdrops and be taken up by conventional osmosis. Indeed, the charged and highly hydrophilic nature of molluscan mucus is ideally suited to enhance osmotic uptake, by spreading droplets contacted by the animal over a wide area of its surface (Machin, 1964a). There is some evidence that the structure of epidermal cells has become slightly modified in terrestrial and semiterrestrial species. In contrast to their aquatic relatives, the epidermal cells of terrestrial species are extensively interdigitated on their lateral borders. Newell (1975) has suggested that this modification might help the epidermis maintain its structural integrity when the body is distorted. Epidermal cilia are comparatively rare in terrestrial snails, except on the undersurface of the foot and around the mouth and pneumostome (Rogers, 1971; Zylstra, 1972a; Newell, 1975). In terrestrial pulmonates the exposed surfaces are well supplied with groups of large subepidermal mucus glands. These are particularly concentrated on lateral margins of the foot and on the mantle collar which is responsible for producing the epiphragm. It has been suggested (Machin, 1964a) that in the absence of cilia, mucus is conveyed from the glands to all areas of the surface by muscular undulations of the body wall. To ensure that there is always an abundant supply of mucus, the network of grooves which is characteristic of the skin of land pulmonates, serves as a reservoir from which mucus is conveyed to the more exposed areas. A survey of the illustrations in Fretter and Graham (1962) suggests that exposed areas of the integument in intertidal prosobranchs are similarly equipped with large mucus glands.

B. Transepithelial Diffusion and Evaporation

One of the most useful general equations governing the diffusion of materials across a complex barrier expresses the relation between overall permeability (P) and that of the constituent layers in the following way:

$$\frac{1}{P} = \frac{1}{P_1} + \frac{1}{P_2} + \frac{1}{P_3} \cdots + \frac{1}{P_n} \tag{1}$$

Permeability constants in G.G.S. units are cm sec^{-1} atm^{-1} (cm^3 cm^{-2} sec^{-1} atm^{-1}), however in evaporation studies, units of mg H$_2$O cm^{-2} h^{-1} mm Hg^{-1} (Leighly, 1937) are more readily applied to direct measurement and interpretation, and will be used below.

When the outer constituent layer is air, the diffusion of water vapour is determined by the appropriate diffusion coefficient D_a, the effective thickness of the layer in which there is a significant vapour pressure gradient, δ, and the membrane permeability, P_m. The equation becomes:

$$\frac{1}{P} = \frac{1}{P_m} + \frac{\delta}{D_a} \tag{2}$$

The units of D_a are mg H$_2$O cm^{-2} h^{-1} mm Hg^{-1} cm^{-1} and δ is in cm.

When a steady-state is reached, concentration gradients become stabilized, and net fluxes through the constituent layers are equal. Under these conditions, assuming that the layers are sufficiently homogeneous to establish linear gradients, it is possible to substitute the Fick relationship for each of the permeability terms:

$$\frac{A \cdot \Delta C}{J} = \frac{A \cdot \Delta C_m}{J} + \frac{A \cdot \Delta C_a}{J} \tag{3}$$

Where J is the net water flux through area A, and ΔC values are corresponding concentration differences.

It can be seen from equation (3) that at steady-state, concentration differences across each constituent layer are inversely proportional to its permeability. The principal resistance to diffusion through a complex barrier is the layer with the lowest permeability and consequently with the greatest change in concentration.

Beament (1961) applied this analysis to membrane-air systems in order to define the conditions under which membrane permeability determines evaporation rate without significant errors due to atmospheric diffusion. According to his criteria "membrane limited" conditions existed when the rate of evaporation into stationary air from a membrane was at least two orders of magnitude lower than the rate of evaporation from an identical free water surface. Evaporation rates from free surfaces which are limited by atmospheric diffusion depend on D_a/δ. Although it is impossible to predict precise values for D_a and δ for any given situation, approximate figures are available. The value for D_a is of the order of 1 mg H_2O cm^{-2} h^{-1} mm Hg^{-1} cm^{-1} for biological temperatures and pressures (Leighly, 1937; Machin, 1970), and δ is of the order of 1 cm (Shiba and Ueda, 1965). It follows from equation (2) that the discrepancy between $1/P$ and $1/P_m$, due to δ/D_a, is 1% for evaporation rates of the order of 0.01 cm^{-2} h^{-1} mm Hg^{-1}. Evaporation from more permeable membranes is partially "vapor limited", leading to 10% error when the rate of loss is 0.1 mg cm^{-2} h^{-1} mm Hg^{-1} and 50% at 1 mg cm^{-2} h^{-1} mm Hg^{-1} (Table I).

There is some evidence that evaporation from land molluscs represents the most extreme form of "vapor limited" conditions, where the loss of water is entirely determined by vapour gradients in air. Experimental and observational evidence (Machin, 1964a) suggests that high surface vapour pressures are maintained in active snails by the extrusion of water-rich mucus. Since epidermal mucus was found to be iso-osmotic to the haemolymph, diffusion gradients would be abolished in favour of a mechanical method of transporting water to the surface. Evidence that water is lost from the haemolymph in this process was obtained by Burton (1965) who showed a rise in blood sodium in snails artificially stimulated to produce mucus.

Analyses of evaporation under "vapor limited" conditions have been made by Machin (1964c) using the snail *Helix aspersa* and Spight (1968) using a series of urodele amphibians. Their approach was to determine empirically various constants of Leighly's (1937) evaporation formula for moving air. Machin (1964c) found that evaporation from *Helix* was affected by obvious variables such as

TABLE I

Water permeability constants for representative biological barriers in terrestrial species

Barrier	Permeability constant (P)[a] $(mg\ H_2O\ cm^{-2}\ h^{-1}\ mm\ Hg^{-1})$	1/P	Authority
Vapour limited			
Air (effective)	1	1	—
Helix aspersa,			
dorsal integument	83	0.012	Machin (1966)
Thin toad skin	30—90	0.01—0.03	Machin (1969)
Thick toad skin	12—37	0.03—0.08	Machin (1969)
H. aspersa,			
epiphragm	2.6	0.38	Machin (1968)
H. pomatia,			
hibernating epiphragm	0.17	5.9	Machin (1968)
Membrane limited			
H. aspersa,			
mantle collar epidermis	0.039	25	Machin (1966)
Otala lactea,			
mantle collar epidermis	0.016	63	Machin (1972)
Cockroach cuticle	0.017	59	Machin (1972)
Mealworm cuticle	0.0015	667	Machin (1975b)

[a] Values for approximately 20°C, atmospheric pressure.

humidity, animal dimensions and wind speed but also by more subtle factors such as orientation of the animal with respect to the wind and the position and streamlinedness of nonevaporating surfaces of the shell. Attempts (Machin, 1964b; 1970) to relate these complex factors directly to the nature of air flow over the snail's body and the formation of a vapour boundary layer have had little success.

Although the conclusion that there should be very little difference between evaporative water loss from a snail and from an identical nonliving free water surface still remains true, the evidence for supposing that mucus extrusion is entirely responsible for net water flux through the integument could be profitably re-examined. The reason for this doubt is that measurements of evaporation rate in moderate to low humidities are insensitive even to large changes in the level of hydration of the epidermis and to alterations in the pattern of water flow across it (Fig. 4). Definitive experiments which could accurately gauge the success of the mucus extrusion mechanism in abolishing diffusion gradients and keeping the skin fully hydrated, especially in extreme or changing ambient conditions, have not yet been performed.

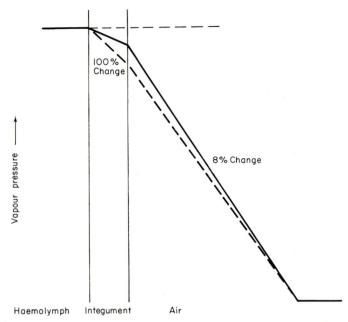

Fig. 4. Hypothetical vapour pressure profiles in and above land snail integument. The diagram shows that doubling the gradient in the integument leads to much smaller changes in the atmosphere.

A number of techniques are potentially available for studying the role of membrane diffusion in "vapor limited" systems. In a study of water flux through isolated amphibian skin in air, Machin (1969) was able to calculate the vapour pressure gradients in air and hence determine the true permeability of the skin. This was done by establishing δ/D_a values for the apparatus with known atmospheric gradients, using a wet Millipore filter (0.1 μm pore size) as a free water surface. Although not used in the amphibian skin study, an experimentally soluble equation for the permeability of a "vapor limited" membrane can be derived as follows. From equations (2) and (3):

$$\frac{A \cdot \Delta C}{J} = \frac{1}{P_m} + \frac{A \cdot \Delta C_a}{J} \qquad (4)$$

If J and J_o are the rates of evaporation from the membrane and free water surface, respectively:

$$\Delta C_a = \frac{\Delta C \cdot J}{J_o} \qquad (5)$$

By substitution:

$$\frac{1}{P_m} = \frac{A \cdot \Delta C}{J} - \frac{A \cdot \Delta C}{J_o} \tag{6}$$

An alternative method for determining surface vapour pressure in a vapour limited system is suggested by Ramsay *et al.*, (1938) early attempts to measure the vapour pressure profile close to the surface of a transpiring leaf. A workable and sensitive dew-point probe (Machin, 1970) which is capable of making a detailed survey of vapour boundary layers could be used to obtain surface vapour pressure directly by extrapolation.

It is possible to estimate the protective value of an epidermal mucus extrusion mechanism and the physiological consequences if such a mechanism were inhibited or lost. The osmotic gradients necessary to sustain a given rate of evaporation, assuming that the net flux of water through the epidermis occurred entirely by diffusion, may be calculated from its permeability. Machin (1966) has obtained such values for the integument of *Helix aspersa* in aqueous media of differing osmotic pressure, using isolated preparations to keep mucus extrusion to a minimum. Representative epidermal diffusion gradients in Table II have been calculated using the highest permeability value obtained (83 mg cm^{-2}. h^{-1} mm Hg^{-1}), since measurements without correction for "unstirred layers" (Dainty and House, 1966) underestimate true permeability. It can be seen that without mucus extrusion, even modest evaporation rates in high humidities would bring about marked dehydration of the

TABLE II

Calculated vapour pressure gradients and surface dehydration necessary to sustain representative evaporation rates in *Helix aspersa* by transepidermal diffusion[a]

Ambient humidity (% R.H. at 20°C)	Expected evaporation rate (mg cm^{-2} h^{-1})	Calculated transepidermal v.p. gradient (mm Hg)	Calculated surface osmotic pressure[b] (mOsm kg^{-1})
99	0.12	0.0014	204
97	0.45	0.0055	217
90	1.67	0.020	262
80	3.41	0.041	327
50	8.04	0.097	523

[a] Epidermal permeability used in calculations was 83 mg cm^{-2} h^{-1} mm Hg^{-1}
[b] Normal haemolymph concentration 200 mOsm, kg^{-1}. Calculated from v.p. from equation given in Machin (1969).

integumentary surface. Other experiments have shown that the extent of dehydration is further aggravated by the tendency in snail integument, along with other living and nonliving organic membranes (Manton and Ramsay, 1937; Machin, 1964a, 1969), for the permeability to fall with decreasing water content. This results in a progressive decline in water content and water flux through the membrane. Clearly, one way of breaking this positive feedback is to have an independent parallel water transport mechanism in the epidermis, such as mucus extrusion.

C. Solute Transport in Land Molluscs; does it exist?

It has already been shown that the change to a terrestrial way of life, even a temporary one, profoundly alters the forces governing water exchange at the animal's surface. Why many soft bodied invertebrates make so little change in epidermal structure in their transition from aqueous to gaseous media is difficult to understand. However it has been possible to show how permeable, actively metabolizing cells are able to function close to a severely desiccating environment. In this respect it is useful to think of the layer of viscous water-rich integumental mucus as a sort of private pond by which each animal is surrounded.

The concept may be usefully extended to include solute transport in land molluscs. Burton (1965) has shown that epidermal mucus of *Helix pomatia* contains twice as much potassium and magnesium as the haemolymph and slightly less sodium. This distribution, he proposed, was best explained by the mucus receiving during its formation, two fluid contributions, one from the blood and the other from within the gland cells. Since far higher ion concentrations exist in this mucus than in any body of fresh water, it is tempting to extend the idea of extrarenal absorption of salts to terrestrial species. Indeed it would seem almost essential with rapid rates of evaporation and correspondingly high potential salt losses, for epidermal reabsorption to occur to preserve the ion balance of the animal. Evaporation from the mucus would in fact, assist this process by increasing concentration gradients favouring ion diffusion back into the snail. It is also tempting to extend the idea of macromolecular recycling, already established in freshwater species to land snails, where epidermal cells also contain lysosomes and small vesicles at the bases of the microvilli.

In some terrestrial members of the prosobranch gastropod

families, Neritacea (Little, 1972) and Pomatiasidae (Rumsey, 1972) there is convincing evidence that the epithelial surfaces of the mantle cavity are used as an additional site of ion reabsorption from the urine. In several species in both groups, fluid collected at the mouth of the mantle cavity, which serves as a lung, was considerably more dilute than urine in its final stages of formation in the kidney. Differences in ion concentrations in the urine and mantle fluid in *Pomatias elegans* (Rumsey, 1972) indicate that all ions measured (Na, K, Ca, Mg) are further reabsorbed by the mantle. There is some evidence that terrestrial members of some pulmonate gastropod families may also use the mantle cavity for a urine modification. Blinn (1964) has observed fluid in the lungs of *Mesodon thyroidus* and *Allogona profunda*, both members of the Polygyridae, but unfortunately no solute concentration measurements were made. There is also evidence that the lung epithelia and possibly other epidermal surfaces in *Helix aspersa* and *Otala lactea* are involved in the elimination of gaseous ammonia (Speeg and Campbell, 1968). Transport processes in the epidermis are terrestrial molluscs are summarized in Fig. 5.

D. Regulation of Water Loss by the Mantle Collar Epidermis in Pulmonates

In the last several years there has been a growing realization that the mantle collar epidermis and not the epiphragm of helicid snails (Machin, 1968; 1975a) was remarkable in its ability to reduce evaporative water loss during periods of inactivity. Recent studies (Machin, unpublished) suggest that this property extends to shelled members of all stylommatophoran pulmonate families so far investigated. Evaporation from inactive basommatophoran pulmonates, a predominantly aquatic group, or from stylommatophoran shell-less slugs which do not have a mantle collar, is not sufficiently low to suggest that epidermal water retarding properties are found in these animals (Machin, 1975a). It can be seen from the low mantle permeabilities of the mantle collar epidermis (Table I.) that this mechanism must have profound effect on the water economy of inactive land snails and greatly prolonging their survival in adverse environmental conditions.

One of the more intriguing aspects of investigating the capacity to control evaporation in a living epidermis is that previous arguments in favour of efficient mucus extrusion, do not seem to hold. Evidence has been put forward (Machin, 1972) that glandular extrusion is severely restricted during inactivity. Without mucus extrusion the diffusional permeability of the epidermis is clearly unable to sustain normal levels of evaporation and rates of water loss decrease exponentially to markedly lower levels (Machin, 1965; 1966). Two pieces of experimental evidence indicate that the principal resistance to water diffusion in the mantle collar lies close to or within the epidermis. The most direct of these comes from the melting behaviour of frozen sections of mantle (Machin, 1974) which indicate the existence of an osmotic gradient across the epidermal cells particularly in a zone close to the mantle surface. It should be pointed out that the presence of an intracellular osmotic gradient and a diffusion barrier within a cell is at odds with currently accepted theories of epithelial water transport and of known properties of cytoplasm. This difficulty is compounded by the fact that the low resolution of frozen section analysis makes it impossible at present to identify the precise site of the barrier or even determine whether it is really within the cells. Yet, the existence of a diffusion barrier, unique to the mantle collar is convincing. Further work (Machin, unpublished) has shown that gradients of the magnitude already described occur only across the mantle epidermis and not in other regions of the integument. Furthermore it takes between four and five minutes to abolish the dehydrated zone by adding haemolymph iso-osmotic Ringer to the mantle surface or by restimulating the mucus glands. Rehydration of "conventional" epidermal cells of the same size would take a matter of seconds. Further evidence in support of a diffusion barrier located close to, but not precisely at the mantle surface, comes from weight analyses of water exchanges across the mantle (Machin, 1972). These studies indicate the presence of a passively exchanging superficial layer of dehydrated mucus, which overlies the barrier itself. Frozen sections indicate this layer to be one or two μm thick. Unfortunately experimental estimates of the water content of this layer in the humidities in which the sections were prepared, suggest a layer at least ten times deeper. However, there are explanations for this discrepancy, such as irregularities in the mantle surface or a more complex relationship between mucus water content and humidity, which would not weaken arguments in favour of the barrier.

It is as yet unknown why or how a single epithelial layer can have such a low permeability to water without the aid of nonliving extracellular structures. It is not known for example, whether low permeability is a constant characteristic of the mantle collar epidermis, merely masked during activity by mucus extrusion, or whether low permeability develops as surface dehydration proceeds, collapsing cell structure and water channels. Dehydration of the superficial mucus layer poses questions about the active or passive solute transport. Does the suggested transport between mucus and cell continue or does partial dehydration prevent exchange and the normal functioning of the mantle epidermal cell?

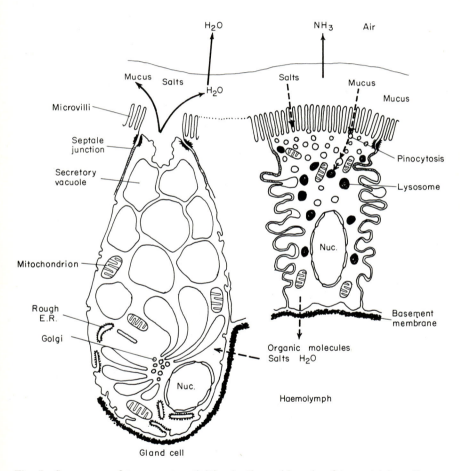

Fig. 5. Summary of transport activities in the epidermis of terrestrial molluscs. Continuous arrows indicate established mechanisms, dotted arrows, hypothetical exchanges. After Wondrak, 1967, 1968a,b; Zylstra, 1972a.

VII. SUMMARY AND CONCLUSIONS

1. The molluscan epidermis consists of a single layer of cells having the structural characteristics of transporting epithelia. Nonglandular cells, bearing apical microvilli, are found on all external surfaces throughout the phylum.

2. The molluscan integument is generously supplied with gland cells which maintain a superficial layer of mucus, consisting principally of sulphated polysaccharides and glycoproteins.

3. The extrapallial epidermis of the mantle readily permits intercellular diffusion of calcium and other ions into the extrapallial space. Polysaccharides and proteins, associated with shell formation, are also transported into this space. Organically bound calcium is probably absorbed pinocytotically and subsequently digested by epidermal lysosomes.

4. Other regions of the body surface in molluscs are also capable of active and passive ion transport and the pinocytotic absorption of organic particles.

5. The epidermis of marine molluscs is probably involved in potassium, chloride and amino acid uptake as well as water absorption to balance urine production. Mucus secretion may contribute to the extrarenal elimination of calcium and sulphate.

6. Water is passively gained and excretory ammonia lost through the permeable integument of fresh water species. Active ion uptake may be aided by the pinocytotic absorption of mucus by the epidermis.

7. Water is lost from the integument of active land molluscs by evaporation, governed by complex physical factors in the atmosphere. Mucus secretion and extrusion transports water through the epidermis, keeping its cells hydrated. Previous conclusions that mucus production completely eliminates epidermal diffusion are re-examined.

8. Methods for the study of epidermal diffusion and vapour-limited evaporation in land molluscs are described.

9. The possibility that organic and ionic constituents of mucus are recycled in terrestrial species, is discussed.

10. The mantle collar in inactive land pulmonates retards evaporation by eliminating mucus extrusion and permitting partial dehydration of the epidermis. Evidence for an epidermal water barrier is critically discussed.

REFERENCES

Avens, A. C. and Sleigh, M. A. (1965). *Comp. Biochem. Physiol.* 16, 121—141.

Barry, P. H. and Diamond, J. M. (1971). *J. memb. Biol.* 4, 295—330.

Beament, J. W. L. (1961). *Biol. Rev.* 36, 281—320.

Bethe, A. (1934). *Pflügers Arch. ges. Physiol.* 234, 629—644.

Bevelander, G. (1952). *Biol. Bull. mar. biol. Lab. Woods Hole* 102, 9—15.

Bevelander, G. and Nakahara, H. (1966). *Biol. Bull. mar. biol. Lab. Woods Hole* 131, 76—82.

Blinn, W. C. (1964). *Physiol. Zool.* 37, 329—337.

Burton, R. F. (1965). *Comp. Biochem. Physiol.* 15, 339—345.

Campion, M. (1961). *Q. Jl. microsc. Sci.* 102, 195—216.

Chaisemartin, C., Martin, P. N. and Bernard, M. (1968). *C. r. Séanc. Soc. Biol.* 162, 523—526.

Chien, P. K., Stephens, G. C. and Healey, P. L. (1972). *Biol. Bull. mar. biol. Lab. Woods Hole* 142, 219—235.

Cloney, R. A. and Florey, E. (1968). *Z. Zellforsch. mikrosk. Anat.* 89, 250—280.

Coggeshall, R. E. (1966). *J. Cell Biol.* 28, 95—108.

Crenshaw, M. A. (1972). *Biol. Bull. mar. biol. Lab. Woods Hole* 143, 506—512.

Crenshaw, M. A. and Neff, J. M. (1969). *Am. Zool.* 9, 881—885.

Crisp, M. (1971). *J. mar. biol. Ass. U.K.* 51, 865—890.

Dainty, J. and House, C. R. (1966). *J. Physiol.* 182, 66—78.

Farquhar, M. G. and Palade, G. E. (1963). *J. Cell Biol.* 17, 375—412.

Fawcett, D. W. and Porter, K. R. (1954). *J. Morph.* 94, 221—281.

Fretter, V. and Graham, A. (1962). "British prosobranch molluscs." Ray Society, London.

Frömter, E. and Diamond, J. (1972). *Nature Lond.* 235, 9—13.

Graf, F. (1971). *C. r. hebd. Séanc. Acad. Sci. Paris* 273, 1828—1831.

Greenaway, P. (1970). *J. exp. Biol.* 53, 147—163.

Greenaway, P. (1971a). *J. exp. Biol.* 54, 199—214.

Greenaway, P. (1971b). *J. exp. Biol.* 54, 609—620.

Gupta, B. L. and Little, C. (1970). *Tissue and Cell* 2, 637—696.

Hanisch, M. E. and Lawrence, A. L. (1972). *Comp. Biochem. Physiol.* 42A, 601—610.

Harrison, F. M. (1962). *J. exp. Biol.* 39, 179—192.

Harrison, F. M. and Martin, A. W. (1965). *J. exp. Biol.* 42, 71—98.

Hiscock, I. D. (1953). *Aust. J. mar. Freswat. Res.* 4, 317—342.

Hubendick, B. (1958). *Ark. Zool.* 11, 31—36.

Hunt, S. (1970). "Polysaccharide — protein complexes in invertebrates." Academic Press, New York and London.

Hunt, S. (1973). *J. mar. biol. Ass. U.K.* 53, 59—71.

Istin, M. and Kirschner, L. B. (1968). *J. gen. Physiol.* 51, 478—496.

Istin, M. and Maetz, J. (1964). *Biochim. biophys. Acta* 88, 225—227.

Kawaguti, S. and Ikemoto, N. (1962a). *Biol. J. Okayama Univ.* 8, 21—30.

Kawaguti, S. and Ikemoto, N. (1962b). *Biol. J. Okayama Univ.* 8, 31—42.

Kirschner, L. B. (1962). *J. gen. Physiol.* **46**, 362A–363A.

Kirschner, L. B., Sorensen, A. L. and Kriebel, M. (1960). *Science* **131**, 735.

Kobayashi, S. (1964). *Biol. Bull. mar. biol. Lab. Woods Hole* **126**, 414–422.

Krogh, A. (1939). "Osmotic regulation in aquatic animals." Cambridge University Press.

Lane, N. J. (1963). *Q. J. microsc. Sci.* **104**, 495–504.

Lawrence, A. L. and Lawrence, D. C. (1967). *Comp. Biochem. Physiol.* **22**, 341–357.

Leighly, J. (1937). *Ecology* **18**, 180–198.

Little, C. (1965a). *J. exp. Biol.* **43**, 23–37.

Little, C. (1965b). *J. exp. Biol.* **43**, 39–54.

Little, C. (1967). *J. exp. Biol.* **46**, 459–474.

Little, C. (1968). *J. exp. Biol.* **48**, 569–585.

Little, C. (1972). *J. exp. Biol.* **56**, 249–261.

Lloyd, D. C. (1969). *Protoplasma* **68**, 327–339.

Locke, M. (1965). *J. Cell Biol.* **25**, 166–169.

Lumsden, R. D. (1975). *Exp. Parasitol.* **37**, 267–339.

Lyons, K. (1973). *Adv. Parasit.* **11**, 193–232.

Machin, J. (1964a). *J. exp. Biol.* **41**, 759–769.

Machin, J. (1964b). *J. exp. Biol.* **41**, 771–781.

Machin, J. (1964c). *J. exp. Biol.* **41**, 783–792.

Machin, J. (1965). *Naturwissenschaften* **52**, 18.

Machin, J. (1966). *J. exp. Biol.* **45**, 269–278.

Machin, J. (1968). *Biol. Bull. mar. biol. Lab. Woods Hole* **134**, 87–95.

Machin, J. (1969). *Am. J. Physiol.* **216**, 1562–1568.

Machin, J. (1970). *J. exp. Biol.* **53**, 753–762.

Machin, J. (1972). *J. exp. Biol.* **57**, 103–111.

Machin, J. (1974). *Science* **183**, 759–760.

Machin, J. (1975a). *In* "Pulmonates." (Eds V. Fretter and J. Peake), Vol. I, Functional anatomy and physiology. pp. 105–163. Academic Press, New York and London.

Machin, J. (1975b). *J. comp. Physiol.* **101**, 121–132.

Manton, S. M. and Ramsay, J. A. (1937). *J. exp. Biol.* **14**, 470–472.

McAlister, R. O. and Fisher, F. M. (1968). *Biol. Bull. mar. biol. Lab. Woods Hole* **134**, 96–117.

Nakahara, H. and Bevelander, G. (1967). *J. Morph.* **122**, 139–145.

Neff, J. M. (1972). *Tissue and Cell* **4**, 591–600.

Newell, P. F. (1977). *Malacologia* **16**, 183–195.

Newell, P. F. and Skelding, J. M. (1973). *Z. Zellforsch. mikrosk. Anat.* **147**, 31–39.

Palade, G. (1975). *Science* **189**, 347–358.

Pancake, S. J. and Karnovsky, M. L. (1971). *J. biol. Chem.* **246**, 253–262.

Pasteels, J. J. (1961). *Z. Zellforsch. mikrosk. Anat.* **92**, 339–359.

Pierce, S. K., Jr. (1971). *Comp. Biochem. Physiol.* **39A**, 103–117.

Pomeroy, L. R. and Haskin, H. H. (1954). *Biol. Bull. mar. biol. Lab. Woods Hole* **107**, 123–129.

Potts, W. T. W. (1953). *J. exp. Biol.* **31**, 376—385.

Potts, W. T. W. (1954). *J. exp. Biol.* **31**, 614—617.

Potts, W. T. W. (1967). *Biol. Rev.* **42**, 1—41.

Pullin, R. S. V. (1971). *Comp. Biochem. Physiol.* **40A**, 617—626.

Ramsay, J. A. (1935a). *J. exp. Biol.* **12**, 355—372.

Ramsay, J. A. (1935b). *J. exp. Biol.* **12**, 373—383.

Ramsay, J. A. (1964). *Phil. Trans. R. Soc. Lond.* B **248**, 279—314.

Ramsay, J. A., Butler, C. G. and Sang, J. H. (1938). *J. exp. Biol.* **15**, 255—265.

Robertson, J. D. (1953). *J. exp. Biol.* **30**, 277—296.

Robertson, J. D. (1964). *In* "Physiology of the Mollusca." (Eds K. M. Wilbur and C. M. Younge), Vol. I, pp. 283—311. Academic Press, New York and London.

Rogers, D. C. (1971). *Z. Zellforsch. mikrosk. Anat.* **144**, 106—116.

Ronkin, R. R. (1950). *J. cell. comp. Physiol.* **35**, 241—250.

Rumsey, T. J. (1972). *J. exp. Biol.* **57**, 205—215.

Saleuddin, A. S. M. (1970). *Canad. J. Zool.* **48**, 409—416.

Satir, P. and Gilula, N. B. (1970). *J. Cell Biol.* **47**, 468—487.

Schmekel, L. and Wechsler, W. (1967). *Z. Zellforsch. mikrosk. Anat.* **77**, 95—114.

Schmidt-Nielsen, B. and Davis, L. E. (1968). *Science* **159**, 1105—1108.

Schmidt-Nielsen, K. (1969). *Q. Rev. Biophys.* **2**, 283—304.

Schoffeniels, E. (1951). *Arch. int. Physiol.* **59**, 49—52.

Schoffeniels, E. and Gilles, R. (1972). *In* "Chemical Zoology." (Eds M. Florkin and B. T. Scheer), Vol. VII, Mollusca, pp. 393—420. Academic Press, New York and London.

Schwalbach, G. and Lickfield, K. G. (1962). *Z. Zellforsch. mikrosk. Anat.* **58**, 277—288.

Scott, J. E. (1968). *In* "The chemical physiology of mucopolysaccharides." (Ed. Quintarelli, G.), pp. 171—186. Little, Brown and Company, Boston.

Shiba, K. and Ueda, M. (1965). *In* "Humidity and Moisture." (Eds A. Wexler and E. J. Amdur), Vol. II, pp. 349—356. Reinhold, New York.

Simkiss, K. (1963). *Comp. Biochem. Physiol.* **16**, 427—435.

Speeg, K. V., Jr. and Campbell, J. W. (1968). *Am. J. Physiol.* **214**, 1392—1402.

Spight, T. M. (1968). *Physiol. Zool.* **41**, 195—203.

Stephens, G. C. (1968). *Am. Zool.* **8**, 95—106.

Storch, V. and Abraham, R. (1972). *Z. Zellforsch. mikrosk. Anat.* **133**, 267—275.

Storch, V. and Welsch, U. (1970). *Z. Morph. Ökol. Tiere* **66**, 310—322.

Storch, V. and Welsch, U. (1972). *Marine Biol.* **17**, 137—144.

Todd, M. E. (1964). *Physiol. Zool.* **37**, 33—44.

Tormey, J. McD. and Diamond, J. M. (1967). *J. gen. Physiol.* **50**, 2031—2060.

Tucker, L. E. (1970). *Comp. Biochem. Physiol.* **36**, 301—319.

Van Aardt (1968). *Neth. J. Zool.* **18**, 253—312.

Van der Borght, O. and van Puymbroeck, S. (1964). *Nature Lond.* **204**, 533—534.

Webber, H. H. and Dehnel, P. A. (1968). *Comp. Biochem. Physiol.* **25**, 49—64.

Weiner, S. and Hood, L. (1975). *Science* **190**, 987—989.

Welsch, U. (1968). *Z. Zellforsch. mikrosk. Anat.* **88**, 565—575.

Welsch, U. and Storch, V. (1969). *Z. Zellforsch. mikrosk. Anat.* **95**, 317—330.

Wilbur, K. M. (1972). *In* "Chemical Zoology." (Eds M. Florkin and B. T. Scheer), Vol. VII, Mollusca, pp. 103—145. Academic Press, New York and London.

Wilson, R. A. (1968). *Comp. Biochem. Physiol.* **24**, 629—633.

Wondrak, G. (1967). *Z. Zellforsch. mikrosk. Anat.* **76**, 287—294.

Wondrak, G. (1968a). *Protoplasma* **66**, 151—171.

Wondrak, G. (1968b). *Z. Zellforsch. mikrosk. Anat.* **80**, 17—40.

Zylstra, U. (1972a). *Z. Zellforsch. mikrosk. Anat.* **130**, 93—134.

Zylstra, U. (1972b). *Neth. J. Zool.* **22**, 299—306.

30. The Function of Nephridia in Annelids

I. Zerbst-Boroffka

Institut für Tierphysiologie und Angwandte Zoologie, Berlin-West

I. INTRODUCTION

In 1946, Ramsay reported that the nephridial urine of *Lumbricus terrestris* is hypo-osmotic to the coelomic fluid. That study, and two subsequent ones (Ramsay, 1949a,c) laid the foundation for subsequent research on the salt and water economy of annelids and on the physiology of the nephridia. First, he described the changes in osmotic pressure and chloride concentration in the blood, coelomic fluid, and urine during adaptation of the earthworm to media of different concentration (Ramsay, 1949a). He concluded that: (a) water penetrates through the skin passively because of the osmotic gradient; (b) salt is absorbed passively from concentrated media and actively from dilute media; and (c) a hypotonic urine is formed in the nephridia by active reabsorption of salts. Secondly, Ramsay (1949c) collected and analysed minute quantities of fluid from different sections of the nephridial canal. The osmotic pressure was reduced in the wide tube and possibly also in the middle, ciliated, tube.

The microanalytical techniques Ramsay developed during his studies on earthworms have been widely used in the study of excretory organs. Specifically, he introduced a microfreezing point method for determining the osmolality of small fluid volumes (Ramsay, 1949b; Ramsay and Brown, 1955) and methods for measuring Na^+, K^+ (Ramsay, 1950; Ramsay *et al.*, 1953) and Cl^- (Ramsay *et al.*, 1955).

At present, research on nephridia of annelids still aims to clarify the mechanism of urine formation and the homeostatic role of the nephridia. These two topics will be discussed in the next two sections.

II. MECHANISM OF URINE FORMATION IN NEPHRIDIA

The functioning of annelid metanephridia has been studied in two species, *Lumbricus terrestris* and *Hirudo medicinalis*. The studies have been described in detail at a symposium (Zerbst-Boroffka and Haupt, 1975). Figure 2 summarizes the results and shows the structural and functional differences between the "open" metanephridium of *Lumbricus* and the secondary "closed" metanephridia of *Hirudo*.

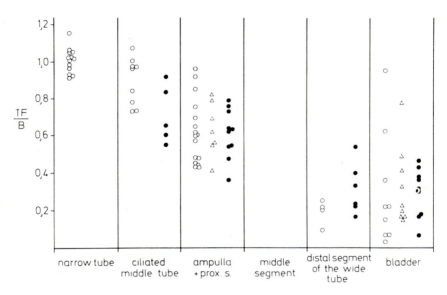

Fig. 1. Osmotic and chloride concentrations in the different canal sections of the nephridium of Lumbricus terrestris. TF = tubule fluid, B = Medium. ○ O.K. (Ramsay, 1949c); △ O.K., ● Cl⁻ (Boroffka, 1965).

In *Lumbricus* the coelomic fluid is probably formed by filtration of the blood into the coelomic cavity, as indicated by the ionic compositions of the two fluids. Recently, filtration structures have been demonstrated in the ventral vessel of *Tubifex* (Peters, 1977). The coelomic fluid is propelled into the nephridial canal through the ciliated funnel. Using micropuncture techniques, Ramsay (1949c) and Boroffka (1965) found that reabsorption of solutes produces a hypotonic urine (Fig. 1). In the ampulla, sodium is reabsorbed actively while chloride follows passively (Boroffka, 1965). One can conclude from the ultrastructure (Graszynski, 1963) that the whole nephridial canal may be involved in reabsorption. There are several possible routes for reabsorption: directly into the coelomic cavity, directly into the blood, or into the blood via the sheath cells.

In contrast to *Lumbricus*, the ultrastructure and micropuncture experiments suggest that in *Hirudo* the canalicular cells secrete NaCl and KCl into the canaliculi to form the primary urine. The secreted salts are derived directly from the blood capillaries, and by reabsorption through the cells lining the nephridial canal adjacent to the canalicular cells (Fig. 2). It was calculated that more than 94% of the salt secreted is reabsorbed, only 6% or less being excreted in the final urine. The salts are recycled by the canalicular cells. In the primary urine, potassium accounts for about one-third of the cation concentration and sodium for two-thirds, whereas, in the blood, potassium makes up only about 8% of the total cation concentration (Boroffka *et al.*, 1970; Haupt, 1974; Zerbst-Boroffka, 1975; Zerbst-Boroffka and Haupt, 1975). In many insects, the potassium concentration is also higher in the urine than in the blood (Maddrell,

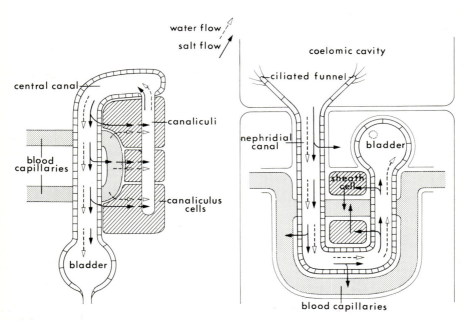

Fig. 2. Functional diagram of the open metanephridium of *Lumbricus* (right) and of the closed metanephridium of *Hirudo* (left) (Zerbst-Boroffka and Haupt, 1975).

Chapter 21). More recent work on the nephridia of different annelids has concentrated on the ultrastructure (e.g. Brandenburg, 1970; Holborrow, 1971; Koechlin, 1970, 1971; Wessing and Polenz, 1974).

III. THE HOMEOSTATIC FUNCTION OF NEPHRIDIA

The literature dealing with osmotic, ionic and volume control has become extensive since Ramsay's early publications. Oglesby's interesting and informative survey (1969) gives a comprehensive summary. Since then additional information has been reported by Brown *et al.* (1972), Dietz and Alvarado (1970), Fletcher (1974a,b), Foster (1974), Oglesby (1970, 1972, 1973), Skaer (1974a,b) and Zerbst-Boroffka (1973). On the other hand, there is relatively little data available on nephridial function in connection with homeostasis.

Research on the homeostatic function of the nephridia seeks answers to two questions.

1. What quantitative part do the nephridia play in the excretion of solutes and water? How great is the nephridial excretion relative to the total output?

2. How does the osmotic pressure of the environment affect the concentration and volume of the urine? How are the nephridia controlled?

Measurements of the excretory fractions were carried out by measuring:

1. the discharge of salt and water from animals in the steady-state (i.e. with constant blood concentration);

2. the discharge of water or salt immediately after a change of medium;

3. the rate of inulin excretion after injection into the coelomic cavity.

All these methods have provided information about osmotic, ionic and volume control during adaptation. However, the results are not directly comparable with one another since each method measures something different. Quantitative data on nephridial excretion can be obtained only by direct measurement of the concentration and flow of the final urine. In annelids such data are rare because of the extraordinary difficulty of collecting urine directly and continuously.

With some annelids it is possible to extract just sufficient urine from the bladder and from the nephridial pores for a measurement of the concentration (Fig. 3). In some sipunculids osmotic pressure and

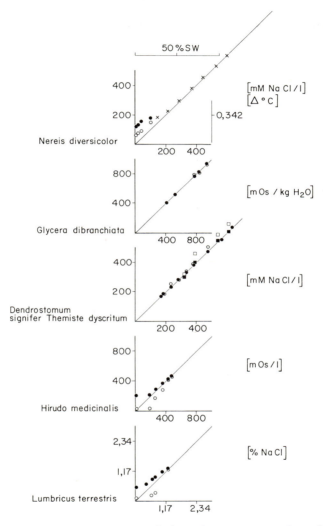

Fig. 3. Change in the osmotic, or total electrolyte, concentration of coelomic fluid and urine corresponding to the concentration of the medium. In each case the duration of adaptation to the given medium and the temperature are given: *Lumbricus* (Ramsay, 1949a) 1 week, 17°C. *Hirudo* (Boroffka, 1968) ½ to 1 month, 20°C. *Themiste* (Oglesby, 1968) after gradual adaptation in the experimental medium 1 week, 10—12°C. *Dendrostomum* (Kamemoto and Larson, 1964) 120 hrs, room temperature. *Glycera* (Machin, 1975) 48 hrs, 14—16°C. *Nereis* (Smith, 1970)?, 18—20°C; (Hohendorf, 1963) 3 days, 10°C.

chloride concentration have been determined using specimens adapted to various degrees of salinity. Both in *Dendrostomum* (Kamemoto and Larson, 1964) and in *Themiste* (Oglesby, 1968) an almost isotonic urine is formed over the whole range of salinity in which the animals are able to survive. The sedentary polychaete *Glycera dibranchiata* also produces urine which is isotonic, at least within the range in which measurements were made (Machin, 1975).

Nereis diversicolor produces a hypotonic urine in media below 400 mOsm (Smith, 1970). However, in media above 400 mOsm the urine is nearly isotonic, even though salinities up to 200% seawater are tolerated (Hohendorf, 1963; Oglesby, 1970).

In *Lumbricus* and *Hirudo* a strongly hypotonic urine is formed in media below 200 mOsm. In media above 200 mOsm the urine is less hypotonic. For both animals, the degree of salinity at which an almost isotonic urine is formed is also the highest which they are able to tolerate (Ramsay, 1949a; Boroffka, 1968).

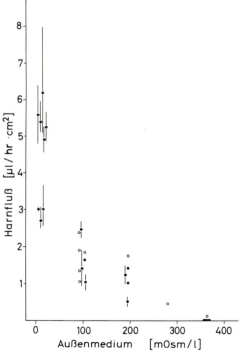

Fig. 4. Urine flow of *Hirudo* after 8 weeks adaptation to various salinities; (Boroffka, 1968).

Only in *Hirudo* are direct measurements of urine flow rate available. The results show clearly that urine flow is greatest in fresh water and diminishes rapidly with increasing salinity of the medium (Fig. 4; Boroffka, 1968).

Perhaps the significance of nephridia for homeostasis can best be understood from a knowledge of the mechanism which controls the excretion rates under different conditions. As pointed out by Ramsay (1949a), the situation is extremely complex. Research on this problem is presently in progress in our laboratory on the leech, *Hirudo*. This animal, being a bloodsucker with the concomitant facilities for osmotic and volume control, presents a very appropriate object for such investigations.

After a meal consisting of the hyperosmotic blood of a vertebrate, the leech absorbs the hyperosmotic salt solution from its stomach and excretes it again via the nephridia. Its blood concentration thus increases temporarily, but decreases to normal again after a few hours or days. At the same time, the concentration and ionic composition of the stomach contents also equilibrates with that of the blood. When the animal is placed in a concentrated medium, water is lost and the blood concentration increases. The animal becomes antidiuretic (Boroffka, 1968). However, after feeding, the leech has a comparable increase in blood concentration, but diuresis occurs, with up to nine times the normal urine flow (Zerbst-Boroffka, 1973). Apparently it is an increase in body volume that triggers diuresis after feeding. A similar phenomenon occurs in the blood-sucking bug, Rhodnius (Maddrell, 1964; Gee, in this volume). In *Rhodnius* the increase in body volume is detected by stretch receptors in the abdominal body wall. However, in *Hirudo* we find that dilation of the stomach or stretching of the dermal muscle sheath do not cause diuresis (Zerbst-Boroffka, 1973). Instead, it appears that it is stretch receptors in the blood vessels that control the rate of urine flow (Zerbst-Boroffka, in preparation). This is comparable to the situation in mammals in which stretch receptors in the left auricle are involved in control of diuresis (Gauer *et al.*, 1970; Goetz *et al.*, 1975).

REFERENCES

Boroffka, I. (1965). *Z. vergl. Physiol.* **51**, 25—48.
Boroffka, I. (1968). *Z. vergl. Physiol.* **57**, 348—375.
Boroffka, I., Altner, H. und Haupt, J. (1970). *Z. vergl. Physiol.* **66**, 421—438.

Brandenburg, J. (1970). *Z. Morph. Tiere* **68**, 83—92.

Brown, S. C., Bdzil, J. B. and Frisch, H. L. (1972). *Biol. Bul. mar. biol. Lab. Woods Hole* **143**, 278—295.

Dietz, T. H. and Alvarado, R. H. (1970). *Biol. Bull. mar. biol. Lab. Woods Hole* **138**, 247—261.

Fletcher, C. R. (1974a). *Comp. Biochem. Physiol.* **47A**, 1199—1214.

Fletcher, C. R. (1974b). *Comp. Biochem. Physiol.* **47A**, 1221—1234.

Foster, R. C. (1974). *Comp. Biochem. Physiol.* **47A**, 855—866.

Gauer, O. H., Henry, J. P. and Behn, C. (1970). *Ann. Rev. Physiol.* **32**, 547—595.

Goetz, K. L., Bond, G. C. and Bloxham, D. D. (1975). *Physiol. Rev.* **55**, 157—205.

Graszynski, K. (1963). *Zool. Beitr. N.F.* **8**, 189—296.

Haupt, J. (1974). *Cell Tiss. Res.* **152**, 385—401.

Hohendorf, K. (1963). *Kieler Meeresforsch* **19**, 196—218.

Holborrow, P. L. (1971). *In* "The Fourth European Marine Biology Symposium." pp. 327—246. Cambridge University Press, Cambridge.

Kamemoto, F. I. and Larson, E. J. (1964). *Comp. Biochem. Physiol.* **13**, 477—480.

Koechlin, N. (1970). *Arch. Anat. Micr. Morph. exp.* **59**, 331—360.

Koechlin, N. (1971). *Arch. Anat. Micr. Morph. exp.* **60**, 37—48.

Machin, J. (1975). *Comp. Biochem. Physiol.* **52A**, 49—54.

Maddrell, S. H. P. (1964). *J. exp. Biol.* **41**, 459—472.

Oglesby, L. C. (1968). *Comp. Biochem. Physiol.* **26**, 155—177.

Oglesby, L. C. (1969). *In* "Chemical Zoology." (Ed. Florkin and Scheer). Vol. IV, pp. 211—310. Academic Press, New York.

Oglesby, L. C. (1970). *Comp. Biochem. Physiol.* **36**, 449—466.

Oglesby, L. C. (1972). *Comp. Biochem. Physiol.* **41A**, 765—790.

Oglesby, L. C. (1973). *Biol. Bull. mar. biol. Lab. Woods Hole* **145**, 180—199.

Peters, W. (1977). *Cell Tiss. Res.* **179**, 367—375.

Ramsay, J. A. (1946). *Nature, Lond.* **158**, 665.

Ramsay, J. A. (1949a). *J. exp. Biol.* **26**, 46—56.

Ramsay, J. A. (1949b). *J. exp. Biol.* **26**, 57—64.

Ramsay, J. A. (1949c). *J. exp. Biol.* **26**, 65—75.

Ramsay, J. A. (1950). *J. exp. Biol.* **27**, 407—419.

Ramsay, J. A. and Brown, R. H. J. (1955). *J. sci. Instrum.* **32**, 372—375.

Ramsay, J. A., Brown, R. H. J. and Falloon, S. W. (1953). *J. exp. Biol.* **30**, 1—17.

Ramsay, J. A., Brown, R. H. J. and Croghan, P. C. (1955). *J. exp. Biol.* **32**, 822—829.

Skaer, H. Le B. (1974a). *J. exp. Biol.* **60**, 321—330.

Skaer, H. Le B. (1974b). *J. exp. Biol.* **60**, 331—338.

Smith, R. I. (1970). *J. exp. Biol.* **53**, 101—108.

Wessing, A. and Polenz, A. (1974). *Cell Tiss. Res.* **156**, 21—33.

Zerbst-Boroffka, I. (1973). *J. comp. Physiol.* **84**, 185—204.

Zerbst-Boroffka, I. (1975). *J. comp. Physiol.* **100**, 307—315.

Zerbst-Boroffka, I. and Haupt, J. (1975). *Fortschr. Zool.* **23**, 33—47.

Author Index

771

Glosson, P. S., 315, 318, *329*
Goertemiller, C. C., 430, 431, *450*
Goetz, K. L., 769, *770*
Goldbard, G. A., 278, *280*
Goldgraben, J. R., 196, *213*
Goldring, S., 498, *509*
Goldstein, D. A., 147, *165*, 187, 194, *213*
Goldsworthy, G. J., 268, 269, 270, 271, 278, *280*
Gomez, D. G., 492, *507*
Gomme, J., 240, *264*
Gomori, G., 350, *359*
Gomperts, B., 235, *238*
Goode, S. R., 18, *27*
Goodenough, D. A., 334, *359*
Goodhall, M. C., 180, *182*
Goodman, D. S., 356, *359*
Gordon, M. S., 321, 323, *326*, 436, *451*
Gorman, A. L. F., 482, 483, *508*
Gottschalk, C. W., 30, 32, 33, 34, *53*, 68, *81*
Graf, F., 742, *759*
Graham, A., 738, 739, 749, *759*
Grandchamp, A., 48, *53*
Grantham, J. J., 45, 46, *53*, 57, 58, 59, 60, 61, 64, 65, 67, 71, 73, 74, 75, 79, *81, 82*, 180, *182*, 275, *280*, 309
Graszynski, K., 764, *770*
Green, K., 428, *451*
Green, N., 41, 43, 45, 46, *52*, 59, 61, 66, 67, 68, 69, *81*
Green, R., 44, *53*
Greenaway, P., 740, 742, 745, 746, *759*
Greenberg, M. J., 323, *329*
Greengard, P., 351, *360*
Greenwald, L., 436, 437, 444, *451*, 461, *476*

Greenwell, J. R., 646, 648, *651*
Gregg, E. C., 310, 313,
Gregor, H. P., 184, *213*
Gregory, A., 288, *332*
Grill, G., 39, *54*
Grim, E., 210, *213*
Grimstone, A. V., 5, *6*, 145, *164*, 571, 581, 588, 589, 590, *595*, 607, *611*
Grisar, T., 503, 504, *508*
Grobben, C., 616, *630*
Gross, J. B., 59, *81*
Gross, W. J., 678, 680, 682, *705*
Grover, N. B., 310, *325*
Gruber, W. D., 641, *651*
Grundfest, H., 304, 305, *330*
Guillery, R. W., 349, *359, 361*
Gullasch, J., 114, 119, *141*
Gunn, D. L., 660, *672*
Gupta, B. L., *27*, 83, 84, 95, 98, 104, 105, 106, 109, 112, 113, 114, 116, 118, 120, 123, 124, 127, 129, 130, 131, 132, 133, 134, 136, *140, 141, 142*, 232, 237, *237, 238*, 276, *280*, 289, *325*, 329, 482, 485, 495, 508, *510*, 545, 551, 552, 567, *568*, 587, 593, *595*, 600, 607, 608, *611*, 633, 646, 649, 650, *651*, 664, *672*, 721, *733*, 737, *759*

H

Hackman, R. H., 665, *672*
Hadley, N. F., 670, *672*
Hagiwara, S., 333, *359*
Hajjar, J. J., 315, 317, *331*
Haljamäe, H., 19, 21, *27*, 499, *508*
Hall, H. H., 246, *264*

P

Subject Index